Robert B. Heimann (Ed.)
Materials for Medical Application

Also of interest

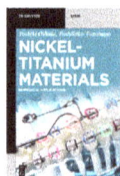

Nickel-Titanium Materials
Biomedical Applications
Yoshiki Oshida, Toshihiko Tominaga, 2020
ISBN 978-3-11-066603-8, e-ISBN 978-3-11-066611-3

Advanced Materials
Theo van de Ven, Armand Soldera (Eds.), 2020
ISBN 978-3-11-053765-9, e-ISBN (PDF) 978-3-11-053773-4

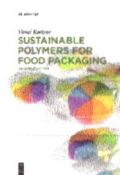

Sustainable Polymers for Food Packaging
Vimal Katiyar, 2020
ISBN 978-3-11-064453-1, e-ISBN 978-3-11-064803-4

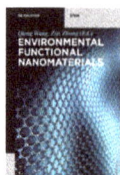

Environmental Functional Nanomaterials
Qiang Wang, Ziyi Zhong (Eds.), 2020
ISBN 978-3-11-054405-3, e-ISBN 978-3-11-054418-3

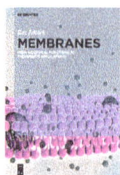

Membranes
From Biological Functions to Therapeutic Applications
Raz Jelinek, 2018
ISBN 978-3-11-045368-3, e-ISBN 978-3-11-045369-0

Materials for Medical Application

Edited by Robert B. Heimann

DE GRUYTER

Editor
Prof. em. Dr. Robert B. Heimann
Am Stadtpark 2A
D-02826 Görlitz, Germany
robert.heimann@ocean-gate.de

ISBN 978-3-11-061919-5
e-ISBN (PDF) 978-3-11-061924-9
e-ISBN (EPUB) 978-3-11-061931-7

Library of Congress Control Number: 2020934825

Bibliographic information published by the Deutsche Nationalbibliothek
The Deutsche Nationalbibliothek lists this publication in the Deutsche Nationalbibliografie;
detailed bibliographic data are available on the Internet at http://dnb.dnb.de.

© 2020 Walter de Gruyter GmbH, Berlin/Boston
Cover image: A laser-structured titanium surface promotes cell spreading via an actin-based
filopodium. SEM image courtesy Dr. Susanne Stählke, Department of Cell Biology, Rostock
University Medical Center. Laser structuring provided by Dr. Rigo Peters and Paul Ordorf, Welding
Training and Research Institute Mecklenburg-Vorpommern GmbH, SLV, Rostock, Germany.
Typesetting: Integra Software Services Pvt. Ltd.
Printing and binding: CPI books GmbH, Leck

www.degruyter.com

Preface

During the past several decades, research into biomaterials has emerged as a hot topic among material scientists and medical researchers alike. Virtually 10,000s of papers on more or less novel or improved biomaterials and biomedical devices can now be found in relevant scientific journals, textbooks, the patent literature, and on the Internet. Attempting to treat this vast field of current endeavor in an encyclopedic fashion is clearly impossible as new contributions are being published daily with ever-increasing speed and regularity. Hence, trying to keep abreast with these developments is akin to shooting at a very fast moving target. The best one can do is to provide snapshots of currently available salient information pertaining to biomaterials and biomedical devices and attempt to chart a course of their future development. For those readers who desire to delve more deeply into the fine details of biomaterials science, many comprehensive specialist accounts are available. A collection of recent treatises on biomaterials science and bioengineering can be found in the reference section.

To say it clearly: it is not our intention to compete with these excellent and comprehensive texts. This would indeed be a tough act to follow. So, why do we write another text on biomaterials science? What is so special about it and which aspects does it cover that are not adequately treated or even missing in previous accounts? Many textbooks on biomaterials and biomedical devices are mere collections of reviews that appear disjointed, and thus, do not convey full appreciation of the highly interdisciplinary nature of the subject matter owing to their lack of cohesion. As they frequently dwell on special cases, they do not allow a comprehensive view at the whole complex and variegated landscape of biomaterial science. Other accounts are merely collated results of conference presentations and symposia proceedings that lack substance since they frequently discuss isolated peculiarities of the research portfolio of the authors involved, and thus, fail to convey the bigger picture.

What are we starting with, then? We realize that many metallic, ceramic, polymeric, and composite materials today are being used successfully in the clinical practice to alleviate many health-related problems. A very substantial part of this effort relates to implants designed to repair congenital defects, replace diseased or missing body parts, or attempt to restore lost body functions. These applications range from artificial dental roots, to alveolar ridge, iliac crest and cheek augmentation, to spinal implants, to scaffolds for tissue engineering, and to restorative hip and knee endoprostheses to repair the compromised ambulatory knee–hip kinematic. Applications also include bone replacement parts in cranial and maxillary-mandibular areas, ocular implants, the ossicular chain of the inner ear, periodontal pocket obliteration, osteosynthetic musculoskeletal fixation devices for improved bone healing, heart pacemakers, and cardiovascular stents to prevent occlusion and stenosis of blood vessels. Much research is being performed on drug and gene

https://doi.org/10.1515/9783110619249-202

delivery vehicles to find nontoxic, biocompatible, nonimmunogenic, biodegradable material combinations that avoid recognition by the host's defense mechanisms by targeting the very locus of disease rather than providing a systemic therapy. In addition, increasingly applications emerge that use biomaterials in the medical device industry as diverse as percutaneous access devices, intramedullary nails and bone screws, antibacterial coatings for surgical instruments and clinical equipment, as well as medical textiles. In the future, the so-called fourth-generation biomaterials based on integrating electronic systems with the human body are poised to manipulate and monitor cellular bioelectrical responses for tissue regeneration aimed to communicating with the host tissue.

In the preface of Brian Love's 2017 book on *Biomaterials: A Systems Approach to Engineering Concepts*, he outlined succinctly the problems a text on biomaterials faces as follows: "Many believe that it is impossible to write a book relating to materials used in medicine. There are several reasons for this dilemma. Books that emphasize the materials' concepts are founded on volumetric replacement and biological response and less on a mechanical or other analysis of the design environment . . . Other monographs that are focused on tissue engineering are also hard to gauge, because there is so much detail needed to define a scaffold structure, optimize the culturing environment, identify an implantation protocol, test in appropriate animal models, to establish when and how to assay, and to confirm a successful outcome . . . The clinical books are also focused more closely on the aspects of biocompatibility and the subtleties associated with materials, compositional variations, or/and end mechanical function are sometimes lacking."

Indeed, these are serious aspects to consider. However, in spite of those warnings, the authors of this book have endeavored to take up the challenge to add their own account on biomaterials to the existing literature by attempting to discuss issues of significance of the history, structure, and mechanical, chemical, and biological behavior of materials used in medical applications, and their interaction with living tissue by reviewing clinical applications in joint replacement, bone grafts, dentistry, and tissue engineering. Information has been appended on existing and emerging markets including sales forecasts as well as standard specifications and regulatory affairs of biomaterials and biomedical devices that underscore the important economic aspect of today's health-related research and development. This foray into the world of economics adds a unique viewpoint that is missing in most competing texts available today. Finally, a host of future developments and challenges have been outlined that need to be addressed in the years to come. Such developments range from biomimetically designed drug and gene delivery vehicles, bioinspired computers based on DNA, biosensors, and hybrid inorganic/organic nanocomposites to 3D-printed scaffolds for regenerative medicine to novel biodegradable metals, bioceramics and carbon materials to surface-active bioglasses and hydrogels, and to the new class of fourth-generation biomaterials based on integrating electronic systems with the human body to provide powerful diagnostic and therapeutic tools for basic research and clinical use. Salient aspects of the role of artificial intelligence in the field of research and development of novel biomaterials and biomedical devices have been briefly outlined.

Alas, this textbook is not designed to satisfy the curiosity of the specialist by disseminating very specific information but, instead, attempts to provide a basic account on the present state of biomaterials for students of materials science and medicine as well as young professionals of both disjointed disciplines at the beginning of their career. Hence, the text is predominantly aimed to those students who may want to inform themselves quickly on the general subject of biomaterials, their properties, synthesis and preparation technologies, analytical characterization, *in vitro* and *in vivo* behavior as well as clinical uses, and standardization and regulatory affairs. The authors are hopeful that their treatise will provide information sought by those students without requiring them to delve into the abyss of specialized literature that threatens to overwhelm even the most patient reader. Consequently, students may benefit from our concise and condensed approach inasmuch as it will relieve them from plowing through piles of original papers scattered over scores of scientific journals. Further, we hope that readers trained in the realms of medical and biological sciences will appreciate the materials science aspects of biomaterials whereas those educated and trained in materials science may find the rather challenging medical content of the book enlightening and useful. We hold it evident that our potential audience in their perception of the biomaterials' world is akin to the much quoted Snowian concept of the "two cultures," here of materials science and the medical profession. Satisfying the expectations of these very diverse groups is intrinsically difficult. Should we have failed here and there in this endeavor, we are begging the gentle reader for pardon.

Robert B. Heimann, editor
Görlitz, Germany

Contents

Contributing authors

Prof. emer. Dr. Robert B. Heimann
Am Stadtpark 2A, D-02826 Görlitz, Germany
robert.heimann@ocean-gate.de
Principal author and editor

Dr. Doaa Adel-Khattab
Department of Experimental and Orofacial Medicine
Philipps University Marburg, Georg-Voigt-Strasse 3, D-35039
Marburg, Germany.

Prof. Dr. Christine Knabe-Ducheyne
Philipps University Marburg, Georg-Voigt-Strasse 3, D-35039
Marburg, Germany.
Department of Experimental and Orofacial Medicine
knabec@med.uni-marburg.de

Prof. Dr. Hans-Jürgen Kock
MEDIAN Hohenfeld-Klinik Bad Camberg
Hohenfeldstrasse 12-14, D-65520 Bad Camberg, Germany
hans-juergen.kock@median-kliniken.de

Privatdozent Dr. Christian Müller-Mai
Katholisches Klinikum Lünen/Werne GmbH
St.-Marien-Hospital, Altstadtstrasse 23, D-44534 Lünen,
Germany
mueller-mai.christian@klinikum-luenen.de

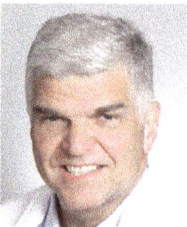

https://doi.org/10.1515/9783110619249-204

Prof. Dr. J. Barbara Nebe
Department of Cell Biology, Universitätsmedizin Rostock
Universität Rostock, Schillingallee 68, D-18057 Rostock, Germany
barbara.nebe@med.uni-rostock.de

Prof. emer. Dr. Mitsuo Niinomi
Institute of Materials Research, Tohoku University
981-0933 Sendai 2-1-1, Katahira, Aobo-ku, Japan
niinomi@imr.tohoku.ac.jp

Privatdozent Dr. Michael Schlosser
Universitätsmedizin Greifswald, Klinik und Poliklinik für Chirurgie
Ferdinand-Sauerbruch-Strasse, D-17489 Greifswald, Germany
michael.schlosser@med.uni-greifswald.de

Dr. Matthias Schnabelrauch
Department of Biomaterials, INNOVENT e.V.
Prüssingstrasse 27b, D-07745 Jena, Germany
ms@innovent-jena.de

List of Abbreviations

AAGR	average annual growth rate
AC	air cooling
ACC	amorphous calcium carbonate
ACP	amorphous calcium phosphate
AD	aerosol deposition
AgNP	silver nanoparticles
AI	artificial intelligence
AISI	American Iron and Steel Institute
ALP	alkaline phosphatase
AM	additive manufacturing
AO	acridine orange
AOD	argon oxygen decarburization
APS	atmospheric (air) plasma spraying
ASTM	American Society for Testing and Materials
AT	aging time
ATCC	American Type Culture Collection
ATMP	Advanced Therapy Medicinal Product
ATZ	alumina-toughened zirconia
AVB	arteriovenous bundle
BAR	bone area ratio
BCC	body-centered cubic
BCP	biphasic calcium phosphate
BCR	bone contact ratio
BCT	body-centered tetragonal
BEPM	blended elemental powder metallurgy
bFGT	basic fibroblast growth factor
BMP	bone morphogenetic protein
BO	bridging oxygen
BOF	basic oxygen furnace
BP	bisphosphonate
BSA	bovine serum albumin
BSP	bone sialoprotein
CAD	computer-aided design
CAGR	compound annual growth rate
CAOP	calcium alkali orthophosphate
CCD	charge-coupled device
CCIM	cold crucible induction melting
CCLM	cold crucible levitation melting
CDRH	Center for Devices and Radiological Health (USA)
CES	cranial electrotherapy
CFA	complete Freund's adjuvant
CFR	Code of Federal Regulation (USA)
CGDS	cold gas dynamic spraying
CNT	carbon nanotube
CQD	carbon quantum dot
CR	cold rolling
CS	chondroitin sulfate

https://doi.org/10.1515/9783110619249-205

CT	computed tomography
CVD	chemical vapor deposition
DCPA	dicalcium phosphate anhydrate
DCPD	dicalcium phosphate dihydrate
DD	degree of deacetylation
DED	directional energy deposition
DES	drug-eluting stent
DFT	density functional theory
DIN	Deutsches Institut für Normung
DLC	diamond-like carbon
DMTA	dynamic mechanical thermal analysis
DNA	deoxyribonucleic acid
DRA	Device Retrieval and Analysis
DSC	differential scanning calorimetry
EAF	electric arc furnace
EBM	electron beam melting
EC	European Community
ECF	extracellular fluid
ECM	extracellular matrix
ECR-CVD	electron cyclotron-resonance chemical vapor deposition
EHLA	extremely high-velocity laser (deposition)
EMR	electrolytically-mediated reaction
EN	European norm
EPD	electrophoretic deposition
EthB	ethidium bromide
ETO	ethylene oxide
FBGC	foreign body giant cell
FBR	foreign body response (reaction)
FC	furnace cooling
FCC	face-centered cubic
FDA	Food and Drug Administration (USA)
FDM	fused deposition molding
FEA	finite element analysis
FESEM	field emission scanning electron microscope
FFC	Fray-Farthing-Chen (Cambridge process)
FFF	fused filament fabrication
GAG	glycosaminoglycan
GMP	Good Manufacturing Practice
GNP	gold nanoparticle
GPa	gigapascal
GPC	gel permeation chromatography
HA	hyaluronic acid
HAp	hydroxylapatite
HB	hydrogen bond
HCA	hydroxylcarbonate apatite
HDH	hydrido-dehydride (process)
HEA	high entropy alloy
HFI	high-frequency induction
HIP	hot isostatic pressing

hMSC	human mesenchymal stem cell
HPSN	hot-pressed silicon nitride
HREE	heavy rare earth element
HVSFS	high-velocity suspension flame spraying
IBAD	ion beam-assisted deposition
ICD	International Classification of Diseases
ICF	intracellular fluid
ICRP	International Commission on Radiological Protection
IgF	insulin-like growth factor
IOL	intraocular lens
ISM	induction skull melting
ISO	International Standard Organization
LB	Langmuir-Blodgett (film)
LCA	life cycle assessment
LDPE	low-density polyethylene
LENS	laser-engineered net shaping
LEPS	low-energy plasma spraying
LLDPE	linear low-density polyethylene
LPPS	low-pressure plasma spraying
LPSO	long-period stacking order
LREE	light rare earth element
LRO	long-range order
LTIC	low-temperature isotropic pyrolytic carbon
MA	marketing authorization
MAO	micro-arc oxidation
MBG	mesoporous bioactive glass
MD	molecular dynamics (simulation)
MDA	Medical Device Agency
MDD	Medical Device Directive
MDR	Medical Device Regulation
MET	micro-current electrical therapy
MHRA	Medicine and Healthcare Products Regulatory Agency (USA)
MIM	metal injection molding
MIM	monovalent ion model (Mg)
MMA	methylmethacrylate
MMC	metal matrix composite
MoM	metal-on-metal (implant)
MPa	megapascal
MPS	micro-plasma spraying
MRI	magnetic resonance imaging
MRSA	multiresistant *Staphylococcus aureus*
MSC	mesenchymal stem cell
MSE	molten salt electrolysis
MTT	3-[4,5-dimethylthiazole-2-yl]-2,5-diphenyltetrazolium bromide (dye)
NASA	National Aeronautics and Space Administration (USA)
NaSiCon	**Na s**uper**i**onic **con**ductor
NBO	non-bridging oxygen
NCP	noncollagenous protein
NDE	negative difference effect (Mg)

NIH	National Institutes of Health (USA)
NK	natural killer (cell)
NMR	nuclear magnetic resonance
OAp	oxyapatite
OC	osteocalcin
OCP	octacalcium phosphate
OFZ	optical float zone
OHAp	oxyhydroxylapatite
ON	osteonectin
OPN	osteopontin
OS	Ono-Suzuki (process)
PACVD	plasma-assisted chemical vapor deposition
PAM	plasma-arc melting
PAMAM	poly(amidoamine)
PBR	pebble-bed reactor (nuclear)
PCL	poly(caprolactone)
PDGF	platelet-derived growth factor
PDLA	poly(D-lactide)
PDMS	poly(dimethylsiloxane)
PECVD	plasma-enhanced chemical vapor deposition
PEEK	poly(ether etherketone)
PEG	poly(ethylene glycol)
PEI	poly(ethylene imine)
PEO	poly(ethylene oxide)
PEO	plasma electrolytic oxidation
PET	poly(ethylene terephthalate)
PGA	poly(glutamic acid)
PHB	poly(hydroxylbutyrate)
PIII	plasma immersion ion implantation
PLA	poly(lactic acid)
PLD	pulsed laser deposition
PLGA	poly(lactide-co-glycolic acid)
PLLA	poly(L-lactide)
PMMA	poly(methylmethacrylate)
PMN	premarket notification
POE	poly(orthoester)
PPAAm	plasma-polymerized allylamine
PREP	plasma rotating electrode process
PSU	polysulfone
PTFE	poly(tetrafluoroethylene)
PU	polyurethane
PVA	poly(vinyl alcohol)
PVC	poly(vinyl chloride)
PVD	physical vapor deposition
QA	quality assurance
QC	quality control
QD	quantum dot
QMS	quality management system
RBSN	reaction-bonded silicon nitride

RCT	randomized control trial
REP	rotating electrode process
rhBMP	recombinant human bone morphogenetic protein
RIM	reaction injection molding
RNA	ribonucleic acid
ROI	return on investment
ROP	ring-opening polymerization
RP	rapid prototyping
RRR	replacement, reduction, refinement (strategy)
RTM	resin transfer molding
RT-PCR	real time polymerase chain reaction
SBF	simulated body fluid
SDE	statistical design of experiments
SED	selected electron diffraction
SFA	sinus floor augmentation
SFF	solid free form (fabrication)
SLM	selective laser melting
SLS	selective laser sintering
S-N	cyclic stress-failure (curve)
SPC	statistical process control
SPPS	solution precursor plasma spraying
SPR	surface plasmon resonance
SPRM	surface plasmon resonance microscopy
SPS	suspension plasma spraying
SQC	statistical quality control
SRO	short-range order
SSM	Schwartzwalder-Somers method
SSN	sintered silicon nitride
ST	solution treatment
SUS	steel use stainless (Japan)
TCP	tricalcium phosphate
TCP	thermochemical processing
TEOS	tetraethylorthosilicate
TEP	triethylphosphate
THA	total hip arthroplasty
THR	total hip replacement
TII	therapeutic inorganic ion
TKA	total knee arthroplasty
TNF	tumor necrosis factor
TSZ	travelling solvent zone
TTCP	tetracalcium phosphate
TZP	tetragonal zirconia polycrystal
UHMWPE	ultra-high molecular weight polyethylene
VAR	vacuum arc (melting)
VCR	vacuum decarburization reactor
VEC	valence electron concentration
VIM	vacuum induction melting
VPS	vacuum plasma spraying
WHO	World Health Organization

WQ	water quenching
XRD	X-ray diffraction
Y-PSZ	Y-partially stabilized zirconia
Y-TZP	Y-stabilized tetragonal zirconia polycrystal
ZTA	zirconia-toughened alumina

Chapter 1
Biomaterials – characteristics, history, applications

Robert B. Heimann

In this introductory chapter, general characteristics of selected metallic, ceramic, polymeric, and composite biomaterials applied to medical use will be discussed. Major historical milestones relating to the discovery or invention, development, and improvement of biomaterials will be addressed. In addition, types and properties of typical biomedical devices that are being introduced into the human body as implants to remedy congenital ailments, heal diseases, or restore lost parts or functions of the human body will be reviewed. Basic information is also provided on typical interaction of biomaterials with living tissue, required to establish the biological safety of a biomaterial or biomedical design by a plethora of biocompatibility tests. These issues are dealt with in much detail in Chapter 4.

Overall, this chapter is meant to be a synoptic approach to the content of this book and thus may serve as a condensed preview useful for quick reference.

Details of the structure–property–application relations of typical and ubiquitously used biomaterials, their synthesis and processing, analyses and testing, quality assurance and regulatory affairs, specifics of their *in vivo* interaction with living tissue and an account on possible future developments and salient challenges will be provided in the remaining five chapters of this treatise. Information on standard specifications of biomaterials and current world market situation and sales forecasts are collected in Appendices A–C.

Definition of biomaterials

Biomaterials are preferentially synthetic substances designed and engineered to be in contact with biological systems. Consequently, the 1986 Consensus Conference of the European Society for Biomaterials defined biomaterials as "nonviable materials used in a medical device, intended to interact with biological systems" (Williams, 1987). A more recent extended definition states that "a biomaterial is a substance that has been engineered to take a form which, alone or as part of a complex system, is used to direct, by control of interaction with components of living systems, the course of any therapeutic or diagnostic procedure, in human or veterinary medicine" (Williams, 2009).

As implied by the above definitions, biomaterials science is highly interdisciplinary as it encompasses elements of materials science, chemistry, physics, biology, medicine, and tissue engineering. Since *biomaterials* are, by definition, synthetic inorganic metallic or ceramic, or organic polymeric structures, they do not include renewable *biological materials* obtained from natural sources such as wood, plant fibers, hides, sinew, bone, and ivory, including hyaluronic acid, collagen, chitosan, gelatin, fibrin, cellulose, alginate, and silk fibroin. However, increasingly hybrid biomaterials/biological materials are being developed, preferably by biomimetic routes that contain natural biological materials as well as biologically derived polymers such as poly(lactic acid) (PLA) and poly(glutamic acid) (PGA) and their copolymers. Hence, the distinction

https://doi.org/10.1515/9783110619249-001

between biomaterials and biological materials, so clear and logical in the past, becomes progressively blurred.

Care should be exercised when defining a biomaterial as biocompatible, since this labeling term is application-specific (see below). This means that a biomaterial that behaves biocompatible or is otherwise suitable in one specific medical application may not be useful in others.

1.1 Biomaterials and the concept of biocompatibility

1.1.1 Role of biomaterials

Biomaterials are designed to interact with biological systems to replace a part or a function of the human body in a safe, reliable, economic, and physiologically and esthetically acceptable manner.

Modern biomaterials are destined to continue to make a large impact as building blocks of prosthetic devices in cardiovascular, orthopedic, dental, ophthalmological, and reconstructive surgery, including tissue engineering scaffolds, surgical sutures, bioadhesives, and drug and gene delivery vehicles. As aptly expressed by Ratner et al. (2013), "the compelling human side to biomaterials is that millions of lives have been saved, and the quality of life improved for millions more" by applying biomaterial to alleviate or remedy a plethora of medical problems. This is an ongoing process as new materials, devices, and applications are emerging with an ever-increasing regularity and rate. Conceivably, there is no end in sight of these developments that have emerged as a substantial economic driving force in our modern world.

1.1.2 A universe of biomaterials

Today, research and development of biomaterials have reached a level of involvement and sophistication comparable only to electronic materials. The reason is obvious since a large proportion of an aging population worldwide relies on replacement or repair of body parts ranging from artificial dental roots and teeth, intraocular lenses (IOLs), and alveolar ridge augmentation to hip and knee endoprostheses. Osseoconductive ceramics and coatings comprising calcium phosphates and bioglasses not only interact with the body by stimulating osseointegration but will be resorbed and, in time, transformed to calcified osseous tissue. In this case, the dominating biorelevant mechanism is *bonding osteogenesis* characterized by a chemical bond between implant and bone. In contrast, bioinert materials such as alumina, zirconia, and carbon and also some metals such as titanium or tantalum react to the host bone by *contact osteogenesis* characterized by a direct contact between implant and bone. Biotolerant materials such as bone cement [Poly(methylmethacrylate) (PMMA)] and stainless steels and CoCrMo alloys will be accepted by the body but develop an implant interface characterized by a layer of connective tissue between implant and bone, thus resulting in *distance osteogenesis*. A plethora of polymeric biomaterials that exist have been designed and engineered to act as low-friction insets of acetabular cups used in total hip of knee arthroplasty, as materials for biodegradable medical devices, as surgical suture materials, or as carriers for drug or gene delivery.

Table 1.1 provides an impression of the multifarious universe of biomaterials used in clinical application today.

Table 1.1: Examples of metallic, ceramic, and polymeric biomaterials, their biological behavior, and typical medical applications.

Material	Biological behavior	Typical medical application
Metals		
Commercially pure titanium	Bioinert but triggers FBR	Acetabular cups
Ti6Al4V, Ti13Nb13Zr, Ti6Al7Nb, Ti12Cr, β-Ti alloys such as Ti12Mo6Zr2Fe	Bioinert but triggers FBR	Shafts for hip implants, spinal fixation rods, many others
CoCrMo alloy	Biotolerant	Femoral balls and shafts, knee implants, cardiovascular stents, others
Austenitic stainless steel (AISI 316L, AISI 304)	Biotolerant	Osteosynthetic devices, hip and knee implants, cardiovascular stents, others
Mg alloys (AZ31, AZ91, AM50, ZK60, WE43)	Bioresorbable	Osteosynthetic devices, cardiovascular stents
Tantalum	Biotolerant/ bioactive (?)	Bearing post of dental prostheses; as porous tantalum for tumor reconstructive surgery, spine fusion
Zirconium	Bioinert (?)	As Ti13Zr and Zr2.5Nb for joint prostheses
NiTi	Biotolerant but potentially allergenic	Spinal correctors, internal fixator for long bones, surgical staples; dental infibulation wires, guide wires for catheters
Ceramics		
Alumina	Bioinert	Femoral balls, inserts of acetabular cups, dental roots, heart valves, bone screws

Table 1.1 (continued)

Material	Biological behavior	Typical medical application
Zirconia (partially stabilized zirconia; tetragonal zirconia polycrystal)	Bioinert	Femoral balls, dental veneers, cell carriers, tooth inlays and crowns
Dental porcelain	Bioinert (?)	Artificial teeth, crowns, inlays, onlays, tooth veneers
Titanium oxide	Bioactive	Heart valve components, anti-microbial surfaces, self-cleaning surfaces, dental implants
Hydroxylapatite	Bioactive	Bone cavity filler, ossicular chain implants, hip implant stem coatings, dental root coating, bone scaffolds
Tricalcium phosphate (β-TCP)	Bioresorbable	Bone replacement, cavity fillings
Tetracalcium phosphate	Bioresorbable	Bone and dental cements
Bioglasses	Bioactive	Bone replacement, ear implants, dental implants, pseudoarthrosis repair, maxillofacial reconstruction, others
Pyrolytic carbon (turbostratic graphite)	Bioinert (?)	Heart valve components
Diamond-like carbon	Bioinert	Antiwear coatings, micro-electromechanical systems
Calcium carbonate (calcite, aragonite, vaterite)	Bioresorbable	Bone tissue scaffolds, bone graft replacement, drug delivery
Calcium sulfate (plaster of Paris, hemihydrate)	Biotolerant/ bioactive	Treatment of osseous defects and nonunion fractures (pseudoarthrosis)
Polymers and composites		
Poly(methylmethacrylate)	Biotolerant	Implant fixation for cemented prostheses, bone cement, hard IOLs
Ultra-high molecular weight polyethylene	Biotolerant	Insets of acetabular cups
Cross-linked polyethylene	Biotolerant	Insets of acetabular cups
PLA	Biodegradable	Biodegradable medical devices (e.g., screws, pins, rods, and plates)
PGA	Biodegradable	Tissue engineering, suture material

Table 1.1 (continued)

Material	Biological behavior	Typical medical application
PLA/PGA copolymer	Biodegradable	Surgical suture material, drug delivery systems, tissue engineering
Poly(methacrylate)/ poly(ethylene glycol) copolymer	Biodegradable	Controlled delivery of insulin, intelligent material
Poly(ethylene oxide)/ poly(butylene terephthalate) copolymer	Biodegradable	Artificial skin, scaffold for tissue engineering
Acrylate/methacrylate copolymers	Biotolerant	Foldable IOLs
Acrylonitrile/butadiene/styrene terpolymer	Biotolerant	Parts for infusion systems and respiratory devices
Polyurethanes	Biotolerant	Heart ventricles, vascular access devices, dialysis devices, surgical tubings, wound dressing, drains
Silicones	Biotolerant	Mammary prostheses, hand and foot joint implants, soft IOLs, tubing, catheters, drains
Poly(glutamic acid)	Biodegradable	Drug and gene delivery vehicle

FBR, foreign body reaction; IOLs, intraocular lenses;
PLA, poly(lactic acid); PGA, poly(glycolic acid).

1.1.3 Hybrid nanocomposites

Since most of the history of natural sciences, inorganic materials and organic substances have been neatly separated and researched in discrete domains of chemistry, physics, material science, and engineering, today the interface between nonbiological and biological materials appears to become increasingly blurred. An entirely novel discipline has emerged to account for this merger: biomimetics (see box) that attempt to develop and describe synthetic assemblies aimed at mimicking biological systems, that is, living systems.

Biomimetics
The term "biomimetics" derives from the Greek words βιος (life) and μιμηση (to imitate). It has been defined as the examination of Nature, her models, systems, processes, and elements to emulate or take inspiration from in order to solve human problems (Benyus, 1997). This and attending to future needs are major driving forces for biomimetics, in particular with regards to biomedical and environmental applications, and diagnostic and therapeutic ("theragnostic")

avenues to treat diseases that have borrowed concepts, materials, and physiological processes from Nature (Jelinek 2013). Novel bioinspired technologies include the development of structural and self-healing materials, design of powerful adhesives and optically active films, deployment of advanced biosensor and – actuators, as well as utilization of biomorphic mineralization routes to deposit quantum dot semiconductor nanoparticles for targeted drug delivery. Self-assembled monolayers produced by bioinspired mineralization processes promise to revolutionize future biomedical applications.

During the last few decades, there has been substantially increased interest in nanocomposite materials for biomedical application. Nanocomposites consist of a combination of two or more nanomaterials, frequently inorganic biomaterials in combination with biological materials including proteins, peptides, lipids, and bacterial and viral matter. By this approach, it is possible to manipulate mechanical properties, such as strength and elastic modulus of the composites, to become closer to those of natural tissue including bone. Hybrid nanocomposites are, for example, hydroxylapatite-chitosan-carbon nanotubes, hydroxylapatite-collagen coatings, or hardystonite-chitosan scaffolds deposited by low-temperature deposition processes such as electrophoretic deposition or sol–gel spin coating. The combination of nanosized bioactive glass particles and biopolymers leads to nanocomposites with enhanced osseoconductive, osteogenic, antibacterial, and angiogenic properties and salient characteristics required for bone tissue engineering.

Hybrid biological–inorganic materials are being researched for their advantageous functionality including their ability to mimic typical properties of natural biological materials (Ivanova et al., 2014a). The burgeoning field of biomimetics (see box) abounds with examples how to engineer the properties and applications of simulated natural tissue to fill the conceptual gap between chemistry, materials science, and biology (Jelinek, 2013). For example, a comprehensive account on hybrid calcium orthophosphate–polymer biomaterials has been published by Sergey Dorozhkin (2018) recently.

An instructive example of such developments is presently accelerated research into protein- and peptide-assisted biomineralization aiming to dental enamel repair and reconstruction, bone formation by collagen-mimicking peptide/amphiphile fiber-induced mineralization of hydroxylapatite (Luk and Abbott, 2002), or design of silica-based scaffolds using collagen fibers as template. These research activities include the synthesis of exceptionally strong and tough nacre ("Mother of Pearl") analogs from alumina and PMMA to simulate the function of the molecular structure of aragonite that in natural nacre is interspaced with "shock-absorbing" polysaccharide layers (Launey et al., 2009). Furthermore, bottom-up synthesis of artificial cell membranes (Ruiz-Hitzky et al., 2010) provides a bioinspired route toward emulating essential natural biological structures. Monomeric silica can be covalently coupled to lipids, forming self-assembled vesicles in water akin to natural liposomes. Polymerization of such silica-lipid structures results in spherical silica constructs, the so-called cerasomes (*cera*mic lipo*somes*). These stable rigid spherical capsules may find application in separation technology, drug or gene delivery, and development of primary cells responsible for skeletal homeostasis, and expression of osteoblasts and osteoclasts (Ha et al., 2018).

Other examples pertain to the synthesis, by a biomimetic route, of biological–inorganic virus-like nanoparticles that are being bioengineered as vehicles for high-contrast functional imaging, such as particles comprising iron oxide cores applicable for magnetic resonance imaging (MRI) (Jelinek, 2013).

As a consequence of these very recent developments, the definition of biomaterials "*nonviable* materials used in a medical device, intended to interact with biological systems" stated by the 1986 Consensus Conference of the European Society for Biomaterials (Williams, 1987) may need to be redefined to account for the hybrid nature of novel bioinspired biomaterials.

Bioceramic materials are playing an increasingly important role in the quest to develop biomedical devices with improved properties and functionalities. Nature has provided ceramic biomaterials such as calcium carbonates or calcium phosphates with structures essential to their host organism, for which they provide varying functions and features. By evolutionary optimization over geological time-scales, many organisms have developed specific material properties as yet unparalleled in man-made materials. Wolf et al. (2019) have identified three key challenges that must be overcome to successfully emulate advanced biomimesis of ceramic materials: (i) temporal control of mineralization, (ii) spatial control of mineralization by self-assembling material scaffolds and, eventually, (iii) control of mineralization by nonclassical crystallization.

Along these lines, the processes of biomineralization are grounded on complex interaction of inorganic solids and organic templates. Following Stephen Mann (1996, 2001), biomineralization, that is, formation of hydroxylapatite-based bones and teeth of vertebrates, calcitic egg shells of birds, and calcitic or aragonitic shells of mussels, can be viewed as a sequence of four consecutive steps. Initially, the organism provides an appropriate reaction volume during a *supramolecular preorganization* step. Frequently, mineralization takes place in a macromolecular network with properties akin to a gel-like structure. Within this network, there exist specific areas the molecular properties of which trigger, in a second step, the deposition of biominerals via *molecular surface recognition*. The third step is characterized by crystal growth during which size and orientation of the crystals are being controlled by the structure and molecular organization of the organic matrix template (*vectorial regulation*). The simplest form of vectorial regulation is provided by the spatial limitation of the reaction volume within which the process of biomineralization occurs. More precisely, vectorial regulation is the result of structuring of the extracellular matrix (ECM). Finally, mesoscopic and macroscopic structural developments of the products of biomineralization lead to more complex, hierarchically organized solids such as bone, teeth, and shells (*cellular construction*).

1.1.4 Four generations of biomaterials

Historically, the general function of biomaterials had been the replacement of diseased or damaged tissues. About 40 years ago, researchers suggested fully crystalline synthetic hydroxylapatite as a suitably biocompatible, but essentially bioinert, material for incorporation in the human body, together with other bioinert ceramics such as high-purity alumina and zirconia. This was the advent of *first-generation biomaterials*, selected to be as bioinert as possible, thereby minimizing the risk of rejection and formation of scar tissue at the interface with host tissues In the next step, bone-like, that is, Ca-deficient, hydroxyl ion-depleted defect hydroxylapatite was introduced as a *second-generation* bioactive, that is, bone growth–supporting (osseoconductive) biomaterial. Among its first application were plasma-sprayed coatings for dental implants, followed by coatings of the stem of hip endoprostheses to improve implant integration with the surrounding bone. These applications are still standard today. Bioactive glasses provided, for the first time, the valid alternative of *second-generation biomaterials* that revealed tight interfacial bonding of an implant with host tissues. *Third-generation biomaterials* such as Bioglass® designed for tissue regeneration and repair use pronounced gene activation properties (Hench and Polak, 2002). By biomimetic techniques, layers of calcium phosphate can be generated and applied to implant coatings biologically functionalized with the osteogenic agent BMP-2. The ultimate aim is to develop a slow-release delivery system for osteogenic agents to promote local bone formation around dental or orthopedic implants and thereby speeding up their osseointegration.

Based on the electrophysical behavior of cells, *fourth-generation biomaterials* that are able to monitor extra- and intracellular electric processes and to tailor them to control the cell network's response to external stimuli have been recently proposed (Ning et al., 2016). Bioelectric signals originating in the activity of ion channels (Figure 1.1) are known to regulate cell behavior in terms of cell numbers (proliferation and apoptosis), positions (orientation and migration), and identities (differentiation) (Cohen et al., 2014).

Suitable fourth-generation biomaterials may include conductive polymers such as poly(pyrrole), poly(vinylidene fluoride), piezoelectric ceramics such as barium titanate or lithium tantalate, transition metal-substituted calcium hexaorthophosphates (Heimann, 2017) and carbon-based materials such as graphene (Kuzum et al., 2014). Conceivably, electronic monitoring and control systems based on electrically active biomaterials could be integrated with the human body to act as powerful diagnostic and therapeutic agents that are able to (i) manipulate cellular

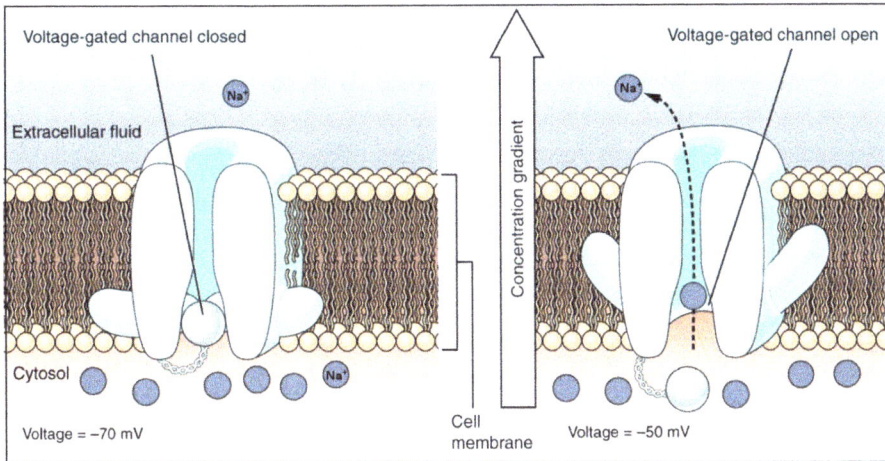

Figure 1.1: Voltage-gated ion channels are pore-forming proteins embedded in the phospholipid bilayer cell membrane. The functioning of ion channels includes establishing a resting membrane potential, shaping action potentials, and other electrical signals by gating the flow of ions across the cell membrane, controlling the flow of ions across secretory and epithelial cells, and regulating cell volume. © Permission granted under Creative Commons Attribution License (CC BY 3.0).

bioelectric signals required for tissue regeneration and (ii) monitor cellular responses to external cues and allow for communication with the host tissues (Ning et al., 2016).

1.1.5 Biocompatibility

Regardless of its origin and chemical and physical nature, any foreign material incorporated into a living organism has to abide by certain properties that will guarantee that there is no negative interaction with the living tissue. Hence, one of the most important properties of any biomaterial is its so-called *biocompatibility*. Over the last decades, the definition of this term has undergone several paradigmatic shifts. Today, biocompatibility is widely seen in a holistic context based on an understanding that it is not meant to be an individual material property per se but relates to various complex interactions on the cell and tissue levels the material is subjected to. Hence, a systemic approach is required that leads to the modern view that biocompatibility refers to the ability of a material to perform with an appropriate host response and is thus confined to a specific application. Accordingly, biocompatibility is neither a single specific material property nor a single biological phenomenon but is meant to describe a collection of properties and processes that involve different but interdependent interaction mechanisms between inorganic material and living tissue. Consequently, the notion of biocompatibility encompasses not only the required biological safety of a material as assessed and certified, for example, by the ISO 10993:1-12 norms (see Appendix A), but also the physicochemical characteristics, the design, sterilization procedures, and even packaging of a biomedical device. The historical development of understanding of biocompatibility and an account on its perceived meaning today has been described in much detail in Chapter 4.3.1.

The multifarious nature of biocompatibility discussed earlier requires ordering. This is commonly being done by defining the character and the severity of interaction of biomaterials with living tissue, using a scale of increasing biocompatibility as follows (Figure 1.2 and Table 1.1):

- *Incompatible materials* are those materials that release to the body substances in undesirable, that is, generally toxic concentrations and/or trigger the release of antigens that may cause strong immune reactions ranging from simple allergies to inflammation to septic rejection associated with severe health consequences.
- *Biotolerant materials*, in contrast, release substances by a slow dissolution mechanism in nontoxic concentrations that may lead to only benign tissue reactions or weak immune responses that cause formation of giant cells or phagocytes. These materials are neither bioactive nor bioinert during long-term residence in the body and include surgical austenitic stainless steels (AISI 316L, AISI 304) or bone cement consisting of PMMA.

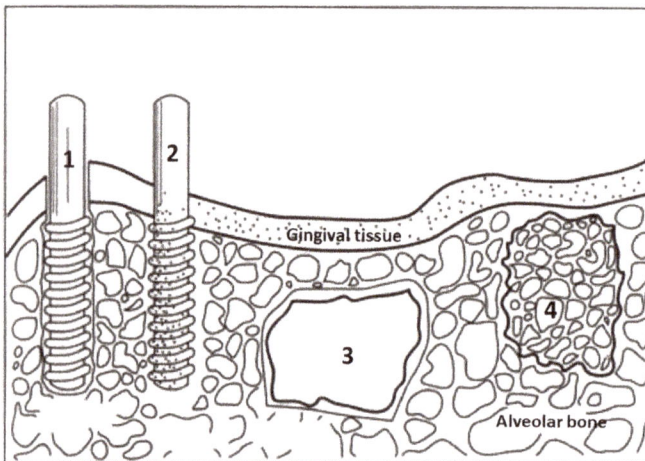

Figure 1.2: Classification of biomaterials according to their degree of biocompatibility. Bioinert (1), bioactive (2), surface-active (3), and bioresorbable (4). (1) Single crystal alumina dental implant; (2) hydroxylapatite coating of dental implant; (3) surface-active Dioglass®, (4) bioresorbable tricalcium phosphate (TCP). Image adapted from Heness and Ben-Nissan (2004). © Permission granted under Creative Commons Attribution 3.0 License Australia.

– *Bioinert materials*, although not releasing any toxic constituents, do not show positive interaction with the living tissue either. As a response of the body to these materials, usually a nonadherent capsule of connective tissue is formed around the bioinert material that in the case of bone remodeling manifests itself by a shape-mediated *contact osteogenesis*. Through the bone–material interface, only compressive forces will be transmitted ("bony on-growth"). Typical bioinert materials are titanium and its alloys, ceramics such as alumina (Figure 1.2, 1), zirconia, and some polymers such as ultra-high-molecular-weight polyethylene (UHMWPE), as well as carbon including highly bioinert diamond-like carbon (DLC).

– *Bioactive materials* show positive interaction with living tissue that in some cases also includes differentiation of immature stem cells toward bone cells. In contrast to bioinert materials, there is strong chemical bonding to the bone tissue along the interface, which is thought to be triggered by the adsorption of bone growth-mediating proteins at the biomaterials surface. Hence, there will be a biochemically mediated strong *bonding osteogenesis*. In addition to compressive forces, to some degree, tensile and even moderate shear forces can also be transmitted through the interface ("bony in-growth"). Typical bioactive materials include calcium phosphates such as hydroxylapatite (Figure 1.2, 2) as well as surface-active bioglasses (Figure 1.2, 3).

– *Bioresorbable materials* dissolve more or less completely within the human body and are slowly replaced by advancing tissue such as bone or cartilage. Common examples of bioresorbable materials are magnesium, iron, zinc, tricalcium phosphate (Figure 1.2, 4), and several polymers including PLA and PGA and their copolymers. Soluble materials such as calcium carbonate (calcite, aragonite, vaterite) and calcium sulfates such as gypsum and hemihydrate are other bioresorbable materials that are being utilized in special applications.

Within the context of continuously evolving understanding of the cellular aspects of complex biomaterials–tissue interaction (see Chapter 4), it should be emphasized that the somewhat vague term *biocompatibility* as defined earlier is slowly making way for the notion of *bioacceptance* that shifts the attention from the material to the biological environment. Accordingly, the term bioacceptance involves two diametrically opposed end points: *biointegration*, meaning that the implant material and the surrounding living structures form a continuum with stable, low-grade interaction, and *biopassivation*, meaning that the material is hardly recognized as foreign by surrounding extracellular fluid (ECF; see box) or tissue, and that its clandestine behavior can persist for clinically meaningful periods of use.

Fluids in the human body
There are two types of fluid in the human body, intracellular (ICF, ~67% of total) and extracellular (ECF, ~ 33% of total) fluids. The volume of ECF is roughly 20% of the body mass. The ECF can be further subdivided into *intercellular* fluid (~26% of total fluid) comprising lymph and transcellular fluid (aqueous humor of the eye, synovial fluid of the joints, cerebrospinal fluid) and *intravascular* fluid (blood plasma, ~7% of total fluid). The ECF provides the medium for the exchange of substances between the ECF and the cells. Thus, the ECF consists of two functional components, plasma and lymph as a delivery system, and interstitial fluid as a medium for water and solute exchange with cells. Main cations in the ECF are Na^+ (136–151 mmol), K^+ (3.4–5.2 mmol), and Ca^{2+} (1.4–1.5 mmol). Main anions are Cl^- (99–110 mmol), HCO_3^- (22–28 mmol), and HPO_4^{2-} (0.8–1.4 mmol). The composition of the ECF is maintained by several homeostatic mechanisms that regulate the pH and the sodium, potassium, and calcium concentrations, as well as the volume of body fluid, and blood glucose, oxygen, and carbon dioxide levels.

Figure 1.2 shows a sketch of the various interactions of biomaterials with living tissue and the type of response allowing different means of achieving attachment of implants to the muscular–skeletal system (Heness and Ben-Nissan, 2004).

1.1.6 Osseointegration

As with biocompatibility, the related term of *osseointegration*, coined by Per-Ingvar Brånemark in 1976, has undergone a conceptual change from the notion of a direct contact between implant materials and bone without any intermediate soft connective tissue layer separating implant and bone (Albrektsson et al., 1981), to osseointegration seen as a foreign body reaction (FBR) whereby interfacial bone is formed as a defense reaction to shield off the foreign material implant from the tissue (Donath et al., 1991; Albrektsson et al., 2017). This paradigmatic shift opens up new avenues for better understanding the complex mechanisms guiding the integration of an implant into the human body. It also involves the notion that titanium is not bioinert but in reality is identified as a foreign body material by the defense mechanisms of the human body. Thus, evolution of metal implants progressively shifted the focus from adequate mechanical strength to improved biocompatibility and absence of toxicity, bacterial infection, and biofilm formation, as well as modulation of inflammation and, finally, fast and complete osseointegration (Spriano et al., 2018).

Fixation of an implant in the human body is a dynamic process that remodels the interfacial zone between the implant and living tissue at all dimensional levels, from the molecular up to the cell and tissue morphology level, and at any time scale from the first second up to several years after implantation. This situation is schematically represented in Figure 1.3. The logarithmic length and time scales indicate the nonlinear nature of this complex dynamic process. While immediately following the implantation, a space filled with biofluid exists adjacent to the

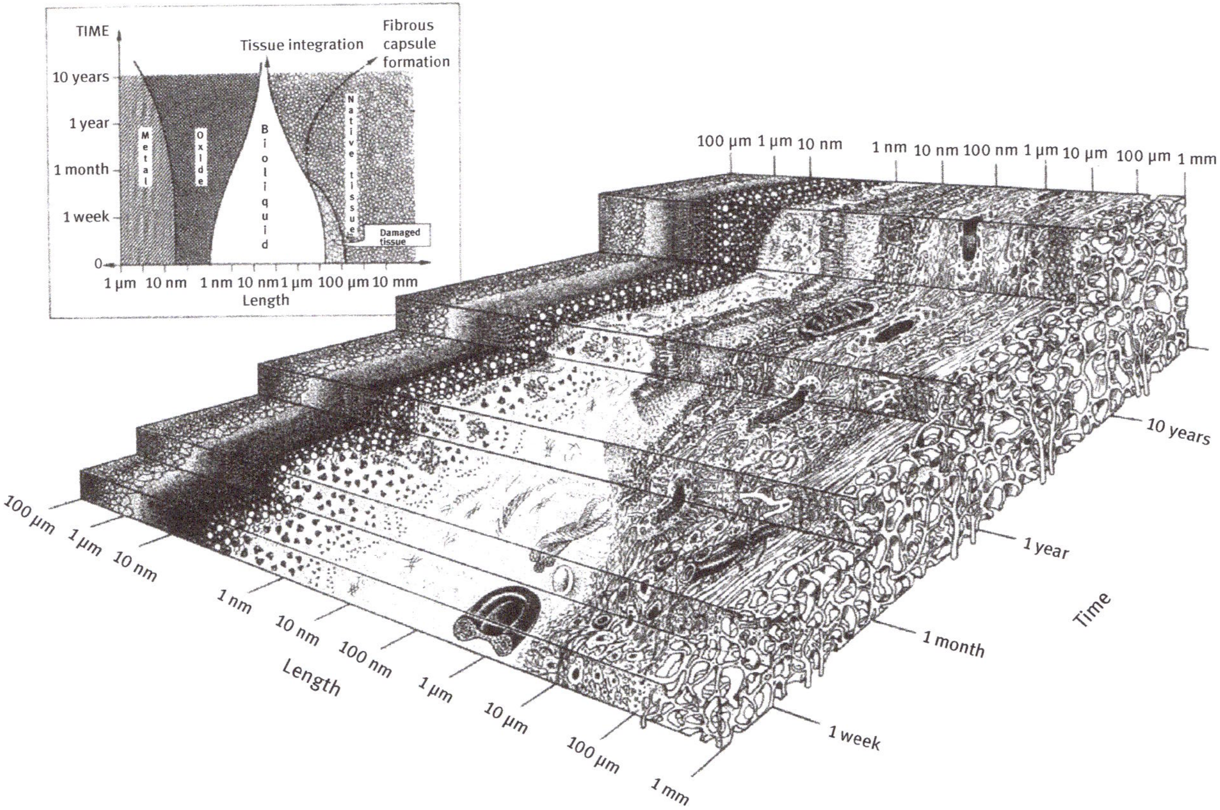

Figure 1.3: Dynamic behavior of the interface between a metallic implant (left) and bony tissue (right) (Kasemo and Lausmaa, 1991). © With permission of Toronto University Press.

implant surface, with time a layer of proteins only several nanometers thick will be adsorbed at the native titanium oxide surface covering the titanium alloy surface. Ideally, a strong bond will be created between implant and tissue. However, sometimes connective tissue is being formed at the interface resulting in a fibrous tissue capsule that prevents osseointegration (see inset) and will cause implant loosening. To prevent this undesirable situation, calcium phosphate coatings several tens or hundreds of micrometers thick can be applied by a plethora of surface coating techniques (Heimann and Lehmann, 2015).

The composition of the bioliquid shown in the inset of Figure 1.3 changes with time in response to chemical adsorption processes of molecules that mediate bone formation and also to transport reactions by outward diffusion of titanium atoms or ions through the thin oxide layer and concomitant inward diffusion of oxygen to the metal–oxide interface.

This diffusion process is aided by lattice defects in the titanium oxide such as grain boundaries, isolated vacancies, or vacancy clusters, and interstitial atoms indicated at the right side of Figure 1.4. This cartoon traces the various steps of ingrowth of alveolar bone into a dental root screw made from titanium (Brånemark, 2016). On the right side of the cartoon, the polycrystalline titanium surface is shown, covered by a thin layer of native titanium oxide, and a layer of adsorbed bone-growth supporting proteins, for example, glycosaminoglycans. These proteins will support osseoinduction that is characterized by proliferation of cells and their differentiation toward bone cells, revascularization, and eventual gap closing. The

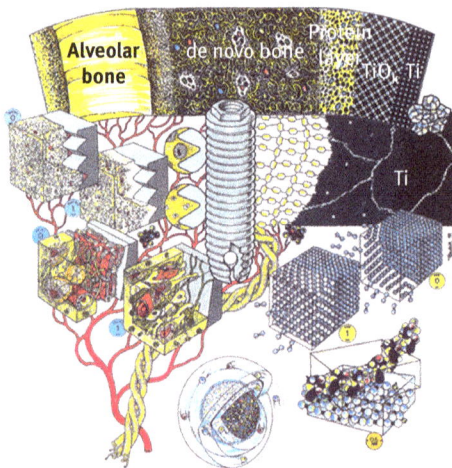

Figure 1.4: Cartoon showing the complexity of osseointegration of a dental root screw made from titanium alloy. The compartments of anchorage are composed of continuous remodeling and reorganizing interfacial tissue and titanium at the cellular as well as at the molecular levels. © 2016 Associated Brånemark Osseointegration Centers.

left side of Figure 1.4 shows how newly formed bone matter grows continuously into the threads of the dental root screw without any mediating interlayer, thus attesting to the bioactive nature of native titanium oxide film. Below this, the triple-helical strands of collagen are shown as well as the revascularization of bone tightly conforming to the geometry of the screw. The top part of the cartoon displays, from the right, the undisturbed titanium implant, the (native) titanium oxide layer, the adsorbed bone-growth mediating protein layer, the de novo bone matter, and the existing alveolar bone bed (yellow, left). The newly grown bone fills the operation-induced gap from both sides, that is, gap closure happens from the site of cortical bone as well as from the implant surface.

Definition of biomedical terms related to bone formation

- *Osseointegration* involves an FBR, whereby interfacial bone between an inorganic implant and living bone is formed as a defense reaction to shield off the implant from tissue (Albrektsson et al., 2017).
- *Biocompatibility* refers to the ability of a material to perform with an appropriate host response in a specific application. Biocompatibility is neither a single specific property of materials nor a single phenomenon but is meant to describe a collection of properties and processes that involve different but interdependent interaction mechanisms between inorganic materials and living human or animal tissue.
- *Osseoconductivity* refers to the ability of biomaterials such as hydroxylapatite to support the in-growth of bone cells, blood capillaries, and perivascular tissue into the operation-induced gap between implant body and existing (cortical) bone bed.
- *Osseoinductivity* refers to the ability to transform undifferentiated mesenchymal precursor stem cells into osseoprogenitor cells that precede endochondral ossification. The process relies crucially on the osseoinductive action of noncollagenous proteins. Silicon and calcium ions released from bioglasses can stimulate the production of proangiogenic and osteogenic factors such as insulin-like growth factor (IGF)-2, thereby providing a potentially promising strategy to enhance neovascularization and resultant bone formation.
- *Osteosynthesis* is the reduction and internal fixation of a bone fracture with implantable devices that usually consist of a metal. It aims at bringing the fractured bone ends together and immobilizing the fracture site until healing has taken place.
- *Osteolysis* is the process of resorption of bone matrix by osteoclasts. The term specifically refers to a pathological process occurring close to a prosthesis, caused either by an immunological response or changes in the structural load of bone (see Wolff's law).

A material is considered osseoinductive if (i) it mineralizes *in vivo*, that is, forms a hydroxylapatite-like layer at the surface of the material, (ii) it is sufficiently porous with (iii) pores large enough to allow the in-growth of blood vessels and cell transport into the core of the material, and (iv) the blood supply is insufficient to keep physiological levels of calcium and/or phosphate ion concentrations (Bohner and Miron, 2019). These authors ventured to propose a mechanism to explain why materials devoid of calcium phosphates such as metals and polymers are sometimes osseoinductive, why intrinsic osseoinduction is so slow, and why ectopic bone formation happens first in the core of implanted materials. They demonstrated that intrinsic osseoinduction is the result of calcium and/or phosphate depletion, and proposed reasons why not only the material surface composition but also the material volume and geometric architecture (e.g., roughness, porosity, pore size, fractal nature) play decisive roles in the process of ossification. They also pointed out similarities between intrinsic, that is, material-driven osseoinduction and trauma-related, that is, material-free heterotopic ossification.

1.1.7 Medical devices and medical equipment: definitions

As far as the clinical application of biomaterials in medically useful structures is concerned, generally accepted WHO and FDA definitions distinguish between medical devices and medical equipment. A *medical device* comprises "an article, instrument, apparatus or machine that is used in the prevention, diagnosis or treatment of illness or disease, or for detecting, measuring, restoring, correcting or modifying the structure and function of the body for some health purposes. Typically, the purpose of a medical device is not achieved by pharmacological, immunological or metabolic means." Following the FDA classification, it has become a standard for registration in the European Community (EC) to classify all medical devices according to the Medical Device Directive (MDD) into three classes according to their inherent risk for patients. The MDD (Council Directive 93/42/EEC of 14 June 1993 concerning medical devices, OJ No L 169/1 of 1993-07-12) was intended to harmonize the laws relating to medical devices within the European Union. In 2016, it has been reissued under the label Medical Device Regulation.

Long-term implants such as hip and knee prostheses, heart valves, vascular stents, pacemakers, breast implants and many others are to be registered into class III. Whereas class I devices present little or no risk to their users, class IIa and IIb devices present some limited risks and thus are subject to performance standards. In contrast, long-term implants are to be registered into class III. They involve some unreasonable risks and consequently require premarket approval prior to their widespread distribution and extensive safety and effectiveness testing prior to commercial sale. Figure 1.5 shows the medical device classification.

In contrast, *medical equipment* involves "medical devices requiring calibration, maintenance, repair, user training and decommissioning – activities usually managed by clinical engineers. Medical equipment is used for the specific purposes of diagnosis and treatment of disease or rehabilitation following disease or injury; it can be used alone or in combination with any accessory, consumable or other piece of medical equipment. Medical equipment excludes implantable, disposable or single-use medical devices" (WHO, 2018). Many aspects of medical equipment and products are described, standardized, and regulated by ISO 14971:2012, and the issue of the third edition EN ISO 14971:2019 is imminent.

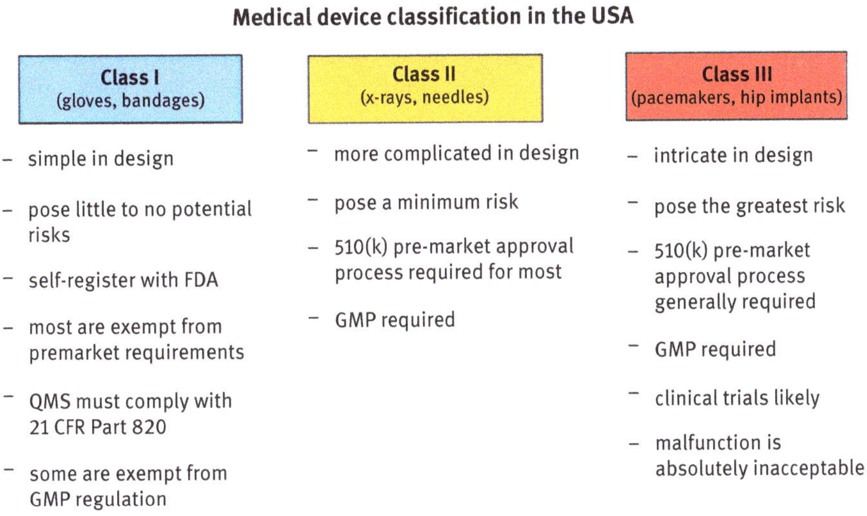

Medical device classification in the USA

Class I (gloves, bandages)	Class II (x-rays, needles)	Class III (pacemakers, hip implants)
– simple in design	– more complicated in design	– intricate in design
– pose little to no potential risks	– pose a minimum risk	– pose the greatest risk
– self-register with FDA	– 510(k) pre-market approval process required for most	– 510(k) pre-market approval process generally required
– most are exempt from premarket requirements	– GMP required	– GMP required
– QMS must comply with 21 CFR Part 820		– clinical trials likely
– some are exempt from GMP regulation		– malfunction is absolutely inacceptable

Figure 1.5: Medical device classification in the USA. FDA, Food and Drug Administration; QMS, Quality Management System; CFR, Code of Federal Regulation; GMP, Good Manufacturing Practice.

1.1.8 Bioinspired materials

Novel biomaterials beyond those mentioned earlier are expected to sprout from the extremely wide range of biomimetic functional materials. However, as stated by Jelinek (2013), although the scope and potential of practical application of such materials are broad, examples of implementation of actual and tangible clinical applications with near-term commercial horizons are rare on account of many hurdles that still need to be tackled. This is manifest in the reluctance of biomedically oriented companies to invest large sums in designing and developing materials at a time when stringent safety and standardization requirements, validation regulations, increasing number of litigation cases, and compatibility issues are important factors counteracting the growth of the biomaterials market. This situation has severe repercussions on return of investment expectations (see Chapter 5, Appendix B).

While some applications of biomimetic materials in tissue engineering, catalysis, and biosensing and bioimaging have already been deployed or are rather close to commercialization, other more ambitious applications such as bioelectronic fourth-generation biomaterials, photonic devices, or magnetic data storage systems and biologically inspired computers based on artificial neural networks are much farther off. For example, attempts are being made to design novel logic computing system of digital comparators. Such state-of-the-art logic systems enable their functionality with large-scale input signals, providing new directions toward prototypical deoxyribonucleic acid (DNA)-based logic Boolean operations and thus promoting the development of advanced logic molecular computing (Adleman, 1998; Benenson et al., 2004; Bonnet et al., 2006; Zhou et al., 2018), DNA-based digital information storage capability (Church et al., 2012; Dahlman, 2019), and DNA bar coding (Savolainen et al., 2005).

Recently, the concept of biomimicry has been extended to nonmedical realms to cover bioinspired construction technologies, kinematic structures, geopolymers, earth construction, as well as complexity and adaptability science (Bezerra et al., 2019).

1.2 A brief history of biomaterials

Although during the entire human history, medical treatments and surgical operations occasionally involved the insertion of foreign materials into the human body, for example as artificial teeth made from bone or ivory, Plaster of Paris for filling bone cavities caused by tuberculosis or dressing of wounds inflicted on the battlefield, biomaterials as we know them today have evolved largely only since the 1920s (Ratner et al., 2013). Biomaterials are employed for fracture fixation of bone, articulation of moving joints, oral implantology, trauma management of any kind, treatment of osteosarcoma and multiple myeloma, neural, and cardiovascular engineering, as well as transport matrices for modern drug and gene delivery, and scaffolds for tissue engineering.

It is a truism that most biomaterials in use today have either originated from entirely different fields of application in aerospace, manufacturing, automotive, construction, or consumer industries or have been discovered serendipitously while searching for materials that could withstand the demanding while highly aggressive environment within the human body. A telling example relates to the invention and further development of artificial heart valves made from pyrolytic graphite, a material that had been originally researched as confining matrix to encapsulate UO_2 nuclear fuel pellets for use in advanced gas-cooled nuclear pebble-bed reactors (PBR) (see Chapter 1.2.2.1).

Whereas the content of the present treatise relates predominantly to synthetic inorganic materials used in medical applications, some notion will be devoted to hybrid inorganic–organic nanocomposite structures synthesized preferentially by innovative biomimetic routes. However, although much of current research into biomaterials is focused on such bioinspired and biohybrid structures designed to emulate natural pathways toward useful medical devices, the examples quoted below are not intended to be exhaustive but just aim to broaden the view and to indicate the directions future research will likely take (see Chapter 6).

1.2.1 Metallic biomaterials

Metallic biomaterials have found many applications as medical implants and functional biomedical devices, with austenitic surgical stainless steels, titanium alloys, and CoCrMo alloys employed in permanent biostable metallic implants (Ivanova et al., 2014b). Whereas these traditional metallic materials enjoy favorable mechanical properties, in particular high compressive and tensile strengths, fracture toughness, and corrosion and fatigue resistance, today biodegradable Mg and Zn alloys as well as NiTi shape-memory materials are gaining increasing attention (Niinomi, 2010; Chen and Thouas, 2015).

1.2.1.1 Bioinert/biotolerant metals

Stainless steels

Over 100 years ago, *low alloyed steel* plates were first used as implanted osseosynthetic fixtures to mend broken bones. Among the first devices was the 1886 Hansman bone plate that was bent at the end to protrude through the skin. The plate was attached to the bone by long-shanked screws that projected outside the soft tissues. This was followed by steel plates developed in 1895 by Sir William A. Lane (1856–1942), and in short succession similar internal fracture fixation devices were introduced by Albin Lambotte (1909) and William O. Sherman (1912), who used vanadium steel, the first metal alloy specifically developed as a biomaterial.

However, in all these cases, strong corrosion and insufficient strength of the material rendered approaches toward longer lasting devices largely unsuccessful. Only the introduction, in 1924, of 18% chromium–8% nickel stainless steel (today's AISI 304), and subsequently, addition of small amounts of Mo to yield AISI 316 *austenitic stainless steel* changed the situation for the better. These novel metal alloys were noticeably more corrosion resistant in contact with aggressive body fluid and mechanically stronger than the vanadium steel used hitherto.

Metallic biomaterials
Metallic biomaterials are engineered systems designed to provide internal support to biological tissues. Based on their high strength, fracture toughness, and resilience against corrosion, they are copiously used for replacement of large and small joints, dental implants, orthopedic osseosynthetic fixation, and cardiovascular stents. Metallic biomaterials may have bioinert (Ti, Ti alloys, CoCrMo alloy), biotolerant (austenitic stainless steels) or biodegradable (Mg, Zn, and Fe alloys) functionality.

The discovery and basic property development of stainless steel (see box) happened essentially between 1900 and 1915. However, as with many other important discoveries, it was the accumulated effort of several individuals that already began in the 1820s when the famous British all-round scientist Michael Faraday (1791–1867) observed that iron–chromium alloys showed remarkable resistance to attack by several acids. In 1906, the Frenchman Leon Guillet (1873–1946) published a study on Fe–Ni–Cr alloys in which the basic metallurgical structure of today's 300 series of austenitic stainless steels was foreshadowed. In 1912, Benno Strauss and Eduard Maurer, working for Friedrich Krupp AG in Germany, filed a patent for an austenitic stainless steel with the tradename Nirosta®. In 1924, William H. Hatfield (1882–1943) invented what today is AISI 304, then called Staybrite 18/8. Soon the potential of stainless steel for medical application was realized, and in the 1940s, the German surgeon Gerhard Küntscher (1900–1972) developed the intramedullary steel nail based on AISI 316. In the 1950s, reduction of the carbon content of AISI 316 from 0.08% to 0.03% was achieved, thus marking the advent of AISI 316L stainless steel with even higher corrosion resistance. This metal alloy is still one of the mainstays in orthopedic fracture devices as well as hip and knee endoprostheses.

Stainless steels
Stainless steels are, respectively, binary iron–chromium and ternary iron–chromium–nickel alloys. There are ferritic [body-centered cubic (bcc) stacking, α-iron type; 12–30% Cr], martensitic (body-centered tetragonal stacking, 12–17% Cr, 0.15–1.0% C) and austenitic [face-centered cubic (fcc) stacking, γ-iron type; 16–25% Cr, 7–20% Ni] stainless steels. Austenitic steels are defined by their fcc crystal structure, that is, a cubic unit cell with one atom at each corner and one atom located on each face of the cube. The unit cell contains close-packed slip planes {111} that, together with the slip direction <110>, account on permutation for 12 slip systems, causing high mechanical formability of these steels. Austenitic steels receive their stainless characteristics and thus, high corrosion resistance from a thin, self-healing chromium oxide surface film. Cold working of austenitic steels can improve their hardness, stress resistance, and strength. The most widely used standard type of the 300 series is all-purpose AISI 304, containing 18% Cr and 8% Ni, followed by AISI 316, containing on average 17% Cr, 12% Ni, 2% Mo and 0.03–0.07% C. So-called surgical steels are austenitic AISI 316 steel for bone fixation screws and hip endoprostheses, and martensitic AISI 440 and AISI 420 steels used for surgical instruments.

CoCr alloys

In addition to stainless steels described earlier, other metal solutions exist such as application of CoCr alloy used in arthroplasty for stems and femoral balls of endoprosthetic hip implants. This alloy was first synthesized by Elwood Haynes (1857–1925) in the early 1900s by fusing together cobalt and chromium metals. Under the name Stellite®, CoCr alloys with a nominal composition of Co30Cr5W (Stellite®6) have been used in various fields where high wear resistance was required. In the early twentieth century, the alloy was first used in medical tool manufacturing, and in 1960, the first CoCr prosthetic mitral heart valve was implanted that happened to last over 30 years on account of its high wear resistance. The high corrosion resistance of CoCrMo is related to formation of a thin surface layer consisting mainly of chromium(III) oxide, Cr_2O_3, akin to the protection mechanism prevalent in stainless steels.

In the 1940s and 1950s, orthopedic implants were predominantly fashioned from first-generation CoCrMo cast alloy ISO 5832–4, austenitic stainless steel AISI 316L (ISO 5832–1B), as well as the commercially pure titanium forge materials (ISO 5832–2) and Ti6Al4V alloy (ISO 5832–3). At the beginning of the 1970s, these materials were supplemented by the CoNiCrMo forge alloy ISO 5832–6. Furthermore, orthopedic implants as well as material for reconstruction of skull and jaw bones have seen the application of additional cobalt-based alloys such as wrought Vitallium (ISO 5832–5) and Phynox (40Co20Cr20Fe14Ni6Mo; ISO 5832–7), now recognized as a high entropy alloy (see Chapter 2.1.1.2) At the end of the 1970s, several CoCrMo alloys were developed using powder metallurgical techniques that in their chemical composition were identical to the CoCrMo cast alloy ISO 5832–4 (Semlitsch, 1984; Mani, 2016; Antunes and de Lima, 2018).

Whereas basic CoCr alloys were applied in biomedical application since the 1930s, it was only after 1990 that second-generation CoCrMo alloys were used for metal-on-metal hip joint replacements owing to their excellent corrosion resistance, biocompatibility, and superior strength. Today, CoCrMo alloys are used to manufacture artificial joints including hip and knee implants, dental partial bridge work, and several other applications including cardiovascular stents. However, today CoCrMo hip replacements have suffered a sharp decline due to reduced biocompatibility related to excessive wear and concurrent accumulation of nanosized cytotoxic corrosion products in tissues surrounding the implant site.

Titanium

In the 1940s, Richard T. Bothe et al. (1940) suggested the application of titanium as suitable for surgical implants. Previously, titanium had shown superior inertness in seawater and thus, expectations rose that its corrosion resistance could also prevail in the chloride-rich ECF environment of the human body. Subsequently, Maurice Down introduced several titanium orthopedic fracture healing devices such as bone plates and fastening screws that stood up well to their task. Further developments culminated in the application of Ti6Al4V for endoprosthetic devices (see Chapter 2.1.1.4). Ti6Al4V is a superior mechanically strong and corrosion resistant alloy that had previously widely been utilized in the aerospace, marine, power generation, and offshore industries. Ti6Al4V derives its resistance to corrosion from formation of a nanometer-thin dense protective surface layer of titanium dioxide that, in addition, confers pronounced osseointegrative properties owing to its perceived osseoinductivity. However, despite its high corrosion resistance, release of alloying metal ions from the Ti6Al4V substrate to the surrounding living tissue was and still is a potential risk since such ion release has been found to cause hepatic degeneration in animals as well as impaired development of human osteoblasts. Heavy metal ions, in particular vanadium, are thought to affect negatively the transcription of ribonucleic acid (RNA) in cell nuclei and, in addition, influence the activity of enzymes by replacing Ca or Mg ions at binding sites. In addition, aluminum ions were found to be neurotoxic. Hence, the metabolic action of aluminum and vanadium ions released even in small quantities from the Ti6Al4V implant and their interference with normal biochemical functions of the human body are important intervention to consider. Another drawback relates to the high elastic modulus of Ti6Al4V beyond 100 GPa that may cause pronounced stress shielding (see box), that is, causes loading stress during walking to be transmitted through the implant rather than the surrounding cortical bone. Consequently, at present, developments are on the way to replace high modulus ($\alpha + \beta$)-type Ti6Al4V by low modulus ($E < 50$ GPa) β-type Zr-containing alloys such as Ti13Nb13Zr (TNZ) or Ti29Nb13Ta4.6Zr (TNTZ), thus avoiding suspected neurotoxic aluminum and, in particular, confirmed cytotoxic vanadium release (Niinomi et al., 2016). Such novel β-type Ti alloys with elastic moduli approaching that of cortical bone (10–30 GPa) tend to prevent or at least reduce stress shielding that may lead to bone resorption by osteolysis and impaired bone remodeling. A thorough account on these activities can be found in Chapter 2.1.1.4.

Stress shielding
Stress shielding happens when metallic implants such as bone plates, screws, and, in particular, endoprosthetic hip implants are used for stabilizing bone fractures and for arthroplastic surgery, respectively. The much higher modulus of elasticity of the implant material (~114 GPa for Ti6Al4V) compared to that of cortical bone (~18 GPa in compression) results in osteolytic bone mass loss in response to decreased physiologic loading of the bone. This means that the load during movement is transmitted through the high modulus metallic implant instead of the low modulus bone. Indeed, bone must be loaded in tension to remain healthy and strong. According to Wolff's law, bone in a healthy person will remodel in response to the load it is placed under. Therefore, if the loading on a bone decreases, it will become less dense and weaker because there is no stimulus for continued remodeling required to maintain bone mass.

Raabe et al. (2007) have provided a promising strategy for a theory-guided bottom-up design of β-Ti alloys for biomedical applications using a quantum mechanical approach in conjunction with supporting experiments. Parameter-free density functional theory calculations (see box) are used to provide theoretical guidance in selecting and optimizing Ti-based alloys with respect to three constraints: (i) the use of nontoxic alloying elements, (ii) the stabilization of the bcc β-phase at room temperature, and (iii) the reduction of the elastic modulus compared to the existing Ti-based alloys (Raabe, 2019).

Density functional theory (DFT)
First principles (ab initio) DFT calculations allow to predict and calculate material behavior on the basis of quantum mechanical considerations, without requiring higher order parameters such as fundamental material properties. In contemporary DFT techniques, the electronic structure is

evaluated using a potential acting on the system's electrons. This DFT potential is constructed as the sum of external potentials V_{ext} that are determined only by the structure and the elemental composition of the system, and an effective potential V_{eff} that represents interelectronic interactions. Thus, a problem for a representative supercell of a material with n electrons can be studied as an n-member set of Schrödinger-like equations that are also known as Kohn–Sham equations (Kohn and Sham, 1965; https://wikipedia.org/wiki/Density_functional_theory).

Reduction of stress shielding by designing implants with low modulus of elasticity as achieved in materials with high porosity has been accomplished by additive manufacturing of titanium implants using thermal spraying technology (Fousová et al., 2017). Bulk titanium with properties akin to human bone to be used for bone augmentations was manufactured from titanium rods sprayed onto a thin substrate wire. The resulting material showed a porosity of about 15% that caused a significant decrease of Young's modulus to the bone range and provided rugged topography for enhanced biological fixation. The compressive strength yielded 628 MPa and the elastic modulus was as low as 11.6 GPa in compression and 5.4 GPa in bending mode, sufficiently low to prevent stress shielding.

Historically, before the advent of titanium, other more or less biocompatible metals were utilized to fashion large joint implants, predominantly the stems and femoral balls of hip endoprostheses as well as parts of knee prostheses. The earliest recorded attempts at hip and knee replacements were carried out in Germany in 1891 by Themistocles Gluck (1853–1942) who used ivory to replace the femoral head and attached it with nickel-plated screws, plaster of Paris, and glue (Wessinghage, 1995; Brand et al., 2011; Figure 1.6).

Marius N. Smith-Petersen (1886–1953) used, in 1939, a hollow hemisphere of Vitallium, a 65Cr30Co5Mo alloy (see above) that could fit over the damaged femoral head of the hip joint with the objective to stimulate cartilage regeneration on both sides of the Vitallium joint. In 1940, the US surgeon Austin T. Moore (1899–1963) performed the first metallic hip replacement likewise based on the cobalt–chromium alloy Vitallium. Other more successful designs of hip endoprosthetic implants were introduced, in 1946, by the French surgeon Robert Judet (1909–1980), followed, among others, by the advanced Austin-Moore (Chillag, 2016) and Thompson prostheses. However, the modern artificial hip joint owes everything to the 1962 invention of Sir John Charnley (1911–1982). The groundbreaking-cemented Charnley design consisted of three parts: (i) a one-piece stainless steel femoral stem and head, (ii) an acetabular cup with polyethylene (originally Teflon®) inset, both of which were fixed to the bone and (iii) PMMA (plexiglass®) bone cement (Figure 1.7). With the advent of the Charnley hip endoprosthesis, the road was opened for further improved implant designs based on Ti6Al4V alloy that make total hip arthroplasty (THA) one of the most often performed and in their clinical outcome highly successful operations today (Figure 1.14).

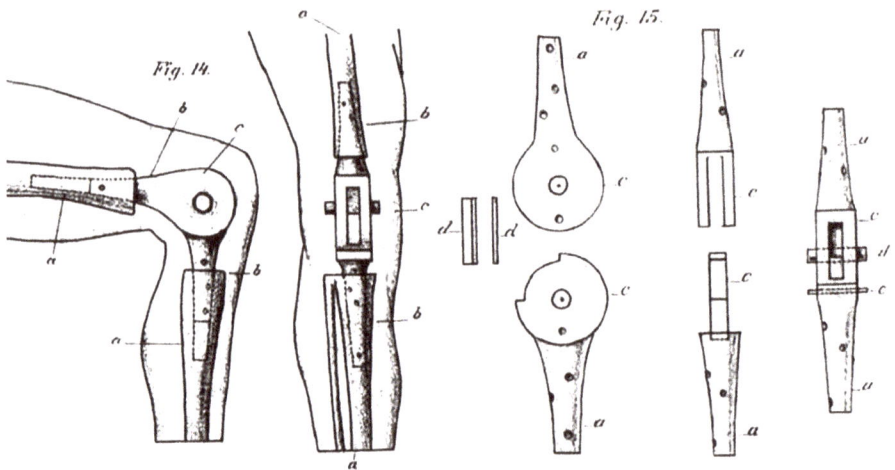

Figure 1.6: Themistocles Gluck's knee joint prostheses based on ivory (Gluck, 1891; Wessinghage, 1995).

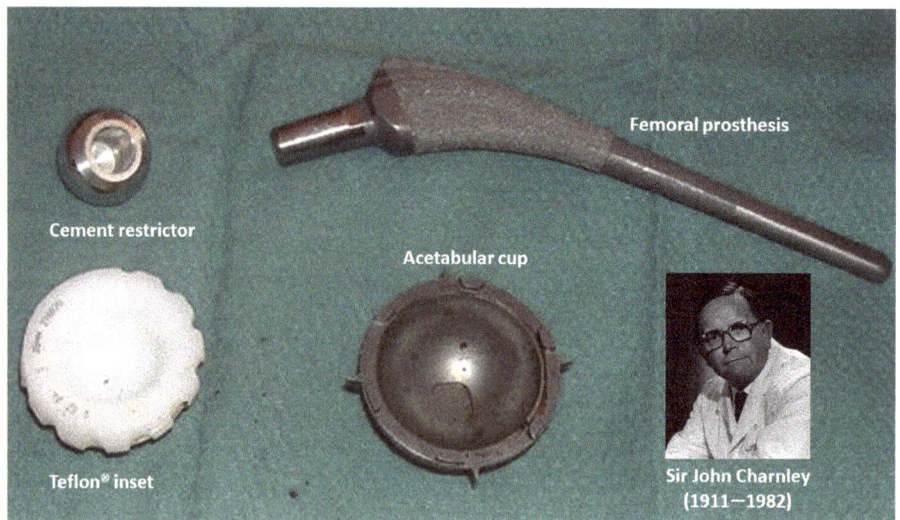

Figure 1.7: Sir John Charnley's hip prosthesis. © Permission granted under Creative Commons Attribution License (CC BY 3.0).

Tantalum

In 1997, porous tantalum (Cohen, 2002) has been approved for use in orthopedic surgeries by the US Food and Drug Administration (FDA) on account of its demonstrated excellent biocompatibility with physical, mechanical, and tissue in-growth properties conducive for enhanced osseointegration and superior structural integrity. Indeed, there are suggestions that porous tantalum not only has remarkable osseoconductive properties but, in some orthopedic domains, it also outperforms the remainder of metallic biomaterials. However, although tantalum, based on animal studies, has previously been considered an excellent biomaterial for soft tissue reattachment surgery, this soft tissue in-growth does not reliably tolerate the high mechanical loads that are generated in the clinical setting. Furthermore, recent laboratory evidence suggested that tantalum may in fact directly inhibit fibroblast development, limiting the potential for mature collagen fibrillogenesis (Gee et al., 2016).

Noble metals

Over 5,000 years ago, Egyptians ingested gold for bodily, mental, and spiritual purification, and well-being as gold leaf was considered the skin of the gods. They also used it in early dental applications such as dental bridges and prosthetic appliances. During the Renaissance, the Father of Iatrochemistry, Paracelsus (1493–1541) appeared to have designed medication ("aurum potabile") based on gold to relief arthritic pain, a treatment later rediscovered by implanting gold particles subcutaneously near an inflamed joint to ease pain of rheumatoid arthritis (for an early account, see Abraham and Himmel, 1997).

Currently, there are a number of direct applications of gold in medical devices including lead wires for pacemakers and gold-plated stents used to inflate and support arteries in the treatment of heart disease. In addition, gold finds application in novel drug delivery microchips that contain drug-filled reservoirs covered, sealed, and protected by thin gold membranes. Catheters and stents incorporating platinum components such as marker bands and guidewires assist the surgeon during guiding the catheter to the treatment site. Recently, the properties of gold compounds have been of interest as potential cancer treatments.

Prior to the introduction of antibiotics, colloidal silver was used widely in hospitals and has been known as a bactericide for at least 1,200 years. The ancient Greeks used silver vessels to keep water and other liquids fresh, and the Romans stored wine in silver urns to prevent spoilage. Before the advent of modern germicides and antibiotics, it was known that disease-causing pathogens could not survive in the presence of silver. Consequently, silver was used in dishware, drinking vessels, and eating utensils. In the early 1800s, doctors used silver sutures in surgical wounds with very successful results, and silver leaf was used to combat infection in wounds sustained by troops during World War I. Silver nitrate eye drops traditionally have been used to prevent eye infections in newborns. However, because these drops often cause the baby's eyes to be irritated, today most hospitals use erythromycin ointment instead. Although the antimicrobial properties of silver have been known since ancient times, today there has been a renewed interest in using silver as an antibacterial agent to fight nosocomial infections, replacing antibiotic drugs in many medical indications.

1.2.1.2 Biodegradable metals

Whereas traditionally metals with sufficient mechanical strengths, corrosion resistance and biocompatibility have been widely used in orthopedic fields, their limitations include stress shielding (see box above) that induces refracturing due to bone mass loss by osteolysis and, thus, may compromise the bioefficacy of these bioinert metallic devices, especially in patients suffering from osteoporotic conditions. In addition, a second surgery will often be required to remove the implanted hardware to avoid potentially adverse effects after fracture healing. Hence, novel biomaterials for a new generation of orthopedic devices are required that are less prone to stress shielding, and are biodegradable, biocompatible, and do not inhibit bone in-growth after surgical implantation.

Magnesium

Increasingly, magnesium alloys such as AZ31, AZ91, AM50, ZK60, or WE43 have come into focus that are considered materials suitable for biomedical applications, owing to their specific mechanical properties as well as their excellent biocompatibility and confirmed osseoconductivity, that is, their ability to support growth of bone cells to anchor solidly an implant to the surrounding bone (Witte, 2010; Poinern et al., 2012). A very early clinical report was released in 1878 by Edward C. Huse, a British physician who used magnesium wires as ligature for bleeding vessels. As early as 1907, a magnesium alloy was used to secure a bone fracture in the lower leg (Lambotte, 1932). Subsequently, magnesium alloys were explored on and off for various cardiovascular, musculoskeletal, and general surgery applications. Although most patients experienced the occurrence of subcutaneous gas cavities caused by rapid implant corrosion and associated release of hydrogen gas, few patients reported any pain, and almost no infections were observed during postoperative follow-up.

Today, biodegradable magnesium-based metal implants are breaking the long cherished paradigm in biomaterial science to apply only highly corrosion resistant metals in biomedical contexts (Sezer et al., 2018; Han et al., 2019). Although all magnesium alloys release corrosion products in contact with ECF, these products do not seriously affect human metabolic pathways. On the contrary, magnesium is an essential trace element in the human body that is necessary as a cofactor in many different enzymatic reactions, thus playing an important role in energy metabolism. Serum magnesium levels can be maintained normal due to the dynamic absorption and excretion equilibrium of magnesium ions in the human body, through gastrointestinal absorption and renal excretion. In addition, magnesium ions are required for effective heart, muscle, nerve, bone, and kidney function, and play an important role in the regulation of ion channels, DNA stabilization, enzyme activation, and stimulation of cell development and proliferation. Indeed, many studies have indicated that magnesium corrosion products cause only minimal, if any, changes to the composition of blood. Most importantly, they are not known to cause damage to excretory organs, that is, liver or kidneys.

However, despite these favorable mechanical (see box) and biological properties, the *in vivo* performance of magnesium is compromised by its high corrosion rate in contact with chloride-rich aqueous solutions such as the ECF. Historically, this rapid corrosion and the associated hydrogen gas evolution were powerful show stoppers of past attempts to use Mg for biomedical applications. Nevertheless, an improved understanding of the complex corrosion mechanisms, combined with development and utilization of well-adhering protective coatings (e.g., Surmeneva et al., 2019), or formation of an reasonably corrosion resistant intermetallic $Mg_{17}Al_{12}$ layer by processes akin to pack cementation (Lu et al., 2019) are presently reviving interest and associated research effort in biodegradable Mg alloys. In parallel, research into developing novel, less corrosion-prone alloy compositions with uniform corrosion behavior are promising approaches toward a new generation of resorbable biomaterials. The corrosion mechanism of such novel alloys would yield uniform and global layer-by-layer removal of material instead of localized pitting or crevice corrosion experienced by common Mg alloys. Increasingly, suitable magnesium alloys are being utilized for specific biomedical implants such as osteosynthetic musculoskeletal bone healing devices and cardiovascular stents (Heimann and Lehmann, 2017). Recently, a novel Mg2Y1Zn0.4Zr quaternary alloy was developed that, in addition to improved corrosion resistance, during MTT (see glossary) assaying revealed suitable cytocompatibility toward murine fibroblast L-929 cell line (Li et al., 2019). Nevertheless, owing to the high electrochemical activity of Mg, numerous environmental factors including temperature, pH, and surrounding ion composition influence its corrosion behavior, thus making it unpredictable. Hence, the need of reliable *in vitro* model(s) to predict *in vivo* implant degradation is increasing (Sanchez et al., 2014).

Magnesium alloys

Favorable mechanical properties of bioresorbable magnesium alloys include low density between 1.74 and 1.81 Mg/m^3 depending on composition as well as elastic moduli (17 GPa in shear, 45 GPa in tension) that are close to those of human cortical bone (10–30 GPa). This is in stark contrast to other metals commonly used in implantology. Moreover, the fracture toughness of the most studied and used AZ31 Mg alloy is, with 18–21 MPa ·√m, roughly comparable to that of cortical bone (2–12 MPa ·√m). Again, this is much closer than that of Ti6Al4V (84–107 MPa ·√m) or stainless steel (AISI 316L: 112–278 MPa ·√m; AISI 304: 119–228 MPa ·√m). In addition, the high specific stiffness, that is, the stiffness-to-mass ratio of magnesium is between $9 \cdot 10^6$ and $26 \cdot 10^6$ $m^2 \, s^{-2}$, and thus, compares favorably to both Ti alloys and stainless steels.

Zinc alloys

There is a disadvantage of current biodegradable Mg alloys that they release hydrogen gas to the human body environment that is prone to trigger gas embolism and inflammation. Hence, other solutions are being sought including the recent development of zinc and zinc–iron alloy compositions (Mostead et al., 2018). Similar to magnesium, zinc is an important metabolic element that plays a crucial role in several regulatory pathways in physiological systems, and is bioresorbable and biocompatible, but, in contrast to magnesium, it does not produce appreciable amounts of gaseous hydrogen *in vivo* (Kafri et al., 2018). Instead, protective $Zn(OH)_2$ and ZnO corrosion products are likely to form on the metal surface not accompanied by gas evolution (Levy et al., 2017).

The efficacy, biological safety, and degradation kinetics of zinc alloyed with up to 2.5 mass% Mg and 2.5 mass% Fe were evaluated *in vivo* in a canine mandibular fracture model. The zinc alloys revealed favorable mechanical properties and biocompatibility, as well as uniform and slow degradation behavior, supporting the notion of potential clinical use of zinc alloys in osteosynthetic applications (Wang et al., 2019b).

Hot extruded Zn alloyed with 0.02% Mg showed low hemolysis rates of 0.74 ± 0.15% and strong inhibitory effect on blood coagulation, platelet adhesion, and aggregation, as well as excellent cytocompatibility and proliferation of endothelial cells (Lin et al., 2019). This suggests a potential for biodegradable cardiovascular stents.

Iron alloys

Research into biodegradable iron-based alloys is still limited even though their corrosion behavior and biocompatibility appear to be promising. Chung (2016) studied a series of 70Fe30Mg, 60Fe20Mg20Mn, and 60Fe20Mg20Zr alloys, synthesized by high-energy mechanical alloying, and deposited as thin films by pulsed laser deposition (PLD). The coatings revealed uniform corrosion properties when immersed in DMEM and were found to support growth and proliferation of murine preosteoblast cells (MC3TC-E1), human mesenchymal stem cells (hMSCs), human umbilical vein endothelial cells, and murine fibroblast cells (NIH3T3). Although iron-based materials for degradable implants are not in clinical practice yet, the research results achieved so far appear to be promising for future applications (Zivic et al., 2018) and thus deserve further study.

1.2.2 Ceramics

Ironically, the earliest evidence of incorporation of what may be called biocompatible ceramics into the human body comes from the field of forensic anthropology. A

79-mm siliceous grey stone spear point of cascade point-style embedded in the pelvic bone of the so-called Kennewick Man (about 9000 BP; Chatters, 2000) attests to the fact that this unintended ceramic "implant" had healed in and thus vividly illustrates the unique capacity of the human body to incorporate and tolerate, for a long time, foreign materials without deleterious effect on health and metabolism.

Bioceramic materials
The main types of bioceramics in clinical use today are bioinert ceramics such as alumina, partially stabilized zirconia and carbon, and bioactive ceramics based on calcium phosphates or bioglasses. On the one hand, *bioinert ceramics* are predominately used as mechanically strong, wear- and corrosion-resistant materials with excellent tribological properties and high reliability designed for femoral heads and acetabular cups of hip endoprosthetic devices but used also as spacers in revision surgery, in knee arthroplasty, and in polycrystalline and single crystal form in dental applications such as tooth root implants. On the other hand, *bioactive ceramics* such as hydroxylapatite, resorbable tri- and tetracalcium phosphates as well as surface-active bioglasses are used for their osseoconductive and, in combination with bone growth-supporting proteins, osseoinductive properties. Hydroxylapatite is predominately used as coatings for endoprosthetic hip and dental root implants, bioresorbable tricalcium phosphates find use as bone cavity filler and bone restoration material, and tetracalcium phosphate is applied in bone and dental cements. Increasingly, bioactive ceramics are used in conjunction with polymers as hybrid biological-inorganic nanocomposites for their ability to mimic salient properties of natural biological materials

1.2.2.1 Bioinert ceramics
Bioinert ceramics excel by high mechanical strength, as well as excellent corrosion, scratch and wear resistance. These properties have earned alumina, zirconia, titania, silicon nitride, graphite, and DLC important roles in designing, manufacturing, and deployment of various load-bearing implants (Ivanova et al., 2014c). However, the relatively high brittleness and limited ability to be integrated with soft and hard tissues *in vivo* are limiting factors affecting their clinical performance.

Aluminum oxide
The favorable properties of alumina have spurred its utilization as a mechanically strong and heat-, wear-, and corrosion-resistant ceramics. Besides many other specialized industrial applications, the potential and utilization of alumina as a suitable biomaterial were first suggested by Max Rock in a Deutsches Reichspatent of 1933, followed in 1966 by a patent issued to Sami Sandhaus for the use of monolithic alumina for dental and jaw implants. However, it was only after the groundbreaking contribution by Pierre M. Boutin (1972) that alumina began to be used as a suitable ceramic material for femoral balls of hip endoprostheses. Alumina was subsequently much improved in terms of its compressive strength and fracture toughness (see Table C.2, Appendix C) by painstakingly engineering its purity and ever decreasing grain size down to the nano-scale level. This development has led to orthopedic

structural ceramics products such as Ceraver-Osteal®, Keramed®, Frialit®, and finally the family of BIOLOX® ceramics by Feldmühle, later CeramTec GmbH as well as BIONIT® manufactured by Mathys Orthopädie GmbH in Bettlach, Switzerland (Willmann, 1996). The current high-end product of CeramTec GmbH is BIOLOX® delta, a chromium-containing zirconia-toughened alumina (ZTA) alloy reinforced with platelets of strontium hexaluminate ($SrAl_{12}O_{19}$) with magnetoplumbite structure as potent crack arresters and deflectors (Palmero et al., 2017).

Zirconium dioxide

Zirconium dioxide was first extracted from the mineral zircon (zirconium silicate, $ZrSiO_4$) by the German chemist Martin Heinrich Klaproth (1743–1817) in 1787, using as starting material the yellowish orange-colored, transparent gemstone jacinth (hyacinth) from Ceylon.

During the following 200 years, zirconia was considered a mere scientific curiosity without any substantial technological merits apart from limited utilization in bricks for high-temperature applications as well as special glasses with a high index of refraction. It was only in 1969 that the first scientific study of the outstanding biomedical properties of zirconia emerged, published by Corning's Helmer and Driskell. Subsequently, it was discovered by Ronald C. Garvie and Patrick S. Nicholson (1972) that by alloying zirconia with oxides such as yttria, calcia, magnesia, and others it was possible to stabilize its tetragonal modification by halting the structurally and mechanically deleterious phase transition from the tetragonal to the monoclinic phase. This discovery allowed utilizing the so-called transformation toughening of zirconia to produce ceramics with unsurpassed crack resistance, hence, somewhat misleadingly dubbed "ceramic steel" (see box). These partially stabilized tetragonal zirconia polycrystalline ceramics excel by high density, small grain size, and high purity that jointly elicit strength and fracture toughness unusually high for a ceramic material. Consequently, such ceramics were employed to fashion femoral ball heads starting by the mid-1980s (Figure 1.8) and, later, to make dental parts of all kinds including dental roots, inlays, onlays, and color-matched tooth veneers (Abd El-Ghany and Sherief, 2016).

Starting in the 1980s, besides structural and mechanical investigations of zirconia, studies on its biocompatibility moved into the limelight. This work triggered a virtual avalanche of research that used increasingly sophisticated evaluation techniques of material properties. In addition, studying the *in vitro* and *in vivo* biomedical performance of zirconia in contact with biofluid and tissues established zirconia as a viable bioceramics. Later, several other applications emerged including the use of zirconia as bond coats as well as reinforcing particles for hydroxylapatite coatings for implants. Today, a large segment of utilization of zirconia exists as color-adapted tooth veneers in dental restoration whereas its application for femoral heads of hip endoprostheses has been abandoned, owing to studies and clinical

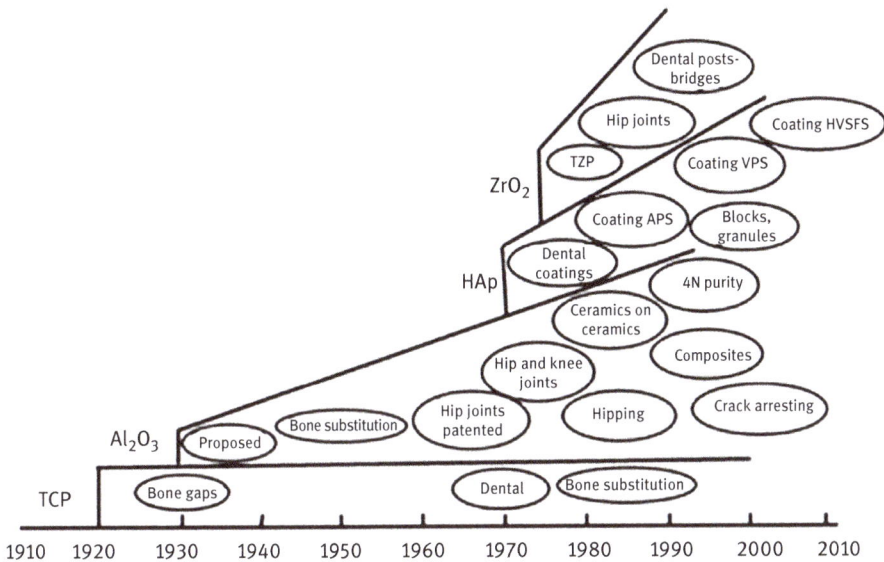

Figure 1.8: Application of bioceramics in medical devices: 100 years of history (adapted from Rieger 2001, and adjusted to current developments). TZP, tetragonal zirconia polycrystal; VPS, vacuum plasma spraying; HVSFS, high-velocity suspension flame spraying; APS, atmospheric plasma spraying (Heimann and Lehmann, 2015).

reports of adverse biological effects that include warnings issued by the British Medical Device Agency (MDA) and the FDA that steam sterilization of stabilized zirconia femoral heads may result in hydrolytic surface degradation, manifest in surface roughening, and eventual mechanical failure.

Ceramic steel?

In partially stabilized zirconia, that is, zirconia containing a certain amount of oxides such as yttria, scandia, ceria, calcia or magnesia that are known to stabilize the tetragonal modification of zirconia, fracture toughness will be increased by the so-called *transformation toughening mechanism*. A crack of critical length is able to move through the stress field but will be deflected by the particles that transform spontaneously from the tetragonal to the monoclinic modification in the vicinity of the propagating crack. The crack energy will be absorbed and the volume change will prevent the advancing crack from spreading further. This strengthening mechanisms leads to a type of ceramic material that has been dubbed "ceramic steel" on account of its drastically improved strength, resilience, and toughness.

Carbon materials

Pyrolytic graphite for biomedical use appeared on the scene in 1966, being a remarkable example of how a chance meeting of two individuals working in very different fields of science and engineering synergistically resulted in novel developments.

Originally, this material had been developed for a new generation of inherently safe nuclear fuel pellet configuration for a very high-temperature (VHT) gas-cooled PBR. Among many other researchers was Jack C. Bokros, working on encasing a 0.5-mm fissionable enriched UO_2 fuel core with a series of spherical layers consisting of a succession of porous carbon, silicon carbide, and finally, pyrolytic graphite. As it happened, Bokros read an article written by Vincent L. Gott, Cardiac Surgeon-in-Charge at Johns Hopkins Medical Institution who was working on a carbon-based paint for use in cardiovascular applications. Subsequently, the two met and eventually developed the Bokros Carbomedics On-X Bileaflet Valve based on pyrolytic graphite. This material quickly became the gold standard for heart valves owing to its superior hemocompatibility, its mechanical strength and durability, as well as its desirable low coefficient of friction. Prior to this time, artificial heart valves had been fashioned from metals such as CoCrMo and/or durable polymers such as PMMA or Nylon® that, however, were rather short-lived owing to their low hemocompatibility that resulted in blood clotting, as well as intolerable wear. Today, pyrolytic carbon still constitutes the most popular material available for artificial mechanical heart valves (see Chapter 1.3.3.3).

DLC is being investigated for use in medical applications owing to its excellent wear properties and biocompatibility. In 1953, Heinz Schmellenmeier at University College of Education, Potsdam, German Democratic Republic, first observed the formation of very thin, very hard surface films on the surface of W–Co alloy that was subjected to a gaseous discharge in an acetylene atmosphere. Subsequently, in 1971, Sol Aisenberg and Ronald Chabot published a landmark paper on ion beam deposition of "DLC" (as they coined it) in an argon atmosphere using graphite electrodes. This was, in fast succession, followed by the 1976 paper by L. Holland and Sankalpa Ojha in which the authors described the formation of DLC in butane subjected to an r.f. glow discharge, and, in 1980, by a contribution by Knut Enke et al. on films with an extremely low friction coefficient. Today, coating techniques applied to deposit DLC films are manifold and include microwave or r.f. plasma-assisted chemical vapor deposition (PACVD) as well as direct ion beam-assisted deposition (IBAD), filtered cathodic arc deposition, electron beam deposition, unbalanced magnetron sputtering, and PLD techniques (Bewilogua and Hofmann, 2014; Moriguchi et al., 2016).

In one of the first studies related to the exceptional biocompatibility of DLC films, Thomson et al. (1991) cultured mouse peritoneal macrophages and mouse fibroblasts on DLC prepared by ion beam deposition and monitored the level of lactate dehydrogenase activity as a marker of cell viability. They concluded that there was no indication of cytotoxicity whatsoever. This verdict was confirmed many times over by subsequent studies (e.g., Love et al., 2013), and today, it can be said with confidence that DCL is a superior biocompatible and, in particular, hemocompatible biomaterial.

Recently, ultra-smooth nanostructured diamond-like coatings deposited by microwave PACVD have been developed for titanium- and cobalt-based alloy implant surfaces. These coatings are characterized by a mixture of sp^3- and sp^2-bonded carbon atoms, with nanometer-sized diamond particles embedded in a quasi-amorphous carbon matrix. They combine ultrahigh hardness, low surface roughness, a very low coefficient of friction, and excellent fracture toughness and adhesion strength to titanium alloys.

Other biomedical applications of DLC include coatings for percutaneous coronary intervention devices to benefit from the unique electrical properties of DLC and hard coatings for metallic cardiovascular stents that reduce the incidence of thrombosis. And finally, the implantable human heart pump can be considered the ultimate biomedical application where coating with DLC is used for blood-contacting surfaces on account of its superior hemocompatibility.

Graphene is a single-layer, 2D-structured nanomaterial with unique physicochemical properties that include a high surface area, excellent electrical conductivity, superior mechanical strength, unparalleled thermal conductivity, remarkable biocompatibility, and ease of functionalization. Owing to these exceptional properties, graphene is receiving increasing attention in biomedical fields. Novel graphene-based developments may include oxidase-based biosensors for small biomolecules such as glucose or dopamine, proteins and DNA detection, bioimaging, drug and gene delivery, and photothermal therapy (Yang et al., 2013; Justino et al., 2017).

1.2.2.2 Bioactive ceramics

Bioactive ceramics are capable of forming direct chemical bonds with hard and soft tissues. Calcium phosphates, in particular hydroxylapatite, are the quintessential bioceramic materials, being similar to the inorganic part of the gravity-defying skeleton of all vertebrates. Bioactive ceramics do not only physically mimic the composition and structure of osseous tissue but, in addition, are able to assist various biological processes associated with osteogenesis (Ivanova et al., 2014d; Prado da Silva, 2016).

Calcium orthophosphates

For at least 250 years, calcium phosphates have been known to be associated with organic tissue such as bone, and, on account of this, were diligently researched, and eventually applied for biomedical purposes. As early as 1769, the Swedish chemists Johan Gottlieb Gahn (1745–1818) and Carl Wilhelm Scheele (1742–1786) discovered that tricalcium phosphate $Ca_3(PO_4)_2$ could be obtained by incinerating cattle bone, and they continued to isolate elemental phosphorus by reducing acid-treated bone ash with charcoal, and distilling off and condensing the escaping phosphorus vapor. The preparation of pure tricalcium orthophosphate by an alternate route was already described 200 years ago in chemical encyclopedia by Good et al. (1813) and Bache (1819). As it turned out, by the end of the eighteenth century, much research had been performed on calcium phosphates that involved the names of many renowned scientists of the time including Klaproth, Proust, Lavoisier, Vauquelin, and de Fourcroy. The nineteenth century saw increasingly important research on calcium phosphates, culminating in a series of contributions by such scientific luminaries as Mitscherlich (1844), Berzelius (1845), Fresenius (1867), Warington Jr. (1871), and Church (1873).

By the early nineteenth century, the knowledge of the presence of calcium phosphates in bone, teeth, blood, milk, and urine as well as urinary and renal calculi were solidly established. There is also faint indication that several calcium phosphate phases, now known to be important intermediaries of biomineralization, were already known, suspected or suggested early, including amorphous calcium phosphate (ACP), octacalcium phosphate ($Ca_8(HPO_4)_2(PO_4)_4 \cdot 5H_2O$) as well as dicalcium phosphate dihydrate (brushite, $CaHPO_4 \cdot 2H_2O$), and dicalcium phosphate anhydrate (monetite, $CaHPO_4$). In 1920, tricalcium phosphate (β-TCP, $Ca_3(PO_4)_2$) was suggested as a bioresorbable ceramic material for filling of bone gaps (Figure 1.2) that, however, was unable to bear extended loads.

During the first half of the twentieth century, a series of studies emerged using X-ray diffraction as a rather new important and versatile tool to assess the structural chemistry of calcium phosphates in general, and hydroxylapatite in particular. Willem F. de Jong (1926) was first to identify the structure of the calcium phosphate phase in bone as being structurally akin to geological apatite that had long been known as an important phosphate mineral. Based on their X-ray diffraction studies, Hendricks et al. (1931) concluded that animal bone consisted of carbonate apatite, $Ca_{10}[CO_3/(PO_4)_6] \cdot H_2O$, a compound thought to be isomorphous with fluorapatite. They also reported on the existence of oxyapatite, $Ca_{10}O(PO_4)_6$ that could presumably be prepared by prolonged heating of hydroxylapatite or bone at 900 °C. However, the stable existence of oxyapatite was later disputed for thermodynamic reason (De Leeuw et al., 2007; Heimann, 2015). The systematic progress of the knowledge on the chemical composition and structure of bone mineral, that is, Ca-deficient hydroxylapatite, was recently reviewed by Christian Rey et al. (2010) and Jill D. Pasteris (2016).

Considering the importance of the structure of bone as a biocomposite of Ca-deficient defect hydroxylapatite intergrown with triple helical strands of collagen I, it is not surprising that as early as about 40 years ago synthetic hydroxylapatite has been suggested by Michael Jarcho (1976, 1981) as a biocompatible artificial material for incorporation in the human body. In the next step, hydroxylapatite was introduced as a bioactive, that is, osseoconductive coating (Zhang et al., 2014; Heimann and Lehmann, 2015). Its first application was in plasma sprayed coatings for dental implants, followed by coatings of the stem of hip endoprostheses to improve implant integration with the surrounding bone (Figure 1.2). Whereas the commercially preferred deposition technique was and still is atmospheric plasma spraying (Heimann, 2008), other techniques abound including low-pressure (vacuum) plasma spraying, and most recently high-velocity suspension flame spraying as well as suspension plasma spraying, solution precursor plasma spraying, and even cold-gas dynamic spraying and aerosol deposition (Heimann and Lehmann, 2015). Hydroxylapatite is also utilized in the form of densified implants for dental root replacement, as a suitable material for filling bone cavities, as well as for fashioning skeletal prostheses and bone growth-supporting scaffolds. Biomimetic pathways toward bone tissue formation include the use of tubular amphiphilic protein fibers consisting of an arginine-glycine-aspartate tripeptide motif sequence acting as a collagen replacement that supports the oriented growth of hydroxylapatite nanocrystals and thus mimic the typical bone structure (Luk and Abbott, 2002).

Comprehensive accounts on the history of calcium orthophosphates as quintessential bioceramic materials have been provided by Sergey Dorozhkin (2012, 2013).

Bioglasses

The deepest and most interesting unsolved problem in solid state theory is probably the theory of the nature of glass and the glass transition (Anderson, 1995)

Glasses are inorganic noncrystalline solids in which "the short range-order (SRO) is preserved in the immediate vicinity of any selected atom, *i.e.* the first coordination sphere; the long range-order (LRO) characteristic of the ideal crystal is dissipated in a way characterized by diversity among different systems and by difficulty in precise description" (Kingery et al., 1976). Clearly, the structure of glass is as complicated as this definition.

Surface-active bioglasses are important to the field of biomaterials as they are the first completely synthetic materials that were found to bond seamlessly and strongly to living bone. Bioglass materials were almost single-handedly developed by the late Larry L. Hench (1971, 1972), then at the University of Florida at Gainesville, FL. In November 1969, the first member of a series of related composition was synthesized, called 45S5 Bioglass®, to indicate its composition (in mass%) of 45% SiO_2, 24.5% CaO, 24.5% NaO_2, and 6% P_2O_5 by mass (Figure 1.9; see box), whereby "45" signifies the SiO_2 mass content and "5" the 5:1 molar ratio of Ca to P.

The extraordinary story of the discovery of bioglasses has been told in Clark et al. (1976) and Hench (2006) as follows:

Constituent	45S5	45S10	45S5F	53SP4	58S
SiO_2	45	45	45	53	60
Na_2O	24.5	27.5	23	23	–
CaO	24.5	24.5	12	20	30
CaF_2	–	–	16	–	–
P_2O_5	6.0	3.0	6.0	4.0	4.0

Figure 1.9: Selected bioglass compositions.

The story of bioglass

In 1967, Hench was an assistant professor at the University of Florida (at Gainesville). At that time his work focused on glass materials and their interaction with nuclear radiation. In August of that year, he shared a bus ride to an Army Materials Conference in Sagamore, New York, with U.S. Army Colonel Klinker who had just returned from Vietnam where he was in charge of supplies to 15 M.A. S.H. units. He was not terribly interested in the radiation resistance of glass. Rather, he challenged Hench with the following: hundreds of limbs a week in Vietnam were being amputated because the body was found to reject the metals and polymer materials used to repair the body. "If you can make a material that will resist gamma rays, why not make a material the body won't resist?" Hench returned from the conference and wrote a proposal to the U.S. Army Medical R & D Command. In October 1969, the project was funded to test the hypothesis that silicate-based glasses and glass-ceramics containing critical amounts of Ca and P ions would not be rejected by bone. In November 1969, Hench made small rectangles of what he called 45S5 glass . . . and Ted Greenlee, Assistant Professor of Orthopedic Surgery at the University of Florida, implanted them in rat femurs at the VA Hospital in Gainesville. Six weeks later, Greenlee called – "Larry, what are those samples you gave me? They will not come out of the bone. I have pulled on them, I have pushed on them, I have cracked the bone and they are still bonded in place." Bioglass was born, and with that the first compositions studied. Later studies by Hench using surface analysis equipment showed that the surface of the bioglass, in biological fluids, transformed from a silicate-rich composition to a phosphate-rich structure, possibly with resemblance to hydroxyapatite (Source: Bioglass. In: World Heritage Encyclopedia™. World Public Library Association. Article ID: WHEBN0002161298).

In a landmark paper by Clark et al. (1976), the theory was advanced that an ideal implant material must have a dynamic surface chemistry that induces histological changes at the implant interface. This challenged the hitherto prevailing opinion of the function of biomaterials as a bioinert component that does not interact with tissue. Since then, several bioactive glass-based products, including small solid blocks, fine powder particles, granules, porous scaffolds, and injectable putties, have been implanted into millions of patients to repair congenital and acquired bone defects and dental deficiencies. Bioactive glasses have also shown promise in emerging applications such as angiogenesis, design of cancellous-like bone scaffolds, interfacial hard–soft tissue engineering, and controlled drug and gene delivery (Baino, 2018).

The original 45S5 Bioglass® composition has been in clinical use since 1985. Its first successful surgical application was to replace the ossicles of the middle ear to treat conductive hearing loss. Since then, bioglasses continue to be used in bone reconstruction. In addition, other applications are widespread and include dental implants, periodontal pocket obliteration, alveolar ridge augmentation, maxillofacial reconstruction, otolaryngologic applications, percutaneous access devices, spinal fusion, coatings for dialysis catheters made from silicone tubing, coatings for surgical screws and wires, cochlear implants, bone graft substitutes, bone tissue engineering scaffolds, antibacterial and antifungal application as wound healing agent, and granular filler for jaw defects following tooth extraction (e.g., NovaBone®).

Comprehensive research starting in the 1990s elucidated the complex mechanisms underlying the bioactive behavior of bioactive glasses, characterized by surface-mediated as well as solution-mediated effects (Ducheyne and Qui, 1999; Knabe et al., 2017; for details see Chapters 2.2.2.6 and 4.1.4). Both of these effects appear to enhance osteoblast differentiation. A key element of bioactive behavior of bioglass is the development of a carbonated apatitic surface layer after immersion in biological fluids rich in HPO_4^{2-} ions (Kokubo, 1993; Radin et al., 1993; Ducheyne and Qui, 1999). Previous review papers and book chapters have summarized events occurring at the bioceramic-tissue interface (Ducheyne et al., 1994; Ducheyne and Qui, 1999; Hench, 2008; Knabe et al., 2017). Collectively, the surface-mediated effects result in enhanced osteoblastic differentiation, bone matrix mineralization and thus bone tissue formation (Schepers et al., 1991; Ducheyne and Qui, 1999). Additional studies have shown that the release of ionic dissolution products of bioactive glasses such as biologically active Ca^{2+} and SiO_4^{4-} also enhances osteoblast differentiation (Xynos et al., 2000, 2001; Radin et al., 2005; Yao et al., 2005; Hench, 2008), emphasizing the synergistic nature of the solution- and surface-mediated effects of bioactive glass on osteoblast function and stimulation of osteogenesis. These processes result in rapid bone regeneration and can also be used to induce angiogenesis, offering the tantalizing potential for novel gene-activating bioglasses for soft tissue regeneration. In addition, research continues into designing and testing of thermally sprayed hydroxylapatite/Bioglass® coatings for improved cell attachment and bone regeneration.

In 2006, Larry Hench provided a comprehensive historical account on the steps toward discovery, characterization, *in vivo* and *in vitro* evaluation, clinical studies, and product development along with the technology transfer processes required to market bioglasses. In 2015, this was followed by a contribution highlighting frontiers and challenges of bioglass research and development (Hench and Jones, 2015).

Titanium dioxide

In contrast to bioinert alumina and zirconia ceramics, titanium dioxide owes its biomedical importance not to useful mechanical properties but largely to its photocatalytic and semiconducting nature (Diebold, 2003, Fujishima et al., 2008), although advantageous biological properties of titanium dioxide include hemocompatibility, superhydrophilicity, and some osseoconductivity. Hence, it may be considered bioactive to a limited extent as the very thin native oxide film formed spontaneously on titanium surfaces used for endoprosthetic applications is thought to confer, together with efficient corrosion protection, osseoconductive properties that are beneficial for bone in-growth, vascularization, and eventual, osseointegration.

In 1972, Akira Fujishima and Kenichi Honda of Yokohama National University discovered that water molecules were split on TiO_2 semiconductor electrodes under strong ultraviolet irradiation. Since this seminal discovery, the photocatalytic properties of TiO_2, in particular of its anatase modification, have been investigated extensively and developed for many successful applications, such as conversion of water to hydrogen and oxygen, conversion of CO_2 to fuel-like hydrocarbons, decontamination, and disinfection of water and air (e.g., Burlacov et al., 2007), self-cleaning surfaces of wall and floor tiles in clinical operation theaters and public swimming pools utilizing the so-called lotus effect, and antimicrobial and bactericidal biomedical materials including medical textiles.

1.2.3 Polymers

Polymers are materials most widely applied as biomaterials and thus copiously used in biomedical devices. Details of their structure, synthesis routes, processing, and applications will be reviewed in much detail in Chapters 2.3 and 3.3.

1.2.3.1 Natural polymers

The basic structural principle of polymers consists of the multiple assemblage of simple structural units (monomers) to form a three-dimensional (3D) polymeric network. This principle can be found in all biological systems, ranging from the intracellular filaments and cytoskeleton to structural proteins of the soft ECM and materials with mechanical function in ligaments or cartilage to keratin present in skin and hair. Natural polymers such as animal horn, hair, cellulose, or latex have been utilized by humans for millennia, and now have found their application in modern medicine (Ivanova et al., 2014e; Maitz, 2015). Today, nanocellulose materials are considered promising precursors of a variety of innovative applications including biomedical uses as wound dressing and materials for some medical implants.

As it were, Nature has fashioned the overwhelming majority of biological materials needed as basic building blocks for living organisms from natural polymeric structures. Indeed, proteins are natural polyamides composed of up to 20 left-handed amino acids, and the genetic molecules DNA and RNA can be considered natural polyesters of orthophosphoric acid. Proteins differ from one another primarily in their sequence of amino acids that is dictated by the nucleotide sequence of their genes, and usually results in protein folding into specific 3D conformations such as α-helices, β-sheets, and β-spirals that determine the biological activity of individual proteins. DNA and RNA are poly(nucleic acids) that alongside proteins, lipids, and complex carbohydrates (polysaccharides) are one of the four major types of polymeric macromolecules that are essential building blocks of all known forms of life, making DNA arguably the most ubiquitous biopolymer on Earth.

1.2.3.2 Semisynthetic polymers

Xyloidine and celluloid

The first semisynthetic polymers were developed during the late eighteenth century by treating organic materials such as wood or cotton with strong acids. Among the earliest developments count the work of Henri Braconnot (1777) who invented xyloidine, a precursor of collodion and nitrocellulose, by treating wood or cotton with strong nitric acid. This polymeric material may be considered the first polymer ever synthesized. Some 70 years later, Christian Schönbein (1846) developed derivatives of the natural polymer cellulose such as the semisynthetic materials celluloid (nitrocellulose treated with camphor), now considered the first thermoplastic. In the same year, Louis-Nicolas Ménard and Florès Domonto discovered that cellulose nitrate, when dissolved in ether or acetone, forms a clear, gelatinous liquid called collodion that initially was applied as surgical wound dressing. Subsequently, in 1851, Frederic Scott Archer discovered that collodion could be used as an alternative to egg white (albumen) on photographic glass plates. In the 1860s, John Wesley Hyatt commenced to produce celluloid products and, in 1878, was able to obtain a patent for injection molding of thermoplastics, although it took another 50 years before it could be commercially realized. During later years, celluloid was used as the base for photographic film that, however, was highly flammable and caused a considerable number of accidents, in particular in early movie theaters.

Subsequently, in 1865, Paul Schützenberger discovered that cellulose reacts with acetic anhydride to form cellulose acetate. In 1903, the German chemists Arthur Eichengrün and Theodor Becker invented the first soluble forms of cellulose acetate (Cellon) that was later copiously used for photographic "safety" films since it was considerably less flammable compared to its celluloid predecessor.

In 1892, a patent was awarded to the British chemists Charles Frederick Cross, Edward John Bevan, and Clayton Beadle for "artificial silk," a product based on cellulose produced and marketed later under the tradename Viscose™. The process was built on the reaction of cellulose with a strong base followed by treatment of the solution with carbon disulfide to obtain a xanthate derivative. Subsequently, the xanthate is then converted back to cellulose fibers. Earlier comparable semipolymeric fiber materials based on natural cellulose are associated with the names of Georges Audemars (1855) and Hilaire de Chardonnet (1839–1924).

1.2.3.3 Fully synthetic polymers

Historically, with a few notable exceptions, all synthetic polymers known as "plastics" were initially developed and applied in nonbiomedical fields. Consequently, their processing technology (see box; Chapter 3.3) and performance requirements reflect characteristics of so-called engineering polymers that include low density, low hardness, high flexibility, electrical insulation, and suitable corrosion resistance against many acids and alkalis but rarely their biological and biomedical functions.

The chemistry of making polymers

Step growth occurs by either *addition polymerization* or *condensation polymerization*. Addition polymerization produces polymers by simple linking of monomers *without* the cogeneration of other products. Examples for addition polymerization are, for example, polyurethanes. Condensation polymerization differs from addition polymerization in that small molecules such as water, ammonia or hydrogen chloride are cogenerated during the polymerization process. Examples are, for example, polyesters.

Chain-growth polymerization, including *ring-opening polymerization* (ROP) involves the bonding of unsaturated or cyclic monomers by opening of a double bond or a ring, respectively. This allows to synthesize polymers of the same or lower density than the monomers. This is important for applications that require constant volume after polymerization such as tooth fillings, coatings, and the molding of electrical and electronic components. In addition, radical ROP is useful to produce polymers with functional groups incorporated in the backbone chain that cannot otherwise be synthesized via conventional addition polymerization of vinyl monomers. For instance, radical ROP can make polymers with ethers, esters, amides, and carbonates as functional groups along the main chain (Billmeyer, 1984).

The mechanism of chain-growth polymerization requires catalysts that can act as radical, cationic, anionic, or coordinative (so-called Ziegler–Natta catalysts) initiators. Examples are polyethylene and polymethacrylate.

The molecular structures of the major classes of synthetic polymers are shown in Figure 1.10.

Bakelite

In 1907, the first fully synthetic polymer was invented by the Belgian-born American chemist Leo Hendrik Baekeland (1863–1944) and named bakelite, 35 years after Adolf von Baeyer (1835–1917) had already experimented with polymerization of pyrogallol and benzaldehyde. However, the "black goop" obtained by von Baeyer was discarded as useless at this time. Bakelite is a poly(oxybenzylmethylene)glycol anhydride, a thermosetting resin, formed by a condensation reaction of phenol with formaldehyde. Its electrically insulating property and thermal resistance made it a breakthrough

Polyolefins

Poly(ethylene) PE Poly(propylene) PP Poly(vinyl chloride) PVC Poly(acrylonitrile) PAN Poly(tetrafluoroethylene) PTFE Poly(ethylene glycol) PEG

Polyacrylate

Poly(methyl methacrylate) PMMA 2-Poly(hydroxyethylacrylate)

Polyamide

Nylon 6.6

Polyester

Poly(ethylene terephthalate) PET

Polyurethane

Poly(caprolactone) PCL Polyurethane PU

Polyether

Poly(ether etherketone) PEEK

Silicone

Poly(dimethylsiloxane) PDMS

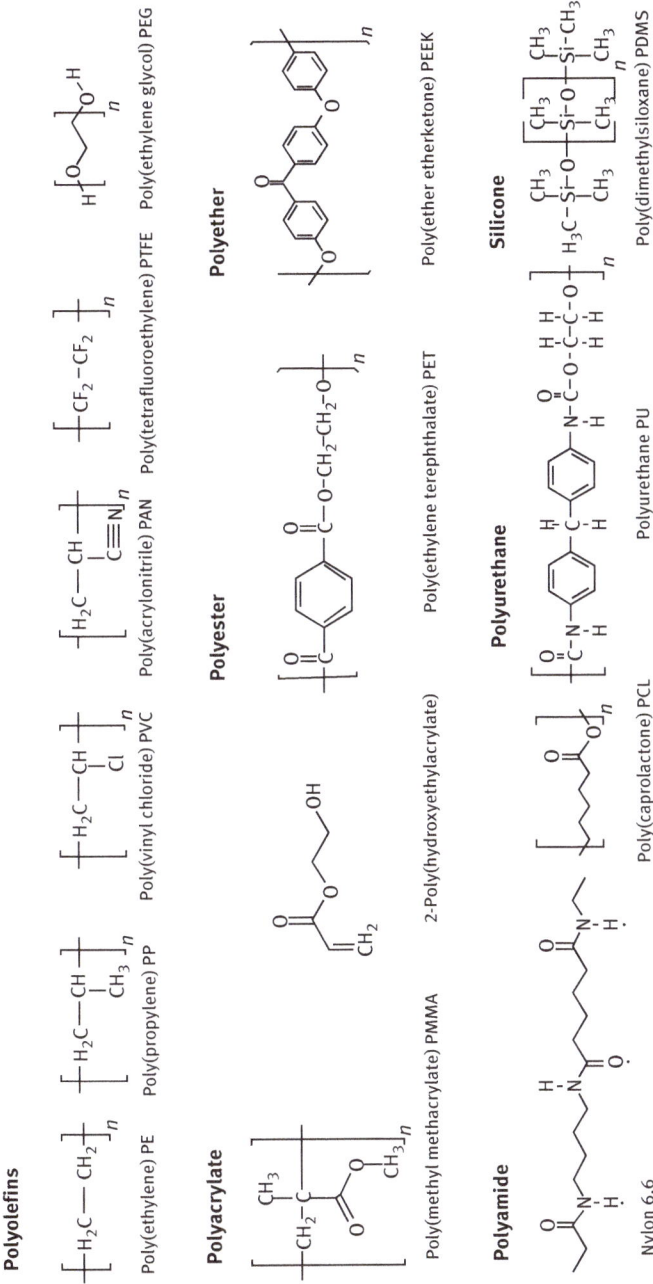

Figure 1.10: Structural formulae of the major classes of synthetic biopolymers.

technology for radio and telephone appliances, firearms, and toys, and at this time was considered "the material of a thousand years," the economic impact of which could not be overestimated as a virtual avalanche of products based on bakelite appeared on the market.

Structure and conformation of polymers

Up to this point, the new materials created were thought to be a collection of colloidal aggregates with small molecular mass, held together by a yet unspecified intermolecular force (Graham, 1861) called secondary valences (Alfred Werner, 1893) or partial valences (Knoevenagel, 1900). Despite significant advances in polymer synthesis, the real nature of the structure of polymers was not understood until the seminal work of Hermann Staudinger (1881–1965). In 1922, Staudinger proposed that polymers consisted of long chains of macromolecules held together by covalent bond within the chains. Staudinger could show that such macromolecular materials, though consisting of several thousand molecular segments, are neither aggregates nor colloids that would be held together by much weaker hydrogen bonds or van der Waals forces. Indeed, the high viscosity of macromolecules even when highly diluted as well as their stability during hydrogenation and dehydrogenation suggested the presence of strong intramolecular covalent bonds instead of weak intermolecular forces. Staudinger's theory met initially with strong resistance among the leading authorities in the field and was only gradually accepted. In 1953, Hermann Staudinger received the Nobel Prize in Chemistry for his essential contributions to the understanding of macromolecular chemistry.

Polyesters and polyamides

The 1930s saw the invention and technological development of several important polymers, their creation being associated with the name of Wallace Hume Carothers (1896–1937) who worked as a chemist for DuPont Company. He and his research group are credited with the inventions of chloroprene (neoprene, 1930), several polyester materials formed by polycondensation of diols such as ethylene glycol (1934), as well as polyamides formed by polycondensation of diamine (1935). Polyesters and polyamides were among the first polymers that have found their way into the biomedical materials toolbox. From here on, synthetic polymers were copiously applied in industry including biomedical industry. The polyesters Trevira® and Diolen® were used for artificial tendons, vascular prostheses, and antimicrobial textiles. Poly(ethyleneterephthalate), invented by J. Rex Winfield and James T. Dickson in 1940, was synthesized by polycondensation of ethylene glycol and terephthalic acid. Subsequently, this polymer was marketed under the brand names Terylene® and Dacron®, and applied as material for artificial heart valves, vascular grafts, and anterior cruciate ligament replacement. The polyamide Nylon family, originally formed by polycondensation of hexamethylenediamine and adipic acid (Nylon 6,6), is used today in a plethora of biomedical applications, including surgical suture material, for ligament and tendon repair, as balloon for catheters, and for dialysis membranes.

Poly(methylmethacrylate)

In the early 1930s, PMMA was introduced by Rowland Hill and John Crawford at Imperial Chemical Industries in England and marketed under the brand name of Perspex®. Some years later, in 1933, Otto Röhm (1876–1939) and Walter H. Bauer (1893–1968) invented and marketed PMMA under the tradename Plexiglas®. Subsequently, the polymer was commercialized by DuPont in the USA as Lucite®. PMMA found many industrial applications including its copious use as bone cement for implant fixation in various orthopedic and trauma surgeries. The first use of bone cement in orthopedics is credited to the famous British surgeon John Charnley, who, in 1958, used it in THA to attach an acrylic cup to the femoral head and to seat a metallic prosthesis within the femoral bone. This was a significant milestone in the advancement of orthopedic surgical procedures designed to repair the ambulatory hip–knee kinematics. Also, Charnley was first to realize that PMMA could be used to fill the medullary canal of long bones and thus, could easily blend with the bone morphology.

Poly(caprolactone)

Poly(caprolactone) (a.k.a. PA6, Perlon®) was developed, in 1938, by Paul Schlack at I.G. Farben corporate group in Germany to emulate the advantageous properties of Nylon 6.6 without violating DuPont's patent on its production. Being a semicrystalline polyamide, unlike most other polymers of the Nylon family, it is not a condensation polymer but is formed by ROP (see box above). Its competition with Nylon 6.6 has shaped the economics of the synthetic fiber industry worldwide. Today, poly(caprolactone) is emerging into biomedical applications because of its easy biodegradability, high biocompatibility, chemical and thermal stability, and sufficient mechanical properties. Recently, porous nanocomposite scaffolds of hydroxylapatite/poly(caprolactone) have been synthesized by thermal cross-linking of poly(caprolactone diacrylate) in the presence of hydroxylapatite and leached with sodium chloride as a porogen for bone tissue engineering application (Koupaei and Karkhaneh, 2016).

Polyurethanes

In 1937, Otto Bayer (1902–1982) of I.G. Farben corporate group in Germany invented a flexible, thermally insulating, and mechanically resilient class of polymers, called polyurethanes. Polyurethanes are produced by reacting an isocyanate containing two or more isocyanate groups per molecule ($R-(N=C=O)_n$) with a polyol containing on average two or more hydroxyl groups per molecule ($R'-(OH)_n$) in the presence of a catalyst or by activation with ultraviolet light.

First widespread use of polyurethanes in Germany was seen during World War II, when these materials were utilized to replace natural rubber that at the time was expensive and hard to obtain. During the war, other applications were developed, largely involving coatings of different kinds, from airplane finishes to resistant insulating clothing. By the 1950s, polyurethanes were being used in adhesives, elastomers, and rigid foams and, in the latter part of the same decade, for flexible cushioning foams. This development culminated, in 1959, with the utilization of polyurethane by NASA for lining of space suits.

Only in the 1970, polyurethanes made their mark in biomedical applications as general purpose tubing, surgical drapes, as well as in a variety of injection molded devices. Their most common use was and is in short-term implants. Novel applications include insulating sheaths for electric leads of artificial hearts and pacemakers, catheter tubing, feeding tubes, surgical drains, intraaortic balloon pumps, dialysis devices, nonallergenic gloves, medical garments, hospital bedding including memory foam mattresses, wound dressings, and others.

Polyurethane elastomers have molecular structures similar to those of human proteins. Protein adsorption initiating the blood coagulation cascade was found to be slower or less than found for other materials. This makes polyurethanes ideal candidates for a variety of medical applications that require adhesive strength as well as unique biomimetic and antithrombogenic properties.

With the advent of new surgical implants, biomedical polyurethanes can lead the way to eliminate some acute and chronic health challenges. Polyurethanes are popularly used in cardiovascular and other biomedical fields due to their suitable biocompatibility as well as their mechanical properties. Many of these polyurethanes have elastomeric properties that are accompanied by toughness, and tear and abrasion resistance. In 2004, the SynCardia total artificial heart with polyurethane ventricles was approved by FDA for use after 10 years of clinical trials.

Today, polyurethanes are synthesized with multiple chemistries and thus properties, including polyester-, polyether-, and polycarbonate-based polyurethanes with aromatic or aliphatic components, whereby the aromatic formulations have better stability in a biological environment. Thermoplastic polyurethanes maintain their elasticity by the mixture of hard and soft segments. Although polycarbonate-based polyurethanes have excellent stability against oxidation and biodegradation similar to poly(vinyl chloride), there are environmental concerns about release of bisphenol A with estrogen-like activity. Polyether-based polyurethanes, in particular aliphatic formulations, show rapid softening in the body, making them more comfortable for the patient. They can be used in many soft elastomeric medical applications such as indwelling catheters and vascular access devices including cuffs for artificial heart valves (Figure 1.16).

Polysiloxanes

Polysiloxanes a.k.a. silicones are elastomeric materials widely used in medical devices. In 1824, the Swedish chemist Jöns Jakob Berzelius (1779–1848) discovered elemental silicon by reducing potassium fluorosilicate with metallic potassium and, by reacting the product with chlorine, obtained tetrachlorosilane ($SiCl_4$), today still one of the precursor products of silicones. In 1863, Charles Friedel (1832–1899) and James Mason Craft (1839–1917) synthesized tetramethylsilane, $Si(CH_3)_4$ as the first silico-organic compound ever. Later, in 1871, Albert Ladenburg (1842–1911) observed that diethyldiethoxysilane, $(C_2H_5)_2(OC_2H_5)_2Si$, reacted with diluted acids to form an oily fluid with remarkable thermal stability. Frederic Stanley Kipping (1863–1949) devoted a large part of his scientific career to silane chemistry and, between 1901 and the 1930s, synthesized various silanes by Grignard-type reactions. He found that hydrolysis of chlorosilanes yielded large molecules, the polymeric nature of which was subsequently confirmed by Alfred Stock (1876–1946) who named the entire group of compounds "silanes" and synthesized various halogen silanes and siloxanes of the general formula $R_3Si\text{-}[O\text{-}SiR_2]_n\text{-}O\text{-}SiR_3$, for example, trimethylsilyloxy-terminated poly (dimethyl siloxanes) $(CH_3)_3Si\text{-}O\text{-}[(CH_3)_2SiO]_n\text{-}O\text{-}Si(CH_3)_3$.

In the 1940s, silicones were commercialized by Dow Corning in the USA as thermally highly stable and electrically resistant materials. Eugene George Rochow (1909–2002) of General Electric invented a direct method to prepare silicones from elemental silicon and methychloride. Today, silicones can be synthesized by several routes, including ring-opening synthesis of cyclosiloxanes $(R_2SiO)_n$ and subsequent polymerization to form long linear chains. These polymer chains can be easily cross-linked at ambient conditions or at elevated temperature to yield elastomers, for example, by reacting vinyl end-blocked polymers with Si–H groups in the presence of a Pt or Rh catalyst (Noll, 1968; Colas and Curtis, 2004). The biomedical uses of silicones are multifarious and include mammary prostheses, testicular implants, hand and foot joint implants, soft and foldable IOLs, and a large variety of catheters, tubing, drains, cannulas, and shunts.

1.2.3.4 Bioresorbable polymers

Among biomedically relevant biologically derived polymers that include collagen, chitosan, alginate, and hyaluronic acid, bioresorbable plastics such as PLA and PGA are high up on the agenda of research and development effort (Perale and Hilborn, 2017). Biodegradable polymers are being used as suture materials, bone fixation materials, adhesives, stents, temporary tissue substitutes, drug and gene delivery carriers, membranes, and artificial skin (Suzuki and Ikada, 2004).

Polylactides

PLA is a biodegradable thermoplastic aliphatic polyester derived from renewable resources, such as corn starch, cassava roots, or sugarcane. In 2010, PLA had enjoyed the second highest consumption volume of any bioplastics worldwide. Owing to the chiral nature (see box) of lactic acid, several distinct forms of polylactide exist including poly-L-lactic acid (PLLA), the product resulting from polymerization of L,L-lactic acid. Being able to degrade into innocuous lactic acid, PLA is used for medical implants in the form of anchors, screws, plates, pins, rods, and as a mesh. Depending on the exact type used, PLA breaks down inside the body within 6 months to 2 years. This slow degradation is desirable for orthopedic support structures, because it gradually transfers the load to the body (e.g., the bone) as that area heals and the polymer degrades.

Chirality

A chiral molecule is a type of molecule that possesses "handedness," that is, it has a nonsuperposable mirror image. The cause of chirality in molecules is frequently an asymmetric carbon atom (see below). Two mirror images of a chiral molecule are called enantiomers or optical isomers. Pairs of enantiomers are designated as right (D)- or left (L)-handed as shown in the formulae of lactic acid:

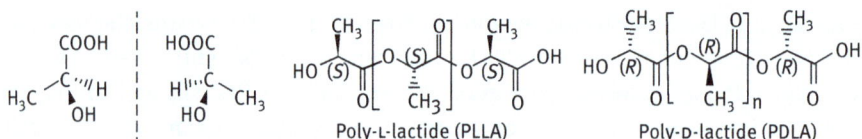

Poly-L-lactide (PLLA)   Poly-D-lactide (PDLA)

Chirality is based on molecular symmetry elements. Specifically, a chiral compound does not contain a proper axis of rotation but may contain planes of symmetry and an inversion center. Chiral molecules are always dissymmetric (lacking a rotational axis) but not always asymmetric (lacking all symmetry elements safe the trivial identity). Asymmetric molecules are always chiral. Owing to the chiral nature of lactic acid, several distinct forms of polylactides exist such as PLLA or PDLA (poly-D-lactide) (see above). The former is used as medical implants in the form of anchors, screws, plates, pins, rods, and as a mesh. PDLA is used for therapeutic drug delivery vehicles or the preparation of microparticles and resorbable polylactide scaffolds.

Recently, "empty" poly(lactide-*co*-glycolide) scaffolds were found to decrease obesity and improve glucose tolerance in mice fed a high fat diet. Histological studies indicated increased cellularity and tissue remodeling around the scaffold as well as increased expression of glucose transporter 1 and IGF-1, proteins involved in wound healing that are also able to modulate blood glucose levels. This work confirms that biomaterial implants have systemic physiological effects and suggests that there may be implications for obesity therapy in humans (Hendley et al., 2019).

Polyglutamic acid

PGA is a polymer of the amino acid glutamic acid (GA). Its poly-y-GA isomer is formed by bacterial fermentation and has a wide number of potential uses ranging from food and medicine to water treatment. Another synthetically produced isomer, poly-α-GA, is applied as a drug delivery carrier in cancer treatment and research is underway for treatment of type I diabetes and its potential use in the production of an AIDS vaccine. Among other novel applications, PGA has the potential to be used for protein crystallization, as a soft tissue adhesive, and a nonviral vector for safe gene delivery (see Chapter 1.3.6). Many medical applications, especially drug delivery, have exploited the advantageous properties of α-PGA. However, y-PGA is essentially different from α-PGA as it does not involve a chemical modification step and is not susceptible to denaturalization by proteases (Ogunleye et al., 2015).

Besides these polymers, copolymers are being developed and used in a variety of applications. For example, poly(lactide-co-glycolic acid) (PLGA) copolymers are being investigated as suitable material for tissue scaffolds to construct artificial skin or intervertebral disks (D'Este and Alini, 2018) and hybrid polyvinyl alcohol/nylon copolymer fibers warrant further preclinical investigation for the development of durable, high-performance bone tissue scaffolds. Polyamide/polyurethane block copolymers combine the flexibility of polyurethanes with the mechanical strength of Nylon®, and therefore, became the material of choice for the balloon of catheters used in angioplasty. Many novel formulations of co- and terpolymers are in the wings that promise to address salient requirements in biomedical application by synergistically combining advantageous properties of their individual components. Other important classes of synthetic polymers such as polyethers, poly(vinyl-chloride), and poly(tetrafluoroethylene) are being discussed in a comprehensive recent review by Maitz (2015). Figure 1.11 shows a selection of bioresorbable natural and synthetic biopolymers.

1.3 Selected types and applications of medical devices

The types and individual expressions of the biomedical devices fashioned from biomaterials discussed above highlight only a very small part of the wide range of devices available today even though their fabrication commands a sizeable portion of the industrial spectrum of developed and developing nations worldwide (see Table B.1, Appendix B) and thus contributes substantially to the gross domestic product. There are in excess of 5,000 different types of implant known to be incorporated in human bodies, and this number is still growing.

Although all biomaterials and biomedical products have undergone scientific research and most of them clinical testing, some of these materials may cause problems. Implant registers help to find explanations and associations of implant solutions (see Chapter 4.5). Producers may find information to help further developments and avoid

Bioresorbable polymers

Poly(glycolic acid) PGA   Chiral structure of lactic acid

Poly-l-lactide PLLA

Poly-d-lactide PDLA

Poly(Lactide-co-glycolide PLGA

Polysaccharide (cellulose)

Hyaluronic acid

Poly(vinyl alcohol) PVA

Primary structure of silk fibroin

Gly   Ser   Gly   Ala   Gly   Ala

Poly(glutamic acid)

Polyhydroxyalkanoate PHA

Polysaccharide (alginate)

Poly(aminosaccharide) (chitosan)

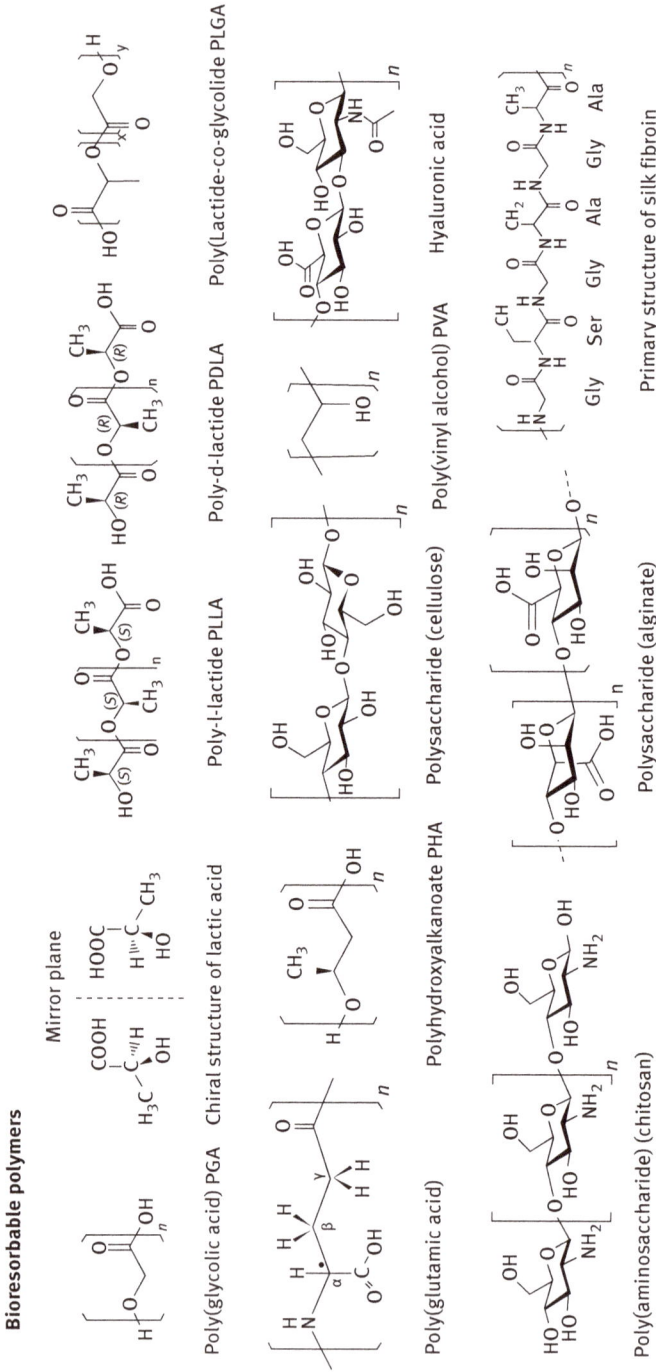

Figure 1.11: Bioresorbable natural and synthetic biopolymers.

double developments. The rules for certification differ in some parts of the world. The Implant-Register® database is the only global and interdisciplinary provider of implant information. It assists surgeons and radiologists to *identify implants*. Within certain limits, the database enables to create *statistics* that may help to survey body implants. Finally, Implant-Register® publishes *findings* that are important for basic implant evaluation grounded on scientific facts (https://implant-register.com).

Recently, in Germany, legislation has been brought on its way to establish an implant register (EIRD, 2019) to provide increased safety for patients undergoing surgical treatment. Patients will be rapidly informed should there be problems with implant quality, postoperative malfunctions, or general concepts and access to new therapeutic solutions.

According to the type and application of biomedical devices, they can be divided into

- Externally worn biomimetic devices such as artificial eye lenses
- Implantable biomimetic devices with physical replacement function such as dental root implants and artificial limb prostheses
- Mechanically active devices such as knee and hip endoprosthetic implants, spine implants, small joint implants, osseosynthetic devices, cerebroperitoneal shunts, and heart valves
- Biologically interacting devices such as cardiovascular stents, vascular prostheses, and blood glucose biosensors
- Electrically active devices such as heart pacemakers, cochlear implants, deep brain stimulators, bone growth stimulators, implantable cardioverter defibrillators, and bionic retinal chips
- Esthetical replacement devices such as silicone breast implants and tooth veneers
- Drug and gene delivery vehicles
- 3D-printed scaffolds for tissue engineering

In the following section, properties and design specifics of selected biomedical devices will be briefly reviewed.

1.3.1 Externally worn biomimetic devices

1.3.1.1 Intraocular lenses

The term "pseudophakia" refers to implanting an artificial eye lens after the natural lens has been surgically removed. During a lifetime of exposure to ionizing ultraviolet radiation coming from the sun, the natural polymeric material of the eye lens deteriorates, resulting in yellowing and clouding and thus severely obscuring vision (cataract). Cataract surgery involves removal of the natural but cloudy lens and its replacement by a pseudophakia IOL. Materials that have been used or are being researched to manufacture intraocular lens implants include PMMA, silicone, collagen, apolar hydrophobic, or polar hydrophilic acrylates, and protein-based hydrogels. Historically, PMMA was the first material to be used successfully in first-generation hard IOLs. This material owes its prominent use in the serendipitous observation of the British ophthalmologist Sir Harold Ridley that Royal Air Force pilots who sustained eye injuries during World War II caused by shattered PMMA windshield material did not show any rejection or FBR. Deducing that the transparent material was inert and useful for implantation in the eye, Ridley designed and implanted the first intraocular lens in a human eye in 1949. Subsequently, soft foldable materials such as silicone or hydrophobic acrylics allowed the insertion of the folded lens through a small incision of the cornea that heals rapidly after implantation, constituting state-of-the-art today.

1.3.2 Implantable devices with physical replacement function

1.3.2.1 Dental implants

For a long time, there was keen desire to replace human teeth lost by disease, old age, or accident. However, the path to modern oral implantology was rocky and fraught with disappointments and failure. This was the outcome of several factors, including the lack of appropriate materials, insufficient understanding of the biological mechanisms involved, and deficient hygienic procedures during oral treatment. As succinctly expressed by Ratner et al. (2013), the occasional success of historical dental implantation attests to "the forgiving nature of the human body and the pressing drive to address the loss of physiologic/anatomic function with an implant." Detailed information on the history of oral implantology can be found elsewhere (Pasqualini and Pasqualini, 2009). Recently, materials used in dental restoration and as dental implants and their history have been reviewed by Monica Saini et al. (2015).

The history of dentistry is almost as ancient as the history of humanity and civilization. In a Sumerian text dated from 5000 BC, the action of a "tooth worm" was mentioned as the cause of dental decay. Remains from the early Harappan periods of the Indus Valley Civilization (c. 3300 BC) show evidence of teeth having been drilled, and in the Egyptian *Ebers* papyrus (1700–1550 BC) various diseases of teeth and remedies for toothache were described.

The earliest attempts to replace lost teeth are attributed to the ancient Greek physician Hippocrates of Kos (460–370 BC), the "Father of Medicine," who described in his *Corpus Hippocraticum* how artificial teeth made from bone or ivory could be fixed with gold or silk threads. Around 100 CE, Aulus C. Celsus, a Roman encyclopedic writer, wrote extensively in his important compendium of medicine on oral hygiene that included stabilization of loose teeth, among others by allogenic tooth replacement. The Etruscans (second century CE) practiced dental prosthetics using gold crowns, gold splint ligatures, and fixed bridgework (Hofmann-Axhelm, 1985).

The Maya (eighth century CE) used a large variety of shaped minerals, including rock crystal, serpentine, jade, cinnabar, turquoise, as well as *Tridacna* shells for dental inlays serving entirely esthetical purpose (Figure 1.12). They used bow drills to perform filling of natural teeth on living individuals, tooth shaping that varied according to regions and tribes, as well as inlaying well-carved stones in meticulously prepared cavities on the labial surface of the front teeth and sometimes in premolars.

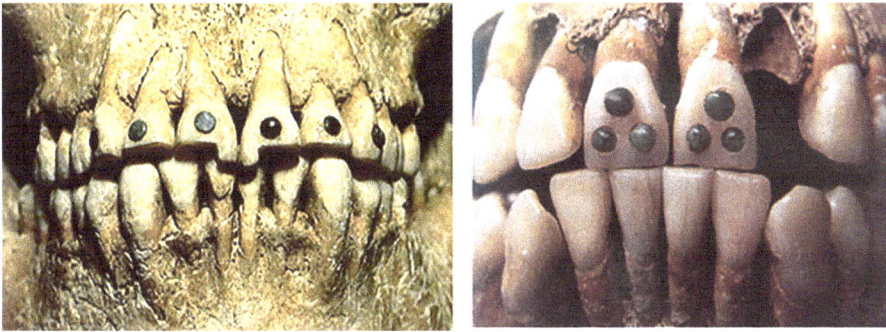

Figure 1.12: Ancient Maya dental inlays with jade buttons discovered in a tomb at Playa de los Muertos (Beaches of the Dead) archaeological site, Honduras. © Permission granted under Creative Commons Attribution License (CC BY 3.0).

Surprisingly, incontrovertible evidence was found to prove the formation of compact bone osteogenesis around the implanted teeth that were highly stable and were probably inserted by Maya medical experts with a technique very similar to modern practice. In 1932, the lower jawbone of a young Mayan woman turned up at the Playa de los Muertos (Beach of the Dead) archaeological site, Honduras. It revealed that three missing teeth were replaced by shaped shell pieces implanted into her jaw. The jawbone, dated back to around 600 CE, showed bone growth around two of the three implants as well as tartar build-up at the implants, making it one of the earliest successful and apparently functional dental implants.

In the Middle Ages, the Arabian scholar and physician Abu al-Qasim (936–1013) described in his comprehensive medical work *Kitāb at-Taṣrīf* the replacement of lost teeth by bony fragments from large animals that were fixed to the remaining teeth with gold ligatures. Similar attempts were attributed to the French doctor Guy de Chauliac (1300–1367) and the Italian Michele Savonarola (1384–1461). During the Renaissance, Ambroise Paré (1510–1590), considered one of the fathers of surgery and modern forensic pathology and a pioneer in surgical techniques, noted that it was possible to replant teeth that had been extracted from their sockets accidentally, tying them to the remaining teeth with gold, silver or linen threads, and keeping them tied until stabilization. Similarly, Gabriel Fallopius (1523–1562) fixed extracted or artificial teeth with gold or other metal wire ligatures.

The 1700s saw the advent of modern dentistry through the eminent early dental surgeon Pierre Fauchard (1678–1761), credited as being the "Father of modern dentistry," who introduced dental fillings as treatment for dental cavities, and suggested the use of lead and tin amalgams and sometimes, gold. In addition, he has been credited with recognizing the potential of porcelain enamels and initiating research with porcelain to imitate natural color and sheen of teeth and gingival tissue. In 1789, the Frenchman Nicolas Dubois de Chémant (1753–1824) received the first patent for porcelain teeth, a feat that was widely publicized in England, the country to which he was forced to flee after the French Revolution. The contemporary Italian dentist Giuseppangelo Fonzi (1768–1840) went a step further by inventing mineral teeth synthesized from kaolinite-bearing argillaceous clay, silica, zinc oxide, titanium oxide, and several other ingredient including metal oxides to provide different color shades. By using different metallic oxides and adjusting their respective proportions, Fonzi was first to obtain 26 different shades, notably the transparent tint that allowed him to make the appearance of artificial teeth similar to that of natural teeth. These "mineral teeth" were made into two parts: (i) the internal opaque part of a mineral tooth was fashioned from kaolinitic argillaceous clay with the optical appearance of dentin; and (ii) the outer transparent part deposited on the surface was made of silica, ensuring brightness and sheen of the prosthetic teeth.

The first documented endosseous metal implant, invented in 1809 by the Italian-born dentist M. Maggiolo working in Paris, involved inserting a gold tube directly in place where a tooth had been recently removed. While the implantation site was permitted to heal prior to the addition of a crown, the process was ultimately a failure due to major inflammation of the gums. Other implants were made of porcelain, silver, and iridium tubes but suffered equally low success rates.

Subsequently, American dental doctors Chapin A. Harris (1806–1888) and Horace H. Hayden (1769–1844) experimented with endosseous implants employing iron teeth of their own design. Harris was first to place a lead-coated platinum post in an artificial socket to resemble the root of a natural tooth. He roughened the lead to provide retention for the *de novo* tissue that was supposed to form inside the artificial cavity. However, today we know that lead is not biocompatible but highly cytotoxic and may trigger the formation of reactive and inflamed hypertrophic tissue that presumably must have formed around that implant, thus, giving the illusion of temporary stability.

In 1931, Alvin E. Strock (1911–1996) inserted a Vitallium dental screw implant into the socket of an extracted tooth and, after the alveolar bone tissue had tightened around the screw, cemented a porcelain crown to its head. Vitallium, a CoCrMo alloy, was the first really successful biocompatible implant metal, and its use had been developed a year earlier by Charles Scott Venable (1877–1961), an orthopedic surgeon.

This development was followed, in 1947, by the invention of the direct endoalveolar infibulation of a hollow spiral screw by Manlio Formiggini (1883–1959). The screw was made from stainless steel or tantalum wire and acted as a bearing post for a future prosthesis. This invention marks the definite transition to the era of successful endosseous implants.

During the late 1970s and early 1980s, Per Ingvar Brånemark (1929–2014) developed truly osseointegrated dental implants by positioning screw-shaped titanium implants with small threads into the empty sockets (see Figure 1.4). The surface of the screw was micro-grooved to permit integration with bone tissue by providing surface structures amenable for easy bone tissue in-growth. In addition, he coined the term "osseointegration," the direct anchorage of an implant by formation of bony tissue around it without development of fibrous connective tissue at the bone–implant interface (see Chapter 1.1.5). This finding contradicted earlier thoughts that implants relied on mechanical retention since it was unknown at this time that metal could be fused directly to bone. However, with the advent of current understanding of osseointegration, rootform endosteal implants became the new standard in implant technology. Hence, Brånemark may be rightly considered the "Father of modern dental implantology."

1.3.3 Mechanically active devices

1.3.3.1 Hip endoprostheses

Modern types of hip implant encompass all three major classes of biomaterials: a metallic stem composed of titanium alloy, stainless steel, or CoCrMo alloy, a femoral ball frequently consisting of advanced alumina or partially stabilized zirconia ceramics, as well as an inset of the acetabular cup actuating against the femoral ball that may be composed of tough polymers such as cross-linked polyethylene or UHMWPE. In addition, in the majority of cases, the metallic stem of the implant is being coated, by atmospheric plasma spraying (see box), with a thin layer of osseoconductive hydroxylapatite, designed to assist in the in-growth of bone cells and to prevent the formation of an acellular fibrous connective tissue capsule, formed as a response of the body to the introduced foreign body.

Plasma spraying

Plasma spraying is a rapid solidification technology during which material introduced predominantly as powder but also as rod or wire into a plasma jet is melted and propelled with high velocity against a surface to be coated. This technology is versatile: any thermally reasonably stable metallic, ceramic or even polymeric material (such as polyamide) with a well-defined congruent melting point can be coated onto nearly any surface. However, the requirement of congruent melting must be relaxed in the case of hydroxylapatite that melts incongruently, that is, decomposes during melting into tricalcium phosphate ($Ca_3(PO_4)_2$, α- and β-TCP) and tetracalcium phosphate ($Ca_4O(PO_4)_2$) or even cytotoxic calcium oxide, CaO following loss of P_2O_5 by evaporation or reduction of phosphate to volatile phosphorus by hydrogen used as auxiliary plasma gas. Moreover, a large portion of the molten material rapidly quenched (cooling rate 10^6 K/s or higher) on contact with the relatively cool implant surface forms easily soluble ACP of variable composition. However, in practice many limitations persist related to coating porosity, thickness, adhesion to the substrate, the presence of residual stresses, and line-of-sight restriction (Heimann, 2008).

Figure 1.13 shows a high-end hip endoprosthesis marketed by CeramTec GmbH with an alumina femoral ball designed to articulate against an alumina acetabular cup. This ceramic–ceramic wear couple excels by much reduced friction, a requirement

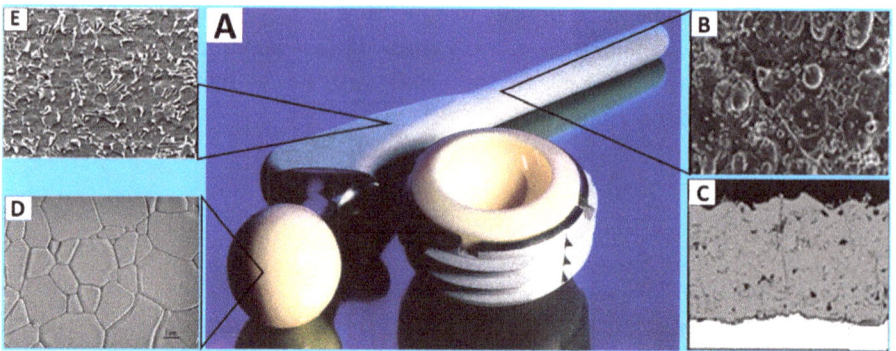

Figure 1.13: (**A**) Endoprosthetic hip implant showing an Ti6Al4V stem (see **E**) with a BIOLOX®forte alumina femoral ball attached (left) and coated with a thin layer of plasma-sprayed hydroxylapatite and the likewise coated titanium acetabular cup (right) with an alumina inset designed to actuate against the ball under very low friction conditions (Image courtesy Professor Gerd Willmann†, CeramTec GmbH, Plochingen, Germany; Clarke and Willmann, 1994). Today, state-of-the-art material is BIOLOX®delta. (**B**) Surface of a typical plasma-sprayed hydroxylapatite coating with well-developed pancake-like molten particle splats and some loosely adhering incompletely melted spherical particles (Heimann and Lehmann, 2015). © With permission by John Wiley and Sons. (**C**) Cross section of a plasma-sprayed hydroxylapatite coating on a Ti6Al4V substrate, showing some pores (dark patches) and radial cracks. The splat-like nature of the coating is apparent (Heimann et al., 2001). © With permission by John Wiley and Sons.
(**D**) Microstructure of the BIOLOX®forte femoral ball. © With permission by CeramTec GmbH, Plochingen, Germany. (**E**) Duplex ($\alpha + \beta$) microstructure of Ti6Al4V. The α-phase is shown in grey, the β-phase in white (Moussaoui et al., 2015). © Permission granted under Creative Commons Attribution license.

necessary to counter for the lack of synovial fluid that in a healthy hip joint provides a very low friction coefficient. This synovial fluid contains high levels of hyaluronic acid, an anionic, non-sulfated glycosaminoglycan that helps to maintain high fluid viscosity and supports the normal integrity of the joint by attenuating inflammation and preserving the cartilaginous matrix covering the femoral head.

Figure 1.14 shows a part of the operational procedure adhered to during an endoprosthetic joint arthroplasty as well as the implant used. Note that in this case the prosthetic stem is not coated but instead is structured to provide a suitable template for bone ingrowth. The patterning of the femoral stem is designed to replicate the gradations of density found in a real femur by using hollowed-out tetrahedra. This structure is thought to reduce the elastic modulus and hence to reduce stress shielding (Gombay-McGill, 2016).

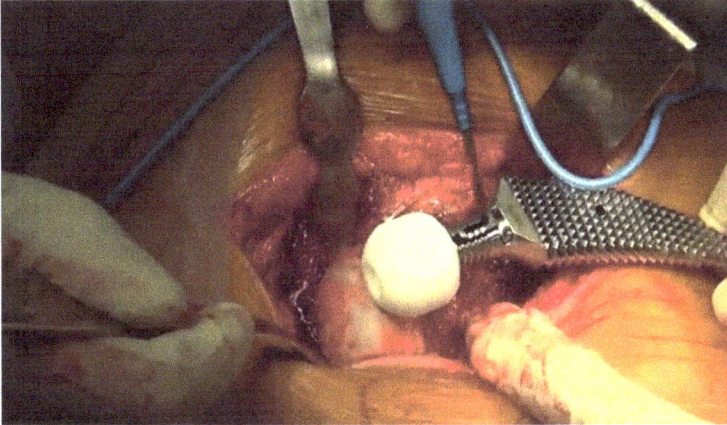

Figure 1.14: Hip joint arthroplasty. © Permission granted under Creative Commons Attribution license.

From its humble beginning at the end of the nineteenth century (Themistocles Gluck, 1853–1942), large joint replacement surgery has expanded greatly, with more than a million hip and knee joints being implanted worldwide each year (see Chapter 1.4; Quinn and Brachmann, 2014). The industry related to this particular field of medical development has substantially grown to supply these joints that reveal ever increasing sophistication and specialization, and constitutes a vigorous, responsible, and research-driven sector of the biomedical industry. Technical principles, design, and safety issues of joint implants were addressed by Gottfried H. Buchhorn and Hans-Georg Willert (1994). Much information can be gained from the Proceedings of the International BIOLOX® Symposium (1995 to the present).

1.3.3.2 Knee endoprostheses

Total knee arthroplasty (TKA) is indicated for patients suffering from a severely painful and/or disabled knee joint as a result of osteoarthritis, traumatic arthritis, rheumatoid arthritis, an accident, or a failed previous implant (Scott, 2015). TKA is intended to provide increased patient mobility and reduce pain by replacing the damaged knee joint articulation, provided there is evidence of sufficiently sound bone to seat and support the components. The knee is one of the most complex joints in the human body. As the joint that is furthest removed both proximally and distally from the next adjacent, that is, hip and ankle joint articulation, it is subjected to major stresses. Deficiencies in mobility cannot easily be compensated for by the hip or the ankle joints. This incongruity of the opposing joint surfaces with the additional complexity of the patella–femoral articulation has presented a surgical challenge for over a century (Murray, 1991).

Figure 1.15 shows on the left a schematic rendering of a typical modern total knee endoprosthesis (1) consisting of a lower tibial part (2) with a shank (3) to secure the part in the existing bone, and an upper femoral part (5) with two curved condylar surfaces (6,7). Between these parts consisting typically of a Ti or CoCr alloy, there is located a meniscus part (8) with a bearing surface (4) typically consisting of low-density polyethylene (LDPE) with two bearing shells (9, 10) in which the condylar surfaces (6, 7) engage (Hagen, 2012). The right side of Figure 1.15

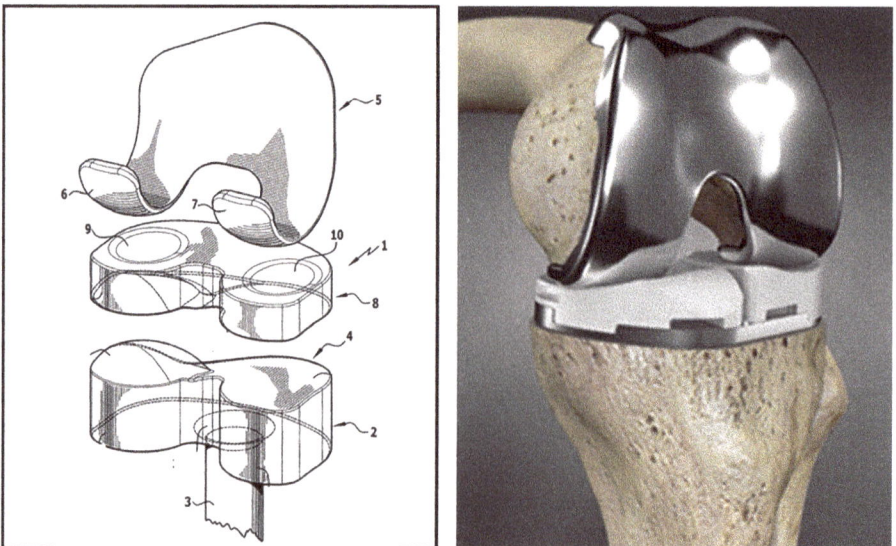

Figure 1.15: Total knee endoprosthesis. (A) Schematic rendering (Hagen, 2012). (B) Attachment of the knee prosthesis to the femoral and tibial bones. © Permission granted under Creative Commons Attribution license.

shows how the lower tibial and the upper femoral parts of a total knee endoprosthesis are attached to the bones.

As early as in 1891, the German surgeon Theophilus Gluck (Gluck, 1891) implanted one of the first total knee replacements, using a hinged ivory TKA fixed with plaster of Paris and colophony glue (Figure 1.6). Although a revolutionary concept at the time, these implants inevitably failed due to high infection rates, poor metallurgy, and inadequate fixation ability. Much later, in 1951, Börje Karl-Olof Walldius developed a hinge made of acrylic resin and later, in 1958, manufactured it from CoCr alloy (Phillips and Taylor, 1975). This prosthesis was in use until the early 1970s along with several other hinged designs including the Shiers design (Watson et al., 1976) in the UK and the Guepar® prosthesis (Aubriot et al., 2014) from France. After that, in 1974, the total condylar knee prosthesis was invented, a plastic component resembling the natural knee structure and allowing for total joint replacement with, however, only limited motion (Vince and Insall, 1991). In the 1980s, the novel concept of mobile-bearing knee replacement was developed in which the component was not completely fixed to allow more motion and a higher angle of knee flexion. Subsequently, in the early 1990s, the LCS® Mobile-Bearing Knee System with low contact stress articular geometry and the Sigma® Cruciate Retaining Knee System, both developed by DePuy Synthes Companies, were introduced, focusing on elimination of undue strain on the patella but without removing the knee flexion ability. Finally, in 2011, the Attune® Primary Total Knee System appeared on the market, one of the largest research and development efforts in the history of DePuy Synthes Joint Reconstruction Company that combined the latest achievements in design, kinematics, engineering, and materials to deliver stability and motion (Clatworthy, 2015).

In conclusion, the advent of TKA was an important milestone in the history of orthopedic surgery. Many surgeons and design engineers were involved in this process that has now become a multibillion dollar industry. TKA has joined THA as a very successful surgical intervention that provides pain relief and much improved mobility to millions of patients worldwide.

Recently, a breakthrough has been achieved in mimicking the complex anisotropic structure of the knee meniscus (Zhang et al., 2019). The natural meniscus is made up of outer and inner regions. The outer region contains fibroblast-like cells within an ECM consisting preferentially of type I collagen that lends resistance to tensile loads. The inner region contains chondrocyte-like cells embedded within an ECM mainly made up of type II collagen and glycosaminoglycans, making the tissue resistant to compressional loads. Instead of LDPE customarily used in knee endoprostheses (Figure 1.15, 8), the authors cultured bone marrow-derived mesenchymal stem cells on a biomimetic scaffold and applied two synergistic biochemical growth factors (cytokines) to the tissue and a loading system that simultaneously exerts both tensile and compressive stresses. The method induces the stem cells to differentiate into separate layers of type I and type II collagen, thus mimicking the anisotropic nature of knee cartilage.

1.3.3.3 Mechanical heart valves

The most widely used mechanical valves to treat valvular heart disease are made from low-temperature isotropic pyrolytic carbon (LTIC) with turbostratic structure, a material that has been used for over 30 years in heart valve design (see Chapter 2.2.1.3). Most mechanical heart valves are of bileaflet type, meaning that they have two carbon "leaflets" to control the blood flow to a single direction (Figure 1.16).

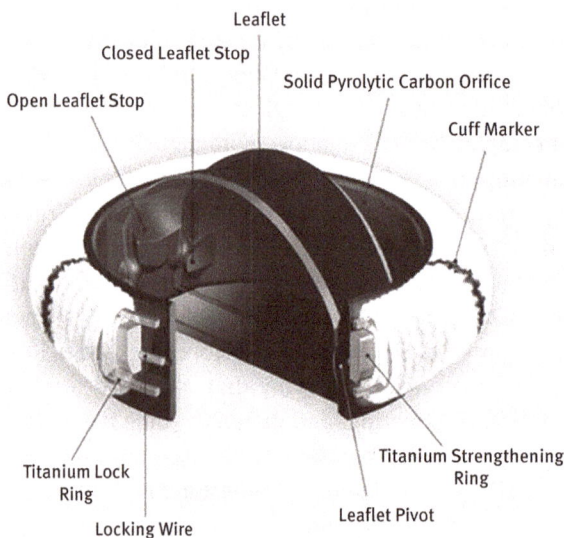

Figure 1.16: Typical artificial heart valve made of a titanium strengthening ring, pyrolytic carbon orifice and leaflets, and a polyurethane cuff. © Permission granted under Creative Commons Attribution license.

The primary advantage of mechanical valves is the likelihood of lasting a patient's lifetime without the need for another valve operation because the valve has worn out. Mechanical valves have excellent blood flow performance that will benefit patient's quality of life and ability for exercise. However, the material is brittle and its hemocompatibility is still not sufficient for prolonged clinical use. A drawback of mechanical valves is the risk associated with the requirement for life-long blood thinner therapy. Hence, the risk of thrombus formation forces patients to take anticoagulation medication, frequently for the remainder of their lives. However, when properly managed, rates are relatively low for both bleeding and clotting.

Owing to the brittleness and limited hemocompatibility of LTIC-based heart valves, the development of materials with better surface properties and much improved hemocompatibility and thus, suppressed thrombogenicity is considered an urgent requirement among cardiologists. Indeed, despite of the general success these heart valves enjoy, there are intrinsic problems related to induction of blood clots, wear-type degeneration of the leaflets, mechanical failure of weld joints of the device, and not in the least, microbial infection.

1.3.3.4 Osteosynthetic devices

Osteosynthesis is the reduction and internal fixation of a bone fracture with implantable devices that are usually made of metal, most often surgical stainless steels such as AISI 316L and AISI 304, or titanium alloy. Osteosynthetic musculo-skeletal fixation devices include fracture fixation screws and plate systems for use in extremity locations, interference screws, surgical clips, screws for femoral head fixation, spinal implants and cages, and several other applications. These surgical devices aim to bring fractured bone ends together and to immobilize the fracture site while healing takes place by intermembranous ossification.

The evolution of bone fracture management started in antiquity with the stabilization of fractures by external splints made from wood, bamboo or reed, and fixed with linen bandages or sometimes already with plaster of Paris. Hence, before the 1950s, fractures were largely externally fixed although pioneering approaches of internal fixation were known such as the 1886 Hansman bone plate that was bend at the end to protrude through the skin. The plate was attached to the bone by screws with long shanks that projected outside the soft tissues. This was followed by a steel plate developed in 1895 by Sir William A. Lane (1856–1942), and in short succession similar internal fracture fixation devices were introduced by Albin Lambotte (1909) and William O. Sherman (1912). However, these internal fixation devices proved to be unstable and suffered from loss of reduction and fracture disease. In a next step, rigid internal fixation was introduced that achieved absolute mechanical stability but tended to promote bone devascularization, delayed union, infection, and in addition led to mechanical metal failure. Hence, in the 1990s, mechanical stabilization was replaced by *biological osteosynthesis* that limited the surgical trauma owing to its less invasive surgical technique. The principle of biological osteosynthesis implies the indirect (closed) reduction and elastic fixation, without intervening on fracture area, and in addition, maximally preserved vascularization of intermediate fragments. In other words, these methods attempt to preserve the blood supply to improve the rate of fracture healing, decrease the need for bone grafting, and decrease the incidence of infection and refracture. Plate osteosynthesis is still recognized as the treatment of choice for most articular fractures, many metaphyseal fractures, and certain diaphyseal fractures such as in the forearm. Since the 1960s, the operational techniques, materials, and implants used for internal fixation with plates have evolved to provide for improved healing.

Recently, attention has shifted from corrosion-resistant bioinert metals such as titanium alloys and austenitic surgical steels to biodegradable magnesium alloys. The main advantage of using magnesium over conventional bone plates, screws, and pins made of environmentally stable metals is that no second operation is required to remove the device after healing of the bone. Instead, magnesium-based fixation devices disappear within 1 year at most by metabolizing without the need for further surgical intervention. There is compelling evidence that osteosynthetic devices based on magnesium alloys such as WE43 are good candidates for replacing conventional devices fashioned from polymers or other, more environmentally stable metals. However, a high evolution rate of hydrogen gas due to rapid Mg degradation can damage the bone growth plates substantially. WZ21 implants were found to maintain their integrity for 4 weeks and corrode subsequently with 0.5% volume loss per day, promote enhanced bone neoformation around the implant, and thus, show evidence of sufficient osteoconductivity and osteoinductivity of magnesium (Kraus et al., 2018).

Mg-Nd-Zn–based alloys were developed at Shanghai's Jiao Tong University (Mao et al., 2012) with lower biodegradation rates and homogeneous nanophasic degradation pattern compared to other biodegradable Mg alloys. This alloy shows an average corrosion rate of 0.34 mm/year in SBF, half of the corrosion rate of 0.71 mm/year of WE43 tested under identical conditions. *In vitro* cytotoxicity tests using various types of cell indicate excellent biocompatibility. Bone implants and cardiovascular stents (see below) have been successfully fabricated using this alloy and *in vivo* long-term assessment via implantation in animal model has been performed. The results confirmed a reduced degradation rate *in vivo*, excellent tissue compatibility, and long-term structural and mechanical durability. Thus, this novel Mg-alloy series with uniform nanophasic biodegradation represents a major breakthrough in the field and may be a promising candidate for manufacturing next-generation biodegradable implants based on magnesium.

1.3.4 Biologically interacting devices

1.3.4.1 Cardiovascular stents

Typical examples of biologically interacting devices are cardiovascular stents (Figure 1.17, left).

Cardiovascular stents are implants designed to provide short-term supporting structures for vascular implants to combat coronary heart and peripheral artery diseases. Conventional permanent drug-eluting stents (DES) fashioned from traditional biocompatible metals such as tantalum, titanium, cobalt–chromium alloys, or the shape-memory alloy NiTi show several disadvantages. These disadvantages include an increased risk of thrombosis, loss of vascular flexibility, inhibition of revascularization, as well as the difficulty to evaluate computer tomographic and MRI. In

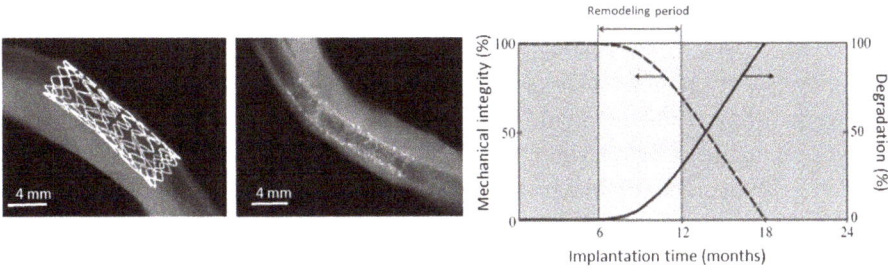

Figure 1.17: Resorption behavior of an Mg alloy stent implanted into a porcine coronary artery. Left: as-implanted. Center: after 56 days (Waksman, 2006). Right: Idealized compromise between mechanical stability and degradation of a biodegradable Mg-based cardiovascular stent (Hermawan et al., 2010). © With permission by Elsevier.

contrast, novel biodegradable stents made from magnesium or iron (Moravej and Mantovani, 2011) allow free physiological vasomotion, foster-improved clinical long-term behavior by inhibiting chronic inflammation, delay thrombosis complications, prevent occurrence of reocclusion and restenosis, and even enjoy better acceptance by patients owing to their nonpermanent nature. Several studies have shown that the critical period of vascular healing normally ends about 3 months after implantation (Figure 1.17, right). Consequently, development effort aims at the survival of Mg-based stents within this time frame.

A second type of bioresorbable cardiovascular stents is polymer based, and most frequently is made of PLLA, a biomaterial that is able to maintain a radially strong scaffold that breaks down over time into innocuous lactic acid, a naturally occurring polymer that the body can easily metabolize. Other polymers include analogs of the amino acid tyrosine (desaminotyrosine) and biocompatible short-chain PLA marketed under the brand name Tyrocore™ (Leibundgut, 2018). Beyond that, biodegradable polymers such as poly(lactide-*co*-glycolide) are extensively studied to optimize their properties and biocompatibility. Due to the degradation of the polymeric coatings and the subsequent transformation into a bare-metal stent, DES are expected to reduce the risk of stent thrombosis (Strohbach and Busch, 2015).

Despite impressive advances in the field of angioplasty, an aggravating problem consists in the necessity to control neointimal hyperplasia, the proliferation and migration of vascular smooth muscle cells located primarily in the *tunica intima*, resulting in the thickening of arterial walls and hence, decreased arterial lumen space. Neointimal hyperplasia is the major cause of restenosis after percutaneous coronary interventions such as stenting or angioplasty. Drug-eluting coatings deposited on the stent surface normally achieve control of neointima proliferation for permanent stents. Antiproliferative drugs are being embedded in carrier substances made from either corrosion-resistant polymers (such as polysulfone or styrene–isoprene butadiene block polymer) or biodegradable polymers (polyesters, such as PLLA, PCL, or PLGA) (Leng et al., 2019).

However, this treatment cannot be used for stents made of magnesium alloy. In this case, resistant polymers are being inflated and lifted off and removed from magnesium surfaces by the evolution of hydrogen bubbles. Degradable polyesters accelerate unfavorably the magnesium degradation process by their acidic decomposition products. Hence, other solutions must be found, for example, through developing and applying of protective coatings that limit the Mg corrosion rate and the associated hydrogen evolution.

Several other novel methods have surfaced attempting to minimize the corrosion rate and thus the hydrogen evolution rate. One of the most successful approaches has involved the use of bioresorbable metallic glasses produced by rapid solidification. Alternative solutions include the development of Mg rare-earth element alloys. Coatings and sophisticated material processing routes are currently being developed to further decrease the corrosion rate (Heimann and Lehmann, 2017). Recently, zinc alloys were shown to exhibit outstanding physiological corrosion behavior, meeting a benchmark penetration rate of 20 µm per year (Bowen et al., 2013). In addition, zinc alloys meet or even exceed mechanical behavior benchmarks for ductility and tensile strength. Although promising, these materials are relatively new to biomaterial science, thus requiring further research and finally, certification to confirm that zinc alloys are feasible base materials for cardiovascular stents.

1.3.5 Electrically active devices

1.3.5.1 Heart pacemaker
A heart pacemaker is a small device placed in the chest or abdomen to help control abnormal heart rhythms. This device produces low-energy electrical pulses to prompt the heart to beat at a normal rate. A pacemaker consists of a battery, a computerized generator, and insulated wires with electrode sensors on one end (Figure 1.18). The electrodes detect the electrical activity of the heart and send data through the wires to the computer in the generator. If the heart rhythm is abnormal, the computer will direct the generator to send electrical pulses to the heart. A modern pacemaker is a marvel of engineering prowess. However, its humble beginning started, in 1952, with the invention of Paul Zoll (1911–1999) who designed a portable version of an existing bulky cardiac resuscitator. While moderately effective, this early pacemaker was primarily used in emergency situations. Through 1957 and 1960, significant improvements were made to Zoll's original invention. In an attempt to reduce the amount of voltage needed to restart the heart and increase the length of time electronic pacing could be accomplished, Clarence W. Lillehei (1918–1999) designed a pacemaker the leads of which were attached directly to the outer wall of the

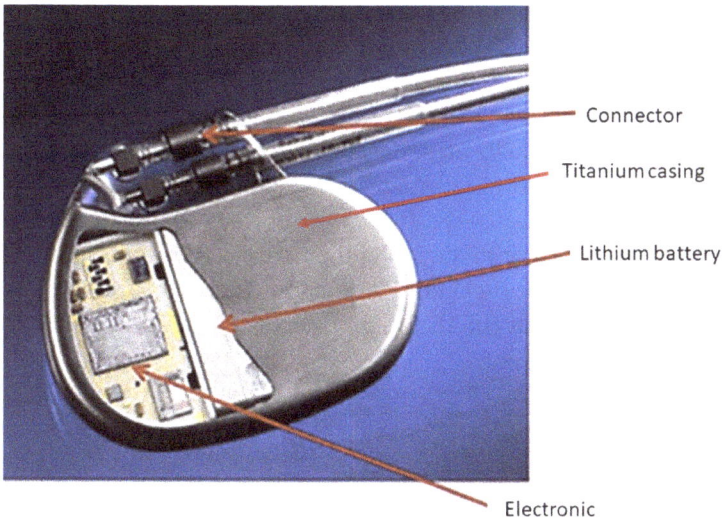

Connector

Titanium casing

Lithium battery

Electronic

Figure 1.18: Typical design of a heart pacemaker. © Permission granted under Creative Commons Attribution License (CC BY 3.0).

heart. Later, in 1958, a battery was added as the power source, making the pacemaker truly portable, thus allowing patients to be mobile. This also enabled patients to use the pacemaker continuously instead of only for emergencies. Whereas Lillehei's pacemaker was still externally worn, in 1960 William Chardack (1915–2006) and Wilson Greatbatch (1919–2011) invented the first implantable pacemaker, together with a corrosion-free lithium battery to power it. Subsequently, Seymour Furman (1932–2006) developed the modern technique to implant an endocardial pacemaker. Instead of cutting open the patient's chest cavity, he inserted the electrical leads into a vein and threaded them up into the ventricles. With the leads inside the heart, even lower voltages were needed to regulate the heartbeat. Since then improvements have been made in their design, including smaller pacemaker devices, longer lasting batteries, and ever-more sophisticated computer controls.

The casing of a heart pacemaker is usually made of corrosion-resistant titanium or titanium alloy, the electrical leads are likewise made from metal alloy, frequently from Pt–Ir alloy and sheathed by biocompatible polyurethane for electrical insulation (Figure 1.18).

The electronic circuitry is that of appropriately modified silicon semiconductor boards. The power source is a long-life lithium battery. The lithium iodide cell invented and manufactured by Greatbatch is now the standard cell for pacemakers, providing the energy density, low self-discharge characteristic, small size, and reliability required. Its function is based on a lithium anode and a cathode made of a proprietary composition of iodine and poly-2-vinyl pyridine.

An alternative energy source was a thermoelectric battery powered by radioactive ^{238}Pu with a half-life of 88 years. The heat from the decaying plutonium is used to generate thermoelectricity to drive the pacemaker. As of 2003, there have been only between 50 and 100 people in the USA who have nuclear-powered pacemakers. When one of these individuals passes away, the pacemaker is supposed to be surgically removed and shipped to Los Alamos National Laboratory where the plutonium will be recovered in a safe way. It is rather obvious why the Pu-based thermoelectric battery never made it to the top of the line. Major manufacturers of heart pacemakers are Medtronic plc (Ireland), Vitatron (The Netherlands), Biotronik (Germany), St. Jude Medical/Abbott (USA), Boston Scientific (USA), and LivaNova plc (UK).

1.3.5.2 Cochlear implants

Cochlear implants function by bypassing most of the peripheral auditory system. In a healthy auditory situation, the cochlea receives sound transmitted via the ear drum and the ossicular chain to the cochlea. The sound wave is converted into movements of tiny hair cells within the cochlea. These hair cells release potassium ions in response to their movement, and the potassium ions in turn stimulate other cells to release the neurotransmitter glutamate that cause the cochlea nerve to relay signals to the brain, creating the impression of sound.

A cochlear implant ("bionic ear") attempts to replicate the functions of the outer, middle, and inner ear by collecting and transmitting sound signals to the auditory nerve that in turn sends the signal to the brain. The individual parts of sound processing system are shown in Figure 1.19.

The microphone is worn behind the ear and picks up sound waves, sending them to the speech processor that analyses the signal and transforms it into a coded electrical signal that in turn is sent to the transmitter. The externally worn transmitter relays the coded signal across the skin to the implanted receiver in form of a special radiofrequency signal. The implanted receiver decodes the signal and sends a corresponding pattern of tiny electrical pulses to the electrodes embedded in the cochlea. Then, the electrical pulse travels along the auditory nerve to the brain. The brain categorizes the sound and assigns meaning.

In 1957, the original single-channel cochlear implant was invented by the French doctors André Djourno and Charles Eyriès, followed, in 1961, by an improved design by William F. House. In 1964, Blair Simmons and Robert J. White implanted a six-channel electrode in a patient's cochlea. The modern microelectronic multichannel cochlear implant was independently developed and commercialized by Graeme Clark in Australia and Ingeborg Hochmair-Desoyer and her husband Erwin Hochmair in Austria. The seminal contribution of French doctors and scientists toward development of multichannel cochlear implants has recently been reviewed by Claude-Henri Chouard (2014).

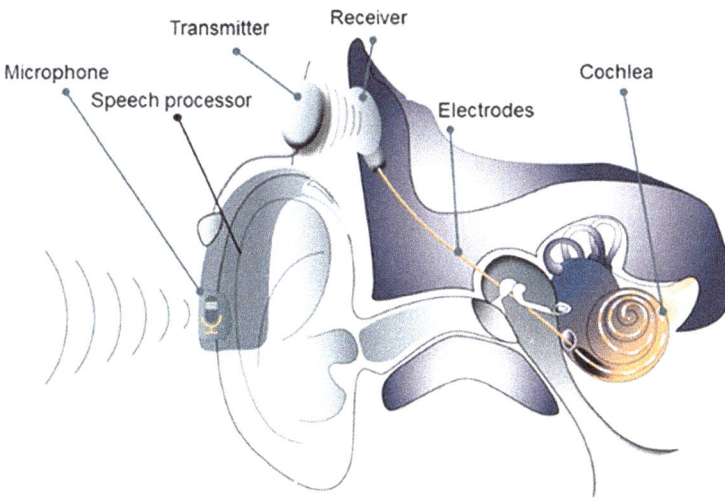

Figure 1.19: Schematic of a typical cochlea implant (Source: Med-El Co. International, Innsbruck, Austria).

1.3.5.3 Retina implants

Retina implants are presently designed to allow blind people suffering from *retinitis pigmentosa* or dry age-related *macula* degeneration to gain rudimentary vision. A small microchip is implanted either epi- or subretinally inside the patient's eye (Figure 1.20). In a healthy eye, photoreceptors (rods and cones) in the retina convert light into nerve signals by changing the polarization state of the cell membrane. These electrical signals are propagated to the intermediate layer mainly composed of bipolar cells that are connected to photoreceptors as well as cell types such as horizontal cells and amacrine cells. The bipolar cells further pass on the electrical signal to the retinal ganglion cells. When a patient is fitted with a retina implant, the implant attempts to carry out this missing function. The microchip consists of 1,500 elements on a surface area of 3×3 mm. Each of these elements contains a silicon photodiode, a differential amplifier, and an electrode. Since the inner retina still functions correctly, light is then converted by the photodiodes into electrical signals that are sent via electrodes to the bipolar cells of the retina, stimulating the retinal nerve cells. These signals in turn stimulate the brain and will be translated into sight.

Presently, the technology allows patients to orientate themselves in space and to recognize faces (Chader et al., 2009). This electrical prosthetic device appears to offer hope in replacing the function of degenerated or dead photoreceptor neurons. Devices with new, sophisticated designs and increasing numbers of electrodes could allow for long-term restoration of functional sight in patients with improvement in object recognition, mobility, independent living, and general quality of life (Daschner et al., 2018).

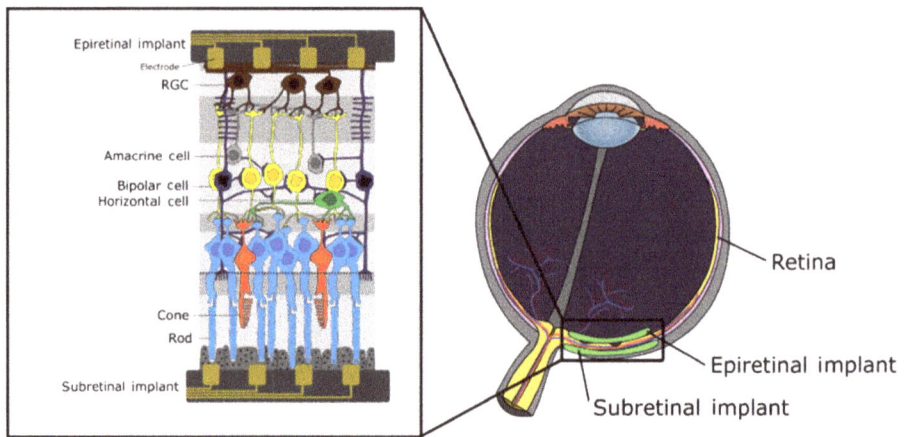

Figure 1.20: Schematic overview of the nervous structure of the retina of the human eye and the location of retinal prostheses. © Permission granted under Creative Commons Attribution-ShareAlike License.

1.3.5.4 Bone growth stimulators

Stimulation of bone growth serves to promote bone healing in difficult to heal fractures including nonunion (pseudoarthrosis) or spinal fusions by applying electrical current or ultrasound to the fracture/fusion site. Electrical and electromagnetic fields are thought to play a role in bone healing through the same principles as mechanical stress applications. When mechanical load is applied to bone, a strain gradient develops. Subsequent pressure gradients in the interstitial fluid drive fluid through the bone canaliculi from regions of high-to-low pressure, and thus, expose osteocyte membranes to flow-related shear stress, as well as to electrical streaming potentials. Application of electrical or electromagnetic fields to the fracture site is supposed to mimic the effect of mechanical stress on bone, leading to enhanced bone growth, and hence assists in healing (Galkowski et al., 2009).

If the recovery time after a THR operation can be reduced by even a small margin due to accelerated healing of the operational trauma site as well as faster bone in-growth, the overall positive effect on the well-being of the patient, as well as the economy, would be substantial (see Chapter 1.3.5.5). Such technology could also be applied to construct novel osteosynthetic devices, that is, surgical fixtures that stabilize and join the ends of fractured bones such as metal plates, pins, or screws. They are designed to increase the healing rate of fresh fractures and osteotomies, that is, surgical procedures of cutting a bone to either lengthen, shorten, or straighten it, spinal fusions, and delayed or nonunion (pseudoarthrosis) fractures (Griffin and Bayat, 2011) that presently account for at least 10% of all clinically treated bone fractures.

Electrical stimulation can be applied either from the outside of the body (noninvasive) or from the inside of the body (invasive). Noninvasive (external) electrical bone growth stimulators are devices worn on the outside of the skin and are of three types: (i) capacitive coupling devices using metal electrodes that are applied to the skin to deliver the current; (ii) pulsed electromagnetic field devices; and (iii) combined magnetic field devices.

Surgically implanted or invasive electrical bone growth stimulators utilize a direct current applied to the nonhealing fracture or bone fusion site. Although multiple randomized trials exist to support the variety of bone stimulation modalities, all are limited to primarily radiologic endpoints. There remains a need to conduct large and definitive trials that use patient-important outcomes before universal acceptance of such modalities will occur (Galkowski et al., 2009).

Advanced osseoconductive implant coatings based on transition metal-substituted calcium hexaorthophosphates (see Chapter 2.2.2.3) with solid-state ionic conductivity may be utilized to construct a capacity-coupled invasive bone growth stimulator (Heimann, 2017). The advantage of such a novel device over already existing bone growth stimulators can be seen in providing an intimate contact of a capacity-coupled electric field with the growing bone tissue as opposed to an externally applied inductivity-coupled electromagnetic field that suffers substantial attenuation when transmitted through soft tissue covering the locus of bone growth. To achieve higher ionic conductivity in Ca(Ti,Zr) hexaorthophosphates, aliovalent doping with highly mobile Na or Li ions appears to be a suitable route.

1.3.5.5 Electroceuticals

Electroceuticals are electronic devices designed to affect biological systems to alleviate or mitigate symptomatological and/or pathological processes in the body. They utilize specific electric waveforms to stimulate electronically the nervous system, and thus, modify bodily functions. Microcurrent electrical therapy (MET) and cranial electrotherapy stimulation (CES) are being used to stimulate cell regeneration, treat injuries, and are thought to relieve acute and chronic pain, anxiety, depression, and even insomnia. Thus, these nerve-stimulating therapies are destined to replace drugs to treat many chronic conditions. In particular, non-invasive handheld vagal nerve stimulators recently approved by the FDA can ease cluster headaches and migraine (Ling and Latham, 2018).

Research into electroceuticals started in 1970, when the orthopedic surgeon Robert O. Becker demonstrated that electric current signals are triggers able to stimulate healing, cell growth, and cell regeneration. These signals originate from an electric control system he called the "current of injury" that is conducted through Schwann cells and the myelin sheaths surrounding neurons (Becker and Selden, 1985). In the 1970s, Candace B. Pert discovered the presence of opiate receptors in the human brain (Pert and Snyder, 1973) and later published her research on ligand–receptor binding, thereby providing a new understanding of cellular physiology and the connection between mind and body, culminating in her famous quip "God is a neuropeptide." Subsequently, Bjorn Nordenstrom proposed a model of biological electric closed circuits analogous to closed circuits in electronic technology. He postulated that mechanical blood circulation is closely integrated anatomically and physiologically with a controlling bioelectric system (Nordenstrom, 1998). Finally, neurobiologist Daniel L. Kirsch patented the Alpha-Stim technology that was the first-dual electromedical device (MET and CES) to be cleared by the US-FDA to market (Kirsch and Marksberry, 2015).

At the heart of electroceutical's working is the fact that cells manufacture peptides acting as ligands attached to surface receptors on other remote cells, thus communicating via the ECF and the circulatory system. Cells within a specific organ or tissue system are thought to communicate through specific frequencies in the microamperage range, activating Becker's "current of injury" and causing the system to develop toward homeostasis. Neuromodulation imparts an external electric signal with a frequency that perfectly matches that of receptors in the body, causing them to resonate and activate intracellular responses.

Although much research is still required to detect, confirm, and utilize these subtle bioelectric regulatory systems to their full capacity, the future of electroceuticals looks promising despite the fact that somewhat mysterious and esoteric ideas such as acupuncture points, Chinese meridians, and the spiritual life force Chi appear to confound the scientific issue.

1.3.5.6 Electrical stimulation to promote osteogenic differentiation

The discovery of the existence of electric fields in biological tissues has led to a variety of efforts to develop techniques to use electrical stimulation for therapeutic applications, for example, to stimulate human mesenchymal stem cell (hMSC) differentiation and corresponding extracellular matrix (ECM)/tissue formation under physiological loading conditions. Several piezoelectric materials have been employed for different tissue repair applications, particularly in bone repair, where charges induced by mechanical stress were found to enhance bone tissue formation, as well as in neural tissue engineering, in which electric pulses can stimulate neurite directional outgrowth to fill gaps in nervous tissue injuries (Rajabi et al., 2015).

During mechanical deformation, piezoelectric scaffolds generate low-voltage outputs or streaming potentials (see box) that were found to promote chondrogenic differentiation. In contrast, piezoelectric scaffolds with a high-voltage output appear to promote osteogenic differentiation. Hence, piezoelectric "smart" materials are attractive as scaffolds for regenerative medicine strategies due to their inherent electrical properties without the need for external power sources for electrical stimulation (Damaraju et al., 2017; Tandon et al., 2018).

Streaming potential

A streaming current and streaming potential are two interrelated electrokinetic phenomena studied in the areas of surface chemistry and electrochemistry. They are an electric current or potential which originates when an electrolyte is driven by a pressure gradient through a channel or porous plug with charged walls.

1.3.6 Drug and gene delivery vehicles

The systemic administration of drugs to treat bone diseases is often associated with poor uptake of the drug in the targeted tissue, potential systemic toxicity, and suboptimal efficacy. To overcome these limitations, microsized and nanosized drug carriers that exhibit specific affinity for bone are being developed to treat bone pathologies (Rotman et al., 2018).

Targeted drug delivery is directed toward delivering medication to a patient so that the local concentration of this medication increases in places where it is required, that is, increases in some parts of the body relative to others that do not require medication. This strategy of delivery is largely grounded on nanomedicine that attempts to employ nanoparticle-mediated drug delivery to combat known disadvantages of conventional oral administration or gavage, or intravenous and intraperitoneal systemic drug delivery. Specific nanoparticles are being loaded with appropriate drugs and directed to those parts of the body where there is diseased tissue. This must be done by avoiding interaction with healthy tissue as well as the host's defense mechanisms. There are two kinds of targeted drug delivery: *active targeted drug delivery* by antibody medications, and *passive targeted drug delivery* based on the (controversial) enhanced permeability and retention effect.

The different types of drug delivery vehicle include a wide variety of polymeric micelles, liposomes, lipoprotein-based drug carriers, amyloid peptides, virus-like particles, nanoparticle drug carriers, dendrimers, and several others (Jelinek, 2013). An ideal drug delivery vehicle should be nontoxic, biocompatible, nonimmunogenic, biodegradable, and, most importantly, able to avoid recognition by the host's defense mechanisms. Hence, since stealth and camouflage are required, approaches akin to a "Trojan horse" have been developed. Apart from biomolecule-based drug delivery vehicles, there are entirely synthetic polymeric micelles prepared from amphiphilic block co-polymers such as methoxypolyethylene glycol-poly(D,L-lactic acid) micelles that consist of both hydrophilic and hydrophobic monomer units (hence, amphiphilic) able to carry drugs with low solubility (Saltzman and Torchilin, 2018).

PEG–PLLA-grafted copolymers are being developed for future somatic gene therapy. At present, local and controlled viral gene DNA release is still plagued by severe safety problems. Hence, non-viral DNA delivery vehicles are used instead such as polymers, proteins, glycosaccharides, and motile cells such as red blood cells, macrophages, and phagocytes. Unfortunately, their activity frequently affects cell viability and shows poor transfection efficiency. PEG-*y*-PLLA-DNA condensates appear to be promising candidates for future local and controlled DNA release for gene therapy. For other potential polymer carriers, refer to Chapter 2.3.

Dendrimers are polymer-based drug- or gene-delivery vehicles with a core that branches out tree-like in regular intervals to form a small, spherical, and very dense nanocarrier structure (Figure 1.21, left). Dendrimer molecules are characterized by their structural perfection, constituting highly symmetrical spherical compounds. Their physical characteristics include monodispersity, water solubility, encapsulation ability, and the capability to attach a large number of functionalizable peripheral groups that make these macromolecules appropriate candidates for drug and gene delivery vehicles (Gupta and Perumal, 2014). Poly(amidoamine) (PAMAM) is among the most well-known dendrimers. The core of PAMAM is a diamine (commonly ethylene diamine) that is reacted with methyl acrylate, and then another ethylene diamine molecules to make the generation 0 (G-0) PAMAM (Figure 1.21, left).

Much research is on the way to use dendrimers as gene delivery vehicles to traffic genes into cells without damaging or deactivating the host's DNA. PAMAM dendrimer/DNA complexes were used to encapsulate functional biodegradable polymer films for substrate-mediated gene delivery. Figure 1.21 (right) shows schematically a cell-specific targeting dendrimer A connected by a strand of DNA to a second dendrimer B. DNA–dendrimer conjugates are potential cancer targeting and imaging agents or therapeutics. Differentially functionalized dendrimers covalently conjugated to complementary deoxyoligonucleotides can readily form duplex combinatorial nanoclusters that constitute cancer cell-specific ligands hybridized to an imaging agent or drug. Cell-specific target ligands (e.g., folic acid) may be appended to dendrimers A and B (Sampathkumar and Yarema, 2005).

In other developments, biomaterials have been engineered to bridge the extracellular barriers in gene delivery and to support direct gene delivery to target tissues. Substrate-mediated transfection that sustains the release of "naked" DNA or vector/DNA complexes also supports cell growth (Fu et al., 2008). In this context, DNA as a biomaterial has evoked great interest as a potential platform for therapeutics and diagnostics, and as hydrogel scaffolds, owing to the relative ease of programming its robust and uniform shape, site-specific functionality, and controlled responsive behavior (Bila et al., 2019).

Other biomimetically engineered gene delivery vehicles, acting as "Trojan horses" to ferry therapeutic nanoparticles for gene repair, include viral particles, synthetic viruses, non-viral gene carriers such as red blood cells, macrophages, and phagocytes as well as mesenchymal stem cells (Jelinek, 2013). Nanoparticles loaded with specific cancer-fighting drugs are tagged with DNA bar codes to identify those nanoparticles that most efficiently deliver their cargo to those tissues or organs where they are therapeutically required (Dahlman, 2019). Novel zwitterionic so-called

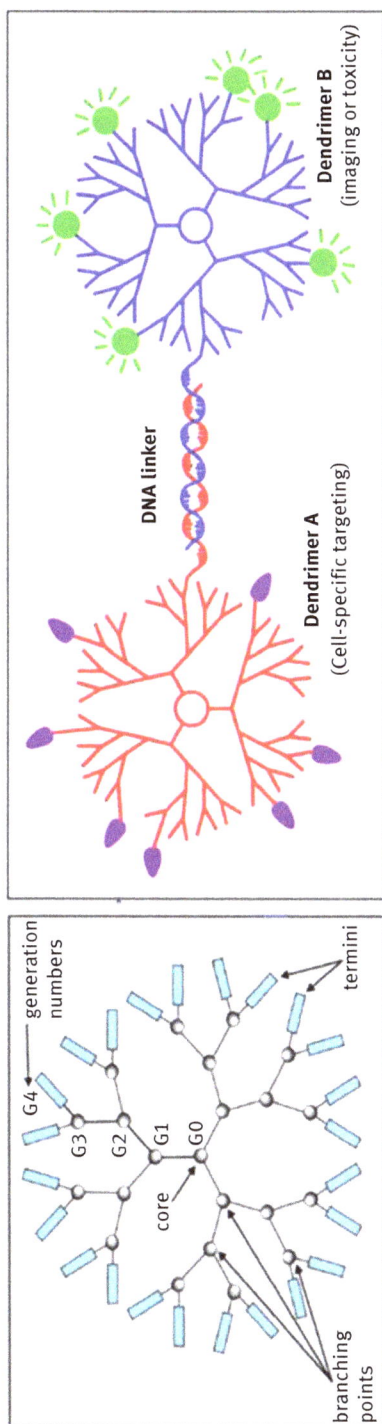

Figure 1.21: Left: Typical spherical three-dimensional morphology of a dendrimer showing, for example, addition of further PAMAM generations at branching points. Right: DNA–dendrimer conjugates are potential cancer targeting and imaging agents or therapeutics (Sampathkumar and Yarema, 2005). © With permission by Elsevier.

Janus dendrimer systems have been developed that are composed of two distinct dendrons with superior protein binding and protein repelling properties, respectively, for efficient spontaneous *in vivo* protein loading and enhanced protein unloading during delivery (Wang et al, 2019a). Amorphous calcium carbonate (ACC) with addition of polyacrylic acid, citric acid, adipic acid, 6-aminocaproic acid, 4-aminobutyric acid, and hexanoic acid shows extremely high porosity and excellent long-term stability that make these ACC-additives promising candidates for drug delivery application (Sun et al., 2019). Although proteins, such as monoclonal antibodies, have shown promise as bone-targeting molecules, they suffer from several limitations based on their generally large molecular size, high production cost, and undesirable immune responses. A viable alternative associated with significantly less side effects is utilization of small molecule-based targeting moieties (Carbone et al., 2017).

1.3.7 Three-dimensional-printed scaffolds for tissue engineering

One of the current milestones in tissue engineering has been the development of 3D scaffolds that guide cells to form functional tissue. Today, 3D-printed scaffolds are being used for osseochondral tissue engineering, bone tumor therapy, general bone tissue engineering, and a plethora of other medical applications. Numerous scaffolds have been designed, tested, and clinically evaluated and applied in orthopedics aimed to improving cell viability, attachment, proliferation and homing, osteogenic differentiation, vascularization, host integration, and load bearing (O'Brien, 2011; Jelinek, 2013; Roseti et al., 2017; Bishop et al., 2017; Meißner et al., 2019). Scaffolds can be constructed by conventional and rapid prototyping techniques. Conventional manufacturing approaches involve subtractive methods whereby sections of the material are removed from an initial block to achieve the desired shape. In contrast, rapid prototyping a.k.a. additive manufacturing (AM, see Chapter 3.1.3.4) is an additive fabrication process that yields the final 3D object through layer-by-layer deposition. An important improvement is the possibility to create custom-made products by means of computer-aided design (CAD) technologies, starting from patient's medical images in conjunction with computer-generated models. Figure 1.22 shows common physicomechanical and biological strategies that need to be considered when applied to manufacture and functionalize bone tissue scaffolds.

To achieve simultaneous addition of cells during the scaffold fabrication, novel robotic assembly and automated 3D-cell encapsulation techniques are being developed. As a result of these technologies, tissue-engineered constructs can be prepared that contain a controlled spatial distribution of cells and growth factors, as well as engineered gradients of scaffold materials with a predictable microstructure (Hutmacher et al., 2004; Wubneh et al., 2018).

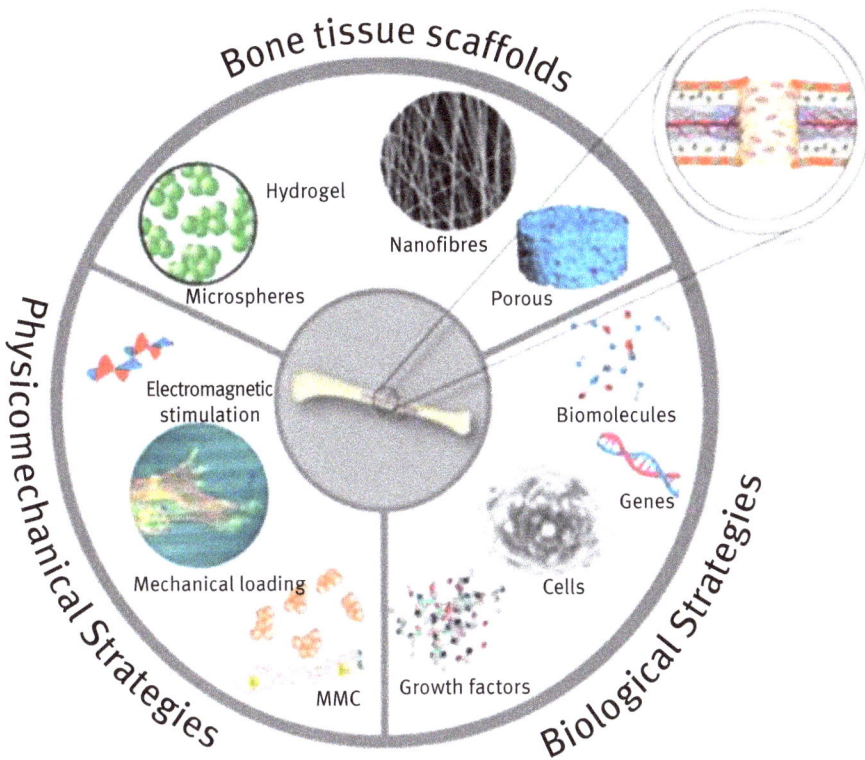

Figure 1.22: Different types of scaffolds (porous matrix, nano-fiber mesh, hydrogels and microspheres) are used to deliver bioactive molecules. This can be combined with a number of physicomechanical strategies to enhance treatment of various bone tissue defects and diseases (adapted from Fernandez-Yague et al., 2015, (MMC) metal matrix composite). © With permission by Elsevier.

Additive manufacturing of scaffolds is generally performed by either solid free-form fabrication (SFF) or laser-engineered net shaping (LENS™). SFF involves 3D layer-by-layer printing from a CAD file. This process has been successfully used to fabricate polymer, ceramic, metal, and composite scaffolds for bone tissue engineering. LENS™ is a layer-by-layer process using a high power laser up to 2 kW to melt metal powders to shape 3D structures based on CAD data. The laser beam is focused onto a metal substrate to create a pool of molten metal whereby metal powder is externally fed into the metal pool in controlled, that is, inert environment. Moving the substrate in the X–Y direction creates a pattern and adding material in the desired area forms a layer that, in time, extends in the Z direction. The next layer is built on top of the previous layer, and the procedure is repeated until the complete 3D body is produced. Although the LENS™ process is similar to other rapid prototyping technologies in its approach to fabricate a solid component by layer additive methods, it is unique in that fully dense metal components are fabricated directly from raw materials, bypassing initial forming operations such as casting, forging, and rough machining (see Chapter 3.1.3.4).

Not only the macrostructure, but also the pore structure can be controlled during LENS™ processing. Through this route, the interconnected porous architecture of trabecular bone can be simulated. Its importance relies on the fact that such architecture facilitates osteogenic cell recruitment, efficient vascularization, and unimpeded flow of oxygen, nutrients, and waste products.

In orthopedic trauma repair, treating full-layer injury of bone and cartilage is still a significant challenge when joint damage includes chondral defects difficult to repair. Hence, multilayer composite scaffolds are designed that contain collagen, bone, and calcified layer intended to simulate full-thickness bone-cartilage structure. In particular, bone and calcified layers are synthesized by 3D-printing technology (Li et al., 2018).

While 3D-printed scaffolds have been successfully applied in various field of medicine, one of the drawbacks of porous scaffolds is that they are, independent of composition, mechanically weak, unlike their natural paragons. As a result of their production process, porosity within most scaffolds is uniformly distributed throughout the scaffold dimension. However, the scaffold may not need to be uniformly porous. Indeed, natural bone does not show a uniform distribution of porosity but instead has a highly porous core (trabecular bone) surrounded by a strong and dense outer shell (cortical bone). With the goal to replicate this bone structure, a porosity gradient from the center to the periphery of the scaffold must be built up through complex design and manufacturing that will ensure mechanical integrity and scaffold interconnectivity. Another limitation of current biodegradable polymer-ceramic scaffolds is grounded in the different rates of degradation of their constituents *in vivo*. Usually, the polymer matrix of composite scaffold degrades much faster than its ceramic component that, in many cases, is hydroxylapatite (e.g., Koupaei and Karkhaneh, 2016). This causes the scaffold structure to degrade unevenly, and thus, may trigger osteolysis. To remedy this issue, ACP with higher solubility could be used instead of hydroxylapatite with the added benefit of generation of a calcium-supersaturated environment for enhanced mineralization. Another route may be based on developing polymeric matrices with higher *in vivo* stability so that the degradation rates of polymer and ceramics match (Bose et al., 2012).

Chapter 2
Types and properties of biomaterials

Robert B. Heimann, Mitsuo Niinomi, Matthias Schnabelrauch

In this chapter, the crystallographic structure, chemical composition, mechanical and thermal properties, biomedically relevant characteristics, as well as current and potential clinical applications of typical biomaterials will be reviewed. This essential information will provide the physicochemical underpinning of a deeper understanding of the eminent role biomaterials are playing today in the complex landscape of our modern world. The structure–property–application profiles of typical metallic, ceramic, polymeric, and composite biomaterials are being described, showing impressively their growing medical importance and socioeconomic impact.

2.1 Metals (Robert B. Heimann, Mitsuo Niinomi)

Today, stainless steels, CoCr-based alloys, titanium and titanium alloys, tantalum, and biodegradable magnesium alloys are the metallic materials used successfully in orthopedic, dental, and in a variety of other biomedical applications. Detailed descriptions of structure, properties, and biomedical application of metals can be found in the comprehensive reviews by Helgen and Breme (1998), Narushima (2010), Ratner et al. (2013), Williams (2014), Weinans and Zadpoor (2016), Love (2017), Ghosh et al. (2018), Thomas et al. (2018), and Niinomi (2018, 2019).

2.1.1 Corrosion-resistant metals

2.1.1.1 Stainless steels
High-alloyed stainless steels are used as biomaterials predominantly for their excellent corrosion resistance in the aggressive body environment as well as their high strengths in load-bearing applications. Their corrosion resistance is based on the high chromium content of at least 12% that enables stainless steels to form a well-adhering and dense chromium oxide surface film that provides a chemical barrier to protect the underlying iron–chromium–(nickel) alloy from further corrosion.

Structure
There are four basic types of stainless steels: ferritic (α-iron) with body-centered cubic (bcc) structure, martensitic (σ-iron) with body-centered tetragonal (bct) structure, austenitic (γ-iron) with face-centered cubic (fcc) structure (Figure 2.1), and two-phasic

https://doi.org/10.1515/9783110619249-002

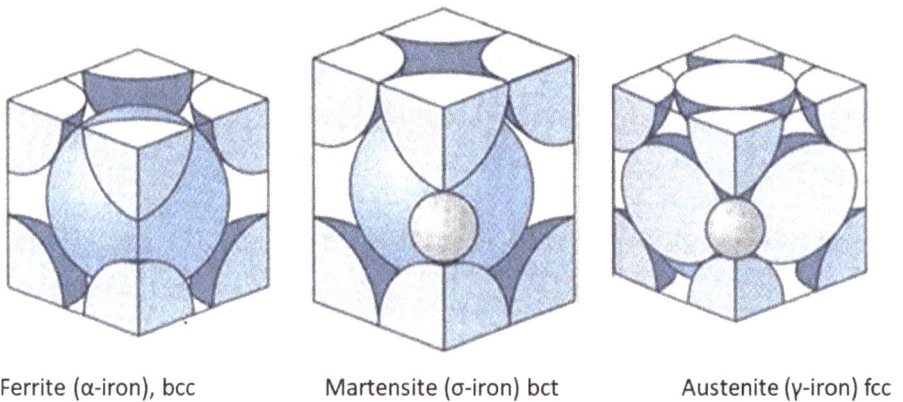

Ferrite (α-iron), bcc   Martensite (σ-iron) bct   Austenite (γ-iron) fcc

Figure 2.1: Schematic rendering of the crystallographic structure of stainless steels. The small spheres denote possible interstitial positions of carbon atoms in martensitic and austenitic stainless steels. The bct structure of the martensitic phase is stabilized by the distortion of the ferritic (bcc) structure by the interstitial carbon atom. © Permission granted under Creative Commons Attribution License (CC BY 3.0).

austenitic–ferritic duplex steels. The iron-rich section of the metastable Fe–Fe$_3$C phase diagram is shown in a simplified version in Figure 2.2.

At the peritectic reaction point, melt L with 0.53% C combines with ferritic δ-iron with 0.09% C to form austenitic γ-iron containing 0.17% C. This reaction occurs at 1,495 °C and can be written as L (0.53% C) + δ-iron (0.09% C) ⟷ γ-iron (0.17% C). In the carbon-free composition, at 912 °C, ferritic α-iron transforms allotropically to austenitic γ-iron. At a carbon content of about 0.8 mass%, there is an invariant (eutectoid) point at which α-iron (ferrite), γ-iron (austenite), and carbon (as cementite, Fe$_3$C) are in equilibrium according to γ-iron (0.8% C) ⟷ α-iron (0.02% C) + Fe$_3$C (6.67% C). On increasing the carbon content to 2 mass%, γ-iron is in a eutectic equilibrium with Fe$_3$C and melt (L) according to L (2.0%C) ⟷ γ-iron (2.08% C) + Fe$_3$C (6.67% C). It should be emphasized that the maximum solubility of carbon in γ-iron (austenite) is 2.08 mass% at 1,148 °C. Contrariwise, the maximum solubility of carbon in α-iron is only 0.02% C at 723 °C.

Finally, at 4.3 mass% C, the phase association γ-iron, cementite (Fe$_3$C), and melt (L) is stable, characterized by the eutectic reaction L (4.3% C) ⟷ γ-iron (2.08% C) + Fe$_3$C (6.67% C). Cast iron exists between 2.0 and 6.7 mass% C. The steel portion of the metastable Fe–C phase diagram can be further subdivided into three regions: the hypoeutectoid region between 0 and 0.8 mass% C, the eutectoid region at 0.8 mass% C, and the hypereutectoid region between 0.8 and 2.0 mass% C.

Ferritic steels are binary iron–chromium alloys with Cr contents between about 12 and 30 mass%. Since these steels contain more than 12 mass% Cr, they do not undergo the (fcc)-to-(bcc) transformation and, thus, cool from high temperature as solid

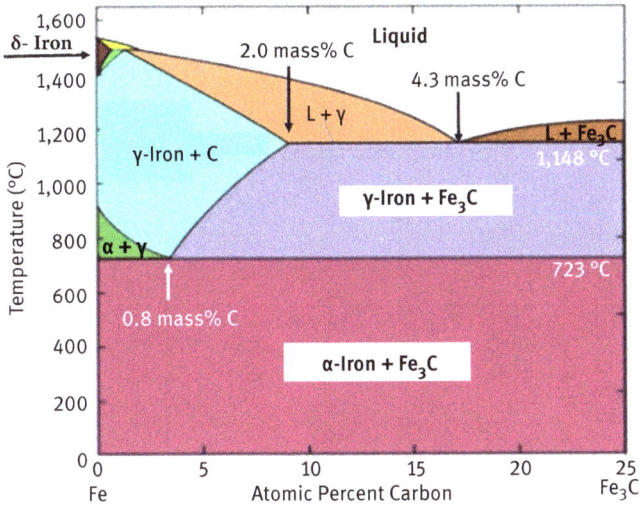

Figure 2.2: Fe-rich portion of the binary Fe–Fe₃C (cementite) phase diagram showing the stability fields of (bcc) α-iron (ferritic), and (fcc) γ-iron (austenitic). Critical carbon concentrations, and eutectic and subsolidus (eutectoid) transformation temperatures have been indicated.
© Permission granted under Creative Commons Attribution License (CC BY 3.0).

solution of Cr in α-iron. Chromium crystallizes in the $Im\bar{3}m$ (229) space group identical to that of α-iron, and an almost identical cubic unit cell length of $a_0 = 288$ pm. Hence, it extends the α-iron phase region by forming a solid solution with iron, and in turn, reduces the γ-iron phase region (Figure 2.3). As a result, a so-called γ-loop is formed in the Fe–Cr phase diagram that divides the phase space into an austenitic γ-iron (fcc) region at Cr contents below about 12% and a ferritic α-iron (bcc) region beyond. In the subsolidus region below about 800 °C, the martensitic σ-iron phase is stable.

Most *martensitic steels* contain between 11 and 18 mass% Cr with sufficient carbon content (0.2–1.0%; Table 2.1) so that a martensitic structure can be produced by quenching from the austenitic phase region. Hence, martensite is a metastable phase consisting of a supersaturated interstitial solid solution of carbon in ferritic iron with bct structure (Figure 2.1). In carbon-containing steels, the appearance of the martensite changes with carbon in the interstitial sites. Low-carbon steels produce lath-type martensite (Figure 2.4B), whereas high-carbon steels produce plate-type martensite when carbon is completely dissolved into the austenite matrix.

Color etching of stainless steels
Certain color etchants reveal residual deformation, segregation, and crystallographic orientation of grains in ferritic, martensitic, and austenitic stainless steels (Figures 2.4C and 2.6C). The color etching reagents deposit a thin transparent film on the steel surface that produces characteristic interference colors, dependent on film thickness and grain orientation. Some color etchants are

phase-specific. For example, electrolytic etching of 7 Mo-duplex steel in 20% NaOH at 3 V d.c. for 10 s reveals tan-colored ferrite (δ-iron), orange-colored martensite (σ-iron), and colorless austenite (γ-iron) (Vander Voort, 2004). This effect is very useful for image analysis work where one needs to detect one particular phase and measure its amount, grain size, and spacing. A typical reagent for color etching of stainless steels contains 100 mL stock solution (1,000 mL water, 200 mL HCl, 24 g $NH_4F \cdot HF$) plus 0.1–0.2 g $K_2S_2O_5$ for martensite stainless steels and 0.3–0.6 g $K_2S_2O_5$ for ferritic and austenitic stainless steels (Beraha's BI reagent; Beraha and Sphigler, 1977).

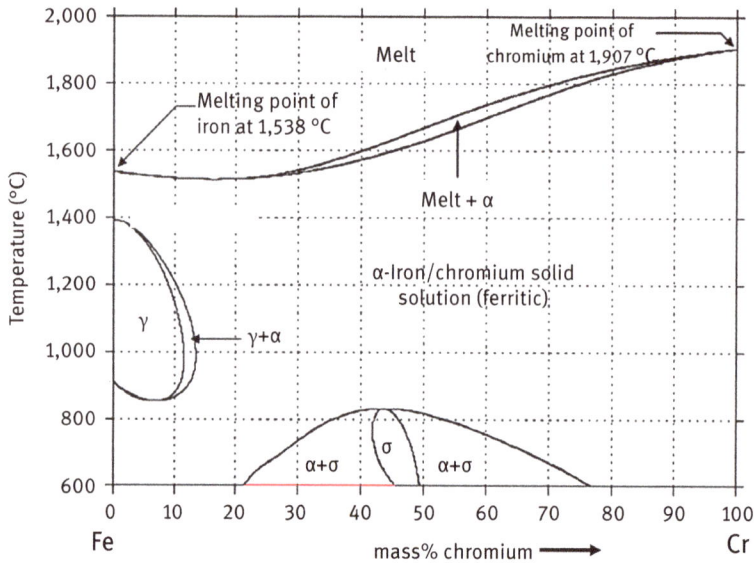

Figure 2.3: Iron–chromium phase diagram. © Permission granted under Creative Commons Attribution License (CC BY 3.0).

Table 2.1: Average compositions (balance: Fe) and mechanical strengths of typical stainless steels (Smith, 1996).

AISI designation	Cr (mass%)	Ni (mass%)	C (mass%)	Tensile strength (MPa)	Yield strength (MPa)
Ferritic stainless steels					
403 (group 1)	12	Mn 1.0	0.15, Si 0.5	485	310
405 (group 1)	13	Mn 1.0	0.08, Si 1.0	470	200
430 (group 2)	17		0.012	517	345
446	25	Al 1.5	0.20	552	345

Table 2.1 (continued)

AISI designation	Cr (mass%)	Ni (mass%)	C (mass%)	Tensile strength (MPa)	Yield strength (MPa)
Martensitic stainless steels					
410	12.5		0.15	517	276
440A	17		0.70	724	414
440A (Q + T)*	17		0.70	1,828	1,690
440C	17		1.10	759	276
440C (Q + T)*	17		1.10	1,966	1,897
Austenitic stainless steels					
301	17	7		759	276
304	19	10		580	290
304L	19	10	0.03	559	269
316	17	12, Mo 2	<0.07	580	290
316L	17	13, Mo 2	<0.03	560	290
321	18	10	Ti = 5× %C	621	241
347	18	10	Nb = 10× %C	655	276
Duplex stainless steels					
182 (2205)	22	5–6; Mo 3	<0.03, N 0.18	621	448
Precipitation-hardened duplex stainless steels					
630 (17-4PH)	16	4, Cu 4	0.03	1,311	1,200
632 (17-7PH)	17	7, Al 1	<0.09	1,500	1,200

Q + T, quenched and tempered.

Austenitic steels with (fcc) structure are ternary iron-chromium-nickel alloys containing about 16–26 mass% Cr and 6–12 mass% nickel that stabilize the structure down to ambient temperature. Figure 2.5 shows the phase relations in the ternary Fe–Cr–Ni compositional space.

Figure 2.6A shows the typical microstructure of an austenitic stainless steel (AISI type 316L) with annealing twins, whereas Figure 2.6B shows lamellar ferrite (α-Fe) and austenite (γ-Fe) in a 2,205 duplex stainless steel (Vander Voort, 2011).

Figure 2.4: (A) Microstructure of a single-phase ferritic (α-iron) stainless steel, etched in 2% nital etch. (B) Lath-type martensite (σ-iron) in AISI type 8620 low alloy tool steel, etched in 2% nital etch (Vander Voort, 2011). (C) Equiaxed ferrite grains, color-etched in Beraha's BI reagent (see box) (Vander Voort, 2005). © Images courtesy of Dr George Vander Voort.

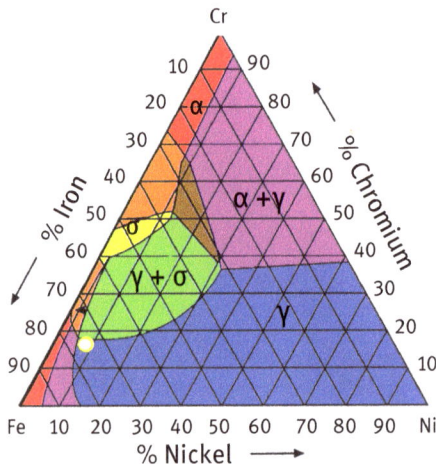

Figure 2.5: Ternary phase diagram Fe–Cr–Ni with stability fields of different stainless steels (α Cr, γ austenitic s.s, σ martensitic s.s.). The white dot represents the composition of the prototypical 18Cr8Ni s.s. © Permission granted under Creative Commons Attribution License (CC BY 3.0).

Properties

Table 2.1 lists the average compositions and mechanical properties of selected stainless steels.

The different types of stainless steel show remarkably variable properties. Ferritic steels cannot be hardened by heat treatment and also cannot be welded to high standards. AISI 430 steel is a general purpose ferritic steel with 17 mass% Cr and very low

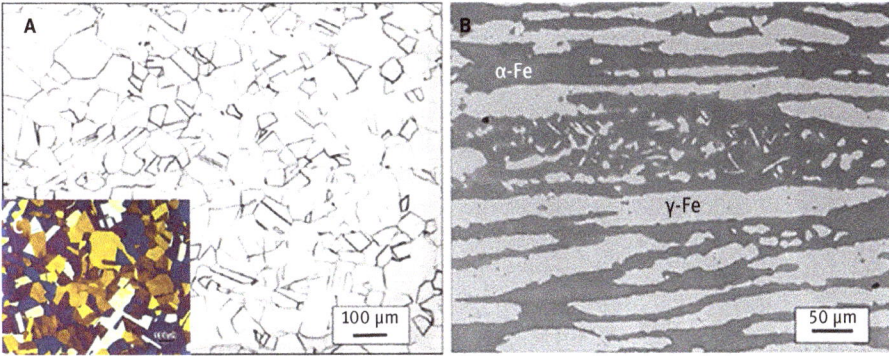

Figure 2.6: (A) Austenitic grains with annealing twins in AISI-type 316L stainless steel, etched in a solution of cupric oxide in ethanolic hydrochloric acid. (B) Ferrite (dark grey) and austenite (light gray) grains in two-phasic precipitation-hardened 2205 duplex stainless steel (Table 2.1), electrochemically etched in 20% NaOH at 3 V d.c. for 12 s. (C) AISI 316 austenitic stainless steel, color-etched with Beraha's BI etch (see box) (Vander Voort, 2011). © Images courtesy of Dr George Vander Voort.

carbon content of 0.012 mass% C distributed as carbide particles that are precipitated among equiaxed ferrite grains (Figure 2.4, C). To some extent, this reduces the corrosion resistance but even though novel ferritic steels have been developed with very low carbon and nitrogen levels, their limited corrosion resistance does not allow application in biomedical devices. In contrast, AISI 410 and 440, and X39Cr13 martensitic steels can be tempered and hardened, and used in applications where strength and toughness are more important than corrosion resistance, e.g., for surgical instruments.

Austenitic stainless steels retain their (fcc) structure at all common heat treatment temperatures, even though they cannot be heat-hardened but only strengthened by work-hardening. The (fcc) structure of these steels is highly deformable and provides a large number of slip systems since the unit cell contains close-packed slip planes {111} that, together with the slip direction <110>, account on permutation for 12 slip systems. Typical austenitic steels include AISI 304L with 19 mass% Cr, 10 mass% Ni, and 0.03 mass% C as well as AISI 316 containing on average 17% Cr, 12% Ni, 2% Mo, and 0.03–0.07% C (Table 2.1). Although these steels normally have better corrosion resistance than ferritic and martensitic steels on account of the fact that carbides can be retained in solid solution by rapid cooling from high temperatures, they become susceptible to intergranular corrosion when chromium-bearing carbides (e.g., $Cr_{23}C_6$) exsolve from the steel matrix and precipitate along grain boundaries during welding or slow cooling from high temperatures through the 870–800 °C range. However, this disadvantage can be somewhat remedied either by lowering the maximum carbon content in the alloy to about 0.03 mass% C as in AISI 316L, or by adding niobium (10× minimum % C) as in AISI 347 or titanium (5 × % C) as in AISI 321 (Table 2.1). The alloying elements Nb and Ti combine preferentially with carbon and stabilize the austenitic crystal structure.

Intergranular corrosion

Intergranular corrosion involves a localized corrosive attack at or adjacent to the grain boundaries. Normally, the metal of an alloy corrodes uniformly with the grain boundaries being only slightly more reactive than the surrounding matrix. Under certain conditions, for example, in the presence of precipitated carbides, the grain boundary region can become very reactive, resulting in intergranular corrosion and, thus, causing loss of strength or even disintegration along grain boundaries.

Another strengthening process exists that, after welding, involves a high-temperature solution treatment (ST) at 500–800 °C to redissolve the chromium carbide into a solid solution, followed by water quenching (WQ). These measures turn the steels from a so-called *sensitized* condition into a *stabilized* condition.

Biomedical AISI 316L stainless steel

In the prototypical biomedical stainless steel AISI (or SUS) 316L, the alloying elements combine synergistically to yield a metallic biomaterial uniquely suited for a large variety of medical applications, that is, Ni stabilizes the austenitic microstructure, Cr produces a thin and relatively durable passivating oxide film at the material's surface, and Mo reduces the risk of pitting and crevice corrosion in Cl-containing solutions including the extracellular fluid (ECF).

Biomedical applications

Typical biomedical applications of biotolerant surgical austenitic stainless steels are in osteosynthetic devices, bone fixation screws and plates, cardiovascular stents, suture and staple materials, and surgical instruments. Starting in 1926, these high-alloyed steels were initially used for osteosynthetic devices, followed in the 1940s by intramedullary Küntscher steel nails based on AISI 316. The first hip endoprosthetic implants developed by Sir John Charnley in 1962 contained a one-piece stainless steel femoral stem and head. With this groundbreaking development, the road was opened for further improved implant designs based on Ti6Al4V alloy that today make total hip arthroplasty one of the most often performed and in their clinical outcome highly successful operations.

Stainless steels designed for medical implant application must be suitable for close and prolonged contact with human tissue, that is, be reasonably stable under warm and saline conditions provided by the ECF. While today the clinical requirements are drivers for the development of special implant steels, originally AISI 316 steel was used for this purpose. Resistance to pitting corrosion, coupled with the need to control the quantity and size of nonmetallic, that is, carbidic inclusions, have spawned special processing routes including vacuum melting or electroslag refining. ISO 5832-1 (2007) and ISO 5832-9 (2007) standards, respectively, specify wrought and high-nitrogen austenitic stainless steels for medical implants, such as aneurysm clips, bone plates and screws, femoral fixation devices, intramedullary nails and pins, and joints for ankles, elbows, fingers, shoulders, wrists, as well as knee and hip endoprostheses.

Cardiovascular stents are commonly fashioned from medical-grade AISI 316L. To prevent attachment of neointimal hyperplasia-causing cells, that is, the proliferation and migration of vascular smooth muscle cells that result in thickening of arterial walls and, hence, decreased arterial lumen space, antiproliferative drugs are used. Neointimal hyperplasia is the major cause of restenosis after percutaneous coronary interventions, such as stenting or angioplasty. Drug-eluting coatings deposited on the stent surface normally achieve control of neointimal proliferation for permanent stents. For example, a bilayer consisting of TiO_2 and poly(lactide-*co*-glycolide) as carrier of an antiproliferative sirolimus (rapamycin) drug was deposited on AISI 316L vascular stents to reduce the incidence of late stent thrombosis (Leng et al., 2019). Despite this, novel developments have earmarked other compositions for stents. For example, the need to reduce the strut thickness of cardiovascular stents has triggered the development of a new high strength radiopaque alloy, based on additions of platinum to a chromium-rich iron-based matrix (O'Brien et al., 2010). The addition of platinum appears to increase the typical yield strength by solid solution strengthening, resulting in average yield strength of 480 MPa. The surface of the material consists of chromium oxide, which contributes to the high corrosion resistance observed. Cell assays suggest that surfaces of this Pt-doped stainless steel endothelialize in a manner comparable to that of undoped stainless steel.

Stainless steels are also used as suture and staples material, applied to close large abdominal incisions (tummy tucks, C-section deliveries) and as orthodontic bonding appliances. Today, the surgical stapling procedure is preferentially used, allowing for rapid wound closure and facilitate an effective binding of wound sites for a period sufficient to create innate wound strength.

In spite of this, the use of martensitic steels for surgical instruments dwarfs the use for implants and suture material because of their ability to be resterilized, and hence, reused. As far as surgical instruments are concerned, designation ISO 7153-1 (2016) contains a survey and a selection of stainless steels available for use in the manufacture of surgical, dental, and specific instruments for orthopedic surgery. Dental and surgical instruments are produced extensively from martensitic stainless steels such as AISI 420 and AISI 440, owing to their hardness being tailored to application, that is, higher hardness for surgical cutting instruments and lower hardness but increased toughness for load-bearing applications. Devices made from these steels include bone curettes, chisels and gouges, dental burs, dental chisels, explorers, dental root elevators, forceps, hemostats, retractors, orthodontic pliers, and a wide range of scalpels. A precipitation-hardening stainless steel has been developed for the shaft and gripping mechanism of forceps used in modern endoscopic keyhole surgery a such special steels are less prone to distortion.

In conclusion, owing to their biocompatibility, proven history, well-known processing technology, and cost-effectiveness, stainless steels have long been a dominant force in the biomaterials market. They combine excellent corrosion resistance in short-term applications and easy machinability with well-known properties, processing routes, and established markets. On the downside, there are risks of pitting and intergranular corrosion as well as stress corrosion cracking in long-term applications in a biological environment rich in chloride, of stress shielding in endoprosthetic implant application owing to the high modulus of stainless steels, and of adverse effects of potentially allergenic and cytotoxic Ni and Cr contents in austenitic stainless steels.

Indeed, corrosion prior to *in vivo* passivation will likely create cytotoxic Ni and Cr metal ions released from the steel that are conveyed to the surrounding tissue, with potentially deleterious effect. To prevent this, novel nickel-free stainless steel grades are currently entering the market that will likely further secure stainless steel's standing in the biomedical industry. Nitrogen-alloyed, Ni-free, austenitic stainless steels with more than 1 wt% nitrogen are a new group of alloys with promising properties that exhibit an interesting combination of high strength and toughness with a high corrosion resistance in various environments (Yang et al., 2019). For example, a Fe25Mn20Cr3 Mo1.3 N alloy shows excellent corrosion resistance in Ringer's solution, artificial saliva, and artificial sweat. This combination of properties makes this alloy not only suitable for fasteners but also for medical and dental implant applications.

As a parting glance, while stainless steels are in no danger of becoming obsolete as a biomaterial, they are increasingly being bypassed in favor of alternative materials. Owing to the passivation of stainless steels, that is, the controlled oxidation of chromium to form a strongly adherent oxide film, there is no obvious mechanism for biodegradation, today an essential property of metallic biomaterials in many applications.

2.1.1.2 CoCr alloys

Owing to their excellent mechanical properties, high corrosion resistance, high wear resistance, and remarkable biotolerance, CoCr alloys have been recognized as effective metallic biomaterials and, thus, have been used for dental and medical devices ever since the first cast CoCrMo alloy, named Vitallium, was developed and applied in the 1930s (see chapter 1.2.1.1).

Structure

Depending on the ratio of cobalt to chromium, CoCr alloys show different crystallographic structures as shown in Figure 2.7. An fcc crystal structure is found in the β-Co phase (formerly called α-Co) with comparatively high strength and ductility, at least when compared to the intermetallic σ-phase stable between about 54 and 67 at% chromium. This σ-phase with a composition close to Co_2Cr_3 tends to be brittle and,

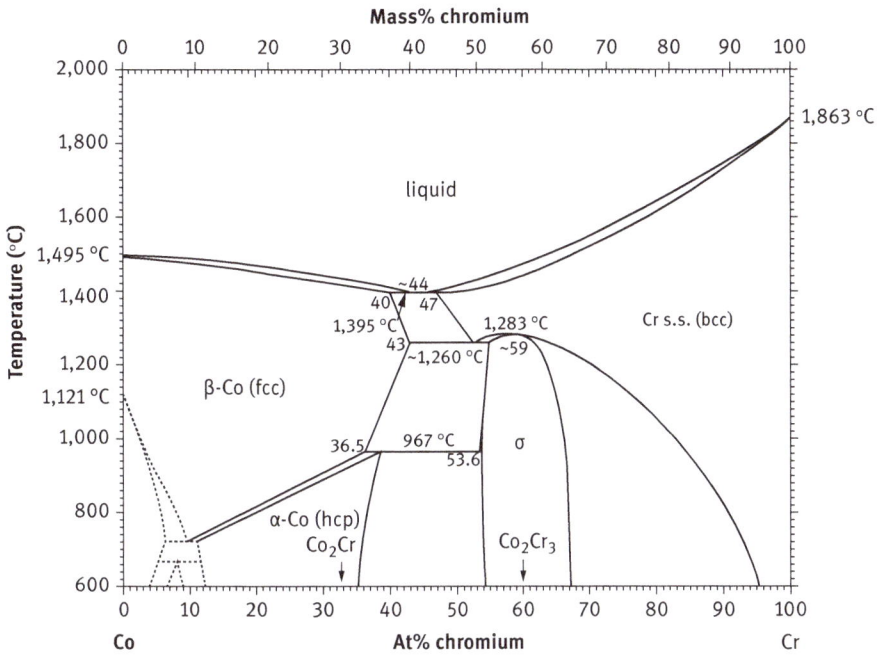

Figure 2.7: Binary phase diagram Co–Cr (modified from Ishida and Nishikawa, 1990). For explanation of phase reactions see Table 2.2. © With permission by Springer Nature.

hence, subject to fracture when subjected to high mechanical loads. Consequently, this compositional range should be avoided when designing devices intended for use in arthroplastic interventions.

The (fcc) crystal structure of β-Co (a_0 = 354.4 pm) is stable in the cobalt-rich part of the phase diagram (Figure 2.7), whereas chromium-rich solid solutions tend to have a (bcc) structure. The β-Co phase can transform to the α-Co phase (formerly called ε-Co) by a peritectoid reaction at high pressure, showing hexagonal close-packed (hcp) crystal structure with a space group of P6$_3$/mmc (194) (Table 2.2). There is a eutectic point at 1,395 °C and a composition of about 40 at% Cr where the melt is in equilibrium with the (fcc) and the (bcc) phases. Unalloyed chromium crystallizes in a bcc structure with a space group Im$\bar{3}$m (229) and a_0 = 288 pm.

In the complex binary Co–Cr system the following reactions occur as shown in Table 2.2.

Table 2.2: Phase reactions in the binary Co–Cr system (Ishida and Nishizawa, 1990; Love, 2017).

Phase reaction	Compositional range (at% Cr)	Temperature (°C)	Reaction type
L ↔ β-Co	0	1,495	Melting point of Co
β-Co ↔ α-Co	0	422	Allotropic reaction
L ↔ β-Co + Cr s.s.	40–47	1,395	Eutectic reaction
Cr s.s. ↔ β-Co + σ	43–55	1,260	Eutectoid reaction
β-Co + σ ↔ α-Co	38.3–53.6	967	Peritectoid reaction
Cr s.s. ↔ σ	59	1,283	Congruent transformation
L ↔ Cr	100	1,863	Melting point of Cr

Phase reactions in heterogeneous equilibrium phase diagrams
When a liquid of eutectic (from Greek *eu*, easy and *teksis*, melting) composition is cooled to the so-called eutectic temperature, the single phase "liquid" crystallizes to form two solid phases (solid solutions α and β) according to L ↔ α solid solution + β solid solution (*eutectic reaction*). According to the Gibbs phase rule, in a binary component system this is an invariant reaction, whereby three phases are in equilibrium with zero degrees of freedom. Another type of reactions frequently occurring is a *peritectic reaction* if the melting point of the two endmember components is quite different. During a peritectic reaction, a liquid phase reacts with a solid phase to form a new and compositionally different solid phase according to L + α ↔ β. Other invariant reactions involve only solid phases that transform into each other by eutectoid or peritectoid reactions. In a *eutectoid reaction*, a decomposing solid phase reacts toward two solid phases according to α ↔ β + γ, whereas in a *peritectoid reaction* two solid phases react to form another, under the prevailing conditions more stable, solid phase according to α + β ↔ γ. Hence, peritectic and peritectoid reactions are the inverse of the corresponding eutectic and eutectoid reactions. As the temperatures and compositions of the reacting phases are fixed, there are zero degrees of freedom. Any deviation in temperature or composition would remove the system from the equilibrium point into a two-phase region. More detailed information can be retrieved from the vast literature on phase transformation (e.g., Hillert, 2007).

The ternary Co–Cr–Mo diagram is shown in Figure 2.8 with the composition of ASTM F799 alloy indicated by a red dot. Details on the Co–Cr–Mo system have been provided by Gupta (2005) and Kareva et al. (2017).

High entropy alloys
Recent developments in the area of a novel class of multicomponent metallic alloys (Lei et al., 2019) may have important ramifications on the development of new high-strength CoCr alloys. The HEA ASTM F562-13 (Table 2.3) with an average

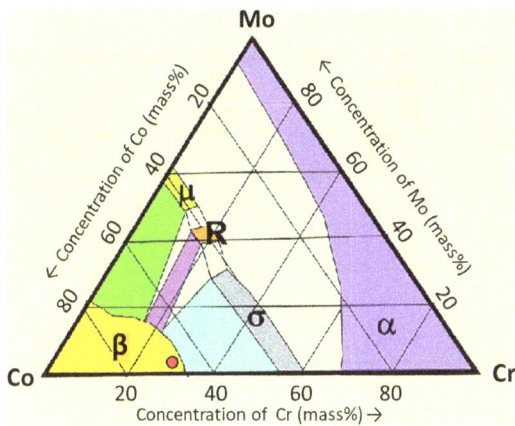

Figure 2.8: Ternary phase diagram of Co–Cr–Mo alloy. The composition of the advanced ASTM F799-11 alloy has been indicated by the red dot. © Permission granted under Creative Commons Attribution License (CC BY 3.0).

composition 33Co35Ni20Cr10Mo1Ti1Fe has shown substantial promise as an alternative material to AISI 316L stainless steel in application as cardiovascular stents. The material has noticeably higher yield strength and fracture toughness compared to AISI 316L, allowing for thinner struts without decreasing radial strength. Its ion release kinetics *in vivo* is comparable to that of stainless steel. Hence, the surfaces of both the Driver® (ASTM F562) and S7 (AISI 316L) coronary stents show a chromium-rich oxide layer that acts as a barrier to stave off the release of ions from the bulk material underneath the surface. Likewise, the HEA Phynox (ISO 5832-7) with composition 40Co20Cr20Fe14Ni6Mo has found application as material for the stem of hip endoprostheses (Semlitsch, 1984).

Other HEAs such as equiatomic CoCrNiFeMn alloy (Cantor et al., 2004; Wang, 2013) show exceptional strength and fracture toughness owing to their high configurational entropy that acts as a potent mechanism to stabilize the solid solution phase. Even though the alloy has not found biomedical uses yet, it might become a promising member of the large family of biomedically relevant CoCr alloys once its biocompatibility has been solidly established.

Recently, a novel Cu-HEA with composition 38.5Fe20Mn20Co15Cr5Si1.5Cu has been explored that showed high corrosion resistance as well as high strength and ductility (Nene et al., 2019).

Table 2.3: Compositions of medical-grade CoCrMo alloys (modified after Black, 1999).

Element	ASTM F75-12 (mass%)	ASTM F799-11 (mass%)	ASTM F90-14 (mass%)	ASTM F562-13 (mass%)
Co	59.9–69.5	58–59	45–56.2	29–38.8
Cr	27–30	26–30	19–21	19–21
Mo	5–7	5–7	–	9–10.5
Mn	1 (max.)	1 (max.)	1–2	0.15 (max.)
Ni	1 (max.)	1 (max.)	9–11	33–37
Fe	0.75 (max.)	1.5 (max.)	3 (max.)	1 (max.)
Si	1 (max.)	1 (max.)	0.4 (max.)	0.15 (max.)
C	0.25 (max.)	0.35 (max.)	0.1 (nomin.)	–
Others		N: 0.25 (max.)	P: 0.04 (max.)	Ti: 1 (max.), S: 0.01 (max.)
W	–	–	14–16	–

High entropy alloys

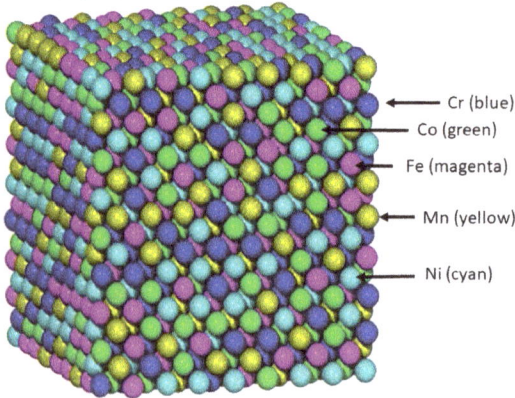

An early definition stipulated that HEAs are materials composed of equal or nearly equal concentrations of at least five metals between 5 and 35 at%. However, subsequent research has suggested that a more critical issue is the requirement of the constituents to form a solid solution without the presence of intermetallic phases since formation of ordered phases would decrease the entropy of the system. Such HEAs have exceptional mechanical (hardness, yield strength, fracture toughness, fatigue life, malleability, oxidation and corrosion resistance, endurance limit), electrical (superconductivity), and magnetic (superparamagnetism) properties, and may even be radiation-hardened materials. HEAs crystallize in fcc lattices when their average valence electron concentration (VEC) is ≥8, and in an (bcc) lattice when VEC is <6.87 (Guo et al., 2011). The atomic structure model of the fcc lattice of CoCrNiFeMn is shown in the figure at right (Wang, 2013). Note the statistical distribution of the metal ions, the disordered nature of which provides the high entropy property. Recently, TiZrNbTaFe (Popescu et al., 2018) and other Ti-based HEAs such as TiNbTaZrMo (Todai et al., 2017) with biomedical relevance are being developed with low modulus of elasticity but retained high mechanical strength (Ghiban et al., 2017).

Properties of CoCr alloys

CoCr alloys show high resistance to corrosion due to the spontaneous formation of a protective passive surface film composed mostly of Cr_2O_3 with minor amounts of cobalt and other metal oxides. CoCr alloys are well known for their excellent biocompatibility owing to the positive interacting of the oxidized surface with the physiological environment. The mechanical properties are similar to those of stainless steel described earlier. In general, they are a result of the multiphase structure and precipitation of carbides acting as reinforcement that strongly increase the hardness and the tensile strength of CoCr alloys. The microindentation hardness of CoCr alloys was found to be around 5 GPa (Sahin et al., 2015), and the tensile strength to be up to 655 MPa in ASTM F75-12 alloy, reaching a value of 900 MPa in ASTM F1537 alloy. The compositional ranges of FDA-approved cast and forged medical-grade CoCrMo alloys are shown in Table 2.3.

Whereas tensile and fatigue strengths of CoCr alloys increase strongly, they show decreased ductility with an associated risk of component fracture when heat-treated. This is a concern when the alloys are intended to be used as stems and/or femoral balls of hip endoprostheses. In an attempt to overcome the low ductility, molybdenum, nickel, tungsten, carbon, and/or nitrogen were added (Table 2.3). This resulted in the creation of advanced biostable alloys such as Co28Cr6Mo (ASTM F799-11), Co28Cr8Mo (ASTM F75) or Co20Cr15W10Ni (ASTM F90-14) (Narushima et al., 2015). In particular, the solubility of nitrogen in the CoCrMo alloys increases with increasing Cr content. By adding nitrogen, the pronounced β-Co phase stabilization and, in turn, the suppression of the brittle σ-phase renders the alloys more ductile, resulting in significant improvements in the mechanical properties of these CoCrMo alloys (Lee et al., 2008).

Biomedical application

Whereas basic CoCr alloys were applied in biomedical application since the 1930s, it was only after 1990 that second-generation CoCrMo alloys were developed and used for metal-on-metal (MoM) hip joint replacements. However, since then, CoCrMo hip replacements have suffered a sharp decline due to reduced biocompatibility related to excessive wear and concurrent accumulation of cytotoxic corrosion products in tissues surrounding the implant site. Despite excellent clinical results, the accelerated release of wear products and the associated adverse tissue response have been close to being show stoppers. Hence, these problems are being addressed by numerous research activities today (e.g., Liao et al., 2013). CoCrMo alloys such as Vitallium

(Co30Cr5Mo, ISO 58342-4) have been used as a biomaterial for MoM hip implants because of their high corrosion resistance and significantly lower volumetric wear loss compared to metal-on-polyethylene implants. This renders them less likely to fail by osteolysis and aseptic implant loosening, the most common hip implant failure mechanisms. However, even CoCrMo-based MoM hip implants suffered a comparatively high failure rate, with many patients experiencing unexplained pain, a development that eventually resulted in the need for widespread revision surgeries. Consequently, the regulatory agencies of the UK (MHRA) and the US (FDA) issued alerts for all MoM hip replacements in 2010 and 2011, respectively.

Further studies on hip simulators have revealed that despite the numerically lower wear debris volume, CoCrMo alloys produce a dramatically higher number of nanometer-sized wear particles up to one trillion particles per year that are able to trigger an inflammatory immune response cascade thought to be linked to implant failure, although the mechanistic details still remain poorly understood. This finding was rather surprising since CoCrMo alloy is nominally an extremely stable material, with high Cr content and associated formation of a protective chromium oxide surface film providing excellent passivity. However, chemical analyses of wear particles from periprosthetic tissue explanted from patients have revealed that the incriminated particles are composed predominately of Cr species, with only trace amounts of Co remaining. The dissolution of these wear particles leads to potential release of highly cytotoxic and even genotoxic Co^{2+} and Cr^{6+} species (Koronfel et al., 2018). Hence, it may be concluded that the presence of these ions in higher concentration could be triggers for other diseases, particularly for long-term implant recipients.

It is not surprising that recently the risk of possible adverse health consequences has caused a flurry of regulatory activity related to MoM total hip replacements using CoCr alloys. These concerns pertain predominately to the significant metal ion release during fretting wear when metal surfaces articulate against one another (Love, 2017). Particles eroded from either side of the MoM implant can invade surrounding areas of the joints, thus, creating adverse tissue reactions. Such particles can be identified by staining with Solochrome cyanine (Papadimitriou-Olivgeri et al., 2019). Although CoCr alloys can be competitively substituted for stainless steels in many applications, wider usage cannot be properly predicted since potential concerns remain. Particulate wear debris from orthopedic implants has been found to accumulate in tissue close to the implant, but also in distant lymph nodes, bone marrow, liver, and spleen. Furthermore, Co and Cr metal ions have been shown to be excreted in urine and through the bile. The large amount of nanosized metal particles and cytotoxic metal ions, and the increasing use of MoM hip and knee endoprostheses have led to serious concerns about carcinogenicity, hypersensitivity, local tissue toxicity, inflammation, and even genotoxicity and teratogenic effects (Jakobsen, 2008).

In addition to implant failure due to excessive production of wear particles, mechanical problems of MoM–CoCrMo alloys have occurred, linked to crack formation within a very fine-grained mixed hard phase (Liao et al., 2012) in high carbon CoCrMo alloy of type ASTM F75-12 (see Table 2.3). The composition of the mixed hard phase was found to consist of $M_{23}C_6$ (M = Cr, Co, Mo) structures, β-Co (fcc) phase, and Co_9Mo_{15} σ-phase (Liao et al., 2013; Stemmer et al., 2013). Coarser (>30 μm) types of mixed hard phases frequently reveal microcracks already below the articulating surfaces (Figure 2.9). Such subsurface microcracks are known to destabilize the gradient below the surface and the balance between tribochemical reactions and surface fatigue.

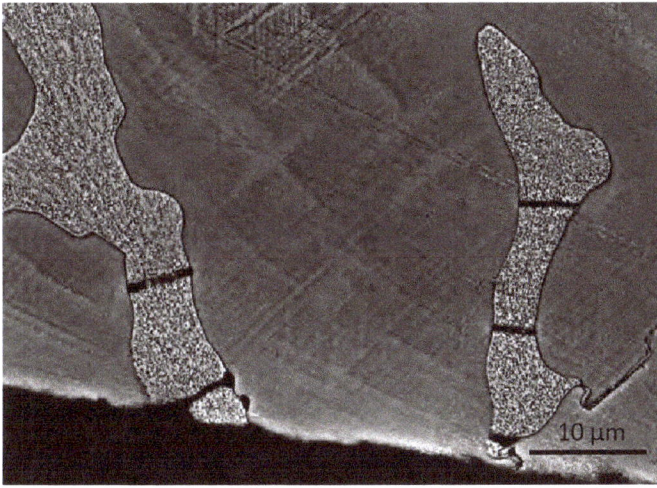

Figure 2.9: Cross section of a retrieved femoral head composed of ASTM F75 CoCrMo alloy showing microcracks within the mixed hard phase underneath the primary articulating areas (Stemmer et al., 2013). © With permission by ASTM International.

Attempts to reduce wear and erosion rates included reinforcement of CoCrMo alloy with up to 3 mass% of hydroxylapatite using Laser Engineered Net Shaping (LENS™). The addition of hydroxylapatite was found to result in stabilizing the α-Co (hcp) phase along with the more common β-Co (fcc) phase of the CoCrMo alloy. Whereas the hydroxylapatite-reinforced alloy showed microhardness comparable to that of the base material, its wear resistance was significantly increased on account of a tribofilm developing at the surface (Sahasrabudha et al., 2018). Other attempts to reduce the wear and corrosion rate of CoCrMo alloy such as ASTM 799–11 (Table 2.3) revolve around implantation of nitrogen ions that appear to form nanocrystalline CrN embedded in the CoCrMo matrix as well as at grain boundaries that enhance hardness, as well as ductility, and corrosion and tribocorrosion resistance (Guo et al., 2015).

2.1.1.3 NiTi alloys (Nitinol®)

Although Nitinol® was initially developed during the 1960s for military purposes (the suffix "nol" refers to Naval Ordnance Laboratory), subsequently it was realized that its properties may be useful for medical applications such as orthodontic wires, dental burs, self-expanding cardiovascular stents, and a plethora of other potential uses. Though being a stoichiometric titanium alloy, it is singled out here on account of its unique shape-memory property.

Structure

Although martensite-type NiTi was reported to have a distorted orthorhombic B19 (space group Pmna (53)) structure, presumably B19′ with the monoclinic space group P2$_1$/m (11) (Figure 2.10, left), first-principles density functional studies suggested that this type of structure is unstable relative to a base-centered orthorhombic (bco) B2 structure that, however, cannot store shape memory at the atomic level.

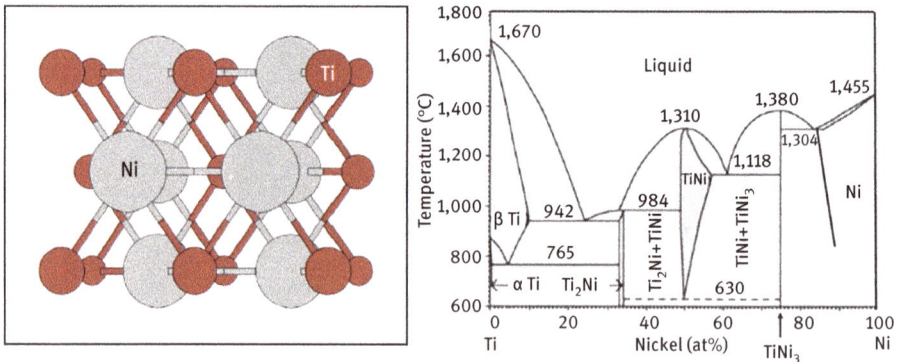

Figure 2.10: Crystal structure of martensitic NiTi with B19′ structure (left) and phase diagram of the binary system Ti–Ni (right). © Permission granted under the Creative Commons Attribution Share Alike 3.0 license.

Thus, the reported B19′ structure may be stabilized by residual internal stresses remaining after cooling (Huang et al., 2003). Experimental lattice constants of the B19′ martensitic structure of NiTi were found to be $a_0 = 464.6$ pm, $b_0 = 410.8$ pm, $c_0 = 289.8$ pm, $\beta = 97.8°$ (Kudoh et al., 1985). Elastic constants were reported by Šesták et al., (2010). Figure 2.10 (right) shows the Ti–Ni phase diagram, with the field of stability of NiTi indicated in shading.

Properties

NiTi is an intermetallic compound that exhibits a thermoelastic martensitic phase transformation around room temperature when cooled from the stronger, high-temperature austenitic phase to the weaker, low-temperature martensitic phase (see box). This inherent phase transformation is the basis for the unique properties of this alloy, in particular its shape memory (Yoneyama and Miyazaki, 2009), pseudoelasticity, and high damping capability (Liu and van Humbeeck, 1997). The martensitic phase of shape-memory alloys can easily deform to a new shape. However, when the alloy is heated through its transformation temperatures, it reverts to the austenitic phase and recovers its previous shape with great force (Figure 2.11). The temperature at which the alloy remembers its high-temperature shape when heated can be adjusted by slight changes in alloy composition and through heat treatment. The shape recovery process occurs over a range of just a few degrees, and the start or finish of the transformation can be controlled within a degree or two.

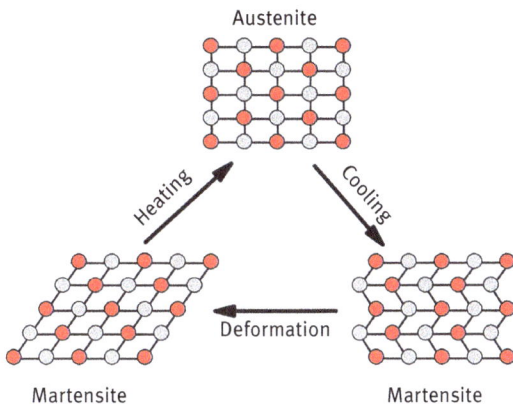

Figure 2.11: Schematics of shape-memory effect of TiNi. © Permission granted under Creative Commons Attribution License (CC BY 3.0).

Shape-memory effect

"Shape memory" refers to the ability of a material to remember its original shape after being plastically deformed while in the low-temperature martensitic form (Figure 2.11). The complex martensitic transformation involved in the shape-memory effect of NiTi is an instantaneous, thermoelastic diffusionless first-order displacive (umklapp) process during which atoms move cooperatively, and often by a shear-like twinning mechanism. During transformation, the bcc NiTi parent phase (austenite; B2 (CsCl) type; space group $Pm\bar{3}m$) shears and forms twinned martensite with bco (B19 (AuCd) type; space group $Pmna$) or monoclinic (B19′; space group $P2_1/m$) arrangements. During transformation, frequently, a rhombohedral R-phase with $P\bar{3}$ symmetry appears as an intermediate phase. Consequently, there are three transformation routes involving (i) a direct transformation without evidence of an R-phase during cooling or heating, (ii) a symmetric R-phase transformation with an R-phase intervening between the austenitic and the martensitic phases during both heating and cooling, and (iii) the most common asymmetric R-phase transformation route in which the R-phase appears during cooling but not upon heating. These changes in solid state allow properties such as single (one-way asymmetric R-phase transformation), double (two-way symmetric R-phase transformation) shape-memory effect, as well as superelasticity.

Figure 2.12 shows the color-etched austenitic NiTi (A) and the martensitic structure obtained after shape-memory deformation (B) (Vander Voort, 2009).

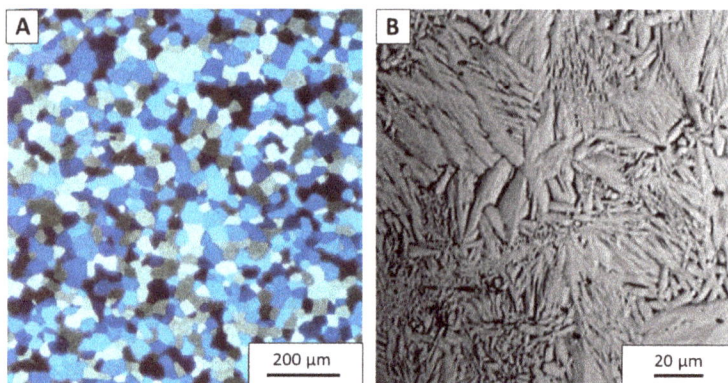

Figure 2.12: (A) Equiaxed austenitic microstructure of Nitinol® color-etched in 90 mL water + 10 mL HCl + 1 g NH_4 F · HF + 28 g $K_2S_2O_5$. Polarized light. (B) Martensitic structure of NiTi alloy after going through the shape-memory effect phase transformation. Etched with equal parts of HNO_3 + acetic acid + HF (Vander Voort, 2009). © Images courtesy of Dr George Vander Voort.

In addition, NiTi alloys show superelastic (pseudoelastic) behavior when deformed at a temperature slightly above their transformation temperature. In contrast to (mechanical) elasticity that relies on bond stretching or the introduction of defects such as dislocations in the crystal lattice, pseudoelasticity results from the reversible motion of domain boundaries during the martensitic phase transformation. Owing to the metastability of the martensitic state, the material reverts back instantaneously to the nondeformed austenitic state as soon as the stress is removed. This also means that in contrast to the shape-memory effect, no temperature change is involved. This process provides a very springy, that is, rubber-like elasticity. The corrosion resistance decreases slightly when the stoichiometry deviates by more than 2% Ni. In contrast, while alloying with elements of the Pt group or Mo improves the corrosion resistance, alloying with Cu, Fe, Mn, or Al decreases it (Shabalovskaya and van Humbeeck, 2009).

Biomedical applications

In addition to their superior mechanical properties, NiTi shape-memory alloys are very biocompatible and behave, in contact with simulated body fluid (SBF), in a bioinert way through passivation by formation of a thin TiO_2 surface film. NiTi alloys have found numerous biomedical applications: in the orthodontic field as wires, palatal arches, and distractors, in the orthopedic field as intraspinal implants and intramedullary nails, in the vascular field as venous filters, self-expanding vascular stents, and stent grafts, in the neurosurgical field as coils for intercranial angioplasty and microguidewires, and in the field of general surgery as mini-invasive surgical instruments such as endodontic files for root canal surgery, steerable biliary guidewires, and urologic graspers (Petrini and Migliavacca, 2011). It is also frequently used as intrauterine device for female contraception. These multifarious applications underline an important aspect of shape-memory alloys: the complexity of the material and its applications requires a close collaboration among clinicians, engineers, physicists, and chemists to accurately define the medical problem, to find the optimum solution in terms of device design, and, accordingly, to optimize the NiTi alloy properties.

Although NiTi shape-memory alloys, introduced to the market already during the 1960s, at this time promised to open an entirely new range of applications owing to their special mechanical properties, the impact of the unsolved allergenic effect of Ni including its cytotoxicity and potential carcinogenicity has limited the use of NiTi in Europe and the USA, even though the material has been successfully used in Russia and China for many novel biomedical devices (Navarro et al., 2008).

2.1.1.4 Titanium and Ti alloys

Since Ti-based biomaterials are of overwhelming importance to the contemporary biomedical device industry, their structures, mechanical properties, and biologically relevant characteristics as well as biomedical applications are being described in much detail below. Among many other pertinent issues, the chapter revolves around the quest for low-modulus β-type Ti alloys designed to counteract stress shielding of endoprosthetic implants while maintaining high strength and workability.

Structure

Titanium alloys are generally grouped into the allotropic crystalline forms α-, (α + β) – and β-Ti (Figure 2.13). The α- and β-phases obey (hcp) and (bcc) structures, respectively, with the latter having a lower atomic density. Since the elastic modulus (Young's modulus) is typically lower in β-type Ti alloys than in α- or (α + β)-type Ti alloys, these β-type Ti alloys have been the focus of much research to obtain Ti alloys with minimum modulus, required to reduce or even prevent stress shielding (see Chapter 1.2.1.1). At ambient temperature, the β-type Ti alloys are composed of a single

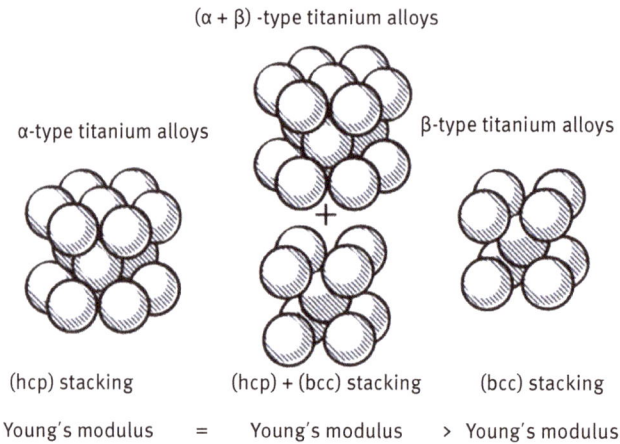

Figure 2.13: Stacking order of α-, (α + β)-, and β-type Ti alloys and order of magnitude of Young's modulus (Niinomi and Boehlert, 2015). © With permission by Springer Nature.

metastable β-phase, formed by rapid cooling from the β-phase region present at high temperature. An illustration of the equilibrium phase diagram of these alloys is presented in Figure 2.14 (Niinomi, 2019).

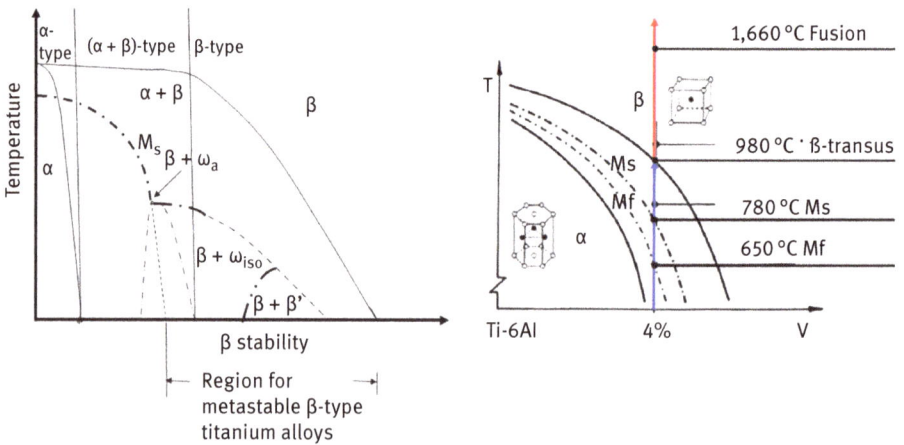

Figure 2.14: Left: Equilibrium phase diagram of Ti alloys. M_S: starting temperature for martensitic transformation (Niinomi, 2019). Right: Phase transformation in Ti6Al4V, showing the starting (M_s) and final (M_f) temperature curves (chain lines) for the martensitic transformation from (hcp) α-phase to (bcc) β-phase. The β-transus for this particular alloy is at 980 °C.

Complexity of phase relations of titanium alloys

Given that all room-temperature β-type titanium alloys at equilibrium are composed of both α- and β-phases, β-type titanium alloys for practical, that is, industrial applications are strictly meta-stable. Specifically, near-α-type and near-β-type titanium alloys exist in the vicinity of the α-type Ti alloy region between the α- and (α + β)-type Ti alloy regions, and in the vicinity of the β-type Ti region between (α + β)- and β-type titanium alloys regions, respectively (see Figure 2.14). Based on the stability of the β-phase, that is, the amount of the β-stabilizing elements present, the cooling rate from high temperature, heat treatment conditions, and many alloy phases can form including a martensitic α′-phase with a (hcp) structure, a martensitic α″-phase with an orthorhombic structure, an athermal hexagonal ω-phase (ω_a), an isothermal ω-phase (ω_{iso}), an α-phase, and a β′-phase with (bcc) structure that appears in β-type titanium alloys owing to a miscibility gap between the two phases of β_{lean} (β′) and β_{rich} (β-matrix). The martensitic α′- and α″-phases appear at relatively lower and higher β stability, respectively, after rapid cooling from high temperature. When the metastable parent β-phase with relatively lower β-stability is deformed, it transforms to the martensitic α″-phase (referred to as deformation-induced martensite phase) or ω-phase (referred to as deformation-induced ω-phase) (see Figure 2.18). The crystallographic nature of martensitic phase transformations in Ti alloys has been elucidated by Zhang et al. (2017). On the general mechanism of martensitic transformation, see Chapter 2.2.1.2.

The near α-, (α + β)-, and β-type titanium alloys are heat-treatable, whereas pure α-type titanium alloys are not. The mechanical strength and ductility of the former are generally microstructurally controlled and can be adjusted by heat treatment including aging followed by ST (see box) and thermomechanical grain boundary engineering including combinations of cold working and heat treatment.

Solution treatment (ST) versus annealing versus tempering

ST is a broad term referring to heating of a material to temperatures sufficiently high to form solid solutions. The temperature is then held for a time until the material is being quenched, causing it to retain the properties of the solid solution. It is a process that softens the alloy and is commonly used as a step in producing alloys that are intended to be used in further manufacturing applications. *Annealing* is a process during which a metal is heated at a temperature slightly *above* the point at which recrystallization occurs. Annealing can increase ductility, alleviate internal stresses introduced by cold working that contribute to brittleness, and increase fracture toughness and homogeneity. During *tempering*, a metal is heated to *below* its critical recrystallization temperature. Tempering increases ductility and toughness, minimizes cracking, and increases general workability. However, it can decrease hardness and strength since ductility and strength are inversely related.

Alloying elements for Ti are generally grouped into α, β, or neutral elements according to which of the α- or β-phases they stabilize, that is, whether they increase or decrease the α/β transformation temperature of pure Ti. Figure 2.15 shows schematically the effect of α-stabilizing, β-isothermal, β-eutectoid, and neutral alloying elements on the equilibrium phase diagram of Ti alloys (Narushima, 2010). Substitutional aluminum and interstitial oxygen, nitrogen, and carbon stabilize the α-phase. Isomorphous vanadium, molybdenum, niobium, and tantalum, as well as eutectoid iron, manganese,

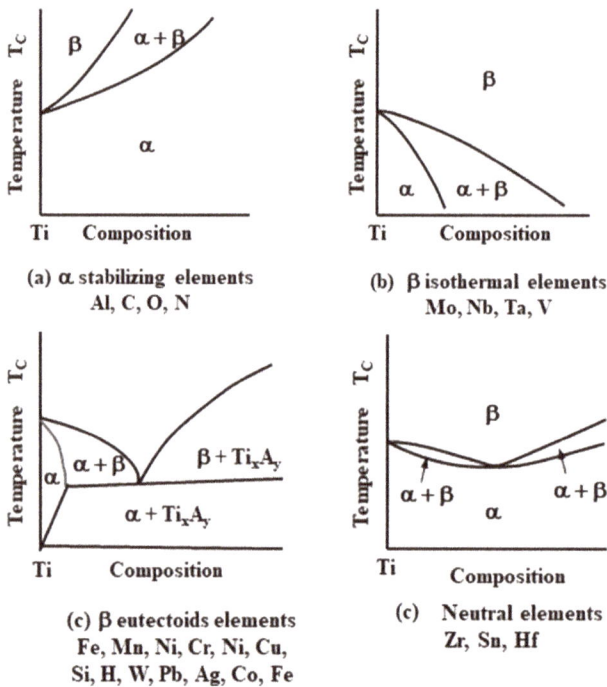

Figure 2.15: Schematic grouping of binary phase diagram of Ti alloys, showing the effect of addition of alloying elements (Narushima, 2010).

chromium, nickel, copper, silicon, hydrogen, tungsten, lead, silver, and cobalt all stabilize the β-phase. Zirconium, tin, and hafnium are neutral elements. However, zirconium may act as a β-stabilizing element when it is present with other stabilizing β-elements (Lütjering and Williams, 2003). To explain the formation and stabilization of α-, β-, and ω-Ti phases, a bottom-up approach based on ab initio calculations has been recently provided by Huang et al. (2016).

Figures 2.16–2.18 display typical microstructures of three representative types of Ti and Ti alloys: α-type commercially pure titanium (referred to as cp-Ti), (α + β)-type Ti6Al4V (referred to as Ti64), and β-type Ti29Nb13Ta4.6Zr (referred to as TNTZ) used for biomedical applications.

The microstructure of cp-Ti subjected to heat treatment below the α/β transformation temperature (referred to as the β-transus temperature; see Figure 2.14, right) exhibits an equiaxed α-structure (Figure 2.16A, A'), whereas an acicular, that is, needle-like α-structure (so-called Thomson–Widmanstätten structure) is observed after heat treatment above the β-transus temperature as shown in Figure 2.16B.

The significance of the existence of the β-transus lies in the behavior of the cast metal: the initially solidified metal is necessarily β-phase, because this is the stable phase at high temperature that must transform on cooling. This transformation occurs

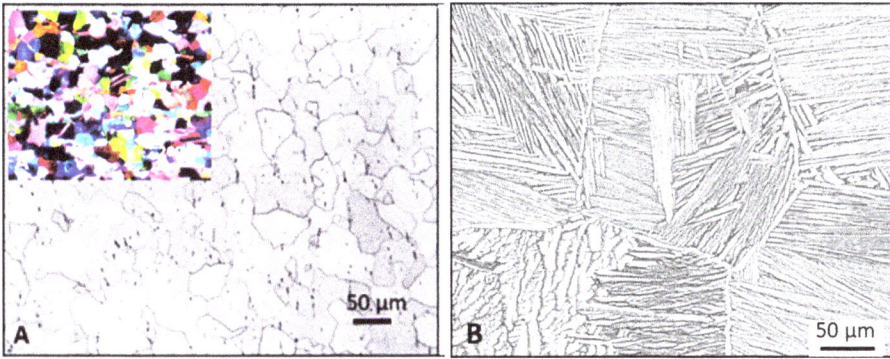

Figure 2.16: Light optical micrographs of polished and etched cp-Ti surfaces. (A) Equiaxed α-structure (Vander Voort, 2014); (A′) Equiaxed α-structured cp-Ti color-etched with modified Weck's reagent (Vander Voort, 2011a); (B) Widmanstätten α-structure of heat-treated Ti6Al4V (Meyer et al., 2008). © With permission by Springer Nature.

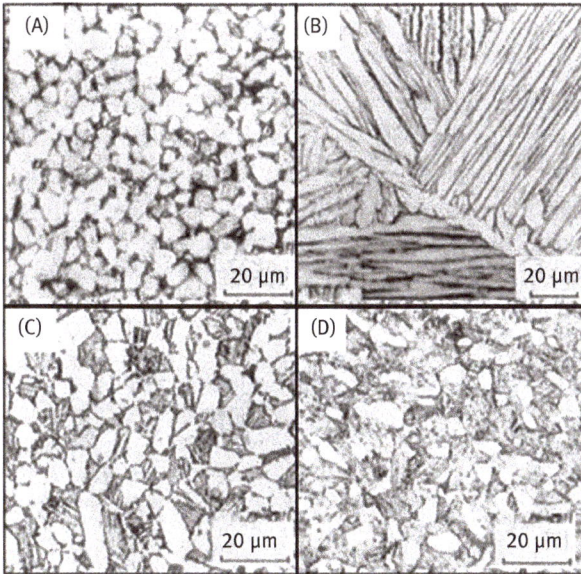

Figure 2.17: Typical light-optical micrographs of Ti6Al4V subjected to various heat treatment: (A) equiaxed α-structure (heated at 920 °C for 2 h, followed by furnace cooling (FC)), (B) needle-like α-structure (Widmanstätten α-structure) (heated at 1,100 °C for 2 h, followed by FC), (C) equiaxed α-structure: bimodal structure (heated at 950 °C for 1.5 h, followed by air cooling (AC) and then heated at 720 °C for 2 h, followed by AC), (D) equiaxed α-structure (heated at 950 °C for 1.5 h), followed by water quenching: solution treatment and then heating at 538 °C for 8 h followed by AC: aging treatment (Narushima, 2010).

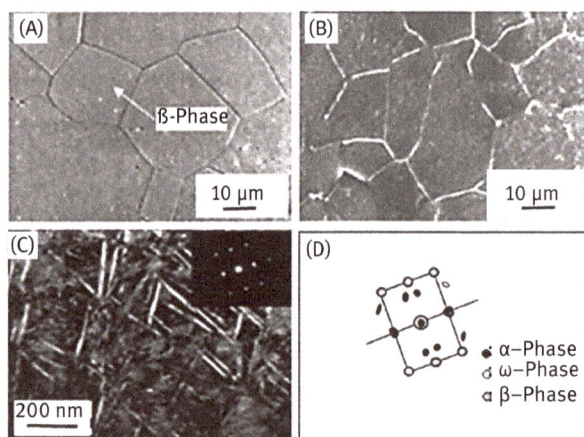

Figure 2.18: SEM and TEM micrographs of Ti29Nb13Ta4.6Zr (TNTZ) subjected to (A) solution treatment (ST) whereby TNTZ$_{ST}$ is composed of single β-phase. (B) and (C) SEM and TEM images, respectively, of TNTZ$_{ST}$ composed of (β + α + ω)-phases after aging treatment at 673 K following ST. The inset of (C) shows a selected electron diffraction (SED) pattern. (D) An index diagram of the crystallographic relationship between the parent β-Ti phase and the martensite α-phase and the ω-phase (Narushima, 2010).

by nucleation and diffusional growth. Thus, at a temperature depending on the alloy composition and the rate of cooling, first the α-phase nucleates at β-grain boundaries, producing a layer over all β-grains. Subsequently, further nucleation allows formation of plate-like crystals of α to grow into the β-grains. Nucleation completely within β-grains may also occur when the cooling rate is increased (see Figure 2.18, panel A). This produces a Widmanstätten structure (Figure 2.16B, Figure 2.17B) and eventually would, in pure Ti, be expected to result in a complete conversion to the α-phase.

Thomson–Widmanstätten structure
A structure characterized by a geometrical pattern resulting from the formation of a new phase along certain crystallographic planes of the parent solid solution. The orientation of the lattice in the new phase is crystallographically related to the orientation of the lattice in the parent phase. This structure was originally observed on polished and etched Ni–Fe solid solution crystallites of octahedrite iron meteorites, but is readily produced in many other alloys, such as titanium by appropriate heat treatment (ASM Metals Handbook).

The (α + β)-type Ti alloys such as Ti6Al4V exhibit a so-called equiaxed α-structure composed of equiaxed primary α-grains and a β-phase when the alloy is processed at a temperature in the (α + β)-region (i.e., below the β-transus temperature) and a needle-like α-structure composed of primary acicular or Widmanstätten α- and β-phase after processing at a temperature in the β-region (i.e., above the β-transus temperature) as shown in Figure 2.17. The β-phase remains metastable at room temperature when

Ti6Al4V is subjected to ST whereby the alloy is heated at a high temperature followed by rapid WQ. Therefore, when the metastable β-phase exists, the alloy is age-hardened, forming a secondary precipitate phase and in general, a fine secondary α-phase (Figure 2.17D). In case of the equiaxed α-structure, a bimodal structure is formed composed of a primary equiaxed α-structure surrounded by the β-phase with a relatively coarse α-phase precipitate. This microstructure leads to incremental mechanical properties. When the alloy is cooled from a temperature above the β-transus temperature, slow cooling such as furnace cooling (FC), relatively slow cooling such as air cooling (AC), and rapid cooling such as WQ result in Widmanstätten α (Figure 2.18A), acicular α, and martensite structures, respectively. The microstructures and the mechanical properties of the (α + β)-type Ti alloys change depending on the processing route including heat and thermomechanical treatments.

The β-type Ti alloys such as TNTZ exhibit a single equiaxed β-phase when heated to a temperature above the β-transus temperature and followed by rapid cooling such as WQ (ST) as shown in Figure 2.18A. Secondary precipitate phases appear in the β-phase due to aging treatment. The α-phase precipitates when the aging temperature is relatively high as shown in Figure 2.18B, whereas the ω-phase precipitates when the aging temperature is relatively low. Although precipitation of the ω-phase significantly increases the strength, the ductility decreases drastically. Hence, ω-phase precipitation should be generally avoided.

The β-type Ti alloys with (bcc) structure are more easily plastically deformed compared to α-type Ti alloys with (hcp) structure, owing to the limited slip systems of the latter. The main slip planes and slip directions in α-type Ti alloys with (hcp) structure are schematically shown in Figure 2.19 (Hagiwara, 2009). There are three main slip systems in the (hcp) structure, but there is a total of 12 in the (bcc) structure (Weinberger et al., 2013). In addition to slip deformation, twinning deformation also occurs in α-type Ti alloys. Twinning enhances plastic deformation of the α-type Ti alloys. Although this deformation also occurs in β-type Ti alloys, it is limited.

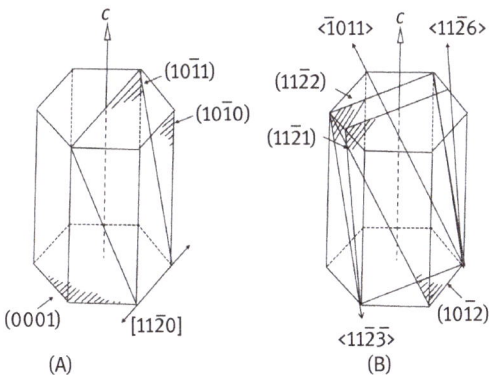

Figure 2.19: (A) Slip system and (B) twinning system of the (hcp) structure (Hagiwara, 2009).

Properties

Tensile strength

The specific strength, that is, the tensile strength divided by the material's density is at a maximum among other metallic materials up to approximately 600 °C as shown in Figure 2.20 (top) (Suzuki and Moriguchi, 1995). This means that Ti alloys are lightweight and strong. The relationship between the fracture toughness and the proof stress of α-, (α + β)- and β-type Ti alloys subjected to various processing routes is shown in Figure 2.20 (bottom) (Niinomi, 2019). High strength and high toughness can be obtained in β-type Ti alloys by applying a variety of processing techniques including heat treatment and thermomechanical treatments. A similar trend is observed in Ti alloys designed for biomedical applications.

Figure 2.21 compares the relationship between tensile strength and ductility (elongation) of SUS 316L stainless steel, CoCrMo alloys, Ti6Al4V, Ti6Al7Nb, and representative β-type Ti alloys for biomedical applications (Hanawa, 2019).

There are four grades of cp-Ti determined by their impurity contents of nitrogen, iron, and oxygen. Evidently, although the tensile strength of cp-Ti grades increases with increasing amounts of impurities (grade number), their elongation tendency decreases. The tensile strength of the β-type Ti alloy TNTZ is, in general, minimized when subjected to ST that generates a single β-phase (open star designation). This is desirable to counteract stress shielding as previously stated. The tensile strength increases with increasing amounts of the α- or ω-phase that precipitate by aging treatment (solid star designation). However, the elongation decreases and Young's modulus increases since ductility and stiffness are inversely related.

The relationships between the tensile strength, Young's modulus and elongation, and the cold working ratio of TNTZ subjected to general cold swaging (see box) are shown in Figure 2.22 (Niinomi et al., 2002a).

Swaging
Swaging is a metallurgical process used to reduce or increase the diameter of tubes and/or rods. This is achieved by placing the tube or rod inside a die that applies compressive force by radial hammering. This can be further expanded by placing a mandrel inside the tube and applying radial compressive forces on the outer diameter. Thus, the inner diameter can have a different shape, for example, a hexagon, while the outer diameter is still circular.

Tensile strength and the 0.2% proof strength both increase with increasing cold working ratio. For TNTZ, they reach the levels of Ti6Al4V ELI (tensile strength approximately 800 MPa) with sufficient elongation at a working ratio in excess of 60%. Nearly the same trend has been reported for TNTZ subjected to general cold rolling (CR). However, in this case, sample elongation decreases with increasing cold working ratio. As further shown in Figure 2.22, Young's modulus of TNTZ

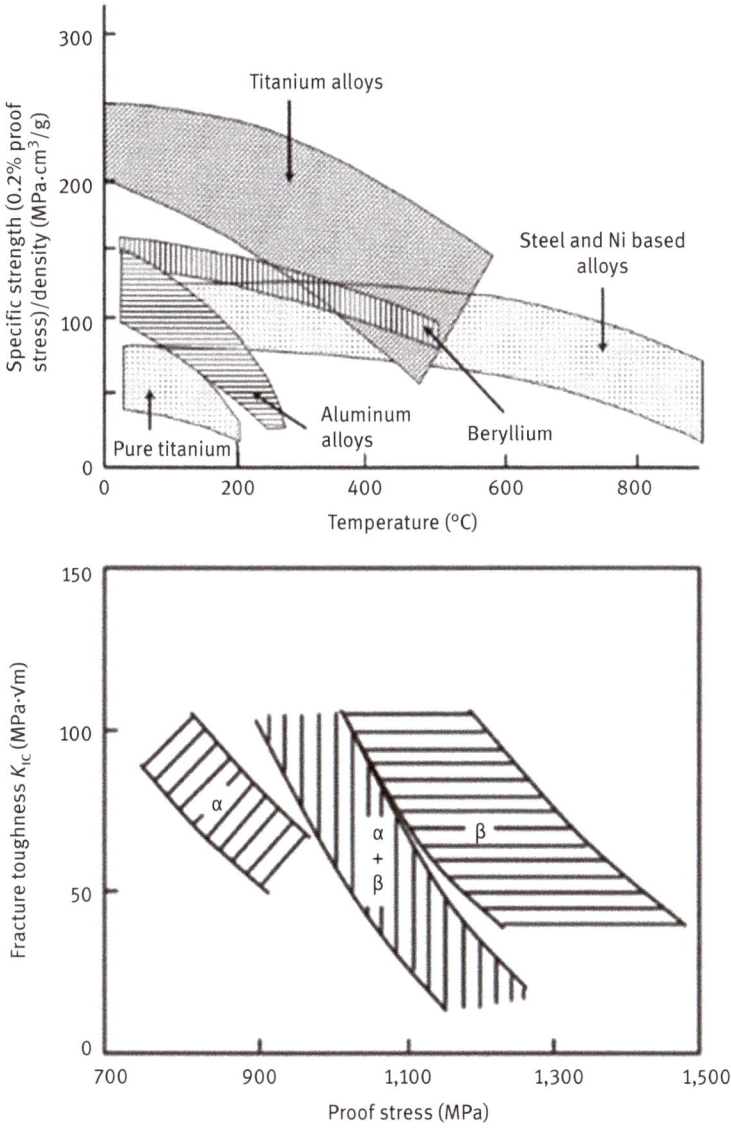

Figure 2.20: Top: Relationship between specific strength (0.2% proof stress/density) and temperature of various metallic materials (Suzuki and Moriguchi, 1995). Bottom: Relationship between fracture toughness and proof stress of α-, (a + β)-, and β-type Ti alloys (Niinomi, 2019).

subjected to cold swaging is close to independent of the working ratio. In case of CR, the modulus of TNTZ was found to decreases at high working ratios owing to formation of texture.

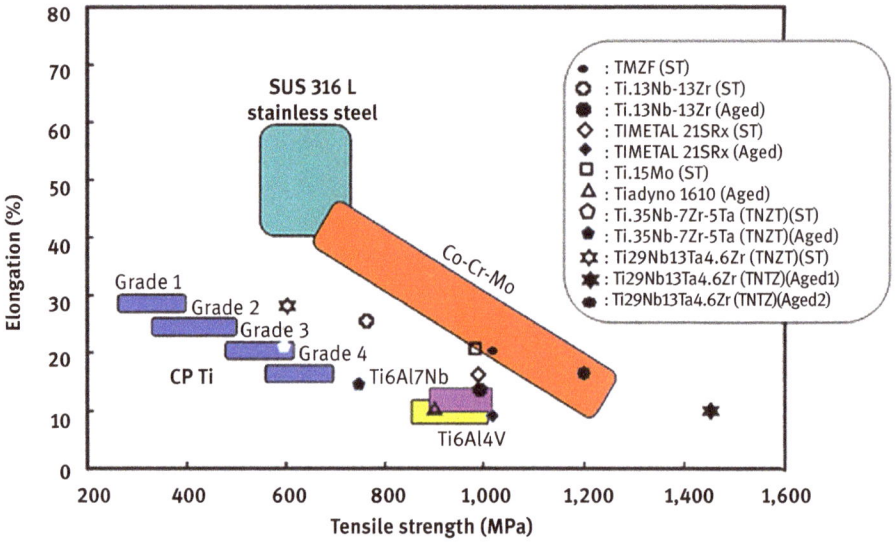

Figure 2.21: Tensile strength versus elongation of SUS 316L (green), CoCrMo alloys (orange), cp-Ti grades (blue), Ti6Al4V (yellow), Ti6Al7Nb (pink), and representative β-type TNTZ alloys (Hanawa, 2019).

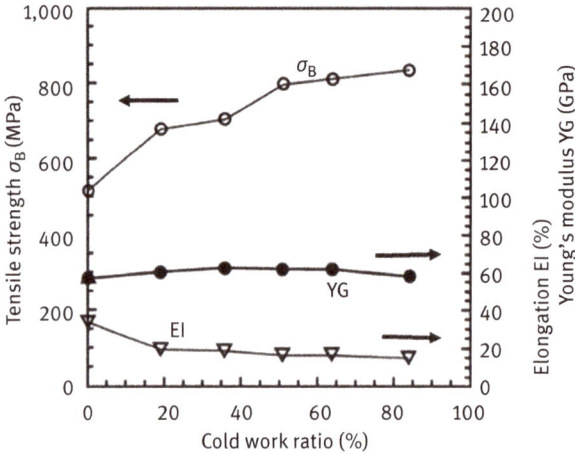

Figure 2.22: Tensile properties of cold-swaged TNTZ bars as a function of cold working ratio (Niinomi et al., 2002a).

Rigidity (Young's modulus)

Figure 2.23 displays Young's elastic moduli of representative α- and (α + β)-Ti alloys for biomedical applications subjected to annealing, and β-type Ti alloys subjected to ST (Niinomi, 2018). As shown, the elastic moduli, that is, the stiffness of the β-type

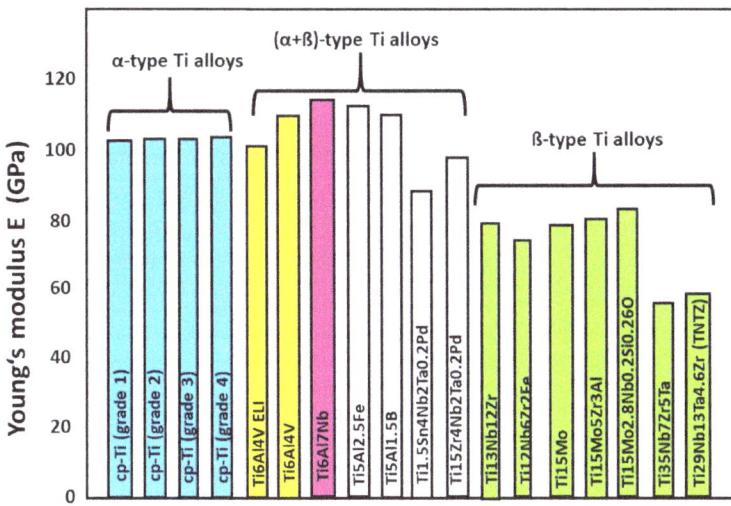

Figure 2.23: Young's moduli of representative α-type, (α + β)-type, and β-type titanium alloys designed for biomedical applications (Niinomi, 2018). Color coding corresponds to that shown in Figure 2.21.

Ti alloys subjected to ST are much lower than those of the α- and (α + β)-Ti alloys subjected to simple annealing. On average, Young's moduli of β-type Ti alloys subjected to ST are around 80 GPa. The Young's moduli of β-type TNZT and TNTZ are lower than those of the other β-type Ti alloys that increase due to aging treatment. For example, the modulus of TNTZ increases to become similar by aging treatment to that of Ti6Al4V, owing to precipitation of α- and ω-phases. In case of the ω-phase, the presence of both athermal and isothermal ω-phases increase the rigidity. The deformation-induced ω-phase also increases Young's modulus. It decreases when α'- and α''-martensite phases are formed, including deformation-induced martensite.

The lowest Young's moduli of β-type titanium alloys are generally obtained after ST. However, since the strength of those alloys is generally poor, an increase in strength is desirable while maintaining a low modulus. Static strength characteristics, such as tensile strength, can be improved by several means including cold work hardening (Figure 2.24), grain refinement strengthening, precipitation strengthening, and distribution strengthening. One of the best ways to increase tensile strength while maintaining a low modulus is through introducing numerous dislocations by classic severe cold working such as CR and cold swaging, or severe plastic deformation processes such as high-pressure torsion, accumulated rolling bonding, and equal channel angular pressing (Yilmazer et al., 2009).

Fatigue strength

Figure 2.24 shows fatigue strengths at 10^7 cycles of representative biomedical stainless steels, Co alloys, Ti alloys, and bone for comparison (Niinomi, 1998). Empty bars refer to cp-Ti data obtained from uniaxial fatigue tests. The fatigue limits (fatigue strength) of all alloys vary widely, dependent on various factors such as the fabrication process, surface condition, microstructure, and fatigue test conditions. The fatigue strength decreases in the following order: Co alloy $\geqq$ Ti6Al4V $\geqq$ 316L stainless steel. Among the Ti alloys, the fatigue strengths of ($\alpha+\beta$)-type Ti alloys such as Ti6Al4V ELI and Ti6A7Nb are greater than those of β-type Ti alloys such as Ti13Nb13Zr, TMZF, and TNTZ. The fatigue strengths of β-type Ti alloys increase by aging treatment after ST. Moreover, the fatigue strength of β-type Ti alloy such as TNTZ has been reported to increase and become comparable to that of Ti6Al4V ELI, while Young's modulus increases. However, the fatigue strength of each metallic biomaterial shows a fairly large scatter due to the abovementioned factors and is dramatically higher than that of bovine bone.

The dynamic strength, that is, the fatigue strength of severe cold-worked TNTZ does not improve compared to TNTZ subjected to ST (Niinomi, 2003). However, the fatigue strength improved considerably when aging treatment was performed following ST or thermomechanical processing including severe cold working. This could be attributed to precipitation of α- or ω-phases in the β-matrix phase during aging treatment, leading to an increase of Young's modulus. Hence, the fatigue strength could be improved by introducing a secondary phase, such as the α-phase or the ω-phase while still minimizing Young's modulus.

However, although both strength and Young's modulus increase significantly due to precipitation of the ω-phase instead of the α-phase, the presence of the ω-phase enhances the brittleness of the alloy. Therefore, the fatigue strength of TNTZ is expected to improve by short-time aging at relatively low temperatures that produces only a small amount of precipitated ω-phase while maintaining a low Young's modulus.

Figure 2.25 (top) displays Young's moduli of TNTZ subjected to solution treatment (ST), severe cold rolling (CR), and aging after CR at 573 K as a function of aging time (AT) (Nakai et al., 2012). Young's modulus remains below 80 GPa, the tentative target value for a low modulus alloy, up to an AT of approximately 10.8 ks (180 h). The fatigue properties in the form of S–N curves for TNTZ subjected to ST, severe CR, and aging for 3.6 ks (60 h; lower curve), and 10.8 ks (180 h; upper curve) at 573 K is shown in Figure 2.25 (bottom) (Nakai et al., 2012). Whereas the fatigue strength of TNTZ improves by aging treatment for 10.8 ks (180 h), the modulus still remains below 80 GPa.

The addition of small amounts of reinforcing ceramic particles to the metallic matrix is also effective in improving the fatigue strength of β-type titanium alloys for biomedical applications while maintaining a low modulus. Figure 2.26 (top) shows Young's moduli of TNTZ with TiB_2 or Y_2O_3 additions subjected to severe CR as a function of B or Y concentration (Song et al., 2011). Young's modulus remains

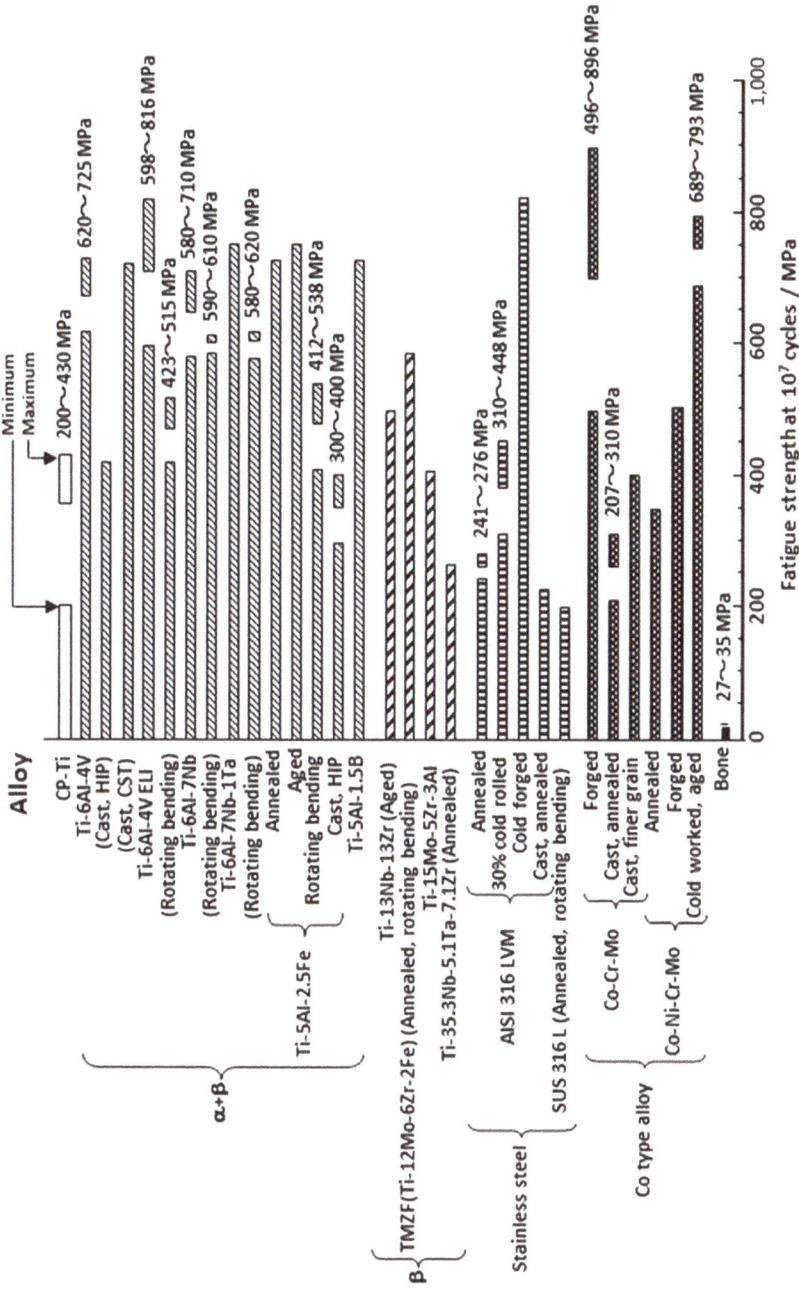

Figure 2.24: Fatigue strengths at 10^7 cycles of biomedical stainless steels, Co alloys, Ti alloys, and bone for comparison. Plain bars refer to data obtained from uniaxial fatigue tests (Niinomi, 1998). © With permission by Elsevier.

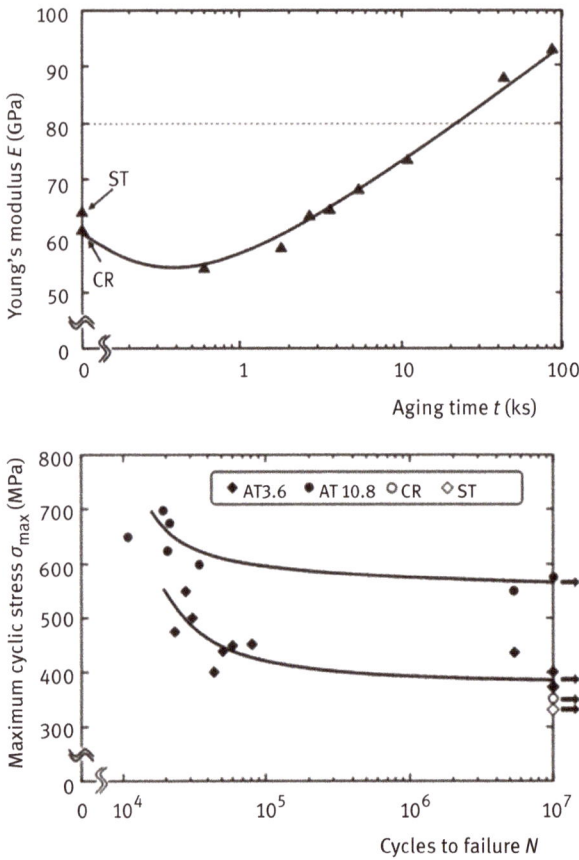

Figure 2.25: Top: Young's moduli of TNTZ subjected to ST, CR and AT treatments as a function of aging time (Nakai et al., 2012). © With permission by Springer Nature. Bottom: Fatigue properties of TNTZ subjected to ST, severe cold rolling (CR), and aging after cold rolling at 573 K as a function of aging time (AT). The lower curve corresponds to an aging time of 60 h, the upper curve to an aging time of 180 h (Nakai et al., 2012). © With permission by Springer Nature.

nearly constant at approximately 60 GPa with increasing B or Y concentrations. Figure 2.26 (bottom) shows the S-N curves for TNTZ with 0.1% and 0.2% B concentration as well as 0.2% and 0.5% Y concentrations subjected to CR after ST (upper curves) along with those TNTZ samples subjected to ST or CR followed by ST (lower curves) (Song et al., 2011). It is evident that the fatigue strength (maximum cyclic stress) of TNTZ is substantially improved by adding TiB_2 or Y_2O_3.

Corrosion resistance

The corrosion resistance of Ti and Ti alloys is generally very high because their surfaces react instantaneously with water molecules in aqueous solutions and atmospheric

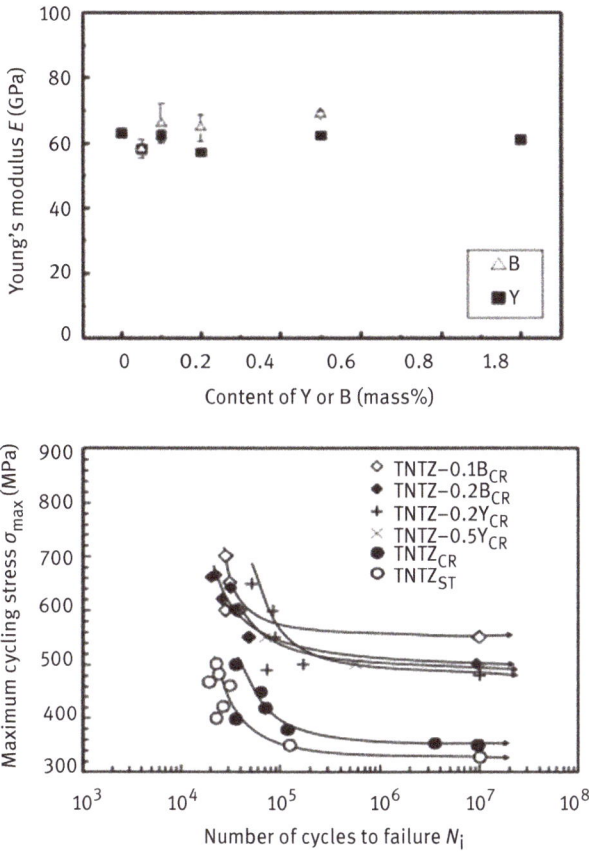

Figure 2.26: Top: Young's moduli of cold-rolled TNTZ as a function of Y or B concentration (Song et al., 2011). © With permission by Elsevier. Bottom: S-N curves of TNTZ with TiB_2 or Y_2O_3 additions subjected to cold rolling (CR) following solution treatment (ST) (upper curves) along with S-N curves of TNTZ subjected to ST or CR followed by ST (lower curves) (Song et al., 2011). © With permission by Elsevier.

moisture to form a dense, homogeneous, several nanometers thin, significantly corrosion-resistant Ti oxide film that acts as a protective passive barrier (Narushima and Niinomi, 2010; Hanawa, 2019). When fractured or sheared-off, this protective Ti oxide film is rapidly restored. Hence, the corrosion rate of Ti and Ti alloys is significantly low, comparable to that of gold and platinum.

Wear properties

Mechanical wear of Ti alloys is, in general, lower than that of stainless steel and CoCrMo alloys. The wear mechanism mode of titanium alloys can be roughly grouped into abrasive and adhesive wear. Abrasive wear is predominant in $(\alpha + \beta)$-type Ti alloys such as Ti6Al4V, whereas adhesive wear is predominant in β-type Ti alloys. Given that the wear losses of biomedical titanium alloys are lower in SBF than in air (Niinomi et al., 2007), adhesive wear is considered to be the main wear mechanism *in vivo*.

The wear resistance of biomedical Ti alloys depends on the type of alloy and its microstructure. It is also dependent on the testing conditions. Figure 2.27 (left) (Niinomi et al., 2007) shows the relationships between mass loss and applied load in Ringer's solution of TNTZ subjected to ST (TNTZ$_{ST}$), thermal aging treatments at 598 K (TNTZ$_{598 K}$), 673 K (TNTZ$_{673 K}$), and 723 K (TNTZ$_{723 K}$), as well as Ti6Al4V subjected to aging treatment following ST (Ti64$_{STA}$). The order of the wear resistance of each material changes before and after application of a certain load. Figure 2.27 (right) (Niinomi et al., 2007) explains this trend schematically.

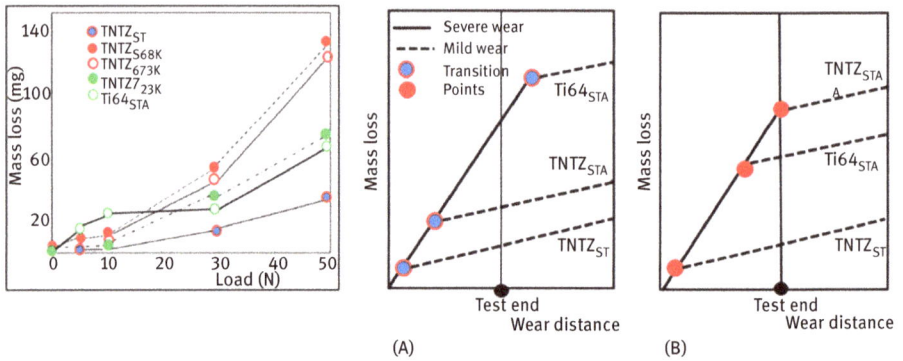

Figure 2.27: Left: Relationship between mass loss and load in Ringer's solution of TNTZ$_{ST}$, TNTZ$_{598 K}$, TNTZ$_{673 K}$ and TNTZ$_{723 K}$, as well as Ti64$_{STA}$ obtained from friction wear tests at various loads (Niinomi et al., 2007). Center and right: Schematic of the relationships between mass loss and wear distance for TNTZ$_{ST}$, TNTZ$_{STA}$, and Ti6Al4V$_{STA}$ under (A) low and (B) high loading conditions (Niinomi et al., 2007). The colored points indicate the transition between severe and mild wear losses. © With permission by IOP Publishing Ltd.

Furthermore, the wear processes can be categorized as severe wear or mild wear. Severe wear changes to mild wear beyond a certain wear distance. Initially, severe wear takes place whereby large wear particles form, consequently leading to a large wear losses. Mild wear is a steady wear process following severe wear whereby fine wear oxide particles are formed, leading to only small wear losses. The wear of Ti6Al4V$_{STA}$ (Ti64$_{STA}$) is in the severe wear region, but the wear of TNTZ$_{STA}$ (TNTZ subjected to aging treatment following ST) as well as TNTZ$_{ST}$ is characterized by mild wear at low-loading condition (Figure 2.27 right, A). However, the transfer from severe wear to mild wear is more strongly delayed in case of TNTZ$_{STA}$ than in the cases of TNTZ$_{ST}$ and Ti6Al4V$_{STA}$ at the high-loading condition (Figure 2.27 right, B).

Surface hardening is one of the most effective methods in improving the wear resistance of titanium alloys. Several techniques have been investigated for this purpose, including oxidizing, nitriding, electroplating, physical vapor deposition (PVD)/chemical vapor deposition (CVD) coating, thermal spraying, and others. Among these techniques, oxidation and gas nitriding are advantageous owing to the simplicity of the technology (Niinomi et al., 2002b; Nakai et al., 2008).

Biocompatibility

Biocompatibility is among the most important material properties of Ti and Ti alloys and can be subdivided into biological and mechanical biocompatibility. The former term refers to the suitability of biomaterials for applications in living tissue from the perspective of biological safety and functionality, while mechanical biocompatibility refers to the ability of a biomaterial to match mechanical properties such as the Young's modulus, static (tensile) strength, dynamic (fatigue) strength, fracture toughness, and fretting fatigue strength with those of living tissues, particularly the cortical bone. For additional information on biocompatibility, see Chapters 1.1.4 and 4.3.

Biological biocompatibility

Indirect information on the biological biocompatibility of a biomaterial can be obtained from cell viability testing. Figure 2.28 (left) shows the result of viability of fibroblast cells and murine L-929 cell line in contact with various pure metals (Kawahara et al., 1963; Niinomi et al., 2002c).

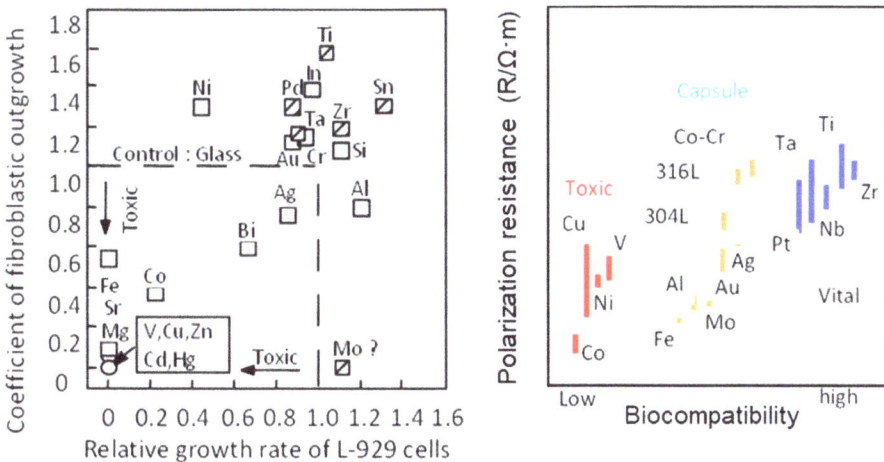

Figure 2.28: Left: Cytotoxicity of pure metals tested for two kinds of cell line (Kawahara et al., 1963; Niinomi et al., 2002b). © With permission by Springer Nature. Right: Relation between polarization resistance and biocompatibility of pure metals and biomedical alloys (Steinemann, 1980; Niinomi et al., 2002b). © With permission by Springer Nature.

In Figure 2.28 (right) the cytotoxicity of pure metals and the relationship between biocompatibility and polarization resistance of typical pure metals and surgical implant materials are displayed (Steinemann, 1980; Niinomi et al., 2002c).

The allergenic potential of metals is a serious problem. The allergy potential of pure metals is rated in Figure 2.29 (top) (Niinomi, 2000). Hg, Ni, Co, Sn, Pd, and Cr are high-risk elements from the perspective of allergy problems. The amount of dissolved Ni ion is strictly limited. As shown in Figures 2.28 (right) and 2.29 (top), Ti, Nb, and Zr are preferable elements for designing low-modulus Ti alloys for biomedical applications. Mo is also a low-toxic element although there is limited data available on its cytotoxicity. Furthermore, cell viability evaluations of L-929, MC3T3-E1, or V-79 cell lines were conducted using extract media (Okazaki et al., 1998) or various metal salts (Yamamoto et al., 1998; Niinomi, 2018). For example, Figure 2.29 (bottom) shows the cytotoxicity of 21 different metal salts tested against L-929 murine fibroblast cells (Yamamoto et al., 1998; Niinomi, 2018). The metal salts are grouped into low, intermediate, and high toxic categories. Low toxicity salts are $FeSO_4$, $FeCl_3$, $SnCl_4$, $TiCl_4$, $ZrCl_4$, $NbCl_5$, $MoCl_5$, $TaCl_5$, WCl_6, and $Al(NO_3)_3$, medium toxicity salts are $CuCl$, $CoCl_2$, $CuCl_2$, $SnCl_2$, $MnCl_2$, $NiCl_2$, $PdCl_2$, $ZnCl_2$, and $Cr(NO_3)_3$, and high toxicity salts are VCl_3 and $K_2Cr_2O_7$.

Results of cytotoxicity testing of TNTZ, cp-Ti, and Ti6Al4V based on the NR (neutral red, toluylene red) absorption method (Borenfreund and Puerner, 1985) using L-929 cells are shown in Figure 2.30 (left) (Niinomi, 2000). The cell viability of TNTZ is nearly equal to that of pure Ti and somewhat greater than that of conventional biomedical Ti6Al4V in both as-extracted and filtered extracted solutions. Therefore, the biocompatibility of TNTZ is considered to be excellent. The trend of cell viability among TNTZ, pure (cp-)Ti and Ti6Al4V is nearly identical regardless of the extracted period, although the cell viability of both as-extracted (nonfiltered) and the filtered extracted solutions decreases slightly with increasing extraction period.

Many studies of biological biocompatibility of titanium alloys were performed using animal models, such as mice, rabbits, dogs, sheep, or cattle. The bonding ability and bone affinity of rods made of TNTZ subjected to ST ($TNTZ_{ST}$) and aging treatment ($TNTZ_{AT}$), cp-Ti, and Ti6Al4V ELI (Ti64) were evaluated by implanting these rods in canine mandibular implant beds of 18-month-old male beagle dogs (Edamatsu et al., 2015). The failure load (maximum load) of these rods implanted for 3 and 6 months was obtained based on punch tests (see box) as shown in Figure 2.30 (right) (Edamatsu et al., 2015). The failure load was found to decrease in the order of $TNTZ_{AT} \geq TNTZ_{ST} \geq$ cp-Ti $\geq$ Ti64 at both observation periods. However, for all types of implants the failure loads after 6 months are smaller than those observed at 3 months.

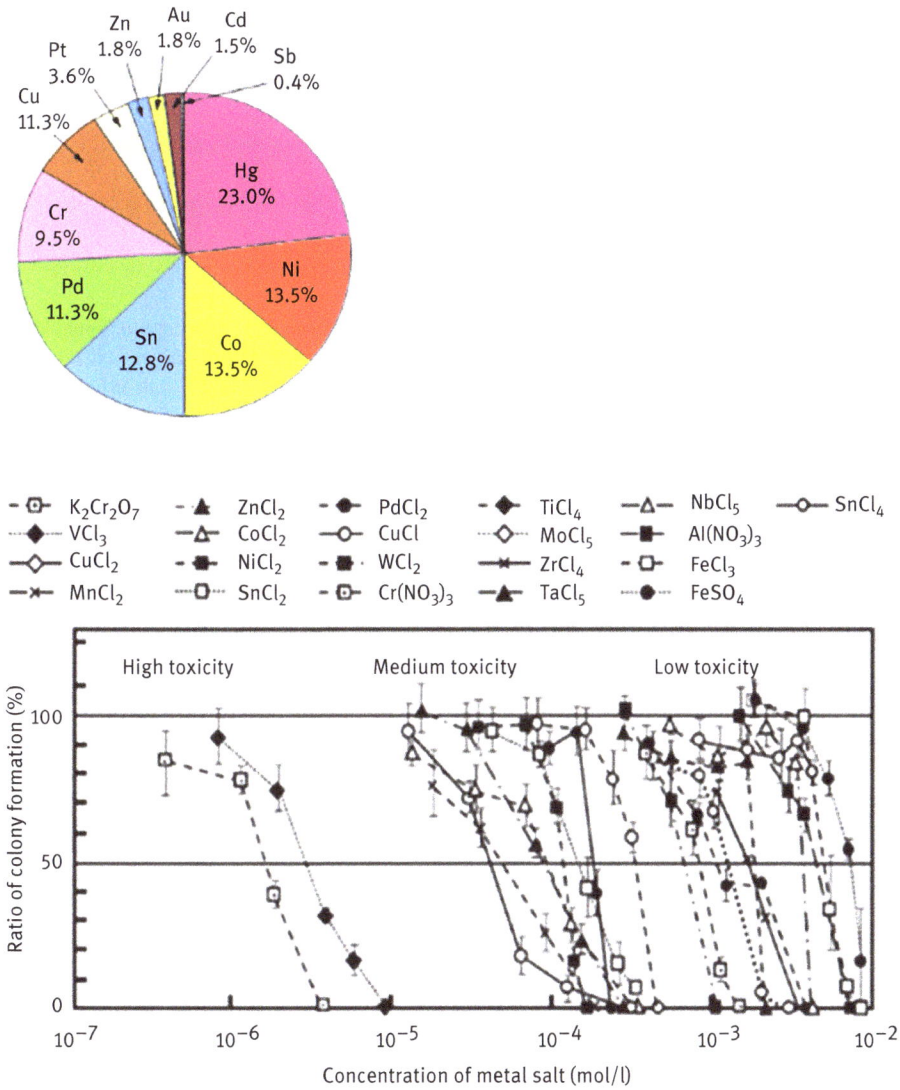

Figure 2.29: Top: Rate of metallic allergy of pure metals (Niinomi, 2000). © With permission by Springer Nature. Bottom: Biocompatibility of metal salts toward L-929 murine fibroblast cells (Yamamoto et al., 1998; Niinomi, 2018).

Punch test

The punch test is a materials testing procedure during which thin disks confined in a clamping installation are centrally loaded by a punch with rounded head until failure. The test results in a force–distance curve that contains information on deformation characteristic and strength of the materials under investigation (Abendroth, 2017).

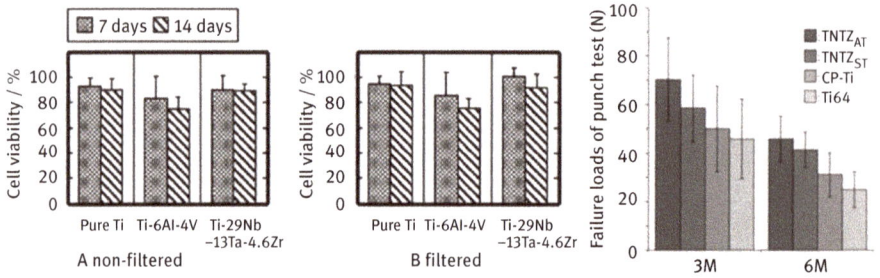

Figure 2.30: Left: Cell viability of L-929 in (A) nonfiltered and (B) filtered culture solutions evaluated by the NR method for cp-Ti, Ti6Al4V and Ti29Nb13Ta4.6Zr (Niinomi, 2000). © With permission by Elsevier. Right: Failure loads of punch tests of $TNTZ_{AT}$, $TNTZ_{ST}$, cp-Ti, and Ti64 rods implanted for 3 and 6 months (3M and 6M, respectively) in canine mandibles (Edamatsu et al., 2015).

The bone area ratio (BAR) and bone contact ratio (BCR) in the cancellous bone region after implantation of $TNTZ_{ST}$, $TNTZ_{AT}$, cp-Ti, and Ti64 rods are shown in Figure 2.31 (Edamatsu et al., 2015). The BAR and the BCR of TNTZ exceeded significantly those of cp-Ti and Ti6Al4V.

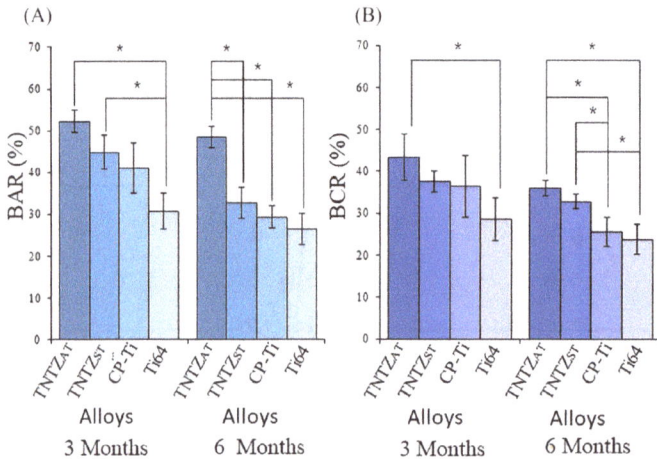

Figure 2.31: Bone area ratio (BAR) (a) and bone contact ratio (BCR) (b) in the cancellous bone area at 3 and 6 months after implantation of $TNTZ_{AT}$, $TNTZ_{ST}$, cp-Ti, and Ti64 (Ti6Al4V) (Edamatsu et al., 2015).

Recent studies show that among several Ti alloys tested, Ti13Nb13Zr alloy was the most effective for inhibition of biofilm formation, cell differentiation, and stimulation for release of immune mediators (Mello et al., 2019).

Mechanical biocompatibility

Mechanical properties such as tensile and fatigue strengths as well as Young's moduli of Ti alloys designed for biomedical applications were discussed above. This paragraph considers the effect of Young's modulus on stress shielding. This includes the requirement of improving the static (tensile strength) and dynamic strengths (fatigue strength) while maintaining a low Young's modulus to inhibit stress shielding and, thus, to avoid bone resorption. Since Young's moduli of metallic biomaterials should be as similar as possible to that of bone, β-type Ti alloys with low Young's modulus have been or are currently being evaluated for biomedical applications. Prior to the regulatory approval of low Young's modulus material as effective stress shielding inhibitors, studies of the healing process were performed in a fracture model of tibiae of rabbits implanted with intramedullary rods or bone plates.

Figure 2.32 shows X-ray images of the fracture models at 24 weeks after implantation into rabbit tibiae of intramedullary TNTZ (A), Ti6Al4V ELI (B), and SUS316 (C) rods with Young's moduli of 58, 110, and 161 GPa, respectively, measured using three-point bending tests (Niinomi et al., 2002d). Bone resorption can be observed in the upper back portion of the tibia after implantation of an SUS stainless steel intramedullary rod (Figure 2.32C), but no bone resorption could be observed after implantation of TNTZ and Ti6Al4V ELI intramedullary rods (Figure 2.32A and B). Substantial bone neoformation could be observed at the frontal portion of the tibia in the presence of the SUS 316L stainless steel intramedullary rods, but very small bone formation was observed at the frontal portion of the tibia for the TNTZ and Ti6Al4V ELI intramedullary rods.

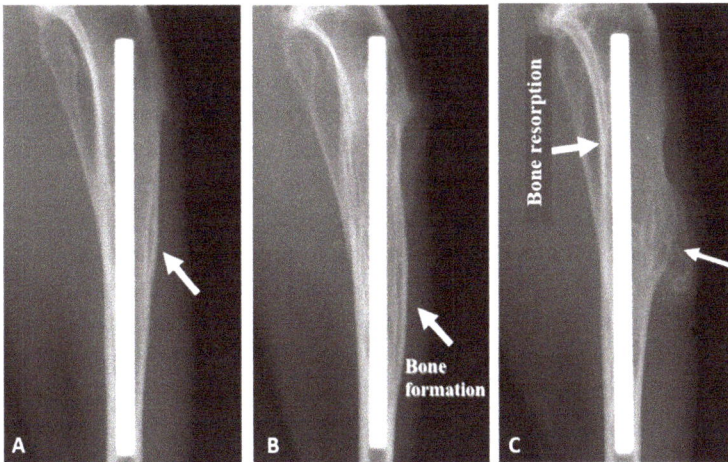

Figure 2.32: Remodeling of tibial bones of rabbits at 24 weeks after implantation of (A) TNTZ, (B) Ti6Al4V ELI, and (C) SUS 316L stainless steel (Niinomi et al., 2002d).

Figure 2.33 shows contact microradiographs of the cross section of tibiae implanted with TNTZ (left), Ti6Al4V ELI (center), and SUS 316L (right) rods at 24 weeks postimplantation (Niinomi, 2008). Evidently, remodeling of bones is best in case of TNTZ. This indicates that Ti alloys with low moduli effectively reduce bone atrophy and, hence, support excellent bone remodeling. The low modulus β-alloy, Ti29Nb13Ta4.6Zr, was found to inhibit stress shielding and improve load transmission compared to other metallic biomaterials with high Young's moduli.

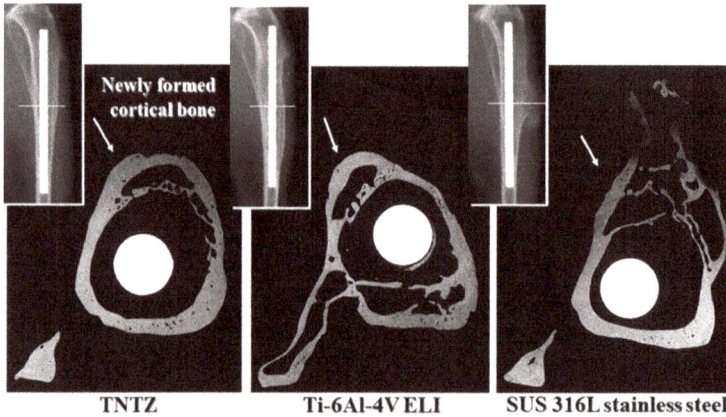

Figure 2.33: Contact microradiographs of cross-sections of rabbit tibiae implanted with intramedullary rods made of TNTZ (left), Ti6Al4V ELI (center) and SUS 316L (right) at 24 weeks after implantation (Niinomi, 2008). © With permission by Springer Nature.

Biomedical applications

The first titanium utilized as implant material was cp-Ti (see Chapter 1.2.1.1), followed by Ti6Al4V. These two alloys, in particular Ti6Al4V, were originally developed for aerospace applications, and subsequently widely used in other applications owing to their high corrosion resistance and excellent mechanical properties. Subsequently, reduced levels of oxygen, nitrogen, carbon, and iron led to the advent of Ti6Al4V ELI as a biomaterial, whereby ELI refers to "extra low interstitials." However, although to date there are no reports of serious health issues regarding their biomedical applications, vanadium in Ti6Al4V ELI has been found to be cytotoxic. Hence, V-free Ti alloys for biomedical applications using Nb and Fe as β-phase stabilizing elements have been developed (Niinomi et al., 2012a). These alloys included Ti6Al7Nb (Semlitsch, 1986) that has been registered in both ASTM (ASTM F1295-05) and ISO (ISO 5832-11:1994) standards and Ti5Al2.5Fe (Brow and Kramer, 1985) that was only registered as ISO (ISO 5832/X) standard. Subsequently, based on the same concept, other Ti alloys for biomedical applications such as Ti6Al6Nb1Ta and Ti6Al2Nb1Ta (Sasaki et al., 1996) were developed. However, they are said to be used predominantly as structural Ti alloys for military rather than for biomedical applications.

Aluminum in Ti6Al4V has been identified as a neurotoxin and there is a debate whether it may be etiologically related to Alzheimer's disease. However, there is no strong evidence yet to confirm such as correlation. Nevertheless, several other V- and/or Al-free Ti alloys for biomedical applications were developed. They include Ti15Sn4Nb2Ta0.2Pd (Okazaki et al., 1995), Ti15Zr4Nb2Ta0.2Pd (JIS T 7401–5) (Okazaki et al., 1995), and eventually Ti4.5Al6Nb2Fe2Mo (Hirano et al., 2007) that has been modified for biomedical applications by reengineering superplastic Ti4.5Al3V2Fe2Mo (SP-700) (Gunawarman et al., 2005). The aforementioned V- and/or Al-free Ti alloys are all (α + β)-type Ti alloys.

Stress shielding and allergic problems are considered in the design of Ti alloys for biomedical applications. Stress is transferred preferentially through the implant because, generally, it has much higher rigidity than that of the bone. Under these conditions, bone resorption may occur because the bone does not receive the appropriate tensile stress required for maintenance. This phenomenon is schematically shown in Figure 2.34 (left) (Niinomi, 2013). This inhomogeneous stress transfer between the implant and the bone is called stress shielding.

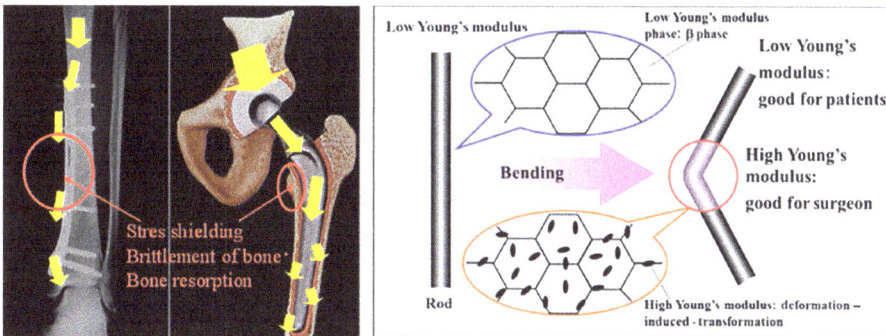

Figure 2.34: Left image: Stress shielding in a bone plate (left) and an artificial hip joint (right). Yellow arrows indicate the stress transfer route (Niinomi, 2013). Right image: Concept of a spinal fixation rod with self-adjustable Young's modulus (Nakai et al., 2011).© With permission by Elsevier.

To inhibit stress shielding, the rigidity (Young's modulus) of the implant should be as similar as possible to that of (cortical) bone. Young's moduli of the major metallic biomaterials for use in implants such as SUS 316L stainless steels (180 GPa), CoCrMo alloys (210 GPa), and Ti6Al4V ELI (110 GPa) are much higher than that of bone (10–30 GPa) (Niinomi et al., 2012a). As a result, strong stress shielding is to be expected (Figure 2.34, left). As previously indicated, Young's moduli of the β-type Ti alloys are expected to be lower than those of α- and (α + β)-type Ti alloys owing to their different crystal structures. Therefore, research and development of β-type Ti alloys without cytotoxic and allergenic elements with low Young's moduli have been initiated. Many alloys such as Ti13Nb13Zr (Mishra et al., 1996; ASTM F1713), Ti12Mo6Zr2Fe (TMZF) (Wang et al., 1996; ASTM F1813), Ti15Mo2.8Nb0.2Si0.28O (TIMETAL 21SRx) (Wang et al., 1996), Ti15Mo (Zardiackas et al., 1996; ASTM F2066), Ti16Nb10Hf (Tiadyne 1610) (Wang et al., 1996), Ti35.3Nb7.1Zr5.1Ta (TNZT) (Ahmed et al., 1996; ASTM designation draft #3), and Ti29Nb13Ta4.6Zr (Kuroda et al., 1998) are in their early development stages. Among them, TIMETAL 21SRx is a derivative of TIMETAL 21S used for aerospace applications. Most of these alloys are roughly categorized into Ti-Nb, Ti-Mo, Ti-Ta, and Ti-Zr (with β-stabilizing elements) system alloys.

Some β-type Ti alloys with low Young's moduli using low cost elements such as Fe, Cr, Mn, Sn, Al, and O have also been developed for biomedical applications because, in general, traditional Ti alloys contain high-cost elements, such as Nb, Ta, Mo, and Zr. Examples of such alloys include Ti10CrAl (Hatanaka et al., 2010), Ti-Mn (Ikeda et al., 2009), Ti-Mn-Fe (Ikeda et al., 2012), Ti-Mn-Al (Ikeda et al., 2010), Ti-Mn-Mo (Santos et al., 2017), Ti-Cr-Al (Ikeda and Sugano, 2005), Ti-Sn-Cr (Ashida et al., 2012), Ti-Cr-Sn-Zr (Murayama and Sasaki, 2009), Ti-(Cr, Mn)-Sn (Kasano et al., 2010), and Ti12Cr (Nakai et al., 2011).

With a focus on the development of support rods for spinal fixation devices to treat juvenile scoliosis, β-type Ti alloys with self-adjustable and low Young's moduli have been introduced. The rod is being bent by the operating surgeon to obtain the proper physiological shape of the spine. The shape of the rod should be maintained after bending, but ought to return back to its original shape, that is, show spring-back. The degree of spring-back depends mainly on the material's strength and Young's modulus. A lower Young's modulus leads to a higher degree of spring-back. As such, the modulus of Ti alloys for spinal fixation devices must be sufficiently low to avoid stress shielding, yet high enough to suppress spring-back (Nakai et al., 2011). New types of titanium alloys the modulus of which prior to deformation is low but increases during deformation (i.e., is self-adjustable), can be achieved and have been proposed based on the concept schematically shown in Figure 2.34 (right) (Nakai et al., 2011). Of these alloys, TiCr was the first to be developed, subsequently followed by Ti17Mo (Zhao et al., 2012), Ti30Zr7Mo (Zhao et al., 2011a), Ti30Zr5Cr (Zhao et al., 2011b), and Ti30Zr3Mo3Cr (Zhao et al., 2011b). Other alloys developed for the same purpose are Ti30Zr5Cr and Ti30Zr3Mo3Cr.

Ni-free shape-memory and superelastic Ti alloys

TiNi alloys were developed as shape-memory alloys and applied as guide wires for catheters and orthodontic wires (see Chapter 2.1.1.3). Since Ni is a high-risk element known to trigger allergic reactions, many novel Ni-free shape-memory β-type Ti alloys have been developed for biomedical applications (Table 2.4; Niinomi, 2012a), grouped into Ti–Nb, Ti–Mo, Ti–Ta, and Ti–Cr system alloys, all with low Young's moduli.

Table 2.4: Representative low modulus Ni-free shape-memory and superelastic alloys (Niinomi, 2012a). © With permission by Elsevier.

Alloy system	Shape-memory alloy
Ti–Nb	Ti-Nb, Ti-Nb-O, Ti-Nb-Sn, Ti-Nb-Al, Ti22Nb(0.5–2.0)O (at%), Ti-Nb-Zr, Ti-Nb-Zr-Ta, Ti-Nb-Zr-Ta-O, Ti-Nb-Ta-Zr-N, Ti-Nb-Mo, Ti22Nb6Ta (at%), Ti-Nb-Au, Ti-Nb-Pt, Ti-Nb-Ta, Ti-Nb-Pd
Ti–Mo	Ti-Mo-Ga, Ti-Mo-Ge, Ti-Mo-Sn, Ti-Mo-Ag, Ti5Mo(2–5)Ag (mol%), Ti5Mo(1–3)Sn (mol%), Ti-Mo-Sc
Ti–Ta	Ti50Ta (mass%), Ti50Ta4Sn (mass%), Ti50Ta10Zr (mass%)
Others	Ti7Cr(1.5, 3.0, 4.5)Al (mass%)
	Superelastic alloys showing only elasticity
	$Ti_3(Ta + Nb + V) + (Zr, Hf) + O$ (at%), Ti29Nb13Ta4.6Zr (mass%)

2.1.1.5 Tantalum

Structure

Tantalum exists in two crystalline phases. The α-Ta phase is relatively ductile and soft, with (bcc) structure (space group $Im\bar{3}m$ (229); $a_0 = 330.58$ pm). The β-Ta phase is hard and brittle, with tetragonal symmetry (space group $P4_2/mnm$ (136), $a_0 = 1{,}019.4$ pm, $c_0 = 531.3$ pm). The β-phase is metastable and converts allotropically to the α-phase upon heating to 750–775 °C. Whereas bulk tantalum consists almost entirely of the α-phase, the β-phase is present in thin films obtained by magnetron sputtering, chemical vapor deposition, or electrochemical deposition from the eutectic molten salt solution.

Properties

Tantalum appears to be inert *in vitro* and *in vivo*. Both the pure metal and its principle oxide possess low solubility and cytotoxicity, with halide compounds being biologically more active. The high modulus of elasticity of tantalum of up to 190 GPa makes it necessary to use it as a highly porous foam or scaffold with lower stiffness for hip replacement implants to avoid stress shielding. Since tantalum is a nonferrous, nonmagnetic metal, these implants are considered acceptable for patients undergoing magnetic resonance imaging (MRI) procedures.

Biomedical applications

The benign local host response to tantalum is characterized by vital encapsulation in soft tissue and frequent osseointegration in hard tissue, similar to titanium. Although tantalum has been in clinical use since the 1940s, it now has found a wide range of diagnostic and implant applications, with apparently overall excellent results (Ferreira de Sousa et al., 2013). Tantalum and tantalum oxide coatings are promising materials for surface modification of Ti or stainless steel for endodontic or endosseous implants, thus reducing the risk of developing postoperation infection and/or peri-implantitis.

The modulus of elasticity of porous tantalum with a high volumetric porosity of 75–80% is as low as 3 GPa and, thus, compares favorably to that of cancellous or subchondral bone (Paganias et al., 2012; George and Nair, 2018). Today, tantalum is already being used in a wide array of orthopedic clinical applications that include primary and revision joint replacement, tumor reconstructive surgery, spine fusion, management of osteonecrosis of the femoral head, foot and ankle surgery, and in the form of thin wires as skin suture material. Although recent studies have demonstrated excellent clinical and radiographic outcomes, even in the presence of extensive bone loss in hip and knee reconstructive surgeries, the use of tantalum in spine surgeries and osteonecrosis of the hip has revealed mixed clinical success (Patil and Goodman, 2015). Hence, further clinical studies are still required to establish its role in specific orthopedic applications and to assess whether the demonstrated advantages of porous tantalum can provide long-term biological fixation and stability (Paganias et al., 2012). With this in mind, selective laser melting technology was used to manufacture highly porous pure tantalum implants with fully interconnected open pores (Wauthle et al., 2015). However, admittedly, the high price of tantalum may be an impediment for future widespread use in biomedical applications.

Owing to its high atomic mass and associated high X-ray energy attenuation, tantalum has also been applied as an X-ray-sensitive admixture to thermally sprayed surface layers of titanium on a nonmetallic substrate for orthopedic implants. The tantalum admixture is supposed to guarantee localization of the substrate in the human or animal body by X-ray mapping with high spatial resolution (Zimmermann, 2013). Thin tantalum as well as tantalum nitride films (<1 μm) may be used to control corrosion of Ti6Al4V alloy in contact with biofluid (Hee et al., 2019).

In addition, tantalum was used since the early days of cardiovascular stent development in both coronary and peripheral stent applications. These types of stents were helically wound wire and knitted wire structures, respectively. One of the factors that led to its use as a stent material is radiopacity, useful for visualizing the device *in vivo*. It is thought that the bioinert nature of the tantalum oxide surface is responsible for improved vascular compatibility and, in particular, reduced thrombogenicity. Early wire-coiled and woven structures were eventually replaced by slotted tube/laser-cut designs (O'Brien and Carroll, 2009).

2.1.1.6 Noble metals

Among corrosion-resistant metals such as stainless steels, cobalt chromium alloys, titanium, and tantalum discussed above that rely on the surface barrier effect of insoluble bioinert oxide films, noble metals such as copper, gold, and silver are highly stable on account of their electronic structure that involve completely filled d-like electron orbitals that do not cross the Fermi level. In particular, gold has been dubbed "the noblest of noble metals," its nobleness being related to the degree of filling of the antibonding states on adsorption of hydrogen molecules, and the degree of orbital overlap with the H adsorbate as ascertained by self-consistent density-functional calculations (Hammer and Norskov, 1995). Although the elements of the platinum group are normally considered noble, according to the strict physical meaning of the term they are not as their d-hybridized bands do cross the Fermi level to a small extent. Instead, when judged by other factors, such as resistance to corrosion and not readily mixing with other elements, then the list of noble metals expands to include rhenium, ruthenium, rhodium, palladium, osmium, iridium, and platinum.

Gold

Gold, platinum, silver, palladium, ruthenium, rhodium, iridium, and osmium are used predominantly in dental application (Roach et al., 2012) as direct gold and silver dental filling materials, traditional amalgam alloys, high-copper amalgam alloys, and gallium alloys.

The excellent biocompatibility, malleability, and resistance to corrosion of gold provide benefits for use inside the human body. For example, it was used as wires in dental applications by the ancient Etruscans as far back as the 7th century BC (see Chapter 1.3.2.1) to hold substitute teeth in place. The ideal dental alloy is one that is easy for the dentist/dental laboratory to manipulate but is strong, stiff, durable, and resistant to tarnish and corrosion. A typical crown and bridge dental alloy may contain 60–75% gold, with silver, platinum, and palladium added to make at least 75% noble metals, plus some copper and zinc. Additional elements are also added to improve castability or to control thermal expansion, important if a porcelain veneer is used.

Currently, there are a number of direct applications of gold in medical devices, including lead wires for pacemakers and gold-plated stents used to inflate and support arteries in the treatment of heart disease (Steegmüller et al., 2002). As gold is highly radiopaque to X-rays, it aids in positioning of the stent. Gold possesses a high degree of resistance to bacterial colonization and because of this it is the material of choice for implants that are at risk of infection, such as the inner ear. For example, gold-plated myringotomy tubes are used for implantation in the tympanic membrane to drain and temporarily aerate the tympanic cavity.

In addition, gold finds application as part of novel drug delivery microchips that contain drug-filled reservoirs covered, sealed, and protected by thin gold membranes. The microchips are implanted, swallowed, or integrated into an intravenous delivery system. A predetermined dose of the drug is administered to the patient by applying a small electric voltage to the gold reservoir cap, causing it to dissolve and allowing the drug to be released. The timing of each dose can be controlled through the use of microprocessors or by remote control or biosensors.

Historically, gold salts have been used in drugs to treat a broad range of ailments (chrysotherapy) such as rheumatoid arthritis, introduced and spearheaded by Jacques Forestier in the 1930s (Forestier, 1929). Today, injectable gold salts such as gold sodium thiomalate and aurothioglucose are considered by many to be the most effective treatment for arthritic ailments.

Recently, the properties of gold compounds have been of interest as potential cancer treatments. Gold nanoparticles (GNPs) are being investigated and used as drug carriers, for thermal therapy to induce apoptotic cell death (see below), as contrast agent in CT and MRI imaging, and as radiosensitizer (Jain et al., 2012). Novel phosphine-supported gold(I) and gold(III) complexes are found to have excellent antitumor activity and clinical trials are presently underway to establish their efficacy for antineoplastic therapy (Reddy et al., 2018).

GNPs have the potential to be used in biosensors by utilizing the surface plasmon resonance (SPR) effect, based on the interaction of an electromagnetic field, provided by laser light, with free electrons at the surface of GNPs. The characteristic reflective shine of an extended metallic surface is caused by light reflected by free electrons. However, for noble metal nanoparticles, the free electrons are confined to a very small space, thus limiting their vibrational frequency. When irradiated with light, the plasmon, that is, the discrete number of oscillations of the free electron gas density, absorbs only that fraction of incoming light that resonates with the characteristic plasmon frequency. When a plasmonic nanomaterials such as gold or silver is coated with a substance that binds to a biomolecule of interest, the frequency of the surface plasmon changes on irradiation with laser light and, hence, the angle of the reflected light. This allows detecting even trace amounts of a biomolecule. In advanced photothermal cancer therapy, plasmonic light-activated GNPs can be infused into the blood of a patient and tend to concentrate inside a tumor. When an external laser light beam of the same frequency as that of the surface plasmon is shone into the tumor mass, the nanoparticles will heat up by resonance, selectively killing the cancer cells without affecting surrounding healthy tissue.

Platinum

The chemical, physical, and mechanical properties of platinum and its alloys include low corrosivity, high biocompatibility, and good mechanical resistance that make the material uniquely suitable for a variety of medical applications. For more than 40 years, platinum alloys have been employed extensively in treatments for coronary artery disease, such as balloon angioplasty and stenting, where inertness and radiopacity under X-ray irradiation are crucial. In the field of cardiac rhythm disorders, platinum's durability, inertness, and electrical conductivity make it an ideal electrode material for devices such as pacemakers, implantable cardioverter defibrillators, and electrophysiology catheters. More recently, its unique properties have been exploited in neuromodulation devices such as brain pacemakers used to treat some movement disorders, and cochlear implants (see Chapter 1.3.5.2) to restore hearing, as well as in coils and catheters to treat brain aneurysms. In 2010, globally some 5.5 kg of Pt were used in biomedical devices, approximately 80% used in established technologies such as guidewires and cardiac rhythm devices, while the remaining 20% were used in more novel technologies such as neuromodulation devices and stents (Cowley and Woodward, 2011).

Catheters and stents incorporating Pt components such as marker bands, tiny metal rings placed on either side of the balloon to keep track of its position in the body, and guidewires assist the surgeon during guiding the catheter to the treatment site, or placing electrodes that are used to diagnose and treat some cardiac rhythm disorders (arrhythmias). The role of platinum in percutaneous transluminal coronary angioplasty is to help ensure that the balloon is correctly located. The guidewire used to direct the balloon to the treatment site is made of base metal such as stainless steel for most of its length but has a coiled Pt–W wire tip. This makes it easier to steer and ensures that it is visible under X-ray irradiation.

Although stents are typically made of AISI 316L stainless steel or cobalt–chromium alloy, in 2009, however, Boston Scientific™ introduced a cardiovascular stent made of a Pt–Cr alloy. The REBEL™ Platinum Chromium Coronary Stent System consists of 37% Fe, 33% Pt, 18% Cr, 9% Ni, and 3% Mo, and provides superior levels of strength, flexibility, and radiopacity. This type of stent has been approved by the European Union, and has in 2014 received approval from the FDA.

The widespread use of platinum agents in the treatment of cancer began with the discovery of the antineoplastic activity of cisplatin, $Pt(NH_3)_2Cl_2$ by Barnett Rosenberg in the 1960s (Rosenberg et al., 1969) that binds to the DNA and inhibits its replication (Kelland and Farrell, 2000). Currently, the most widely used treatments for many types of cancer are platinum-based antineoplastic drugs, although they carry the risk of causing a combination of more than 40 specific side effects which include neurotoxicity, manifested by peripheral neuropathies including polyneuropathy. Nevertheless, highly toxic platinum-based drugs are still a mainstay of cancer chemotherapy as approximately half of all patients undergoing therapeutic treatment receive a platinum drug (Johnstone et al., 2014).

Silver

Owing to their pronounced antimicrobial properties, excellent biocompatibility, and low toxicity, silver ions have been used for a long time in different fields of medicine as wound dressings, catheters, and antibacterial coatings for dental implants (Corrêa et al., 2015). With the advent of nanotechnology, silver nanoparticles (AgNPs) have demonstrated unique interactions with bacterial and fungal species and, hence, are widely used in the medical arena, such as in wound sutures, endotracheal tubes, and surgical instruments. AgNPs have also been applied in several areas of dentistry, as endodontics, coatings for dental prostheses, and in restorative dentistry. AgNPs incorporation aims to avoid or at least to mitigate microbial colonization of dental materials, thus increasing oral health levels and improving quality of life.

Although the antimicrobial mechanism of AgNPs has been extensively investigated in the past, its fine details remain unclear and several explanations have been proposed. Presumably, silver ions interact with the peptidoglycan cell wall (Figure 1.1), causing structural changes, increased membrane permeability and, finally, apoptotic cell death. It is thought that silver atoms bind to thiol groups (-SH) in enzymes and subsequently cause the deactivation of enzymes. Silver forms stable S–Ag bonds with thiol-containing compounds in the cell membrane that are involved in transmembrane energy generation and ion transport (Klueh et al., 2000). It is also believed that silver can take part in catalytic oxidation reactions that result in the formation of disulfide bonds (R-S-S-R). Since silver ions catalyze the reaction between oxygen molecules in the cell and hydrogen atoms of thiol groups, water is released as a product and two thiol groups become covalently bonded to one another through a disulfide bridge (Davies and Etris, 1997). Possibly, the silver-catalyzed formation of disulfide bonds changes the conformation of cellular enzymes and subsequently, affects their function.

Further, AgNPs can interact with the exposed thiol groups in bacterial proteins, thereby inhibiting DNA replication (Seth et al., 2011). It was also proposed that Ag^+ enters the cell and intercalates between the purine (adenine, guanine) and pyrimidine (cytosine, thymine) base pairs, disrupting the hydrogen bonding between the two antiparallel DNA strands and, thus, denaturing the molecule (Klueh et al., 2000).

In conclusion, with the ever-increasing number of antibiotic-resistant strains of bacteria and silver's comparatively low toxicity to humans, its use as an antimicrobial agent is an exciting topic with great relevance to many fields of biomedical research and related industry. Since the overuse of antibiotics has caused the emergence of antibiotic-resistant strains, such as community-acquired multiresistant *Staphylococcus aureus* and healthcare-acquired multiresistant *S. aureus*, there has recently been a renewed interest in using silver as an antibacterial agent to fight nosocomial infections, thereby replacing antibiotic drugs in many medical indications.

2.1.2 Biodegradable metals

2.1.2.1 Magnesium

Medium-entropy magnesium alloys such as AZ31, AZ91, ZK60, or WE43 are currently considered suitable candidate materials for biomedical applications based on their favorable mechanical properties, and their confirmed biocompatibility and osseoconductivity. Hence, these alloys could potentially be utilized for osteosynthetic musculoskeletal fixation devices that include fracture fixation screws and plate systems for use in extremity locations, interference screws, surgical clips, screws for femoral head fixation, spinal implants and cages, and others. Moreover, substantial research is underway to improve bioresorbable cardiovascular stents based on Mg alloys that are being designed to provide short-term supporting structures for vascular implants to combat coronary heart and peripheral artery diseases.

Although magnesium alloys release corrosion products in contact with body fluid, these products do not deleteriously affect human metabolic pathways. Indeed, many studies have revealed that implanted magnesium devices cause only minimal changes to blood composition, and, in particular, do not cause damage to excretory organs, that is, liver or kidneys. During the past decade, many reviews on using magnesium as an emerging biomaterial have been published. Among them, the most authoritative contribution is a review of the status of clinical application of magnesium-based orthopedic implants by Zhao et al. (2016a). A review on more recent research and patents aimed at controlling corrosion of bioresorbable Mg alloy implants and steering their development toward next-generation biomaterials was provided by Heimann and Lehmann (2017).

Structure and properties

Magnesium adopts an hcp structure (see Figure 2.13) that can be described as an ABAB layer packing where each Mg atom is surrounded by six nearest neighbors. Suitable mechanical properties of magnesium and its alloys include low density of 1.74–1.81 g/cm^3, a high strength-to-mass ratio, and elastic modulus (17 GPa in shear, 45 GPa in tension) close to that of cortical bone (7–30 GPa), in contrast to Ti6Al4V or surgical stainless austenitic steels. The fracture toughness of AZ31 alloy (18–21 $MPa/m^{1/2}$) is almost in the range of that of cortical bone (2–12 $MPa/m^{1/2}$), much closer than that of Ti6Al4V (84–107 $MPa/m^{1/2}$) and stainless steels (AISI 316L: 112–278 $MPa/m^{1/2}$; AISI 304: 119–228 $MPa/m^{1/2}$). The high stiffness-to-mass ratio (specific stiffness) of (9–26)$\cdot10^6$ m^2/s^2 compares favorably to that of both steel and titanium.

However, in contrast to these advantageous mechanical properties, magnesium exhibits low corrosion resistance in contact with aqueous solutions, such as Cl^- ion-rich ECF. Indeed, rapid corrosion and associated hydrogen gas evolution have been "show stoppers" of early attempts to use Mg for biomedical applications (Witte, 2010). Nevertheless, today an improved understanding of corrosion mechanisms, and development and application of innovative corrosion protection techniques, most notably deposition of dense coatings with high adhesion strength, are reviving research interest in biodegradable Mg alloy (Song, 2011; Heimann and

Lehmann, 2017). Deposition strategies of such protective coatings are dealt with in Chapter 3.1.4.4. In parallel, research into developing novel, less corrosion-prone alloy compositions with uniform corrosion behavior will provide promising approaches toward a new generation of resorbable biomaterials.

The electrochemical mechanisms governing corrosion of magnesium in aqueous media are complex. In particular, they are unlike those that control corrosion of ordinary metals, such as titanium, iron, or steel, the corrosion mechanism of which is consistent with Tafel theory.

Tafel theory

This theory stipulates that the rates of anodic and cathodic partial reactions become equal at the corrosion potential E_{corr}. Hence, the electrochemical parameters of ordinary metals can be obtained from the so-called Tafel slopes of a potential-current density plot provided by potentiodynamic polarization experiments. In this case, an increase of the applied potential affects an increase in anodic potential but a decrease of the cathodic potential. The current density i can be expressed as a function of the overpotential η ($\eta = E_{applied} - E_{open\ circuit}$) by the Tafel equation as $\eta = \beta \cdot \log(i/i_0)$, where β is the so-called Tafel slope, i is the applied current density, and i_0 is the exchange current density, that is, the corrosion reaction rate at the reversible potential.

In contrast, magnesium and its alloys behave differently as they show an increase of both anodic and cathodic partial potentials with increasing applied potential. This behavior is called the negative difference effect (NDE) (Thomaz et al., 2010). To account for the complex corrosion mechanism of magnesium, three basic models have been invoked: the NDE model, the monovalent ion model (MIM), and a combination of both (Persaud-Sharma and McGoron, 2012) as shown in Figure 2.35.

The naturally formed passive oxide film that partially protect the magnesium surface from aqueous corrosion breaks down in the presence of an electrolyte precisely as the result of the NDE. This breakdown process releases Mg^{2+} ions through the oxidation half-cell reaction (eq. 2.1a) and gaseous H_2 through the reduction half-cell reaction (eq. 2.1b) under low potential or current density conditions (Figure 2.35, A1). In the process, less localized corrosion pitting occurs compared to a magnesium surface exposed to higher voltage or current density (Figure 2.35, A2).

Accordingly, the corrosion process is governed by the following contributing half-cell reactions:

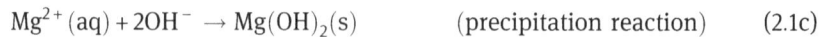

$$Mg(s) \;\rightarrow\; Mg^{2+}(aq) + 2e \qquad\qquad \text{(oxidation reaction)} \qquad (2.1a)$$

$$2H_2O(aq) + 2e \;\rightarrow\; H_2(g) + 2OH^-(aq) \qquad \text{(reduction reaction)} \qquad (2.1b)$$

$$Mg^{2+}(aq) + 2OH^- \;\rightarrow\; Mg(OH)_2(s) \qquad \text{(precipitation reaction)} \qquad (2.1c)$$

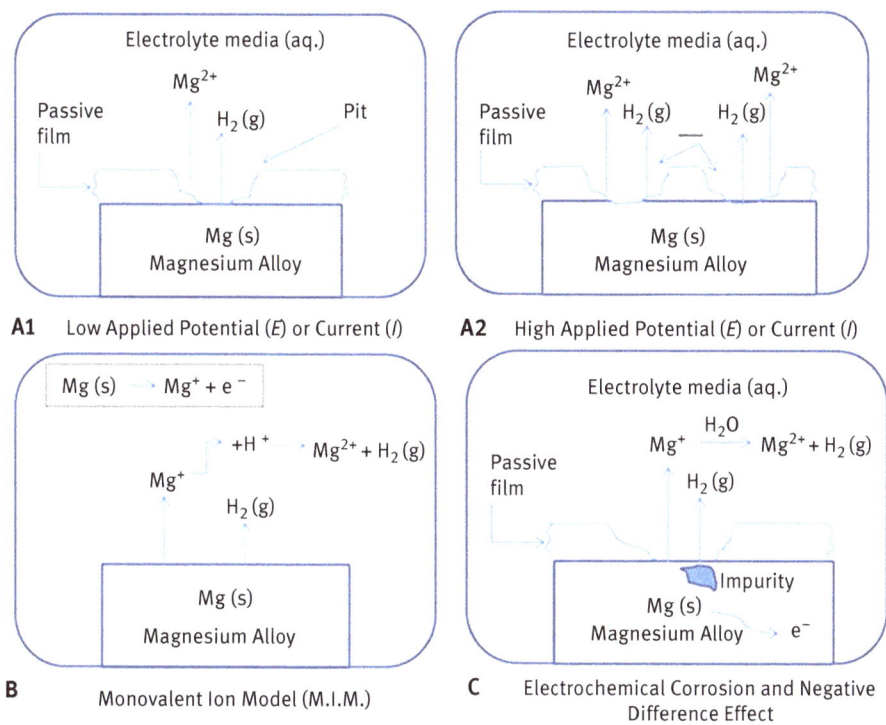

Figure 2.35: Schematic rendering of different corrosion mechanisms for magnesium. (A1) Mg^{2+} formation and $H_2(g)$ evolution under low E or I. (A2) Mg^{2+} formation and $H_2(g)$ evolution at high E or I, with multiple pit formation. (B) Monovalent ion model (MIM). (C) Combined model (Persaud-Sharma and McGoron, 2012). © With permission by PubMed Central.

However, the NDE model has been questioned since investigation of corrosion of AZ91 alloy showed the presence of a protective film consisting of hydromagnesite, $Mg_5(CO_3)_4(OH)_2 \cdot 8H_2O$, and MgO.

The MIM model (Figure 2.35B) proposes the formation of monovalent transient Mg^+ ions according to

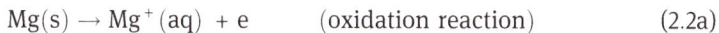

$$Mg(s) \rightarrow Mg^+(aq) + e \qquad \text{(oxidation reaction)} \qquad (2.2a)$$

that react with two proton to form gaseous hydrogen by

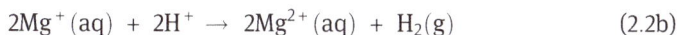

$$2Mg^+(aq) + 2H^+ \rightarrow 2Mg^{2+}(aq) + H_2(g) \qquad (2.2b)$$

Alternatively, Song et al. (1997) proposed an electrochemical corrosion model that combines the MIM model with the protective film NDE model. This combined model is based on the assumption that with increasing applied potential or current density, more and more surface area looses its protective film. In these film-depleted areas, corrosion produces Mg^+ ions that are still able to react with water according to eq. (2.3) to generate gaseous hydrogen and hydroxyl ions that drive up the solution pH (Figure 2.35C):

$$Mg^+ + 2H_2O \rightarrow Mg^{2+} + OH^- + H_2(g) \tag{2.3}$$

The model further proposes that a negative applied potential would result in an intact protective film covering the metal surface. Cathodic hydrogen evolution is supposed to decrease with increasing potential until the pitting potential has been attained.

Biomedical application

One of the unique advantages of magnesium, and also its limiting disadvantage, is its pronounced bioresorbability. On the one hand, this property enables implants to gradually dissolve within the body when healing has completed without the need of a secondary operation to remove the implant. On the other hand, rapid corrosion may destabilize an implant before its medical purpose has been achieved. Recent reviews (e.g., Agarwal et al., 2016; Kamrani and Fleck, 2019) report on corrosion properties, biocompatibility, and surface modifications of Mg-base alloys applied in biomedical context.

Nevertheless, suitable magnesium alloys are increasingly being utilized for specific biomedical implants, most notably *osteosynthetic musculoskeletal bone fracture fixation devices* (Chaya et al., 2015; Lee et al., 2016; Marukawa et al., 2016; Torroni et al., 2017) and cardiovascular stents (Moravey and Mantovani, 2011). Osteosynthetic applications include fracture fixation screws and plate systems for use in extremity locations, interference screws, surgical clips, screws for femoral head fixation, spinal implants and cages, and others.

Research employing animal models suggests that osteosynthetic plates and screws made from polyethylene oxide (PEO)-coated Mg cast alloy WE43 (93.6Mg4Y2.25Nd0.15Zr) are promising materials for further development of implants for osteosynthesis of the human facial skeleton (Schaller et al., 2016). Animals showed a good tolerance of the plate/screw system without wound healing disturbance, attesting to excellent biocompatibility of the Mg alloy. Although radiology revealed formation of hydrogen gas pockets about four weeks after surgery, micro-CT, and histological analyses of PEO-coated screws revealed significantly lower corrosion rates, and increased bone density and bone–implant contact (BIC) area compared to uncoated screws.

Substantial research is underway to improve bioresorbable *cardiovascular stents* based on Mg alloys designed to provide short-term supporting structures for vascular implants to combat coronary heart and peripheral artery diseases (see Chapter 1.3.4.1). This development has been instigated by several disadvantages of permanent drug-eluting stents fashioned from traditional biocompatible metals, such as tantalum, titanium, CoCrMo alloys, or the shape-memory alloy NiTi that present increased risks of thrombosis, loss of vascular flexibility, inhibition of revascularization, as well as difficult to evaluate CT and MRI images. In contrast, novel biodegradable stents made from magnesium alloys allow free physiological vasomotion, foster improved clinical long-term behavior by inhibiting chronic inflammation, delay thrombosis complications, prevent occurrence of re-occlusion and restenosis, and even enjoy better acceptance by patients. Several studies have shown that the critical period of vascular healing normally ends three months after implantation. Hence, developments aim to providing survival of Mg-based stents within this timeframe.

A plethora of studies underscore the notion that research into developing adequate protective coatings (see Chapter 3.1.4; Heimann and Lehmann, 2017) to limit magnesium corrosion in the highly aggressive Cl ion-containing body environment is high up on the agenda of contemporary research and development effort to incorporate magnesium alloys into the toolbox of biomedical engineers (Hornberger et al., 2012; Wu et al., 2013). Rapid degradation of Mg leads to undesirable outcomes for most medical implants, including accumulation of hydrogen gas (see eq. (2.2b)) that leads to cavities, establishment of local alkaline pH due to the release of hydroxide (OH^-) ions during degradation (see eqs. (2.1b) and (2.3)), and premature mechanical failure by pitting, crevice, or intergranular corrosion before bone tissue heals. This interferes with the clinical requirement for Mg-based implants to retain their mechanical stability during tissue healing (typically 12 weeks for bone) and then gradually degrade afterwards (see Figure 1.17).

Magnesium implant devices are found to effect only minimal changes to blood composition and, in particular, are not known to cause damage to excretory organs, that is, liver or kidneys. Magnesium is an essential trace element in the human body and is necessary as a cofactor in many different enzymatic reactions, playing an important role in energy metabolism. Serum magnesium levels can be maintained normal due to the dynamic absorption and excretion equilibrium of magnesium in the human body, through gastrointestinal absorption and renal excretion. In addition, magnesium is needed for effective heart, muscle, nerve, bone, and kidney function. Hence, magnesium ions play important parts in the regulation of ion channels, DNA stabilization, enzyme activation, and stimulation of cell growth and proliferation as detailed in a recent review on different magnesium-based bioceramics (Nabiyouni et al., 2018). Magnesium ions have been shown to increase cell adhesion and osteoblastic activity *in vitro*, and many *in vivo* material implantation studies confirm improved bone regeneration and healing in their presence. In addition, recent studies have reported potential antimicrobial properties associated with Mg degradation products.

Related to other applications, *in vitro* studies using human bone marrow-derived mesenchymal stem cells (hBMSCs) showed that the presence of biodegradable Mg alloys positively affect the proliferation of stems cells as well as the progress of osteogenic development by inducing stem cell differentiation toward osteoblasts (Berglund et al., 2018). When AZ31B surfaces were coated with calcium phosphate (CaP), proliferation, adhesion, and viability of hBMSCs increased as ascertained by MTT assay and calcein-AM/EthD-1 staining. As far as osteogenic differentiation is concerned, real-time polymerase chain reaction (RT-PCR; see box) analysis showed that a CaP coating significantly upregulated the expression of collagen 1 (Col-1), alkaline phosphatase (ALP), osteocalcin (OC), and osteopontin (OPN) in hBMSCs, suggesting that calcium phosphates strongly promote proliferation and differentiation of mesenchymal stem cells (MSCs).

Polymerase chain reaction
PCR is a modern method widely used in molecular biology to make several copies of a specific DNA segment. Using PCR, copies of DNA sequences are exponentially amplified to generate thousands to millions of more copies of that particular DNA segment. PCR is now a common and often indispensable technique used in medical laboratory and clinical laboratory research for a broad variety of applications including biomedical research and criminal forensics. The majority of PCR methods rely on thermal cycling. Thermal cycling exposes reactants to repeated cycles of heating and cooling to permit different temperature-dependent reactions – specifically, DNA melting (denaturation), and enzyme-driven DNA replication (elongation). PCR employs two main reagents – primers (which are short single strand DNA fragments known as oligonucleotides that are a complementary sequence to the target DNA region) and a DNA polymerase. In the first step of PCR, the two strands of the DNA double helix are physically separated (denaturated) at high temperature (94–96 °C) in a process called DNA melting. In the second annealing step, the temperature is lowered to 68 °C and the primers bind to the complementary sequences of DNA. The two DNA strands then become templates for DNA polymerase to enzymatically assemble at 72 °C a new DNA strand from free nucleotides, the building blocks of DNA. As PCR progresses, the DNA generated is itself used as a template for replication, setting in motion a chain reaction in which the original DNA template is exponentially amplified (https://en.wikipedia.org/wiki/Polymerase_chain-reaction).

Morphological studies in animal models showed that no major detrimental systemic effects were associated with the implantation of WE43 alloy. However, a tendency was observed toward lysis of red blood cells and a decrease in the index of survival of mesenchymal stromal cells (Lukyanova et al., 2018). In another contribution, the presence of chemical conversion coatings was found to decrease the accumulation of Ca and P around the implant, thus avoiding ectopic calcification within muscle tissue (Chen et al., 2017).

Electrochemical monitoring of hydrogen, the product of corrosion of magnesium during contact with body fluid *in vivo* has been used to record the degree of interaction between implant and living body environment. Zhao et al. (2016b, 2017) developed a visual sensor to monitor transdermally the amount of hydrogen evolved during biodegradation of magnesium. The sensor, subcutaneously implanted in mice, consisted of a thin, flexible polymer sheet coated with hydrogen-sensitive material (MoO_3 and Pt catalyst) that changes color in the presence of H_2 gas. This noninvasive method has the capability to measure in real time the corrosion rate of Mg implants.

In conclusion, it is crucial to control the degradation rate of Mg in the harsh physiological environment of the human body. On the one hand, implant materials made of magnesium are required to have sufficient mechanical strength and integrity during their residence time in the human body, that is, until the affected part of the body is healed. On the other hand, reasonable resistance is required to corrosion in the body environment during the initial periods of implantation, as well as establishment of controlled, uniform and thus predictable corrosion rate during subsequent stages. Furthermore, the released corrosion products should not exceed the acceptable tolerance level of the human body. It is fortuitous that magnesium alloys excel in most of these departments.

During the past decade, biodegradable magnesium alloys have advanced significantly toward a variety of medical applications, including osseosynthetic bone healing devices and cardio-vascular stents (see Figure 1.17). However, many still unanswered questions are waiting to be tackled, and there is a long way to go before Mg alloy-based implants will become a routine implant option in the clinical operation theater (Willumeit-Römer and Müller, 2017). An ultimate challenge pertains to the application of Mg for load-bearing endoprosthetic hip and knee implants. The favorable mechanical properties of Mg alloy in terms of a high stiffness-to-mass ratio and, in particular, an elastic modulus close to that of cortical bone suggest, in the long run, the emergence of a biomaterial able to mitigate or even alleviate stress shielding (Poinern et al., 2012). This would enormously benefit the outcome of total hip replacement operations that are frequently plagued by diverting the load-imposed forces away from low-modulus bone toward the high-modulus metal implant, a process that eventually results in osteoclastic bone resorption and hence, atrophy. Today, truly isoelastic endoprosthetic implants possibly based on magnesium might be considered the holy grail of implantology. However, to reach this lofty goal there are still very substantial barriers to overcome. First among the challenges is the need to improve the corrosion resistance of Mg-based alloys by orders of magnitude, either by discovering novel biocompatible alloying elements or by developing appropriate biocompatible protective surface coatings that synergistically will guarantee long-term *in vivo* stability of the implant. Indeed, recently developed MgNdZn alloys (Mao et al., 2012) show not only much reduced biodegradation rates but also homogenous nanophasic degradation pattern. This invention is considered a substantial step toward emergence of Mg alloys with controlled layer-by-layer surface removal that will allow more accurate prediction of implant survival time *in vivo*. All this requires a tremendous paradigmatic shift from magnesium being an established *biodegradable* material at present toward a future reasonably *biostable* material that will survive in the human body for a long time.

In the light of this challenge, the progress achieved so far and related to improving the corrosion rate, that is, the *in vivo* survival time of Mg-based implants in osteosynthetic and cardio-vascular stent applications may appear rather paltry. This perception, however, is far from reality. Mg alloys have made substantial inroads into the biomedical market as confirmed by the avalanche of original research papers and patent applications that attest to the importance that is being bestowed on this area of biomedical research. What at this point is needed most are more targeted approaches toward a common discussion among the international research community. Otherwise, progress in this important field of biomedical research will be too slow to meet the expectations and actual needs of our present society and of future generations

2.2 Ceramics (Robert B. Heimann)

During the last few decades, bioceramic materials have evolved into a powerful driver of advanced ceramics research and development. For many years, *bioinert* ceramic materials such as alumina, zirconia, and carbon as well as *bioactive* ceramic materials such as hydroxylapatite, tricalcium phosphate (TCP) and tetracalcium phosphates (TTCP), silica-based bioglasses, calcium carbonate, and, to a limited extent, titanium dioxide have been used successfully in the clinical practice. An excellent early account on the wide range of bioceramics available today has been given in a comprehensive book edited by Tadashi Kokubo (2008) that discusses issues of significance of their structure, mechanical properties and biological interaction, and reviews their clinical applications in joint replacement, bone grafts, tissue engineering, and dentistry. Additional information is available from Heimann (2008, 2010) and Heimann and Lehmann (2015), the latter dealing specifically with bioceramic coatings for medical implants.

Despite all this elaborate and cost- and manpower-intensive research, to date no ceramic biomaterial is known that is, at the same time, mechanically and chemically stable, and sufficiently osseoconductive to fulfill its expected biomedical functions. Indeed, conventional bioceramics such as alumina or zirconia are mechanically strong but bioinert, and osseoconductive hydroxylapatite is mechanically weak and essentially nonresorbable when highly crystalline, whereas the mechanically even weaker osseoconductive TCP and TTCP are bioresorbable.

Figure 2.36 shows this apparent dichotomy as well as the trend that is being followed today toward the development of composite biomaterials consisting of a mechanically stable scaffold seeded with biochemically signaling growth factors. Scaffolding materials are considered either bioinert metals (titanium, tantalum) or hard ceramics (alumina, zirconia), or bioactive ceramics such as hydroxylapatite and bioglass as well as calcium carbonates derived from organic templates (bovine spongiosa, corals, sea urchin spines).

Figure 2.37 and Table 2.5 compare important mechanical, thermal, and elastic properties of oxides ceramics used in biomedical applications. The figure of merit

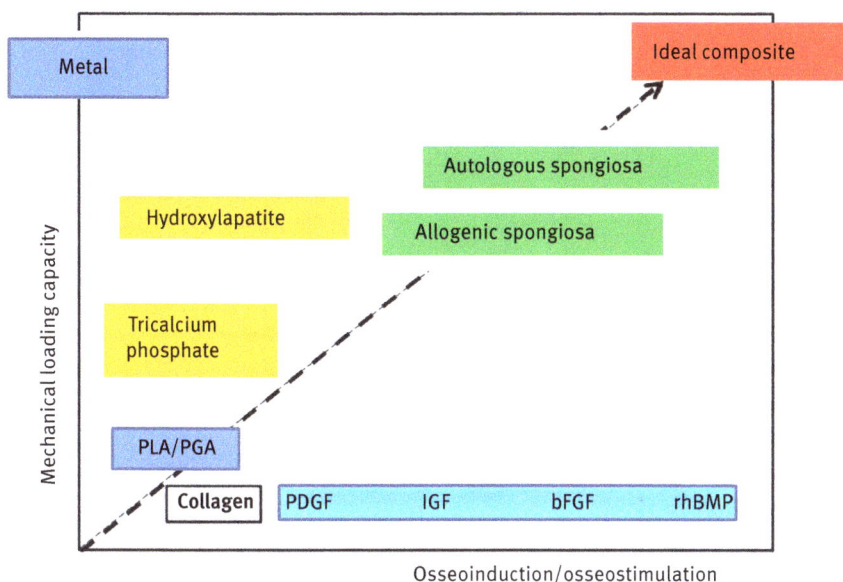

Figure 2.36: Mechanical loading capacity versus osseoinductive/osseostimulating ability of several bone replacement materials. Yellow: bioactive/bioresorbable ceramics. Green: biological bone replacement materials. Dark blue: metals or polymers (PLA, polylactic acid; PGA, poly(glutamic acid)). Light blue: biological signaling factors (PDGF, platelet-derived growth factor; IGF, insulin-like growth factor; bFGF, basic fibroblast growth factor; rhBMP, recombinant human bone morphogenetic protein) (modified after Niedhart and Niethard, 1998).

values reported in the scientific literature, ASTM and ISO standards, granted or applied for patents, and promotional brochures vary widely, depending on purity, grain size, and processing routes of the ceramic materials. In the remainder of this chapter, structure–property–application relations of prominent bioceramics materials will be evaluated.

2.2.1 Bioinert/biotolerant ceramics

By definition, during incorporation into the human body, bioinert bioceramics do not release any of their constituents in toxic concentrations that may negatively interfere with body functions or human metabolism. However, these materials do not show positive interaction with living tissue, meaning that they neither support formation of de novo tissue nor participate in cell differentiation. As a response of the body to these materials, usually a nonadherent fibrous capsule of connective tissue is formed around the bioinert ceramic material that in the case of bone remodeling

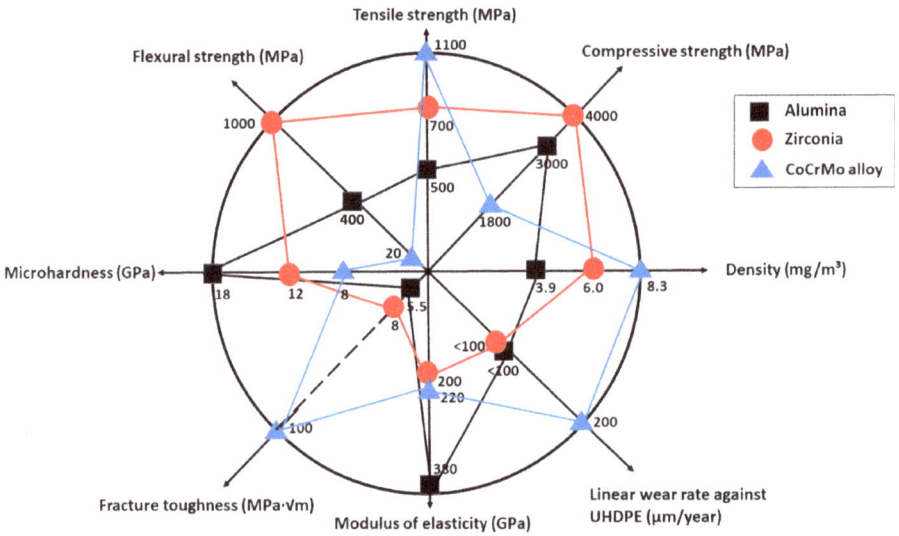

Figure 2.37: Mechanical, thermal, and elastic performance-application diagram of selected bioceramic materials compared to CoCrMo alloy. The data are averages of the ranges given in Table 2.5 and shown here only for sake of comparison.

Table 2.5: Selected property ranges of oxide ceramics for biomedical application.

Property	Dimension	Alumina	Zirconia	Titania
Density	Mg/m^3	3.7–3.9	5.6–6.1	3.9–4.1
Compressive strength	MPa	2,000–4,000	1,200–5,000	800–1,000
Tensile strength	MPa	300–630	120–700	250–300
Flexural strength	MPa	280–420	900–1,400	30–100
Fracture toughness	MPa · Vm	3–5	7–10	2–3
Elastic modulus	GPa	350–400	200–220	230–290
Shear modulus	GPa	140–160	50–90	90–110
Microhardness	GPa	15–20	12–13	8–10
Thermal expansion coefficient	$10^{-6}/K$	7.5–8.5	10–12	8–12
Thermal conductivity	W/m · K	30–40	2.5–3.0	5–11

manifests itself by a shape-mediated *contact osteogenesis*. Through the bone–material interface, only compressive forces will be transmitted (bony on-growth). Typical bioinert bioceramic materials are alumina (α-Al$_2$O$_3$), zirconia (partially stabilized zirconia, PSZ; tetragonal zirconia polycrystal, TZP), as well as pyrolytic graphitic carbon and highly biocompatible diamond-like carbon (DLC).

2.2.1.1 Alumina

Structure
The structure of the thermodynamically most stable form of crystalline aluminum oxide (corundum, α-Al_2O_3) consists of oxygen ions forming a slightly distorted hcp lattice with the aluminum ions filling two-thirds of the octahedral interstices (Figure 2.38, left).

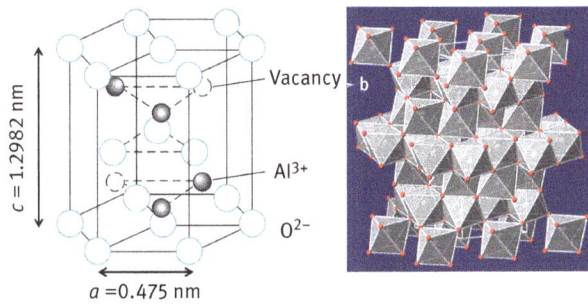

Figure 2.38: Hexagonal closed-packed (hcp) structure of α-Al_2O_3. Left: The primitive unit cell of Al_2O_3. Right: Each aluminum atom is octahedrally surrounded by oxygen atoms (red dots), forming AlO_6 octahedra. © Permission granted under Creative Commons Attribution License (CC BY 3.0).

Corundum obeys a hexagonal-scalenohedral ($\bar{3}m$) crystal class with a $R\bar{3}c(167)$ space group. The primitive cell with lattice constants $a_0 = 475.1$ pm and $c_0 = 1{,}298.2$ pm contains $Z = 6$ formula units of aluminum oxide. Apart from α-Al_2O_3, several other transitional alumina polymorphs exist such as γ-, η-, δ-, and θ-alumina with fcc structure as well as χ- and κ-alumina with hcp structure. These polymorphic modifications are formed during stepwise dehydroxylation of aluminum hydroxides such as gibbsite, bayerite, boehmite, or diaspor.

Properties
Alumina (corundum, sapphire, α-Al_2O_3) is considered the workhorse material of the structural ceramics industry. It excels by very high hardness (Mohs scratch hardness scale 9; Vickers microhardness up to 20 GPa), high abrasion resistance, and chemical inertness against most acids and alkalis, making alumina an ideal material to perform well in a variety of aggressive environments ranging from mining industry to chemical industry to metal manufacturing and processing to ceramic

armor and biomedical applications. Mechanically very strong, abrasion resistant, and thoroughly biocompatible corundum single crystals are being successfully used as dental root implants.

However, these advantageous properties are partially offset by rather low tensile and flexural strengths, and fracture toughness, and also low thermal shock resistance. While these properties may affect the wear performance of alumina in biomedical devices such as the femoral ball of hip endoprostheses, the decision to include alumina in bioengineering designs must be judiciously assessed to avoid catastrophic failure in service under harsh conditions as prevailing in the human body environment. Table 2.6 provides an overview of important properties of alumina ceramics. Most of these properties depend on processing parameters including sintering temperature, sintering atmosphere, impurity content, grain size, and several other factors. Extremely pure and fine-grained alumina used in femoral

Table 2.6: Mechanical, thermal, elastic, and electrical properties of high alumina ceramics. Data of high purity 99.9% Al_2O_3 were taken from information provided by Kyocera Global (http://global. kyocera.com/prdct list/material/alumina/alumina.html). Data of 99.5% Al_2O_3 refer to https://accu ratus.com/alumox.html.

Property	Dimension	Average value	Comment
Density	Mg/m^3	3.98	99.9% Al_2O_3
Flexural strength	MPa	400	3-Point bending strength
Compressive strength	MPa	2,600	99.5% Al_2O_3
Tensile strength	MPa	660	99.5% Al_2O_3
Elastic modulus	GPa	380	99.9% Al_2O_3
Shear modulus	GPa	152	99.5% Al_2O_3
Bulk modulus	GPa	228	99.5% Al_2O_3
Vickers hardness	GPa	17.5	HV 9.8 N, 99.9% Al_2O_3
Fracture toughness	MPa · $\sqrt{m}$	4.5	99.9% Al_2O_3
Thermal conductivity	W/m · K	34	@ 20 °C, 99.9% Al_2O_3
Thermal expansion	10^{-6}/K	7.2	(40–400 °C), 99.9% Al_2O_3
Thermal expansion	10^{-6}/K	8.0	(40–800 °C), 99.9% Al_2O_3
Specific heat capacity	J/kg · K	780	99.9% Al_2O_3
Dielectric strength	kV/mm	15	99.9% Al_2O_3
Dielectric constant		9.9	@ 1 MHz, 99.9% Al_2O_3

heads for hip endoprostheses show noticeable higher strength, fracture toughness, and hardness as shown in Table 2.7.

Table 2.7: ISO 6474 norm requirements and properties of clinically utilized alumina-based ceramics.

Property	High alumina	ISO 6474–1:2010	ISO 6474–2:2012	BIOLOX® forte	BIOLOX® delta
Al_2O_3 content	>99.7	>99.5			~82
ZrO_2 content					~17
$SiO_2 + Na_2O$ (%)	<0.02	0.1			
$SiO_2 + Na_2O + CaO$ (%)			<0.1		
Cr_2O_3 (%)					0.5
SrO (%)					0.5
Density (Mg/m^3)	3.98	>3.90	>3.94	3.98	4.37
Average grain size of Al_2O_3 (μm)	3.6	<7	<4.5	1.7	0.6
Average grain size of ZrO_2 (μm)					0.3
4-point bending strength (MPa)	280–420	400–500	> 450	630	1,380
Fracture toughness (MPa·√m)	3–4	4–6	–	3.2	6.5
Modulus of elasticity (GPa)	350–400	380–420	–	410	360
Vickers hardness (GPa)	24	>20		20	19
Average wear rate (mm^3/10^6 cycles)				8.3	0.6

Biomedical applications

Pure, fine-grained polycrystalline alumina materials have been used for almost 50 years (Boutin, 1972) for femoral heads of hip endoprostheses. Today, there exist a large variety of clinical options to combine femoral heads and acetabular cups (Ben-Nissan et al., 2008).

In Germany, medical alumina products are being marketed by CeramTec GmbH under the brand names BIOLOX® and BIOLOX® forte. The properties and required purity of alumina used in biomedical applications are summarized in Table 2.7. The ISO 6474-1:2010 norm specifies the characteristics of, and corresponding test methods for, a biocompatible and biostable ceramic bone substitute material based on high purity alumina for use as bone spacers, bone replacements, and components of orthopedic joint prostheses. This norm has been reviewed and confirmed in 2015.

Grain boundary engineering

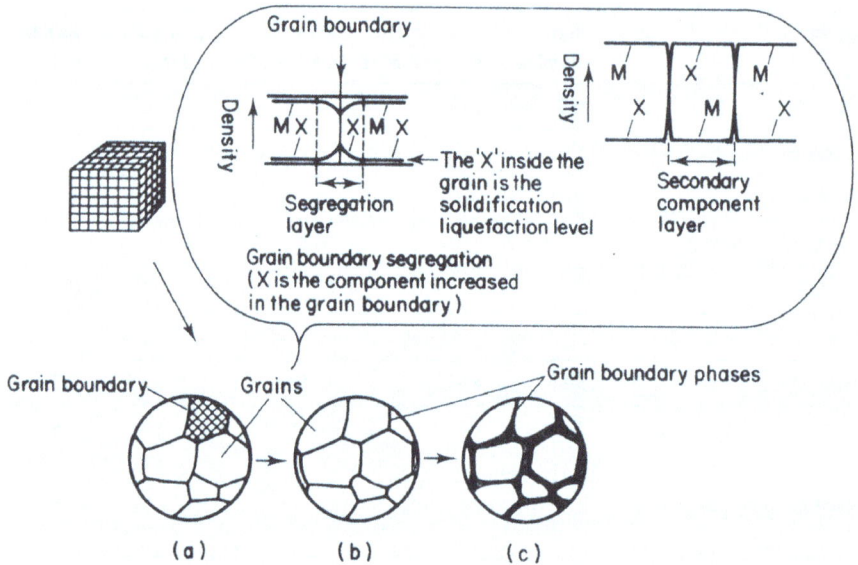

Grain boundary engineering involves the intentional addition of impurities to block grain boundary movement during sintering at high temperature (Watanabe, 2011). In the grain boundary region, free surface energy is increased so that impurities, in this case MgO, tend to concentrate along the grain boundaries, suppressing grain growth by recrystallization. Hence, the impurities such as MgO exist as a second phase of $MgAl_2O_4$ among the alumina constituent particles. With increasing amount of MgO impurities, the microstructure shifts from (a) to (c). In such case, the shapes of the grain boundaries depend on the material, its constituents, and the temperature and time of the sintering process. The figure shows microstructural pattern (a) to (c) of segregation of impurities along energetically preferred grain boundaries, resulting in suppression of grain growth by recrystallization (Ichinose, 1987). © With permission by John Wiley & Sons Ltd.

High strength alumina ceramics for medical use can be achieved by grain boundary engineering (see box) during which suppression of grain growth at high sintering temperatures is achieved by adding minor amounts of magnesium oxide in the range of a few tenths of a percent. Accumulation of magnesium oxide along the grain boundaries of alumina will result in a thin surface layer consisting of spinel ($MgAl_2O_4$) that acts as a barrier toward the grain boundary movement normally associated with the process of recrystallization. Hence, formation of large grains by recrystallization will be effectively suppressed as utilized in third-generation BIOLOX® forte material.

Figure 2.39 shows the surface of thermally etched BIOLOX® forte. The etching technique used involved heating of the material at temperatures around 1,500 °C for a prolonged time (Chen and Tuan, 2000). This heat treatment causes formation of facets that act to minimize the surface free energy by a so called "hill-and-valley" structure, involving thermally activated movement of surface atoms (Herring, 1951; Heimann, 1975). Thus, thermal etching increases the topographical contrast of otherwise featureless polished surfaces by delineating small angle grain boundaries and thus, is an important ceramographic tool for quality assessment.

(A)   (B)

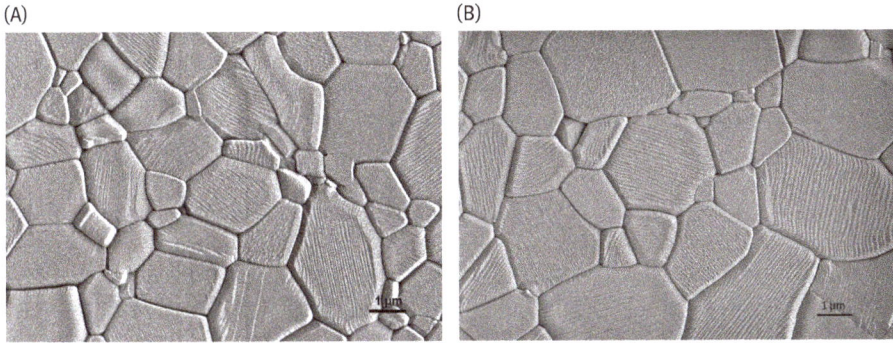

Figure 2.39: Microstructure of thermally etched BIOLOX® forte, showing "hill-and-valley" structure of alumina grains (A) and enhanced grain boundary etching (B). © Image courtesy of CeramTec GmbH, Plochingen, Germany.

The mechanical behavior of alumina ceramics in simulated physiological environment has led to long-term survival predictions for these materials when subjected to subcritical stresses. Figure 2.40 shows a cumulative Weibull plot (see box) as representative of the failure probability of typical femoral heads made from high-purity alumina such as BIOLOX® forte. The very low data scatter and the high average bending strength of 641 MPa (d_{50}) attest to both the strength and the extraordinary reliability as indicated by a high value of m, that is, the slope of the straight line. Considering the tensile stresses encountered in many implants, an alumina ceramic femoral ball can be reliably employed (Piconi, 2017).

Wear of the sliding pair alumina/UHMWPE (ultra-high-molecular-weight PE) of hip endoprostheses was found to have decreased by 25–30% compared to that of a metal/UHMWPE pair in hip simulator tests and clinical results. Wear of alumina/alumina pairs was observed to be close to zero in a similar hip simulator test.

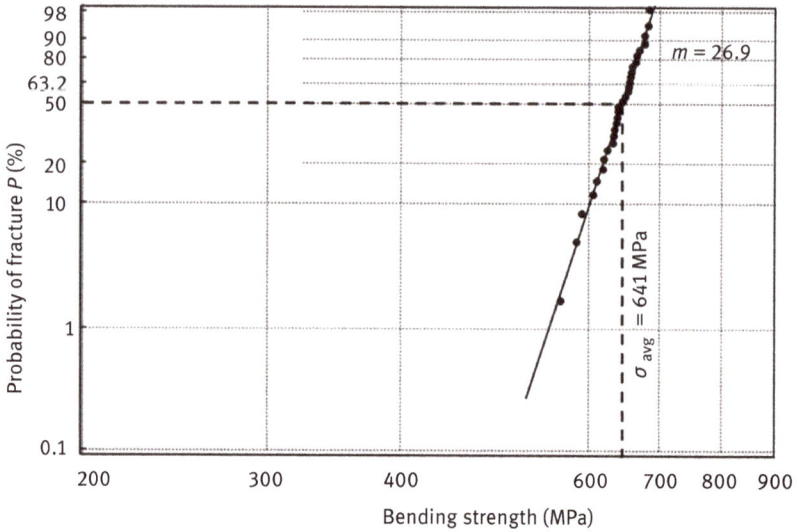

Figure 2.40: Weibull fracture probability plot of 30 BIOLOX® forte bars subjected to a 4-point bending test. At the mean strength value of 641 MPa, there is a probability that 50% of the investigated specimens will fail. The high Weibull modulus m attests to the high engineering reliability of the material © Courtesy of CeramTec GmbH, Plochingen, Germany.

Weibull distribution

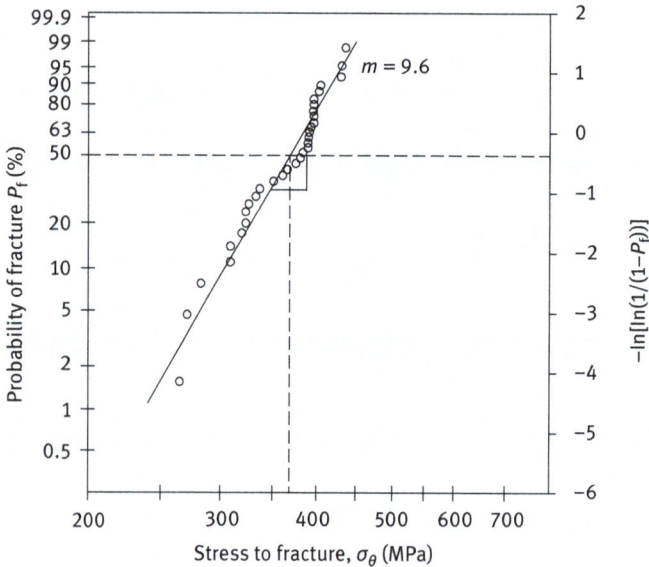

The stochastic Weibull distribution is frequently used in reliability engineering and failure analysis, including life data analysis of alumina femoral heads. The probability P_f of failure during a 3- or 4-point flexural test is given by $P_f = 1 - \exp[-(\sigma-\sigma_\theta)]^m$, whereby σ is the applied stress, σ_θ

is the stress to failure, and m is the Weibull modulus. Linearization yields $\ln(1-P_f) = -(\sigma-\sigma_\theta)^m$ and finally $\ln[\ln(1/(1-P_f))] = m \cdot \ln\sigma - m \cdot \ln\sigma_\theta$. Then, m is the slope of the linear plot of $x = \ln\sigma$ versus $y = \ln[\ln(1/(1-P_f))]$. The graph shows a typical Weibull plot of feldspathic dental porcelain subjected to a flexural test. The mean strength value is the 50% probability of failure, that is, half of the collection of specimens investigated will fail at the flexural strength of 365 MPa. A large slope $m = 9.6$ shown in the graph indicates increased engineering reliability. More information can be gained from Quinn and Quinn (2010).

Duplex Al_2O_3–ZrO_2 ZTA ceramics

The ISO 6474-2:2012 norm relates to a high-purity alumina matrix materials reinforced with zirconia to form a compound that is considered zirconia-toughened alumina (ZTA). It specifies a much lower average grain size with a concurrent increase in the flexural strength to well beyond 450 MPa (Table 2.7). Novel developments in the field of femoral heads for hip endoprostheses rely on high purity alumina (~82 vol%) with additions of about 17 vol% tetragonal Y-stabilized zirconia, 0.5 vol% chromia (Cr_2O_3), as well as 0.5 vol% strontium oxide (SrO) additives (BIOLOX® delta). During sintering at temperature as high as 1,800 °C, the ZTA matrix takes in the chromium oxide in solid solution coloring it pink, and tiny platelets of strontium hexaluminate ($SrAl_{12}O_{19}$) form acting as potent crack arresters (Figure 2.41, left).

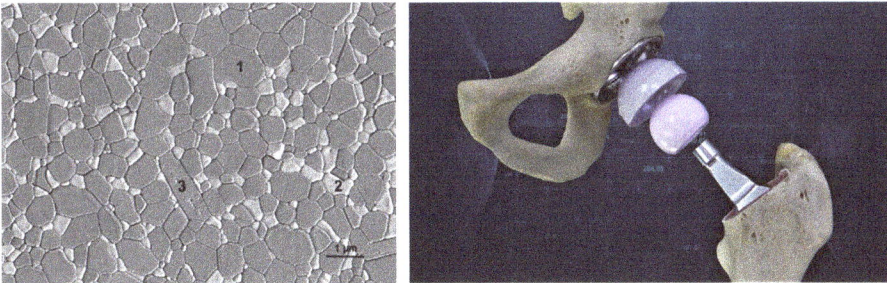

Figure 2.41: Left: Microstructure of thermally etched BIOLOX®delta. 1, Alumina grain; 2, zirconia reinforcing grain; 3, strontium hexaluminate platelet acting as crack arrester. © With permission of CeramTec GmbH, Plochingen, Germany. Right: Articulating parts (femoral ball and acetabular cup) of a hip endoprosthesis consisting of BIOLOX®delta designed by Conformis Hip System, Conformis, Inc. © 2019 GlobeNewswire, Inc.

This ZTA composite material is mechanically strengthened by two synergistic mechanisms: (i) zirconia transformation toughening and (ii) reinforcement by strontium hexaluminate platelets that dissipate crack energy via deflecting of crack paths. By these mechanisms, the four-point bending strength increases almost threefold compared to that of unalloyed alumina (Table 2.6) to reach about 1,400 MPa, and, in parallel, the fracture toughness increases to 6.5 MPa·√m. Initially, it was thought that the addition of chromium oxide would cause hardening of the ZTA matrix material, an assumption that was subsequently disproved (Kuntz, 2015). At this point in time, ZTA-based BIOLOX®delta arthroplastic implants (Figure 2.41, right) are considered state-of-the-art technology, with unsurpassed mechanical, tribological, and biocompatible properties. In particular, in a hip simulator test this design shows dramatically reduced wear rates of about 0.6 mm³/Mc (millions of cycles) compared to 8.3 mm³/Mc of its BIOLOX®forte predecessor (Clarke et al., 2006). As of 2017, some 6.6 million BIOLOX®delta femoral ball heads and over 2 million acetabular inserts have been sold worldwide, making it one of the most successful biomedical products ever.

Addition of polycrystalline, metastable tetragonal Y-partially stabilized zirconia (Y-PSZ) particles to a fine-grained alumina matrix results in a so-called duplex structure (Figure 2.42) with high fracture strength exceeding 700 MPa and fracture toughness up to 12 MPa·√m. Since Y-PSZ has lower elastic modulus (around 210 GPa) compared to alumina (380 GPa; Table 2.6), cracks of critical size introduced by external loads tend to move toward the zirconia particle aggregates. The crack energy will then be dissipated by forcing tetragonal t-ZrO_2 particles to transform to monoclinic m-ZrO_2. This mechanism will be augmented by the compressive stress generated that tends to counteract crack movement. This mechanism accounts for the extraordinary strength of modern bioceramics materials such as BIOLOX®delta discussed above. Figure 2.42 shows how an advancing crack will trigger transformation of t-ZrO_2 to m-ZrO_2 around the crack tip by releasing its crack energy.

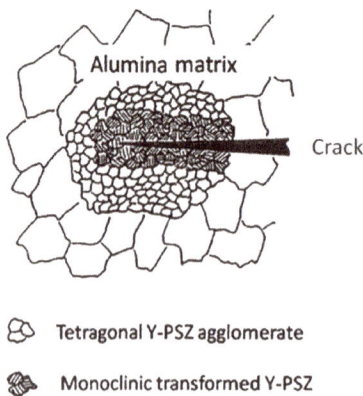

Alumina matrix

Crack

⬡ Tetragonal Y-PSZ agglomerate

🐾 Monoclinic transformed Y-PSZ

Figure 2.42: Schematic rendering of the reinforcement mechanism of an alumina–zirconia duplex ZTA ceramics with agglomerated Y-PSZ particles (Heimann, 2010).

Biological safety considerations

While alumina as the prototypical bioinert ceramics is considered extremely stable against corrosion even in the aggressive body environment, over the years some concern has been voiced that high alumina-bearing implants could lead to elevated concentration of potentially cytotoxic aluminum in the body. Such enhanced aluminum levels are considered etiological agents in dialysis osteomalacia, encephalopathy, and some forms of anemia. Under normal physiologic conditions, the usual daily dietary intake of aluminum of 5–10 mg is completely eliminated via the kidneys and the renal pathway. However, if aluminum is not removed by renal filtration, it will accumulate in the blood where it binds to proteins such as albumin. It is then rapidly distributed throughout the body, and concentrates in the brain and in bone. Brain deposition has been implicated as a possible cause of the so-called dialysis dementia or even autism (Mold et al., 2018) and Alzheimer's disease (Bhattacharjee et al., 2014). When incorporated in bone, aluminum is known to replace calcium in biological apatite at the mineralization front, thus disrupting osteoid formation and normal calcium exchange. Serum aluminum concentrations are likely to be increased above the reference range in patients with metallic joint prosthesis whereby aluminum release can occur from either the Ti6Al4V alloy or, to a much lower degree, from the alumina-based femoral head. Hence, even though to date there are no reliable clinical reports on upper safety levels of aluminum, the effect of even minute quantities of aluminum released from joint prostheses needs further investigation.

2.2.1.2 Zirconia

Since zirconia is produced from naturally occurring zirconium silicate (zircon, $ZrSiO_4$) or from rare baddeleyite (monoclinic β-ZrO_2), trace amounts of [226]Ra (daughter decay product of [235]U) and [228]Th, replacing the isovalent zirconium ion in the crystal lattice may remain in the processed material, rendering it slightly radioactive. This indeed was a major concern that in the past had hampered the development of otherwise mechanically superior zirconia ceramics for biomedical applications. However, novel sophisticated processing routes (see Chapter 3.2.5.1) have decreased the content of potentially dangerous radioactivity down to virtually zero as the body dose received from Y-stabilized zirconia today is less than 200 times the limit recommended by the International Commission on Radiological Protection (ICRP, 1991).

Today, zirconia has found various applications in biomedical devices, most importantly as hard and tough structural ceramic material for femoral balls in hip endoprostheses and material for restorative dentistry, including inlays, onlays, and tooth veneers. Since stabilized zirconia shows substantially higher fracture toughness compared to alumina, it is advantageously be applied in endoprosthetic devices (Chevalier and Gremillard, 2017).

Structure

Crystallographic data of the three polymorphic structures of zirconia are shown in Table 2.8.

Table 2.8: Crystallographic data of zirconia.

Modification	Monoclinic	Tetragonal	Cubic
Space group	$P2_1/c$ (14)	$P4_2/nmc$ (137)	$Fm\bar{3}m$ (225)
Lattice constants (pm)	$a = 516.9$	$a = 509$	$a = 512$
	$b = 523.2$	$c = 518$	
	$c = 534.2$		
	$\beta = 99°15'$		
Density (Mg/m³)	5.83	6.10	6.09

On heating, the monoclinic m-ZrO_2 phase (natural mineral baddeleyite) is stable up to 1,170 °C (Figure 2.45). Beyond this temperature, it transform by a so-called martensitic (diffusionless) shear mechanism (see box) to the tetragonal t-ZrO_2 form that is stable up to 2,370 °C. On further heating, cubic c-ZrO_2 forms that is stable up to the melting point at 2,700 °C. In the *monoclinic structure*, Zr^{4+} ions have the unusual coordination number of 7, whereby four O_{II} oxygen ions form with the central zirconium an almost symmetrical tetrahedral arrangement (Figure 2.43 A).

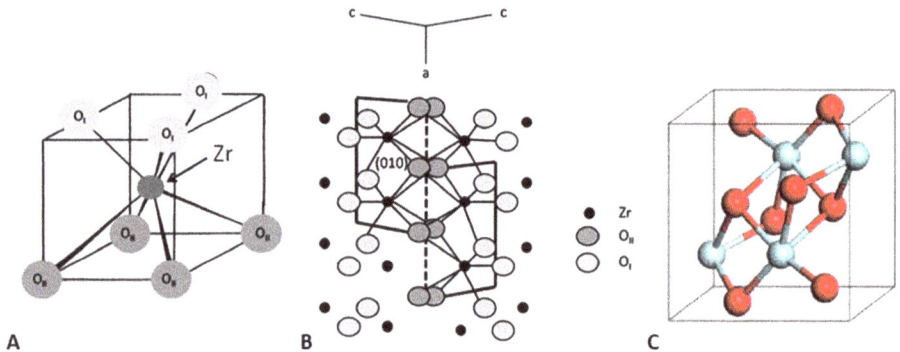

Figure 2.43: Structure of monoclinic zirconia. (A) Idealized ZrO_7 polyhedron with coordination number [7], showing the two types of oxygen ions (O_I and O_{II}). (B) Projection onto the (010) plane, showing the plane of twinning (dotted line). (C) Unit cell of m-ZrO_2.

The three remaining O_I oxygen ions are strongly disordered. This is the reason why m-ZrO_2 shows a strong tendency for deformation twinning to relieve lattice strain. The twin plane (001) is composed of O_{II} ions that are rather mobile so that they are easily moved out of their equilibrium positions. In the projection of the (010) plane (Figure 2.43 B), the boundary of the twin plane (001) is shown as the dotted line, formed by rotation around a twofold axis through $x = \frac{1}{2}$ and $y = \frac{1}{2}$. This displaces one half of the twin against the other by $\frac{1}{2} a$.

Even though the *tetragonal form* has a high coordination number of [8], its structure is rotationally distorted. Four oxygen ions surround the central zirconium ion in form of a flattened tetrahedron at a distance of 206.5 pm. The other four oxygen ions are found at a distance of 245.5 pm at the vertices of an elongated tetrahedron rotated against the former by 90° (Figure 2.44A, B). The *cubic zirconia* modification (Figure 2.44C, D) exists in the fcc fluorite-type structure whereby each Zr^{4+} ion is octahedrally surrounded by eight oxygen ions that are arranged in ideal tetrahedral conformation.

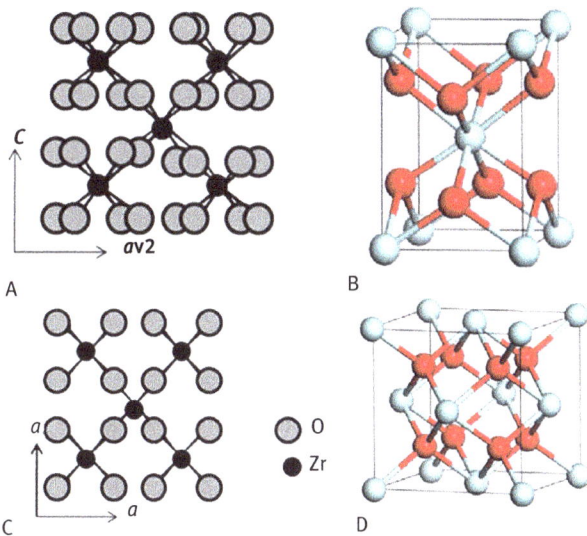

Figure 2.44: Structures of tetragonal (A, B) and cubic (C, D) zirconia.

Transformation toughening of zirconia ceramics

The tetragonal-monoclinic phase transformation of zirconia is of martensitic type, that is, it is a diffusionless shear or "umklapp" process progressing at near the velocity of sound (see box). In its formal crystallographic mechanism, the process is akin to the transformation of α-iron with (bcc) structure to γ-iron with (fcc) structure on which the excellent malleability of heated steel is based.

Martensitic phase transformation of tetragonal zirconia
The mechanism of the martensitic monoclinic-tetragonal phase transformation of zirconia is analogous to that encountered during transformation of α-iron to γ-iron shown schematically in the left figure. It also bears resemblance to the martensitic transformation of NiTi that results in its shape-memory effect (see above).

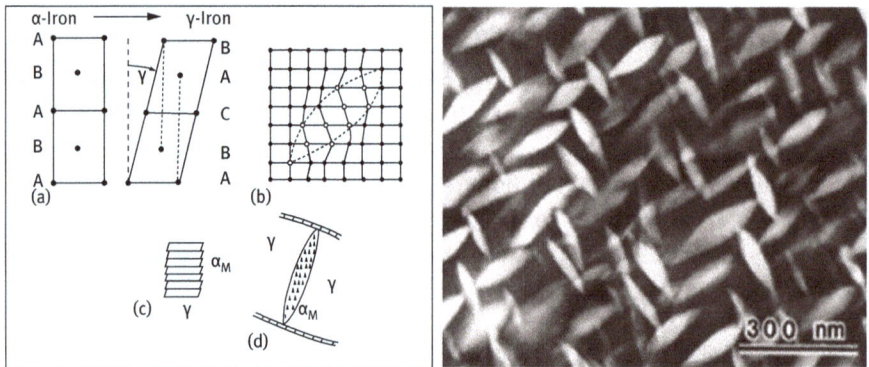

Left: Mechanism of martensitic transformation of α- to γ-iron. (a) Shearing during which the (bcc) stacking order ABABA . . . is transformed to the fcc stacking order BACBA . . . (b) Shearing within the matrix lattice. (c) Inner plastic deformation by sideway sliding of small crystalline slabs to maintain the original volume of the martensitic σ_M phase within the matrix γ-iron phase. (d) Lamella of σ_M phase with a γ-iron crystal at a grain boundary (after Hornbogen, 2019). Right: TEM micrograph showing spindle-shaped tetragonal zirconia precipitates within a monoclinic zirconia matrix phase in Mg-partially stabilized zirconia, akin to the martensitic lamella shown in (d) (Birkby and Stevens, 1996). Martensitic transformation of leucite in dental porcelain is shown in Figure 2.51.

Since the transformation process results in a volume change of about 5 vol%, in pure, that is, unalloyed zirconia the elastic limit and yield strength will be exceeded on cooling, causing formation of cracks and, eventually, complete structural disintegration of monolithic zirconia bodies. However, addition of oxides such as yttria, calcia, magnesia, ceria, or scandia delays the deleterious phase transformation by stabilizing the tetragonal structure, thus inhibiting its transformation to monoclinic zirconia. This process is at the very heart of the important mechanism of *transformation toughening of zirconia* ceramics for biomedical application. As shown in the box above, nucleation and growth of spindle-shaped tetragonal particles occur that are stable against transformation even at room temperature. The metastable tetragonal particles induce tangential stresses and hence produce microcracks in the surrounding matrix that are subcritical in terms of the Griffith–Orowan fracture theory (see box).

Basics of Griffith–Orowan fracture theory

The classic Griffith–Orowan theory describes the relationship between strength and toughness of brittle materials such as ceramics. In the simple basic approximation of the theory, the stress to fracture σ_f is related to the modulus of elasticity E, the fracture energy γ, and the critical crack length c by the equation

$$\sigma_f = \sqrt{\left(\frac{2E\gamma}{\pi c}\right)}$$

Since the critical stress intensity factor (= fracture toughness) is $K_{Ic} = \sqrt{2E \cdot \gamma}$, the fracture stress becomes

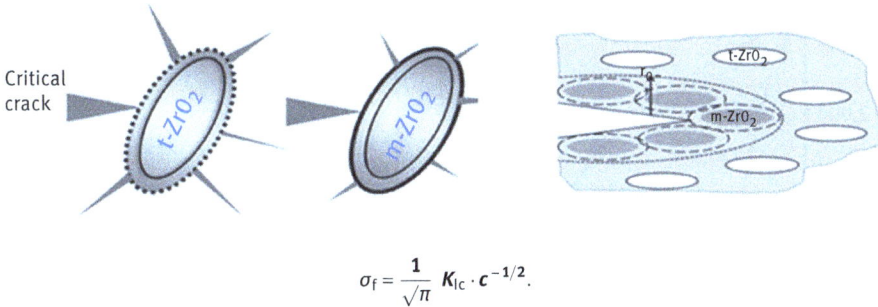

$$\sigma_f = \frac{1}{\sqrt{\pi}}\, K_{Ic} \cdot c^{-1/2}.$$

In the case of zirconia, a crack of critical length is able to move through the stress field but will be deflected by the particles that transform spontaneously to the monoclinic form in the vicinity of the propagating crack (left panel). The crack energy will be adsorbed and the volume change will prevent the advancing crack from spreading further. This is expressed in an increase of the fracture toughness of the PSZ by the so-called *transformation toughening mechanism* (right panel). A crack penetrating monoclinic zirconia with embedded untransformed tetragonal (t-ZrO_2) particles generates around its tip a stress field with radius r_0 that triggers the martensitic transformation t → m by releasing the matrix pressure on the t-phase. The volume expansion and existing shear stresses cause compressive stresses around the advancing crack tip and thus, halt its further propagation.

Yttria-stabilized tetragonal zirconia polycrystal (Y-TZP) material

Alloying of zirconia with yttrium oxide strengthens the ceramic material by the mechanisms of transformation toughening. Figure 2.45 shows the binary phase diagram ZrO_2–Y_2O_3.

The most significant characteristic of the system is the decrease of the temperature of the tetragonal (t) to monoclinic (m) phase transformation with increasing yttria content. Whereas at high temperature there exists a small phase field in which, above the transformation temperature, the two-phase assembly t + m is stable, beyond this area the transformable t-phase is stabilized between 0 and about 4 mol% Y_2O_3. On increase of the yttria content, the system enters the stability field of t + c such that, eventually, a homogeneous cubic solid solution will be formed that exists between the melting point of pure ZrO_2 around 2,700 °C and ambient temperature. This is the field of existence of fully stabilized (cubic) zirconia. During rapid quenching of molten Y-PSZ as occurring in plasma spraying, the system enters the field of metastable (nontransformable) tetragonal (t′) zirconia between about 3 and 12 mol% Y_2O_3 that, on annealing, lead to exsolution of yttria and thus, transformation of t′-zirconia to yttria-poor t-zirconia with 4–5 mol% Y_2O_3 and yttria-rich c-zirconia (>14 mol% Y_2O_3).

For biomedical application of Y-stabilized zirconia, for example, as tough ceramic materials for femoral heads or dental restoration material, the mechanical

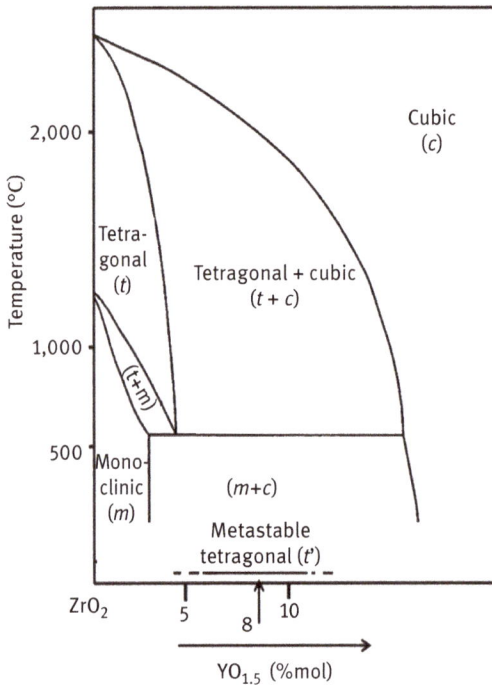

Figure 2.45: High-temperature region of the ZrO_2-rich side of the binary phase diagram ZrO_2–Y_2O_3. © Permission granted under Creative Commons Attribution License (CC BY 3.0).

properties, in particular flexural strength and fracture toughness, have to be maximized. This requires careful adjustment of the relative amounts of ZrO_2-phases present. For example, while the fracture toughness of t′-zirconia was found to be less than about 4 MPa·√m but that of t-zirconia to be beyond 7 MPa·√m, a maximum value of the stress intensity factor K_{Ic} exceeding 10 MPa·√m was observed for a laser-sealed plasma-sprayed zirconia coating partially stabilized with 7.5 mol% yttria and annealed at 1,200 °C for up to 170 h (Jasim et al., 1992). The complex phase composition was found to be 43% t + 34% t′ + 15% c + 8% m.

Table 2.9 shows selected properties of commercially available Y-TZP (ISO 13356:2016) and ZIOLOX® material formerly marketed by CeramTec GmbH. Figure 2.46 shows the uniform microstructure of ZIOLOX® with an average grain size of less than 500 nm.

Table 2.9: Characteristic material properties of Y-stabilized zirconia and ZIOLOX® (CeramTec GmbH).

Propert Property	Dimension	Range of values of Y-stabilized zirconia	ZIOLOX®
Density	Mg/m³	6.05–6.09	6.09
Zr content	%	95–97	
Y content	%	3–5	
Average grain size	μm	0.2–0.4	<0.5
Vickers hardness	GPa	12–13	12.5
Modulus of elasticity	GPa	150–210	210
Compressive strength	MPa	>2,000	
Tensile strength	MPa	>650	
Flexural strength	MPa	900–1,300	>900
Fracture toughness	MPa · √m	7–10	~8
Thermal conductivity	W/m · K	2.5–3.0	

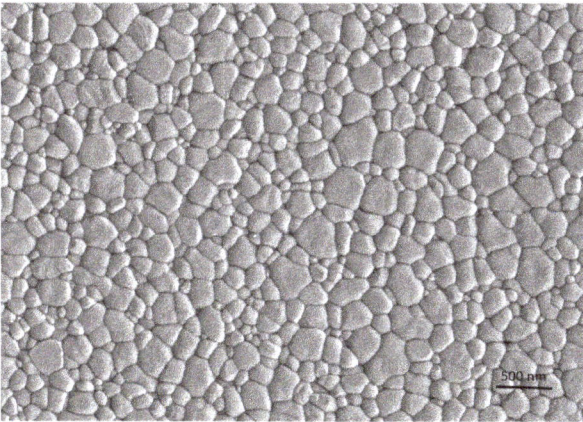

Figure 2.46: Microstructure of ZIOLOX®. © With permission of CeramTec GmbH, Plochingen, Germany.

Other strengthening mechanisms

Apart from transformation toughening, research and development effort have resulted in engineered bioceramics strengthened by several other mechanisms, including reinforcing alumina ceramics by mixing with zirconia, heat treatment of

PSZ, chemical reactions to produce a matrix with dispersed zirconia particles, and particle-reinforced zirconia.

–   Zirconia-reinforced ceramics consist of ceramic matrix materials such as alumina with 15–25 vol% ZrO_2 added as schematically shown in Figure 2.42. The fracture toughness of this ZTA ceramic will be considerably increased from about 4 MPa · √m for pure alumina (Table 2.6) to about 13 MPa · √m for hot-pressed alumina with 16 vol% zirconia added.

–   Replacing Y_2O_3 as stabilizing oxide by CeO_2 yields cerium-stabilized TZP (Ce-TZP) material with fracture toughness as high as 19 MPa ·√m and flexural strength of 1,400 MPa (Tanaka et al., 2002). Addition of alumina results in very tough and strong Ce-TZP/alumina nanocomposites for dental prosthetic application.

–   Heat treatment of PSZ can dramatically increase the strength of the ceramics. For example, annealing of zirconia partially stabilized with 3.3% CaO at 1,300 °C for 10 hours has shown to increase its modulus of rupture from 300 to 650 MPa. The reason for this is related to precipitation of CaO-rich fully stabilized, that is, cubic zirconia clusters at grain boundaries that act as crack-branching centers.

–   *In situ* dispersion of zirconia particles in a mullite matrix can be obtained by reacting zirconium silicate with alumina according to $2ZrSiO_4 + 3Al_2O_3 \rightarrow 2ZrO_2 + 3Al_2O_3 \cdot 2SiO_2$. Hot-pressed mixtures develop strength values up to 400 MPa.

–   Particle-reinforced zirconia has been obtained by adding alumina particles to TZP material. The resulting ATZ (alumina-toughened zirconia) ceramics hot-pressed at 1,450 °C shows substantial increase of strength at 1,000 °C and significantly improved cyclic fatigue strength.

–   Gentle abrading and polishing of the surface causes more t-ZrO_2 grains to transform to m-ZrO_2 so that the thickness of the surface layer under compressive stress can be increased to 100 µm and beyond (Figure 2.47). The maximum strengthening effect is reached if the thickness of the transformation zone subjected to compressive stress is higher than the critical crack length but still small compared to the cross section of the ceramic body. This fact is very important: the stabilized zirconia material reacts much less sensitive to small surface defects as other ceramics and glasses do since each service-imposed wear loss the thickness of which is less than the critical crack length is rendered as compressive stress. The effect is exploited to strengthen the surface of zirconia femoral heads for hip endoprostheses by thorough polishing (Figure 3.36). A corollary of this effect is the observation that measuring the elastic modulus of stabilized zirconia by indentation revealed significant changes of the modulus with indentation depth as result of indentation-induced t → m phase transformation.

Biomedical applications

Historically, there has been some controversy related to the cytocompatibility of zirconia in contact with living tissue. Histomorphologic and morphometric studies of the

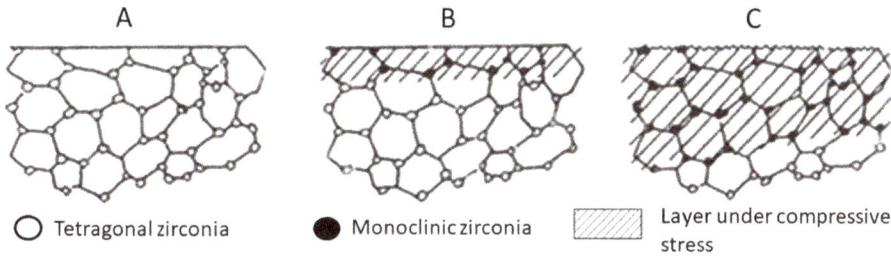

Figure 2.47: Free surface of Y-stabilized zirconia at sintering temperature (A). During cooling, the tetragonal t-ZrO_2 particles near the surface transform to monoclinic m-ZrO_2 and thus induce a compressive stress in the matrix (B). The thickness of the compressive layer can be further increased by abrading/polishing the surface (C) (Heimann, 2010).

interface of zirconia ceramics with a bony implantation bed have shown incomplete transformation of chondroid and osteoid cells to osteoblasts in a Sprague-Dawley rat femoral model. This initially suggested that the presence of zirconia could be counter-productive to bone bonding. Indeed, Zr-containing materials have been suspected to inhibit matrix vesicle development and impairment of their function (Gross et al. 2003, 2004).

In contrast, Ti6Al4V rods coated with $CaTiZr_3(PO_4)_6$ and implanted into the femora of sheep did not show any adverse reactions despite the high zirconia con-tent of the ceramics. On the contrary, the gap-bridging potential of such coatings was found to be excellent and supported bone apposition without development of a connective tissue capsule (Heimann et al., 2004). Much earlier work by Hulbert et al. (1972) had confirmed that calcium zirconate ceramics implanted intramuscu-latory in rabbits promoted the formation of a 100–200 μm thick pseudomembrane that within 6–9 months gradually densified without the presence of inflammatory cells such as phagocytes or macrophages. More recent work by Liu et al. (2006) had shown satisfactory apatite film growth and proliferation of bone marrow MSCs on zirconia substrates, suggesting favorable cytocompatibility of zirconia.

However, several years ago, concerns have been raised on the mechanical sta-bility of steam-sterilized Y-stabilized zirconia femoral ball heads. In 1996, the British Medical Device Agency (MDA) published reports that zirconia femoral heads during autoclaving suffered surface degradation/roughening due to their hydrothermal instability and, consequently, substantially increased mechanical wear rates. Subsequently, a similar warning was issued by FDA. These facts, in conjunction with thermodynamic calculation that suggest disastrous decomposi-tion of Y-TZP in ZIOLOX® after an implantation period exceeding 10 years (Pfaff and Willmann, 1998) have initially put some doubt on the long-term performance of femoral ball heads manufactured from zirconia. An explanation of the mechanism of hydrothermal degradation of stabilized zirconia assumes that water molecules ad-sorbed at the ceramic surface produce Zr-OH moieties that will act as stress concentration sites. Simultaneously, thermally activated dipole reorientation creates localized strain within the crystal lattice. The combination of these stress modes, together with residual stresses generated from thermal expansion anisotropy of the t-ZrO_2 particles, will eventually exceed the critical conditions for tetragonal phase retention, thus fostering nucleation of undesirable monoclinic zirconia. Then, an autocatalytic reaction chain is initiated and t $\rightarrow$ m transformation propagates. If the zirconia grains in Y-ZTP are large enough, micro- and macrocracking occur, opening up pathways for further penetration of water vapor into the bulk of the zirconia ceramic during steam sterilization and inducing formation of new monoclinic nuclei (Deville et al., 2017).

To confirm these pathways and to seek remedial action, additional work is required to alleviate these concerns by developing safe manufacturing, sterilization, and implantation protocols. For example, increased monoclinic content have been previously reported for retrieved zirconia components as well as significantly increased roughness in the bore of fractured femoral heads. This was partially attributed to differences in surface finishing (ground vs. polished surfaces), but appropriate controls are still required to determine if the observed values relate to fracture as a reason for revision (Sakona et al., 2010). In addition, sintering additives such as copper oxide, iron oxide, alumina, or magnesium oxide were found to be beneficial in suppressing hydrothermal aging-induced formation of monoclinic zirconia in Y-TZP (Ramesh et al., 2018).

There is still another issue that may cast doubt on the suitability of zirconia as a bioceramic material (Chevalier, 2006). It was revealed that the failure series of femoral heads sold under the commercial name Prozyr® by Saint Gobain Ceramiques Desmarqueste in 2001–2002 was caused by the fact that the inner core of the ball was not as dense as required. Consequently, a crack that started near the ball and neck boundary quickly propagated into the inner, less compact, core. In addition, the wet and warm environment provided by biofluids under cyclic loading was found to hasten the process. Since different zirconia products from different vendors have different process-related microstructures, there is a need to assess their ageing sensitivity with advanced and accurate techniques, and consequently, ISO standards were modified accordingly to gain confidence from clinicians.

At the end of the past century, Y-TZP femoral heads still had made up about 25% of the total annual number of hip joint implants in Europe and some 8% in the USA. Between 1985 and 2001, more than 400,000 Y-TZP femoral heads were implanted worldwide. However, at present there is a tendency among orthopedic surgeons to abstain from using zirconia femoral heads on account of the perceived deficiencies discussed above. Consequently, utilization of Y-TZP in joint arthroplasty has been largely abandoned today.

Nevertheless, apart from the abandoned use in prosthetic joint application, today zirconia is copiously utilized in dental parts of all kind including implants for dental roots, inlays, onlays, and color-matched tooth veneers. Indeed, zirconia has emerged as a versatile dental material due to its excellent esthetic outcome in terms of color, opacity, and shine, as well as mechanical properties that can mimic the appearance of natural teeth and decrease peri-implant inflammatory reactions. To optimize adhesion, proliferation, and differentiation of osteoblast/odontoblast cells at the dental implant–bone interface during osseointegration, a plethora of techniques have been implemented to modify the implant surfaces and hence, to improve the early and late bone-to-implant integration. This included acid etching, grit blasting, laser treatment, irradiation with UV light, as well as CVD and PVD vapor deposition. In addition, coating of zirconia with silica, magnesium, graphene, dopamine, and bioactive molecules has been assessed. Although such modified zirconia surfaces have clearly demonstrated faster and more complete osseointegration compared to untreated surfaces, there is no general consensus among dentists and clinicians regarding the most advantageous surface treatments and the resulting morphological requirements of the surfaces to enhance osseointegration.

In spite of their undisputed advantages, zirconia dental implants do not last indefinitely. A study on recovered broken zirconia implant parts with their restorative crowns provided not only information on the failure origin using fractography but also knowledge regarding occlusal crown loading with respect to the implant's axis. A mathematical model was applied, helpful in showing how occlusal loading affects the location of the fracture initiation site on clinical zirconia implant fracture cases (Scherrer et al., 2019). A recent review concluded that the good biocompatibility and the favorable physical and mechanical properties render zirconia implants a potential alternative to titanium implants. However, knowledge regarding the implant-restorative complex and related aspects is still immature to recommend its application for general daily practice (Nishihara et al., 2019).

A different field of application of zirconia ceramics has opened up in the past as bond coats for osseoconductive hydroxylapatite coatings applied to the stem of hip endoprosthetic implants, as strengthening particles for hydroxylapatite composite bodies including coatings, or as particle reinforcement material of monolithic hydroxylapatite. While hydroxylapatite coatings deposited on titanium alloy are notoriously weak in cohesion as well as adhesion to the metallic substrate surface, work was performed to strengthen these coatings by adding reinforcing particles of zirconia to form composite coatings deposited by radio-frequency suspension plasma spraying, atmospheric plasma spraying (APS), or other advanced deposition techniques. The idea was to provide a means for mechanical interlocking between bond coat layer and substrate as well as establishing chemical bonding between bond coat and hydroxylapatite by a thin interfacial reaction layer that was found to consist of calcium zirconate.

Monolithic sintered hydroxylapatite bodies show insufficient mechanical strength in terms of bending strength, fracture toughness, modulus of elasticity, and microhardness (see Chapter 2.2.2.1), making them undesirable for load bearing applications. Research has been performed to improve these properties by adding, among others, fine zirconia particulates. Generally, an increase of the bending strength and fracture toughness by a factor of two to three was observed, and has been attributed to the combined action of formation of reaction phases such as calcium zirconate, and the transition of tetragonal to cubic zirconia. Addition of $CaZrO_3$ to hydroxylapatite stabilizes the latter at least up to 1,300 °C and inhibits formation of α- and β-TCP phases. Moreover, the mechanical properties of the hydroxylapatite were improved by addition of 10 mass% of $CaZrO_3$, despite an increase of porosity. *In vitro* assays demonstrated that the calcium zirconate/hydroxylapatite composites are biocompatible, favoring human osteoblast cells adhesion and proliferation (Vassal et al., 2019).

While hydroxylapatite/zirconia composite coatings as well as hydroxylapatite coatings with a zirconia bond coat appeared to provide only a marginal improvement of coating adhesion strength, plasma-sprayed gradient coatings behave much more favorably, increasing tensile adhesion strength of heat-treated coatings to beyond 50 MPa (Ning et al., 2005). Hardness and modulus increased gradually from the Ti6Al4V substrate into the coating perpendicular to the interface, whereas microstructure and composition varied smoothly without showing distinct interfaces among adjacent layers.

2.2.1.3 Carbon materials

All lives on the Earth, and possibly beyond it, are based on the element carbon (Pace, 2001). Two allotropic modifications of carbon play important roles in biomaterials engineering today, pyrolytic graphite and DLC, with nanostructured graphene and carbon nanotubes (CNTs) waiting in the wing destined to be developed for therapeutic and diagnostic (theragnostic) application owing to a set of unparalleled physicochemical properties (Rifai et al., 2019).

Structures

Graphite crystallizes in the hexagonal system and obeys the crystal class 6/mmm with space group $P6_3/mmc$ (194). Lattice parameters are $a = 246$ pm, $c = 671$ pm, with $Z = 4$. As shown in Figure 2.48 (left) graphite has a layered, planar structure. Within each graphene layer, the carbon atoms are arranged in a hexagonal lattice with carbon–carbon atom distances of 142 pm. The separation distance between individual graphene layers in highly ordered graphite is 335.4 pm. Atoms in the plane are bonded covalently whereby only three of the four potential bonding sites are satisfied. The fourth electron is free to migrate within the plane, thus rendering graphite electrically conductive. However, graphite does not conduct in a direction at right angle to the plane. Bonding between layers is via weak van der Waals bonds that allow layers of graphene to be easily separated or to slide past each other, precondition for application as solid lubricant.

Diamond crystallizes in the cubic crystal class m3̄m, with space group Fd3̄m (227). The lattice constant is $a = 356.7$ pm. The diamond lattice can be viewed as a pair of intersecting fcc lattices, with each separated by 1/4 of the width of the unit cell in each dimension (Figure 2.48, right).

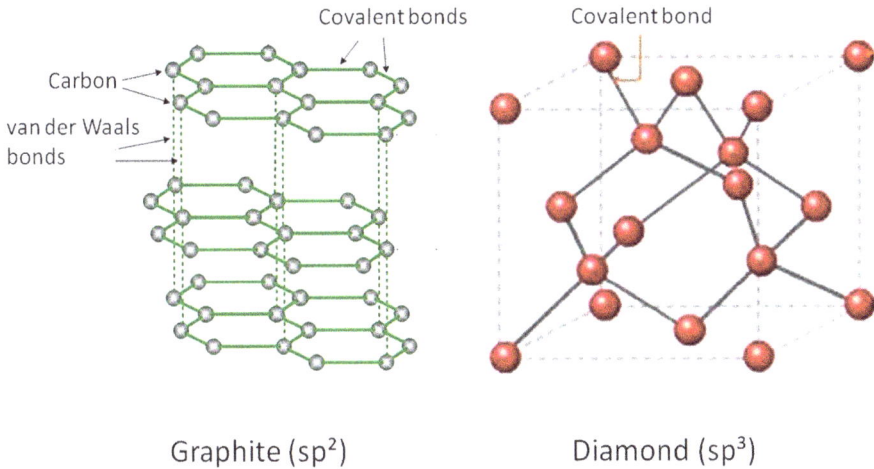

Figure 2.48: Crystallographic structures of graphite and diamond. © Permission granted under Creative Commons Attribution License (CC BY 3.0).

Properties

Pyrolytic graphite belongs to the family of turbostratic carbons that are structurally closely related to graphite but consist of covalently bonded stacks of hexagonal nets of carbon atoms with sp^2-hybridization that are rotationally disordered in the crystallographic *c*-direction, that is, perpendicular to the layer stacks. Whereas in normal graphite the layers can easily slide against each other, in turbostratically disordered graphite wrinkles or distortions exist within the layers that will resist sliding, giving pyrolytic turbostratic graphite improved stabilization and durability compared to ordered graphite. Hence, although turbostratic carbon is very similar to graphite, adjacent planes are out of registry with one another, even though there may be some degree of perfection within the planes. The disorder effects an increase of the interlayer spacing from 335.4 pm for ordered graphite to about 344 pm (Figure 2.49). This turbostratic arrangement is the characteristic structure of the so-called amorphous carbon found in chars. At more than 344 pm layer spacing, the parallel planar carbon layers assume a completely random lateral ordering, causing exfoliation.

Diamond is an ultrahard, wear- and corrosion-resistant, tribologically superior, and electrically insulating crystalline allotrope of carbon with the highest thermal conductivity of all known materials.

However, the most ubiquitous carbon allotrope for biomedical applications is DLC that, in contrast to fully sp^3-hybridized diamond, is an amorphous carbon material with mixed sp^3/sp^2-bonding. Depending on the preparation conditions, a large number of different types of DCL films exist distinguished by the absence or presence of hydrogen in its amorphous matrix: hydrogen-free amorphous carbon

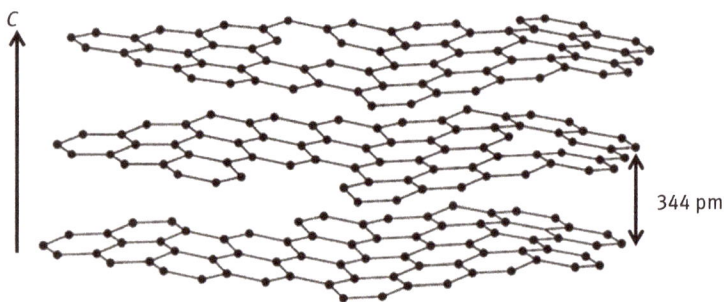

C

344 pm

(a-C), hydrogen-free tetrahedral sp^3-bonded amorphous carbon (ta-C), metal-containing (Ti, W) hydrogen-free amorphous carbon (a-C:Me), hydrogenated amorphous carbon (a-C:H), metal-containing hydrogenated amorphous carbon (a-C:H:Me) as well as modified hydrogenated amorphous carbon (a-C:H:X) where X is related to nonmetallic elements such as silicon, oxygen, nitrogen, fluorine, or boron (Bewilogua and Hofmann, 2014; see also Chapter 3.2.6).

Biomedical applications

Currently, low-temperature isotropic pyrolytic graphitic carbon (LTIC) with turbostratic structure is being generally regarded the material of choice for fabricating artificial heart valves. However, the material is brittle and its hemocompatibility is still not sufficient for prolonged clinical use. As a result, the risk of thrombus formation forces patients to take anti-coagulation medication, frequently for the remainder of their life. This requires development of a material with better surface properties and much improved hemocompatibility and thus, suppressed thrombogenicity. Pyrolytic carbon is also in medical use to coat anatomically correct orthopedic implant and currently marketed under the brand name PyroCarbon® (Ascension Orthopedics, Austin, TX). These pyrolytic carbon implants have been approved by the U.S. Food and Drug Administration (FDA) for metacarpophalangeal and proximal interphalangeal joint replacements. The results of several studies demonstrated that pyrolytic carbon joint implants reduce pain and are functionally superior to arthrodesis (Ritz and Scott, 2015).

DCL coatings are being developed as alternate bearing surfaces for endoprosthetic hip and knee implants, designed to combat wear in articulating implant devices (Thorwarth et al., 2010; Dalibón et al., 2017; Boehm et al., 2017). DLC coatings on Ti6Al4V alloy substrates have shown superior wear resistance, not in the least owing to their tight adherence. In addition, immersion and electrochemical tests have revealed improved stability of DLC coatings on Ti6Al4V substrates compared to 316L stainless steel and CoCrMo alloy substrates (Zhang et al., 2015a; Li et al., 2019b).

On the down side, several studies have shown that DLC coatings on articulating joints failed in service owing to partial coating delamination some years after implantation. This delayed delamination was caused by crevice corrosion of the adhesion-promoting Si interlayer. The rate of interlayer corrosion was measured *in vitro* in solutions containing proteins and was found to exceed 100 µm/year. Although proteins are not directly involved in the corrosion reactions, they can block existing small cracks and crevices underneath the coating, thus hindering the exchange of liquid. This may results in a build-up of crevice corrosion conditions within the crack, causing slow dissolution of the Si interlayer (Hauert et al., 2012). Delamination can also occur by a slowly advancing crack in a thin carbidic interlayer due to stress corrosion cracking. However, when stable coating adhesion can be obtained, DLC-coated articulating implants show essentially no wear of the coating up to 100 million articulations in a hip simulator, corresponding to about 100 years of *in vivo* service (Hauert et al., 2013).

The main goals of recent development interest in DLC coatings revolve around improvement of surface resistance to scratching and biocompatibility. DLC coatings were deposited using four different methods: plasma-activated chemical vapor deposition, physical vapor deposition by magnetron sputtering, and physical vapor deposition via unfiltered and filtered cathodic arc (Gotzmann et al., 2017). The different DLC coatings were characterized for their wettability, morphology, roughness, and coating adhesion, as well as their ability to withstand repeated sterilization cycles. Subsequent electron-beam modification resulted in increased hydrophilicity without changes in surface morphology or roughness. However, the adhesion of human fibroblast cells to electron beam-modified DLC coatings was reduced to 30% of that of untreated DLC coatings, presumably caused by changes in the surface chemistry. Nevertheless, electron-beam treatment of DLC coatings deposited by magnetron sputtering may be a promising tool for modifying and functionalizing DLC coatings, in particular when varying requirements in biomedical functionality and cell adhesion are involved.

CNTs are suitable for therapeutic and diagnostic applications, including tumor-targeted drug delivery, as well as drug delivery across the blood-brain barrier. In addition, CNT-based biosensors were found to be useful for detection of various biomolecules (Raphey et al., 2019).

Owing to their advantageous photoluminescence properties (D'Amora and Giordani, 2018), *carbon quantum dots* (CQDs) replacing toxic semiconducting metal-based quantum dots (QDs) such as CdSe are of increasing interest in a wide range of applications such as fluorescence imaging, nanocarriers for drug and gene delivery (see box; Chandra Ray and Ranjan Jana, 2017), medical diagnostics and theranostics, photodynamic therapy agents, analyte detection, biosensing, optical and electrochemical sensors, light-emitting diodes, energy conversion and storage, as well as electro- and photocatalysis (Sagbar and Sahiner, 2019). CQDs display intriguing properties that include exceptional optical and fluorescence characteristics with high quantum yield, simple and inexpensive preparation methods from renewable sources (Das et al., 2018), high thermal and optical photostability, tunable excitation and emission, easy surface functionalization with a large variety of biomolecules, and nontoxic nature with high biocompatibility.

Quantum dots

Quantum dots are nanosized (semiconductor) particles that will emit light of specific frequencies if electricity or light is applied. These frequencies can be precisely tuned by changing the size, shape, or material of the dots. In emerging biomedical applications, quantum dots can be utilized for highly sensitive cellular imaging, real-time tracking of molecules and cells over extended periods of time, as effective agent against both gram-positive and gram-negative bacteria, active and passive tumor targeting under *in vivo* conditions, and as inorganic fluorophores for intra-operative detection of hidden tumors using fluorescence spectroscopy. Challenges exist regarding the potential cytotoxicity of many quantum dot materials. For example, CdSe nanocrystals are highly toxic to cultured cells under UV illumination, because the particles dissolving by photolysis will release toxic cadmium and selenium ions to the culture medium. However, in the absence of UV irradiation quantum dots with a stable polymer coating have been found to be essentially nontoxic.

2.2.1.4 Dental porcelains

Dental porcelains are used to replace dental superstructures, including crowns, bridges, inlays, onlays, tooth veneers, and artificial teeth. Increased demands for esthetic restorations and the encouraging performance of all-ceramic restorations in the permanent dentition have led to significant advances in dental ceramics. Although dental glass-ceramics achieve highly esthetic results today, their relatively low strength and brittle nature prohibit their predictable use in high-stress applications unless supported by a high-strength substrate (Platt, 2016). Reviews on properties and application of dental ceramics are available from Denry and Holloway (2010), Zarone et al. (2011), and Yin and Stoll (2014).

Composition

Dental porcelain is a feldspathic multiphase ceramic material that consists of silicon dioxide and aluminum oxide, together with a few percent of alkali oxides as fluxes and, sometimes, a few tenths of a percent of transition metal oxides for color matching in dental applications. Most dental porcelains have compositions with SiO_2 contents between 60% and 75 mass%, Al_2O_3 contents between 7 and 15 mass%, and alkali ($K_2O + Na_2O$) contents between 10 and 20 mass%. A *high-fusing dental porcelain* mix may be composed, before sintering, of 70 to 80 mass% of alkali feldspar, 4 mass% of kaolinite, and 15 to 20 mass% of quartz. Figure 2.50 shows an isothermal section of the silica-rich portion of the ternary K_2O-Al_2O_3-SiO_2 phase diagram at

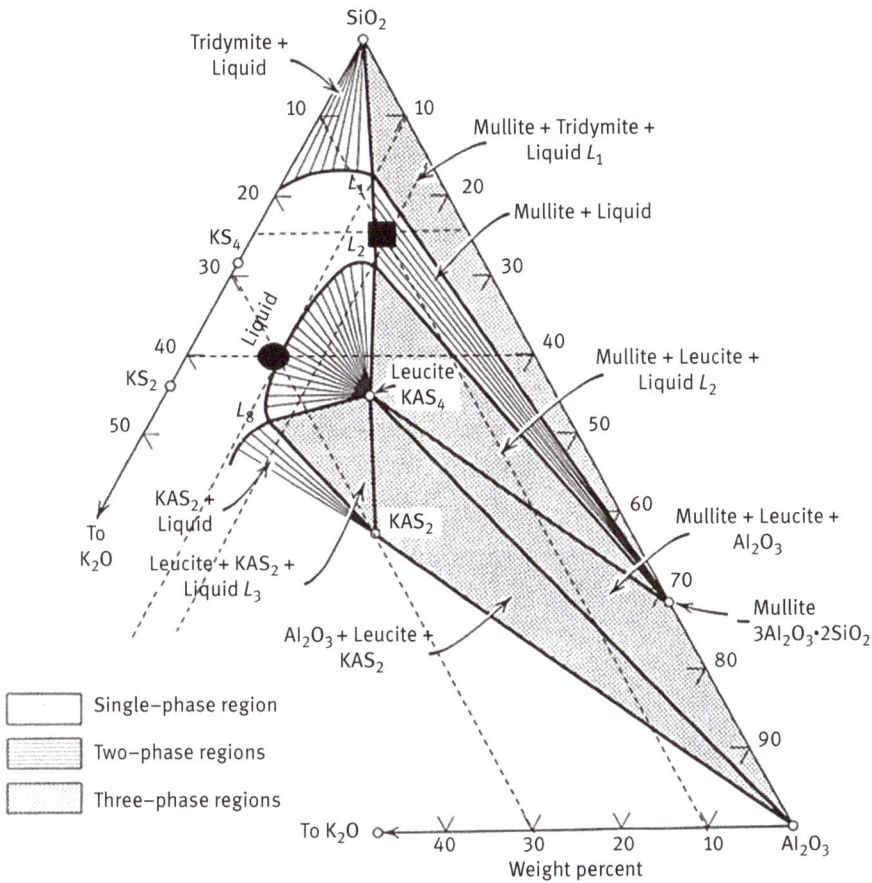

Figure 2.50: Silica-rich isothermal section of the ternary K_2O-Al_2O_3-SiO_2 phase diagram at 1,200 °C (modified after Kingery et al., 1976). Some molar compositions are given in cement-chemical notation (K = K_2O; A = Al_2O_3; S = SiO_2). The black square relates to a typical high-fusing dental porcelain composition, the black dot to a low-fusing one. © With permission by John Wiley & Sons.

1,200 °C. This temperature is the typical sintering temperature of a high-fusing mixture of 75 mass% SiO_2, 15 mass% Al_2O_3, and 10 mass% K_2O (black square). The composition is positioned in the two-phase region mullite + liquid. Compositional fluctuation, for example, by evaporation of K_2O during sintering may move the system into the three-phase region mullite + tridymite (or cristobalite) + liquid. Indeed, high-fusing dental porcelain frequently shows crystalline high-temperature silica phases such as cristobalite or tridymite within the glassy matrix of the porcelain (Love, 2017).

In contrast, *low-fusing dental porcelains* consisting on average of 60 mass% silica, 10 mass % alumina, and 30 mass% potassia are located at the boundary between leucite and liquid (black dot). These dental porcelains contain leucite ($KAlSi_2O_6$) as crystalline component embedded in an aluminum silicate glass phase (Sakaguchi and Powers, 2012). Tetragonal leucite stable at ambient conditions undergoes a martensitic phase transformation to a cubic phase at 625 °C (Shalukho and Kuz'menkov, 2011). This transformation toughens the ceramic matrix in the same way as martensitic transformation toughens zirconia (see Chapter 2.2.1.2). Phase transformation is accompanied by polysynthetic twinning to relieve transformation-induced lattice strain (Figure 2.51) as well as a thermal expansion resulting in a 1.2% increase of the unit cell volume that is associated with the risk of cracking or spalling of the ceramic body (Mackert et al., 1986). This also explains the high thermal expansion coefficient associated with tetragonal leucite ($>20 \cdot 10^{-6}$/°C). Hence, the amount of crystalline leucite, typically between 10 and 20 vol%, controls the thermal expansion coefficient of the dental porcelain so that it can be adequately matched to that of metallic dental materials including gold alloy (Brodkin et al., 2000).

Figure 2.51: Leucite crystallite (center) embedded in the glassy matrix of low-fusing dental porcelain, showing polysynthetic twin lamellae associated with the diffusionless (martensitic) transformation of leucite. Twinning is the result of inner plastic deformation by sideway sliding of small crystalline slabs to maintain the original volume of the low-temperature tetragonal phase during martensitic transformation to the high-temperature cubic phase (see box in Chapter 2.2.1.2) (Denry and Holloway, 2010). © Permission granted under Creative Commons Attribution License (CC BY 3.0).

Properties

Feldspathic porcelains are well suited for metal-ceramic restorations as they fuse at lower temperatures than do many other ceramic materials and thus, reduce the risk of distortion of the metallic overlay. Lower processing temperatures can be achieved by adding alkali oxides (Na_2O and K_2O) to the glassy matrix. These oxides create nonbridging oxygen (NBO) atoms in the glassy network, thereby lowering the fusing temperatures to 930° to 980 °C (Table 2.10). Some porcelains with an even lower fusing temperature (760–780 °C) and high coefficients of thermal expansion ($15.8 \cdot 10^{-6}$/°C) are also available. These porcelains are designed to be compatible for bonding to high-Au alloys that have thermal expansion coefficients between $16.1 \cdot 10^{-6}$/°C and $16.8 \cdot 10^{-6}$/°C. However, owing to their comparatively high hardness, they can be abrasive to opposing teeth. This becomes a significant problem if the porcelain surface is roughened by occlusal adjustments or sensitivity to aging in the oral environment (Sakaguchi and Powers, 2012).

Table 2.10: Fusion temperature ranges of various dental porcelains and their clinical applications (Babu et al., 2015).

Porcelain type	Fusion temperature range (°C)	Clinical application	Typical composition (%)		
			Feldspar	Silica	Kaolin
High fusing	>1,300	Denture teeth	80	15	5
Medium fusing	1,000–1,300	Jacket crowns, bridges, inlays			
Low fusing	850–1,000	Veneers over cast metal crowns	60	25	15
Ultra-low fusing	<850	Used with Ti and Ti alloys			

Since porcelains are commonly sintered, adequate mixing of the different components is required to produce reliable material for crowns for grinding bite surfaces (molars) and porcelain veneers. The powder components are mixed to form gel-like slurries that prior to sintering hold the particles together. The gel can be poured or pressed into molds that are typically oversized relative to the final product. During sintering, the particles densify and shrink, the amount of which can be controlled by adjusting the composition to yield a correctly sized crown for installation.

Biomedical applications

Dental porcelains are playing a vital role in dentistry owing to their natural esthetics and excellent biocompatibility with no known adverse tissue reactions. Despite this, there will always remain a compromise between esthetic appearance and biomechanical strength. To achieve adequate mechanical and optical properties, the amounts of glassy and crystalline phases must be optimized. However, there is a conundrum: superior translucency requires a higher content of the glassy phase but high strength requires a higher content of the crystalline phase. Hence, the two disparate material phases need to be thoroughly balanced. Even though the material is highly abrasion resistant, low fracture toughness and resistance to tensile stresses are inherent disadvantages. Although attempts have been made to overcome these shortcomings, ways to improve fracture toughness and tensile stress resistance need further research (Babu et al., 2015).

2.2.1.5 Silicon nitride

Structure

Silicon nitride crystallizes in two hexagonal (α- and β-Si_3N_4) and a cubic γ-modification with spinel structure stable at high pressure and temperature. The space groups of α-Si_3N_4 is *P31c* (159) with $a = 774.8(1)$ pm and $c = 561.7(1)$ pm. The space group of β-Si_3N_4 is *P6$_3$/m* (176) with $a = 760.8$ pm and $c = 291.1$ pm. The symmetry of γ-Si_3N_4 with diamond structure is *Fd$\bar{3}$m* (227) with $a = 773.8$ pm.

As shown in Figure 2.52, the structure consists of slightly distorted corner-sharing SiN_4 tetrahedra forming distorted hexagonal rings arranged in layers with ABCDABCD . . . stacking (α- Si_3N_4) and ABAB . . . stacking (β-Si_3N_4). In β-Si_3N_4, the bond lengths between Si and N1 are 173.0 and 173.9 pm, respectively, and between Si and N2 174.5 pm. In α-Si_3N_4, the bond lengths vary between 156.6 pm (N1-Si1) and 189.6 pm (N1-Si2).

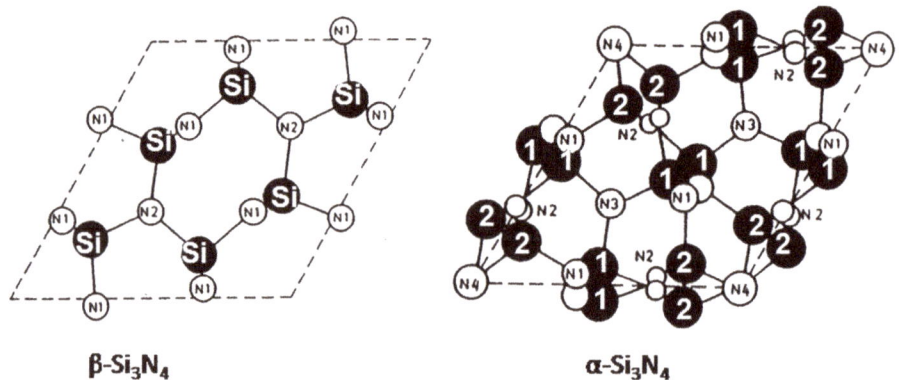

β-Si_3N_4 α-Si_3N_4

Figure 2.52: Structures of hexagonal β- and α-silicon nitride. Projection down the c-axis onto the AB plane. Black circles: silicon atoms, white circles: nitrogen atoms (Heimann, 2010).

The AB layer is identical in both α- and β-phases; the CD layer in the α-phase is related to AB by a c-glide plane. Hence, a double layer in α-Si_3N_4 can be thought of as a superposition of a β-Si_3N_4 layer and its counterpart inverted by 180°. Consequently, there are twice as many atoms per unit cell in α-Si_3N_4 ($Z = 4$) than in β-Si_3N_4 ($Z = 2$). The Si_3N_4 tetrahedra in β-Si_3N_4 are interconnected in such a way that channels are formed extending parallel to the *c*-axis of the unit cell. Owing to the c-glide plane, the α-structure contains two (isolated) interstitial sites per unit cell instead of channels. Since the channel structure of the β-phase supports easy diffusion of ions through the lattice thus promoting greatly sinterability at high temperatures, fabricating load-carrying biomedical devices with required high density should start from β-Si_3N_4 precursor powder.

Properties

Depending on the synthesis route, there are several structural and textural variants of silicon nitride that differ remarkably in their mechanical, elastic, and thermal properties (Table 2.11). The variants most frequently used in engineering applications are porous reaction-bonded silicon nitride (RBSN) and dense sintered silicon nitride (SSN). To form RBSN, direct nitriding of fine compacted silicon powder is employed that leads to formation of a mixture of α- and β-Si_3N_4 by prolonged heating in nitrogen or ammonia atmospheres between 1,000 and 1,450 °C. Since a volume increase of about 22% is completely accommodated by the inter-particle void space of the compacted silicon powder, there is no shrinking during nitriding so that the original dimensions of the green component will be faithfully retained. This is the underlying cause for using RBSN for complex shaped components. Since expensive diamond grinding and machining can be avoided, parts manufactured from RBSN are economically highly competitive. Although the porosity of RBSN is between 20% and 30%, the mechanical performance of RBSN is remarkable as flexural strengths in the range of 200–300 MPa can readily be attained. An additional benefit is that this strength can be retained to about 1,400 °C so that designing with

Table 2.11: Selected properties of RBSN and SSN.

Property	RBSN	SSN
Density (Mg/m^3) (% of theoretical density)	70–88	95–100
Flexural strength (4-point, 25 °C) (MPa)	150–350	500–1,000
Fracture toughness (25 °C) (MPa · Vm)	1.5–3	5–8
Fracture energy (J/m^2)	4–10	~ 60
Modulus of elasticity (25 °C) (GPa)	120–220	300–330
Thermal conductivity (25 °C) (W/mK)	4–30	15–50
Thermal shock resistance R (K)	220–580	300–780
Thermal shock fracture toughness R' (W/m)	500–10,000	7,000–32,000
Coefficient of thermal expansion (10^{-6} K^{-1})	3.2	3.2

RBSN leads to monolithic ceramics with high Weibull moduli and hence, high reliability in service.

To achieve fully dense SSN monolithic bodies, hot pressing or hot isostatic pressing of silicon nitride powder with added metal oxides is being employed. At temperatures above 1,550 °C, these additives form with contaminant silicon dioxide films around individual silicon nitride grains a liquid siliceous phase in which the silicon nitride readily dissolves. This essentially glassy binder phase leads to efficient densification of the sintered ceramic body (Heimann, 2010).

Biomedical applications

Although nonoxide ceramics such as silicon nitride are predominantly used in high-temperature applications, recent research results have encouraged their use in a variety of biomedical applications. Silicon nitride has been used as porous intervertebral spacers for spinal fusion surgery for more than a decade without any undesirable effects (Sorrell et al., 2004; Guedes e Silva et al., 2008; Anderson and Olsen, 2010). Silicon nitride comprises a high strength, long-term resilient, and mechanically reliable ceramic material, based on its high proportion of covalent bonds (see Figure 3.32). In addition, silicon nitride possesses suitable biocompatibility as confirmed by cell culture tests with mouse fibroblasts (L929) and human mesenchymal stem cells (hMSCs) (Cappi et al., 2010). These tests revealed excellent cytocompatibility as demonstrated by live/dead cell staining and differentiation toward osteoblasts. *In vitro* tribological tests in a hip simulator showed an extremely low friction coefficient that was found to decrease to even lower values with increasing articulating time. This behavior may be attributed to hydrodynamic lubrication by a developing hydrogen-terminated silicon oxide boundary film separating the articulating components (Jahanmir et al., 2004).

The strength of silicon nitride with a high degree of covalent bonding was shown to be significantly higher than that of oxide ceramics that possess largely ionic-type bonds (Figure 3.32). A finite element analysis (FEA) was performed to study the effects of silicon nitride HR (hip resurfacing) prostheses on the stress distribution in the femoral neck area. The result showed that the stress distribution within the femoral bone implanted with silicon nitride prostheses was similar to that of a healthy, intact femoral bone. In addition, lifetime predictions revealed that silicon nitride implants are mechanically reliable and, thus, suitable for HR prostheses (Zhang et al., 2010). This indicates that high-strength nonoxide ceramics such as silicon nitride and possibly silicon carbide (SiC) are future biomaterial candidates, in particular for highly loaded, thin-walled implants such as ceramic HR prostheses. Indeed, silicon nitride coatings are currently under investigation as bearing surfaces for joint implants, owing to their low wear rate and good biocompatibility of both coating and its potential wear debris (Olofsson et al., 2012).

The propitious mechanical properties of silicon nitride (Table 2.11) associated with high biocompatibility and promising tribological features including a low friction coefficient, high fracture toughness, and high wear resistance render this ceramic material a possible candidate for replacing joint components (Guedes e Silva, 2012). However, the high modulus of elasticity of both hot-pressed silicon nitride (HPSN) and pressureless sintered SSN in excess of 300 GPa are counter indicative of their use in hip implants, a fact based on the risk of strong stress shielding. As an alternative, RBSN could be used with a considerably lower modulus around 150 GPa that, however, is still much higher than that of cortical bone. On the downside, both the flexural strength (150–350 MPa) and fracture toughness (1.5–3 MPa $\cdot \sqrt{m}$) of RBSN are much lower than those of HPSN and SSN (500–1,000 MPa and 5–8 MPa $\cdot \sqrt{m}$, respectively), a consequence of the high porosity of the former (Heimann, 2010).

In a recent study, silicon nitride (SiN_x) coatings were deposited onto CoCrMo surfaces with an adhesion promoting interlayer. First, reactive d.c. magnetron sputtering was applied to coat a CoCrMo substrate with a 0.5–0.8 μm thick CrN interlayer, followed by a 4 μm SiN_x top layer (N/Si = 1.10–1.25) deposited by reactive high-power impulse magnetron sputtering. The CrN interlayer was deposited using negative bias voltages ranging from 100 to 900 V and various substrate rotation. Coating wear tests performed against Si_3N_4 in a reciprocating ball-on-disk test revealed specific wear rates lower than, or comparable to CoCrMo. The study suggests that low negative bias voltages may improve the adhesion of SiN_x coatings to the metallic substrate (Filho et al., 2019).

2.2.2 Bioactive ceramics

In contrast to bioinert ceramics discussed above, bioactive, that is, osseoconductive ceramics show positive interaction with living tissue that may even include differentiation of immature stem cells toward osteocytes, that is, osseoinductive behavior. These ceramics exhibit chemical bonding to the bone along the interface, thought to be triggered by the adsorption of bone growth-mediating proteins at the biomaterials' surface. Hence, there will be a biochemically mediated strong *bonding osteogenesis* (Yuan et al., 2017). In addition to compressive forces, to some degree tensile and shear forces can also be transmitted through the interface (bony in-growth). Typical bioactive ceramic materials are calcium phosphates and bioglasses. It is generally believed that the bioactivity of calcium phosphates is strongly associated with the formation of hydroxylcarbonate apatite (HCA), chemically similar to biological, that is, bone-like apatite (LeGeros and LeGeros, 1984).

In a recent comprehensive review, Habraken et al. (2016) have highlighted the tremendous improvements achieved in calcium phosphate materials research in the past, in particular in the field of biomineralization, as carrier for drug and gene delivery, as biologically active agent, as coatings for biomedical devices (Ben-Nissan,

2014; Layrolle, 2017), and as bone graft substitute. Evaluation of specifically hollow 1D structures consisting of nanotubes of hydroxylapatite confirmed this material's efficacy for cellular internalization and drug delivery using two distinguished anticancer model drugs: hydrophobic Paclitaxel® and hydrophilic Doxorubicin hydrochloride® (Srivastav et al., 2019).

In addition to their outstanding biological performance, calcium phosphates can be inexpensively produced, are biologically safe, and can relatively easy be certified for clinical use, thus offering great promise for the future. Beyond this, incorporation of metabolically important ions such as magnesium, strontium, silicon, copper, zinc, or cobalt into calcium phosphate-based bone graft materials are destined to provide an economic and feasible solution for bone defect healing (Wang and Yeung, 2017).

2.2.2.1 Hydroxylapatite

In the clinical practice, in noncemented hip endoprostheses an osseoconductive hydroxylapatite layer is provided that will allow a *bonding osteogenesis* that through "bony in-growth" will be able to transmit moderate tensile and shear forces. In this case, two ossification fronts will be present, one growing from the bone toward the implant and another growing from the implant toward the bone. Strong and convincing clinical evidence is mounting that a 150–200 µm thick long-term stable bioactive hydroxylapatite coating will elicit a specific biological response at the interface of the implant material by controlling its surface chemistry through adsorption of bone growth-mediating noncollagenous matrix proteins (see box) such as osteocalcin, osteonectin, silylated glycoproteins, and proteoglycans. This will result of the eventual establishment of a strong and lasting osseoconductive bond between living tissue and biomaterial.

Noncollagenous matrix proteins

Noncollagenous proteins (NCPs) play fundamental roles in promoting, controlling, and regulating fibrillogenesis, nucleation and crystal growth, and mineralization during osseo- and dentinogenesis. Similar to collagen fibrils, NCPs are synthesized and secreted by osteoblasts and odontoblasts, respectively. Many biological functions of NPCs during tissue mineralization are still unknown. Members of this group include osteocalcin, osteonectin, proteoglycan(s), sialoprotein, osteopontin, serum proteins, and ALP. *Osteocalcin* makes up about 20% of the group and is an important biochemical marker of bone formation. *Osteonectin* is thought to promote mineralization of collagen. *Proteoglycans* making up about 10% of NCPs play a significant part in controlling movement of water molecules in cartilage and thus, provide the tissue with resilience under compressive load. *Bone sialoprotein* II constitutes about 7.5% of NCPs with a yet unclear function. *Osteopontin* (sialoprotein I) is presumably involved in calcium binding. *Serum proteins* including serum albumin and several immunoglobulins make up some 25% of NCPs. Finally, *ALP* may be involved in the degradation of phosphate esters to provide a high local concentration of phosphate to kick-start mineralization (Al-Qtaitat and Aldalaen, 2014).

However, there are dissenting views. Several *in vivo* studies showed that neither peri-implant bone volume nor BIC was significantly affected by different calcium phosphate coatings including hydroxylapatite, α-TCP, or TTCP. Hydroxylapatite-coated acetabular cups were observed to have a similar risk of aseptic loosening as uncoated cups. Hence, using such coatings may not confer any added value in terms of implant stability (Lazarini et al., 2017; see also Gotfredsen, 2015; Aldosari et al., 2018; Surmenev and Surmeneva, 2019).

Structure

Hydroxylapatite, $Ca_{10}(PO_4)_6(OH)_2$ is a member of a large group of chemically different but structurally identical compounds with the general formula $\mathbf{M}_{10}(\mathbf{Z}O_4)_6\mathbf{X}_2$ ($\mathbf{M}$ = **Ca**, Pb, Cd, Zn, Sr, La, Ce, K, Na; $\mathbf{Z}$ = **P**, V, As, Cr, Si, C, Al, S; $\mathbf{X}$ = **OH**, Cl, F, CO_3, H_2O, □) that obey the hexagonal space group $P6_3/m$ (176). In stoichiometric hydroxylapatite, Ca polyhedra share faces to form chains parallel to the crystallographic c-axis [0001] that functions as a 6_3 screw axis. These chains are linked into a hexagonal array by sharing edges and corners with $[PO_4]$ tetrahedra. The OH^- ions are located in wide hexagonal channels parallel to [00.1] (Figure 2.53A). Two oxygen ions of the $[PO_4]$ tetrahedra are located on mirror planes through $z = 1/4$ and $z = 3/4$. The other two are symmetrically arranged above and below the mirror plane. The Ca^{2+} ions are situated in two different positions: Ca_I at $z = 0$ and 1/2 along the three-fold axes a_i, and Ca_{II} at $z = 1/4$ and 3/4 along the hexagonal screw axis 6_3 parallel [00.1] (Figures 2.53B and 2.55). This positional difference can be expressed by the formula $[(Ca_I)_4(Ca_{II})_6](PO_4)_6(OH)_2$.

The Ca_I ions are coordinated by nine oxygen ions that belong to six different phosphate tetrahedra, whereas Ca_{II} ions have an irregular sevenfold coordination with six oxygen ions of five phosphate groups in addition to the OH^- ions. Since each of the 16 OH^- positions in the unit cell is statistically occupied to only 50% there exist on average 8 vacancies/unit cell along the c-axis. Hence, there are direction-dependent differences in the mobility of OH^- ions and also the Ca_I ions associated with them that are extremely relevant when considering structural transformation from amorphous calcium phosphate (ACP) to crystalline hydroxylapatite as well as stepwise dehydroxylation of hydroxylapatite to form oxyhydroxylapatite (OHAp), and eventually, oxyapatite (OAp).

The lattice parameters of hexagonal hydroxylapatite are $a = 943.2$ pm and $c = 688.1$ pm (Posner et al., 1958). In addition, there exists a stoichiometric and ordered form of hydroxylapatite crystallizing in the monoclinic space group $P2_1/b$ (14) with $a = 942.1$ pm, $b = 2a$, and $c = 688.1$ pm, $\gamma = 120°$ (Elliot et al., 1973). It is thought that the deviation of the symmetry from the archetypal space group $P6_3/m$ may result from local ordering of OH^- ions in [00.z] anionic columns. The ordered monoclinic form of hydroxylapatite transforms to a hexagonal anion-disordered form at approximately 205 °C. However, since this monoclinic form occurs only under such

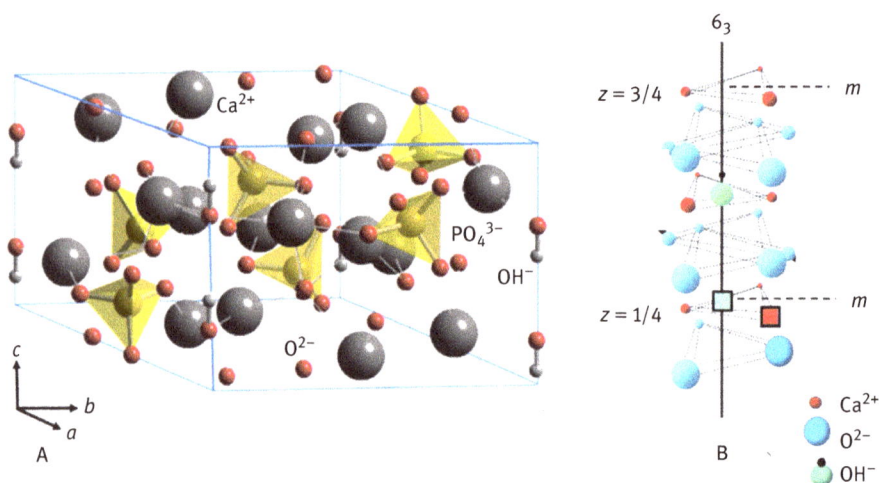

Figure 2.53: (A) Crystallographic structure of hydroxylapatite showing the projection down the c-axis of the hexagonal unit cell with $a = 943.2$ pm, $b = a\sqrt{3}$, $c = 688.1$ pm. Large gray spheres: Ca^{2+}; medium-sized yellow spheres: P^{5+}; small red spheres: O^{2-}, small grey spheres: H^+. (B) Details of the arrangement of the Ca_I ions at $z = \frac{1}{2}$ and Ca_{II} ions at $z = \frac{1}{4}$ and $\frac{3}{4}$ (red) and O^{2-} ions (blue) surrounding the 6_3 screw axis parallel to [00.1]. The OH^- ion (green) is located within the triangular Ca configuration at $z = \frac{1}{2}$. (Rey et al., 2009). © With permission by Springer Nature.

special thermal conditions it can be safely neglected in the present discussion. This lack of relevance notwithstanding, there is some evidence that growth of bone-like hydroxylapatite in an electric field is accelerated by reorientation of the dipole moments between O^{2-} and H^+ of lattice OH^- ions in response to the electric polarization conditions (Yamashita et al., 1996). It is in this way that the ordered alignment of OH^- columns present in monoclinic hydroxylapatite is being attained.

Several studies have shown that the texture of plasma-sprayed hydroxylapatite coatings noticeably influences the stability, crystallinity, phase composition, and surface roughness (Sun et al., 2001). The (00.2)/[11.0]-oriented coatings revealed higher hardness, elastic modulus, and fracture toughness (Kim et al., 2010). In addition, the preferred orientation of crystals in textured hydroxylapatite coatings presumably affects their biological and biomechanical characteristics. Indeed, more recent investigations suggest that textured hydroxylapatite coatings can elicit different cellular behavior owing to anisotropy of protein adsorption on various crystal faces of hexagonal hydroxylapatite crystals (Surmenev et al., 2014). Completely molten in-flight particles and optimum substrate preheating temperatures are necessary requirements to deposit such coatings with strong (00.2)-oriented crystallographic texture (Wang et al., 2016).

It has also been found that the crystal lattice of plasma-sprayed hydroxylapatite appears to be severely distorted (Shamray et al., 2017). As described earlier, the basic elements of the hydroxylapatite crystal structure are two columns parallel to the c-axis formed by the Ca_I atoms with coordinates $x = 0.333$ and $y = 0.666$ and channels formed by triangles of Ca_{II} atoms located in planes with $z = 1/4$ and $z = 3/4$; (Figure 2.53B), with a screw axis 6_3 in the center, where the OH groups are located. Since in plasma-sprayed hydroxylapatite the Ca_I positions are shifted from $z = 0$, there are "large" $(Ca_I–Ca_I)_L$ and "small" $(Ca_I–Ca_I)_S$ distances (Shamray et al., 2019).

In plasma-sprayed hydroxylapatite, the $[PO_4]$ tetrahedra are deformed with an experimentally obtained Baur distortion coefficient (Baur, 1974) of $D_1(TO)$ of ~0.3. The distances $Ca_I–O_{II}$ increase from 246 to 257 pm, and the $Ca_I–O_I$ distances decrease from 240 to 230 pm compared to those in stoichiometric hydroxylapatite, considered to be the result of internal residual stresses introduced by thermal spraying as also demonstrated by the broadening of X-ray diffraction (XRD) peaks. These residual stresses originate from the large temperature gradients that are generated when molten particles strike the cool substrate and are being rapidly quenched. Since their contraction is restricted by adherence to the roughened substrate surface, tensile stresses are generated, commonly referred to as "quenching stresses" (Matejicek and Sampath, 2003).

Furthermore, the substructural $Ca_5(PO_4)_3$ cluster elements (Shamray et al., 2019) are flattened along the c-axis by shortening of the $Ca_I–Ca_I$ distances. Likewise, within the clusters, the Ca–O distances change. These distortions have important consequences for the thermal stability and dissolution behavior of plasma-sprayed hydroxylapatite coatings.

Biomedical activity of hydroxylapatite

The reason why hydroxylapatite is copiously applied to biomedical devices ranging from coatings of the stem of hip endoprostheses to densified monolithic implants for dental root replacement, to suitable material for filling bone cavities, and to bone growth-supporting three-dimensional (3D) scaffolds (Derry and Kuhn, 2016) is grounded on its chemical and structural similarity to biological, that is, bone-like apatite. Emphasis should be put on the word "similarity" as many authors maintain that "the inorganic part of bone consists of hydroxylapatite." This statement is essentially wrong as biological apatite differs in many aspects from (geological) hydroxylapatite in terms of chemical composition, overall stoichiometry, crystalline order, defect density, hydroxyl ion content, and degree of substitution with metabolically important trace elements (Pasteris, 2016).

During the past decade, much work has been performed on biomimetic formation and application of hydroxylapatite/polymer nanocomposites (see Chapter 1.1.3). However, since most of the research has been devoted to *in vitro* characterization of the bioactive performance of hydroxylapatite coatings, mostly for dental and orthopedic implants, this *in vitro* immersion in simple SBF solution does not provide a direct link to evaluate coating success *in vivo* during clinical trials as confirmed by meta-meta-analyses by Surmenev and Surmeneva (2019). This is augmented by the admonition by Drouet (2013) that precipitation of a white solid during immersion in SBF does not automatically indicates the existence of a "bone-like carbonate apatite layer" as is sometimes too hastily concluded: "not all that glitters is gold." Hence, the identification of an apatitic phase should be carefully demonstrated by appropriate characterization techniques, preferably using complementary analytical methods.

Bone is a composite material the water-free substance of which consists of about 70 mass% Ca-depleted defect hydroxylapatite (called biological apatite or bone-like apatite) and some 30 mass% collagen I. Collagen type I constitutes 90–95 mass% of the organic matrix whereas the remaining 5–10 mass%, the so-called ground substance, consists of glycosaminoglycans (GAGs) such as hyaluronic acid and chondroitin sulfate (see Chapter 2.3.8.2), as well as NCPs such as osteocalcin, osteopontin and bone sialoprotein (see box above).

In bone, the apatite nanocrystal platelets (Figure 2.54, left, A) of some $30 \times 50 \times 2$ nm in size are orderly arranged along the triple-helical strands of collagen I (Figure 2.54, left, B). The basic structural unit of collagen is a helix consisting of three polypeptide chains coiled around each other to form a spiral that is stabilized by inter-chain bonds (Figure 2.54, left, C). Hence, the abundant protein collagen I functions as a structural template for crystallization of nanosized hydroxylapatite, a process that is presumably mediated by carboxylate terminal groups and the NCP osteocalcin.

The ubiquitous substitution of Ca by other ions reduces the theoretical Ca/P ratio of 1.67 of stoichiometric hydroxylapatite to substantially below 1.6 for bone-like apatite. This nonstoichiometry of biological apatite has been described by the approximate formula (e.g., Liu et al., 2001)

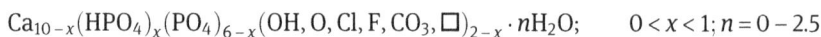

$$Ca_{10-x}(HPO_4)_x(PO_4)_{6-x}(OH, O, Cl, F, CO_3, \square)_{2-x} \cdot nH_2O; \qquad 0 < x < 1; n = 0 - 2.5$$

As suggested by the formula, biological hydroxylapatite should be described as calcium-and hydroxyl-deficient carbonated hydroxylapatite particles in which a fraction of the PO_4^{3-} lattice sites are being replaced by HPO_4^{2-} ions. However, 1D- and 2D-NMR investigation on calcium phosphate nanoparticles precipitated from aqueous solutions by Jäger et al. (2006a, 2006b) and subsequently confirmed by von Euw et al. (2019) have revealed that hydroxylapatite nanoparticles comprise a crystalline stoichiometric core coated by a very thin layer (~1 nm) of disordered, that is, ACP with a Ca/P ratio around 1.5 (Figure 2.54, right). This accounts for the well-known fact that secondary hydroxylapatite formed after incubation of coatings in SBF frequently show Ca/P ratios strongly deviating from stoichiometry. Given the small size of the particles of about 4 nm, this amorphous layer yields a much higher proportion of the overall phosphorus content as previously estimated. Hence, Von Euw et al. (2019) concluded that the dynamics of mineral–mineral and mineral–biomolecule interfaces in bone tissue appear to be driven by metastable hydrated amorphous environments rich in HPO_4^{2-} ions rather than by stable crystalline environments of the

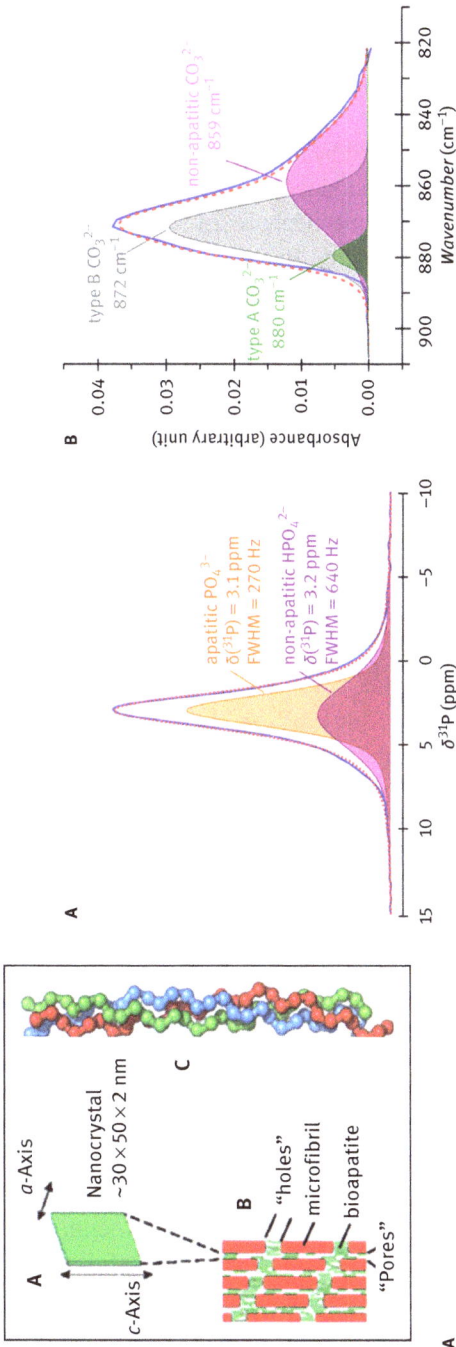

Figure 2.54: Left: Schematics of the hierarchical architecture of bone. (A) Individual nanocrystalline platelet of biological Ca- and OH-deficient hydroxylapatite. (B) Enlargement of a collagen microfibril showing the oriented intergrowth with hydroxylapatite nanocrystals (modified after Pasteris et al., 2008). © With permission by The Mineralogical Society of America. C: Structure of tropocollagen, consisting of triple-helical strands of polypeptide chains. Right: (A) Quantitative ^{31}P magic angle spinning (MAS) solid-state ^{31}P spectrum of a fresh 2-year-old sheep bone tissue sample (blue enveloping line) and its corresponding deconvolution into two peaks corresponding to the PO_4^{3-} containing internal crystalline hydroxylapatite core (orange) and the HPO_4^{2-} containing nonapatitic environments in the form of an amorphous surface layer (purple). (B) FT-IR spectrum of the $v_2(CO_3)$ vibration mode of a 2-year-old sheep bone tissue sample (blue line) and its deconvolution (red dashed line). Type B CO_3^{2-} ions occupy the PO_4^{3-} sites within the crystal lattice; type A CO_3^{2-} ions occupy the OH^- sites within the crystal lattice. The nonapatitic CO_3^{2-} ions are present within the amorphous surface layer (Von Euw et al., 2019). © Permission granted under Creative Commons Attribution 4.0 International License.

hydroxylapatite structure. This is in accord with earlier findings that HPO_4^{2-} ions are known to stimulate nucleation outside the matrix vesicles in the interstitial space (ECM) and on triple-helical collagen I strands (Höhling et al., 1971; Weiner et al., 1999).

The fact that the OH^- positions can be occupied by mobile O^{2-} ions or by vacancies □ is of vital importance for understanding the kinetics of the dehydroxylation reaction of hydroxylapatite to OHAp and OAp, respectively when subjected to high temperature as encountered during thermal spraying. Heat-imposed dehydroxylation of hydroxylapatite presumably results in a symmetry change in that the 6_3 screw axis of hydroxylapatite is being reduced to a polar axis $\bar{6}$ thought to be present in the structure of OAp, $Ca_{10}O(PO_4)_6$ (Figure 2.55; Heimann, 2015; 2018a).

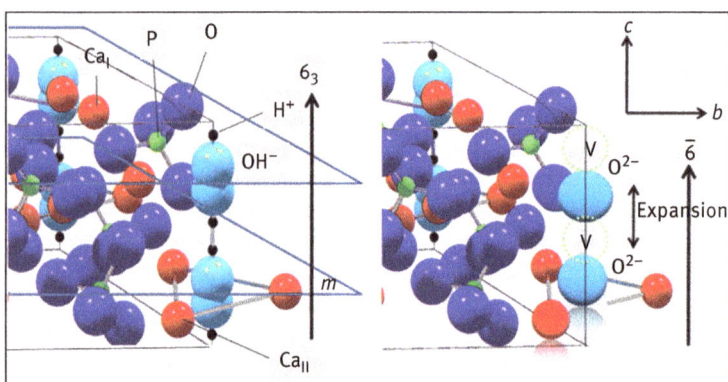

Figure 2.55: Left: Ball-and-spoke model of hydroxylapatite with space group P6$_3$/m. The mirror planes m through z = ¼ and ¾ are shown perpendicular to the screw axis 6$_3$. Right: Hypothetical structure of oxyapatite with lower symmetry $P\bar{6}$ showing a chain of O^{2-} ions parallel to the polar hexagonal axis $\bar{6}$ separated by vacancies V (Heimann, 2018a).

While the occurrence and structural behavior of OH^- ions in the hydroxylapatite lattice has been recognized and widely studied in synthetic material, it came as a complete surprise that natural biological bone-like apatite was found to be essentially free of hydroxyl ions when looking at Raman spectroscopy data (Pasteris et al., 2004; Wopenka and Pasteris, 2005), a fact that was later confirmed by nuclear magnetic resonance (NMR) spectroscopy and inelastic neutron scattering. Hence, contrary to the general medical nomenclature, bone-like apatite appears to be nonhydroxylated. Interestingly, there has been found a strong correlation between the concentration of OH^- ions and the crystallographic degree of atomic order. While it is not obvious at all how charge balance can be maintained within an essentially OH^- free nanocrystalline apatite lattice, there are now suggestions that the specific state of atomic disorder, imposed biochemically by the body, is essential for cell metabolism and the ability of the body to carry out tissue-specific functions. In addition, there are assumptions that the lack of hydroxyl ions in the lattice of nanocrystalline biological Ca-deficient hydroxylapatite causes a high density of vacancies along the 6_3 screw axis (Figure 2.55, left) and hence, a high mobility of Schottky-type defects that in turn, influences its solubility and in doing so provides a mechanism for fast and efficient bone reorganization by dissolution (osteoclastesis) and reprecipitation (osteoblastesis) during bone reconstruction in response to changing stress and load levels according to Wolff's law (see box).

Wolff's law

Wolff's law (Julius Wolff, 1836–1902) states that load-carrying bone will adapt to the load to which it is subjected. Increase of loading on a particular bone results in remodeling over time so that the bone becomes stronger. The internal architecture of the trabeculae undergoes adaptive changes that results in thickening. Contrariwise, if the loading on a bone decreases, the bone will atrophy and become less dense and, thus, weaker. This reduction in bone density (osteopenia) is known as stress shielding and can occur as a result of a hip arthroplasty. The normal stress on a bone is shielded from load and the stress is relocated to the prosthetic implant with higher modulus of elasticity.

Substituted hydroxylapatite

Because of its open channel structure, hydroxylapatite is able to incorporate a wide range of other ions by substituting Ca^{2+} cations as well as OH^- and PO_4^{3-} anions without large distortion of the lattice, thus maintaining the crystal structure of pure stoichiometric hydroxylapatite. In biological apatite, Ca^{2+} is being partially substituted by metabolically important cations such as Na^+, Mg^{2+}, Sr^{2+}, K^+, and some trace elements such as Pb^{2+}, Ba^{2+}, Zn^{2+}, and Fe^{2+}. Replacement of PO_4^{3-} with CO_3^{2-}, SiO_4^{4-}, and SO_4^{2-} anions as well as OH^- with Cl^-, F^-, O^{2-}, and CO_3^{2-} contributes to a host of biochemical pathways in which bone matter is involved.

As result of its chemical similarity to the mineral component of bone, hydroxylapatite has a wide range of applications both in bone grafts and for the coating of metallic implants. However, to more accurately mirror the chemistry of biological hydroxyapatite, various substitutions, both cationic (substituting for calcium ions) and anionic (substituting for phosphate or hydroxyl groups) have been synthesized. Shepherd et al. (2012) have summarized key effects of substitutions including Mg, Zn, Sr, Si, and carbonate on physical and biological characteristics. They showed that even small substitutions may have very significant effects on thermal stability, solubility, osteoclastic, and osteoblastic response *in vitro*, and degradation and bone regeneration *in vivo*.

Hence, there is a wide-open field of research into synthesis and characterization of various substituted hydroxylapatites that has led to a host of publications in the recent literature (e.g., Laskus and Kolmas, 2017; Graziani et al., 2018). Indeed, for some researchers it is tempting to use a shotgun approach, employing the entire

PSE for substitutional modification of hydroxylapatite, regardless whether the resulting materials may possess any biomedical merits. A review on antibacterial properties of hydroxylapatite substituted with ions of silver, copper, zinc, selenium, strontium, cerium, europium, gallium, and titanium has been provided by Kolmas et al. (2014).

Carbonate substitution

The PO_4^{3-} groups in hydroxylapatite can be replaced partially by CO_3^{2-}, forming carbonate-apatite, dahlite; Brophy and Nash, 1968), whereas OH^- can be substituted by CO_3^{2-}, Cl^-, and, in particular, F^- as found in tooth enamel and dentin. The carbonate substitutional defects are either located in the hydroxyl-occupied channel parallel [001] (type-A defect) or at the position of an orthophosphate group (type-B defect) (Figure 2.54, right). Computer modelling has shown that the lowest energy configuration exists for type-A defects when two hydroxyl groups are replaced by one carbonate group in such a way that the O–C–O axis is aligned with the c-axis channel of the apatite lattice and the third oxygen atom of the CO_3^{2-} group lying in the a/b plane (Peroos et al., 2006). This compositional variability is prominent among the causes of the high biocompatibility and osseoconductivity of bone-like apatite as well as its crystallite morphology. However, although it has been known for some time that nanosize and morphology of bioapatite crystallites are critical to the proper mechanical and physiological functioning of bone, the exact mechanism that controls these properties is still unclear.

Recently, it has been found that carbonate ion substitution for orthophosphate groups accounts for profound changes of the physicochemical and mechanical properties of bone-like apatite. With increasing CO_3^{2-} replacement between 3 and 8 mass%, the crystallite size, degree of crystallinity, and stiffness decrease, whereas the degree of aggregation and solubility increase (Deymier et al., 2017). It is remarkable that no proteins or other organic materials are required to form biological apatite with properties conducive for easy and effective bone remodeling but that an inorganic substitute such as carbonate ion suffices to achieve this task.

Strontium substitution

Among the plethora of cations that can substitute for calcium in the structure of hydroxylapatite, strontium has provoked an increasing interest because of its perceived beneficial effect on bone formation (Bussola Tovani et al., 2019), including prevention of bone resorption. Strontium can be quantitatively incorporated into hydroxylapatite whereby its substitution for Ca^{2+} (Shannon ionic radius 114 pm) causes a linear increase in the lattice constants and an associated linear shift of the infrared absorption bands of the hydroxyl and phosphate groups, coherent with the greater ionic radius of Sr^{2+} (Shannon ionic radius 132 pm). Rietveld refinement indicates that in most of the range of concentration strontium displays a slight preference for the $Ca_{(II)}$ cation site, consistent with its larger ionic radius. However, somewhat surprisingly, at very low concentrations Sr^{2+} ions tend to occupy the smaller $Ca_{(I)}$ site (Bigi et al., 2007).

Strontium addition to hydroxylapatite is responsible for a significant variation of the shape and dimension of nanopowder particles as well as a preferential growth along the c-axis direction at higher strontium loads. Modifications in the local chemical environment of phosphate and hydroxyl groups in the apatite lattice have been also observed (Frasnelli et al., 2017).

Strontium has a strong affinity for biological hydroxylapatite, and 98% of its distribution in the human body is in the skeleton. Strontium was found to regulate osteoblast-related gene expression, to enhance ALP activity in MSCs, and to reduce osteoclast differentiation. Animal studies and clinical trials have also shown that Sr stimulates bone formation and reduces bone resorption.

In this context, it is noteworthy to recall the grave concern that the 1986 Chernobyl nuclear accident has caused among the European population when roughly 10 PBq of radioactive ^{90}Sr, about 5% of the core inventory, were released into the environment and contaminated more than 200,000 km^2 of Eastern and Central Europe. This worry has repeated itself after the 2011 Fukushima nuclear disaster in Japan. There was widespread fear that incorporation of radioactive strontium ions in the biological hydroxylapatite matter of growing bone of children may result in severe health consequences among future generations. This was reminiscent of the St. Louis baby tooth survey (1959–1970) that concluded that children born in 1964 had about 50 times more ^{90}Sr in their baby teeth than those born prior to atmospheric testing of nuclear weapons, a finding that eventually led to a ban on such testing (Alley and Alley, 2013).

Since ^{90}Sr is metabolized similar to calcium, it can easily be incorporated into bone, and can remain there for many decades. Within bone, the β-emitter ^{90}Sr irradiates the sensible blood-producing bone marrow cells and may cause leukemia and other malignant diseases of the blood (Rosen, 2012). However, during extensive monitoring of the ecosystems affected by the Chernobyl fallout over the past three decades it became clear that most of the strontium (and plutonium) isotopes were deposited within only 100 km of the damaged reactor, thus, restricting its deleterious health effect to a comparatively small area and little population. While radioactive ^{131}I, of great concern after the accident owing to its risk of causing thyroid cancer in children, has a short half-life, and has now decayed away nearly completely, ^{90}Sr and ^{137}Cs, with half-lives around 30 years, persist and will remain a concern for decades to come. Although plutonium isotopes and ^{241}Am will remain radioactive for thousands of years, their contribution to human radioactive exposure is considered to be low (WHO, 2005).

Mg substitution

Owing to the lower Shannon ionic radius of 89 pm, Mg^{2+} ions tend to occupy the smaller $Ca_{(I)}$ cation sites. Mg influences osteoblast and osteoclast activity, and its deficiency affects bone metabolism, causing osteopenia and cessation of bone growth. Electrochemically deposited Mg-substituted hydroxylapatite increased the proliferation of MC3T3-E1 cells, ALP activity, and osteocalcin secretion compared with a group without Mg substitution (Zhao et al., 2013). Mg substitution was found to promote osteogenic differentiation of pre-osteoblast cells and, thus, may accelerate early bone healing with improved implant osseointegration. Moreover, based on the underlying signaling pathway, Mg has been proposed as a factor significantly upregulating Akt phosphorylation, enhancing cell adhesion and viability, cell differentiation, ALP activity, and the expression of osteogenesis-related genes (Wang et al., 2017).

Zn substitution

The use of hydroxylapatite is frequently limited owing to its slow osseointegration rate and low antibacterial activity, in particular when it is being used for long-term biomedical applications. Substitution of Zn (Shannon ionic radius 90 pm) into hydroxylapatite results in decrease in the *a*- and *c*-axes lengths of the unit cell, thereby causing the crystal structure to be altered. Bond order calculations show a preference of Zn for the $Ca_{(II)}$ positions with fourfold coordination. When occupying the octahedral $Ca_{(I)}$ position, Zn remains in sixfold coordination in the bulk that decreases to fivefold at the surface (Matos et al., 2010). Site-specific doping of Zn in hydroxylapatite can serve to develop potential bioceramics for bone implant applications with tailored biological properties and controlled antibacterial efficacy depending on the doping site (Bhattacharjee et al., 2019). *In vitro* cell culture investigation of human adipose-derived MSCs revealed enhanced bioactivity of Zn-substituted hydroxylapatite. In addition, the materials were found to possess antibacterial capability against *S. aureus* bacteria (Thian et al., 2013).

Ag substitution

At present, much work is being done on Ag-substituted hydroxylapatite owing to its antimicrobial and fungicidal behavior and its propensity to reduce or even prevent biofilm formation (Miranda et al., 2010; Ghani et al., 2011). Since biofilm-producing bacteria are the principal cause of infections associated with arthroplasty, silver is an attractive element to be incorporated into hydroxylapatite layers deposited on implant substrates by air plasma or suspension plasma spraying (Prentice et al., 2013; Cizek et al., 2018). In particular, silver ions are known to reduce biofilm formation involved in nosocomial infection with methicillin-resistant *S. aureus* (Ueno et al., 2016). Plasma-sprayed hydroxylapatite coatings doped with Ag were found to be highly effective against *Pseudomonas aeruginosa* bacteria but revealed evidence of cytotoxic effects marked by poor cellular morphology, cell death, and almost complete loss of functional ALP activity. Addition of SrO to the Ag-doped coatings served to offset these negative effects and led to performance improvement comparable to pure hydroxylapatite-coated samples (Fielding et al., 2012).

Silicon substitution

Silicon plays an important role in connective tissue metabolism, especially in bone and cartilage (Carlisle, 1981). On the one hand, lack of sufficient amounts of silicon in bone results in a decrease in the number of osteoblasts, osseomatriceal collagen, and GAGs. On the other hand, the addition of Si during hydroxylapatite synthesis leads to an improvement of its bioactive behavior (Gibson et al., 1999) as well as significantly higher bone apposition rate and more rapid remodeling of bone compared to unsubstituted hydroxylapatite (Porter et al., 2004). The enhanced bioactivity of silicon-substituted hydroxylapatite over pure hydroxylapatite has been attributed to the effect of silicate ions to accelerate dissolution (Friederichs et al., 2013). Solid-state NMR provided direct evidence for the isomorphous substitution of PO_4^{3-} by SiO_4^{4-} groups in the hydroxylapatite structure. Since the radius of Si^{4+} with 54 pm is slightly larger than that of P^{5+} (52 pm), and the average length of the Si–O bonds (162 pm) is greater than that of the P–O bonds (152 pm), a slight change in the lattice constants of the Si-substituted hydroxylapatite was observed (Balas et al., 2003).

Sulfate substitution

Sulfate ions have been found to substitute for phosphate anions in naturally occurring apatite (Comodi et al., 1999). This geological apatite has exceptionally high S and relatively high Si, Sr and LREE contents, whereas the HREE content is negligible. A high positive correlation between Ca-site Substitution Index (CSI = 100(10−Ca)/Ca) and Tetrahedral Substitution Index (TSI = 100 (Si + C + S)/P atom/a.p.f.u.) and a systematic parallel increase in REE, S and Si contents indicate two coupled substitution mechanisms, that is, $REE^{3+} + Si^{4+} = Ca^{2+} + P^{5+}$ and $Si^{4+} + S^{6+} = 2\ P^{5+}$. LREE and Sr show a marked preference for the $Ca_{(II)}$ site (see above).

The substitution of sulfate in apatite is of potential importance in synthetic biomaterials used in bone repair and reconstruction. Sulfate ions can be incorporated into synthetic apatite (Toyama et al., 2013) or in composites with calcium sulfate, showing that the design of new apatite compositions and composites could include the use of medically desirable counter cations such as lithium, sodium, potassium, or strontium (Tran et al., 2017). The maximum molar amount of sulfate ions that can be incorporated in hydroxylapatite in the presence of Na^+ as counterion is more than three times lower than the maximum molar amount of carbonate ions that can be incorporated. This difference can be explained by different solubility of the substituted apatites. The unit-cell parameters determined for both sulfated calcium and strontium hydroxylapatites synthesized with sodium counter-ions show a slight increase in the a-axis length but a nearly constant *c*-axis length with increasing sulfate content on account of the difference in size of the anion. These synthetic materials show close relationships with sulfated silicate apatites such as hydroxylellestadite, $Ca_{10}[(OH, Cl, F)_2/(SO_4)_3/(SiO_4)_3]$ (Harada et al., 1971).

Other substitutions

Cerium-doped hydroxylapatite/collagen coatings with suitable antibacterial efficacy against *E. coli* and *S. aureus* bacteria were deposited by a biomimetic method on Ti substrates that were previously treated by an alkali + thermal oxidation process (Ciobanu and Harja, 2019).

Selenium-substituted hydroxylapatite particles showed increased ALP activity of bone marrow MSCs, presumably owing to their nanosized surface morphology. Selenium released from particles effectively inhibited the growth of human osteosarcoma cells by inducing apoptosis. Hence, this material is potentially useful for bone defect repair after surgical tumor removal, offering safety, increased osteogenic differentiation of stem cells, and inhibition of tumor cell growth (He et al., 2019).

2.2.2.2 Bioresorbable calcium phosphate ceramics

According to Bohner (2010), the term "bioresorbable" is used to describe materials that will disappear from the implantation site over time, regardless of the mechanism leading to the material removal. These ceramic materials with very variable resorption mechanisms include Plaster of Paris (calcium sulfate hemihydrate), gypsum (calcium sulfate dihydrate), calcium carbonate (calcite, aragonite, vaterite), as well as a host of calcium phosphates such as dicalcium phosphate anhydrate (monetite, DCPA), dicalcium phosphate dihydrate (brushite, DCPD), octacalcium phosphate (OCP), TCP (α-TCP, α'-TCP, β-TCP), TTCP, and biphasic calcium phosphate (BCP).

State-of-the-art materials for resorbable calcium orthophosphate ceramics are TCPs, either as the low-temperature modification (β-TCP) or one of its high-temperature modifications (α-TCP, α'-TCP). Even more rapidly resorbable calcium alkali orthophosphates such as $Ca_2KNa(PO_4)_2$ (CAOP, see Chapter 4.1.4.1) were synthesized (Berger et al., 1995a,b) and evaluated for special biomedical applications including enhanced osteoblast differentiation and bone formation (see Chapter 4.1.4.3; Knabe et al., 2017a).

Structures

α-TCP crystallizes in the monoclinic space group P2$_1$/a (14) with a large unit cell with parameters $a = 1,287.3$ pm, $b = 2,728.0$ pm, $c = 1,521.3$ pm, $\beta = 126.2°$, and $Z = 24$ (Figure 2.56, left). Since the space group is akin to that of monoclinic hydroxylapatite with space group P2$_1$/b, there exist close structural relations. Consequently, α-TCP can

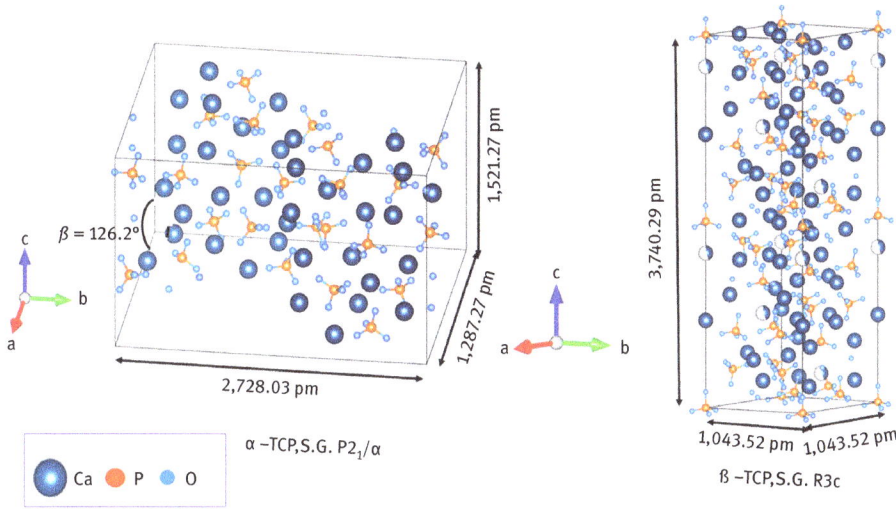

Figure 2.56: Crystallographic unit cells of α- and β-tricalcium phosphate (modified after Laskus and Kolmas, 2017). The lattice dimensions of α-TCP were taken from Yashima et al., (2007), those of β-TCP from Yashima et al., 2003). © Permission granted under the Creative Commons Attribution (CC BY) license by MDPI, Basel, Switzerland.

change readily to the apatite structure by replacing the cation-cation columns at the edges of the unit cell with anionic OH$^-$ columns. Synthetic β-TCP crystallizes in the rhombohedral space group R3c (161) with $a = b = 1043.5$ pm, $c = 3740.3$ pm, $\alpha = \beta = 90.0°$, $y = 120°$, $Z = 21$ (Figure 2.56, right). There is a close structural relationship with the natural mineral whitlockite, $Ca_9(Mg,Fe)[(HPO_4)(PO_4)_6]$ (Calvo and Gopal, 1975).

Results of molecular-dynamic calculations indicate that β-TCP is much more stable than α-TCP, confirming experimental results. This is thought to be related to different distributions of Ca atoms that have a pronounced effect on the stability and the electronic properties of the different modifications of TCP. In particular, the uniformly distributed Ca vacancies stabilize the β-TCP structure so that its solubility is much lower than that of the α-TCP modification.

TTCP, $Ca_4O(PO_4)_2$ is the least stable calcium orthophosphate. It crystallizes in the monoclinic space group P2$_1$ (4) with unit cell parameters $a = 702.3$ pm, $b = 1,198.6$ pm, $c = 947.3$ pm, $\beta = 90.90(1)°$. The Ca^{2+} and PO_4^{3-} ions are located in four sheets perpendicular to the b-axis. Each sheet contains two $Ca^{2+}-PO_4^{3-}$ columns and one $Ca^{2+}-Ca^{2+}$ column, similar to the structure of glaserite, $K_3Na(SO_4)_2$ with surplus oxygen ions. Significantly, two adjacent sheets in TTCP form a layer that is closely related to the structure of hydroxylapatite. Structural models of TTCP have been given by Goto and Katsui (2015).

In the presence of DCPD (brushite, $CaHPO_4 \cdot 2H_2O$) or DCPA (monetite, $CaHPO_4$), and water, TTCP forms easily hydroxylapatite according to

$$2\,Ca_4O(PO_4)_2 + 2\,CaHPO_4 \rightarrow Ca_{10}(PO_4)_6(OH)_2.$$

This reaction describes the formation of self-setting bone cement used ubiquitously in dentistry (Posset et al., 1998). With diammonium hydrogen phosphate, TTCP forms whisker-like structures of hydroxylapatite (Wang et al., 2005).

Figure 2.57 shows the pH dependence of the solubility isotherms of the biologically most important calcium orthophosphates at 25 °C. From this diagram, the following order of solubility results:

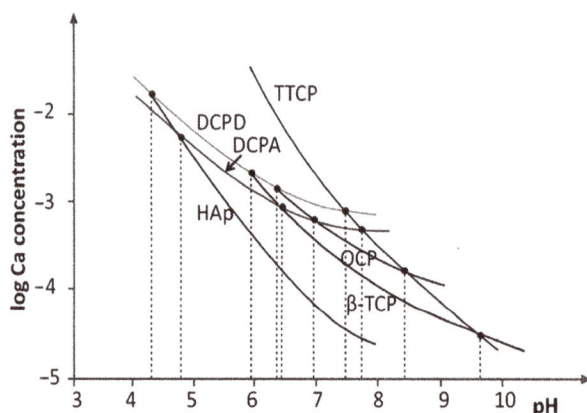

$$HAp < DCPD < DCPA < \beta - TCP < OCP < TTCP$$

Figure 2.57: Solubility isotherms of crystalline calcium orthophosphates at 25 °C. HAp, hydroxylapatite; DCPD, brushite; DCPA, monetite; β-TCP, β-tricalcium phosphate; OCP, octacalcium phosphate; TTCP, tetracalcium phosphate (after de Groot et al., 1990). © With permission by John Wiley and Sons.

Biomedical uses of bioresorbable calcium orthophosphates

Phase-pure β-TCP is being used either in the form of sintered ceramic monolithic shapes (e.g., Cerasorb®; KSI tricalcium phosphate®; chrOs™; BioResorb®) or as coatings for implants. Fast resorbable alkali-bearing calcium phosphates are supplied to the medical community predominately as granular materials, but increasingly also as porous spongiosa-like bodies or pastes. Porosity in monolithic ceramic bodies can be introduced by bloating agents, organic spacers or by utilization of naturally occurring cellular structures such as sponges (Schwartzwalder and Somers, 1963; see box). However, owing to the large variability of pore sizes and pore size distributions it is difficult to obtain products conforming to the stringent quality control requirements imposed by governmental regulations. Here the field is open for development of process engineering tools to obtain materials with highly reproducible properties. Such properties will include control of the size of microstructural surface features that were found to act as an osteogenic factor (Zhang et al., 2014).

Doping of TCP with Zn ions was found to stimulate a faster osteogenic differentiation of MSCs to osteoblasts and significantly improved cell proliferation as revealed by enhanced ALP activity (Chou et al., 2014).

Schwartzwalder–Somers process
A polymeric sponge is covered with a ceramic material in suspension, and after rolling to ensure that all pores have been filled, the ceramic-coated sponge is dried and heated to decompose the polymer, leaving only the porous ceramic structure in place. The foam must then be sintered for final densification. This method is widely used because it is effective with any ceramic able to be suspended. However, large amounts of gaseous byproducts are released and cracking of the foam due to differences in thermal expansion coefficients is common.

Amorphous calcium phosphates

In the field of bone regeneration, some clinical conditions require highly resorbable, reactive bone substitutes in addition to TCP preparations, needed to initiate rapid de novo tissue formation (Gelli et al., 2019). The thermodynamic instability of ACPs and their easy transformation to crystalline phases, most often hydroxylapatite, are of great biological relevance since ACP is thought to play an initiating role in matrix vesicle biomineralization. In addition, due to significant chemical and structural similarities with calcified mammalian tissues, as well as excellent biocompatibility and bioresorbability, all types of ACPs are promising candidates when efficient artificial bone grafts are desired (Dorozhkin, 2010; Combes and Rey, 2010).

Although ACPs appear to be well suited for this task owing to their high metastability, this desirable property leads to difficulties as sintering will result in transformation into crystalline, and thus, more stable compounds, defeating the purpose. Recently, various calcium phosphate samples codoped with carbonate (CO_3^{2-}) and magnesium ions were synthesized using ammonium and potassium hydroxide solutions. Spark plasma sintering at very low temperature (150 °C) leads to consolidation of the initial powders under maintaining their amorphous character. Such consolidated ACP compounds may be considered a new family of bioceramics with high metastability, allowing for fast release of bioactive ions upon resorption (Luginina et al., 2019).

More information on synthesis, structure, and thermal behavior of ACP can be found in recent papers by Vecstaudza et al. (2017, 2019).

Biphasic calcium phosphates

BCP composed of different ratios of stable hydroxylapatite and more soluble β-TCP have significant advantages over other calcium phosphate ceramics due to their controllable bioactivity and resorption kinetics that guarantee the stability of the biomaterial while promoting bone in-growth (Ebrahimi et al., 2017). Depending upon the concentration of the more stable and soluble phases, a ceramic material can be obtained that can be applied to large bone defects, in load bearing areas, and as customized monolithic bodies. Such monolithic bodies will maintain their shape and integrity over long periods of time. BCP bioceramics have been shown to be biocompatible, osseoconductive, biologically safe, mechanically predictable, and capable of inducing osteogenic differentiation of stem cells. These characteristics associated with their advantageous cost, biomedical effectiveness, unlimited supply, and absence of disease transmission make BCPs a viable alternative to autologous bone grafts, allografts, and others implants (Lobo and Arinzeh, 2010; Bouler et al., 2017). Hence, BCP bioceramics are recommended for use as an alternative or additive to autologous bone for orthopedic and dental applications. It is available in the form of particulates, blocks, customized designs for specific applications and as an injectable biomaterial in a polymer carrier. BCP ceramic can be used also as an abrasive for grit-blasting to modify biomedical implant surfaces. Exploratory studies demonstrated the potential uses of BCP ceramic as scaffold for tissue engineering, drug delivery systems, and carrier of growth factors (LeGeros et al., 2003).

In addition to bone grafts, biphasic hydroxylapatite/TCP can be produced as implant coatings from mixtures of high purity DCPD (brushite) and calcium carbonate powders with a Ca/P molar ratio of approximately 1.6 (Jinawath et al., 2004). Powder mixtures were formed into free-flowing agglomerates of $-300 + 105$ μm and $-500 + 300$ μm size, and air plasma-sprayed onto grit-blasted Ti6Al4V substrates at plasma powers between 26 and 34 kW and a stand-off distance of 90 mm to obtain coatings of 160–200 μm thickness. After deposition, the coatings were heat-treated at 900 °C. The average composition of heat-treated biphasic coatings was found to be around 62 mass% TCP, 34 mass% hydroxylapatite, and residual 4 mass% CaO. The reaction proceeded according to

$$12 \ [CaHPO_4 \cdot 2H_2O] \ + 7 \ CaCO_3 \ \rightarrow \ 3 \ Ca_3(PO_4)_2 + Ca_{10}(PO_4)_6(OH)_2 + 29 \ H_2O + 7 \ CO_2.$$

However, the large amounts of residual CaO render the coatings unfit for producing suitable biphasic coatings. Hence, immersing the coatings in revised SBF (r-SBF) (Kim et al., 2001) for 28 days converted the residual CaO to $CaCO_3$ by reaction with the HEPES-stabilized HCO_3^- ions contained in the r-SBF. In addition, remaining α-TCP was transformed to β-TCP. The average composition of the incubated coatings was found to be 52 mass% β-TCP, 35 mass% hydroxylapatite, and 13 mass% $CaCO_3$. Unexpectedly, after the postdepositional heat treatment the coatings showed very high adhesion strength to the Ti6Al4V substrate. Presumably, the merely mechanical adhesion typical for as-sprayed coatings changed after heat treatment to a chemical bond developed at a reactive interlayer between coating and substrate. This interlayer, presumably consisting of a calcium titanate, is preventing effectively the oxidation of the metal at the coated surface and is keeping the entire coating layer mechanically intact as well (Jinawath et al., 2004).

The generally positive effect of BCP toward enhanced bone in-growth can be augmented by incubation in SBF containing bovine serum albumin (BSA). It was found that this treatment upregulated ALP activity and osteogenically related genes and proteins, thus confirming the positive effect of SBF and BSA-SBF on cell growth (Huang et al., 2017).

D'Arros et al. (2020) showed the crystallographic evolution in BCP of an initial rhombohedral β-TCP structure to a micro-sized needle-like layer corresponding to an hexagonal "apatitic" TCP form. This phenomenon leads to an increase of the hydroxylapatite/TCP ratio, since hexagonal apatitic TCP is structurally similar to hexagonal hydroxylapatite. However, the Ca/P ratio (reflecting the chemical composition hydroxylapatite/TCP) remains unchanged. Thus, the high reactivity of BCP involves a dynamic evolution from a rhombohedral to a hexagonal structure, but not a chemical change.

Today, in addition to BCPs, there are bioceramic materials that combine the positive properties of several calcium phosphate phases. The electrochemically deposited biphasic BONIT® (DOT GmbH, Rostock, Germany) consists of about 70 mass% brushite and 30 mass% of crystalline hydroxylapatite (Ca/P ~ 1.1). The material contains chloride ions in a concentration similar to that in natural bone (39 mg/kg) and has a high porosity (60–70%), corresponding to a solid density of 0.9 Mg/m^3. The more soluble phase (brushite) promotes short-term bone synthesis whereas the microcrystalline hydroxyapatite phase will be more slowly resorbed and, thus, releases their ions to the surrounding tissue over a relatively long period. Owing to its high porosity and easy bioresorbability, BONIT® can be loaded with antibiotics such as gentamycin or bone growth-supporting factors such as rhBMPs or NCPs. The company also provides dental root implants consisting of a hydroxylapatite grit-blasted, double acid-etched surface that, in addition, contains a thin resorbable calcium phosphate coating (BONITex®). This creates an implant surface able to reduce healing time and to facilitate early loading.

Highly soluble alkali-containing calcium orthophosphates

Even though TCP and TTCP are rather easily resorbed in a physiological environment there is the quest for products even more soluble. There is a development trend concerned with alkali-containing calcium orthophosphates in the system $Ca_3(PO_4)_2$-CaNaPO$_4$-CaKPO$_4$ (Berger et al., 1995a,b) that when used as temporary bone replacement material show high resorbability and biodegradation without imposing acute or chronic damage to the surrounding tissue. Addition of alkali metal ions (Na, K) to TCP leads to formation of easily resorbable bone replacement ceramics containing $Ca_{10}Na(PO_4)_7$ and $Ca_{10}K(PO_4)_7$, respectively the structures of which appear to be very close to that of β-TCP so that it is difficult to distinguish it from the parent structure. Further addition of alkali ions lead to crystalline phase of the type $Ca_2 M_2(PO_4)_2$ with M = Na, K. In particular, $Ca_2KNa(PO_4)_2$ facilitates enhanced osteoblast growth and extracellular matrix (ECM) elaboration (Berger et al., 1995a; Knabe et al., 1998; see Chapter 4.1.4). Finally $CaMPO_4$ (β-rhenanite) will be formed. However, its solubility is so high that application of this compound for bone substitution is not feasible. This property notwithstanding composite hydroxyapatite/β-rhenanite biomaterials were developed (Suchanek et al., 1998) in which β-rhenanite acts as a weak interphase with high bioactivity. Moreover, it significantly enhances the sinterability of hydroxyapatite at 1,000 °C without forming undesirable secondary phases.

2.2.2.3 Calcium (Ti, Zr) hexaorthophosphates

Transition metal-substituted calcium hexaorthophosphate ceramics may be good candidate materials for orthopedic implant application (Heimann, 2012, 2017a, 2019). However, only scant information on this novel class of bioceramic materials is available in the relevant literature.

Structure

Calcium (titanium, zirconium) hexaorthophosphates obey the NaSiCon (**Na** *superionic con*ductor) structure (Alamo, 1993). The composition of the NASICONs can be described by the general formula $\mathbf{AM_2X_3O_{12}}$, where **A** = Ti, Zr, Hf, Nb or other transition metals of appropriate size, **M** = Na, K, Ca or lattice vacancies and **X** = P or Si. All members of this family show a low coefficient of linear thermal expansion and hence, high thermal shock resistance. Figure 2.58 (left) shows the geometry of the $A_2(XO_4)_3$ structural unit, Figure 2.58 (right) the unit cell of $CaTi_4(PO_4)_6$. The $Ti_2(PO_4)_3$ groups form a 3D network of two TiO_6 octahedra that are connected through their vertices to three PO_4 tetrahedra. These basic units appear as $-O_3TiO_3-O_3TiO_3-$ bands along the c-axis of the hexagonal unit cell. Along the ab plane these bands are connected by PO_4 tetrahedra.

The structural formula is $[M_1][M_2][A^{VI}_2][B^{IV}_3]O_{12}$, whereby M_1 and M_2 are interstitial vacancy sites, either partially or fully occupied by cations. Small highly charged ions such as Zr or Ti occupy the octahedral A-sites, and Si or P fill the tetrahedral B-sites. Many members of the NaSiCon group crystallize in the rhombohedral space group $R\bar{3}c$ (167). The structure consists of vertex-linked $[TiO_6]$ and/or $[ZrO_6]$ octahedra that form chains parallel to the c-axis (Figure 2.58). These chains are linked by $[PO_4]$ tetrahedra perpendicular to the c-axis, resulting in a 3D network. This configuration allows for two kinds of cavities, M_1 and M_2; mobile Na cations occupy the M_1 cavities, which also align along the c-axis.

When the monovalent Na or K cations are substituted by a divalent cation such as Ca, the symmetry is lowered to $R\bar{3}$ (148). This happens by ordering of cations and vacancies in the M_2 cavities, a configuration that leads to the loss of the c

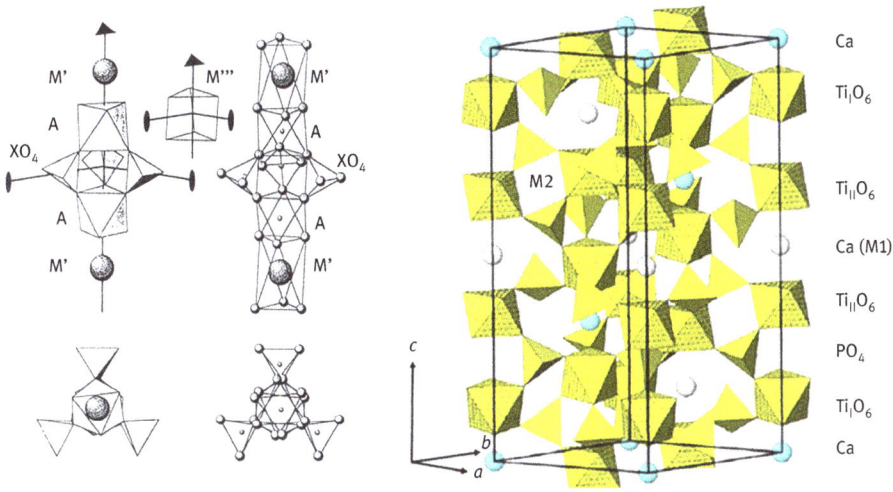

Figure 2.58: Left: Geometry of the $A_2(XO_4)_3$ structural group (Alamo, 1993) © With permission by Elsevier. Right: Unit cell of $CaTi_4(PO_4)_6$ (Heimann, 2017a). © With permission by The Mineralogical Society of America.

glide plane owing to the fact that half the M_1 sites are vacant (Alamo, 1993). The smaller M_2 cavities located between the chains are normally empty and only filled if additional ion contributions are required for charge compensation. The M_1 site can be either completely empty as in the case of $\square Nb^{5+}Nb^{4+}(PO_4)_3$, or partially filled as in $\square_{0.5}Ca^{2+}_{0.5}Ti_2(PO_4)_3$ and $\square_{0.66}La^{3+}_{0.33}Ti_2(PO_4)_3$. These vacancies account for the structural variability of the NaSiCon family as well as their substantial ionic conductivity (Senbhagaraman et al., 1993).

Properties

Calcium (Ti, Zr) hexaorthophosphate ceramics show solubility in SBF at least one order of magnitude lower than that of hydroxylapatite and in particular, TCP. Information on this substantially reduced *in vitro* solubility compared to hydroxylapatite can be found in Heimann (2010). In addition, these compounds excel by exceptionally low coefficients of thermal expansion (Alamo, 1993; Senbhagaraman et al., 1993).

Plasma-spraying of powders with $CaTiZr_3(PO_4)_6$ composition results in coatings with strong adhesion to Ti6Al4V substrates (>40 MPa) even though considerable thermal decomposition has been observed that leads to formation of zirconium pyrophosphate (ZrP_2O_7), rutile (TiO_2), and baddeleyite (β-ZrO_2), as well as several phosphorus-depleted structures. This is demonstrated in the cross-sectional images of plasma-sprayed coatings shown in Figure 2.59.

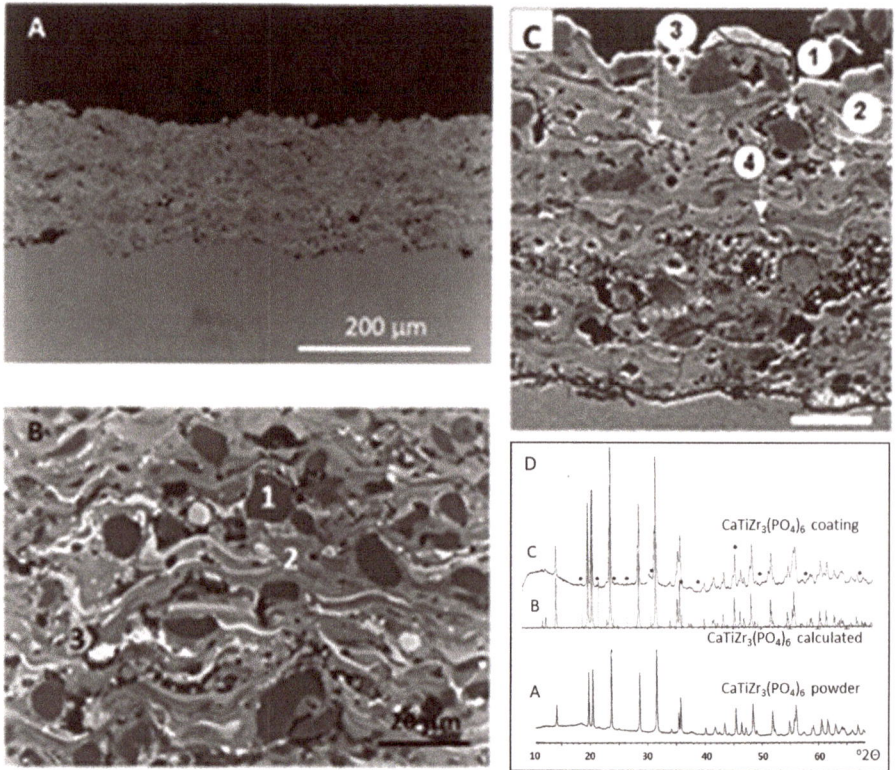

Figure 2.59: Cross-sectional images of plasma-sprayed $CaTi_{4-x}Zr_x(PO_4)_6$ coatings. (A) CZ_4P_3 coating (plasma power 36 kW, spray distance 90 mm, argon gas flow rate 50 slpm, hydrogen gas flow rate 10 slpm). (B) Enlarged image of (A) (Reisel, 1996). © Images courtesy Dr. Guido Reisel, Oerlikon Metco. (C) CTZ_3P_3 coating (plasma power 35 kW, spray distance 120 mm, argon gas flow rate 50 slpm, hydrogen gas flow rate 12 slpm) (Heimann, 2012). (D) XRD pattern of phase-pure CTZ_3P_3 starting powder (A), calculated interplanar spacings (B) and a plasma-sprayed coating (C). The dots indicate the presence of ZrP_2O_7, one of the decomposition products of incongruent melting (Heimann, 2019). © Permission granted under Creative Commons Attribution (CCBY) license.

The compositions of the individual coating phases shown in Figure 2.59B and C are indicated in Table 2.12. Phase 1 is essentially the target phase, whereas phases 2 and 3 are heat-affected calcium hexaorthophosphates strongly depleted in phosphorus, and phase 4 is composed of β-ZrO_2.

There is some evidence that the products of incongruent melting may lead to particle-mediated reinforcement of the coating microstructure. In addition, the substantially lower coefficient of the linear thermal expansion of calcium hexaorthophosphate coatings compared to hydroxylapatite predicts the development of compressive coating stresses when applied to Ti6Al4V implant surfaces, as opposed to strong tensile stresses in the case of hydroxylapatite coatings (see Chapter 3.2.7.3).

Table 2.12: Composition in mol% of the phases identified in Figure 2.59B and C.

Oxide	$CaTiZr_3(PO_4)_6$	Phase 1	Phase 2	Phase 3	Phase 4
CaO	12.5	12.6	13.5	16.5	0
ZrO_2	37.5	39.5	43.6	49.7	100
TiO_2	12.5	12.9	18.0	19.8	0
P_2O_5	37.5	35.0	24.9	14.0	0

This makes a successful enhancement of the adhesive strength of this novel type of coating highly likely when appropriate parameter optimization during plasma spraying will be carried out (Heimann, 2019).

In conclusion, transition metal-substituted calcium hexaorthophosphates were found to provide dense, well-adhering coatings (Schneider et al., 2001) with excellent biocompatibility and osseoconductivity as well as comparatively high thermal stability and low solubility *in vivo* compared to hydroxylapatite (Heimann, 2017a). As an added bonus, such compounds attain substantial ionic conductivity when doped with highly mobile ions such as sodium or lithium that are able to move in response to an outside electric field within the cavities of the crystalline structure. This suggests interesting options to design so-called *fourth-generation biomaterials* that utilize bioelectric properties for theragnostic purposes (Ning et al., 2015) or electrically stimulated bone growth (see Chapter 1.5.5).

Biomedical applications

In vitro biocompatibility tests with primary rat bone marrow cells showed substantial cell proliferation in the presence of fetal bovine serum. Animal tests proved that 150 μm thick coatings based on $CaTiZr_3(PO_4)_6$ applied to Ti6Al4V rods implanted in the femura of sheep led to strong neoformation of dense bone at a stable interface implant–bioceramic coating without coating delamination often observed with hydroxylapatite (Heimann et al., 2004). Build-up of a $Ti6Al4V/TiO_2/NASICON/$(hydroxylapatite) multilayered coating system could potentially lead to a device with the equivalent circuit of a capacitor that by proper poling could store negative electrical charges close to the interface with the growing bone, thus enhancing bone apposition rate and presumably, bone density (Heimann, 2017a).

Based on their multifarious advantageous properties, these NASICON compounds may be considered the "Sleeping Beauty" of osseoconductive coatings for endoprosthetic implants, osteosynthetic fixation devices, and bone growth stimulators. Among the earliest biological studies on this novel class of bioceramics were *in vivo* tests with $CaZr_4(PO_4)_6$ disks implanted into the distal epiphyseal parts of canine femora and tibiae that revealed that a direct and stable contact between implant and bone could be established. After 9 months, extensive remodeling of osteons had occurred in direct contact with the biomaterial without noticeable resorption of the latter. During fracturing associated with specimen preparation, separation occurred within the bone but not at the ceramic–bone interface, suggesting very strong bone apposition (Szmukler-Moncler et al., 1992). This work was indeed a key experiment that established the *in vivo* osseoconductive capacity and stability of these materials sufficiently well. Subsequently, Knabe et al. (2004) grew human bone-derived cells on sintered calcium (Ti,Zr) hexaorthophosphate samples, and tested for expression of various biochemical indicators for cell proliferation and cell vitality, such as levels of osteocalcin, osteonectin, osteopontin, ALP, and bone sialoprotein I. Compositions conforming specifically to $CaTiZr_3(PO_4)_6$ displayed maximum osteoblastic differentiation including sufficient expression of an array of osteogenic markers thus suggesting a high degree of osseoconductive potential. Similar results were reported for the growth of bone marrow stromal cells cultured on calcium titanium phosphate microspheres to be used as potential scaffolds for bone reconstruction (Barrias et al., 2005).

2.2.2.4 Calcium carbonates

The marine environment is rich in mineralizing organisms with porous structures, some of which are currently being used as bone graft materials and others that are in early stages of development (Ben-Nissan et al., 2019). Their exoskeletons are built up by calcium carbonates, either as calcite, aragonite, or magnesian calcite. Organism-templated biomineralization provides biological pathways potentially able to emulate nature's physiological mechanisms for biomedical purposes (Jelinek, 2013; Wolf et al., 2019). Biomimetic strategies have led to the investigation of naturally occurring porous structures as templates for bone growth, in particular cancellous bone (Clarke et al., 2011).

Structures

Calcium carbonate ($CaCO_3$) crystallizes in three polymorphic modifications. The most common and thermodynamically stable form is *calcite*, crystallizing in the trigonal system (crystal class $\bar{3}m$, space group $R\bar{3}m$ (166)) with $a = 49.89$ pm, $c = 170.61$ pm, $Z = 6$. The calcium ions are coordinated by six oxygen ions as shown in Figure 2.60 (left). Carbon ions are threefold coordinated with oxygen ions in planar conformation. Calcite is often the primary constituent of the shells of marine organisms, for example, plankton (such as coccoliths and planktic foraminifera), the hard parts of red algae, some sponges, brachiopods, echinoderms, some serpulids, most bryozoa, and parts of the shells of some bivalves such as oysters and rudists.

Aragonite is the high-pressure polymorph of calcium carbonate. Hence, inorganic aragonite occurs in high-pressure metamorphic rocks such as those formed in subduction zones. However, formation at ambient pressure requires biological control.

Aragonite crystallizes in the orthorhombic system (crystal class mmm, space group Pmcn (62)) with $a = 49.5$ pm, $b = 79.6$ pm, $c = 57.4$ pm, $Z = 4$. As shown in Figure 2.60 (right) aragonite, in contrast to calcite, has calcium ions coordinated by nine oxygen ions whereby each oxygen ion is bonded to three calcium ions. Aragonite occurs in almost all mollusk shells, and in cold-water corals (Scleractinia). Because the mineral deposition in mollusk shells is strongly biologically controlled, some crystal forms are distinctively different from those of inorganic aragonite. Since aragonite is thermodynamically unstable at ambient temperature and pressure, it tends to convert very slowly to calcite within 10^7–10^8 years. However, this transformation is readily achieved within hours at temperatures beyond 400 °C.

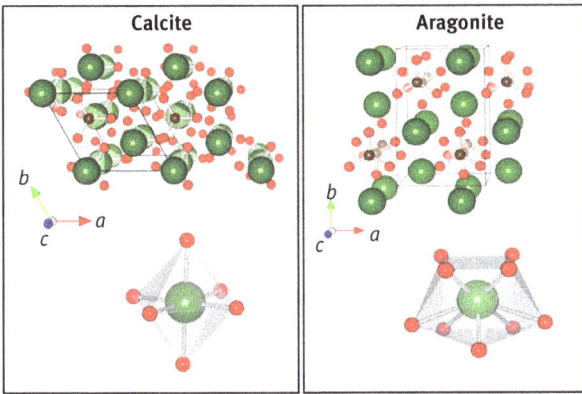

Figure 2.60: Structural models of calcite and aragonite. Large green spheres: calcium ions. Red spheres: oxygen ions. Black spheres: carbon ions. The trigonal unit cell of calcite and the orthorhombic unit cell of aragonite are outlined. © Permission granted under Creative Commons Attribution License (CC BY 3.0).

The third polymorphic form of calcium carbonate, *vaterite* (μ-CaCO$_3$) is metastable under ambient conditions, and decomposes even more readily than aragonite. Vaterite crystallizes in the hexagonal system (crystal class 6/mmm, space group P6$_3$/mmc (194)) with $a = 41.3$ pm, $c = 84.9$ pm, $Z = 6$. Since vaterite has a higher solubility than either calcite or aragonite, it transforms to calcite (at low temperature) or aragonite (at higher temperature: ~ 60 °C) when exposed to water. For example, at physiological temperature solution-mediated transition from vaterite to calcite occurs, whereby the former dissolves and subsequently precipitates as calcite, thermodynamically assisted by an Ostwald ripening process. Vaterite is a common biomineral that is metastably formed from a hydrated amorphous calcium carbonate (ACC) precursor via dehydration, a process inhibited in the presence of PO$_4^{3-}$ ions. Vaterite is frequently the first product to be formed during the repair mechanism of damaged shells of living marine organisms that later converts to either calcite or aragonite. In addition to water-free calcium carbonate modifications, there are also hydrated calcium carbonates such as monohydrocalcite ($CaCO_3 \cdot H_2O$) and ikaite, $CaCO_3 \cdot 6H_2O$ (Hesse et al., 1983).

Biomedical applications

Starting in the early 1970s, research has been performed to utilize corals of the genus *Porites* as potential xenogenic bone graft substitute. As the porous structure and mechanical properties of these stony corals with finger-like morphology somewhat resembles those of trabecular human bone, attempts of their using in biomedical devices were rather obvious. The exoskeleton of these calcium carbonate scaffolds has since been shown to be not only biocompatible, but also osseoconductive and biodegradable at variable rates that depend on porosity and implantation site. Although neither osseoinductive nor osseogenic *per se*, coral grafts can act as adequate carriers for growth factors. Appropriate designs may allow cell attachment, growth, spreading, and differentiation. When selected to match the resorption rate of coral grafts with the bone formation rate at the implantation site, natural coral exoskeletons have been found to be impressive bone graft substitutes (Demers et al., 2002). Indeed, experiments have shown sufficient coral resorption by the Zn-metalloenzyme carbonate dehydratase (carboanhydrase) and subsequent replacement by new tissue in transcortical resections of the canine femur (Patat et al., 1987).

However, there was a problem since the bone-like coral compound failed to biodegrade completely as new bone formed, leading to some unfavorable complications. As a result, coralline grafts were limited to very specific surgeries in certain situations. However, a possible solution was found to this problem by covering a calcium carbonate scaffold with a layer of hydroxylapatite to form a coralline hydroxylapatite/calcium carbonate (CHACC) composite material. Unlike pure hydroxylapatite, CHACC was found to completely biodegrade as bone growth progressed, ideally suited for tissue scaffold functions (Paddock, 2013).

Based on the potential of coralline bone graft materials, there are several commercial products on the market, including CoreBone®, ProOsteon® (Interpore Cross International), Biocoral,® or Vitoss® (Stryker). CoreBone® is the product of an Israeli company that has developed, and is distributing, bioactive, coral-based bone graft material that provides a natural alternative to human and animal-derived (bovine, porcine) bone grafts. The technology embeds bioactive minerals (e.g., silica) into the skeleton of corals during their growth process to accelerate healing and provide better connectivity. The CoreBone® graft is thought to offer the same bone-like qualities as human- or animal-derived bone substitute materials without the associated risks of contamination by bacteria, viruses, or prions. In contrast, ProOsteon® is made from marine coral exoskeletons hydrothermally converted to hydroxylapatite with architecture similar to cancellous bone (Walsh et al., 2006). Vitoss® (Stryker) possess an open, interconnected structure designed to allow for 3D bone regeneration. When combined with bone marrow aspirate, it has been shown to demonstrate high bone healing capacity.

Although these biomaterials have shown great potential owing to their chemical and structural characteristics strikingly similar to those of the human cancellous bone, clinical data presented to date are ambiguous (Pountos and Giannoudis, 2016). Consequently, correct formulation and design of the grafts to ensure adequate biocompatibility and resorption behavior appear intrinsic to a successful clinical performance and thus, require more research. Among the major drawbacks of calcium carbonates is their exceptionally high resorption rate *in vivo* that can be somewhat controlled, for example, by adding phosphate groups to create composite calcite/hydroxylapatite biomaterials (Smirnov et al., 2010). For example, artificial eyeballs have been fashioned from coralline calcium carbonate and superficially converted to hydroxylapatite. Placed into the eye socket, muscle tissue will attach to its biocompatible surface, thus allowing the artificial eye to be tracked along with the healthy one (Colen et al., 2000).

Non-sintered porous aragonite-based bioceramic materials have been investigated for the influence of process parameters, such as mass fraction and particle size of pore-former, and isostatic pressure, on porosity and compressive strength of compacted bodies in uniaxial or isostatic compaction modes (Lucas-Girot et al., 2002). Sintering of calcium carbonate with biodegradable phosphate-based glass as sintering aid resulted in a composite material with enhanced compressive strength, reduced degradation rate, and acceptable biocompatibility (He et al., 2015).

In addition to bone graft materials, calcium carbonate has been proposed as a smooth thin surface film of a porous 3D-poly(caprolactone) scaffold for bone replacement (Sommerdijk et al., 2007). These thin calcium carbonate films can be formed using anionic macromolecules, such as DNA as crystallization inhibitors. Indeed, DNA was found to be a powerful inhibitor of calcium carbonate crystallization and thus, can be used to prepare ACC films that slowly crystallize to form calcite with a preferred (11.0) orientation.

2.2.2.5 Silica

The marine sponges (Porifera) are organisms able to develop skeletons composed of unique rigid multiphase SiO_2- or $CaCO_3$-based composites (Ehrlich et al., 2015). The beneficial properties of the three different phases present, that is, amorphous silica, biopolymers such as collagen or chitin, and crystalline aragonite, extend the biomimetic potential of marine sponges toward bioinspired design of novel multiphase composite materials for biomedical applications (Niu et al., 2013).

Silica–collagen and silica–chitin nanocomposites are found naturally in siliceous tissue of sponges of the classes Hexactinellida, Demospongia, and Homoscleromorpha. These composite materials may be promising candidates for modern biomedicine and bone replacement applications since bioinspired, silicified collagen scaffolds possess the ability to enhance recruitment of progenitor cells and promote osteogenesis and angiogenesis by immunomodulation of monocytes. Hence, these scaffolds may be components of future tissue engineering applications (Wysokowski et al., 2018). While the functional role of silica–collagen composites in bone development is still enigmatic and in many aspects unclear, recent research results indicate that they may trigger precipitation of calcium phosphates. This finding represents an important advance in the translation of biomineralization concepts for the *in situ* mineralization of bone and teeth (Shadjou and Hasanzadeha, 2015).

Making glass is a high-temperature technology. In contrast to human technological effort over many millenia, nature has succeeded to form very complex glass structures at low temperature, an achievement far beyond the reach of current human endeavor (Schoeppler et al., 2017). Hence, it is compelling to use these natural glass structures as templates for a variety of biomedical applications. For example, diatomaceous biosilica has been proposed as an alternative source of mesoporous materials in the field of multifunctional supports for cell growth (Cicco et al., 2016). The biosilica surfaces were functionalized with 3-mercaptopropyl-trimethoxysilane and 3-aminopropyl-triethoxysilane. Fourier-transform infrared spectroscopy and X-ray photoelectron spectroscopy analyses revealed that the $-SH$ or $-NH_2$ functional groups were tightly grafted onto the biosilica surface. Viability of human primary osteosarcoma (Saos-2) and normal human dermal fibroblast cells was established in response to the nanoporosity of cell walls (frustules) of the diatoms. These results show that (i) diatom microparticles are promising natural biomaterials suitable for cell growth, and (ii) their surfaces exhibit good biocompatibility, owing to the presence of the -SH (mercapto) groups.

The complex hierarchical mesoscale structures of sponge or diatomaceous biosilica combines hard and stiff inorganic components with soft and tough organic tissue components to produce composite materials with extraordinary fracture resistance. This also leads to an unusual combination of fracture toughness and optical light propagation properties (Zhang et al., 2015c).

2.2.2.6 Bioglasses

In contrast to osseoconductive hydroxylapatite, bioglasses have been found to be osseoinductive. Osteogenetic cells receive signals from bioglasses that guide them to generate new bone. Although the various intracellular signaling pathways have not been clarified in detail, they appear to be associated with silicon and calcium ions released from the degrading bioglass surface (Carlisle, 1982; Hench and Polak, 2002). Studies have shown that several hormones including insulin growth factor II, known to be involved in the synthesis of osteoblasts on collagen, are stimulated by the inorganic dissolution products of bioglass (Xynos et al., 2001) and that in this way bioglasses determine gene expression by the rate and type of ions released from dissolving bioglass (Jones et al., 2007).

During implantation, reactions occur at the material-tissue interface that lead to time-dependent changes of the surface characteristics of the implant material and the tissues at the interface (Ducheyne and Qui, 1999; Knabe et al., 2017). Bioactive ceramics and glasses undergo solution-mediated surface reactions after immersion in biological fluids. These reactions include dissolution, reprecipitation, and ion-exchange phenomena in combination with protein adsorption events occurring at the bioactive ceramic surface (Kokubo et al, 1992, 1993; Radin et al., 1996, 1997a, 1997b; Ducheyne and Qui, 1999; El-Ghannam et al., 1999; Kaufmann et al., 2000). A key element of bone bioactive behavior is the development of a carbonated apatite surface after immersion in biological fluids containing HCO_3^- ions. Previous review papers summarized these events that were reported to occur at the bioactive ceramic–tissue interface (Ducheyne et al., 1994; Ducheyne and Qui, 1999).

The following enumeration does not imply a ranking in terms of time sequence or importance: (i) leaching/dissolution of the ceramic surface (Hench et al., 1971; Kokubo et al., 1992; Kokubo, 1993; LeGeros et al., 1993; Ducheyne et al., 1994), (ii) precipitation from solution onto the ceramic surface (Hench et al., 1971; Kokubo et al 1992; Kokubo, 1993; LeGeros et al. 1993, Ducheyne et al.,1994; Radin et al., 1997a, 1997b), (iii) ion exchange and structural rearrangement at the ceramic–tissue interface (Hench et al., 1971; Kokubo et al 1992; Kokubo, 1993; Ducheyne et al., 1992; Radin et al., 1993,1997a); (iv) interdiffusion from the surface boundary layer into the ceramic (Ducheyne et al., 1992) (v) solution-mediated effects on cellular activity (El-Ghannam et al., 1997; Yao et al., 2005); (vi) deposition of either the mineral (a) or the organic phase (b) without integration into the ceramic surface (Daculsi et al., 1989; LeGeros et al., 1993; de Bruijn et al., 1993,1994; El-Ghannam et al., 1995); (vii) deposition with integration into the ceramic (Daculsi et al., 1989; LeGeros et al., 1993); (viii) serum protein adsorption and chemotaxis onto the ceramic surface (Schepers et al., 1991); (ix) cell attachment and proliferation (Matsuda and Davies, 1987; El-Ghannam et al,. 1995,1997); (x) cell differentiation (El-Ghannam et al., 1995) and (xi) ECM formation (Matsuda and Davies, 1987; Schepers et al., 1991; de Bruijn et al., 1993; El-Ghannam et al., 1995). Collectively, the surface-mediated effects result in enhanced osteoblast differentiation, bone matrix mineralization and thus, bone tissue formation (Ducheyne and Qui, 1999). In addition, solution-mediated effects require adequate consideration.

Several studies demonstrating the release of ionic dissolution products of bioactive glass such as biologically active Ca^{2+} and SiO_4^{4-} that enhance osteoblast differentiation (Xynos et al., 2000, 2001; Radin et al., 2005; Yao et al., 2005; Hench, 2008) emphasize the synergistic nature of solution- and surface-mediated effects of bioactive glass on osteoblast function and stimulation of osteogenesis. These processes result in rapid bone regeneration and can also be used to induce angiogenesis, hence, offering the tantalizing potential for novel gene-activating glasses for soft tissue regeneration. In addition, research continues into designing and testing of thermally sprayed hydroxylapatite/Bioglass® coatings for improved cell attachment and bone regeneration.

Composition

Bioactive glass compositions are predominately found in the system CaO-Na_2O-P_2O_5-SiO_2. The first of the series of bioglasses introduced to the biomedical industry was 45S5 Bioglass® with a composition of 45% SiO_2, 24.5% CaO, 24.5% NaO_2, and 6% P_2O_5 by mass (Table 2.13).

Table 2.13: Composition of typical bioactive glasses and glass-ceramics used for medical and dental applications (Hench, 2013a).

Composition (mass%)	45S5 Bioglass® (NovaBone)	S53P4 (AbminDent 1)	A-W glass-ceramic (Cerabone)
SiO_2	45	53	34
Na_2O	24.5	23	0
CaO	24.5	20	44.7
P_2O_5	6	4	16.2
MgO	0	0	4.6
CaF_2	0	0	0.5
Phases	Glass	Glass	Apatite/β-wollastonite glass
Class of bioactivity	A	B	B

Class A of bioactivity involves both osseoconduction and osseostimulation owing to rapid reaction of the bioactive glass surface *in vivo*. In contrast, class B materials react much more slowly within several weeks and do rarely lead to differentiation of osseoprogenitor cells into mature osteoblasts (Hench, 2013a).

The clinical outcome of introducing this composition into the human body revealed that 45S5 Bioglass® has greater osteoblastic activity compared to hydroxylapatite (class of bioactivity A). Today, many bioactive silica-based formulations derived from 45S5 Bioglass® exist (see Table 2.14). Glasses with a Ca/P ratio substantially below these compositional values do not bond to bone. The classic ternary diagram of compositional dependence of bone bonding capability of surface-active bioglasses given originally by Hench (1991) is shown in Figure 2.61. This ternary diagram is a section through the quaternary diagram CaO-Na_2O-SiO_2-P_2O_5 so that all glasses in the bone-bonding range a have a constant P_2O_5 concentration of 6 mass%.

Table 2.14: Composition in mass% of selected bioglasses and glass-ceramics used in clinical applications (Hench, 1991; 2013a).

Type	SiO_2	CaO	Na_2O	P_2O_5	MgO	$Ca(PO_3)_2$	CaF_2	Others
30S15B5	30	24.5	24.5	6				15 B_2O_3
45S5 (NovaBone)	45	24.5	24.5	6				
45S10	45	24.5	27.5	3				
45S5.4F	45	14.7	24.5	6			9.8	
40S5B5	40	24.5	24.5	6				5 B_2O_3
52S4.6	52	21	21	6				
53SP4(AbminDent)	53	20	23	4				
55S4.3	55	19.5	19.5	6				
58S	60	36		4				
S45P7	45	22	24	7				2 Ta_2O_5
Ceravital KGC	46	20	5		3	26		0.5 K_2O
Ceravital KGS	46	33	4			16		
Ceravital KGy213	38	31	4			14		7 Al_2O_3, 6 Ta_2O_5/TiO_2
Ceravital bioactive	40–50	30–35	5–10	10–15	2–5			0.5–3 K_2O
A-W G (Cerabone)	34	45	1	16	5			
MB glass-ceramic	19–52	3–9	3–5	4–24	5–15			12–33 Al_2O_3, 3–5 K_2O

Following the invention of the archetypical 45S5 Bioglass®, various kinds of bioactive glass and glass-ceramics have been developed with different functions that include high mechanical strength, sufficient machinability, fast setting ability, and even antibacterial property (53SP4 Bioglass®).

Compared to traditional CNS glasses (CaO-Na_2O-SiO_2), there are critical differences in composition between bioactive glasses and the former. Bioglass®, including 45S5 and 30S15B5 (30 mass% SiO_2, 15 mass% B_2O_5, 24.5 mass Na_2O, 24.5 mass% CaO, 6 mass% P_2O_5) are characterized by less than 60 mol% SiO_2, high Na_2O and CaO contents, and sufficiently high CaO/P_2O_5 ratios to be able to form HCA *in vivo*. Table 2.14 shows a range of clinically used bioglass compositions.

Hydrolytic stability of bioglasses

A common feature of bioactive glasses is a time-dependent kinetic modification of their surfaces during implantation. While they are generally nonresorbable, the

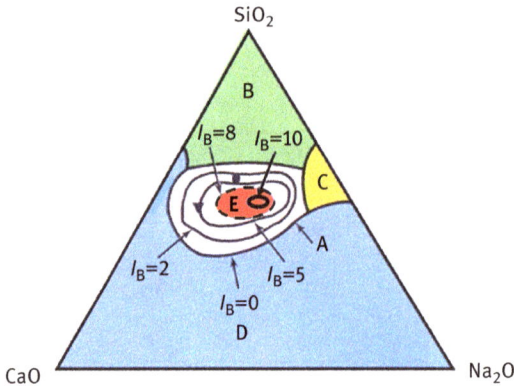

Figure 2.61: Bioglass compositions in the pseudoquaternary system CaO-Na$_2$O-SiO$_2$-(P$_2$O$_5$) (after Hench, 1991). Region A (white): Bone-bonding compositions. Region B (green): Bioinert SiO$_2$-rich compositions. Region C (yellow): Completely resorbable CaO-poor glasses. Region D (blue): Technically impractical compositions. Region E (red): Soft tissue-bonding compositions. The isopleths shown inside the region A (0 < I_B < 10) are a measure of bone-bonding ability whereby $I_B = 100/t_{0.5}$. $t_{0.5}$ is the time required to achieve 50% bone bonding. © With permission by John Wiley & Sons.

release of sodium and calcium ions triggers a cascade of reactions that culminate in nucleating a thin layer of biological HCA (Clark et al., 1976). This layer provides a bonding interface with bony tissue that is mechanically strong, so that in many cases the interfacial strength of adhesion exceeds the cohesive strength of the implant material or the bone tissue bonded to it (Hench, 1991).

The low hydrolytic stability of Bioglass® compositions renders the surface of such implants highly reactive when exposed to ECF *in vivo*. There are five distinct and kinetically separable stages during interaction of a glass with liquid defined as follows (Hench, 1998; see also Heimann, 2018b).

- Stage 1: The dissolution reaction starts with diffusion-controlled, that is, $t^{1/2}$-dependent rapid ion exchange of Na$^+$ and Ca^{2+} with H$_3$O$^+$ from solution (Figure 2.62) to yield

$$\equiv Si - O - Na + H_3O^+ \rightarrow \ \equiv Si - OH + Na^+ + H_2O$$

- Stage 2: The reaction continues with surface-controlled, that is, $t^{1.0}$-dependent slow dissolution of silica attained by breaking bridging oxygen (BO) bonds in Si–O–Si and formation of Si–OH (silanol) groups at the interface bioglass–solution according to

$$\equiv Si - O - Si \equiv \ + H_2O \rightarrow \ \equiv Si - OH + OH - Si \equiv$$

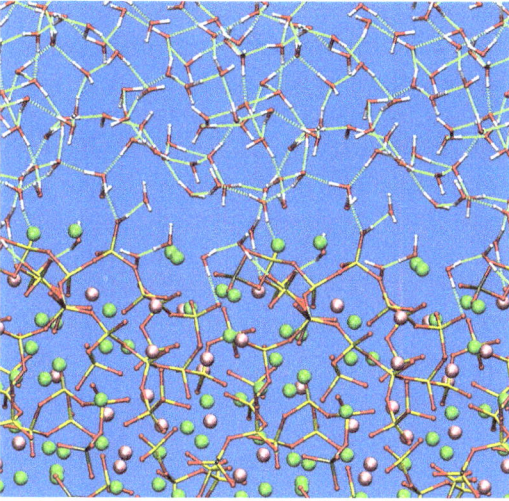

Figure 2.62: A snapshot of the bioglass–water interface, extracted from a molecular dynamic (MD) simulation trajectory. The phosphorus-silicate network (bottom) is represented as yellow/red ball-and-stick, with green spheres denoting Na^+ and brown spheres denoting Ca^{2+} cations. Dashed lines indicate hydrogen bonds (HB) (Tilocca and Cormack, 2011). © Permission granted under the terms of the Creative Commons Attribution License by The Royal Society.

- Stage 3: Subsequently, condensation and repolymerization of the alkali- and alkaline earth-depleted porous silica-rich surface layer happens according to

$$\equiv Si - OH + OH - Si \equiv \ \rightarrow \ \equiv Si - O - Si \equiv\ + H_2O$$

- Stage 4: Migration of Ca^{2+} and HPO_4^{2-} ions to the surface through the porous silica network forms a $CaO–P_2O_5$-rich surface film that subsequently transforms into an ACP layer by incorporation of soluble calcium and phosphate ions from solution.
- Stage 5: The reaction sequence is concluded by nucleation and growth of nano-sized HCA crystals by incorporating OH^- and CO_3^{2-} ions from solution, via transformation of ACP, possibly via OCP.
- Stage 6: Adsorption and chemical fixation of biologically active components such as NCPs within the growing HCA layer that support ingrowth of collagen fibrils.

It should be emphasized that the sequence shown above is applicable in its simplicity only to single-phase glasses. For multiphase glasses, in particular glass–ceramics and bioglasses, a similar sequence must be established for each individual phase.

While this sequence of events has been widely accepted, there were attempts to provide a modified, supplementary view (Hill, 1996) by considering the effect of re-placing SiO_4 by AlO_4 tetrahedral groups on the reactivity and thus, bone-bonding ability. Also, in the presence of network-modifying Ca^{2+} ions, the network connectivity (Holliday, 1977) will change toward a weaker, more ionic bonding between two silicate chains in the glassy network. Indeed, the network connectivity of a glass (see box) can be used to predict its surface reactivity, solubility in aqueous solutions, coefficient of thermal expansion, and tendency of phase separation (Hill, 1996). For example, all bone-bonding bioglasses with network connectivity below 2 show high surface reactivity, including the archetypical 45S5 Bioglass® with connectivity 1.90.

Furthermore, phosphate ions are thought to play a key role in bioactivity and apatite formation of bioglasses, the bioactivity of which increases monotonically with increasing phosphate content as long as P remains predominantly as Q_P^0 tetrahedra (see box; Edén, 2011). This can be explained by the fact that phosphorus does not enter the silicate network, but instead forms a separate orthophosphate phase (Hill and Brauer, 2011).

Network connectivity of glasses

Silicate glasses may be considered inorganic polymers of oxygen atoms cross-linked by silicon atoms (Holliday, 1977). The cross-link density X (Stevels, 1957) is defined as the average number of additional cross-linking bonds above 2 for elements other than oxygen forming the glass network backbone. In other words, X is the average number of NBO ions per polyhedron. In bioglasses, these elements are silicon, phosphorus, boron and, occasionally, aluminum. The network connectivity Y (Stevels, 1957) is the average number of BO ions per polyhedron, and can be calculated by $Y = 8-2\,R$, whereby R is the average number of oxygen ions per network-forming ion, usually the oxygen–silicon ratio (Kingery et al., 1976). The cross-link density is then given by $X = 2\,R-4$. Hence, a glass with $R = 2$ has a cross-link density $X = 0$ and a network connectivity $Y = 4$, corresponding to a pure 3D silica glass. Otherwise, glasses with $R = 3$ ($X = 2$; $Y = 2$) cannot form a 3D network since the tetrahedra have less than two oxygen ions in common with neighboring tetrahedra. Consequently, chains of tetrahedra of various lengths exist as, for example, in simple $Na_2O \cdot SiO_2$ glasses with linear polymer chains.

In a more modern view, the network connectivity can be described by the "Q^x" terminology, where x is the number of BO ions on a tetrahedron. The structure of pure silica glass consists of Q^4 tetrahedra (four bridging, no NBOs, corresponding to $Y = 4$ above). Modifying oxides create lower "Q^x" structures by replacing BOs with NBO: Q^3 tetrahedra (phyllosilicates; 3 bridging, 1 NBO per silicon ion), Q^2 tetrahedra (inosilicates; 2 bridging, 2 NBOs per silicon ion, corresponding to $Y = 2$ above), Q^1 tetrahedra (sorosilicates; 1 bridging, 3 NBOs per silicon ion), and Q^0 tetrahedra (nesosilicates; 0 bridging, 4 NBOs per silicon ion) (Shelby, 2005).

Timeline and mechanism of osseointegration of bioglass

During interaction of Bioglass® with body fluid, alkali ions will be leached out of the glass, leaving a gel-like layer enriched in silica at its surface as discussed above. Three to four weeks after implantation, fibrous collagen strands were found to grow into the gel layer that, together with hydroxylapatite nanocrystals derived from Ca^{2+} and PO_4^{3-} ions of the glass, did generate new bone, thus anchoring the Bioglass® implant solidly to the existing bone bed. *In vivo* strength testing using animal models revealed that the adhesion strength at the Bioglass®–bone interface frequently exceeded the intrinsic yield strength of the bone.

Concurrent with the first four reaction stages, adsorption of proteins and other biological growth factors occurs and is believed to contribute to the biological activity of the HCA layer. The controlled release from bioglass of biologically active Ca^{2+} and SiO_4^{4-} ions leads to up-regulation and activation of genes in osseoprogenitor cells (Hench, 2008). Within approximately 3–6 h, an HCA layer (stage 5) forms at the surface of a bioglass implant that is chemically and structurally nearly identical to natural bone mineral. Consequently, bone cells are able to attach directly to the bioglass surface. As the reaction continues, the HCA surface layer grows in thickness to form a bonding zone of 100–150 µm that constitutes a mechanically compliant interface that is essential to extend the bioactive bonding of the implant to the natural tissue. These surface reactions occur within the first 12–24 h after implantation. By the time osteogenic cells infiltrate the implantation site that normally takes 24–72 h, they will encounter a bone-like surface, complete with adsorbed organic components. It is this sequence of events, in which the bioactive glass participates in the repair process that allows for the creation of a direct bond of the material to tissue. The normal healing and regeneration processes of the body begin in concurrence with the formation of these surface-active layers. Bioactive glasses appear to minimize the duration of the macrophage and inflammatory responses that accompany any trauma, including trauma deliberately imposed by surgery.

More recent research has clarified the biological mechanisms that make bioglasses highly desirable biomaterials. Research continues into designing and testing of thermally sprayed hydroxylapatite/Bioglass® coatings for improved cell attachment and bone regeneration. In addition, boron-containing bioglasses are known to influence the performance of several metabolic enzymes since boron deficiency is associated with impaired growth and abnormal bone development. The addition of network-forming boron in different proportions to bioactive silicate glasses has significant effects on glass structure, glass processing parameters, biodegradability, biocompatibility, bioactivity, and cytotoxicity. Controlled release of boron *in vivo* may be beneficial for enhanced bone and soft tissue growth (Balasubramanian et al., 2018) since borate-containing bioglasses have been shown to convert faster and more completely to hydroxylapatite compared to silicate bioactive glass (Deliormanli, 2013).

However, under certain environmental *in vitro* conditions, undesirable calcium carbonate deposits have been found to form at the surface of bioactive glasses, instead of or in competition with HCA. Major factors controlling calcium carbonate formation include a high concentration of Ca^{2+} ions in the SBF testing solution, an appropriate surface of bioglass-to-volume of solution ratio (S/V) that controls supersaturation toward calcium carbonate precipitation, and the composition and surface texture of the bioglass (Mozafari et al., 2019).

As discussed above, immersion of a surface-active silicate glass in water results in partial dissolution at the glass surface the kinetics of which determines the degree of osseogenicity. Fast initial dissolution leads to reduced time for forming a stable bonding interface with bone (Peitl et al., 2001), whereby the growth rate of new bone depends on the release of a critical amount of dissolved ions in the space surrounding the dissolving bioglass (Xynos et al., 2001). In contrast to completely resorbable bioceramics, silicate-based bioglasses form a self-passivating surface layer that limits further dissolution. It was also found that small silicate structural units such as the cyclic silicate trimer $[Si_3O_9]^{6-}$ group present in pseudowollastonite show unusual kinetic stability compared with those moieties found on the surface of amorphous SiO_2. This confirms their potential role as active sites for heterogeneous, stereochemically promoted nucleation of hydroxylapatite in SBF and, thus, as templates for adsorption of bone-growth stimulating proteins and eventually, growth and strengthening of the bone-bonding interface *in vivo* (Sahai and Anseau, 2005).

The critical initial stages of bioglass dissolution in contact with water (step 1 above) have been thoroughly investigated by Tilocca and Cormack (2011) using *ab initio* finite-temperature molecular-dynamic (MD) simulation (see box).

Molecular dynamic simulation

Molecular dynamic (MD) simulation involves the study of the physical movement of atoms and molecules that are allowed to interact for a fixed period of time to arrive at a view of the dynamic evolution of the system under investigation. The trajectories of atoms or molecules are determined by numerically solving Newton's equation of motion. The forces between interacting particles are calculated using interatomic potentials or MD force fields. Though these equations are mathematically complex, they can easily be solved since their representations are simple, including simulating springs for bond length and angles, periodic functions for bond rotation and Lennard–Jones potentials, and the Coulomb attraction for van der Waals and electrostatic interactions (Hospital et al., 2015).

This work considered details of the mechanism how water molecules are organized near the bioglass surface, and also focused on the hydrogen-bond (HB) connectivity and its effect on the coordination and mechanism of release of Na^+ cations. MD simulations by these authors (Tilocca and Cormack, 2009) have studied the role of exposed Na^+ cations in coordinating water molecules and promoting solvent penetration in the earliest stages of the dissolution mechanism.

Figure 2.62 presents a snapshot of the glass–water interface, extracted from the MD simulation, revealing the open nature of the bioglass surface that reflects the high fragmentation of the bulk silicate network as well as the high amount of exposed Na^+ cations, in direct contact with the water layer. The main water–surface interaction at the initial interface involves water coordination by NBO^--Na^+ pairs, whereas Ca^{2+} ions have a minor role in the initial stages of surface hydration and dissolution.

Biomedical application of bioactive glasses

The application of bioactive glasses for medical use is a relatively recent accomplishment. Hence, there have been only comparatively few but generally successful clinical applications of these materials. Perhaps most important is the absence in the medical literature of any reports of adverse responses to these materials in the body that confirm the antibacterial and antifungal properties of bioglasses. Today, application of bioglasses include (Hench, 2013a)

- Dental implants
- Periodontal pocket obliteration
- Alveolar ridge augmentation
- Maxillofacial reconstruction
- Cervical and lumbar interbody fusion
- Otolaryngologic applications
- Percutaneous access devices
- Spinal fusion for adolescent idiopathic scoliosis
- Femoral nonunion repair
- Cranial-facial reconstruction
- Cochlear implants
- Bone graft substitutes
- Bone tissue engineering scaffolds
- Drug and gene delivery vehicles
- Antibacterial and antifungal application as wound healing agent
- Granular filler for jaw defects following tooth extraction
- Coatings for dialysis catheters made from silicone tubing
- Coatings for surgical screws and wires
- Toothpaste for treatment of dentinal hypersensitivity.

Addition of 45S5 Bioglass® particulate (NovaMin®) to toothpaste was found to support occlusion of dentinal tubules and re-mineralization of the surface of teeth, thereby eliminating the cause of dentinal hypersensitivity. In 2011, Glaxo-Smith-Kline plc acquired this technology and has launched an extremely successful over-the-counter version of "bioactive" toothpaste called "Sensodyne Repair & Protect" that prevents dental pain sensitivity and inhibits gingivitis. As an example of the use of bioactive materials as a preventive treatment, it is considered a revolution in dental healthcare (Hench, 2013b). A similar approach was recently taken by using nanohydroxylapatite particles in toothpastes as an agent to fight dentinal hypersensitivity (Pei et al., 2019; De Melo Alencar et al., 2019).

Further studies suggest that bioactivity of glasses occurs only within certain compositional limits and very specific ratios of oxides in the hexanary **Na_2O**-K_2O-**CaO**-MgO-**P_2O_5**-**SiO_2** system (see Figure 2.61). The extent of these limits, and the physicochemical and biochemical reasons thereof are only poorly understood at present. For example, in subcutaneous installations it was discovered that two adjacent samples of 53SP4 glasses (Table 2.14) in contact with each other can establish a bond across their superficial apatite layers. Hence, it appears that there exists a certain tendency of self-repair if a monolithic bioglass device may accidentally break within the surrounding soft tissue. A similar phenomenon was observed in A/W (apatite/β-wollastonite) glass–ceramics with substantially higher phosphorus content (Table 2.14) but lower I_b compared to Bioglass®. Bioglasses used in periodontics are marketed as PerioGlas® to dentists and oral surgeons as a bone graft substitute material used in ridge augmentation procedures and filling extraction sockets to prevent bone loss and aid regeneration of jaw.

At present, there is also substantial research interest in the use of bioactive glassy materials for tissue scaffolds that mimic the structure of trabecular, that is, spongy bone. The challenge is to prepare bioresorbable scaffolds of suitable geometry and bioactivity to support the growth of artificially seeded tissues that can be tailored to fit specific bone defects. Research is currently being done to fine-tune the architecture and resorption characteristics of sol–gel-derived bioactive glasses. Foaming agents and surfactants are being incorporated into the sol–gel reaction mixture to introduce 3D-interconnected pore geometry with the aim to emulate the porous structure of trabecular bone (Jones, 2013). Along these lines, inorganic materials such as bioglasses that can augment the body's own ability to regenerate are destined to become significant in future clinical approaches to functionally restore damaged tissue. In addition, the possibility of large-scale manufacturing of engineered tissues seeded with the patients' own cells to minimize the risk of rejection is an innovative alternative to some problems currently associated with prosthetic implants and with donor organs. If successful, this approach will dramatically improve the quality of life for millions of people worldwide.

In addition, some attention has been given to the use of mesoporous bioactive glasses for antibacterial strategies, primarily because of their capability of acting as potent carriers for the local release of antimicrobial agents (Kargozar et al., 2018). The incorporation of antibacterial metallic ions including silver (Ag^+), zinc (Zn^{2+}), copper (Cu^+ and Cu^{2+}), cerium (Ce^{3+} and Ce^{4+}), and gallium (Ga^{3+}) ions into the mesoporous glass structure and their controlled release is thought today to be one of the most attractive ways to inhibit bacterial growth and reproduction.

Mesoporous bioactive glasses (MBGs) have been recently proposed as drug delivery vehicles for advanced treatment of bone cancer. Different types of MBGs including granular particles and 3D scaffolds can be applied for cancer therapy (Kargozar et al., 2019).

In addition, silica-based mesoporous bioglasses offer a potential alternative to the systemic delivery of antibiotics for prevention against infections. The antibacterial efficacy of antibiotic-free MBGs is controlled by the composition-dependent dissolution rate and the concentration of the dopant elements (Kaya et al., 2018).

Modern advances in drug delivery applied to regenerative medicine (Zeng et al., 2019) focus on the rational design of scaffolds tailored for specific cargo and engineered to exert distinct biological functions. Among these scaffold materials, bioglasses loaded with therapeutic inorganic ions (TIIs) have been subject to recent research (Mouriño et al., 2019). These TIIs with potential angiogenic and osteogenic effects must be delivered in a controlled and sustained manner from bone tissue-engineering scaffolds. Successful strategies to overcome current challenges will have broad application in bone regeneration with great promise for future clinical therapies.

Biomedical application of nanoscaled magnetic bioglasses

Research into and development of magnetic ceramic nanoparticles, intended for cancer treatment, by a glass ceramic route has been the focus of some research effort (Parveen et al., 2012). Promising biomedical applications of *soft magnetic nanoparticles* are emerging including magnetic drug targeting, magnetic separation of leukemia cells from blood plasma, magnetic hyperthermia to fight cancer cells (Li et al., 2019a), and use as contrast agent in MRI. The magnetic particles attach to the cell membrane and, upon application of a magnetic field, activate the membrane and initiate some biochemical reactions within the cell, not only supporting the growth of functional bone and cartilage cells but also enhancing tissue regeneration (Miola et al., 2019a). In addition, magnetic liposomes, that is, lipid vesicles that incorporate magnetic nanoparticles either in the lipid bilayer or in an aqueous compartment have been suggested as vehicles for controlled drug delivery (Gao et al., 2009). Alternatively, there has been a revival of attention in the field of *hard magnetic nanoparticles* such as barium ferrite by precipitation of nanocrystals with an appropriate narrow crystal size distribution in the glassy matrix. The crystal size can be controlled by the temperature and time of heat treatment. Superparamagnetic magnetite nanoparticles have been incorporated into polymethyl methacrylate (PMMA)-based bone cement to impart bioactive and magnetic properties potentially useful for treatment of bone tumors (Miola et al., 2019b).

Progress of magnetic bioactive glass–ceramics has been made as a thermoseed in hyperthermia treatment of cancer, particularly to combat deep-seated bone tumors. These deep-regional tumors are efficiently heated and destroyed at temperatures around 42–45 °C without damaging neighboring healthy tissue. Although there are still substantial challenges to overcome, there is some indication that magnetic bioactive glass–ceramics and mesoporous bioactive glasses could be effective to treat cancerous cells by a hyperthermia route (Kargozar et al., 2019).

2.2.2.7 Titanium dioxide

In comparison to the bioactive ceramic materials discussed above, titania takes on a somewhat Janus-faced image. On the one hand, titanium dioxide does not owe its importance for biomaterials science to advantageous mechanical properties but predominantly to its photocatalytic and semiconducting behavior. On the other hand, its hemocompatibility, superhydrophilicity, and (limited) osseoconductivity are of biological importance and, hence, exploited in a variety of biomedical applications and devices.

Structure

Under ambient conditions, titanium dioxide occurs in three polymorphic modifications: tetragonal rutile (space group P4$_2$/mnm (136)), tetragonal anatase (space group I4$_1$/amd (141)), and orthorhombic brookite (space group Pbca (61)). The crystal structure of the thermodynamically stable rutile consists of edge-sharing TiO$_6$ octahedra that form straight chains parallel [001]. The chains are linked by sharing corners of the octahedra. As evident from the left of Figure 2.63, this configuration resembles an fcc lattice. However, the close-packed oxygen sheets are slightly kinked by about 20° to result instead of a cubic in a tetragonal symmetry with a 4$_2$ screw axis.

In anatase, the TiO$_6$ octahedra share four edges to form a pseudotetragonal framework structure (Figure 2.63, right), and in the least stable polymorph brookite the octahedra share three edges. Each octahedron is linked with three others to form zig-zag chains with each octahedron sharing one edge with an octahedron of the neighboring chain, thus forming nets parallel (100).

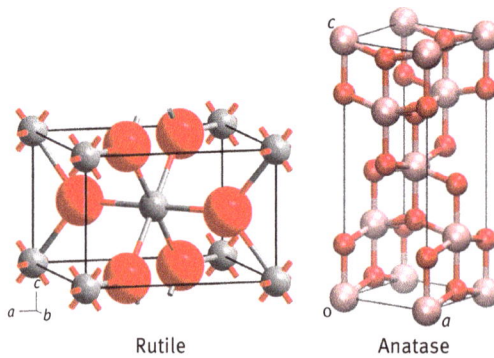

Rutile   Anatase

Figure 2.63: Ball-and-spoke model of the tetragonal rutile (left) and anatase (right) structures of titanium dioxide. The central Ti ion (gray, pink) is octahedrally surrounded by eight oxygen ions (red). © Permission granted under Creative Commons Attribution License (CC BY 3.0).

Besides these modification, at least eight other synthetic crystalline structures of stoichiometric TiO$_2$ are known, five of which are high pressure phases with α-PbO$_2$-like, baddeleyite-like, cotunnite-like, orthorhombic OI, and cubic structures. Nonstoichiometric Ti oxides are known as Andersson-Magnéli phases Ti$_n$O$_{2n-1}$ (Ti$_4$O$_7$ to Ti$_{10}$O$_{19}$).

Properties

Table 2.15 shows salient properties of titanium dioxide (rutile modification) (see also Table 2.5).

Table 2.15: Mechanical and thermal properties of rutile.

Property	Dimension	Value
Density	Mg/m^3	4.26
Compressive strength	MPa	680
Tensile strength	MPa	350
Modulus of rupture	MPa	140
Shear modulus	GPa	90
Modulus of elasticity	GPa	230
Fracture toughness	$MPa \cdot \sqrt{m}$	3.2
Vickers hardness	GPa	8.8
Thermal expansion	$10^{-6}/K$	9.0
Thermal conductivity	$W/m \cdot K$	11.7

Biomedical applications

Titanium oxides play an important role in the complex process of osseointegration of an implant fashioned from Ti or Ti alloy (see Chapter 1.1.5). Fixation of an implant in the human body is a dynamic process that remodels the interface zone between the implant and living tissue at all dimensional levels from the molecular up to the cell and tissue morphology level and at all time scales from the first second up to several years after implantation (Kasemo and Lausmaa, 1991). This is represented in Figure 1.3 in which the logarithmic distance and time scales indicate this complex dynamic process. While immediately following the implantation, a space filled with ECF exists adjacent to the implant surface, with time bone growth-supporting proteins will be adsorbed at the native titanium oxide surface that is only several nanometers thick. This exceedingly thin oxide film covering the titanium alloy surface will give rise to osseoinduction by proliferation of cells and their differentiation toward bone cells, revascularization, and eventual gap closing. Ideally, a strong bond will be formed between implant and tissue. However, sometimes connective tissue is being formed at the interface resulting in a fibrous tissue capsule that prevents osseointegration (see Figure 2.64) and will cause implant loosening. To prevent this undesirable situation, a hydroxylapatite coating of several tens or hundreds of micrometers is being applied by several surface coating techniques (see Chapter 3.2.6.2).

The composition of the ECF will change with time in response to chemical adsorption processes of molecules that mediate bone formation and also to transport reactions by outward diffusion of titanium atoms or ions through the thin oxide layer and concomitant inward diffusion of oxygen to the metal–oxide interface. This diffusion process is aided by lattice defects in the titanium oxide film such as grain boundaries, isolated vacancies or vacancy clusters, and interstitial atoms. Several other reactions may occur, for example, corrosion and partial dissolution/resorption of the oxide layer. This latter process appears to be a limiting factor for some implants even though the resorption rate is slow owing to the high chemical stability of titanium oxide. It ought to be noted that the relations shown in Figure 2.64 are, in essence, identical to those shown in Figure 1.4.

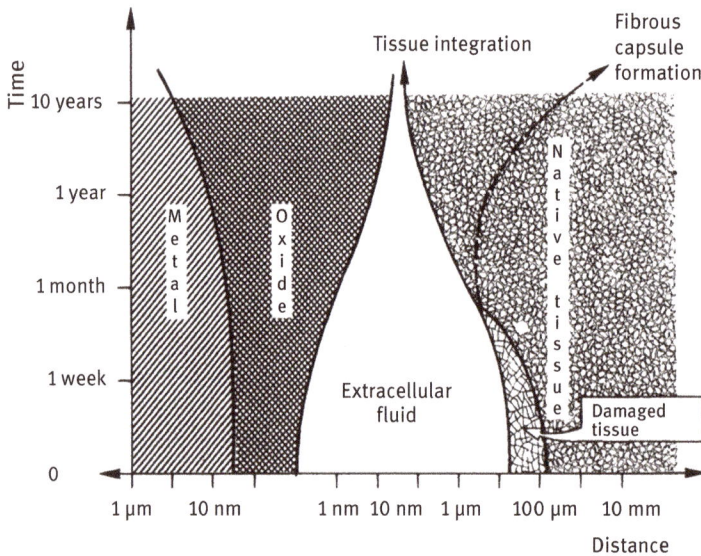

Figure 2.64: Schematic rendering of the double logarithmic distance-time relation of osseointegration to illustrate the dynamic behavior of the interface between a metallic implant (left) and bony tissue (right) (after Kasemo and Lausmaa, 1991). © With permission by John Wiley and Sons.

In clinical application, hydroxylapatite coatings may fail by chipping, spalling, delamination, and dissolution. Such deleterious behavior is being observed on explanted endoprostheses consistently close to the Ti implant/coating interface. This has been frequently attributed to the existence of a layer of ACP at the coating/Ti interface formed during rapid quenching of molten or semimolten droplets of calcium phosphate with high cooling rates exceeding 10^6 °C/s as experienced during thermal spraying. A continuous ACP layer is thought to act as a low energy fracture path and, owing to its comparatively high solubility, will preferentially dissolve *in vivo*, causing further weakening of the mechanical integrity of the interface. To assist improved adhesion of the coating to the metallic substrate, titanium oxide bond coats have been deposited directly onto the roughened metal substrate surface by APS (Heimann and Lehmann, 2015). Even though the feedstock material consists of the rutile modification of titania, the deposited bond coat shows brookite structure (Figure 2.65) on account of rapid quenching of the molten TiO_2 material and solidification of the metastable brookite phase that is presumably controlled by an Ostwald step rule mechanism (Heimann and Wirth, 2006).

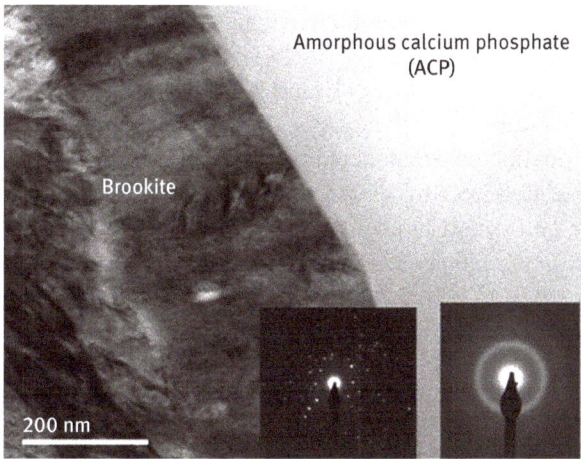

Figure 2.65: Bright field STEM image of the interface between an air plasma-sprayed polycrystalline titanium oxide (brookite) bond coat with columnar morphology (left) adjacent to an amorphous calcium phosphate (ACP) layer (right). Both phases are metastable obtained by quenching during rapid solidification. The insets show selected electron diffraction (SED) pattern of both phases (Heimann and Wirth, 2006).

Owing to its pronounced hemocompatibility, titanium dioxide is considered a suitable material for heart valve components. Failure of heart valve function accounts for a large proportion of heart diseases and approximately 400,000 patients worldwide undergo heart valve replacements annually. Consequently, the hemocompatibility of titanium oxide is widely being studied *in vitro* and *in vivo*, including measurement of clotting time and severity of platelet adhesion. Important findings revealed that the clotting time of rutile-type titanium oxide films and membranes is substantially extended over that of LTIC. In addition, fewer platelets were found to adhere to the titanium oxide films, making them hopeful substitutes for LTIC in artificial heart valves.

This requires solving still critical limitation issues including the observed increase of platelet adhesion with increasing coating thickness and density of negative surface charge.

In addition, titanium dioxide films are ubiquitously applied to coat medical devices and structures owing to their photocatalytic properties. Applications abound in self-cleaning surfaces of wall and floor tiles in clinical operation theatres and public swimming pools, utilizing the so-called lotus effect, and antimicrobial and bactericidal surfaces of biomedical materials and devices (Jackson and Ahmed, 2007). Although reactive radicals formed by nanoparticles of titanium dioxide under UV radiation are advantageous as they are able to decompose deleterious microorganisms, they pose potential risks for humans. Hence, a cytotoxicity assessment of titanium dioxide was performed (Kasper et al., 2010).

At present, TiO_2 films are studied as efficient immobilizing substrates for amino-terminated dendrimers for drug carriers and gene transfection vehicles (see Chapter 1.3.6) (Li et al., 2018).

Also, titanium dioxide in form of its brookite polymorph has been suggested as useful antimicrobial coating for dental root implants (Hotokebuchi and Noda, 2013). Furthermore, anatase formation on titanium surfaces treated with NaOH and H_2O_2 was found to be the key driving force for enhanced bioactivity, thus fostering osseointegration of orthopedic and dental implants (Rajendran et al., 2019). A recent review charts present and future biomedical application routes of titanium dioxide (Raj et al., 2014).

2.2.2.8 Other bioceramic materials

Calcium titanate ($CaTiO_3$) has been implicated in enhanced osteoblastesis *in vitro* on titanium-coated hydroxylapatite surfaces. Hence, it has been concluded that orthopedic coatings that form $CaTiO_3$ could support osseointegration with juxtaposed bone needed for enhanced implant efficacy (Webster et al,. 2003). It has been suggested also that bonding of atmospheric plasma-sprayed hydroxylapatite to a titanium alloy substrate may possibly be mediated by a very thin reaction layer essentially composed of either perovskitic calcium titanate or calcium dititanate, $CaTi_2O_5$ that not only provides increased adhesive bonding but also controls nucleation of apatite in contact with SBF or ECF, respectively. However, experimental evidence of such a reaction layer in as-sprayed coatings is scant or absent, and its visualization by electron microscopy even at high magnification is hampered by its exiguity owing to the very short diffusion paths of Ca^{2+} and Ti^{4+} ions, respectively that render any potential reaction zone extremely thin (Heimann, 2008).

The control of enhanced apatite nucleation by calcium titanate surfaces has been explained by the presence of an epitaxial structural relation between the (022) lattice plane of calcium titanate and the (00.1) lattice plane of hydroxylapatite (Wei et al., 2007; Figure 2.66). The (022) plane of the perovskite structure is defined by

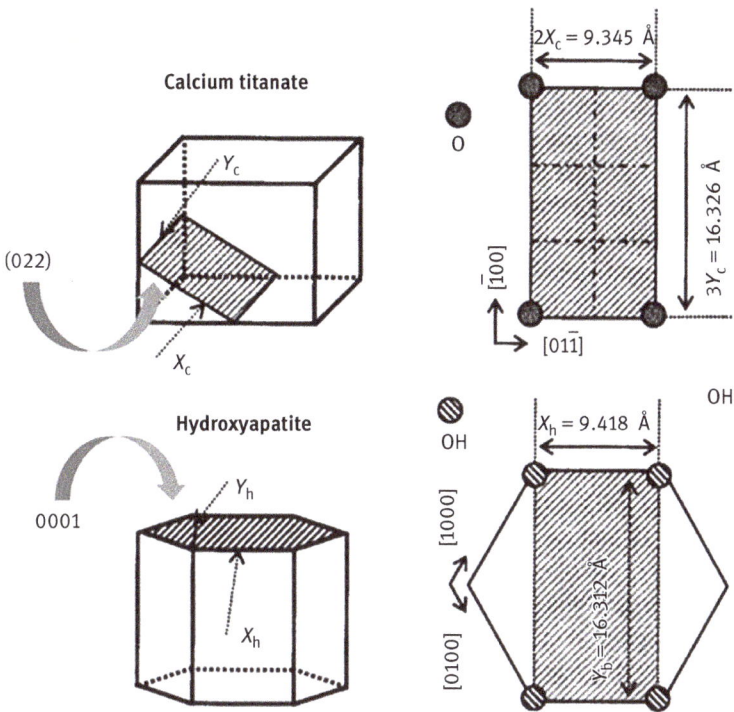

Figure 2.66: Epitaxial relation between (022) of calcium titanate (perovskite) and (00.1) of hydroxylapatite (Wei et al. 2007). © With permission by Elsevier.

the position of the oxygen atoms whereas the (00.1) plane of hydroxylapatite is defined by the position of hydroxyl ions. A 2D lattice match exists that is characterized by the fits of the lattice distances $2X_P \approx X_H$ (mismatch: 0.8%) and $3Y_P \approx Y_H$ (mismatch: 0.09%). Hence, there is a suggestion that biomimetic formation of hydroxylapatite during incubation in SBF and during osseoconductive integration in the bone bed *in vivo* relies on interfacial molecular recognition determined by the degree of epitaxial crystal lattice matching between hydroxylapatite and perovskite.

Bioactive glass and silicate-based ceramic coatings showed sufficient efficiency of the apatite mineralization in SBF exceeding that of hydroxyapatite. The apatite forming ability and dissolution rate mainly depend on the chemical composition and structure of these coatings as well as the coating process that control the total porosity and surface area of the coatings, in contact with SBF (Brunetto et al., 2019).

Some research has been performed aimed to utilizing *sphene* (calcium titanium silicate, $CaTiSiO_5$) as a biocompatible osseoconductive coating on titanium-based implants (Ramaswamy et al., 2009; Biasetto et al., 2016). This novel type of coating has revealed suitable chemical stability, excellent adhesion strength, and reasonable osseoconductivity. Hence, the material appears to be an potential candidate to be used as bioactive coatings for orthopedic and dental implant application (Elsayed et al., 2018) in as much as it may advantageously combine the effect of higher bone apposition rate and more rapid remodeling of bone attributed to released SiO_4^{4-} anions and the accelerating effect of Ca^{2+} cations on precipitation of bone-like apatite *in vivo*.

Occasionally, biocompatible *calcium silicate phases* such as larnite (β-Ca_2SiO_4) and wollastonite (α-$Ca_3[Si_3O_9]$) have been investigated for their application as ceramic biomaterials since calcium silicate glasses and ceramics were found to bond easily to living bone, and to form, by a biomimetic process, an apatitic surface layer when exposed to SBF (Garcia et al., 2018; Buga et al., 2019). In fact, even P_2O_5-free CaO–SiO_2 glasses were shown to form a tight bond with bone mediated by a thin apatite layer formed adjacent to the bone tissue. The cyclic silicate trimer $[Si_3O_9]^{6-}$ group of pseudowollastonite (β-$Ca_3[Si_3O_9]$) possibly acts as an active site for heterogeneous, stereo-chemically promoted nucleation of hydroxylapatite in SBF. In addition, histological investigations provided evidence that silicon may be allied to the initiation of mineralization of preosseous tissue in periosteal or endochondral ossification, presumably through Si–OH functional groups that are known to induce apatite nucleation by providing bonding sites for cation-specific osteonectin attachment complexes on progenitor cells.

It has been further suggested that the specific crystal structure of a bioceramic material controls largely the attachment, viability, and osteogenic differentiation of hMSCs. It was observed that the two calcium silicate polymorphs wollastonite and pseudowollastonite behave differently in a biological environment, presumably based on their different crystal structure that controls their solubility in the ECF. Indeed, the strained silicate ring structure of pseudowollastonite lends itself to higher solubility compared to the more stable, open silicate chain structure of wollastonite, thus proving to be more osseoconductive. This confirms the notion that the crystal structure is among the fundamental parameter to be considered in the intelligent design of pro-osteogenic, partially resorbable bioceramics (Bohner and Miron, 2019). Additional studies have revealed upregulation of the vascular endothelial growth factor as well as BMP-2 and nitric oxide expression. Calcium silicate-based scaffolds, cocultured with hMSCs and human umbilical vein endothelial cells showed significantly enhanced vascularization and osteogenic differentiation *in vitro* and *in vivo*, which appear to indicate that utilization of calcium silicate bioceramics may be a promising way to enhance bone regeneration.

Calcium magnesium silicates such as diopside ($CaMgSi_2O_6$: Sainz et al., 2010), merwinite ($Ca_3MgSi_2O_8$; Ardakani et al., 2011) and monticellite ($CaMgSiO_4$; Kalantari and Naghib, 2019) were explored for biomedical applications (Pouroutzidou et al., 2019). In general, sufficient mechanical stability, adhesion strength, and biocompatibility of coatings for metallic implants were observed. Diopside has shown significant potential to form an apatitic layer during incubation in SBF (Salahinejad and Vahedifard, 2017). The bonding strength of this potential bioceramic material to the implant material was about 350 ± 7 MPa, at least one order of magnitude higher than that reported for hydroxylapatite. The diopside samples revealed a reasonable fracture toughness of 4 ± 0.3 MPa·$\sqrt{m}$ (Kazemi et al., 2017). Doping of diopside with monovalent ions such as Li, Na, and K revealed improved viability and proliferation of osteoblast-like MG-63 cell cultures (Rahmani and Salahinejad, 2019). Nanostructured merwinite coatings not only improved the corrosion resistance of the implant substrate but also enhanced the bioactivity, mechanical stability, and cytocompatibility of AZ91Mg alloy (Razawi et al., 2014). The sorosilicate baghdadite ($Ca_3Zr[(Si_2O_7)O_2]$) has been suggested as suitable biocompatible coating for Ti-based implants (Huang et al., 2011). Other potentially advantageous silicate-based bioceramics such as melilite-type hardystonite ($Ca_2Zn(Si_2O_7)$) (Wu et al., 2005; Gheisari Dehsheikh and Karamian, 2016) and åkermanite ($Ca_2Mg(Si_2O_7)$) (Hoppe et al., 2011; Diba et al., 2014) are under scrutiny, owing to their ability to inhibit bone resorption and promote osteoblast differentiation, osteocalcin secretion, and ALP activity.

Lithium disilicate ($Li_2Si_2O_5$) glass–ceramic was introduced, in 1998, as a material for dental applications by Ivoclar Vivadent Corporation, Lichtenstein. The material was obtained in the shape of hot-pressed ingots and could be processed by a procedure similar to the lost-wax technique used for dental metal alloys by either press or CAD/CAM technology. The material shows optimum distribution of small needle-shaped crystals in a glassy matrix with low porosity. The glass ceramic is ideally suited for the fabrication of monolithic or veneered restorations in the anterior and posterior dental regions owing to its color and sheen resembling those of natural teeth. The outstanding mechanical performance of the material is based on a combination of excellent flexural strength around 500 MPa and comparatively high fracture toughness in the range of 2.5 MPa·$\sqrt{m}$ that can be improved by reinforcing with zirconia particles (Bergamo et al., 2019). Dental indications include veneers (≥ 0.3 mm), inlays and onlays, occlusal veneers, partial crowns, minimally invasive crowns (≥ 1 mm), implant superstructures, hybrid abutment solutions, and three-unit bridges up to the second premolar as the terminal abutment. The survival rate of dental restorations of all kinds based on lithium disilicate was found to be an impressive > 99% over an observation time of 10 years (Malament, 2015). However, other studies yielded less optimistic results. A meta-analysis of 12 studies reported that the existing evidence indicates excellent short-term survival rates for lithium disilicate single crowns, but that the evidence for their medium-term survival is limited (Pieger et al., 2014). For lithium disilicate-fixed dental prostheses, the evidence for short-term survival is fair, although limited, but the evidence for medium-term survival is not promising. The majority of failures in both types of restorations were reported in the posterior region. Details on properties, history, test results, and applications of lithium disilicate glass ceramic can be found in Zarone et al. (2016). The firing regime and cooling protocol was found to influence development of residual stresses during crystallization firing of lithium silicate glass–ceramics (Wendler et al., 2019).

Owing to their hardness, corrosion resistance and intrinsic biocompatibility, *titanium nitride* coatings are used in cardiology in ventricular-assist devices for patients with heart failure and for pacemaker leads. In neurology, TiN-coated electrodes are investigated for the development of permanent implanted devices for the treatment of, for example, spinal cord injury as well as for implantable stimulation electrodes (Canillas et al., 2019). In addition, TiN is used in dentistry to coat CoCrMo dental implants to reduce the release of cytotoxic ions, and for the esthetic effect of its golden color (van Howe et al., 2015). *In vitro* platelet adhesion experiments were performed to examine the interaction between blood and TiN. On coated samples, platelets were seen as isolated entities without any significant spreading (Subramanian et al., 2011) and showed encouraging blood tolerability with a hemolysis percentage close to zero, suggesting application as surface coatings for LTIC-based artificial heart valves.

For a long time, thin surface films of TiN were deposited on endoprosthetic implants to prevent their abrasion and corrosion and to prevent fretting of contacting implant surfaces (Steinemenan, 1972; Subramanian et al., 2011). The load-bearing surfaces are coated with a preferably 8–10 µm thick layer of the biologically inert, abrasion resistant material TiN that is harder than the implant substrate to prevent wear as well as leaching of ions out of the implant material. Besides the suggested beneficial effect of TiN coatings on bearing surfaces of cemented prostheses, the coating might also be beneficial at the bone-implant surface of noncemented prostheses owing to its biological inertness.

Thin oxide films such as Ta_2O_5 (Chang et al., 2014) or Nb_2O_5 deposited by magnetron sputtering or electron beam deposition exhibit high hardness, high resistance against corrosion, and reasonable osseoconductive properties, possibly leading to an increased service life of dental, maxillofacial or orthopedic implants (Pauline and Rajedran, 2014; Pradhan et al., 2016; Rahmati et al., 2016). The as-deposited films are generally amorphous and require annealing at temperatures beyond 400 °C to crystallize. The interaction of osteoblasts was found to be correlated to the roughness of nanosized surface structures of Nb_2O_5 coatings on polished Ti surfaces (Eisenbarth et al., 2006).

2.3 Polymers (Matthias Schnabelrauch)

2.3.1 Introduction

Polymeric materials currently represent the largest class of biomaterials used in a variety of medical applications including surgical sutures, orthopedic bone cements, osteosynthesis materials, dental adhesives and composites, cardiovascular devices, soft tissue implants, wound closures, drug and gene delivery carriers, biosensor components, and many others.

A great advantage of polymers is their wide structural variability achievable by using a multifarious toolbox to generate target-specific polymeric materials. Major instruments of this toolbox are the availability and selection of suitable monomers containing polymerizable groups and diverse other functionalities, the use of co-monomers and compatible fillers, the production of blends with other polymers, the performance of cross-linking surface grafting processes and, last but not least, the adjustment of the molecular mass of the polymer. Varying these instruments of the toolbox in a proper way allows matching the desired application properties of a specific polymeric biomaterial to a wide range of biomedical requirements. A further advantage of polymers is that they can be easily processed into required shapes using manufacturing techniques either known for a long time such as injection molding and solvent casting or recently developed additive manufacturing processes such as 3D printing or stereolithography (see Chapter 3.3). The surface of polymers can be modified by different physical or chemical treatments to improve or impede the attachment of other molecules, cells or tissues, and by this means, controlling the polymer's bioactivity. Furthermore, bioresorbable polymers offer the possibility to manufacture temporary implants that disappear after having fulfilled their function in the human body (e.g., as drug reservoir). In such a case, resorbable or biodegradable polymers are accessible which can be degraded by enzymatic or hydrolytic processes, forming nontoxic degradation products.

In this chapter, a general survey on polymers and their biomaterial-relevant properties will be presented and most widely employed polymers used in medical applications introduced in more detail. Additional information can be found in a recently published volume on functional biopolymers relating to syntheses, properties, and biomedical applications of biopolymers including hemocompatible, ophthalmic, and stimuli-responsive polymers (Mazumder et al., 2019).

2.3.2 Classification of polymers

There exist several ways to classify polymers depending on their source or on various structural features such as type and composition of monomers, chain structure, tacticity, molecular forces or other characteristics (Parisi et al., 2015). An exemplary classification is given in Table 2.16.

2.3.3 Molecular structure

Polymers (polus *Greek* many + merizo *Greek* part) are macromolecules generated by combining a large number of low molecular mass molecules, so called monomers. These monomers contain one or more reactive functional site that react with other monomers to build up the polymer in a chemical or enzymatic polymerization reaction.

Table 2.16: Classification of polymers in terms of source and structural parameters (various numbered polymer classes can also be found in Figure 2.67).

Classification	Explanation	Examples
Source	– *Natural polymers*, occurring in plants or animals – *Semisynthetic polymers*, derived by chemical modification of natural polymers – *Synthetic polymers*, "man-made" artificial polymers synthesized in the laboratory	Collagen Cellulose acetate Polyurethane
Monomer type	– Homopolymers (1): contain only a single type of monomer – Co-/terpolymers (2, 3): comprising two or more monomers – For they are distinguishable by the arrangement of the repeating unit in the chain: – Alternating copolymer (2a) – Block copolymer (2) – Graft copolymer (2c) – Random copolymer (2d)	Polystyrene Poly(L-lactide- *co*-glycolide)
Chain structure	– Linear (4): long straight chains – Branched (5): linear side chains of different length along the main chain – Cross-linked (6): 3D network of linear chains	Polystyrene Low-density polyethylene Melamine
Tacticity	– Isotactic (7): pendant group arranged on the same side of the backbone – Syndiotactic (8): alternative arrangement of pendant group – Atactic (9): pendant group arranged randomly	Poly-L-lactide Poly(meso-lactide)
Molecular forces	– Thermoplastics: intermolecular forces hold polymer chains together; soften on heating and can be molded – Thermosets: their individual chains are bonded covalently, once set, their form cannot be shaped anymore – Elastomers: their chains are elastic; they can be stretched	Polypropylene Melamine Natural rubber

According to its molecular mass, a polymer molecule can contain hundreds up to several millions of repeating units formed by combining a certain number of monomers. Polymeric substances with molecular masses below about 10,000 g mol^{-1} are often called oligomers (oligoi *Greek* few), although there is no strict definition at which molecular mass a molecule is named a polymer or an oligomer.

The repeating units of polymers are predominantly connected by covalent bonds characterized by relative high binding energies, fixed binding angles, and short distances (0.11–0.16 nm). The strength of covalent bonds is affected by the electronegativity of the binding atoms. Atoms with equal electronegativity will form nonpolar covalent bonds such as C–C bonds, whereas from atoms with different electronegativities polar covalent bonds result as in the case of C–N bonds. In general, covalent

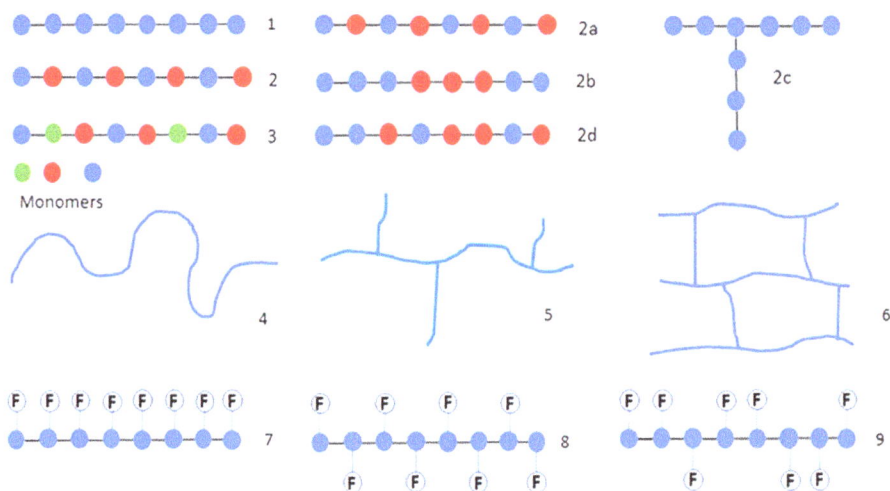

Figure 2.67: Classes of polymers (for numbered polymers, see Table 2.16).

bonding includes several kinds of interaction as σ-bonding in polymers allowing free rotation around the bond axis. In contrast, π-bonding present in many monomers and conjugated polymers leads to double bonds characterized by a smaller bonding length and a loss of free rotatability. In adjacent polymeric chains or different segments of a polymeric chain, intermolecular forces or van der Waals bonds can also occur. In addition, the formation of hydrogen bonds in polymers containing units with heteroatomic O, N, or S functionalities (e.g., carbonyl, amino, or sulfonyl groups) can strongly influence the properties of these macromolecules.

2.3.4 Molecular mass and molecular mass distribution

Most polymers are characterized by high molecular mass often ranging from 10^4 to 10^7 g/mol. In this context, it has to be noted that the molecular mass of a defined polymer is not a unique value as it is for most low-molecular-mass organic compounds. In the overwhelming majority of cases, polymerization reactions result in a mixture of single macromolecules with differing number of containing monomers, that is, a polymer with average molecular mass and a distinct molecular mass distribution is obtained. Based on the different ways to determine the average molecular mass of a polymer, two useful definitions are the number-averaged molecular mass M_n (Figure 2.68; eq. (2.1)) and the mass-averaged molecular mass M_m (Figure 2.68; eq. (2.2)), where N_i is the number of moles of species i and M_i is the molecular mass of this species i (Guarino et al., 2015).

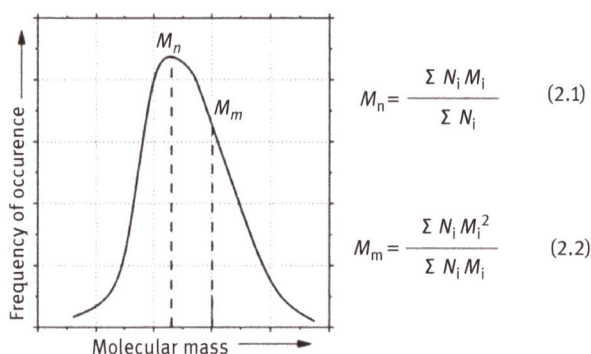

The plot shows Frequency of occurence (y-axis) versus Molecular mass (x-axis), with M_n and M_m marked.

$$M_n = \frac{\Sigma\, N_i\, M_i}{\Sigma\, N_i} \qquad (2.1)$$

$$M_m = \frac{\Sigma\, N_i\, M_i^2}{\Sigma\, N_i\, M_i} \qquad (2.2)$$

Figure 2.68: Molecular mass distribution plot showing number- and mass-averaged molecular masses that can be estimated by eqs. 2.1 and 2.2, respectively.

The M_m/M_n ratio is the so-called dispersity index (more commonly known as polydispersity index, PI), a measure for the width of the molecular mass distribution of a given polymer. The molecular mass of polymers and the PI strongly influence important application properties of a polymer such as brittleness, tensile strength, solution and melt viscosity, and extrudability.

There exist different analytical methods to estimate the molecular mass of a polymer including gel permeation chromatography (GPC), viscosimetry and osmometry techniques, ebullioscopy, cryoscopy, sedimentation or light scattering experiments, or end-group analysis. The most common technique widely used in polymer characterization is GPC, a variant of size exclusion chromatography. This technique uses the passage of the polymer solution over a column of porous beads. Whereas high-molecular mass polymer molecules are excluded from being incorporated into the pores of the beads and thus, elute first, lower molecular mass molecules pass the column with increased elution times. The use of low dispersity polymer standards allows a correlation between the elution (= retention) time of the sample with its molecular mass. Using both multiangle light-scattering and differential refractive index detectors in parallel, the determination of absolute molecular masses is possible.

2.3.5 Solid state of polymers

As large macromolecules, many polymers are able to assume different conformations by rotation of valence bonds. Substituents present in the repeating unit of a polymer (e.g., the phenyl ring in polystyrene) may be arranged in different ways around the extended polymer backbone. This phenomenon is called *tacticity*. In Figure 2.67, the different forms of polymeric arrangements (isotactic, syndiotactic, atactic) are schematically depicted. As a consequence, isotactic and syndiotactic polymers may crystallize under suitable conditions, atactic polymers are usually amorphous.

Polymers can crystallize upon cooling from the melt, mechanical stretching or solvent evaporation. Because the degree of crystallinity typically ranges from 10% to 80%, and both crystallized and amorphous phases coexist, such polymers are often called "semicrystalline." The degree of crystallinity together with the size and orientation of the molecular chains tremendously affects the optical, mechanical, thermal, and chemical properties of a polymer.

Conventional methods to determine the degree of crystallinity of polymers are density measurements, differential scanning calorimetry (DSC), XRD, spectroscopic procedures like infrared spectroscopy, and NMR spectroscopy. In addition, the distribution of crystalline and amorphous regions in polymers can be evaluated by microscopic techniques such as polarized microscopy and transmission electron microscopy.

2.3.6 Thermal and mechanical behavior

Although some polymers are completely amorphous, most are semicrystalline containing crystalline areas surrounded by disordered regions. The presence of crystalline areas in the polymer usually results in enhanced mechanical properties, unique thermal behavior, and increased fatigue strength, making such materials favorable for most medical applications.

During cooling, a polymer melt with a highly random molecular structure generates during crystallization an ordered solid phase. The melting transformation is the reverse process which take place when a polymer is heated and the glass transition (see box, Chapter 3.3.2) occurs with amorphous or noncrystallizable polymers. In semicrystalline polymers, the noncrystalline regions undergo the phenomenon of the glass transition, while the crystalline regions are affected by the melt phenomenon. Although complete crystallization is not achieved in polymers, it is of fundamental importance because it is associated with partial alignment of polymeric molecular chains folding together and forming ordered regions, so called lamellae and subsequently, growing spherulites. Polymers with some crystallinity also exhibit a melting temperature (T_m) owing to melting of the crystalline phase. Usually, melting occurs over a narrow temperature range.

The phenomenon of the glass transition occurs in amorphous and semicrystalline polymers due to a reduction in motion of large segments of molecular chains when the temperature of the polymer melt decreases. Upon cooling, the glass transition corresponds to a gradual transformation from a liquid into a rubbery material, and then into a rigid solid. The last step corresponds to the glass transition characterized by the glass transition temperature (T_g). In this rubber-like amorphous phase, a polymer is not liquid but softens and deforms easily.

The mechanical behavior of polymers can be roughly categorized into the three characteristics brittle, ductile, and rubbery (Guarino et al., 2015):

-   Brittle polymers show high elastic moduli and high ultimate tensile strength but low ductility and toughness. Their T_g is much higher than room temperature. Example: PMMA.
-   Ductile polymers are semicrystalline with T_g values below room temperature. These polymers have generally lower strength and elastic modulus but greater toughness than brittle polymers. Examples: polyethylene (PE), polytetrafluoroethylene (PTFE).
-   Rubbery polymers show low elastic moduli and T_g below room temperature. They can return to their original shape following high extensions because cross-linking prevents significant polymer chain translations. Example: polybutadiene

The three categories roughly correspond to the classification of polymers according to acting molecular forces (see Table 2.16) into thermoplastics, thermosets, and elastomers. This correlation is shown in Figure 2.69.

Investigation of the thermal properties of polymers is normally performed by thermal analysis, particularly DSC. Changes in the compositional and structural parameters of the polymer material usually affect its melting or glass transition temperatures. Further analytical techniques include differential thermal analysis and dynamic mechanical thermal analysis (DMTA). DMTA is the most common technique used to characterize viscoelastic behavior and it is also another important tool to understand the temperature dependence of the mechanical behavior of polymers. Furthermore, it is a characterization technique to measure storage modulus and glass transition temperature, and also to confirm polymer cross-linking.

2.3.7 Biostability and biodegradability

Polymeric materials surrounded by living tissue are subject to a variety of physical, chemical, and biological influences including abrasion, mechanical deformation, dissolution, chemical, and biological agents (water, acids, bases, salts, proteins, or enzymes) resulting in various degradation processes. These changes are in many, especially long-term clinical applications, undesirable and a result of the strongly aggressive medium present in the living medium of the human body. For this reason, even polymers which have been intended to be biostable, normally will show some degree of degradation *in vivo*. Nevertheless, it can be stated that homochain polymers with backbones solely containing carbon atoms or rings are essentially biostable. Examples of such polymers are PE, polypropylene (PP), PTFE, poly

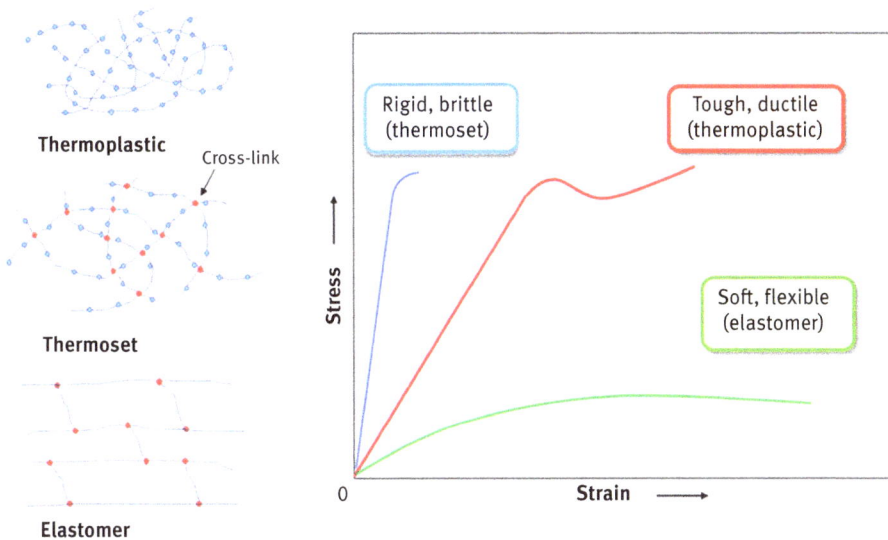

Figure 2.69: Stress versus strain curves of three different polymer classes characterized by different molecular interactions (Guarino et al, 2015). © With permission by Springer Nature.

(etheretherketone) (PEEK), or poly(vinylchloride) (PVC) that survive in the body under normal circumstances for many years.

Nevertheless, during the last decades, the interest in biocompatible polymers designed to degrade has increased dramatically. Biodegradable polymers have several advantages. Owing to their gradual degradation *in vivo* they do not permanently elicit foreign body reactions and all traces of artificial material, which always pose a potential threat in terms of material-related infections, completely disappear. Furthermore, there is no second surgery necessary to remove implants after fulfilling their temporary function, for example, as osteosynthesis devices for bone fixation.

Degradable biomaterials also have a great potential in advanced concepts of regenerative medicine such as tissue engineering. They may serve as temporary scaffolds for the formation of new viable 3D tissue. Biodegradation is an essential factor since scaffolds should preferably be absorbed by the surrounding tissues without the necessity of a surgical removal.

Both natural and synthetic polymers are used in medicine as biodegradable materials. Depending on their structure, three different routes are known (Figure 2.70) to form degraded metabolizable, mainly water-soluble products (Schnabelrauch, 2018).

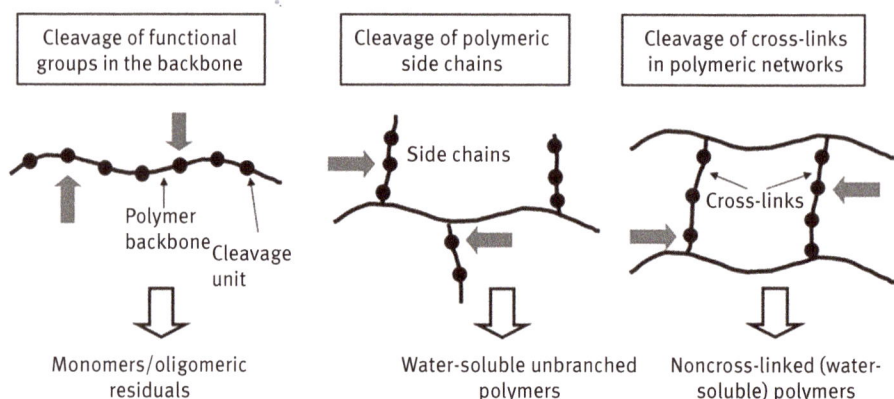

Figure 2.70: Degradation routes of biodegradable polymers (Schnabelrauch, 2018).
© With permission by Springer Nature.

- The polymer has degradable side chains that are hydrolyzed in the body providing residual polymers with hydrating groups (e.g., hydroxyl, carboxyl groups) making the polymer water soluble and excretable from the body.
- Cross-linking of a water-soluble polymer with hydrolyzable cross-linking units. Hydrolysis of the cross-links in the body results in the formation of the water-soluble polymer which can also be excreted.
- Direct incorporation of the hydrolyzable unit in the backbone of the polymer chain leading to slow hydrolysis of the polymer chain with progressing time. Shorter polymer fragments are formed which at a distinct point of time become water-soluble and can be excreted or metabolized from the body.

The main material-related factors that determine the overall rate of the degradation process are the chemical stability of the hydrolytically susceptible groups in the polymer, the hydrophilic/hydrophobic character of the repeating unit, the morphology of the polymer including porosity, the molecular mass, the presence of additives, and the geometry of the device implanted in the body. In addition, there are several media-related factors influencing the degradation rate including pH, temperature, the presence and concentrations of solutes, salts, enzymes, and microbes, and the occurrence of stress or strain.

The chemical composition of the polymer chain, that is, the type of hydrolytically susceptible units within the backbone is probably the key parameter determining the rate of degradation. A survey of common hydrolytic bonds appearing in polymer backbones and corresponding examples used in medical applications are given in Figure 2.71.

Normally, anhydride and orthoester units hydrolyze faster than ester links that in turn hydrolyze faster than urethane or amide groups. Based on the knowledge of hydrolytic susceptibility of the polymer structure, it is possible to adjust the degradation

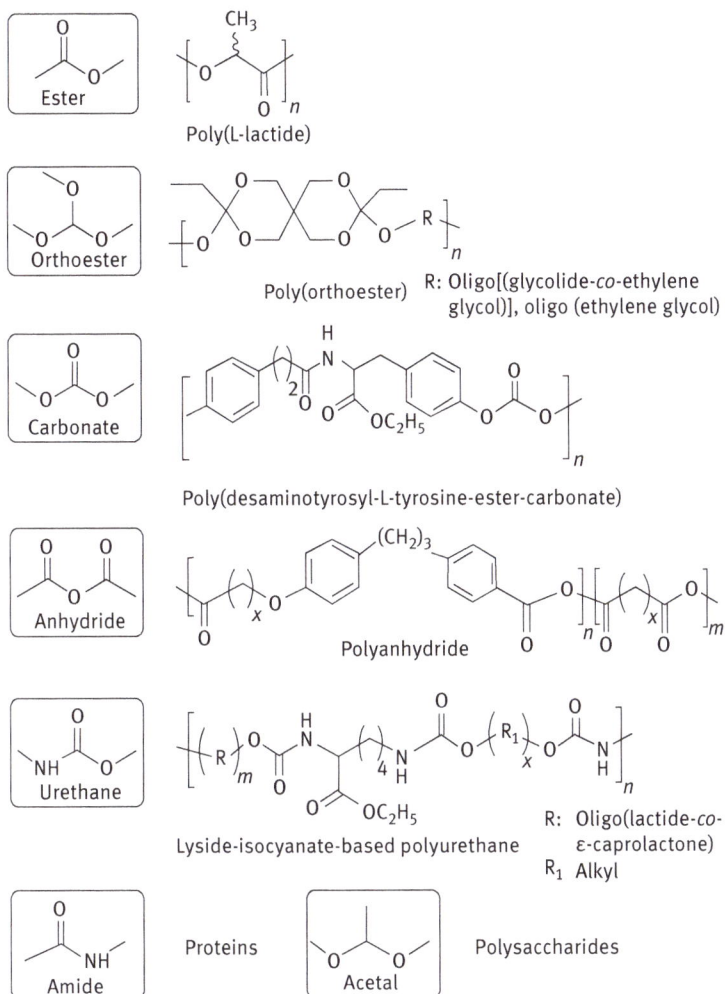

Figure 2.71: Various types of hydrolysis-sensitive bonds within polymer backbones (left) and examples of corresponding biodegradable polymers containing these bonds (right).

rate of a polymer to the requirements of the special medical application (design to degrade). The morphology of the polymer is another important parameter affecting polymer degradation. As already mentioned above (see Chapter 2.3.6), polymers can be either amorphous or semicrystalline. In the crystalline state, the polymeric chains are densely packed, resisting the penetration of water. Consequently, backbone hydrolysis tends to occur in the amorphous regions of a semicrystalline polymer and at the surface of the crystalline domains. Among polylactones, poly(L-lactide) and polyglycolide (PGA) are typical semicrystalline polymers whereas poly(D,L-lactide) is amorphous. This is the reason why poly(L-lactide) and poly(D,L-lactide), both having identical chemical compositions, differ remarkably in their degradation rate and why the amorphous poly(D,L-lactide) normally degrades much faster.

During hydrolytic degradation, low-molecular-mass water-soluble oligomers are formed by cleavage of polymer chains and released into the media, resulting in polymer mass loss. Two different basic mechanisms, surface erosion and bulk erosion, are known to describe the polymer material degradation (Figure 2.72) (Göpferich, 1997). Surface erosion takes place when the hydrolytic degradation rate is much higher than the diffusion rate of water within the polymer. The hydrolytic degradation occurs then solely on the polymer surface. In contrast, bulk erosion takes place when the hydrolytic degradation rate of the polymer is much lower than the diffusion of water within the polymer. As a consequence, hydrolytic degradation occurs nearly homogeneously, irrespectively of the depth from the material surface. This might be an open problem because in the final stage of degradation a collapse of the complete polymer structure may occur, resulting in the release of many polymeric fragments.

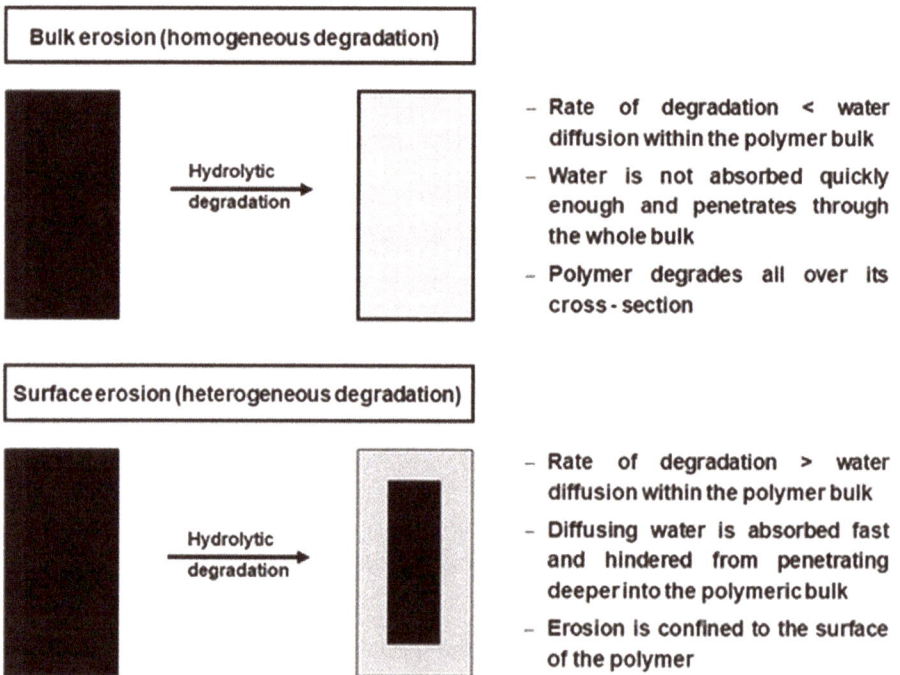

Bulk erosion (homogeneous degradation)

Hydrolytic degradation

– Rate of degradation < water diffusion within the polymer bulk
– Water is not absorbed quickly enough and penetrates through the whole bulk
– Polymer degrades all over its cross-section

Surface erosion (heterogeneous degradation)

Hydrolytic degradation

– Rate of degradation > water diffusion within the polymer bulk
– Diffusing water is absorbed fast and hindered from penetrating deeper into the polymeric bulk
– Erosion is confined to the surface of the polymer

Figure 2.72: Bulk and surface erosion of degradable polymers (Schnabelrauch, 2018). © With permission by Springer Nature.

2.3.8 Natural polymers

Although synthetic polymers have been used extensively in medical applications over the last decades, there is a steadily growing interest in nature-derived polymeric materials (Yannas, 2004), frequently called biopolymers. Natural polymers offer the advantage of being very similar to macromolecular substances of the body, for example, concerning the ECM. Problems of toxicity and stimulation of chronic inflammation and often even the lack of recognition by cells, which will frequently be triggered by synthetic polymers, may be suppressed by natural polymers. Further favorable characteristics of natural polymeric materials include facilitating cell attachment, enhancing the mechanical properties of synthetic polymers, and binding and protecting bioactive endogenous substances. However, natural polymers have also serious disadvantages impairing their extensive use as biomaterials. With regard to their biological activity, it has to be noted that natural polymers, especially proteins, may have a substantial immunogenic potential. Furthermore, due to the fact that natural polymers are produced by living organisms (plants, animals, microorganisms), their availability in a required, always identical batch-to-batch structural composition and purity is often a serious challenge. Finally, processing and other technological manipulations are often more elaborate for natural polymers compared to synthetic ones.

Among the different groups of natural polymers, polysaccharide- and protein-based materials in particular are currently being used as biomaterials. Members of both classes frequently used as biomaterials will be discussed in the next chapters.

2.3.8.1 Polysaccharides

Polysaccharides are polymeric carbohydrates composed of long chains of monosaccharide units bound together by glycosidic linkages. Their structure ranges from linear to highly branched. Polysaccharides are frequently of heterogeneous nature, containing slight modifications of the repeating unit. Polysaccharides containing only one type of monosaccharide as repeating unit are called homopolysaccharides. Heteropolysaccharides are composed of at least two different repeating units. In living organisms, polysaccharides function as storage or structure materials. GAGs are a group of complex, unbranched, negatively charged polysaccharides composed of disaccharide repeating units. GAGs are components of many biological macromolecules, often covalently added to proteins, forming proteoglycans. They are important constituents of the ECMs of many tissues and possess important biological functions in blood coagulation, central nervous system development, wound repair, infection, growth factor signaling, morphogenesis, and cell division, differentiation, and migration.

Polysaccharides including GAGs frequently used as biomaterials are described below.

Cellulose is a high-molecular-mass, linear polysaccharide composed of β-1,4-linked D-glucose units (Figure 2.73). In the cellulose chain, the glucose units form six-membered pyranose rings joined by an acetal (glycosidic) linkage between the C-1 carbon atom of one pyranose ring and the C-4 of the next ring. Since water is eliminated during formation of the acetal linkage from two glucose molecules, the repeating unit is called an anhydroglucose unit.

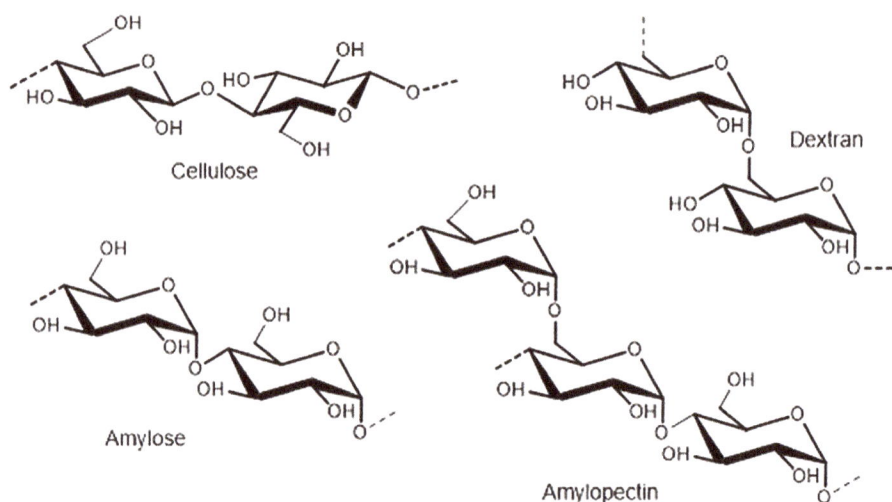

Figure 2.73: Glucose-containing polysaccharides used as biomaterials.

Cellulose is the most abundant organic polymer in nature as the main constituent of the cell walls in green plants. It is also formed by some species of bacteria and algae. In plants, cellulose is synthesized at the plasma membrane of cells by the enzyme cellulose synthase starting from uridine diphosphate-glucose. The main technical sources of cellulose are wood, cotton linters, and other plant fibers. Besides plants, bacteria can serve as a source to produce highly purified cellulose. The molecular mass of cellulose depends on its origin and on the processing treatment (Gilbert and Kadla, 1998).

Cellulose adopts an extended and rather stiff rod-like conformation. The multiple hydroxyl groups on the glucose ring of one chain forms hydrogen bonds with oxygen atoms on the same or on a neighboring chain, holding the chains firmly together side-by-side and forming microfibrils. As a consequence, cellulose can have highly ordered crystalline structures providing cellulosic fibers with high tensile strength. The native polymer exists normally as cellulose I. Cellulose II, the thermodynamically more stable polymorph, is irreversibly formed after regeneration of cellulose I. After special treatments of the polymer, two additional polymorphs are formed.

Although hydrophilic due to numerous hydrogen bonds, cellulose is insoluble in water and most common organic solvents. Cellulose can be dissolved in several rather nonconventional solvents like Schweizer's reagent (Cuprammonium hydroxide $[Cu(NH_3)_4(H_2O)_2](OH)_2$), N-methylmorpholine N-oxide, lithium chloride/N,N-dimethylacetamide mixtures, or various ionic liquids. The former solvents are also used to produce regenerated cellulose from cellulose manufacturing processes (dissolving pulp from sulfite or sulfate processes).

As a polyalcohol, cellulose can be derivatized at the free hydroxyl groups by typical esterification, etherification and also oxidation reactions. Due to the high crystallinity, frequently an adequate preactivation of the cellulosic starting material or homogeneous reaction conditions are needed to obtain reproducible results. Today, several ethers and esters of cellulose are used as high-tonnage products in domestic as well as in construction, textile, and chemical industries. These derivatives include methyl, ethyl, hydroxyethyl or carboxymethyl ethers of cellulose, and cellulose acetate or acetate-butyrate esters. Most of the ethers are water soluble and gel forming whereas ethyl cellulose and the mentioned esters are soluble in common organic solvents and thermally processable thermoplastics.

Cellulose and its derivatives have also found considerable interest in pharmaceutical and medical applications as tablet binder, drug delivery formulations, hemodialysis membranes, wound coverages, or artificial skin in burn therapy or ulcer treatments.

Partial oxidation of cellulose has been found to provide a biomaterial with controlled degradation characteristics. The oxidation of regenerated cellulose using nitrogen dioxide as oxidant selectively transforms primary hydroxyl groups of cellulose into a carboxyl group. The resulting product is biodegradable under *in vivo* conditions in only a few weeks (Stilwell et al., 1997). The oxidized cellulose material is used clinically since the 1940s as hemostatic membrane and absorbable adhesion barrier in a variety of surgical procedures.

A promising future biomaterial is bacterial cellulose (BC), produced by several types of aerobic bacteria (Klemm et al., 2006; Keshk and El-Kott, 2017), mainly belonging to the *Acetobacter* genus. BC is a highly pure form of cellulose. It consists of microfibrils arranged in a 3D web-shaped structure, providing a porous geometry. Compared to common plant cellulose, it has higher crystallinity (80–90%), degree of polymerization (up to 8,000), and water absorption capacity. Future biomedical applications may include wound healing materials, artificial skin and blood vessels, scaffolds for tissue engineering, and drug delivery devices.

Starch, in contrast to cellulose, is composed of D-glucose molecules linked together by α-glycosidic bonds. It consists of two types of macromolecules, the mostly unbranched α-1,4-linked amylose (20–30%) and amylopectin (70–80%) composed of longer α-1,4-linked D-glucose units bearing α-1,6-linked branches of shorter α-1,4-linked D-glucose units (Shogren, 1998) (Figure 2.73). In green plants, starch as a reserve polysaccharide is arranged as semicrystalline granules of different size. Starch becomes soluble in water when heated.

Like cellulose, starch can be derivatized using the same procedures. Hydroxyethyl starch (HES) is used clinically as a volume expander in intravenous therapy to prevent shock following severe blood loss by trauma or surgery. However, it was recently found that the use of HES may be associated with adverse effects such as anaphylactoid reactions and other critical side effects (Perner, 2012). In 2018, the European Medicines Agency recommended to suspend or strongly limit the marketing authorization of HES products in Europe. Other starch products are used in pharmaceutical industry as tablet disintegrant and binder.

Dextran, another D-glucose containing polysaccharide is composed of α-1,6-linked D-glucose units with α-1,2-, α-1,3-, and α-1,4-linked branches. Natural dextrans are reserve polysaccharides produced by microorganisms of the family *Lactobacillus* (species *Leuconostoc*, e.g., *L. mesenteroides* and *L. dextranicum*). Whereas the degree of branching and the molecular mass depend on the individual microbial strain, the production of nearly unbranched dextran is also known.

Depending on its molecular mass, dextran is water soluble forming viscous solutions. It is highly biocompatible, slowly degradable in the body, and excreted from it (only larger dextran molecules (>60 kDa) are excreted poorly from the kidney and remain in the blood for as long as weeks until they are metabolized). Previously, dextran has been used as plasma volume expander but serious anaphylactoid reactions observed have now strongly limited their use. Dextran has an antithrombotic effect and reduces blood viscosity. In microsurgery, dextran can be used to decrease vascular thrombosis by reducing erythrocyte aggregation and platelet adhesiveness. In pharmaceutical applications, it can act as lubricant and solubilizer in drug formulations (Dräger et al., 2011).

Like cellulose or starch, dextran can be chemically modified to obtain esters, ethers or oxidized derivatives (Heinze et al., 2006). Another common reaction is cross-linking of polysaccharides to adjust gel formation or water solubility, and to increase the mechanical stability of the formed materials. The cross-linking of dextran or other dextran derivatives with epichlorohydrin resulted in 3D stable networks which can act as stationary phase in gel filtration (Porath and Flodin, 1959). These materials are originally known under the trade name Sephadex™. Chromatographic applications include gel permeation, affinity and ion exchange chromatography, and immunoprecipitation.

Chitosan is a linear cationic polymer derived from chitin, the second most abundant polysaccharide after cellulose, and found in cell walls of some fungi, cuticles of insects, and in the exoskeletons of crustaceans. Whereas chitin is composed of β-1,4-linked *N*-acetyl-β-D-glucosamine (more precisely 2-acetamido-2-desoxy-β-D-glucose) repeating units, in chitosan more than the half of the *N*-acetyl groups are deacetylated. Therefore, chitosan can be denoted a copolymer of β-1,4-amino-β-D-glucosamine and randomly distributed β-1,4-*N*-acetyl-β-D-glucosamine building units (Dash et al., 2011). Chitosan itself can be isolated together with chitin from some *Morales* fungi such as *Mucor rouxii*. Technically, chitosan is produced by alkaline or enzymatic (chitin deacetylase) hydrolysis of chitin (Scheme 2.1) (Kurita, 2001). The degree of deacetylation (DD) of chitosan, which is a measure of molar fraction of glucosamine to *N*-acetyl glucosamines, normally ranges from 60% to 90%. The molecular mass of commercial chitosans is between 10 and 1,000 kDa.

Chitosan is crystalline with varying degree of crystallinity depending on the DD value. Since the primary amino group in chitosan is protonated below pH 6.0 in contrast to chitin, chitosan is soluble in diluted acids such as acetic, lactic, or

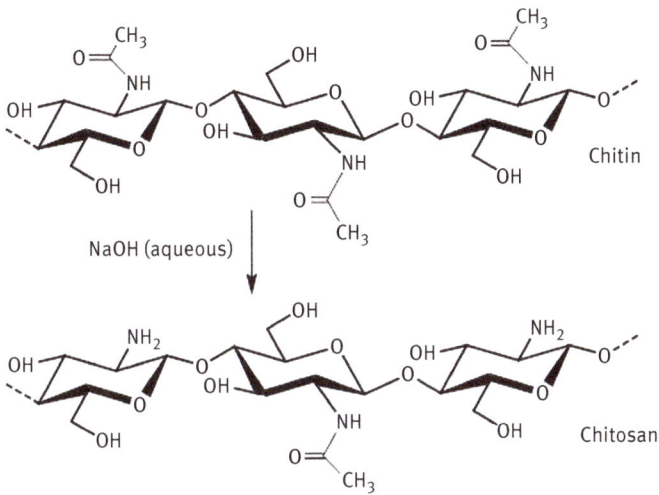

Scheme 2.1: Synthesis of chitosan by alkaline deacetylation of chitin.

hydrochloric acids. At higher pH, the ammonium salt is deprotonated and chitosan becomes insoluble. This solubility transition is strongly influenced by the DD value and chitosan with a DD of 50% might be soluble in water at neutral pH. Chitosan, like chitin, can be enzymatically degraded by cleavage of the glycosidic bond of acetylated repeating units in the presence of lysozyme, which is present in many tissues and in secretions such as blood, saliva, and tears.

The limited solubility of chitosan and its polyfunctional character brings special attention toward chemical modification of chitosan (Prabaharan and Tiwari, 2011). During the last decades, many derivatization reactions of chitosan have been studied to impart specific physicochemical properties, such as solubility, hydrophilicity, hydrophobicity, or charge density. In addition, quaternized chitosan derivatives have been extensively investigated. Despite the amino-group containing building blocks, chitosan has a good biocompatibility and a low immunogenicity. The highly positively charged molecule undergoes rather various ionic and hydrogen bonding interactions, thus qualifying chitosan as a bioadhesive, bacteriostatic, and hemostatic material. Its solubility in aqueous solutions allows processing of chitosan into films, fibers, coatings, and, after cross-linking, into gels, beads or nano- and microsized structures (Anitha et al., 2014).

Current biomedical applications include wound and hemostatic dressings (Chitoderm®, HemCon®, Celox™ Gauze, ChitoSAM™), CARGEL, a chitosan gel bioscaffold for the repair of cartilage lesions, or nerve conduits supporting nerve repair (Reaxon®). Chitosan is further used as a component in dental care products (e.g., Chitodent® toothpaste). Currently, its future use is widely discussed as carrier and permeation enhancer in drug delivery systems, and vaccines, and gene delivery vehicles (Agrawal et al., 2014).

Alginic acid or its salts, *alginates*, are a family of nonbranched, anionic polysaccharides composed of blocks of β-D-mannuronic acid (M) and blocks of its C-5 epimer, α-L-gulur-onic acid (G) linked via β-(1-4)-linked glycosidic bonds (Figure 2.74). The biopolymer can be regarded as true copolymer of homopolymeric M and G blocks interrupted by alternating M-G blocks (Dräger et al., 2011). Alginates can be harvested from ma-rine brown algae and are also produced from some bacteria (e.g., *Pseudomonas* or *Azotobacter* strains). Depending on the source of the alginate, the relative amounts of G and M monomers and their sequential arrangement can considerably vary within the biopolymer chain (Agrawal et al., 2014).

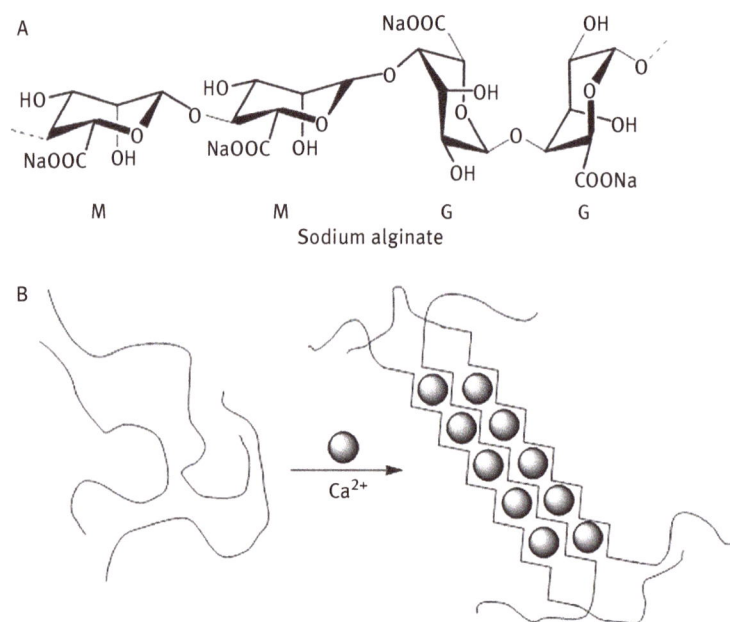

Figure 2.74: Chemical structure of (A) alginate and (B) mechanism and structure of calcium alginate gel formation (egg-carton model).

Alginate is well known for its gelling properties in aqueous solution. The gel forma-tion takes place by complexation of the carboxylic moieties of the polysaccharide and bivalent counter ions such as calcium, lead, and copper. Only low concentrations of divalent ions of >0.1% (w/w) are needed to initiate the gelling reaction. During gelling, alginate chains, preferable G blocks, interact ionically with the bivalent cations induc-ing helical chains generally known as "egg-box" model (Figure 2.74). Unlike di- or trivalent cations, monovalent cations and even magnesium ions do not form gels. The mechanical properties of the formed alginate networks are strongly influ-enced by the G content of alginate (Dräger et al., 2011). Alginate gels are relatively temperature stable and can be developed and set at physiologically relevant temperatures in various forms like pastes, films, meshes or beads.

Alginate is used also as food additive, thickener, emulsifier, and moisturizing agent in cosmetics. In biomedical applications, its importance is still one of the best options for cell encapsulation. Recently, alginate gels have also gained attention as bioactive macromolecules for drug delivery formulations. It is also used in dentistry and prosthetics as pastes or impression-making material. Furthermore, alginate has found attention in wound dressings for larger wounds and burns. As meshes or nonwoven structures, alginate is able to absorb large amounts of exudate from the wound bed. Simultaneously, it keeps the wound moist and can be more easily removed from the wound compared to other materials like, e.g., cellulosics. Currently, there are several alginate-based wound dressings clinically used (e.g., Algisite®M, Restore®, SeaSorb®, Askina® Sorb, or Kaltostat®) (Agrawal et al., 2014).

2.3.8.2 Glycosaminoglycans

GAGs, formerly also called mucopolysaccharides, are natural, negatively charged unbranched heteropolysaccharides composed of disaccharide repeating units. Normally, the repeating disaccharide unit is formed from D-glucuronic acid (GlcUA), L-iduronic acid (IdoUA), or D-galactose (in case of keratan sulfate) linked to a D-N-acetylglucosamine (GlcNAc) or D-N-acetylgalactosamine (GalNAc) residue. GAGs have a high degree of heterogeneity in terms of molecular mass, disaccharide composition, and sulfation pattern. Most GAGs are linked to proteins by N- or O-glycosylation in the form of proteoglycans. They are found throughout the body, often in mucus and in fluids around the joints, and as components of fibrous ECM. GAGs are involved in various biochemical processes such as cell adhesion, growth and proliferation, cell surface binding, wound healing or tumor metastasis. The chemical structures of different important GAGs (hyaluronan, chondroitin sulfate, heparin) are shown in Figure 2.75.

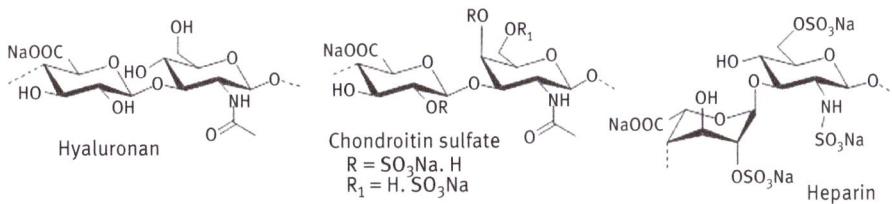

Figure 2.75: Chemical structures of different important GAGs (hyaluronan, chondroitin sulfate, heparin).

Hyaluronic acid, also called hyaluronan (HA), is the only nonsulfated GAG formed from β(1-3)-linked repeating disaccharide units of D-N-acetylglucosamine (GlcNAc) and

D-glucuronic acid (GlcA) linked by alternating β-(1-4) glycosidic bonds (Figure 2.75). Dependent on the biological source, HA may possess extremely high molecular masses up to 10^4 kDa. It is found naturally in the ECM, various tissues (e.g., connective tissue), synovial fluid, vitreous humor, and the umbilical cord. In eukaryotic cells, the biosynthesis occurs at the inner surface of the plasma membrane catalyzed by HA synthases. In the human body, HA plays an important role in several physiological processes including lubricating of articulating joints, water regulation in wound repair, and radical scavenger in inflammation. As component of the ECM, HA also contributes to tissue hydrodynamics, movement and proliferation of cells, and participates in a number of cell surface receptor interactions (Dicker et al., 2014).

Whereas HA was formerly isolated from animal sources (e.g., rooster combs), nearly all commercially available HA is currently being produced from bacteria (Streptococci, Bacilli, *Escherichia coli*) in fermentation processes.

High-molecular mass HA is highly water-soluble. In an aqueous solution, the HA molecule is stiffened due to a combination of the chemical structure, internal hydrogen bonds, and interactions with solvent. Hence, a HA molecule assumes an expanded random coil structure in solutions which occupies a very large domain, and is able to bind large amounts of water (Scott, 2000).

The properties of HA can be engineered by derivatization such as esterification, etherification or cross-linking of the functional hydroxyl, carboxyl, or acetamido groups of the HA molecule. Esterification and etherification reactions mostly result in more hydrophobic, more rigid, and less enzymatically cleavable derivatives. Numerous cross-linking agents including di-epoxides, water-soluble carbodiimides, dialdehydes, or divinyl sulfones are known to prepare swellable but water-insoluble hydrogels of HA. Cross-linked hydrogels can also be prepared by radical polymerization of (meth)acrylated HA derivatives (Prestwich and Kuo, 2008, Schiller et al., 2010).

The application of chemically modified HA materials in tissue regeneration, drug delivery systems or implant coatings has been intensively studied during the last decades. HA products are currently used as lubricants in osteoarthritis therapy (Hyalgan®, Orthovisc®, Synvisc®), gels in dermal wound (Bionect®) or viscoelastic gels in surgical aid (Healon®, DuoVisc®, Z-Hyalin®) (Agrawal et al., 2014).

Chondroitin sulfate (CS), a sulfated GAG, comprises β-(1-3) linked repeating disaccharide units of D-*N*-acetylgalactosamine (GlcNAc) and D-glucuronic acid (GlcA) alternating linked by β-(1-4) glycosidic bonds (Figure 2.75). The regular disaccharide sequence of chondroitin-4-sulfate (formerly called CS A) is sulfated in position C-4 of the GalNAc unit, while chondroitin-6-sulfate (CS C) is sulfated in position C-6 of GalNAc. Disaccharides with varying numbers and positions of the sulfate groups can be located, in different moieties, along the polysaccharide chains (Lamari and Karamanos, 2006). Therefore, no pure isomeric CS is available from biological materials. Compared to HA, the molecular mass of native CS is significantly lower and does not exceed 50 kDa. CS is water-soluble and usually linked via short oligosaccharides to core proteins forming proteoglycans. Technically, CS is isolated from extracts of cartilaginous cow and pig tissues (cow trachea, pig ears and noses) and also marine sources (shark, fish, bird cartilage).

As a major component of the ECM, CS is important to maintain the structural integrity of tissues. Currently, CS is used as dietary supplement for osteoarthritis patients. It is also included in some viscoelastic formulations (Viscoat®) for ophthalmic surgery. Furthermore, CS has also gained attraction for tissue regeneration scaffolds such as the Integra® template, a bilayer membrane system for skin regeneration (Agrawal et al., 2014).

Heparin is a highly sulfated GAG of variably sulfated disaccharide repeating units. The most common disaccharide repeating unit is based on L-iduronic acid (IdoA), β-(1-4) linked to D-*N*-acetylglucosamine and an alternating α-(1-4) linkage (see Figure 2.75). The C-2 position of IdoA and the C-6 position of GlcNAc are sulfated and in addition the *N*-acetyl group of GlcNAc is replaced by a *N*-sulfate moiety. The molecular mass of native heparin ranges between 3 and 30 kDa (Mulloy, 2012). It is a naturally occurring anticoagulant produced by basilophils and mast cells. Pharmaceutical-grade heparin is produced from mucosal tissues of slaughtered animals such as porcine (pig) intestines or bovine (cattle) lungs.

Heparin is generally used as medication for anticoagulation. In addition, numerous surface coatings have been developed to improve blood compatibility of biomaterials (Sakiyama-Elbert, 2014), in particular, coronary stents (CORLINE®), hemodialysis catheters (Astute™), cardiopulmonary bypasses (Astute™), vascular grafts (CARMEDA®, Flowline BIPORE®), and extracorporeal circulation devices (BIOLINE®) (Biran and Pond, 2017).

2.3.8.3 Proteins

A wide variety of proteins and their derivatives are being used as biomaterials. The majority of these biomaterials are found in mammalian tissues as components of the ECM, the structural skeleton, or the blood plasma. Nevertheless, some proteins, not originating from mammalian sources (e.g., silk), have been developed as biomaterials based on their specific, outstanding properties combined with excellent biocompatibility. In this context, a promising future approach is the use of recombinant manufacture technologies to prepare tailor-made proteins.

Collagen (kólla *Greek* glue) has been used as biomaterial for many years. Up to now, 28 different molecular types of collagen have been identified, differing in their composition, structure, function, and tissue specificity. The most abundant collagen type in the human body by far is collagen type I, representing together with types II to V 90% of all collagens in the human body. With the exception of type V, these collagens are able to assemble into cross-striated fibrils (fibril-forming collagens) (Hulmes, 2008) that have been widely investigated for use in biomaterials applications.

Collagen type I is composed of a triple helix, consisting of identical, left-handed polypeptide chains (α1) and an additional chain differing slightly in its chemical composition (α2). The α1 chains contain 1,056, and the α2 chain 1,029 amino acid residues. The most common structural motifs of collagen are the amino acid sequences glycine-proline-X, and glycine-X-hydroxyproline, where X is any amino acid other than glycine, proline or hydroxyproline. Twisting or intertwining of the three α-strains around a molecular axis form a triple-helical structure. The collagen monomers (tropocollagen) assemble, after cleavage of globular domains formed during collagen biosynthesis, to form collagen fibrils (Figure 2.76). In addition, further structural stabilization occurs by covalent cross-linking of fibrils. These fibrils are important structural building blocks in collagen type I forming fibers and fiber bundles which are characterized by a parallel aligned to the direction of load (Agrawal et al., 2014). Structure and occurrence of collagen types II to IV, also present in larger amount in the human body, are explained in more detail in Table 2.17.

Amino acids
~ 1 nm

Tropocollagen
~ 300 nm

Fibrils
~ 1 µm

Fibers
~ 10 µm

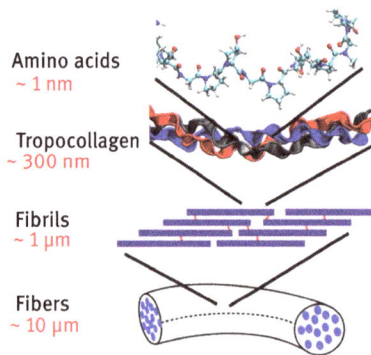

Figure 2.76: The hierarchical design of collagen type I (Buehler, 2006). © Copyright (2006) National Academy of Sciences, U.S.A.

Collagen is widely used as a biomaterial mostly derived from porcine, equine, or bovine sources. Previously, collagen was employed as Catgut sutures made from bovine material to close surgical wounds. Purified collagen has been used as carrier in drug delivery (Infuse™ Bone Graft, a collagen carrier for recombinant human bone morphogenetic protein-2 delivery) and in injectable form as tissue filler for functional (urethra, vocal chords) and also for cosmetic effects. Collagen type I is also being applied in tissue regeneration of skin and bone. For example, the artificial skin product Apligraf® contains collagen in combination with living human cells to heal venous leg and diabetic foot ulcers. For bone tissue regeneration purposes, collagen is used in pure form as membrane (Biomend®) or combined with TCP or hydroxylapatite to form composite materials. Another application is as topical hemostasis material in the form of sheets, fleeces, sponges, foams, or gels to control bleeding. Furthermore, collagen has also been used to fabricate tube allografts for peripheral nerve regeneration and for vascular prostheses.

Table 2.17: Occurrence and structural characterization of collagens type I to IV.

Collagen type	Occurrence	Structural characterization
I	Most abundant collagen (bone, tendon, ligament, skin, dentin, cornea, blood vessels, scar tissue)	– Aggregates into fibrils 50–500 nm long depending on age and kind of tissue – Fibrils form fibers and fiber bundle – In bone, formation of the fibrillar phase of mineralized cortical and trabecular composite
II	Articular cartilage, hyaline cartilage, ribs, nose, trachea, larynx	– Formation of a dense network of thin fibrils (interfibrillar matrix) made of proteoglycans, glycoproteins, other proteins, and water – Provides tensile integrity – Contributes as shock absorber in joints
III	Skin (together with type I), blood vessels, ligaments, internal organs	– Homotrimer of 3 α1 (III) chains, resembles other fibrillar collagens in its structure – Formation of disulfide bond (elastic properties)
IV	Basement membranes (eye lens, blood vessels, kidneys, basal lamina structures of skin)	– Fibers link head-to-head, rather than in parallel – lacks the regular glycine in every third residue (more random overall arrangement with kinks) – Causes the collagen to form in sheets – Acting with noncollagenous components to form meshes and networks

Gelatin is obtained by partial hydrolysis of collagen derived from bone or other connective tissue of animals. Two types of gelatin can generally be obtained, depending on the pre-treatment procedure (prior to a subsequent extraction process). Acidic pretreatment resulting in gelatin type A barely affects the amide groups while the alkaline pretreatment (type B) targets the amide groups of asparagines and glutamine and hydrolyzes them into carboxyl groups, thus converting many of these residues to aspartate and glutamate. This explains the different isoelectric points of type A (pH 8–9) and type B (pH 4.8–5.5) gelatin. During the extraction process, tropocollagen molecules are broken down cleaving stabilizing cross-links and forming chains with lower degree of molecular assembly (Schrieber and Gareis, 2007).

Gelatin readily dissolves in hot water forming colloidal solutions (sols). On cooling, these sols convert to gels, and on warming they revert to sols again. This important property of thermoreversibility is the basis for several technical applications of gelatin. Furthermore, as a denatured product, gelatin is much less antigenic than collagen. Gelatin chains contain abundant motifs such as arginine-glycine-aspartic (RGD) sequences stimulating cell adhesion. These advantages make gelatin attractive for drug delivery (Unigel™ – for fixed dose combinations), tissue engineering, wound management (e.g., SURGIFOAM®, SURGISPON® gelatin absorbable sponges for hemostatic use), and cell therapy (e.g., beMatrix™).

Elastin is an ECM protein of mammals. It is secreted as 60–70 kDa building blocks called tropoelastin. It has a characteristic domain arrangement of hydrophobic sequences (mainly glycine, valine, alanine, proline amino acids) alternating hydrophilic lysine-containing sequences. Catalyzed by lysyl oxidases, elastin is covalently stabilized by interchain cross-linking forming a highly insoluble polymer resistant to further enzymatic, chemical or physical degradation (Agrawal et al., 2014). The cross-linking occurs at the lysine residues via desmosine and isodesmosine molecules. In contrast to collagen, elastin is a highly elastic protein and allows tissues to resume their shape after stretching or contracting. It is therefore found in locations where elasticity and flexibility are of great importance such as skin, bladder, lung, blood vessels, or intervertebral disks (Wise et al., 2014).

Although the insolubility of elastin has limited processing and practical applications, it can be used as biomaterial in different forms including autografts, allografts, xenografts, decellularized ECM, and purified elastin preparations. There are also degraded, solubilized forms of elastin available by acid or basic degradation protocols (e.g., α-elastin) of insoluble elastin. Repeated elastin-like sequences can also be produced by synthetic pathways (Nivison-Smith and Weiss, 2011). A more promising approach, developed during the last years, appears to be by recombinant techniques that allow the preparation of proteins mimicking the elastic properties of elastin (Williams, 2014).

Fibrin is formed during blood coagulation by the action of fibrinogen and thrombin. Fibrinogen is a soluble, complex glycoprotein (molecular mass about 340 kDa) circulating in blood. It is composed of two sets of three polypeptide chains (Aα, Bβ, and γ), including the fibrinopeptides A and B, which are joined together by disulfide bonds within the E domain (Figure 2.77) (Noori et al., 2017).

In the last step of the coagulation cascade, thrombin, a serine protease, cleaves off two sets of peptides, fibrinopeptides A and B, from the amino terminal ends of the Aα and Bβ chains. Then, each E-site interacts with a complementary binding site located on the D domain of adjacent molecules. This "E-D polymerization" result in the spontaneous formation of so-called protofibrils, subsequently yielding a 3D network that is further stabilized by covalent cross-linking of fibrin by factor XIIIa, a transglutaminase, leading to a stable structure (Spotnitz, 2014).

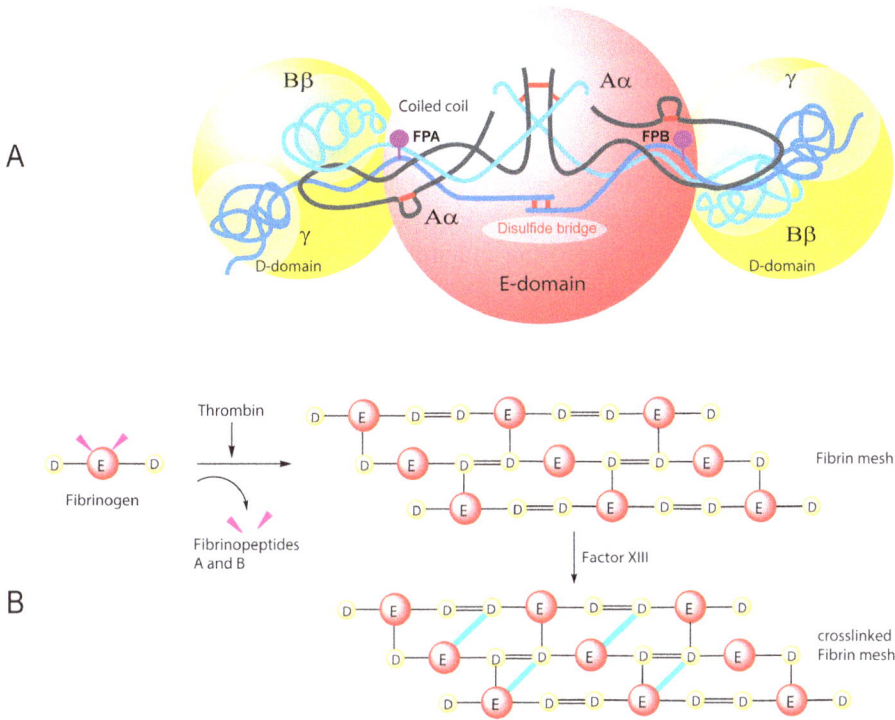

Figure 2.77: Fibrinogen structure (A) and fibrin formation within the cascade of blood clotting (B).

This naturally occurring process is used artificially in biomaterial products known as fibrin glues or sealants. It is accepted for the use as hemostat, sealant, or adhesive in surgery, tissue regeneration, and even drug delivery. Examples for specific clinical use include the repair of tissue dura, bronchial fistulas, peripheral nerves, or achieving hemostasis after spleen or liver trauma. Commercial products are, for example, Tisseel®, Evicel®, or Beriplast®. For the different applications, separate preparations of fibrinogen and thrombin are mixed just before use to employ the *in situ* formed glue. In general, the source for fibrinogen is plasma or heterologous/autologous cryoprecipitate whereby bovine, human, or recombinant materials are used as thrombin source.

Silk is a structural protein produced as fibers from different silkworms (e.g., *Bombyx mori*) and other insects, as well as spiders. It is comprised of two proteins, the long-chained fibroin (~ 70–80%) that is surrounded by the sticky-like sericin (~ 20–30%). Fibroin (molecular mass about 350 kDa) can be described as a block copolymer containing crystalline domains of highly repetitive amino acid sequences such as (glycine-serine-glycine-alanine-glycine-alanine), interrupted by amorphous regions consisting of amino acids with polar, aromatic, and bulkier side chains.

The predominant secondary structure is an antiparallel β-sheet stabilized by an extensive hydrogen bond system and hydrophobic interactions (Rising, 2014). Silk fibers are hygroscopic, but insoluble in water and most organic solvent, nondegradable by many proteases, and characterized by high tensile strengths of up to 700 MPa.

As a biomaterial, silk has gained growing interest during the last decades (Spiess et al., 2010). It can be used as suture material in general (PERMA-HAND®, TRUESILK) and cardiovascular, ophthalmic, and neurological surgery. Further applications include tissue engineering and drug delivery.

2.3.9 Synthetic polymers

2.3.9.1 Polyolefins

PE, the simplest hydrocarbon polymer, is derived from ethylene (C_2H_4). The main properties of PE are determined by the extent of chain branching and the molecular mass and its distribution. There are different types of PE used in medical applications (Table 2.18), namely low-density PE (LDPE), linear low-density PE (LLDPE), high-density PE (HDPE), and UHMWPE.

Table 2.18: Selected characteristics of different PE used for medical applications (Ha et al., 2009).

Characteristics	LDPE	HDPE	UHMWPE
Density (g/cm³)	0.91–0.925	0.941–0.965	0.94–0.99
Molecular mass (kDa)	20–600	<450	2,000–10,000
Crystallinity (%)	40–55	60–80	50–90
Tensile strength (N/mm²)	10	27	41
Young's modulus (N/mm²)	210	1,400	800–2,700
Melting range (°C)	105–110	130–135	135–155
Water uptake (%)[a]	< 0.1	< 0.1	0.01

[a]Measuring conditions: 23 °C, 50% relative humidity.

The manufacture of PE is frequently performed by coordination polymerization of gaseous ethylene in the presence of metal chloride or oxide catalysts. LDPE is produced under high pressure (100–300 MPa, [hypercritical state of ethylene]) in the presence of catalysts. By this process, a highly branched polymer is obtained with less compact packing of macromolecules and a low density. LDPE is used for bottles, tubes, syringes, containers, or other disposables. Polymerization under low pressure (10–80 MPa) in the presence of Ti- or V-based so-called Ziegler–Natta catalysts or employing a Cr-based Phillips catalyst result in HDPE with linear structure. LLDPE without long-chain branches is produced by Ziegler–Natta or Phillips processes in the presence of higher olefins as comonomer, or with a Ti- or V-based metallocene catalyst.

Extending the polymerization time in the coordination-catalyzed process to obtain higher molecular masses leads to UHMWPE used in medicine for surgical implants, for example, as bearing surfaces in total hip and knee prostheses. In the latter case, further improvement of the long-term wear properties of UHMWPE can be obtained by cross-linking that can be initiated by different procedures including exposure of the PE to γ-irradiation, repeated irradiation and annealing cycles, or the incorporation of antioxidants as, for example, vitamin E within the material.

PP is another hydrocarbon-based polymer derived from propylene. It is industrially manufactured from propylene by gas phase, bulk, or slurry polymerization in the presence of different catalysts such as Ziegler–Natta catalysts or metallocene compounds. The use of the catalysts strongly affect the tacticity of the formed PP. The tacticity, that is, the orientation of the methyl groups relative to the methyl groups in neighboring monomer units (see Table 2.16 and Figure 2.67) has a strong influence on the crystallinity of the final PP. PP is widely used as nonabsorbable monofilament surgical suture material. It is further used as meshes for tissue repair, especially for hernia treatment, and also for urological or urogynecological regeneration processes (Williams, 2014).

PVC is hydrocarbon-derived polymer (chemical structure, see Figure 3.56) that has only a very low degree of crystallization (<5%). Although the pure polymer is biocompatible, the use of low molecular plasticizers (e.g., dioctyl phthalate) to improve flexibility of the polymer may cause adverse effects, especially in younger patients. PVC is therefore not suited for long-term applications in the body. The polymer is mainly used as packaging material for medical disposables and containers (Lee et al., 2003).

PTFE, also known as Teflon®, the brand name of Chemours™, is a perfluorinated olefin polymer. It was accidentally discovered by DuPont™ in 1938. It is manufactured by radical polymerization of tetrafluoroethylene with peroxide initiators under pressure in aqueous suspension or emulsion. Molecular masses of highly crystalline PTFE range between 500 and 5,000 kDa. Although thermoplastic like other olefins, PTFE processing is rather difficult. Suitable processing technologies for PTFE are compression molding (at about 380 °C) and powder or paste extrusion (Lee et al., 2003).

PTFE is resistant to many solvents and chemicals, and also stable within a wide range of temperature (−270 to 315 °C). Above 400 °C, it decomposes forming toxic fluorinated products. The surface energy of PTFE is very low and it has one of the lowest coefficients of friction of any solid. The mechanical properties of PTFE are rather poor with a tensile strength of only 15–35 N/mm^2 and a compressive strength below 10 N/mm^2 (Williams, 2014).

Due to its inertness and low coefficient of friction, PTFE is interesting as nonadhesive coating. The main biomedical application of PTFE is as artificial vascular grafts to replace blood vessels, in particular with small diameters down to 5 mm (e.g., Gore-Tex®). For this purpose, PTFE undergoes an expansion process where it is heated to temperatures above 330 °C while being held in a restraining device to prevent shrinkage. The resulting expanded PTFE (ePTFE) has a density as low as 0.1 g/cm^3, and a porosity up to 96% with small pore sizes less than 1 μm. This expansion process is also able to improve the mechanical properties of PTFE (Xue and Greisler, 2003).

2.3.9.2 (Meth)acrylate-based polymers

The term (meth)acrylate is used as a generic for acrylate and methacrylate. (Meth)acrylate-based polymers are derived from acrylic and methacrylic acid monomers. There exist a variety of (meth)acrylate monomers with different chemical structures used today in biomedical applications.

(Meth)acrylate-based units are important building blocks for *in situ*-curable polymeric materials used as bone cements for fixation of endoprostheses, ionomer cements, or photochemically cross-linkable dental adhesives and composites (Figure 2.78) (van Landuyt et al., 2007). Further applications include hydrogel-type soft contact lenses, and drug delivery systems (Lee et al., 2003). Acrylamide-based cross-linked gels are used in gel electrophoresis (polyacrylamide gel electrophoresis, PAGE) to separate biological macromolecules (e.g., proteins, DNA) according to their electrophoretic mobility. Another acrylamide-derived monomer, *N*-isopropylacrylamide, forms thermosensitive hydrogels after radical polymerization to poly(*N*-isopropylacrylamide) (PNIPAm), usable as carrier for several drugs such as ibuprofen or calcitonin.

The main monomer in bone cement compositions is methyl methacrylate (MMA). The bone cement kit consists of two phases, a liquid phase containing MMA, and sometimes other (meth)acrylate comonomers, a polymerization activator (e.g., *N,N*-dimethyl-*p*-toluidine), and a stabilizator (e.g., hydroquinone) and a solid phase. The solid phase comprises polymethylmethacrylate, sometimes a further prepolymerized (meth)acrylate, an opacifier (e.g., barium sulfate), and dibenzoyl peroxide as radical initiator. A stabilizer, a dye, or antibiotics can be additional additives. Once liquid and solid phases are mixed, the radical polymerization of the monomer(s) starts. In most commercial bone cement kits, a mixing ratio of two to three parts of solid phase to one part of liquid phase is used to reduce the shrinkage and the generation of heat. For the surgeon remain 10–12 min for manually modelling the cement mass until further molding is not possible anymore (Kühn, 2014). Commercial trade names of bone cements are PALACOS®, BonOs®, cemSys®, or Synicem®.

In dentistry, (meth)acrylate-based monomers are used as matrix resins in dental composites. Those composites are applied as anterior and posterior filling materials, pit and fissure sealants, luting composites, and for crown build-ups, and the bonding of brackets and orthodontic bands. Monomers such as bisphenol A-glycidyl

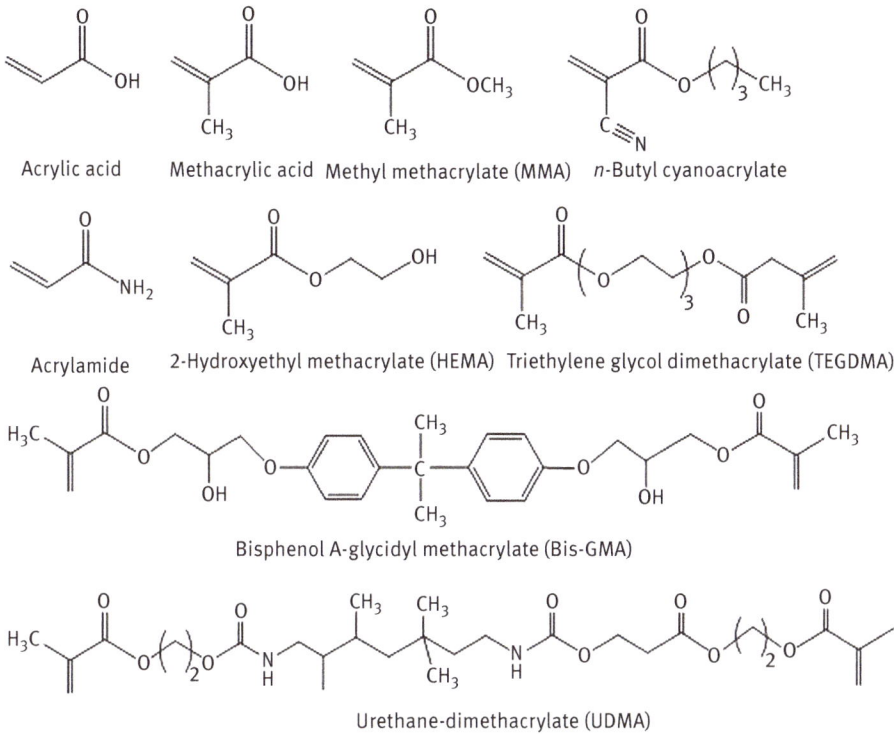

Acrylic acid   Methacrylic acid   Methyl methacrylate (MMA)   *n*-Butyl cyanoacrylate

Acrylamide   2-Hydroxyethyl methacrylate (HEMA)   Triethylene glycol dimethacrylate (TEGDMA)

Bisphenol A-glycidyl methacrylate (Bis-GMA)

Urethane-dimethacrylate (UDMA)

Figure 2.78: (Meth)acrylate monomers used to manufacture biomedical polymers (van Landuyt et al., 2007).

methacrylate or urethane dimethacrylate are often used as resins in addition to triethylene glycol dimethylacrylate (TEGDMA) which can be employed to decrease the viscosity of the matrix, thus allowing increased filler contents. Dental adhesives applied for bonding to the natural substance of teeth, enamel and dentin also contain (meth)acrylate monomers such as hydroxyethyl methacrylate, TEGDMA, or phosphonated acrylates (van Landuyt et al., 2007). The polymerization of resin-based composites will be either chemically initiated (by addition of radical initiators) or, more frequently today, cured by light initiation (light-curing composites).

Cyanoacrylates are a class of monomers derived from cyanoacrylic acid esters. They are able to polymerize anionically in the presence of traces of water to form long-chained poly(alkyl cyanoacrylates). Cyanoacrylates are widely used as adhesives (Duarte et al., 2012; Borie, 2019). For medical use, *n*-butyl and 2-octyl cyanoacrylates (Histoacryl®, DERMABOND, INDERMIL®) are approved as topical glues usable for surgical closure of skin incisions.

2.3.9.3 Polyesters

Polyesters are a class of polymers containing an ester functional group in their backbone. Natural polyesters and a few synthetic ones are biodegradable but the most synthetic polyesters, especially if they contain aromatic building blocks, are long-term biostable. The most common polyesters are thermoplastics.

Polyethylene terephthalate (PET) is among the most common thermoplastic polyesters and as such finds widespread use in biomedical applications, for example, as artificial vascular grafts (trade name: Dacron®), sutures, and meshes. The manufacture of PET occurs either by transesterification of dimethyl terephthalate with ethylene glycol or direct esterification of terephthalic acid with ethylene glycol. PET is hydrophobic, highly crystalline with a high melting temperature (250–260 °C) and resistant to hydrolysis in dilute acids (Lee et al., 2003).

Besides ePTFE, PET is the most important polymer material to generate artificial vascular grafts. Clinically available grafts are fabricated in either woven or knitted forms. The multifilament PET threads in woven grafts are fabricated in an over-and-under pattern resulting in very limited porosity and minimal creep of the finished graft. Knitted grafts are made with a textile technique in which the PET threads are looped to create greater porosity and radial distensibility. A crimping technique is used to increase the flexibility, distensibility, and kink-resistance of textile grafts. The high porosity of the knitted graft necessitates pre-clotting as a means of preventing transmural blood extravasation. Gelatin, collagen, and normally weakly cross-linked albumin, and even heparin, are used to seal knitted PET graft pores. Similar to ePTFE grafts, Dacron® grafts perform well as aortic substitutes, with a 5-year primary patency rate of over 90%. Grafts made of both material types have been shown to perform well at diameters >6 mm, but neither material has been ideally suitable for small-diameter (<4 mm) applications (Xue and Greisler, 2003).

Among polyesters, there is a group of synthetic, biodegradable polyesters, named polylactones. The term polylactones is derived from the starting products of the polymerization reactions in which polyesters are prepared by ring-opening polymerization (ROP) of lactones or cyclic diesters. The most prominent cyclic starting components for producing biodegradable polyesters are depicted in Figure 2.79.

Although synthesis of polylactones can be achieved by direct condensation of α-hydroxy carboxylic acids, this route normally results in low-molecular-mass polymers (Hartmann, 1998). The main route therefore is ROP of mono- or dilactones in the presence of an initiator/catalyst. A typical ROP to convert glycolide or lactide to corresponding high-molecular-mass polymers or copolymers is performed at elevated temperatures in the bulk with tin(II)-bis-(2-ethylhexanoate), also termed tin (II)octanoate, as a catalyst. This catalyst is known to preserve the stereochemistry in case of the dilactide monomer. The reaction mechanism is shown in Scheme 2.2.

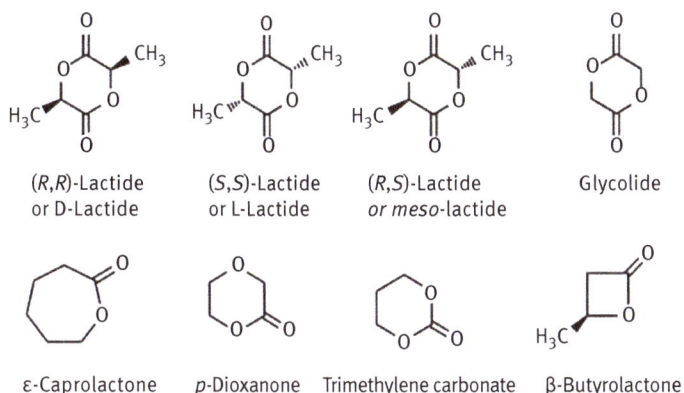

(*R,R*)-Lactide   (*S,S*)-Lactide   (*R,S*)-Lactide   Glycolide
or D-Lactide   or L-Lactide   *or meso*-lactide

ε-Caprolactone   *p*-Dioxanone   Trimethylene carbonate   β-Butyrolactone

Figure 2.79: Cyclic monomers for the synthesis of biodegradable polyesters.

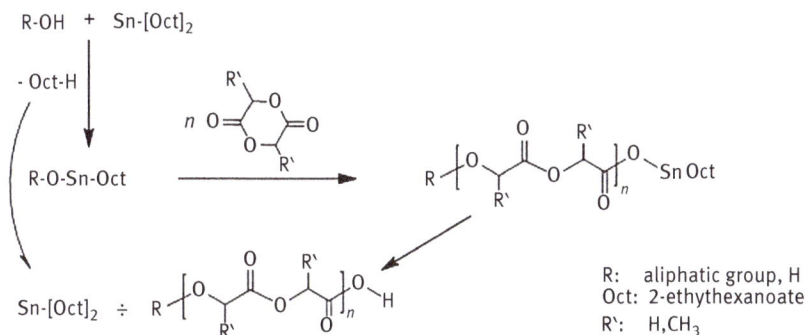

Scheme 2.2: Mechanism of tin(II)octanoate-catalyzed ring-opening polymerization (ROP) of lactones.

Polylactide and *PGA* as well as their copolymers are the most attractive and widely investigated biodegradable polymers (for chemical structures, see Figure 2.80) (Perrin and English, 1997). According to the stereochemistry of the monomers, polylactide is present in the three isomeric forms D(−), L(+), and racemic (D, L). Both poly-L-lactide (PLLA) and PGA homopolymers, and also the poly(L-lactide-glycolide)-based copolymers and stereocopolymers with predominant L-lactide and glycolide portion are semicrystalline polymers. Poly-D,L-lactide (PDLL) is an amorphous polymer (Fambri and Migliaresi, 2010).

Further biodegradable *polylactones*, used for medical applications are poly(ε-caprolactone) (PCL) (Cama et al., 2017), poly(*p*-dioxanone) (PD) (Bezwada et al., 1997), and poly[(*R*)3-hydroxybutyrate] (PHB) (Doi, 1997) as well as various types of corresponding copolymers. In a few applications, copolymers with poly(trimethylene carbonate) (Dobrzynski et al., 2017) segments are also used. Among these

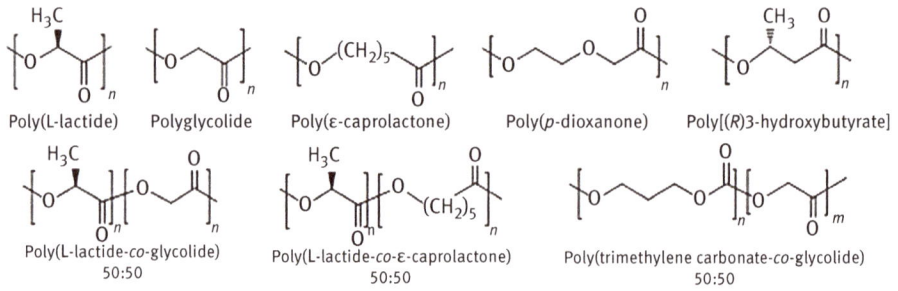

Poly(L-lactide) Polyglycolide Poly(ε-caprolactone) Poly(p-dioxanone) Poly[(R)3-hydroxybutyrate]

Poly(L-lactide-co-glycolide) Poly(L-lactide-co-ε-caprolactone) Poly(trimethylene carbonate-co-glycolide)
50:50 50:50 50:50

Figure 2.80: Biodegradable homo- and copolyesters derived from α-hydroxy carboxylic acid monomers (Perrin and English, 1997).

polymers, poly[(R)3-hydroxybutyrate] can be synthesized both chemically as mentioned above and also from microorganisms by a fermentation process starting from glucose and starch materials.

The homopolyesters are typically strong, stiff materials with high modulus and tensile strength. A concept to improve the elastic properties of the polymers is to prepare co- and terpolymers. Frequently, this approach decreases the degree of crystallinity, resulting in an enhanced rate of degradation.

Table 2.19 presents an overview on thermal, mechanical, and degradation properties of several lactone-based homo- and copolymers.

Table 2.19: Physical properties of selected biodegradable polyesters and related copolyesters (Chu, 2003; Suzuki and Ikada, 2005). T_m, melting temperature; T_g, glass transition temperature.

Polymer	T_m (°C)	T_g (°C)	Modulus (GPa)	Elongation (%)	Degradation
PLLA	170	56	8.5	25	3–5 years
PDLA	–	59	3.2	9	12–16 months
PGA	230	36	8.4	30	2–4 months
PD	106	< 20	2.1	35	6–12 months
PCL	60	−60	0.3–0.4	20–120	> 5 years
PHB	179	10	3.5	6	3 years
PLLA-GA (10:90)	200	40	8.6	24	10 weeks
PLLA-PCL (50:50)	90–120	−17	0.9	600	6–8 months

Evidently, the physical and degradation properties of the polylactone family is strongly affected by their chemical structure, copolymer ratio, molecular packing, and molecular mass. Thus, there exist a wide variety to tailor the material properties with regard to specific clinical requirements.

The earliest and most frequent clinical applications of biodegradable polylactones have been in wound closure materials. This includes absorbable suture materials from polyglycolide such as Dexon®, poly(p-dioxanone) (PDS®), poly(lactide-co-glycolide) (Vicryl®, also termed polyglactin 910), poly(glycolide-co-ε-caprolactone) copolymer (Monocryl®, also termed polyglecapron 25), and poly(glycolide-co-trimethylene carbonate) copolymer (Maxon™, polyglyconate). Some of these materials are also used as surgical meshes for the repair of hernia or the body wall (Chu, 2003). Polylactide and polylactide-based copolymers have been largely applied as osteosynthetic implants such as pins, plates, or interference screws (LactoSorb®, Sysorb® Arthrex®, Bioscrew®, Resorb-X®), as suture anchors (Bio-Statak®), surgical clips and staples (Lactomer®) or bone regeneration membranes (Guidor®). A cardiovascular stent based on polylactide has recently been developed (Absorb GT1), reaching now clinical trials. A further field of application is drug release and so far, a number of drug delivery devices based on polylactides, PCL, and polyglycolide-containing copolymers have been commercialized. Along with classical implant devices (Capronor®, Durin™, Zoladex®), novel injectable microspheres (Lupron® Depot, Eligard®, Trelstar™) or hydrogels (OncoGel™) have been developed (Jain et al., 2011). For this application, poly(ethylene glycol)-containing block-copolymers, often with amphiphilic and stimuli-responsive properties, also have been manufactured.

Poly(orthoesters) (POE) are other interesting biodegradable polymers that degrade by a surface erosion mechanism. Since 1970, four generations of this relatively fast degradable polymer class have been developed (Heller et al., 2002). The fourth generation (see Figure 2.71 for an illustrative chemical structure) that can also be used as injectables, may contain glycolic or lactic structural units in combination with hydrophilic units of ethylene glycol oligomers in the polymer backbone to control hydrolytic degradation of the acid-labile ortho-ester linkage. Due to the surface eroding behavior and the hydrophobicity of the POE unit, incorporated drugs are released concomitantly with the polymer degradation and with a zero-order kinetics, without any burst release effect (Boesel and Reis, 2005). The POE system is manufactured as drug carrier system under the trade name Biochronomer™ (Ottoboni et al., 2014).

Polyanhydrides are another class of surface-eroding degradable polymers. The high hydrolytic reactivity of the anhydride linkage offers an intrinsic advantage in versatility and control of degradation rates. By varying the monomer type and the monomer ratio in copolymers, surface-eroding polymers with degradation times between 1 week and several years are designable (Jain et al., 2005). The long list of investigated polyanhydride structures covers aliphatic, aromatic, and aliphatic-aromatic homopolymers as well as amino acid- and fatty acid-based polymers, and poly(ester anhydrides) (Domb et al., 2011). The most extensively studied polyanhydride is a poly[1,3-bis(p-carboxyphenoxy) propane-co-sebacic anhydride] copolymer (PCPP-SA) that was approved as drug carrier (Gliadel®) in brain cancer therapy.

2.3.9.4 Polycarbonates

Polycarbonates are low crystalline thermoplastic polymers characterized by the -O-CO-O-carbonate group. Their synthesis normally occurs by condensation of a dihydroxy compound with a derivative of "carbonic acid." In a well-known example, 2,2-bis (4-hydroxyphenyl)propane (bisphenol A) is treated with phosgene resulting in linear homopolymers with molecular masses of up to 30 kDa. Alternative routes include a transesterification of diols with diphenyl carbonate or the ROP of cyclic carbonates (see Chapter 2.3.9.3). In medicine, polycarbonates are used as containers, tubes, syringes, bottles, and components for dialysis apparatus (Williams, 2014).

Due to safety concerns of aromatic polycarbonates, current focus is directed toward aliphatic polycarbonates. Recently, biodegradable polycarbonates derived from natural amino acid L-tyrosine, for example, poly(desaminotyrosyl-L-tyrosine-ethylester-carbonate) (Figure 2.71) have been developed (Kemnitzer and Kohn, 1997). These polymers have potential in absorbable implant materials such as cardiovascular stents or drug delivery devices.

2.3.9.5 Polyamides

Polyamides (for general formulae, see Figure 3.55) are synthetic, mostly semicrystalline thermoplastics. The most important members of this polymer class are *poly(caprolactam)*, (Polyamide 6, PA6, Perlon®), and *poly(hexamethylene adipamide)* (PA 66, Nylon). Whereas PA6 is technically produced by ROP of caprolactam, PA66 is manufactured by polycondensation of hexamethylene diamine with adipic acid. Polyamides are suitable for extrusion and injection molding. Owing to its high tensile strength, PA66 is used as nondegradable suture material. Further applications include short-term applications such as catheters and catheter balloons in angioplasty procedures or stent deployment. It is also known to be hemocompatible (Srichana, and Domb, 2009). Aromatic polyamides such as the high-strength *poly(p-phenylene terephthalamide)*, known as Kevlar®, are under investigation as artificial ligament material.

2.3.9.6 Polyurethanes

Polyurethanes (for general formulae, see Figure 3.55) represent an important polymer class, forming, depending on their structure, rigid thermoplastics or thermoplastic elastomers (Zdrahala and Zdrahala, 1999). They are constituted of polymeric chains joined by -NH-COO-urethane links.

Polyurethanes are obtained from the reaction of isocyanates having the $-N=C=O$ functionality with compounds containing active hydrogen atoms such as alcohols or amines. Urethane bonds are formed from the reaction of isocyanate and alcohol groups, while urea bonds result from the reaction of isocyanate with amine ones (see Scheme 2.3). The reaction of isocyanates with water generates unstable carbamic acid that decomposes into an amine and carbon dioxide acting as an *in situ* blowing agent.

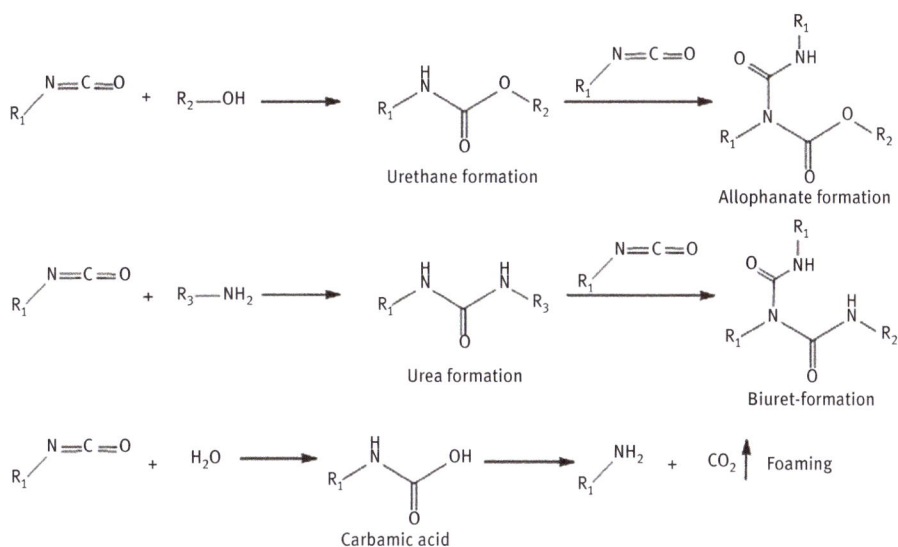

Scheme 2.3: Reactions of isocyanates with alcohol, amine, and water (foam formation).

Di- or polyisocyanates used to synthesize polyurethanes can be of aromatic nature including 4,4′-methylene diphenyl diisocyanate and toluene diisocyanate, whereas aliphatic isocyanates include 1,6-hexamethylene diisocyanate, 1,4-butane diisocyanate, and lysine ethyl ester diisocyanate. In comparison to aliphatic isocyanates, aromatic isocyanates have higher reactivity and the generated polyurethanes have higher mechanical properties, but also a higher toxicity. Therefore, for biomedical polyurethanes, aliphatic isocyanates are preferred starting materials. Di- or polyols reacting with isocyanates to polyurethanes have terminal hydroxyl groups and a variable backbone composed of polyesters, polyethers, polycarbonates, polydimethylsiloxane (PDMS), or polybutadiene. Normally, the design of the polyol backbone is dictated by the desired properties of the polyurethane and its final application (Prieto and Guelcher, 2014).

Today, segmented polyurethanes are used comprising a low-molecular-mass soft segment that is commonly a macrodiol and a semicrystalline hard segment of a diisocyanate and chain extender. In a typical polyurethane synthesis, the macrodiol is reacted with an excess of isocyanate to form a prepolymer that subsequently is treated with a chain extender to build up molecular mass and form a linear block copolymer with alternating blocks of hard and soft segments (Touchet and Cosgriff-Hernandez, 2016). Both urethanes and ureas can react with excess isocyanate groups to allophanate and biuret units, forming thermally labile cross-links that provide additional structural diversity of polyurethanes.

Currently, there are several clinically used polyether-based polyurethanes as Biomer®, Tecoflex™, or Pellethane® on the market. They are used for cardiovascular devices (e.g., heart valves, leads for pacemakers, catheters). Chain extenders with hydrolyzable units such as lactide and oligo(ethylene glycol) units offer an interesting option to generate biodegradable polyurethanes (Guelcher, 2008).

2.3.9.7 Polyaryletherketones and polysulfones

Polyaryletherketones represent a polymer class of high-performance, semicrystalline thermoplastics with high temperature stability and high mechanical strength. The polymer backbone of polyaryletherketones contains alternating ketone and ether groups. PEEK is the most prominent member of this polymer class with regard to clinical applications (Kurtz and Devine, 2007).

PEEK can be manufactured by polycondensation of 4,4'-difluorobenzophenone with the disodium salt of hydroquinone at higher temperatures in polar solvents. The commercially available polymer is semicrystalline (30–35%) and has a molecular mass around 100 kDa with a melting temperature of 343 °C and a glass transition temperature of 143 °C. It can be processed by conventional thermoplastic methods such as extrusion, injection, or compression molding. Due to its molecular structure, PEEK is insoluble in many common organic solvents, highly resistant to thermal degradation and stabile to acids, bases, and salt solutions. It is not degraded in the body and can be sterilized by γ-irradiation. Unfilled PEEK has a Young's modulus of 3–5 GPa. Further increase of the modulus can be achieved by reinforcement of PEEK with chopped or continuous carbon fibers to modify their elastic modulus to approach that of cortical bone (Rahmitasari et al., 2017).

A medical grade PEEK material is provided under trade names such as PEEK-Optima® and available as unfilled PEEK, and also reinforced with hydroxylapatite, or carbon fibers. Main medical applications cover spinal fixation devices, interbody fusion devices, and also CAD-CAM manufactured craniofacial implants.

Polysulfones (PSU) are other high-performance polymers with high temperature stability. To this polymer class also belong *polyethersulfones* (PES).

PSU are prepared by polycondensation of aromatic dihydroxy components (e.g., the Na salt of bisphenol A) and a bis(halophenyl)-sulfone. They are amorphous, transparent thermoplastics characterized by high strength and stiffness, retaining these properties between – 100 °C and + 150 °C. The glass transition temperature of PSU is around 200 °C. They are highly resistant to acids and alkalis at pH values between 2 and 13, but soluble in dichloromethane. The Young's modulus is on the order of 7 GPa. Due to the great resistance to water absorption, PES can be repeatedly steam-sterilized. Similar to PSU, it is used in hollow fiber microfiltration and ultrafiltration membranes, including those for hemodialysis (Williams, 2014; Koga et al., 2018).

2.3.9.8 Polysiloxanes

Polysiloxanes, commonly called silicones, consist of an inorganic silicon-oxygen . . . -Si-O-Si-O- . . . backbone chain with organic side groups attached to the silicon atoms.

In general, polysiloxanes have the chemical formula $[R_2SiO]_n$, where R is an aliphatic (methyl, alkyl, etc.) or aromatic (phenyl) organic group. The usual method to manufacture polysiloxanes is the hydrolysis of single or mixed chlorosilanes (R_3SiCl, R_2SiCl_2, $RSiCl_3$, $SiCl_4$). In this procedure, the intermediately formed silanol groups are rapidly condensed with elimination of water, forming Si–O–Si links.

PDMS is by far the most common silicon-based polymer (Srichana and Domb, 2009). The synthesis route is shown in Scheme 2.4.

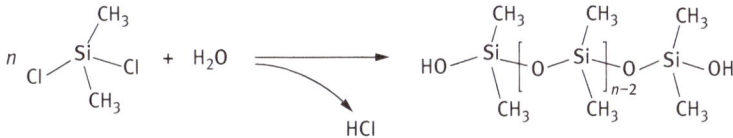

Scheme 2.4: Synthesis of polydimethylsiloxane by hydrolysis of dichloromethylsilane.

The linear polysiloxanes with n between 10 and 20 are known as silicon oils. Under specific conditions, cyclic siloxanes are also formed that are convertible into high-molecular mass linear silicone elastomers. There exist three main types of curing/cross-linking, namely (i) radical cross-linking using radical initiators to generate radical end-groups at silicon chains able to recombine with other chains, (ii) condensation reactions such as the reaction of a Si–OH end-group with an acetate-group at another chain end releasing acetic acid during the curing process, and finally (iii) addition cross-linking using, for example, the addition to Si–H containing silicones of unsaturated end-groups including vinylic units. Silicones are produced as oils, gels, gums, and elastomers. The strength and elasticity of silicone elastomers is a function of polymer chain length, the type of side group, and the degree of cross-linking. In order to enhance the mechanical properties of silicones, different fillers are added such as silica fume, talc, or TiO_2. The glass transition temperatures (T_g) of silicones used in medical applications are below room temperature. There are several techniques available for tailoring of PDMS properties to application-specific requirements by surface and bulk modification techniques (Wolf et al., 2018).

PDMS and other polysiloxanes are biocompatible, and hence, inert to cells and tissues. They are stable to heat and able to retain their properties under different environmental conditions at a wide temperature range. Furthermore, silicones have a high permeability to oxygen, nitrogen, and water vapor. Due to their great hydrophobicity and their low surface tension, polysiloxanes are capable of self-wetting, thus promoting good film formation and surface covering (Colas and Curtis, 2004).

Polysiloxanes are extensively used in medical industry. Due to their good blood compatibility, they are used for many cardiovascular devices including catheters and heart valves. In orthopedics, silicones are used for hand, finger, and foot joint implants. In reconstructive and plastic surgery, silicone implants find application for many parts of the human body including breast, scrotum, nose, cheek, chin, or buttock reconstruction. The most prominent esthetic implant is the silicone breast implant, introduced in the 1960s by Dow Corning Inc. Silicone pressure-sensitive adhesives are in common use, for example, in transdermal drug delivery systems (Curtis and Colas, 2004). Although silicones are known for their high biostability and biocompatibility, there is some concern about the biocompatibility of low molecular mass PDMS gels used in breast implants. It is known that PDMS is not only hydrophobic, but to some extent also lipophilic. For this reason, diffusion of PDMS through the silicone elastomer shell of the implant into the surrounding tissue appears to be possible, causing adverse reactions. This topic is currently a subject of further studies (Williams, 2014).

2.3.9.9 Polyethylene glycols and other water-soluble, synthetic polymers

Polyethylene glycol (PEG) is a nontoxic, water-soluble polyether (chemical structure Scheme 2.5), also known as PEO. Depending on their molecular masses, PEGs are liquids or low-melting solids. PEG is produced by base- catalyzed polymerization of ethylene oxide. Oligomeric PEGs with molecular masses between 400 and 600 Da are liquids at room temperature. At molecular masses above 3 kDa, PEGs are solids. PEGs are commercially available over a wide range of molecular masses from 300 Da to 10,000 kDa. They are also available with different geometries, forming branched or star PEGs, having a different number of PEG chains emanating from a local core, as well as comb PEG with multiple PEG chains grafted onto a polymer backbone.

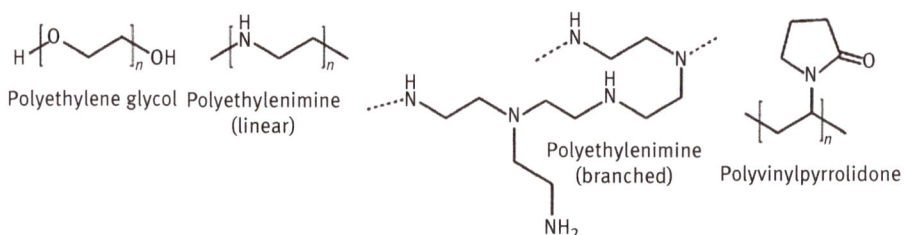

Scheme 2.5: Water-soluble synthetic polymers.

PEG and its monomethoxy ether are manufactured under the tradename Carbowax Sentry™ for numerous applications in foods, cosmetics, pharmaceutics, and in biomedicine, as surfactants, dispersing agents, as solvents, in ointments, suppository bases, tablet excipients, and laxatives.

Oligomeric PEGs are often used as hydrophilic segments in polylactone-based copolymers to enhance hydrophilicity and flexibility of the resulting biodegradable block co- or terpolymers (Kutikov and Song, 2015).

A further important application of PEG in biomedicine is its ability to modify the surfaces of other bioactive molecules such as proteins, polypeptides, or even nanoparticles. As shown in Figure 2.81, there exist several advantages of this procedure, dubbed "PEGylation," to alter the pharmacokinetic and pharmacodynamic properties of the bioactive species.

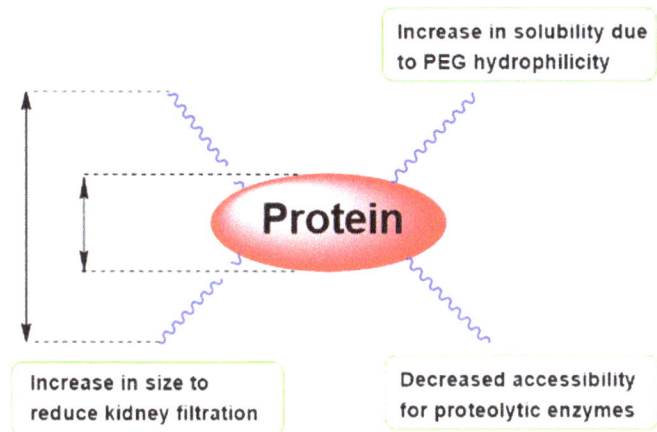

Figure 2.81: Process of PEGylation of proteins (modified after Veronese and Pasut, 2005).

At first, PEGylation makes the bioactive molecule larger, thus reducing renal clearance. Furthermore, PEG acts as a shield to protect proteins from degradation by enzymes or antibodies. Further advantages include increasing of water solubility and minimizing cytotoxicity. With respect to nanoparticles, PEG can serve as hydrophilic outer shell to obtain biological properties. In combination with conjugated ligands, it is also used in gene delivery systems. Furthermore, PEG hydrogels are used in tissue engineering (Peppas, 2004), frequently after incorporation of ECM-derived cell adhesive molecules, and hydrolytically or enzymatically cleavable segments.

Polyethylenimine (PEI) is formally the polymerization product of ethylene imine but it is synthesized by ROP of aziridine. Depending on the reaction conditions, different degree of branching can be achieved (Scheme 2.5). Linear PEI is synthesized by hydrolysis of poly(2-ethyl-2-oxazoline) and sold as jetPEI® transfection reagent. The linear PEIs are solids at room temperature while branched PEIs are liquids at all molecular masses. Linear PEIs are soluble in hot water, at low pH, in methanol, ethanol, or chloroform.

Similar to PEG, PEIs are used as gene carrier materials in biomedicine. As a nonviral vector, the polymer is able to carry genes with reduced risk of toxicity, although with lower efficiency compared to viral vectors (Boussif et al., 1995). Due to the highly cationic nature of PEI under physiological conditions, it has a dose-related cytotoxicity, and there are currently several attempts to reduce cytotoxicity and at the same time, enhance the transfection potency using chemical modification of PEIs (Williams, 2014).

Polyvinylpyrrolidone (PVP) is the polymerization product of *N*-vinylpyrrolidone (Scheme 2.5). It is a water-soluble amorphous polymer with a glass transition temperature between 110 and 180 °C, dependent on molecular mass. It is commercialized under the name Povidone®.

Together with iodine, PVP forms a complex that can be used as a disinfectant (e.g., Braunol®). In addition to being used as blood plasma expander, PVP has numerous ancillary medical applications such as a binder in tablets, and as lubricant in eye drops and packaging solutions for contact lenses. PVP is employed as wetting agent in the so-called MoistureSeal® technique to hydrophilize contact lenses. Presently, the introduction of PVP as hydrophilic component in the production of hydrogels is being explored (Husain et al., 2018).

Chapter 3
Synthesis and process technology of biomaterials

Robert B. Heimann, Mitsuo Niinomi, Matthias Schnabelrauch

3.1 Metals (Robert B. Heimann, Mitsuo Niinomi)

3.1.1 Stainless steels

3.1.1.1 Melting

Prior to the early 1970s, most of the stainless steels were produced by melting together raw materials such as iron or scrap iron, chromium, nickel, silicon, manganese, titanium, and other alloying elements in an electric arc furnace (EAF), and injecting oxygen gas into the melt to amplify the required decarburization rate. This, however, was accompanied by formation of chromium oxide that subsequently accumulated in the slag; hence, the resulting alloy was more or less severely depleted of chromium. This drawback required remedial action that subsequently consisted in a well-defined reduction period during which ferrosilicon was added to deoxidize the steel, thus reducing considerably the chromium oxide losses to the slag.

In the next development step, invented in 1954 by the Linde Division of the Union Carbide Corporation, production of stainless steel was carried out by the *duplex process*, using an argon oxygen decarburization (AOD) converter (Figure 3.1). In this process, iron or steel scrap, ferroalloys, and other raw materials were melted in an EAF or similar furnace to produce liquid low-carbon stainless steel. The metal was then transferred to an AOD vessel and subjected to three refining steps: decarburization, reduction, and desulfurization.

The versatility of the EAF–AOD duplex process led steelmakers to reexamine the use of different converters for melting of stainless steels. This has led to the development of various other types of converter for duplex processes, including Kawasaki Steel Corporation's K-BOP converter, Mannesmann Demag Hüttentechnik's (now SMS Siemag) MPR converter, Creusot-Loire-Uddeholm's CLU converter, and Sumitomo's top and bottom blowing process converter.

Already in the mid-1960s, some steelmakers began modifying existing conventional basic oxygen furnace (BOF) converters for partial decarburization, followed by decarburization in a ladle under vacuum to make low-carbon stainless steels. This process is known as *triplex process* because three process units are involved: an EAF, a converter for preblowing (AOD), and a vacuum converter refining unit. Vacuum refining involves lowering the pressure above the steel bath to promote and accelerate evolution of carbon monoxide gas. The liquid stainless steel treated

https://doi.org/10.1515/9783110619249-003

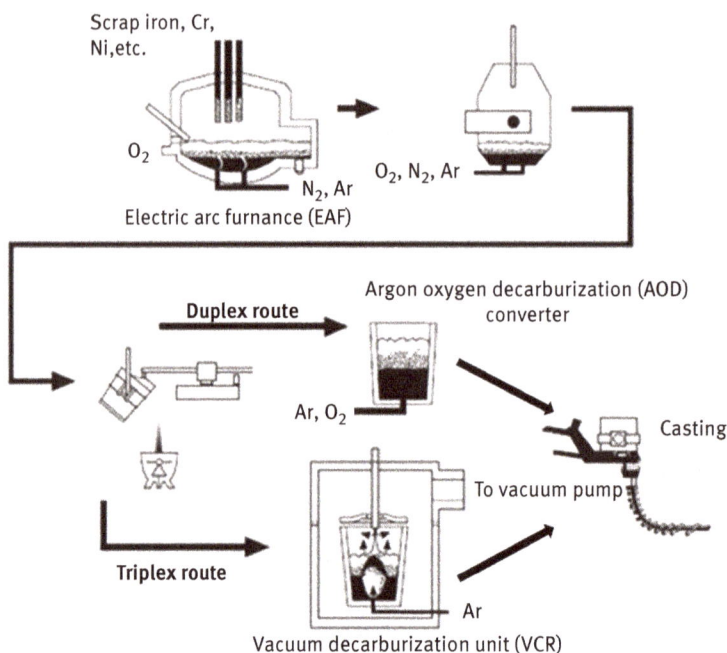

Figure 3.1: Stainless steel production routes (Satyendra, 2014). http://ispatguru.com/stainless-steel-manufacturing-processes © Permission granted under Creative Commons Attribution License (CC BY 3.0).

by the vacuum process generally contains an average of 0.5% carbon. Most vacuum processes are performed in a chamber with a ladle full of metal as opposed to a separate refining vessel used in the dilution/converter processes. In most triplex processes, vacuum processing of steel in the teeming ladle is the final step prior to casting. Figure 3.1 compares the duplex and triplex process routes.

The choice of the process route is influenced by raw material availability, the nature of the desired product, downstream processing, existing shop logistics, and capital economics. The duplex process during which EAF steel making is followed by refining in a converter and tends to be flexible with respect to raw material selection. The triplex process, whereby EAF steel making and converter refining are followed by refining in a vacuum system, is often desirable when the final product has to have very low carbon, sulfur, and nitrogen contents. Overall, the triplex process is technically more involved, and hence, more costly as its cycle times are longer than those required for the duplex process on account that there is an extra transfer needed from process converter to the vacuum unit. It also tends to have slightly higher refractory costs because there are two furnaces required to perform decarburization.

The high temperatures at the tuyère tip and the high bath agitation rate place great demands on the thermal and mechanical stability of the refractory lining.

While typical BOF refractory turnover times are as long as months or even years, replacement cycles for refractory linings for converters in stainless steel manufacturing are only several weeks or days. Since refractory costs comprise a significant fraction of the total operating costs, the choice of refractory type is critical. Options include magnesite-chromite and dolomite refractories (Sadik et al., 2016), the choice of which is dependent on the vessel operation pattern, final product specifications, and process economics.

3.1.1.2 Forming

Semifinished products are manufactured from the basic cast sheet and plate ingots by hot and/or cold rolling. Before hot rolling, the ingots are preheated to high temperature in a furnace called a soaking pit. Subsequently, they are hot-rolled in a reversing breakdown rolling mill and this rolling process is continued until the temperature of the slab drops so low that rolling becomes too difficult. Then the slab is reheated, and rolling is continued until the desired thickness has been reached.

In some cases, hot extrusion technique is applied that requires very powerful extrusion presses and improved lubricants such as glass for manufacturing seamless stainless steel pipe by high-frequency induction (HFI) resistance pressure welding.

3.1.1.3 Heat treatment

After the stainless steel is cast or formed, it usually goes through an annealing step. Annealing involves heat treatment during which the steel is heated and cooled under controlled conditions to relieve internal stresses generated by the buildup of dislocations within the microstructure that results in softening of the metal. Some steels are heat-treated for higher strength. However, such heat treatment, also known as age hardening (see box), requires careful control since even small deviations from the recommended temperature, time, or cooling rate regimes can seriously affect the properties. Lower aging temperatures produce high-strength stainless steel with low fracture toughness, while higher temperature aging produces stainless steels with lower strength but enhanced toughness. Although the heating rate to reach the aging temperature (480–540 °C) does not affect the properties, the cooling rate does. A post-aging quenching, that is, rapid cooling, may increase the toughness without a significant loss in strength. The nature of heat treatment depends on the type of steel: ferritic, austenitic, or martensitic steels (see 2.1.1.1). Austenitic steels are heated to above 1,040 °C for a certain time depending on the thickness of the ingot, sheet or plate. Water quenching (WQ) is used for thick sections, whereas air cooling or air blasting is used for thinner sections. If cooled too slowly, carbide precipitation can occur. Contrariwise, since carbides can be retained in solid solution by rapid cooling from high temperatures, these steels become susceptible to intergranular corrosion when chromium-bearing carbides (e.g., $Cr_{23}C_6$) exsolve from the steel matrix and precipitate along grain boundaries during welding or slow cooling from high temperatures through the 870–800 °C range.

Age hardening

Age hardening or precipitation hardening is a heat treatment technique used to increase the yield strength of malleable metallic materials, including most structural alloys of aluminum, magnesium, nickel, titanium, and some tool and stainless steels. Precipitation hardening relies on changes in solid solubility with temperature to produce fine particles of an impurity phase that impede the movement of dislocations or other defects in a crystalline lattice. Since dislocations are known to be the dominant cause of plasticity, this dislocation arresting serves to harden the material.

3.1.1.4 Descaling

Annealing in ambient atmosphere described above causes the buildup of an oxide layer (scale) on the steel surface. The scale can be removed using several processes. One of the most common methods called pickling uses a nitric acid–hydrofluoric acid bath. Alternatively, electrocleaning is used by applying an electric current to the steel surface in an orthophosphoric acid bath. The annealing and descaling steps occur at different stages depending on the type of steel being worked.

3.1.2 CoCr alloys

Since the early 1900s, a wide range of metals and their alloys have been used for surgically implanted medical devices, prostheses and dental materials, designed to provide enhanced physical and chemical properties, such as strength, durability, and corrosion resistance. Among them, CoCr-based alloys are considered economic alternatives to nickel-based austenitic stainless steels known to be potential allergens, as well as titanium and its alloys (see Chapter 2.1.1.2).

3.1.2.1 Extraction

CoCr alloy manufacturing requires the extraction of cobalt and chromium from naturally occurring oxide ores that need to be reduced to obtain the pure metals. Metallic chromium is usually processed by aluminothermic reduction (see box), and pure cobalt metal can be won by several different processing routes depending on the mineralogy of the specific ore.

Aluminothermic reduction

The process was developed by Hans Goldschmidt (Goldschmidt, 1898) for carbon-free extraction of heavy reducible metals from their oxides such as chromium, vanadium, niobium, and titanium through combustion of aluminum powder. High-purity metals and alloys can be produced in refractive ceramic crucibles. Temperatures of up to 2,500 °C can be obtained by automated continuous exothermic reactions between metal oxides and metallic aluminum. The heavy metal accumulates at the bottom of the crucible, the lighter aluminum oxide floats on top of the liquid metal. Frequently produced alloy types include VAl, NiNb, FeNb, MoAl, CoCr, Cr, and even Ti6Al4V (Hassan-Pour et al., 2015).

The pure cobalt and chromium metals are then fused together under vacuum either by an electric arc or by induction melting under exclusion of oxygen that would otherwise oxidize the reactive metals. For example, ASTM F75 CoCrMo alloy is produced in an inert argon atmosphere by ejecting through a small orifice the molten metal that is immediately cooled to produce fine alloy powder (Ratner et al., 2013). However, this synthesis route is expensive and difficult to control. Recently, a novel electrochemically based solid-state reduction technique known as the Fray–Farthing–Chen (FFC) Cambridge process has been developed that involves the reduction of an oxide precursor cathode preform in a molten calcium chloride electrolyte bath (Figure 3.2; Chen et al., 2004; Hyslob et al., 2010).

3.1.2.2 Forming and shaping

Once the desired CoCrMo alloy has been obtained by techniques described earlier, current processing technologies that are used to shape biomedical devices, in particular dental prosthetic implants, include the following (Reclaru and Ardelan, 2019):

- Traditional casting processes that, however, are laborious and time-consuming methods that frequently introduce porosity when critical flow parameters are not appropriately considered.
- Milling processes involving computer-aided design/computer-aided manufacturing (CAD/CAM) technology to obtain virtual images of dental prostheses based on impressions or 3D images. CAD software is used to design a virtual prosthetic element, followed by milling to manufacture it. However, milling of CoCr blanks may be difficult because of their hardness.
- Thus, high demand is placed on the manufacturing unit (coolant delivery, rigidity of the machine, etc.). Recently, great progress has been achieved by using milling units with four or five axes.
- CAD/CAM sintering of powder: This technology uses metal blanks with a wax-like texture and allows them to be dry milled in a green, that is, nonsintered state. After the required shapes have been milled from the blank, they are debonded and sintered for densification in a special furnace. This results in high-quality products with a homogeneous, distortion-free framework devoid of contraction cavities.

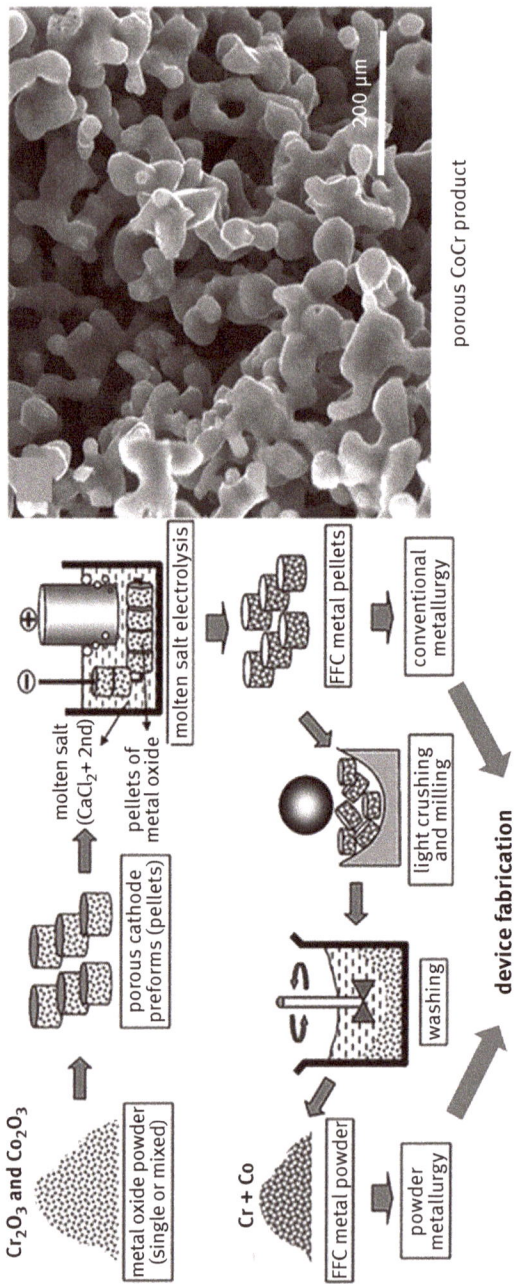

Figure 3.2: Schematic rendering of the FFC Cambridge process to produce CoCr (Chen et al., 2004). © With permission by ASM International & TMS.

- 3D-printing technologies: Modern 3D printers seamlessly integrate with CAD software and other digital files such as magnetic resonance images. Some alternatives are available including selective laser sintering (SLS) using a high power CO_2 laser, direct metal laser sintering, and selective laser melting (SLM; Figure 3.23).
- Electron beam melting (EBM; Figure 3.24): The technology produces dental prosthetic parts by melting metal powder layer by layer with an electron beam under high vacuum condition. Unlike most conventionally sintered products, the parts are fully dense, and thus, extremely strong.

3.1.3 Titanium and titanium alloys

3.1.3.1 Reduction and refining

Titanium is the 10th most abundant element (Clark number 10) in the outer 16 km layer of the Earth's crust; hence, there are sufficient Ti resources available. The ores used to produce metallic Ti are either rutile (tetragonal TiO_2) or ilmenite ($FeTiO_3$; about 50 mass% TiO_2). Liquid $TiCl_4$ is obtained from TiO_2 as an intermediate material to produce metallic Ti as a sponge. $TiCl_4$ is synthesized by reacting TiO_2 with a mixture of carbon and chlorine according to:

$$TiO_2 + 2Cl_2 + 2C \rightarrow TiCl_4 + 2CO \tag{3.1}$$

Subsequent reduction of $TiCl_4$ to Ti can be achieved by several processes including the Kroll (Mg reductant, eq. (3.2)), Hunter (Na reductant, eq. (3.3)), Van Arkel (iodine refining, eq. (3.4)), or electrolytic (eq. (3.5)) processes, with the Kroll process currently being the main practical reduction process for large-scale production of titanium.

The *Kroll process* is schematically shown in Figure 3.3 (Takeda and Okabe, 2017). The reduction reaction of the precursor $TiCl_4$ to Ti sponge can be expressed by

$$TiCl_4(g) + 2Mg(l) \rightarrow Ti(s) + 2MgCl_2(l) \tag{3.2}$$

During the *Hunter process*, $TiCl_4$ is reduced to Ti by reaction with Na according to:

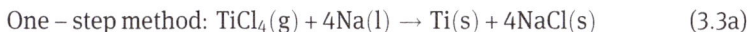

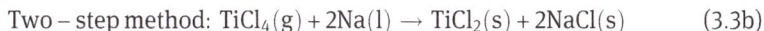

$$One - step\ method:\ TiCl_4(g) + 4Na(l) \rightarrow Ti(s) + 4NaCl(s) \tag{3.3a}$$

$$Two - step\ method:\ TiCl_4(g) + 2Na(l) \rightarrow TiCl_2(s) + 2NaCl(s) \tag{3.3b}$$

$$TiCl_2(l) + 2Na(l) \rightarrow Ti(s) + 2NaCl(s), \tag{3.3c}$$

where g, l, and s refer to gas, liquid, and solid. Increasing attention receives the new low-cost Armstrong Process® likewise using Na as a reducing agent (Araci et al., 2015).

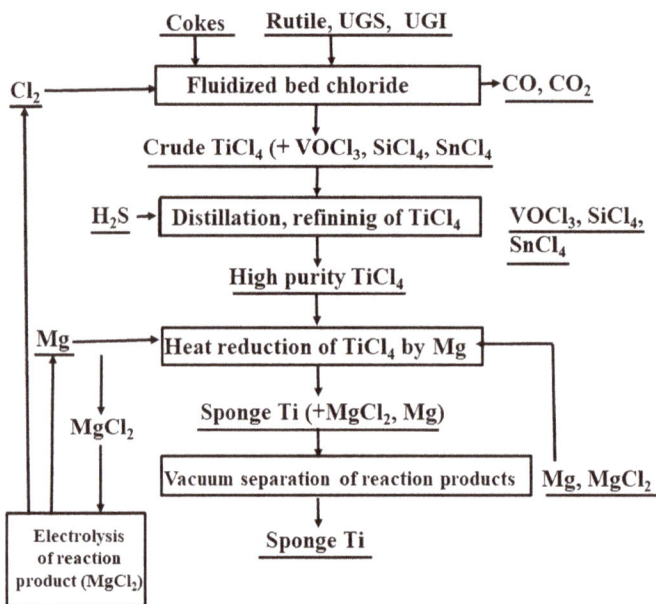

Figure 3.3: Flow sheet of the Kroll process. UGS and UGI refer to upgrade slag and upgrade ilmenite, respectively (Takada and Okabe, 2017). © With permission by The Japan Institute of Light Metals.

During the *Van Arkel iodine process*, gaseous TiI_4 is generated by reacting low-purity Ti with iodine. Subsequently, TiI_4 is thermally decomposed on a white-hot tungsten filament to yield high-purity Ti according to:

$$TiI_4(g) \rightarrow Ti(s) + 2I_2(s, g) \tag{3.4}$$

The *electrolytic process* within the chloride ion-containing molten salt is characterized by the following reactions:

$$\text{Cathode reaction: } Ti^{4+} + 2e^- \rightarrow Ti^{2+} \tag{3.5a}$$

$$Ti^{2+} + 2e^- \rightarrow Ti \tag{3.5b}$$

$$\text{Anode reaction: } 4Cl^- \rightarrow 2Cl_2 + 4e^- \tag{3.5c}$$

Direct Ti reduction processes are also being investigated. Among them, the *FFC process* (see Figure 3.2) based on the direct electrochemical reduction of TiO_2 to metallic Ti in molten $CaCl_2$ has received considerable attention. The FFC process is schematically shown in Figure 3.4a (Okabe, 2005).

The sintered TiO_2 feed electrode acting as a cathode is immersed in molten salt and cathodically polarized to remove oxygen. The electrolysis reactions occurring in this process are as follows:

(a)

(b)

(c)   TiO$_2$ CaCl$_2$ molten salt   Ca–X alloy

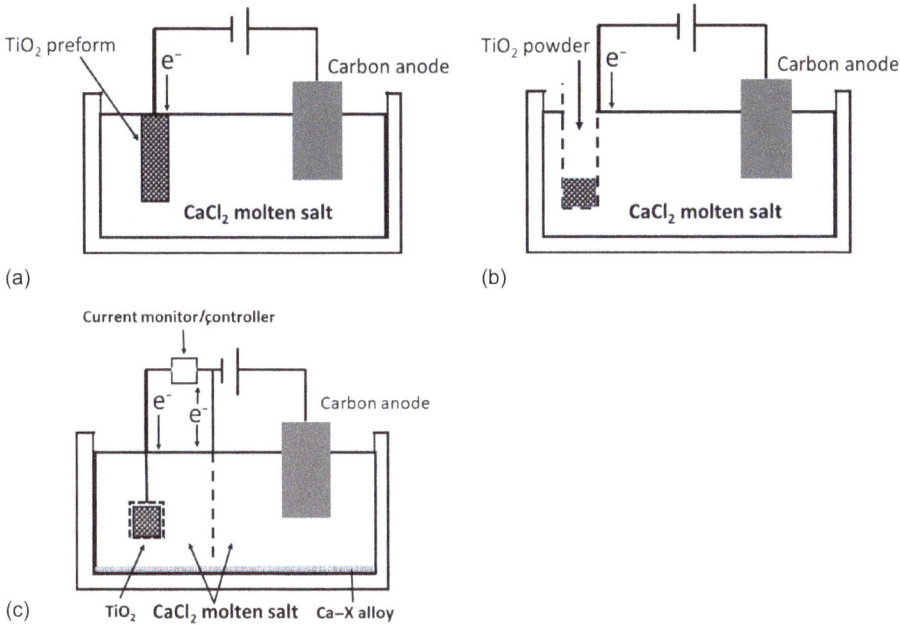

Figure 3.4: Schematic of the FFC (a), OC (b) and EMR/MSE (c) processes (Okabe, 2005). © With permission by The Japan Institute of Light Metals.

$$\text{Cathodic reaction: TiO}_2 + \ 4e^- \ \rightarrow \text{Ti} + 2O^{2-} \qquad (3.6a)$$

$$\text{Anodic reaction: C} + \ xO^{2-} \ \rightarrow CO_x + \ 2xe^- \qquad (3.6b)$$

Although the oxygen content of Ti produced by the FCC process is as low as 10 ppm, the amount of impurities such as C and Fe is relatively high.

In the *Ono and Suzuki* (OS) *process*, Ti oxides are electrochemically reduced in molten CaCl$_2$-CaO, using Ca as the reducing agent as schematically shown in Figure 3.4b (Okabe, 2005).

The reduction of TiO$_2$ or CaTiO$_3$ formed during the process has been reported to be inhibited, thereby leading to a residual high oxygen content in the Ti end product. Since the direct reduction of TiO$_2$ is slower than that of CaTiO$_3$, the reduction route via CaTiO$_3$ is more economical. When CaTiO$_3$ is reduced, the oxygen content in Ti is lower when using a carbon rod electrode as opposed to using a carbon crucible because the surface area is increased which, in turn, increases the cathodic electric current density (Suzuki et al., 2017). The OS process is also sensitive to C and Fe contamination. The reactions occurring during this process are as follows:

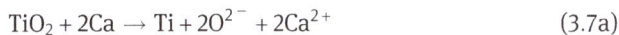

$$\text{TiO}_2 + 2\text{Ca} \rightarrow \text{Ti} + 2O^{2-} + 2\text{Ca}^{2+} \qquad (3.7a)$$

Electrolysis

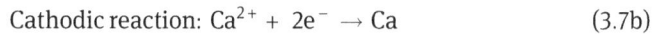

$$\text{Cathodic reaction: } Ca^{2+} + 2e^- \rightarrow Ca \tag{3.7b}$$

$$\text{Anodic reaction: } C + xO^{2-} \rightarrow CO_x + 2xe^- \tag{3.7c}$$

A direct reduction process via electrolysis of molten ($CaCl_2$ + CaO) to produce alloys such as Ti6Al4V and Ti29Nb13Ta4.6Zr for biomedical applications using the OS method has been developed by Osaki et al. (2009). To synthesize Ti6Al4V or Ti29Nb13Ta4.6Zr, mixtures of TiO_2, Al_2O_3, and V_2O_5 or TiO_2, Nb_2O_5, Ta_2O_5, and ZrO_2 are being placed into the titanium net shown in Figure 3.4b.

In the electrolytically mediated reaction (EMR)/molten salt electrolysis (MSE) process, schematically shown in Figure 3.4c (Okabe, 2005), the metallothermic reaction is divided into two electrochemical steps as expressed by eqs. (3.8a) and (3.8b).

Purity is effectively controlled by the EMR of CaO in molten $CaCl_2$, the cathodic and anodic partial reactions of which are expressed by eqs. (3.9a) and (3.9b). The overall reaction of the EMR/MSE process is expressed by eq. (3.9c).

$$\text{Cathodic reaction: } TiO_2 + 4e^- \rightarrow Ti + 2O^{2-} \tag{3.8a}$$

$$\text{Anodic reaction: } 2Ca \rightarrow 2Ca^{2+} + 4e^- \tag{3.8b}$$

Electrolysis

$$\text{Cathode: } Ca^{2+} + 2e^- \rightarrow Ti + Ca \tag{3.9a}$$

$$\text{Anode: } C + xO^{2-} \rightarrow CO_x + 2xe^- \tag{3.9b}$$

$$\text{Overall reaction: } TiO_2 + C \rightarrow Ti + CO_2 \tag{3.9c}$$

The direct electrode position process using a direct current electroslag remelting (DC-ESR) unit as shown in Figure 3.5 (left) (Niinomi, 2012) is applied to produce liquid Ti at the bottom of the unit containing Ti ions. The reactions during this process are expressed by the following equations:

$$\text{Cathodic reaction: } Ti^{4+} + 4e \rightarrow Ti \tag{3.10a}$$

$$\text{Anodic reaction: } xO^{2-} + C \rightarrow CO_x + 2xe^-, \; x = 1 or 2 \tag{3.10b}$$

The DC-ESR process can also be applied to produce Ti alloys such as Ti–Fe and Ti–Al series alloys.

To further reduce the oxygen content of Ti, zone refining processes such as optical floating zone (OFZ; Dabkowska and Dabkowski, 2010) melting or traveling solvent zone (TSZ) refining processes (see box) under an extremely low partial pressure of oxygen (Hagiwara et al., 2009) are effective means to obtain purified

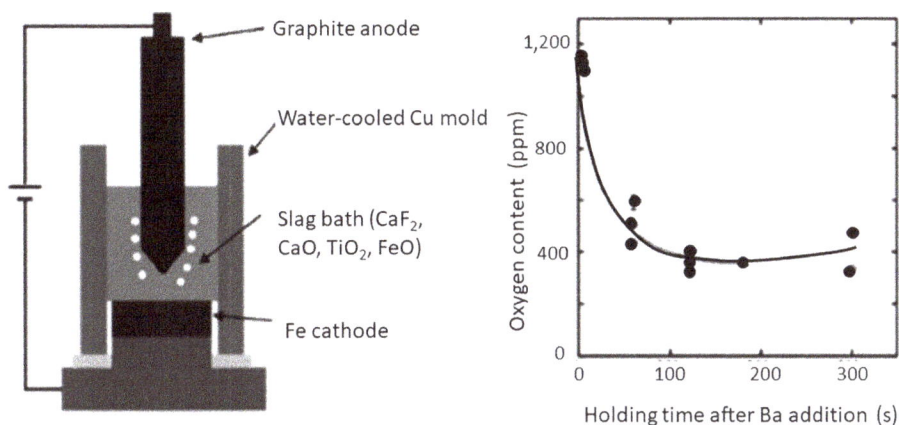

Figure 3.5: Left: Schematic drawing of direct electrode position process (Niinomi, 2012). Right: Change in oxygen content in NiTi alloy melts with holding time after barium addition at 1,673 K (Miyamoto et al., 2008). © With permission by The Japan Institute of Metals and Materials.

titanium. Since the oxygen concentration decreases along the crystal growth direction, oxygen depletion refinement of titanium can be achieved using the OFZ or TSZ processes.

Traveling solvent zone refining process
In the traveling heater zone refining process, a circular mobile induction-coil heater is fixed at one end of a horizontally arranged metal rod which is made up of the impure metal. When the mobile heater is moved slowly along the metal rod, a thin melt zone develops in which the impurities concentrate. This is because the impurities are more soluble in their corresponding melt state. Hence, the impurities in the metal are concentrated at the leading edge by the molten zone and move through the metal rod, leaving the solidified pure metal behind. The pure metal is left to solidify as the heater moves along the rod. Eventually, the impurities accumulate at one end of the metal rod and this small part of the rod can be cut off. The OFZ melting process uses vertically arranged metal rods that are being heated at the focus of four high power halogen lamps.

As an alternative, a deoxidation process using Ba as an oxygen sink has been proposed. Figure 3.5 (right) (Miyamoto et al., 2008; Niinomi, 2012) shows the decrease in oxygen content in NiTi alloy melts with holding time after Ba addition at 1,673 K.

The removal of oxygen has also been reported in an electron beam-melted Ti–Al alloy melt by adding excess Al and in an electroslag-remelted Ti6Al4V alloy melt by adding CaF_2 flux and metallic Ca. In addition, deoxidation routes of Ti and Ti–Al alloy melts by adding Ca or Y (Iizuka et al., 2019) as well as by plasma-arc melting using hydrogen are feasible.

3.1.3.2 Melting

Numerous methods to melt Ti and its alloys have been applied including vacuum arc melting (VAR) with consumable and nonconsumable electrodes, EBM, plasma arc melting (PAM), vacuum induction melting (VIM) using a refractory crucible, cold crucible induction melting (CCIM) using a crucible composed of water-cooled copper segments, and cold crucible levitation melting (CCLM). These methods are schematically shown in Figure 3.6 (Niinomi and Nakai, 2017).

VAR is the main melting process to fabricate large Ti ingots, but EBM and PAM are also receiving considerable attention. The advantages of EBM over VAR are as follows:

– Use of a consumable electrode can be omitted.
– Not only ingots with circular cross sections but also those with rectangular cross sections can be fabricated, and the ingot can be directly subjected to hot-rolling immediately after casting.
– Microstructurally and compositionally homogeneous ingots can be fabricated.
– High-quality products can be produced by removing nonmetallic inclusions.
– The yield of the process is excellent because the number of the melting cycles can be reduced: one-time melting is sufficient whereas two or three times melting is required in case of VAR.

However, when attempting to fabricate a large ingot requiring a large melt pool on top of the ingot, EBM (Figure 3.6B) shows a disadvantage as liquid segregation is likely to occur. To solve this problem, it is necessary to reduce the size of the melt pool by either reducing the applied electric power or the efficiency of the melting process. It is difficult to inhibit or, ideally, prevent evaporative losses from the melt pool of elements with high vapor pressure such as Al, Sn, and Cr when developing the alloy ingot since EBM melting is conducted under high-vacuum conditions.

Figure 3.7 shows a direct cast slab of Ti fabricated by EBM and direct hot rolling immediately after casting (Niinomi and Kagami, 2016). Application of advantageous wrought processes such as blooming or hot forging to fabricate the slab by hot rolling from the cast ingot, followed by full-scale manufacturing of Ti sheet can significantly shorten the manufacturing process.

Table 3.1 compares the characteristic process parameters of Ti using VAR, EBM, and PAM (Niinomi and Nakai, 2017; Report of Research Study on Cost Reduction of Titanium Manufacturing-Overview, 2005, Niinomi, 2017).

VIM, CCIM, and CCLM (Figure 3.6D–F) are induction-heated melting processes that are advantageously utilized to fabricate small ingots. Owing to electromagnetic induction stirring, a homogeneous distribution of the alloying elements can be achieved, including those elements the melting points or specific gravities of which differ substantially from those of Ti. In case of CCLM (Figure 3.6F), impurity contaminations can be reduced since the melt does not contact the crucible walls as the magnetic force causes the melt to float.

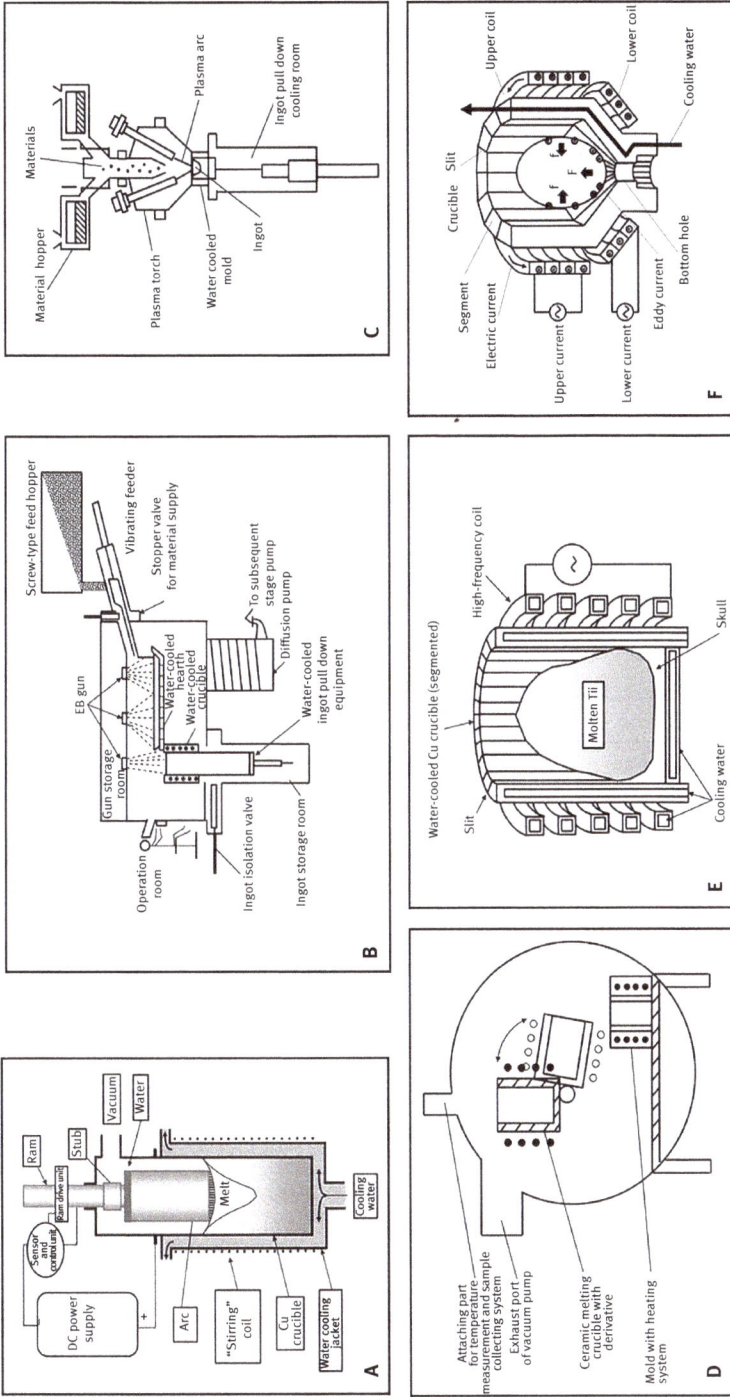

Figure 3.6: Melting methods for Ti and Ti alloys. (A) Vacuum arc melting (VAR); (B) electron beam melting (EBM); (C) plasma arc melting (PAM); (D) vacuum induction melting (VIM) with refractory crucible; (E) cold crucible induction melting (CCIM) with a crucible using water-cooled copper segments; and (F) cold crucible levitation melting (CCLM) (Niinomi and Nakai, 2017). © With permission by The Japan Institute of Light Metals.

Figure 3.7: Direct cast Ti slab (left) and conventional titanium ingot (right) (Courtesy of Nippon Steel Co., Ltd. and TOHO Titanium Co., Ltd.).

Table 3.1: Comparison of characteristics of vacuum arc melting (VAR), electron beam melting (EBM), and plasma arc melting (PAM) (Niinomi and Nakai, 2017).

	VAR	EBM	PAM
Heat source	Electric arc	Electron beam	Plasma arc
Range of vacuum	0.1–1.0 Pa	0.01–0.1 Pa	Argon <0.1 Pa
Compositional control	By homogeneous distribution of elements in the consumer electrode	Special consideration required for elements with high vapor pressure (Al, etc.)	Relatively good
Melt pool depth	Deep	Shallow	Shallow

3.1.3.3 Forming

Forging

Plastic forming of Ti and its alloys can be carried out by various method such as bending, pressing, roll-forming, and forging. Among them, forging is the most important method to fabricate products composed of Ti alloys. According to the operation temperature, forging processes are categorized as cold, warm, and hot forging. Cold forging is conducted at a temperature between room and recovery temperatures, at approximately $0.2T_m$, whereby T_m is the material's melting point. Warm forging is carried out at a temperature between the recovery and recrystallization temperatures at approximately $0.4T_m$. Hot forging is conducted at a temperature above the recrystallization temperature (from $0.4T_m$ to $0.6T_m$) when work hardening disappears.

Typical forging processes are open-die-forging (Figure 3.8A) and closed-die-forging processes (Figure 3.8B) (Chandrasekaran, 2019) whereby open-die forging is limited to producing products with simple geometries. Hot die forging and isothermal forging are appropriate to fabricate biomedical prostheses. During hot forging, the die is heated to a temperature between 200 and 300 °C. Therefore, the temperature of the product surface rapidly decreases, and the temperature difference near the surface and inside the surface becomes large because the temperature of the inside increases as a result of plastic working. Then, the microstructural dehomogenization difference near the surface and the inside of the billet becomes large, too. To solve this problem, the die is heated to a temperature between 925 and 1,260 °C. For $(\alpha + \beta)$-type titanium alloys such as Ti6Al4V, forging starts in the β-region (β-processing) and finishes in the $(\alpha + \beta)$-region to achieve a desired equiaxed α-structure, obtained by annealing at a temperature near 700 °C. The β-processing only results in a needle-like structure (acicular α-structure, Widmannstätten structure) (see Chapter 2.1.1.4).

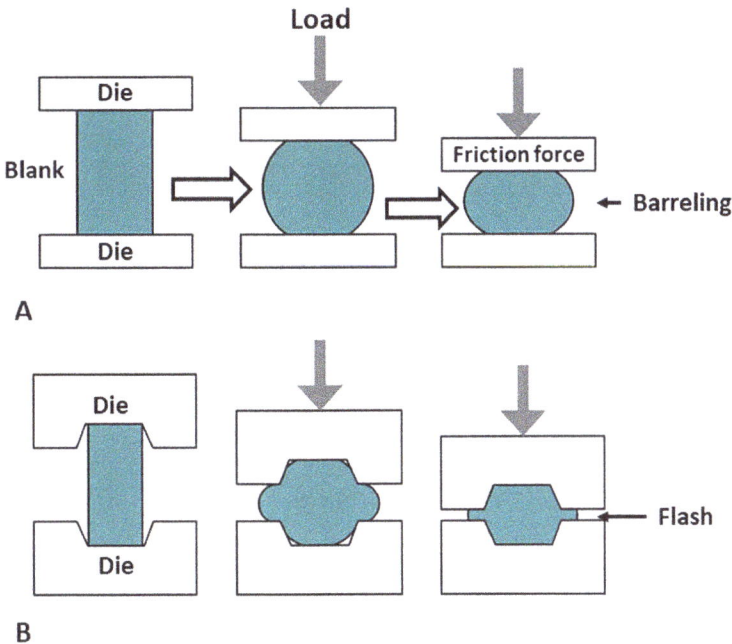

Figure 3.8: Schematic drawings of (A) open-die and (B) closed-die forging processes (modified after Chandrasekaran, 2019). © With permission by Elsevier.

During (α + β) processing, the microstructure changes depending on the forging reduction ratio (see box) as shown in Figure 3.9 (Inagaki, 2008). When the forging reduction ratio is low, the equiaxed α-structure with excellent mechanical properties cannot be obtained because of lack of plastic strain required to achieve a fully recrystallized structure. To obtain a fully recrystallized structure (equiaxed α-structure), an appropriately high forging reduction ratio is required. During isothermal forging, superplastic behavior can be achieved in products composed of titanium alloys such as Ti6Al4V, Ti4.7Al3V2Fe2Mo (SP-700), Ti5Al2.5Fe, and Ti6Al7Nb, which show superplastic behavior when the forging is conducted at appropriate temperature and deformation rate. Most of the final shapes of the products can be obtained using this type of isothermal forging similar to near-net shaping (see below).

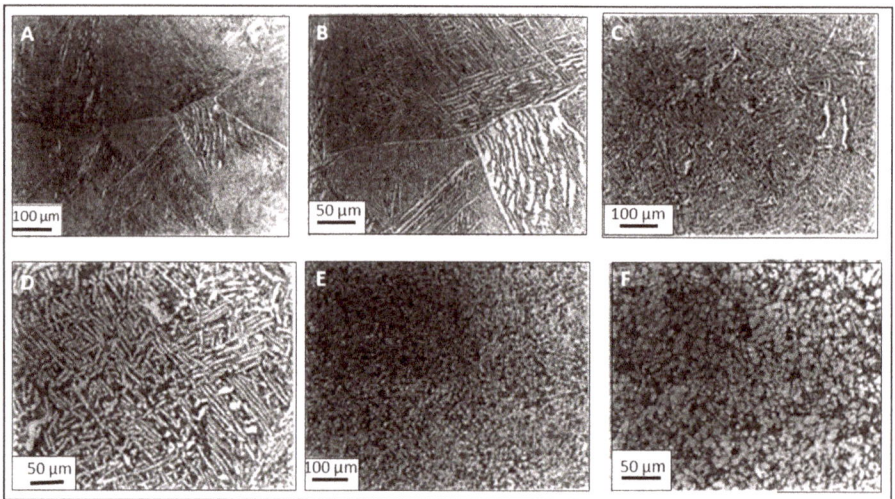

Figure 3.9: Variation in microstructure of (α + β)-type Ti alloys by forging with different reduction ratios. Heat treatment: 700 °C, 2 h. AC. (A) and (B) Forging reduction ratio of 22%; (C) and (D) forging reduction ratio of 62%; and (E) and (F) forging reduction ratio of 91% (modified after Inagaki, 2008).

Forging reduction ratio
The forging reduction ratio is defined as the amount of cross-sectional reduction taking place during drawing out of a bar or billet. The original cross section divided by the final cross section is the forging reduction ratio.

Machining

Machining is generally categorized as turning, milling, end-milling, drilling, and cutting. The materials used for cutting tools are carbon tool steels, alloy tool steels, high-speed tool steels, cemented carbide, cermets, alumina ceramics, cubic boron nitride, and diamond. Because the reactivity of Ti and its alloys with other metallic materials is very high, tool materials exhibiting low reactivity (low adhesive ability) toward Ti and Ti alloys and high thermal conductivity are preferably used. This includes high-speed tool steels and cemented carbides. Appropriate selection of water-soluble or water-immiscible cutting fluids is also of importance. Water-soluble cutting fluids are suitable because sufficiently high cooling rates are required. In general, during machining of Ti and Ti alloys, the cutting speed should be slow and the cutting depth deep.

On the one hand, the cutting resistance of cp-Ti and most Ti alloys is lower than that of carbon and stainless steels. On the other hand, the machinability of cp-Ti is better than that of stainless steels, but that of Ti6Al4V is lower than that of both stainless steels and cp-Ti. The reason for relative poor machinability of Ti and its alloys is based on the high strength of some Ti alloys exceeding 1,500 MPa. Consequently, high forces need to be applied to the cutting edge of the tool that, however, may cause chipping and wear of the blade. Since the heat transfer rate is slow owing to very low heat conductivity, heat accumulates in both the tool and the workpiece, and consequently, tool wear increases. In addition, workpieces of Ti and Ti alloys experience large deformation during machining as a result of their low Young's moduli, chatter vibration may occur, and the machining accuracy tends to be low. Chips sometimes burn when machining is done with a worn cutting tool or if the chips are too thin.

3.1.3.4 Near-net shaping

Casting

Casting is among those near-net shape processes that effectively reduce the cost of products. Precision casting is very effective to fabricate products of complex shape, particularly biomedical prostheses as well as dental parts such as crowns, inlays, bridges, and dentures made via dental precision casting. Castings are subject to casting defects such as pores and shrinkages, and coarse microstructures that lead to reduced ductility and fatigue strength. Therefore, improving the mechanical properties of castings is required.

As molten Ti and Ti alloys are generally cast into molds machined from graphite blocks or ceramic molds composed of various types of ceramic, it is mandatory to avoid chemical reactions between the mold material and the Ti alloy melt with very high reactivity. Therefore, oxides thermodynamically more stable than titanium oxide (TiO_2) are preferentially selected as mold materials. The Ellingham diagram (Ellingham, 1944), that is, the temperature dependence of the standard free energy of formation of oxides, is an effective tool to select the proper oxide to make a mold (Figure 3.10; Niinomi, 2019).

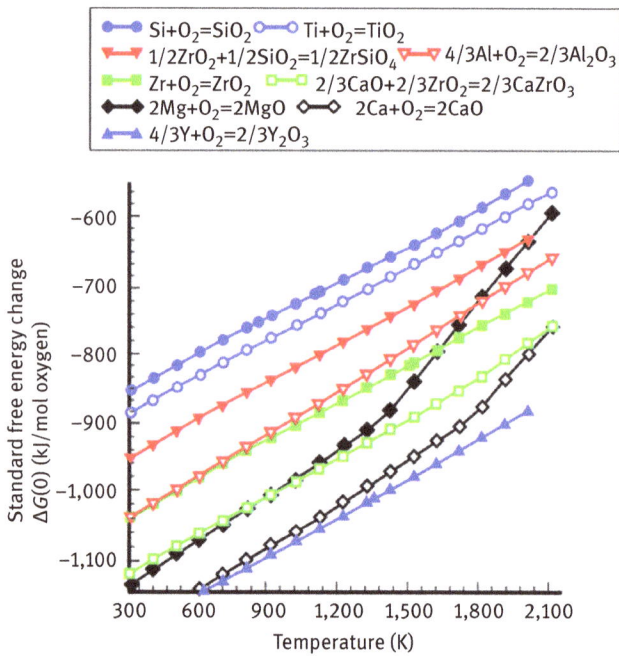

Figure 3.10: Ellingham diagram: standard free energy change during formation of oxides (Niinomi, 2019). © With permission by Elsevier.

Since alumina, magnesia, zirconia, calcia, yttria (Y_2O_3), and other refractory oxides are thermally more stable than TiO_2, they can be used as mold material for Ti casting. Mixtures of these oxides are also effective mold materials when they are of low cost and high operability. Because the solubility of α-stabilizing oxygen in Ti and Ti alloys is very high (see Fig. 2.15), Ti oxide tends to concentrate near the surface of the casting, leading to an α-phase segregation area ("α-case") that results in brittle castings. Therefore, diffusion of oxygen from the mold material into Ti must be inhibited by using a more stable oxide as a mold material compared to TiO_2 as justified above. As the α-case must be removed during fabrication of the final cast product, reducing its formation leads to substantial cost reduction.

Popular mold materials for dental precision casting of Ti and its alloys are silica-, alumina-, or magnesia-based ceramics. Zirconia-, yttria-, and calcia-based mold materials are also frequently used for industrial precision casting of Ti and its alloys.

However, silica- and alumina-based mold materials generally contain phosphate as a binding agent that shows high reactivity with the Ti melt (Niinomi, 2001). Generally employed melting techniques for dental castings include gas melting using combustible gases such as (city gas + air), (propane + air), (city gas + oxygen), (oxygen + acetylene), and (oxygen + hydrogen). Other techniques frequently applied include HFI melting, arc melting, and electric resistance melting.

Fabrication of Ti and Ti products for medical uses, in particular dental applications, is mainly done using precision investment casting methods such as centrifugal precision casting or levicasting. The levicast method (Demukai and Tshuishima, 2005; Niinomi, 2019) is based on a combination of levitation melting and counter-gravity low-pressure casting as schematically shown in Figure 3.11 (Niinomi, 2019). It can be applied to cast implants such as the stems of artificial hip joints. Finally, Figure 3.12 (Niinomi et al., 2001) shows a schematic rendering of the dental precision casting process by the lost-wax (investment casting) technique that is used to fabricate dental prostheses including crowns. Metal mold casting was once attempted for Ti casting, but has not been used in practical applications yet.

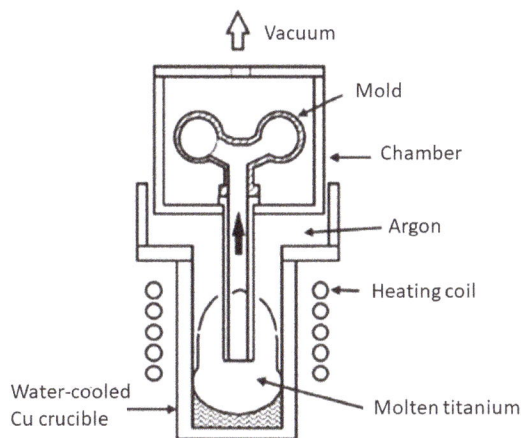

Figure 3.11: Schematic drawing of the LEVICAST process (Niinomi, 2019). © With permission by Elsevier.

Casting dental Ti alloys is commonly performed with general structural Ti alloys. Several titanium alloys for dental applications and their mechanical properties are listed in Table 3.2 (Niinomi, 2019). The most widely used Ti alloys for dental and medical prostheses are cp-Ti and Ti6Al4V ELI. The use of TNTZ and Ti15ZrNbTa for dental castings, originally developed for other biomedical applications, has also been investigated. Ti5Al13Ta (Doi et al., 1998) and Ti40Zr (Niinomi et al., 2001) were also developed for dental castings. Since the high melting points of Ti alloys are one of the causes of the high reactivity of the Ti melt with mold materials, it is

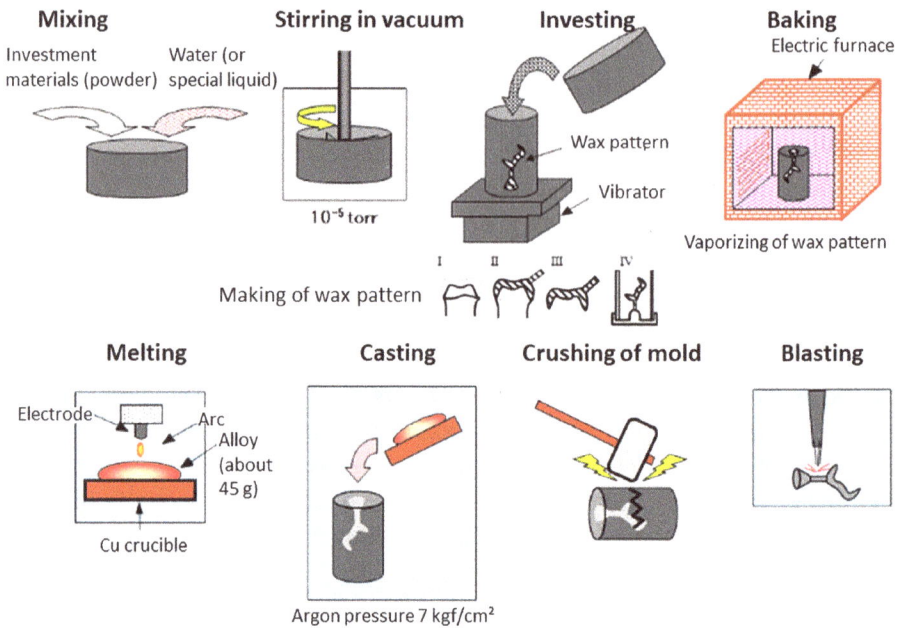

Figure 3.12: Schematic illustration of dental precision casting method (Niinomi et al., 2001).

Table 3.2: Selected titanium alloys for dental applications and their mechanical properties (Niinomi, 2019). © With permission by Elsevier.

Alloy	Process	Tensile strength (MPa)	Yield strength (MPa)	Elongation (%)	Vickers hardness (kgf/mm^2)
Ti20Cr0.2Si	Casting	874	669	6	318
Ti25Pd5Cr	Casting	880	659	5	261
Ti13Cu4.5Ni	Casting	703		2.1	
Ti6Al4V	Casting	976	847	5.1	
Ti6Al4V	Superplastic forming	954	729	10	346
Ti6Al7Nb	Casting	933	817	7.1	
TiNi	Casting	470		8	190

desirable to lower the alloy melting points. To satisfy this purpose, alloying elements with melting points below that of Ti and high solid solubility is effective since they are expected to improve the mechanical properties. Examples include chromium, manganese, palladium, zirconium, and cobalt additions (Niinomi, 2001).

Although TNTZ is useable as a dental casting material, its melting point of approximately 2,300 K is very high (Figure 3.13). Hence, only a CaO mold may be suitable for casting, although CaO is relatively difficult to handle and, consequently, costly. Therefore, the melting point of TNTZ needs to be lowered by replacing the high-melting Ta content with small amounts of Fe, Cr, or Si. Using the d-electron alloy design methodology (see box), the compositions of the alloys are adjusted such that the d-orbital energy level (Md) and bond order (Bo) values of the derived alloys are (nearly) equal to those of TNTZ.

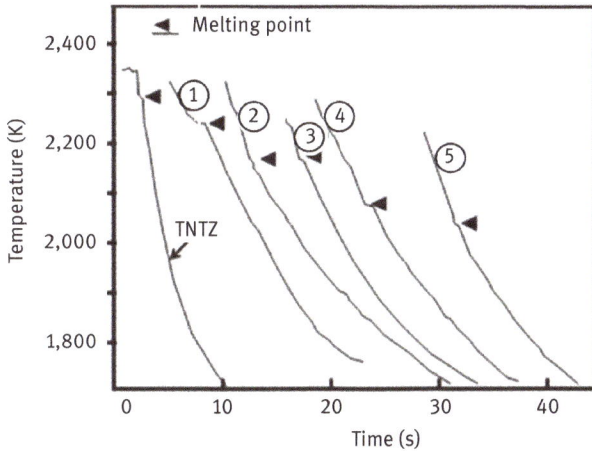

Figure 3.13: Cooling curves of various designed Ti alloys compared to that of the parent TNTZ. 1: Ti29Nb15Zr1.5Fe; 2: Ti29Nb10Zr0.5Si; 3: Ti29Nb18Zr2Cr0.5Si; 4: Ti29Nb10Zr0.5C50.5Fe; 5: Ti29Nb13Zr2Cr (Akahori et al., 2005). © With permission by Elsevier.

d-Electron alloy design
The chemical bonds of transition metal-based Ti alloys are characterized by their d-electron orbitals. Two alloying parameters, the bond order (Bo) and the d-orbital energy level (Md) are used to design such alloys. Bo is a parameter related to the overlapping of the electron clouds between alloying element (M) and mother metal (X), and hence, is a measure of the strength of the covalent bond between M and X atoms. In case when both M and X are transition metals, the d–d covalent bond between them is most important. The more the Bo value increases, the stronger is the chemical bond operating between M and X atoms. The d-orbital energy level Md controls the direction of charge transfer and, hence, is related to the electronegativity. The higher electronegative element has a lower d-orbital energy level. Also, Md correlates with the atomic radius, that is, the larger the atomic radius the higher the d-orbital energy level as well as the (excess) mixing enthalpy. By plotting various alloying elements in a Bo–Md diagram, optimization of alloy properties in terms of strength, elastic modulus or degree of solid solution can be achieved. Following this method, high-temperature α-type Ti alloys, high-strength β-type

Ti alloys, and high corrosion-resistant alloys have been developed. Also, β-type Ti alloys for bio-medical applications with low elastic moduli have been successfully developed (Morinaga et al., 1988; Morinaga, 2016).

As a result of d-electron alloy design, alternative compositions such as (1) Ti29 Nb15Zr1.5**Fe**, (2) Ti29Nb10Zr0.5**Si**, (3) Ti29Nb18Zr2**Cr**0.5**Si**, (4) Ti29Nb10Zr0.5**Cr**0.5**Fe**, and (5) Ti29Nb13Zr2**Cr** have been proposed. Their cooling curves are shown in Figure 3.13 from which the melting points can be extracted (Akahori et al., 2005). Evidently, the melting points of all novel alloys are lower than that of the parent TNTZ, with Ti29Nb13Zr2Cr (5) exhibiting the lowest melting point. Figure 3.14 shows as-cast tensile test specimens consisting of Ti29Nb13Zr2Cr and TNTZ, cast in a modified MgO-based mold material (Akahori et al., 2005). The removability of the mold material is improved for Ti29Nb13Zr2Cr over TNTZ. A similar trend was reported for the other alloy designs.

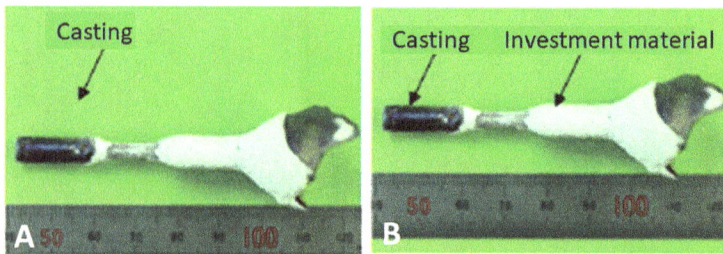

Figure 3.14: Cast tensile test specimens of (A) Ti29Nb13Zr2Cr and (B) Ti29Nb13Ta4.6Zr (TNTZ) (Akahori et al., 2005). © With permission by Elsevier.

Figure 3.15 (Akahori et al., 2005) displays the Vickers hardness of various Ti alloys the cooling curves of which are shown in Figure 3.13 as a function of the distance from the surface of the specimen. The Vickers hardness of each alloy was found to be nearly constant at approximately 200 μm depth, but that of the parent TNTZ was higher than that of all other designed alloys up to approximately 100 μm depth. Therefore, the reactivity of the melt of each designed alloy can considered to be lower than that of TNTZ.

Figure 3.16 (top) (Akahori et al., 2005) shows the average volume fractions of casting defects of various designed Ti alloys and TNTZ in cross sections of tensile test specimens. The average volume fractions of casting defects for Ti29Nb18Zr2Cr0.5Si, Ti29Nb10Zr0.5Cr0.5Fe and Ti29Nb13Zr2Cr are equal or lower than those of TNTZ. As shown in Figure 3.16 (bottom) (Akahori et al., 2005), the mechanical properties of cast Ti29Nb10Zr0.5Cr0.5Fe and Ti29Nb13Zr2Cr alloys reveal a good balance of strength and elongation.

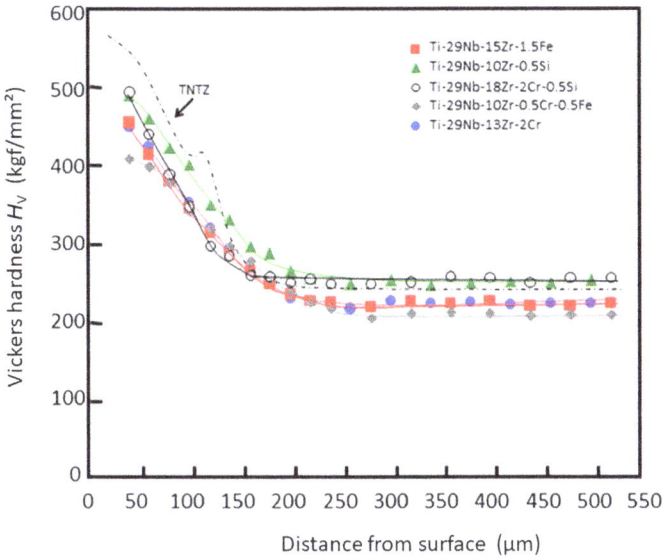

Figure 3.15: Vickers hardness of various Ti alloy designs and TNTZ as a function of the distance from the surface of the specimens (Akahori et al., 2005).

The microstructure of cast metallic dental materials is generally much coarser than that of wrought materials and displays numerous casting defects such as pores and shrinkages as well. Although hot isostatic pressing (HIPing) is an effective way to eliminate microstructural pores of cast $(\alpha + \beta)$-type Ti alloys such as Ti6Al4V and Ti6Al7Nb, their microstructure still remains coarse.

Hence, to refine the microstructures of cast $(\alpha + \beta)$-type Ti alloys such as Ti6Al4V, Ti6Al7Nb, and Ti6.5Al2ZrMoV, thermochemical processing (TCP; see box) is advantageously applied (Akahori et al., 2000; Kou et al., 2015). During TCP, hydrogenation is performed below the β-transus temperature in a hydrogen gas atmosphere. The amount of hydrogen entering the Ti lattice is approximately 1 mass%. Thereafter, solution treatment (ST) is conducted, followed by dehydrogenation under high vacuum conditions. The final hydrogen content is equivalent to or slightly lower that that before TCP since hydrogenation and dehydrogenation are reversible processes.

Thermochemical processing

TCP involves the use of hydrogen as a temporary alloying element of titanium to enhance both processability and final mechanical properties. In this process, hydrogen is added to the titanium alloy by holding the material at a relatively high temperature in a hydrogen environment. The presence of hydrogen allows the titanium alloy to be processed at lower stresses/lower temperatures (because of the increased amount of β-phase) and heat-treated to produce novel microstructures (because the hydrogen allows the material to behave as a eutectoid former system; see Fig. 2.15C) with enhanced mechanical properties after hydrogen removal by vacuum annealing to levels below which no detrimental effects occur (Froes et al., 1990).

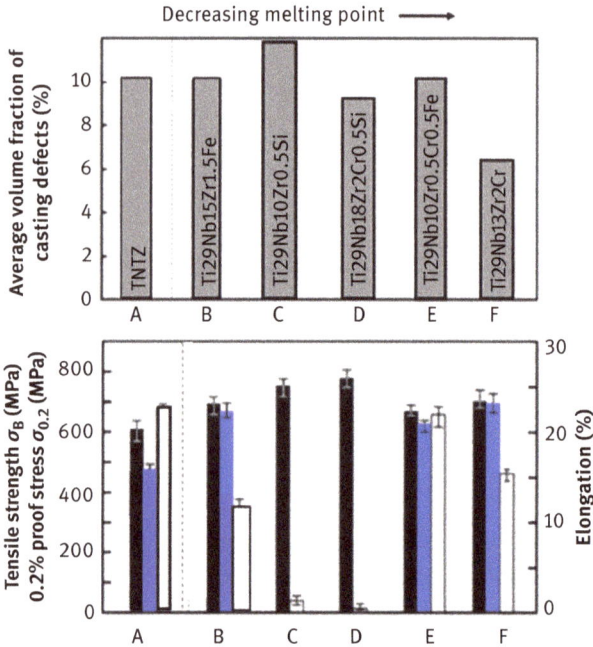

Figure 3.16: Top: Average volume fractions of casting defects of various designed Ti alloys and TNTZ in cross sections of tensile test specimens (Akahori et al., 2005). Bottom: Tensile properties of various Ti alloys. (A) TNTZ; (B) Ti29Nb15Zr1.5Fe; (C) Ti29Nb10Zr0.5Si; (D) Ti29Nb18Zr2Cr0.5Si; € Ti29Nb10Zr0.5C50.5Fe; (F) Ti29Nb13Zr2Cr. Black bars: tensile strength; blue bars: 0.2% proof stress; white bars: elongation (Akahori et al., 2005).

Figure 3.17 (top left) shows SEM micrographs of Ti6Al4V subjected to hot isostatic pressing (HIPing) and to HIPing + TCP (Akahori et al., 2000) whereby the microstructure of the latter is much finer than that of the only HIPed cast Ti6Al4V.

The tensile properties of as-HIPed cast Ti6Al7Nb and HIPed cast Ti6Al7Nb + TCP are shown in Figure 3.17 (right) (Akahori et al., 2000). The tensile strength and 0.2% proof stress of the as-HIPed cast Ti6Al7Nb subjected to TCP are greater than those of the as-HIPed cast Ti6Al7Nb, whereas the elongation of the former is smaller. The S–N curves of these two materials are compared in Figure 3.17 (bottom left), showing that the fatigue strength of as-HIPed cast Ti6Al7Nb has significantly increased through TCP (Akahori et al., 2000).

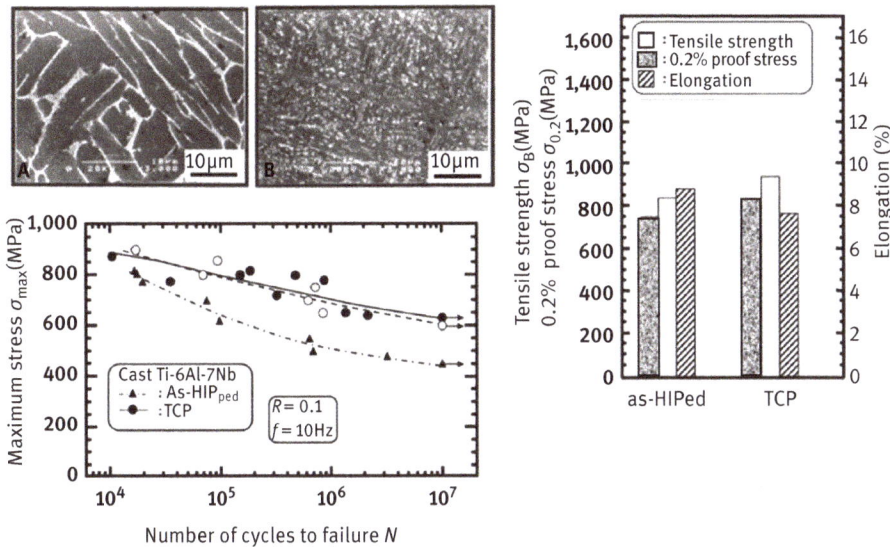

Figure 3.17: Top left: SEM micrographs of (A) HIPed cast Ti6Al4V and (B) HIPed cast Ti6Al4V subjected to TCP. Right: Tensile strength, 0.2% proof stress, and elongated of as-HIPed cast Ti6Al7Nb and HIPed cast Ti6Al7Nb + TCP. Bottom left: S–N curves of as-HIPed cast Ti6Al7Nb and HIPed cast Ti6Al7Nb + TCP (Akahori et al., 2000).

Besides TCP, addition of small amounts of boron, TiB_2, or silicon is an effective way to refine the microstructures of cast Ti alloys. Both boron and TiB_2 are distributed as TiB in the material. The β-grain sizes of cast boron- or TiB_2-added Ti6Al4V fabricated via arc melting or induction skull melting (ISM; see box) are shown in Figure 3.18 (Huang et. al., 2013).

Induction skull melting
The skull crucible process was invented to melt and cast materials with melting points too high for customary crucibles. In essence, by heating only the center of a volume of material, the material forms its own "crucible" from its cooler outer layers. The term "skull" refers to these outer layers that form a shell enclosing the molten volume. Titanium powder is heated, then gradually allowed to cool. Heating is accomplished by radiofrequency (RF) induction using a coil wrapped around the apparatus. The outside of the device is water-cooled in order to keep the RF coil from melting and also to cool the outside of the titanium and thus maintain its shape (see Fig. 3.6E).

Regardless of the melting method applied, the β-grain size of cast Ti6Al4V and the length of the primary precipitated acicular α-phase decrease upon addition of boron or TiB_2. This microstructural refinement is considered to be the result of compositional supercooling. In case of pure Ti or Ti0.5Si, both strength and ductility are improved upon adding an appropriate amount of Si. In case of Ti6Al4V, strength gradually improves by a small addition of Si while the ductility strongly decreases.

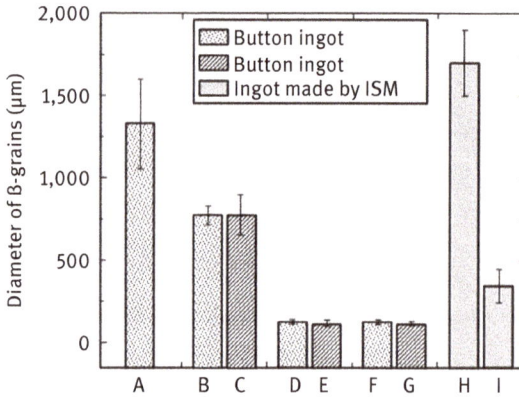

Figure 3.18: Grain diameters of B- or TiB$_2$-added Ti6Al4V (Ti64) fabricated by arc melting (button ingot) and induction skull melting (ISM). (A) Ti6Al4V; (B) A + 0.05B; (C) A + 0.16TiB$_2$; (D) A + 0.1B; (E) A + 0.32 TiB$_2$; (F) A + 0.3B; (G) A + 0.96 TiB$_2$; (H) Ti6Al4V; (I) A + 0.32TiB$_2$ (Huang et al., 2013).

Although it has been reported for many Ti-based castings that small additions of boron or TiB$_2$ refine their microstructure, it is unclear whether ductility increases or decreases. Clearly, more research is required here.

Powder metallurgy

Powder metallurgy (P/M) processes can be generally divided into prealloyed P/M and blended elemental P/M (BEPM) processes. To fabricate prealloyed powder of Ti alloys, atomization processes such as gas atomization, plasma atomization, rotating electrode process (REP), and plasma rotating electrode process (PREP) are utilized. REP and PREP processes are schematically shown in Figure 3.19.

Figure 3.20 shows the spheroidal morphologies of Ti6Al4V powders fabricated by gas atomization, plasma atomization, and the PREP process (Niinomi, 2014). The particle diameters of powders produced via gas atomization, plasma atomization, REP and PREP are up to 500 µm, 25–250 µm, and 50–350 µm, respectively.

Another powder-producing process for Ti and Ti alloys is the hydride-dehydride (HDH) process, mainly used to fabricate commercially pure Ti (cp-Ti). In the hydride-dehydride process, sponge Ti is hydrogenated under atmospheric pressure of hydrogen at approximately 800 °C to produce very brittle TiH$_2$ powder. Then, the TiH$_2$ powder particles are mechanically crushed and dehydrogenated under high vacuum conditions and high temperature to obtain very pure Ti. The powder morphology is irregular blocky as shown in Figure 3.21 (Miura, 2008). Although the oxygen content of the HDH-Ti powders is relatively high compared to that of atomized Ti powders, the HDH-Ti powders are generally less expensive compared to those created by an atomization process.

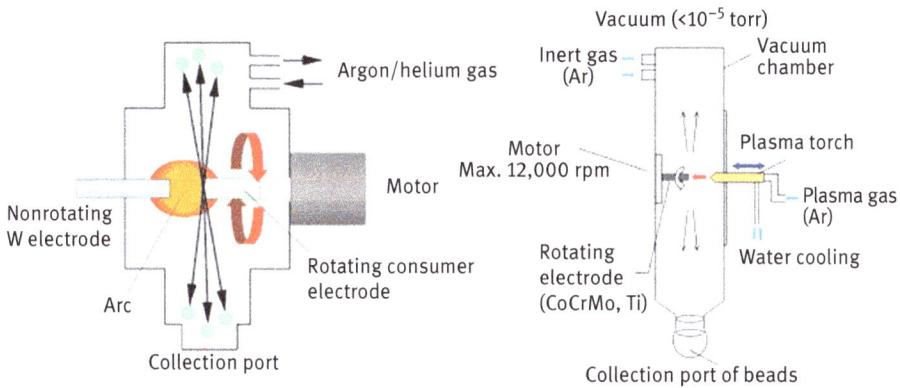

Figure 3.19: Left: Schematic drawing of the rotating electrode process (REP). Right: Schematic drawing of the plasma rotating electrode process (PRE) (Courtesy of Prof. Akihiko Chiba, Inst. of Materials Research, Tohoku University, Sendai, Japan).

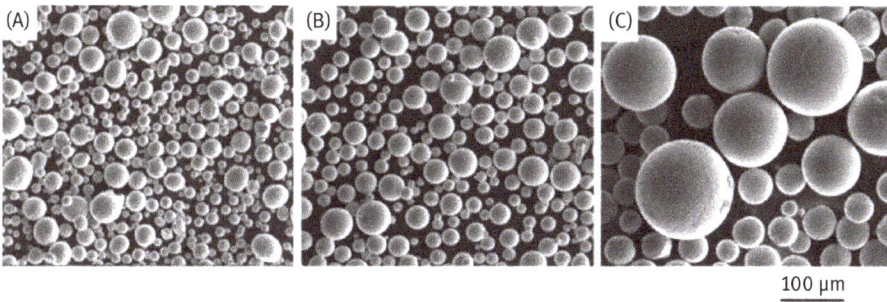

Figure 3.20: SEM micrographs of Ti6Al4V powders produced by (A) gas atomization, (B) plasma atomization, and (C) the plasma rotating electrode process (PREP) (Niinomi, 2014).

Figure 3.21: SEM micrograph of Ti6Al4V powder fabricated by the HDH process (Miura, 2008).

HDH-produced Ti powders are suitable for being processed by mold sintering, during which cold compaction is performed by uniaxial die pressing. However, the inherent strength of the final product is relatively low. Atomized Ti powders are suitable for additive manufacturing (AM) and metal injection molding (MIM). However, HDH-Ti powders can also be processed by MIM.

During the prealloyed P/M process, consolidation of powders is achieved by HIPing at a temperature below the β-transus temperature. Other consolidation processes using prealloyed powders are conventional forging, during which the die is filled with powder and heated, and vacuum hot pressing. The as-HIPed densities of Ti products using PREP powders reach up to 99.99%. However, in the BEPM process, pure Ti powders are first mixed with alloying element powders to obtain the targeted chemical compositions, and then consolidated by cold isostatic pressing (CIPing) in a soft plastic or elastomeric mold (Adams et al., 2007). After CIPing, the powder compacts are sintered by HIPing. The densities of the products using the prealloyed P/M process are in excess of 94%.

Metal injection molding (MIM)

The MIM process is a near net-shape manufacturing process that combines P/M with plastic injection molding. Figure 3.22 shows a schematic rendering of the basic stages of the MIM process.

During processing, metallic Ti powder is initially mixed with approximately 40 vol% of an organic binder whereby its fluidity increases to become similar to that of plastics. The injection molded "green" body is then debonded either by aqueous or solvent treatment or thermally, and sintered under vacuum conditions. By this process, small- and intermediate-sized complex-shaped parts can be mass produced. Powders with an average diameter of 20 μm are used, causing the relative density of the sintered products to exceed 95%. Thus, high-performance Ti alloy parts can be produced using various metallic or alloy precursor powders obtained by gas atomization or the HDH process.

Impurities, in particular, oxygen are important quality factors in Ti products manufactured by the MIM process as well as in other P/M processes. For example, the tensile strength of Ti6Al4V produced by the MIM process increases linearly with increasing oxygen content up to approximately 0.5 mass%, and then, on further increasing the oxygen content decreases drastically. The elongation is maintained up to an oxygen content of about 0.35 mass% and then drastically decreases at still higher oxygen levels.

Additive manufacturing (AM)

AM is a 3D-printing and 3D-prototyping process. During AM processing, 3D-structured bodies with complex geometry can be produced via layer-by-layer addition of material based on CAD strategy. At present, AM processes are receiving

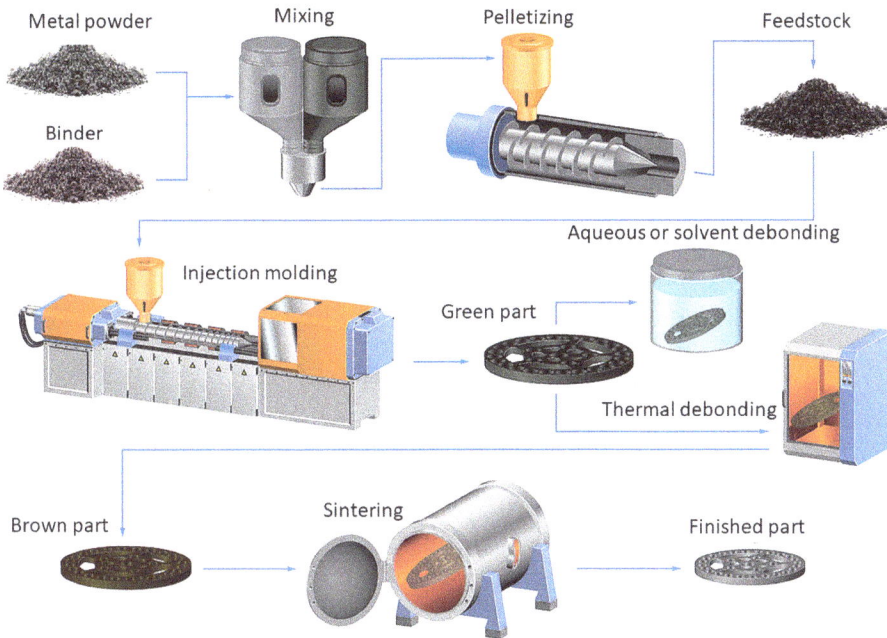

Figure 3.22: Schematics of the basic stages of the metal injection molding (MIM) process (https://en.wikipedia.org/wiki/Metal_injection_molding). © Permission granted under Creative Commons Attribution-ShareAlike License.

considerable attention for 3D near-net shape forming of Ti and its alloys. These processes include powder bed fusion, binder jetting, directional energy deposition (DED), sheet lamination, material extrusion, and material jetting methods. Among these methods, powder bed-fusion methods such as SLM (Figure 3.23) and EBM (Figure 3.24, left) are prominently used.

The 3D geometry data of the final product shape are generated and assisted by a 3D-CAD system using a 3D scanner or computerized tomography (CT) scans. Then, the 3D data are sliced into 2D data using appropriate software. The 3D body of the final product is printed layer-by-layer. The AM process using a laser beam is schematically shown in Figure 3.24 (right). To initiate the AM process, Ti powders are deposited on a substrate using a roller or scraper. Based on the 2D-sliced data, the desired 2D cross section of the first layer is melted using laser or electron beams and solidifies. Subsequently, the substrate is moved downward by the incremental thickness of each layer that is set by the user, and a new layer is deposited. This layer-by-layer process is continued until the final product shape has been obtained.

Whereas the SLM process is carried out under inert argon or helium gas atmospheres, the EBM process uses high vacuum conditions instead. The appropriate powder size is + 45–105 μm for the SLM process and + 10–45 μm for the EBM process

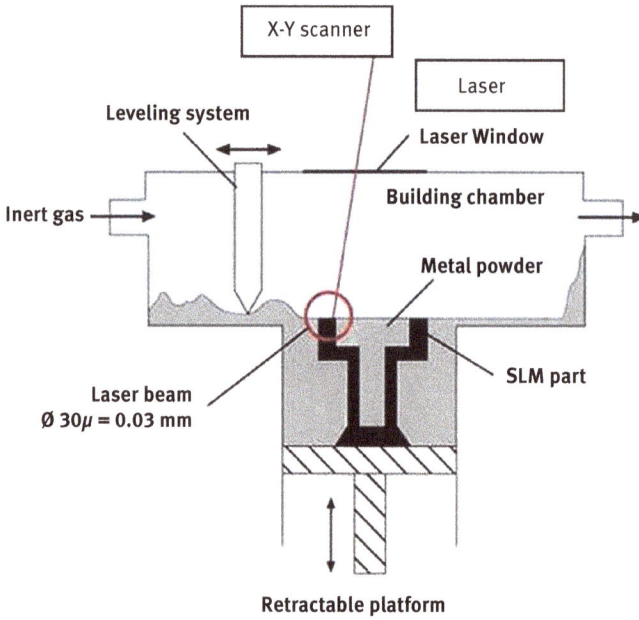

Figure 3.23: Schematic of the selective laser melting (SLM) process. © 2019 Compolight.

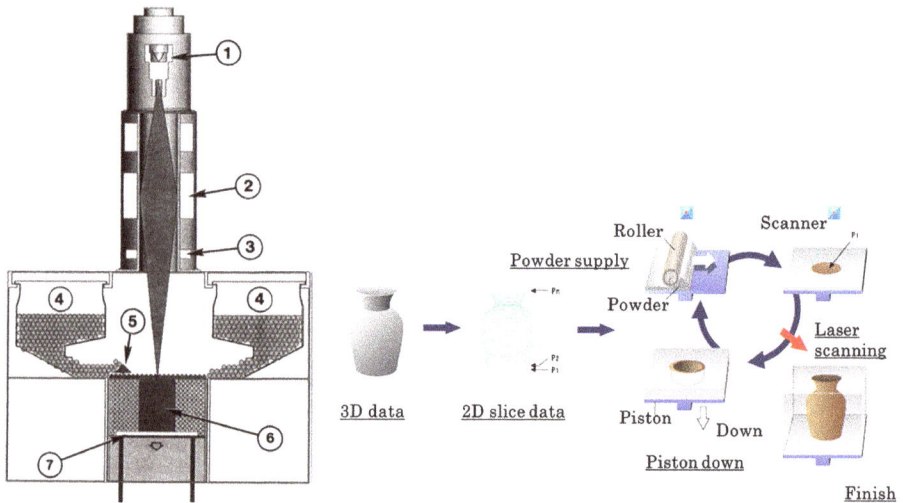

Figure 3.24: Left: Schematic of an electron beam melting (EBM) system. (1) Electron gun assembly; (2) EB focusing lens; (3) EB deflection coils (x–y); (4) powder hopper; (5) powder (layer) rake; (6) cylindrical build; (7) build table. Right: Right: Schematic rendering of 3D near-net shape laser forming (Courtesy of SOKEIZAI Center, Tokyo, Japan).

because the energy absorption rate using a laser beam exceeds 80%, higher than that of an electron beam used in the EBM process.

To fabricate a product with excellent mechanical properties, various process parameters must be optimized including beam power, scan speed, frequency in the case of SLM process, powder layer thickness, and atmosphere. For example, Figure 3.25 shows the effect of the layer thickness on the microstructural and mechanical properties of Ti6Al4V fabricated using an SLM process. Macro-pores are observed when the incremental powder layer thickness exceeds 150 μm, but are absent when the powder layer is only 100 μm. In this case, the tensile strength of the Ti6Al4V product exceeds 1,000 MPa, somewhat higher than that of the ASTM standard value of 950 MPa. Figure 3.26 shows Ti6Al4V implants with different compartment sizes fabricated by the EBM process and designed for regenerating bone with high quality.

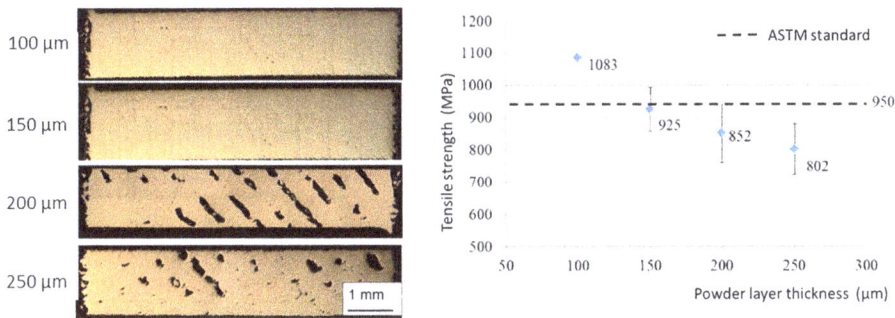

Figure 3.25: Left: Macrographs of cross sections of laser-formed Ti6Al4V with various layer thickness. Right: Effect of powder layer thickness on tensile strength of laser-formed Ti6Al4V (Courtesy of SOKEIZAI Center, Tokyo, Japan).

Superplastic forming

Ti alloys such as Ti6A4V and Ti4.7Al3V2Fe2Mo (SP-700) show abnormally high ductility under low stress conditions. This phenomenon is called superplasticity that can occur in fine-grained materials (see box). The superplastic forming temperature for Ti6Al4V is approximately 850 °C, but for SP-700, it is some 100 °C lower. For example, maxillofacial prostheses, dentures, and blade-type artificial dental roots can be fabricated using a superplastic forming route. Hot-press forming (Figure 3.27, right; Wakabayashi, 1992) or blow-molding under vacuum conditions or using an inert gas such as argon (Figure 3.27, left; Nakahigashi and Yoshimura, 2002) are suitable to form dental parts and dentures.

Figure 3.26: Ti6Al4V implants with various compartment sizes fabricated by EBM for regenerating bone with high quality (Courtesy of Prof. T. Nakano, Graduate School, Osaka University, Osaka, Japan).

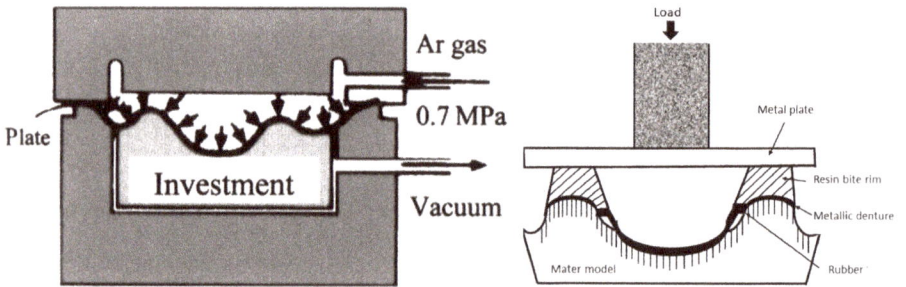

Figure 3.27: Schematics of techniques applied to fabricate dentures. Left: Blow-molding under argon (Nakahigashi and Yoshimura, 2002). © With permission by The Japan Inst. of Metals and Materials. Right: Hot-press forming (Wakabayashi, 1992). © With permission by The Stomatological Society of Japan.

Superplasticity

Superplasticity is a material property by which solid crystalline material is deformed well beyond its fracture point, usually over about 600% during tensile deformation. In metals, but also ceramics, requirements for superplasticity include a fine grain size of less than approximately 20 μm and a fine dispersion of thermally stable particles that pin the grain boundary movement and maintain the fine grain structure at the high processing temperatures as well as the existence of two phases required for superplastic deformation. Although the exact mechanism of superplasticity in metals is still under debate, it appears to rely on atomic diffusion and the sliding of grains past each other during deformation. When metals are cycled around their phase transformation temperature, internal stresses are produced and behavior akin to superplasticity may develop.

3.1.4 Magnesium

3.1.4.1 Production of magnesium

At present, metallic magnesium is produced predominantly by three industrial processes: the silicothermal Pigeon-Bolzano process, the Dow process, and the Norsk Hydro process. Schematics of the process flows are shown in Figure 3.28. Several other processes such as the electrolytic solid oxide reduction of MgO, and the I.G. Farben, the MagCan, and the National Lead Industries processes appear to be less frequently used.

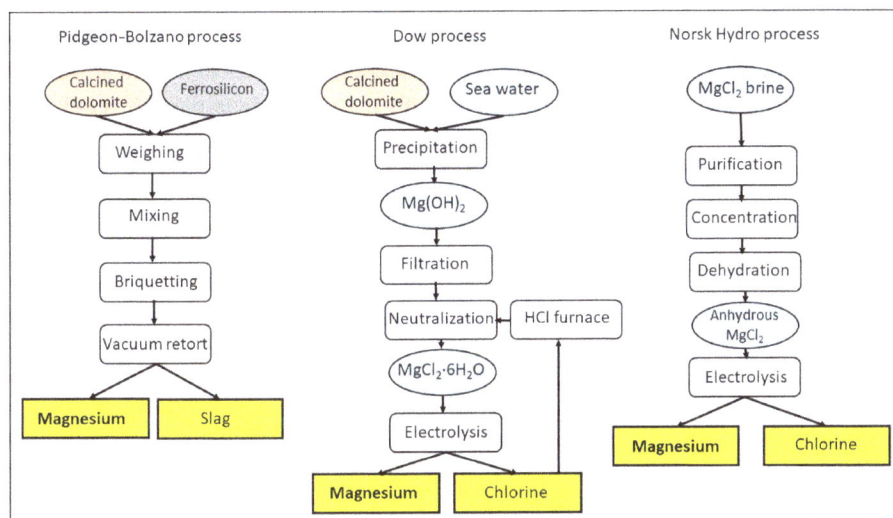

Figure 3.28: Process flow diagrams of common magnesium production routes.

The *Pidgeon-Bolzano process* involves the reduction of magnesium oxide obtained from calcined dolomite at high temperatures with silicon (or carbon) according to

$$2MgO + 2CaO + Si \rightarrow 2Mg + Ca_2SiO_4 \tag{3.11}$$

The related *Magnetherm process* was developed by Pechiney-Ugine-Kuhlman in 1963 in France. Aluminum or bauxite are added to the calcined dolomite charge in a furnace at 1,300–1,700 °C and act to maintain the calcium silicate slag liquid so that it may be tapped and thus separated from magnesium. The process is currently used in France, the USA, and Japan.

Although an energy and labor intensive form of thermal reduction and thus one of the least efficient methods of magnesium production, over the past 20 years, it has become the most prevalent. This is in spite of the fact that, according to Ellingham diagrams, the reaction (eq. (3.11)) is thermodynamically unfavorable, in accordance with Le Chatelier's principle. However, it can still be driven to the right side of the reaction equation by continuous supply of heat, and by removing one of the products, namely either distilling off the magnesium vapor or tapping the siliceous liquid slag. The Pidgeon process is a batch process during which finely powdered calcined dolomite or magnesite and ferrosilicon are mixed, briquetted, and charged in retorts made of stainless steel. The hot zone of the retort is either gas-fired, coal-fired, or electrically heated, whereas the condensing section equipped with removable baffles is water-cooled. By distillation, high purity magnesium is produced that is remelted and cast into ingots.

The *Dow process* produces magnesium by electrolysis of fused magnesium chloride obtained from salt brine or sea water. The magnesium chloride at these sources still contain significant amounts of water and must be dried before it can be electrolyzed to produce metal. Although the first magnesium metal extracted from sea water was produced by Dow Chemicals at their Freeport, Texas plant in 1948 that operated until 1998, presently the only remaining salt water magnesium producer is the Dead Sea Magnesium Ltd. (Israel), a joint venture between Israel Chemicals Ltd. (ICL) and Volkswagen AG.

A saline solution containing Mg^{2+} ions is first treated with lime (calcium oxide) obtained from calcined dolomite and the precipitated magnesium hydroxide is collected according to

$$Mg^{2+} + CaO + H_2O \rightarrow Ca^{2+} + Mg(OH)_2 \qquad (3.12)$$

The magnesium hydroxide is then converted to a partial hydrate of magnesium chloride by treating it with hydrochloric acid and heating of the product according to

$$Mg(OH)_2 + 2HCl \rightarrow MgCl_2 + 2H_2O \qquad (3.13)$$

The salt is then electrolyzed in the molten state. At the cathode, the Mg^{2+} ions are reduced by two electrons to metallic magnesium:

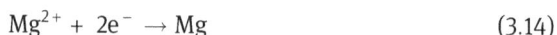

$$Mg^{2+} + 2e^- \rightarrow Mg \qquad (3.14)$$

At the anode, the Cl^- ions are oxidized to chlorine gas, releasing two electrons to complete the electrolysis circuit:

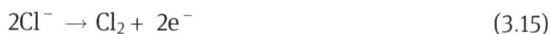

$$2Cl^- \rightarrow Cl_2 + 2e^- \qquad (3.15)$$

The related *Norsk Hydro process* starts with a magnesium chloride brine which could be prepared from a mineral base, sea water, or bittern, dependent on comparative costs. The brine is then purified, concentrated, and dehydrated in prilling towers to pure anhydrous magnesium chloride. The by-product chlorine from the subsequent electrolysis may be sold or recycled to the process.

The *I.G. Farben process* relies on calcination of magnesium hydroxide into magnesium oxide and chlorination in the presence of carbon to yield anhydrous magnesium chloride cell feed. The by-product chlorine from this process is recycled to a chlorinator. First developed by the German I.G. Farben Industries in the 1930s, it is currently used in China, Russia, and the USA.

The *solid oxide membrane* technology involves the electrolytic reduction of MgO (Figure 3.29A). At the cathode, Mg^{2+} ions are reduced to magnesium metal.

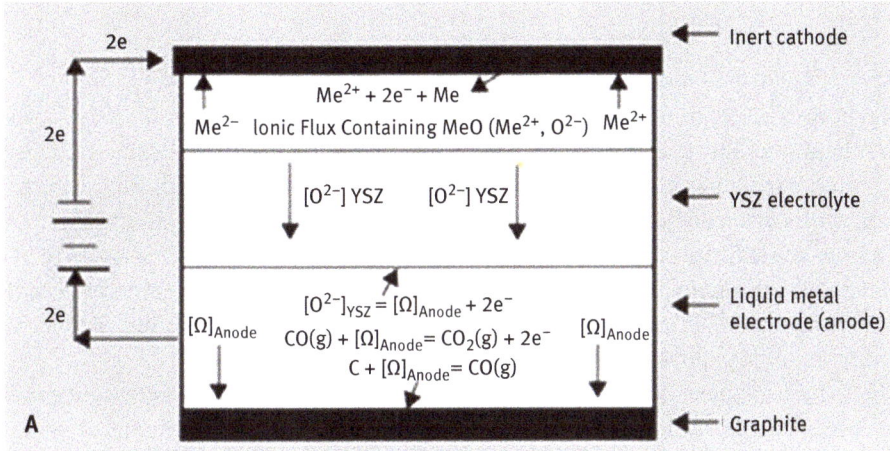

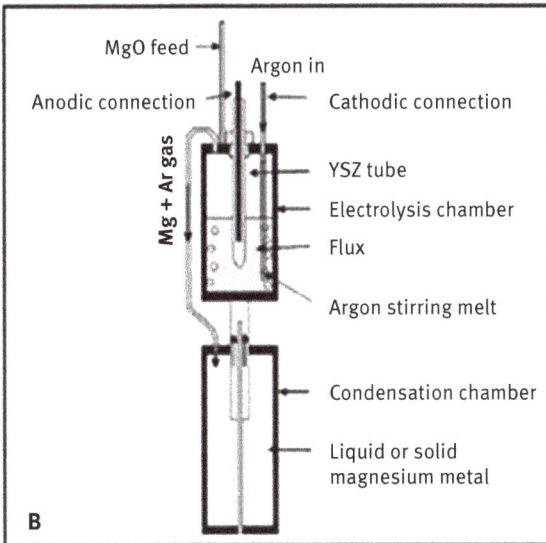

The electrolyte is yttria-stabilized zirconia (YSZ). The anode is a liquid metal. At the YSZ/liquid metal anode O^{2-} is oxidized and the electrons released are used to re-duce MgO (Figure 3.29B). A layer of graphite borders the liquid metal anode, and at this interface carbon and oxygen react to form carbon monoxide. When silver is used as the liquid metal anode, there is no reductant carbon or hydrogen needed, and only oxygen gas is evolved at the anode (Pal and Powell, 2007). It has been reported that this method provides a 40% reduction in cost per pound over the elec-trolytic reduction method. This method is more environmentally sound compared to others because there is much less carbon dioxide emitted.

3.1.4.2 Alloying of magnesium

Magnesium in its pure form is insufficient for biomedical application as it shows low hardness and strength as well as poor corrosion resistance. Hence, alloying with other metals is required such as aluminum, zinc, manganese, silicon, copper, rare earths elements (REE), or zirconium. Alloying with aluminum, zinc and zirco-nium promote the mechanical properties by precipitation hardening, whereas man-ganese is known to improve corrosion resistance. For a comprehensive recent review on the effect of different alloying elements on mechanical and biological properties of Mg alloys, see Riaz et al. (2018).

The designation of magnesium alloys is usually given by two letters followed by two numbers. Letters refer to main alloying elements (A = aluminum, Z = zinc, M = manganese, E = REE). Numbers indicate the nominal compositions of main alloy-ing elements. For example, the common AZ91 alloy is composed of roughly 9 mass% aluminum (A) and 1 mass% zinc (Z) in addition to 90 mass% magnesium. However, real compositions are more complex as shown in Table 3.3. Typical cast Mg alloys are AZ63, AZ81, AZ91, or WE43 (4 mass% Y, 3 mass% REE), typical wrought Mg alloys

Table 3.3: Compositions of selected Mg alloys.

Alloy	Mg	Al	Zn	Si	Mn	Others
AE44	92	4				4% REE (Ce, La)
AJ62A	89.8–91.8	5.6–6.6	0.2	0.08	0.26–0.5	2.1–2.8% Sr
AM60	93.5	6	0.1		0.35	
AZ31B	96	2.5–3.5	0.7–1.3	< 0.05	0.2	
AZ81	~ 92	7.5	0.7			
AZ91D	90.8	8.25	0.63	0.035	0.22	0.003% Cu, 0.014% Fe
WE43	93.6					4% Y, 2.25% Nd, 0.15% Zr

are AZ31, AZ61, or ZK60. Table 3.3 shows the composition of selected Mg alloys frequently used for biomedical application.

There have been many experimental attempts to find "the" Mg super-alloy that satisfies all requirements needed for a quintessential biomedical material. For example, recent developments aim to age hardening of magnesium by adding gadolinium and yttrium. This results in precipitation of fine-grained β phase with DO19 (Ti₃Al type) crystal structure and β phase with bco crystal structure at aging temperatures exceeding 200 °C. On the one hand, both precipitates convey increased peak hardness, tensile strength, and 0.2% proof stress but reveal decreased elongation. On the other hand, higher Y contents increase the elongation of the alloys but result in decreased strength (Black and Kohser, 2012; Xu et al., 2016). Presumably, the strengthening effect is related to formation of long-period stacking order (LPSO) phases as discussed below.

A feasible way to control corrosion behavior is to modify Mg by alloying, thus altering its composition so that addition of certain elements inhibit degradation by biocorrosion and ideally induce a uniform corrosion mode that allow precise prediction of survival rates as opposed to localized pitting, grain boundary, or crevice corrosion. Even though improvement of the corrosion performance of Mg through addition of certain alloying elements is accompanied by microstructural refinement expressed by decreased grain size, it is also subject to concurrent phenomena such as intergranular segregation and formation of secondary phases at grain boundaries that may compromise the mechanical performance and corrosion resistance of the material (Chen et al., 2016). Indeed, higher concentrations of alloying elements tend to increase degradation owing to increased formation of galvanic coupling sites at grain boundaries between the magnesium matrix and newly formed alloy phases.

Frequently, zinc, lithium, calcium, strontium, or rare earth elements (REE) and yttrium were used as alloying constituents. The substantial increase in strength of Mg alloys following addition of alloying elements such as REE (Gd) and/or Y can be related to the formation of lamellar LPSO phases. The LPSO structures comprise long-period stacking derivatives of the hexagonal close-packed (hcp)-Mg structure and the Zn/REE distributions are restricted at the four close-packed atomic layers, thus forming local face-centered cubic, that is, local ABCA stacking. Energy-favored structural relaxations of the initial model cause significant displacements of the Zn/REE positions, implying that strong Zn-REE interaction may play a critical role for phase stability. The LPSO phases appear to tolerate a considerable degree of disorder of atoms at the Zn and REE sites with statistical co-occupation by Mg, extending substantially the nonstoichiometric phase region.

Novel Mg-Y-Ca-Zr-based alloys were developed with alloying elements judiciously selected to impart favorable properties (Chou, 2016). Processing techniques including ST combined with hot extrusion were employed to further enhance the desired properties of the material such as controlled corrosion, high strength and ductility, and minimal toxic response. Increasing the Y content contributed to improved corrosion resistance. Hot extrusion was used to reduce the grain size, thereby improving mechanical properties. Extrusion yielded high strength compared to other Mg alloys, with values approaching those of iron-based alloys. This may be related to the presence of Mg12YZr, a LPSO phase that serves to impede dislocation propagation and thus improves mechanical resilience. Both as-cast and extruded Mg-Y-Ca-Zr alloys demonstrated excellent *in vitro* cytocompatibility, eliciting high viability and proliferation of MC3T3 preosteoblast cells and human mesenchymal stem cells (hMSCs). Finally, implantation of Mg-Y-Ca-Zr–based alloys into subcutaneous tissue of mice and intramedullary cavities of fractured rat femora resulted in normal host response and fracture healing, without eliciting any local or systemic toxicity. Thus, the novel alloys demonstrated great potential for applications as orthopedic and craniofacial implant biomaterials, warranting additional preclinical safety and efficacy trials.

3.1.4.3 Manufacturing of magnesium-based biomedical devices

Besides conventional forming of parts by casting and forging, AM techniques are increasingly developed. By AM, individual tissue regeneration could be obtained. Binder-based sintering technologies such as MIM (see Figure 3.22) and, more recently, fused filament fabrication (FFF) of magnesium powder are useful means to produce patient-specific implants tailored to individual needs. The advantage of FFF over MIM is that although no expensive molds are required, individual prototyping of sophisticated shaped parts at low costs is still possible (Wolff et al., 2019). FFF of metals is a novel technique and processing of Mg powders enables manufacturing of flexible filaments and failure-free green parts by using SF_6-free P/M sintering technique. Test specimens and implant demonstrator parts were successfully produced with mechanical properties of up to 177 MPa UTS, 123 MPa yield strength, and 2.8% elongation at fracture. Hence, as the mechanical properties of FFF-produced magnesium are comparable to cast material, FFF appears to be a promising approach to Mg implant production (Wolff et al., 2019).

3.1.4.4 Corrosion protection coatings

Improvement of magnesium designed for biomedical applications focuses on corrosion control that can be achieved by two means: deposition of protective coatings that isolate the surface from aggressive body environment, and alloying the base magnesium metal with biocompatible elements that stabilize the surface against accelerated corrosive deterioration (Chapter 3.1.4.2). The majority of approaches hitherto are based on application of protective coatings, either *in situ* conversion coatings or *ex situ* coatings with compositions that differ chemically from those of the substrate material.

Coating methods applied include predominantly plasma electrolytic oxidation (PEO), and, to a lesser degree, physical vapor deposition (PVD), electrophoretic deposition (EPD), pulsed laser deposition (PLD), galvanic anodization, sol–gel deposition, cold gas dynamic spraying (CGDS), coincident microblasting (CoBlast) and several other, more elusive and less tested, techniques (Heimann and Lehmann, 2015).

In situ conversion coatings consist of MgO or Mg(OH)$_2$ that are directly and reactively obtained from the magnesium substrate by surface oxidation, for example, by PEO, galvanic anodizing, hydrothermal, or biomimetic, that is, chemical methods.

PEO (Yerokhin et al., 1999) is the method of choice, owing to its technical simplicity, environmentally benign character, and experimental versatility in terms of desired coating composition (see box). Hence, it is preferred over other conversion coatings, for example, those based on MgF$_2$ that require immersion of the magnesium substrate into potentially dangerous hydrofluoric acid. A typical PEO coating consisting of MgO is shown in Figure 3.30A. The inherent porosity has subsequently been sealed with phytic acid + Ce(NO$_3$)$_3$ (Figure 3.30B; Jiang et al., 2018).

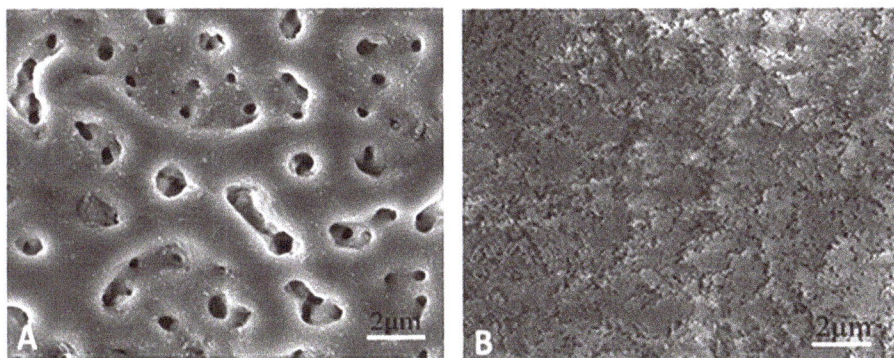

Figure 3.30: Typical surface of a PEO-treated Mg alloy substrate with extensive surface porosity (A) and the same PEO coating subsequently sealed by dip coating with phytic acid + Ce(NO$_3$)$_3$ (B) (Jiang et al., 2018). © With permission by Elsevier.

Plasma electrolytic oxidation

PEO, sometimes called micro-arc oxidation (MAO), is a generic term referring to metal oxidation at potentials where electrical discharge phenomena play a dominant role. The application of this process allows forming an oxide layer with variable porosity on metallic implant materials. In addition, it is feasible to absorb onto these oxide layers some species contained in solutions used for anodizing, yielding chemical surface properties suitable for biomedical applications. The surface discharge process during the PEO process can be divided into three stages. As shown in Figure 3.31A, many small gas bubbles develop at the sample surface immersed in an electrolyte, and a thin oxide film with dopants and defects forms during the conventional oxidation stage (anodizing). During the transition stage shown in Figure 3.31B, an adhering bubble layer is created due to increasing number of gas bubbles. The anions in the electrolyte are collected at the surface of the bubble layer, forming numerous microregions characterized by strong electric fields at the working electrode (anode). Consequently, O_2 and H_2O gases trapped in the bubble layer will be ionized owing to the strong electric field within the microregions. This differs from the traditional view that the initial breakdown process occurs via electron injection into the films, with the subsequent electron avalanche leading to the breakdown. During further increase of the potential, the discharge stage will be reached as shown in Figure 3.31C at which the bubble layer breaks down. The discharge initially occurs at the weakest dielectric points, that is, at locations where dopants and defects exist. During this stage, the evolved gas is mainly oxygen produced under the action of electrochemical and plasma-assisted processes. Near-surface temperatures generated at PEO of magnesium or aluminum can quickly reach 1,200 °C over a wide range of working potentials. The PEO coatings are thus characterized by columnar melt pools with eruption of molten materials at the free surface, and a network of fine cracks, pipes, and pores.

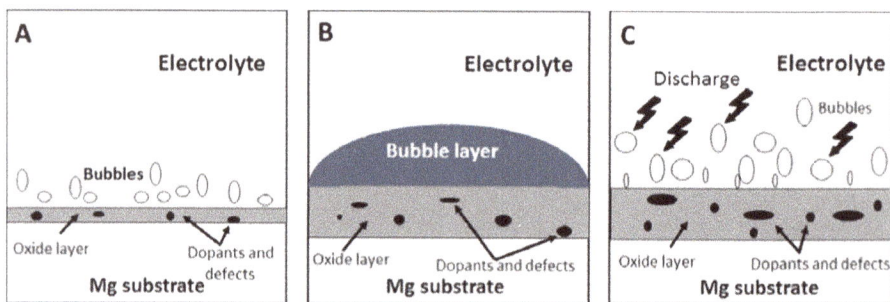

Figure 3.31: Cartoon illustrating the plasma discharge phenomena during different stages of the PEO process. (A) Conventional anodizing, (B) transition, (C) plasma discharge (modified after Wang et al., 2010).

In contrast to *in situ* magnesium conversion coatings that modify the composition of the base metal to yield a coating that predominantly consists of magnesium oxide/magnesium hydroxide, the composition of *ex situ* deposited coatings is independent of that of the substrate. Most of the reported researches on *ex situ* deposited coatings consider calcium phosphates, most prominently hydroxylapatite. Deposition techniques include *physical methods* such as PVD, ion plating/ion implantation, RF magnetron sputtering, EPD, and CGDS as well as blast coating, and *chemical methods* such as chemical vapor deposition (CVD), hydrothermal deposition, pulsed electrodeposition (PED), and sol–gel techniques including dip and spin coatings (Heimann and Lehmann, 2017).

Polymer coatings applied to reduce corrosion of biomedical magnesium implant devices include poly(hydroxyalkanoates), organosilanes, parylene such as poly-*p*-xylene, epoxy resins, phenolics, elastomers such as polyurethanes, poly(lactic acid)

and poly(glycolic acid) and their copolymers, as well as poly(caprolactone)/poly
(lactide-*co*-glycolide) (see also Chapter 3.3.2.3).

3.2 Ceramics (Robert B. Heimann)

3.2.1 General properties of advanced ceramics

Ceramics are, by definition, inorganic, nonmetallic, and mostly polycrystalline materials shaped at room temperature from diverse raw materials by various processes
and subsequently consolidated by sintering at high temperature. In contrast to classic ceramics based on natural raw materials such as clays, advanced ceramics rely
on very fine-grained, high purity, and generally synthetically produced oxide materials such as alumina, zirconia or titania, or nonoxide materials such as carbides,
nitrides, or borides of transition metals.
Advanced ceramic materials are (Heimann, 2010a)
- highly specialized by exploiting unique mechanical, electric, magnetic, optical,
  biological, and environmental properties;
- designed to perform well under extreme conditions such as high temperature,
  high pressure, high stress, high radiation fields, and high corrosive exposure;
- relatively expensive with properties and failure mechanisms not yet fully understood in all their complex details;
- high value-added products owing to sophisticated and, thus, expensive processing technology;
- not presently profitable in terms of return of investment but offering great promises in the future; and
- positioned at the beginning of the development cycle and, hence, not yet widely
  used with respect to potential.

However, in the case of bioceramics, several caveats expressed in the listing above
do not apply anymore as impressive R&D effort in the past two decades has moved
this class of ceramics further up the line toward well-performing, high-value-added
and hence copiously applied advanced materials.
  During shaping of advanced ceramics prior to consolidation by firing at high
temperature, frequently suitable plastic materials are added as binders. These
binder materials will be burned away during the subsequent firing process. The
well-mixed ceramic precursor powders are spray-dried (Bastan et al., 2017) into a
loosely agglomerated, easy flowing powder. After compaction by several techniques
including cold pressing or slip casting, the green ceramics are sintered at high temperature to obtain their typical properties. The strong ionic and covalent bonds
(Figure 3.32) lend advanced ceramics their characteristic mechanical, elastic, thermal, optical, and tribological properties.

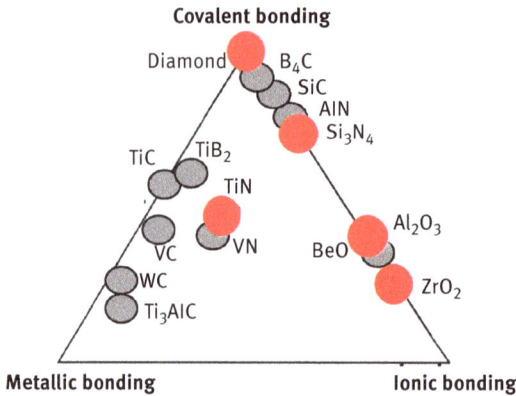

Figure 3.32: Chemical bonding types of different advanced ceramic materials. Ceramics used in biomedical applications are shown in red (Heimann, 2010).

In metals with nonlocalized electrons ("free electron cloud model"), planes of atoms are able to slide past one another by dislocation movement and lend the metallic materials their thermal and electric conductivity, ductility, and high fracture toughness. In contrast, ceramics are electrically insulating and inherently brittle. Consequently, crack formation occurs easily and may cause the material to disintegrate catastrophically. This means that ceramics tend to be strong in compression, but behave very brittle during impact, in tension or under flexural loading as expressed by their generally low fracture toughness. However, the experimentally measured fracture toughness of ceramics tends to be one to two orders of magnitude lower than its theoretical value predicts that can be calculated by considering the strong interatomic forces. Unless ceramics are manufactured under near ideal conditions using precursor powders with optimum properties, they will contain flaws and cracks that act as sites of stress concentration and, eventually, initiate failure in service by catastrophic crack propagation.

Owing to the nature of ceramics, the precursor materials used in their production have a much larger impact on the properties of the final product than in the case of the other groups of materials, that is, metals and polymers. This is because there are no intermediate refinement/purification steps available for ceramics as there are for metals that may include melting, controlled solidification, refining, and eventually plastic deformation designed to improve the properties of the end product (see, e.g., Chapter 3.1.3.3). Not so in the case of ceramics. All imperfections inherent in the green ceramic body such as voids, agglomerates, or inhomogeneous grain size distributions propagate into magnified imperfections within the sintered product. This has been dubbed the "domino effect" that emphasizes the dependence of the final properties of the ceramic product on the characteristics of every processing step, and in particular on the characteristics of the raw material precursor powders. Adding the variability inherent in the manufacturing process such as type of forming, extent of drying, homogeneity of force distribution during pressing, sintering time and temperature, sintering atmosphere, and many other critical parameters, it is evident that the end product of this chain of production steps is uniquely dependent on a myriad of factors and their mutual interactions. This makes the production of advanced ceramic parts intended for use in biomedical applications a highly sophisticated technology, requiring stringent materials and process control, from synthesizing the precursor powder, powder consolidation, and sintering to cutting, shaping, grinding, and polishing the final ceramic biomedical device (see Chapter 3.2.4).

3.2.2 Manufacturing of ceramic powders and their consolidation

As mentioned earlier, the properties of the final ceramic product are critically dependent on the quality of the precursor powders and the unfired, so-called green body. Indeed, many limitations to the performance of ceramics in biomedical use are a direct consequence of nonoptimum physicochemical properties of the powders used during their fabrication. The average particle size, particle size distribution, degree of agglomeration, and dispersion characteristics of the precursor powders combine to limit the macroscopic homogeneity of the ceramics in the green and final sintered states. The microstructural homogeneity of the ceramics can be optimized by processing powders with tightly controlled characteristics. Monodisperse ceramic powders derived from sol–gel processes are prime examples of these materials. Such powders offer many processing advantages over conventional ceramic powders obtained by milling since their near spherical shape and narrow particle size distribution allow for a tight control over the packing density of the powder particles in the green ceramic. Since there are very few agglomerates, the particles pack very densely and uniformly with residual pore sizes in the green ceramic, often being on the order of just two particle diameters, that is, less than about 1 μm. Whereas aqueous dispersions of such powders settle only slowly under ambient gravity conditions, very high green densities are attainable due to statistical ordering of the particles with residual pore sizes often on the order of just one particle diameter. This is a crucial advantage since residual porosity or the presence of particle agglomerates in a sintered structural ceramic such as alumina or zirconia will act as crack initiation sites or may create defects in the material that are decreasing the overall performance and strength of the final product and, hence, their longevity. During processing conventional ceramic powders obtained by mechanical means such as milling, elimination of residual porosity and flaws resulting from poor powder packing and agglomerates is very difficult if not impossible to achieve. Densification is typically accomplished by applying external pressure and/or adding sintering aids to control grain growth during sintering. "HIPing" is extremely cost intensive and therefore undesirable for economic reasons, and also limited to simple geometrical shapes such as cylinders or rectangular blocks. Addition of sintering aids can adversely affect the high temperature properties of the material by lowering its melting/softening point and is therefore also undesirable. However, when using monosized powders, such processing steps are unnecessary to achieve suitable densities, microstructural uniformity, and superior sinterability.

Many different ways to synthesize monodisperse powders by sol–gel technique were described in the literature (see Chapter 3.2.8). The quest is to prepare powders with spherical shape, uniform dimensions, and a low state of aggregation. Powder isolation procedures by centrifuging must be initiated within a short time from the onset of precipitation, otherwise undesirable hard-necked aggregates of individual powder particles may form. The powder are usually washed with ethanol followed by washing with alkaline water to impart a negative charge to the surface of the ceramic particles, thereby providing a net repulsive interparticle potential that effectively inhibits flocculation and, thus, agglomeration (Heimann, 2010a).

3.2.3 Sintering

During firing of the consolidated green ceramic body, changes occur that are related to (i) changes in grain size and shape, (ii) changes in pore geometry, and (iii) changes in pore size and pore size distribution. Specifically, whereas during the onset of sintering the pores become more spherical in shape and smaller in size, subsequently they change shape and shrink to become a 3D network of cylindrical pores (Brinker and Scherer, 1984), presumably by transport of vacancies along these pore channels to the outside of the ceramic body (Iler, 1986). This is schematically shown in Figure 3.33.

During sintering, the density of the ceramic body strongly increases. Driving forces that result in densification are the change in free energy by the decreased surface area and the change in surface free energy by eliminating solid–vapor interfaces. These effects are of substantial magnitude if the powder particles are small and consequently, have a small radius of curvature of only a few micrometers (see box). This is the reason why in pressureless sintering very fine-grained ceramic powders are preferred. During sintering, new low energy solid–solid interfaces develop, triggered by different mechanisms of material transfer (Table 3.4) including evaporation and condensation, viscous flow, surface diffusion, grain boundary or lattice diffusion, plastic deformation, or even transport of vacancies. These processes and their physicochemical underpinnings have been described in much detail by Kingery et al. (1976). See also Iler (1986).

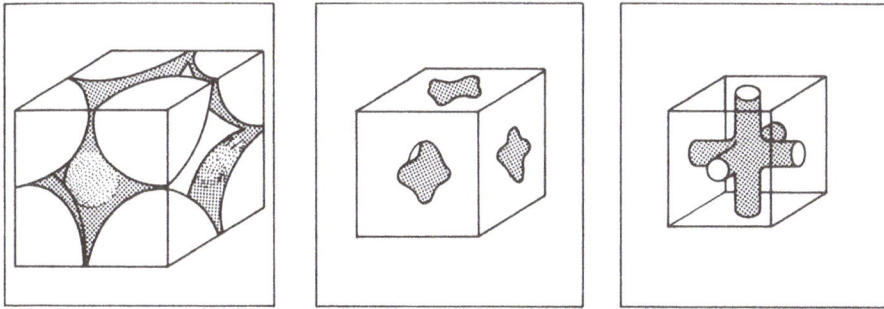

Figure 3.33: Pore cavities change shape and shrink during sintering to become 3D networks of cylindrical pores (Brinker and Scherer, 1984).

Table 3.4: Sintering mechanisms and rates ordered according to decreasing exponent m (Ichinose, 1987; Kingery et al., 1976).

Mechanism	Exponent m	Exponent n	Type (see box)
Viscous flow due to plastic deformation	1/2	1	vi
Evaporation–condensation	1/3	$\Delta L/L = 0$	ii
Grain boundary diffusion controlled by vacancy formation	1/4	1/2	iii
Volume (lattice) diffusion	1/5	2/5	iv
Grain boundary diffusion controlled by vacancy transfer	1/6	1/3	v
Surface diffusion	1/7	$\Delta L/L = 0$	i

On the physics of sintering

Sintering is the thermally activated densification of powder compacts in the solid state to yield a coherent mass without the presence of an intervening liquid, that is, a melt phase. Sintering occurs by diffusion of atoms through the microstructure. The driving force for diffusion and thus, densification is the change in surface free energy caused by (i) the decreasing surface area and (ii) the lowering of the surface free energy resulting in elimination of solid–vapor interfaces (Kingery et al., 1976). Using the so-called equal sphere model (Figure 3.34), this driving force can be described by the equation

$$\sigma = 2\gamma/r$$

where σ is the surface stress, γ the surface free energy, and r the radius of curvature.

Sintering creates new lower energy solid–solid interfaces accompanied by a total decrease in free energy. On an atomic scale, transfer of material is affected by the change in vapor pressure and differences in free energy across curved particle surfaces. This is governed by the Thomson–Freundlich equation,

$$\ln\left(\frac{p_1}{p_0}\right) = \frac{\gamma M}{dRT}\left(\frac{1}{\rho} + \frac{1}{r}\right),$$

where p_1 and p_0 are the vapor pressures over the small radius of curvature and a flat surface, respectively; γ is the surface energy, M is the molecular mass of the vapor, d is its density, and r and ρ are the principal radii of curvature at the surface (see Figure 3.34).

If the size of the particle is small, that is, if its curvature is high, these effects become very large in magnitude. Particles with a radius of curvature below a few micrometers show a high degree of energy change, and thus, sinter more easily. This is one of the main reasons why ceramic technology is generally based on processing materials with suitably fine particle size. There are five common sinter mechanisms, characterized by the different paths the atoms adopt to move from one location to another. They include (i) surface diffusion (diffusion of atoms along the surface of a particle), (ii) vapor transport (evaporation of atoms that condense at a different surface location), (iii) lattice diffusion away from the surface (surface atoms diffuse through the lattice), (iv) lattice diffusion away from grain boundary (grain boundary atoms diffuse through the lattice), and (v) grain boundary diffusion (atoms diffuse along grain boundaries). Another mechanism involves (vi) plastic deformation whereby viscous flow of matter is effectuated by dislocation movement. Mechanisms (i)–(iii) do not lead to decreasing porosity, that is, they merely rearrange matter but do not cause pores to shrink. In contrast, mechanisms (iv)–(vi) result in densification as atoms are moved from the bulk of pores towards their surface and thus, eliminate porosity.

The sintering process can be explained using the simplified equal sphere model shown in Figure 3.34. Into the bonding region, that is, neck area, material is

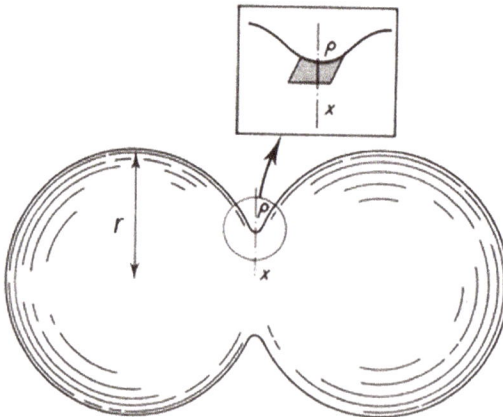

Figure 3.34: Equal sphere model of sintering.

transferred by diffusion. The surface stress across the two surfaces with radii of curvature r and ρ can be expressed by

$$\sigma = \gamma \, (1/r \, - \, 1/\rho) \tag{3.16}$$

Since $r \gg \rho$, the surface stress $\sigma \sim -y/\rho$ is directed outward from the neck area. Depending on the curvature ρ, the lattice vacancy density near the neck increases, thus creating a difference between it and that of the equilibrium area. Consequently, atoms tend to migrate toward the neck area, thus causing sintering.

As mentioned earlier, material transfer takes place by a number of different paths, and sintering processes are divided accordingly as shown by the mechanisms (i)–(vi) indicated in the box above and also shown in Table 3.4. The rates of growth or contraction in the neck area are usually expressed as functions of time as follows:

$$(x/r)^m \infty \, t \quad \text{or} \quad (\Delta L/L)^n \infty \, t \tag{3.17}$$

whereby x/r is the growth rate and $\Delta L/L$ is the shrinkage rate in the neck area. The two exponents m and n characterize the different sintering mechanisms as shown in Table 3.4.

In case of surface diffusion and evaporation–condensation processes, no volume change, that is, no shrinkage occurs ($\Delta L/L = 0$).

In terms of thermodynamics, the sintering process is driven by the reduction of free enthalpy dG as result of the surface reduction dA_s, the reduction of the grain boundary surface dA_b, and the volume reduction dV as expressed by the so-called sintering equation (Aldinger and Weberruß, 2010):

$$dG \, = \gamma_s \cdot dA_s + \gamma_b \cdot dA_b + p \cdot dV \tag{3.18}$$

where y_s is the specific surface energy, y_b the specific grain boundary surface energy, and p the capillary pressure.

Besides the external pressureless mechanisms that depend on the capillary pressure p resulting from surface energy to provide the driving force for densification, another way to achieve densification of a ceramic body involves pressure sintering and hot pressing. Applying an external pressure eliminates the need for very fine-grained powders and removes large pores caused by nonuniform mixing. In some cases, a sintering temperature regime can be chosen such that densification can be achieved without the occurrence of excessive recrystallization. Heat can be applied by three methods: inductive heating, indirect resistance heating, or field-assisted sintering technique. In very special cases, HIPing is used as a densification tool that involves ceramic powder consolidation at both elevated temperatures and pressures provided by isostatically pressurized gases such as argon or helium in an appropriate high-pressure containment vessel (Bocanegra-Bernal et al., 2009). The mechanisms by which densification occurs involve solid-state sintering, vitrification, and liquid-phase sintering. As the grain-growth process is insensitive to pressure, sintering by hot pressing allows the fabrication of high density/small grain size ceramics with near theoretical mechanical, elastic, and thermal properties and close to zero porosity that in some cases even results in optical transparency (Fang et al., 2004). However, the HIPing process is expensive and generally confined to simple geometrical shapes so that near-net shaping technology is excluded for many types of ceramic.

3.2.4 Alumina

The worldwide production of alumina was estimated to be 132.4 million metric tons in 2017, with China (70.7 million tons), Australia (20.8 million tons), Western Europe (20.7 million tons), Brazil (12.7 million tons), and North America (3.0 million tons) being the leading producers (Source: www.world-aluminium.org/statistics/alumina-production). The key players in the high-purity alumina market include Orbite Technologies Inc.; Sumitomo Chemical Co., Ltd; Baikowski SAS; Alcoa Inc.; Nippon Light Metal Holdings Company, Ltd.; Rio Tinto; Chalco; Norsk Hydro; Altech Chemicals Ltd; Zibo Honghe Chemical Co. Ltd.; UC Rusal, Sasol Performance Chemicals, and Xuan Cheng Jing Rui New Material Co. Ltd.

The global high-purity alumina (4N to 6N) market was valued at US$ 987 million in 2017 and is expected to reach US$ 4.1 billion by 2025, registering a CAGR of 19.4% from 2018 to 2025.

It is very significant to note that this recent global production estimate is more than three times that of 1995 (IPAI, 1995), with obvious environmental implications (see Chapter 3.2.4.3). About 95% of the alumina produced worldwide are still won by the Bayer process. While some 94% of this synthesized alumina are used to extract metallic aluminum by electrolysis (Hall–Héroult process), only 6% of the total are nonmetallurgical or chemical alumina used for advanced ceramic products including bioceramic materials.

The raw alumina material is natural bauxite ore containing predominately the aluminum hydroxide gibbsite (γ-Al(OH)$_3$) as well as minor amounts of the aluminum oxyhydroxides boehmite (γ-AlOOH) and diaspore (α-AlOOH). These minerals are selectively extracted from insoluble components such as quartz, clay minerals, and iron and titanium oxides by dissolving the bauxite ore in highly heated (150–250 °C) sodium hydroxide (Bayer process; Habashi, 2005) or moderately heated (40–80 °C) sodium carbonate solutions (Pederson process) (Pederson, 1927). The entire processes can be divided into three steps: extraction, precipitation, and calcination. Figure 3.35 shows the flow diagrams of the two processes.

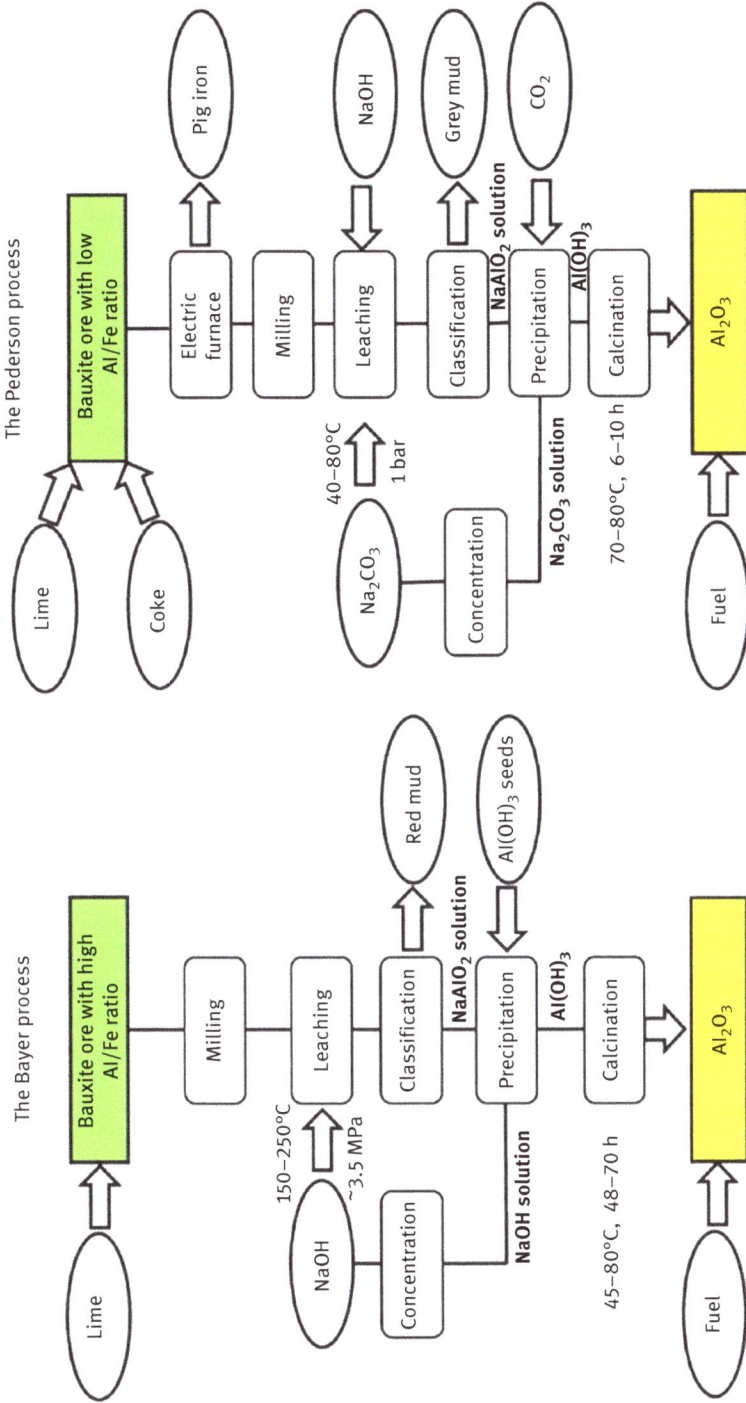

Figure 3.35: Flow diagrams of the Bayer and the Pederson processes (modified after Safarian and Kolbeinsen, 2016).

3.2.4.1 Bayer process

Extraction

The ore is washed, crushed, and milled to reduce the particle size and make the minerals more available for extraction. It is then combined with the process caustic liquor and sent as slurry to a heated pressure digester for extraction. Concentration, temperature, and pressure within the digester are set according to the properties of the bauxite ore, that is, the ratio of hydroxides to oxyhydroxides. Ores with high gibbsite content can be processed at 140 °C. Processing of boehmite requires higher temperatures between 200 and 240 °C. The steam pressure at 240 °C is approximately 3.5 MPa. After the extraction stage, the insoluble oxide residue ("red mud") must be separated from the aluminum-containing liquor by settling. The liquor is purified by filtering before being transferred to the precipitators. The insoluble "red mud" is thickened and washed to recover part of the caustic soda, which is then recycled back into the main process.

Precipitation

Crystalline pure aluminum hydroxide (gibbsite) is precipitated from the so-called pregnant digestion liquor after cooling and diluting, and adding gibbsite seed crystals according to $[Al(OH)_4]^- \rightarrow Al(OH)_3 + OH^-$. This is essentially the reverse of the extraction process. The precipitated gibbsite crystals are classified into size fractions and fed into a rotary or fluidized bed calcination kiln. Undersized particles are returned to the precipitation stage.

Calcination

The gibbsite crystals are calcined to form high-purity alumina (α-Al_2O_3) as the precursor material for alumina bioceramics according to $2\,Al(OH)_3 \rightarrow Al_2O_3 + 3\,H_2O$. The calcination process using rotary kilns, fluid bed kilns, or flash calciners must be carefully controlled since it dictates the properties of the final product. A secondary process stream containing dried gibbsite can be separated to produce aluminates, zeolites, filler material for toothpaste, fire retardants, and others.

Since the stability of aluminum hydroxides shows a pronounced pH dependence, controlled dehydration of aluminum hydroxides leads to alumina products with tailored properties for a large variety of industrial application. In particular, the size and morphology of crystalline alumina particles can be adjusted in such a way that on subsequent sintering ceramics with predetermined and reproducible properties for use in biomedical and other applications can be obtained.

High-purity alumina with 99.99% Al_2O_3 content can be produced by successive activation and washings of the Bayer hydrate or through treatment with chlorine to remove traces of contamination by iron and titanium oxides. Even higher purity alumina can be manufactured by calcining ammonium aluminum sulfate purified by successive recrystallization steps or from high-purity aluminum metal via an alkoxides route. The latter involves reacting the metal with an alcohol to produce aluminum alkoxide, purifying it by distillation, hydrolyzing, and, finally, calcination. However, the substantially higher cost of this processing route limits its application.

3.2.4.2 Pederson process

In contrast to the hydrometallurgical Bayer process, the Pederson process is a combined pyro- and hydrometallurgical process, advantageously used to process lateritic aluminum ore with low Al_2O_3/Fe_2O_3 ratio (Figure 3.35). The process starts with sintering the bauxite ore with lime and coke at temperatures around 1,000 °C in an electric furnace. The hematite or limonite content of the bauxite ore will be reduced to pig iron which is an economically valuable materials that can be recovered. The remaining calcium aluminate slag is milled and digested at temperatures between 40 and 80 °C under ambient pressure condition in sodium carbonate solution, in contrast to the high pressure regime attained during leaching in the Bayer process. Settling separates the $NaAlO_2$ solution from the so-called gray mud that finds use in cement production. Precipitation of aluminum hydroxide from the solution is through injection of carbon dioxide that yields additional sodium carbonate that is recycled to the leaching step of the process. The final calcination step converts the hydroxide to α-Al_2O_3. The advantages of the Pederson process over the classic Bayer process include the lack of environmentally problematic production of red mud (see below), production of consumable by-products such as pig iron and gray mud instead, much reduced digesting temperature, much shorter $Al(OH)_3$ precipitation time, as well as more than 50% lower capital and investment costs (Safarian and Kolbeinsen, 2016).

3.2.4.3 Environmental impact

The currently worldwide leading Bayer process has serious environmental repercussions that involve the production of staggering amounts of toxic red mud that must be collected, moved, neutralized, securely stored in huge holding ponds, and constantly guarded against catastrophic flooding like the one that happened on October 4, 2010, near the western Hungarian city of Ajka, where 0.7 million m^3 of red sludge overflowed the retaining walls of the holding pond. Since it produces no red mud, the Pederson process is, in principle, a more environmentally benign and sustainable process that yields consumable by-products such as pig iron and gray mud, the latter being utilized in cement production. In addition, lower processing temperatures and pressures during the digestion steps and accelerated $Al(OH)_3$ precipitation kinetics, together with the option to collect and recycle CO_2, make the environmental footprint of the Pederson process advantageous over that of the Bayer process.

Environmental challenges involved in mining and processing of aluminum ore to extract metallic aluminum include the extremely high amount of electric energy required to feed the Hall–Héroult electrolysis process, and the release of toxic perfluorocarbons formed by decomposing cryolite and polycyclic aromatic hydrocarbons from consumption of the Söderberg electrodes that consist of anthracite, coke, and hard coal tar pitch.

3.2.4.4 Manufacturing of alumina femoral balls

Femoral balls of hip endoprostheses made from bioinert alumina have to sustain high mechanical stresses, resorption/corrosion by aggressive body fluid, and abrasive wear over the lifetime of the implant in the human body for 15–20 years. Some important properties of alumina femoral ball heads are listed in Table 3.5 (Willmann, 1995).

Table 3.5: Important mechanical and functional properties of alumina femoral balls (after Willmann, 1995).

Materials property	Prerequisite for
High hardness	Wear resistance over many years
High wear resistance	Low risk of particle-induced osteolysis
No plastic deformation under load	
No elastic deformation under load	
No creep	
Finely grained microstructure	Excellent surface finish, low coefficient of friction
Dense (near-zero porosity)	Mechanical stability
High flexural strength	Load-bearing capability
High compressive strength	High fracture strength
Good fatigue resistance	Improved reliability
High Weibull modulus	Reliable product
Extreme corrosion resistance	Biocompatibility
High chemical purity	Bioinert behavior
Electrical insulator	No galvanic or fretting corrosion

The wear performance of the sliding couple ceramic femoral ball/acetabular cup liner (see Fig. 1.13) is of crucial importance since the lubricating synovial fluid present in natural hip joints is absent in the artificial system. Hence, the coefficient of friction must be as low as possible. Some data of linear wear of clinically established wear couples as well as data obtained during wear screening tests are shown in Table 3.6 (Heimann and Willmann, 1998).

Table 3.6: Wear data of materials combinations used in femoral ball/ acetabular cup liner wear pairs (Heimann and Willmann, 1998).

Materials combination	Linear wear (µm/year)
Metal/UHMW-PE [1]	200
Alumina/UHMW-PE	<100
Alumina/CFRP [2]	<4
Y-TZP [3]/UHMW-PE	<100
Biolox™/Biolox™ [4]	<5
Biolox forte™/Biolox forte™ [4]	<1
Biolox delta™/Biolox delta™ [5]	<1
Y-TZP/Y-TZP	Disastrous

[1] UHMW-PE: ultrahigh-molecular-weight poly(ethylene)
[2] CFRP: carbon fiber-reinforced polymer
[3] Y-TZP: yttria-stabilized tetragonal zirconia polycrystal
[4] Biolox™, Biolox forte™: trade names of medical alumina bioceramics produced by CeramTec GmbH, Plochingen, Germany
[5] Biolox delta™: trade name of alumina-zirconia reinforced bioceramics produced by CeramTec GmbH, Plochingen, Germany

Figure 3.36 shows schematically the manufacturing process of ceramic femoral ball heads (Clarke and Willmann, 1994). Starting from high-purity alumina powders (Figure 3.36A), a cylindrical precursor shape is formed by cold uniaxial pressing (Figure 3.36B) that subsequently is being turned on a CNC-controlled lathe to shape the inside taper and the outside spherical surface (Figure 3.36C). After laser-etched engraving (see Figure 3.38) for identification (Figure 3.36D), the ceramic green body will be densified by sintering at temperatures up to 1,600 °C (Figure 3.36E). First, the binder material will be driven off, followed by sintering into a dense, polycrystalline ceramic material at temperatures that allow atoms or ions to become mobile enough to reach their equilibrium arrangement but low enough as to avoid melting. This sintering process is essentially a solid-state reaction characterized by diffusion. During cooling from high temperature, each ceramic grain shrinks anisotropically, and tensile forces will be created between adjacent grains with the potential to initiate microcracking. To avoid this, cooling cycles have to be controlled very carefully. During sintering, compaction occurs associated with linear shrinkage that may be

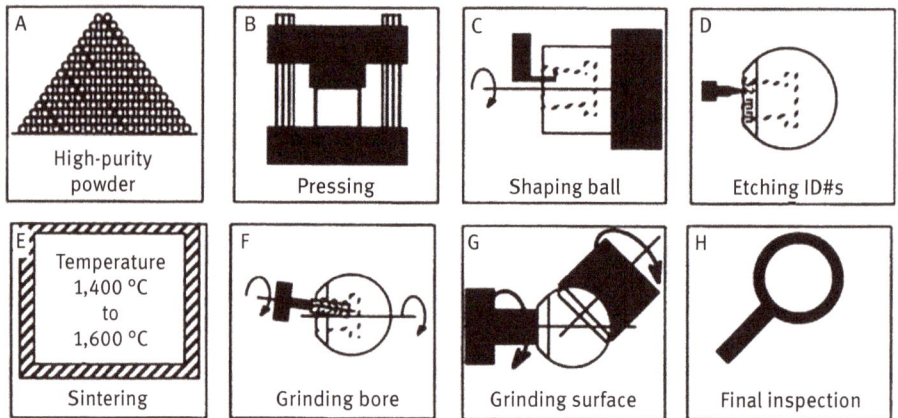

Figure 3.36: Schematic of the manufacturing process of ceramic femoral balls (Clarke and Willmann, 1994).

on the order of 15–25% in any one direction. Hence, the volume shrinkage of the ceramic part may reach some 50% overall.

Finishing is being done by grinding, with diamond tools, the internal bore and the spherical surface to a very low roughness value of <1 μm to achieve a low coefficient of friction (Figure 3.36F, G). Final inspection (Figure 3.36H) will assure an extremely high degree of reliability that is required for long-term survival of the ball in the aggressive body environment.

Stringent quality control measures during production of femoral balls are of the utmost importance. To assess and estimate performance *in vivo*, wear screening and corrosion tests are being performed on simplified specimen geometries (cylindrical pins, annulus) in the presence of lubricating simulated body fluids (SBFs: Ringer's solution, Hanks' balanced salt solution, etc.) or protein-containing solutions such as fetal bovine serum and human serum albumin solutions under physiological conditions. Some of these tests are carried on for up to 2 million cycles that correspond to a service time of 2 years considering the annual average number of load changes during walking.

3.2.5 Zirconia

As with alumina, manufacturing zirconia precursor powders with superior properties are indispensable for long-lived biomedical zirconia devices. Extremely fine-grained powders with average grain size below 1 μm, near spherical shape, and narrow grain size distribution can be produced by several synthesis routes from natural minerals. In contrast to alumina, zirconia powder production is complicated by the facts that the polymorphism of zirconia and the need for stabilizing the tetragonal modification by alloying with other metal oxides requires additional attention.

3.2.5.1 Processing of zirconia

Raw ore for the production of zirconia is the mineral zircon ($ZrSiO_4$), mined in Australia (42%), South Africa (25%), China (10%), Indonesia (8%), Mozambique (5%), and India (3%) from predominantly beach placer deposits. At present, China dominates the downstream zirconium metal business at ~ 90% and has recently declared zirconium to be a strategic metal, presumably owing its major use as corrosion-resistant Zircaloy® cladding for nuclear fuel rods. This move invokes the possibility of a worldwide supply crunch and thus, steeply increased sales price for countries previously dependent on Chinese exports.

The sand and gravel containing zircon grains mixed with other high-density silicates such as garnets or tourmaline, ilmenite ($FeTiO_3$), and rutile (TiO_2) are typically collected from coastal waters by a floating dredge consisting of a large steam shovel fitted on a floating barge. The scooped-up gravel and sand are purified by spiral concentrators that separate the minerals based on their density. Ilmenite and rutile are then removed by magnetic and electrostatic separators. The purest concentrates of zircon are shipped to end-product manufacturers to be used in zirconia and eventually, zirconium metal production. Major producers of high-purity zirconia powders are Tosoh Corp. (Japan), Anhui Jinao Chemical Co., Ltd. (China), Zhimo New Material Technology Co., Ltd (China), Ortech Advanced Ceramics (USA), Sasol Performance Chemicals (Germany), Nippon Light Metal Comp., Ltd. (Japan), and Jyoti Ceramic Industries (India).

The less common zirconium ore baddeleyite (β-ZrO_2) is won from hard-rock deposits, predominantly in South Africa (Phalaborwa) as a byproduct of copper and phosphate-rock mining. At present, it is also mined at the Kovdor magnetite-apatite-baddeleyite deposit, part of a carbonatite intrusion complex of the Kola peninsula, Murmansk Region, Russia (Ivanyuk et al., 2016). In 2012, EuroChem purchased the Kovdor open pit mine and the attached processing facility. The site now produces some 2.5 million tons/year of apatite concentrate, 5.6 million tons/year of iron concentrate, and 8,000 tons/year of baddeleyite concentrate (Dickson, 2015).

There are several chemical routes leading from purified zircon sand concentrate to zirconium oxide for biomedical application. These include (i) carbothermal reduction of zirconium silicate above 1,750 °C, (ii) direct chlorination by a mixture of coke and chlorine below 1,200 °C, (iii) alkali demixing by reacting zirconium silicate with sodium hydroxide or sodium carbonate, (iv) reaction of zircon ore with calcium oxide, and (v) plasma decomposition above 2,000 °C (Figure 3.37) (Heimann, 2010a).

(i) Thermal demixing by *carbothermal reduction* at temperatures >1,750 °C generated by an electric arc is governed by the reaction equation

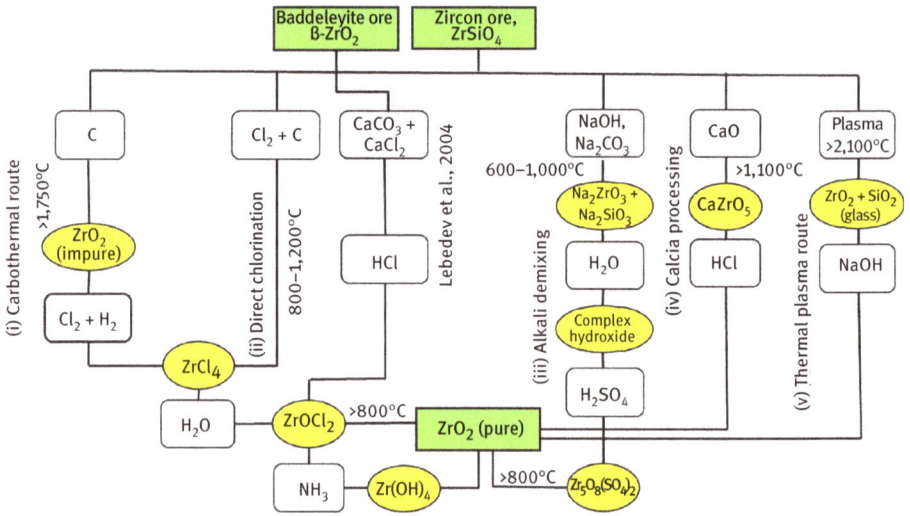

Figure 3.37: Processing routes of zircon and baddeleyite ores to obtain pure zirconia for biomedical applications. Intermediate products are shown in yellow, starting and final products in green.

$$ZrSiO_4 + C \rightarrow ZrO_2 + SiO \uparrow + CO \uparrow \tag{3.19a}$$

In the presence of excess carbon, some SiC will form (Kljajevićet al., 2011). In a second processing step, the zirconia is purified by treatment with chlorine and hydrogen according to

$$ZrO_2 + 2Cl_2 + 2H_2 \rightarrow ZrCl_4 + 2H_2O \tag{3.19b}$$

The zirconium tetrachloride is subsequently converted to zirconia (see ii).

(ii) A variant of this process is the *direct chlorination* (Manieh and Spink, 1973) at temperatures between 800 and 1,200 °C according to

$$ZrSiO_4 + 4C + 4Cl_2 \rightarrow ZrCl_4 \uparrow + SiCl_4 \uparrow + 4CO \uparrow \tag{3.19c}$$

$$ZrCl_4 + H_2O \rightarrow ZrOCl_2 + 2HCl \tag{3.19d}$$

Zirconium tetrachloride evaporates and is subsequently condensed at temperatures between 150 and 180 °C and thus, separated from $SiCl_4$ that condenses only at the much lower temperature of -10 °C. Subsequently, $ZrCl_4$ is hydrolyzed with water and forms a saturated solution of $ZrOCl_2$ (eq. (3.19d)). On cooling between 65° and 20 °C, zirconyl chloride crystallizes. The crystals are separated and dried at 85 °C. Eventually, calcination at >800 °C yields pure zirconia. Alternatively, ammonia can be added to the zirconyl chloride solution to yield zirconium hydroxide:

$$ZrOCl_2 + 2NH_3 + 3H_2O \rightarrow Zr(OH)_4 + 2NH_4Cl \qquad (3.19e)$$

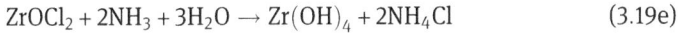

Finally, the zirconium hydroxide is calcined to form zirconia.

(iii) The *alkali demixing process* (Zhang et al., 2012) is based on the following sequence of reactions:

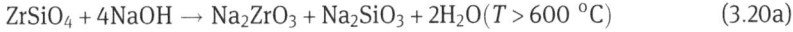

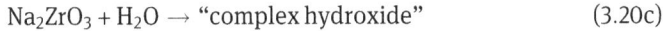

$$ZrSiO_4 + 4NaOH \rightarrow Na_2ZrO_3 + Na_2SiO_3 + 2H_2O(T > 600\ °C) \qquad (3.20a)$$

$$\text{or } ZrSiO_4 + 2Na_2CO_3 \rightarrow Na_2ZrO_3 + Na_2SiO_3 + 2CO_2 \uparrow \quad (T > 1,000\ °C) \qquad (3.20b)$$

$$Na_2ZrO_3 + H_2O \rightarrow \text{"complex hydroxide"} \qquad (3.20c)$$

The resulting "complex hydroxide" is being treated with sulfuric acid to form substoichiometric zirconyl sulfate that will be calcined above 800 °C.

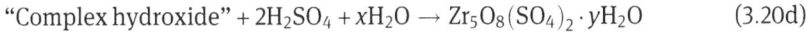

$$\text{"Complex hydroxide"} + 2H_2SO_4 + xH_2O \rightarrow Zr_5O_8(SO_4)_2 \cdot yH_2O \qquad (3.20d)$$

$$Zr_5O_8(SO_4)_2 \cdot yH_2O \rightarrow 5ZrO_2 + 2SO_2 \uparrow + y\ H_2O(T > 800\ °C) \qquad (3.20e)$$

(iv) The *calcia processing* reaction can be formulated as follows:

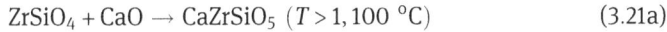

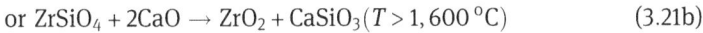

$$ZrSiO_4 + CaO \rightarrow CaZrSiO_5 \quad (T > 1,100\ °C) \qquad (3.21a)$$

$$\text{or } ZrSiO_4 + 2CaO \rightarrow ZrO_2 + CaSiO_3(T > 1,600\ °C) \qquad (3.21b)$$

The calcium silicate byproduct will subsequently be dissolved by hydrochloric acid. The thermal decomposition reaction occurring between 1,200 and 1,500 °C was found to follow a diffusion kinetic model in the form of $x + (1-x)[\ln(1-x)] = k \cdot t$, where x is the extent of solubilized silica in time t and k is the rate constant. The apparent activation energy of the process was calculated to be 205 kJ/mol (El-Barawy et al., 1999).

(v) The *plasma route* can be expressed by a thermal dissociation reaction according to

$$ZrSiO_4 \rightarrow (ZrO_2 + SiO_2)\ (T > 2,100\ °C) \qquad (3.22a)$$

$$SiO_2 + 2NaOH \rightarrow Na_2SiO_3 + H_2O \qquad (3.22b)$$

On cooling, the silica solidifies as a glass, and the crystalline zirconia suspended in the glassy matrix is liberated by dissolving the silica glass in NaOH. This process yields very fine-grained highly pure crystalline zirconia since natural impurities of zircon such as titanium and iron oxides will effectively be removed by incorporation into the glassy silica phase (Rendtorff et al., 2012).

Naturally occurring zirconium oxide (baddeleyite) can be processed by sintering with $CaCO_3$ and $CaCl_2$ and dissolving the product in HCl to form $ZrOCl_2$ that, during thermal hydrolysis, is converted to pure ZrO_2 (Lebedev et al., 2004).

To produce Y_2O_3-partially stabilized zirconia (Y-PSZ), sol–gel processing or coprecipitation routes have been adopted based on hydrolysis of stoichiometric mixtures of zirconyl chloride and yttrium trichloride (e.g., Carter et al., 2009). The hydroxides formed are subjected to azeotropic distillation, drying, calcining between 850 and 950 °C, wet milling, and spray drying.

The various processing routes are schematically shown in Figure 3.37.

3.2.5.2 Environmental impact

The mineral zircon contains significant levels of natural radioisotopes (500–1,000 Bq/kg from ^{232}Th and 1,000–5,000 Bq/kg from ^{238}U). Hence, extraction of zirconia from this mineral requires removal of its radioactive content (Hollitt et al., 1997) as well as specific impact analyses of the health effects and the potential risk for people living near a processing plant. However, comparison between soil samples collected in areas near the plants and soils collected far from them shows a negligible environmental effect. This means that the radioactivity level in the area of maximum particulate deposition is indistinguishable from that of natural background (Righi et al., 2002).

An life cycle assessment (LCA) study to evaluate the global warming, acidification, eutrophization, abiotic depletion, and ozone depletion potential occurring during processing of zircon sand, commissioned by the Zircon Industry Association confirmed that zirconia manufacture has a relatively low overall environmental impact, overwhelmingly associated with local electricity consumption linked to upstream mining processes (extraction, separation and drying), much like the mining of other minerals (LCA Zircon, 2017). Moign et al. (2010) provided an LCA on yttria-stabilized zirconia feedstock used in plasma spraying with powder and suspension feed.

The recent large-scale Dubbo Zirconia Project is located in the Central West region of New South Wales, Australia. The project is based upon rich in-ground resources of zirconium, hafnium, niobium, (tantalum), yttrium, and rare earth elements within the sub-volcanic Jurassic Toongi trachyte rock deposit. The ore will be processed by sulfuric acid leach followed by solvent extraction recovery and refining to produce several products. The environmental impact statement contains a description of the locality and predicted impacts on the surrounding physical, biological, and socioeconomic environment (Dubbo Zirconia Project, 2013).

3.2.5.3 Manufacturing of zirconia femoral balls

The individual processing steps shown above for alumina femoral heads are also applied to their zirconia counterparts. Figure 3.38 shows three standard diameter

sizes (22, 28, and 32 mm) of zirconia femoral balls with laser markings to identify un-
equivocally the individual part, important for documentation in possible litigation
cases. For example, the top ball produced by CeramTec GmbH shown in Figure 3.38
carries the identification code 32-12/14M 96 @ Z 8179 that refers to a ball diameter of
32 mm, an inside taper of 12/14 mm, a medium neck length M, the year of manufacture (1996), the type of ceramics (Z for zirconia), as well as the production number
8179. With this complete identification, it is possible to establish a quality test protocol and trace the history of this particular part that may become very important in
any court procedure associated with litigation cases against the manufacturer, the
hospital, or the surgeon who had performed the operation. It also provides a unique
means of identification in case other identification traits have been obliterated as for
example, of a body charred in a fiery accident including a plane crash.

Figure 3.38: Zirconia femoral balls with laser-engraved identification (Heimann and Willmann, 1998).

3.2.5.4 Color change during sterilization with ionizing radiation

Metallic and ceramic femoral heads must be thoroughly sterilized before surgical
implantation to guarantee a successful integration of the implant by avoiding bacterial colonization through nosocomial infection. However, the hydrolytic stability
of zirconia may be compromised during common steam sterilization at 121 °C since
water absorbed at the ceramic surface produces Zr–OH moieties that will act as
stress concentration sites. Indeed, this has been one of the leading causes of massive failure series of zirconia femoral balls in service (see Chapter 2.2.1.2).
Consequently, other methods must be applied including sterilization with gasses
such as ethylene oxide, ozone, chlorine dioxide, or vaporized hydrogen peroxide as
well as plasmas and ionizing radiation (γ-rays, X-rays, and e-beam exposure).
Sterilization of zirconia femoral heads with ionizing radiation introduces color

centers in Y-stabilized zirconia. Even though this effect does neither compromise the mechanical nor the biomedical performance of the femoral head, in the past it has caused concern among orthopedic surgeons. Indeed, the bluish- or brownish-tinged irradiated zirconia femoral heads shipped by the manufacturer to the hospital are known to have been rejected under suspicion of deleterious contamination. However, the actual cause of the reversible discoloration was found only on an electronic level. These electronic mechanisms responsible for irradiation-induced formation of color centers in zirconia have been probed by electron spin resonance, 3D-thermoluminescence, and residual optical spectroscopy (Dietrich et al., 1996; Heimann and Willmann, 1998; Heimann and Lehmann, 2015).

3.2.6 Carbon

As far as its structure is concerned, *pyrolytic carbon*, although in many ways similar to graphite, contains some covalent bonding between its monoatomic graphene sheets (see Chapter 2.2.1.3). Pyrolytic carbon is generally produced by thermal decomposition (pyrolysis) of gaseous hydrocarbons such as methane, acetylene, as well as ethanol (Li et al., 2010) by CVD. This thermal CVD process differs from plasma CVD, wherein the gaseous hydrocarbons are cracked, that is, decomposed by plasma action that is operational at temperatures as low as room temperature (Hassler, 2012). As a result of thermal CVD processing, the material properties include high density, a high degree of anisotropy, and a smooth surface morphology.

Pyrolytic carbon coatings of heart valve housing and bileaflets are produced by codepositing carbon and silicon carbide using CVD in a fluidized bed furnace, fed with silicon carrier gas, and a suitable hydrocarbon. Most artificial valves are made of titanium, graphite, pyrolytic carbon, and polyester (see Fig. 1.16). The titanium is used for the housing or outer ring, graphite coated with pyrolytic carbon is used for the bileaflets, and 100% pyrolytic carbon is used for the inner ring. The pyrolytic carbon is sometimes impregnated with radio-opaque tungsten or tantalum so that the valve can easily be visualized by radiological means *in situ* following implantation. The pyrolytic carbon coating is produced through a CVD process at temperatures between 1,800 and 2,300 °C. The inner rings are made from 100% pyrolytic carbon using a fluidized bed process.

Diamond-like carbon (DLC) coating deposition typically occurs at low pressure conditions (1–27 kPa) and involves feeding varying amounts of hydrocarbon gases into a deposition chamber, energizing them, and providing conditions for DLC growth on a substrate. Typical energy sources include a hot filament, microwave power, and arc discharges. The energy source generates a plasma in which the gases are decomposed ("cracked") and more complex chemistries occur.

DLC coatings are being deposited either by CVD or PVD techniques. Most of the processes occurring during DLC deposition are of physical nature, and the sp^3

bonds are produced by the impact of carbon (or hydrocarbon) ions on the growing film. The physical processes utilized to deposit DLC thin films include direct ion beam (IB) and IB-assisted deposition (IBAD), filtered cathodic vacuum arc, DC, and RF sputtering, PLD, and plasma immersion ion implantation (PIII). Other techniques involving chemical processes include plasma-enhanced CVD (PE-CVD) and electron cyclotron-resonance plasma CVD (ECR-CVD) (Fedel, 2013). Further details inform the authoritative contribution by Bewilogua and Hofmann (2014).

As shown in the phase diagram of Figure 3.39, in addition to the sp^3-hybridized bonds of diamond type the remainder of the carbon atoms is predominately sp^2-hybridized as typical for graphite. The ratio of sp^3/sp^2 bonds as well as the hydrogen content determines the properties of DLC coatings. Since most DLC coatings are produced using hydrocarbon gases, they always contain a certain amount of hydrogen. However, tetragonal amorphous carbon (ta-C) coatings made from solid graphite do not contain significant amounts of hydrogen. Their sp^3 fraction ranges from 40% to 90%, depending on the fabrication conditions. Hence, ta-C coatings approach the properties of crystalline diamond. They can be deposited with thickness exceeding 10 µm and are composed of nanosized thin layers with periodically varying sp^3 contributions. The individual nanolayers possess different moduli of elasticity and density, and act to alleviate the generally high compressive coating stresses and thus, prevent spalling. Investigation of DCL coatings, deposited by PE-CVD onto PLA foils demonstrated a shift of the dominant binding from sp^3 to sp^2 that indicates a pending stress relief in the layer resulting in cord buckling and cracking and thus, an imminent layer failure. This was also demonstrated by the widely fluctuating contact angles (Schlebowski et al., 2019).

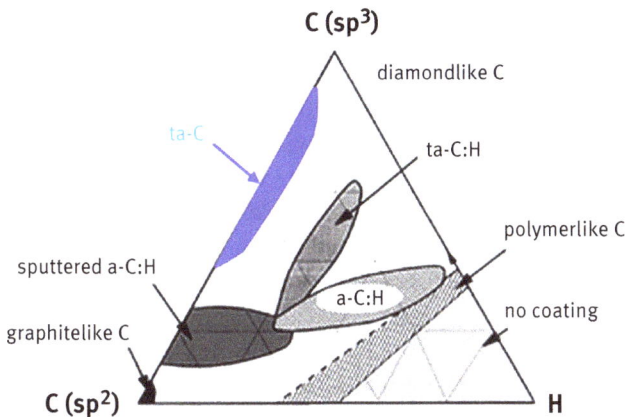

Figure 3.39: Jacob–Möller–Robertson C–H phase diagram of hybridized carbon (Robertson, 2002). © With permission by Elsevier.

3.2.7 Calcium phosphates

Owing to their low mechanical strength and fracture toughness, monolithic parts of calcium phosphates including hydroxylapatite cannot be advantageously used in load-bearing applications. Instead, their chemical, structural, and biological similarity to the inorganic component of the natural composite materials "bone" is being utilized, preferentially in applications such as filler material for bone cavities, ossicular chain implants, bone scaffolds, and, most frequently, coatings for the stem of hip endoprostheses and endosseous dental root implants to support in-growth of bone cells. In addition, biomimetic hydroxylapatite/organic hybrid nanocomposites (see Chapter 1.1.3) are increasingly being developed for a variety of applications including drug and gene delivery vehicles and bone-substituting scaffolds (Wubneh et al., 2018).

3.2.7.1 Powder synthesis

There exist a rather large number of calcium phosphate compositions with Ca/P ratios ranging from 0.5 to 2.0, making the calcium phosphate system highly complex and the outcome of syntheses crucially dependent on the correct choice of composition of the precursor materials, pH, and reaction conditions such as synthesis temperature and pressure. Figure 3.40 shows the binary $CaO–P_2O_5$ phase diagrams in the absence (left) and presence (right) of water.

Given this complexity, it is imperative to consider that the phase purity and particle characteristics of the final synthesized powder affect their bioactivity, as well as their mechanical and dissolution properties. Since these characteristics ultimately determine the medical application of the material, it is mandatory to develop synthesis methods enabling to exercise control over crystal morphology, chemical composition, crystallinity, particle size distribution, and agglomeration tendency. Hence, a variety of synthesis techniques have been developed including wet-chemical precipitation (Yelten and Yilmaz, 2016, 2018), hydro- or solvothermal processing (Earl et al., 2006; Suchanek et al., 2019), sol–gel techniques (Agrawal et al., 2011), solid-state synthesis (Pramanik et al., 2007), spray-drying (Bastan et al., 2017), and electrodeposition (Kar et al., 2006). Less frequently applied methods include biomimetic precipitation (Cengiz et al., 2008), self-propagating combustion synthesis (Tas, 2000), solution combustion technique (Rao et al., 2016), and water-in-oil and oil-in-water emulsion syntheses (Bose and Saha, 2003). Recent reviews by Nayak (2010) and Cox (2014) inform about these techniques and indicate their strengths and weaknesses.

Owing to the importance of the powder characteristics for appropriate biological performance, a complete understanding is mandatory of how process variables impact on hydroxylapatite powder properties. Statistical design of experiments (SDE) strategies can be a very useful and powerful tool to determine which

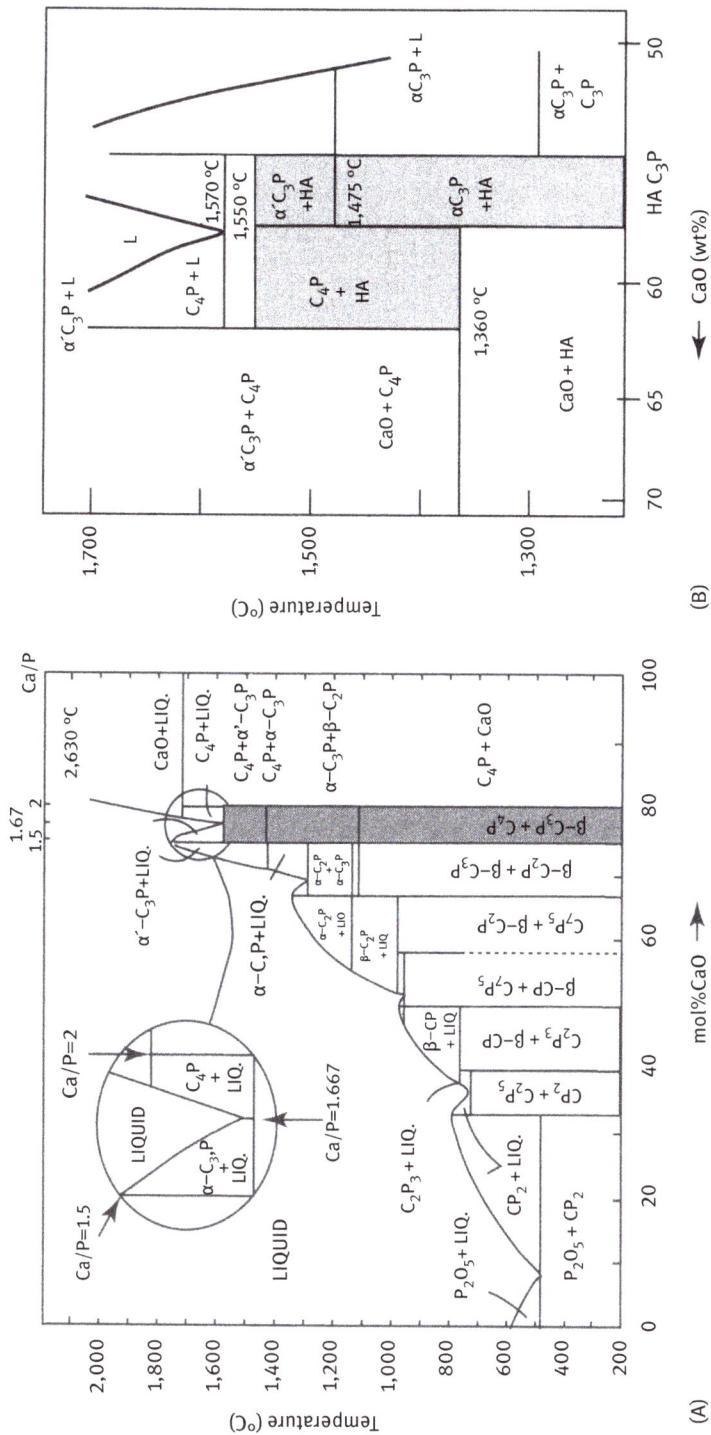

Figure 3.40: Left: Binary phase diagram of CaO-P$_2$O$_5$ in the absence of water (Kreidler and Hummel, 1967). The inset shows the region of interest, that is, the incongruent thermal decomposition of hydroxylapatite (Ca/P = 1.67) to α'-tricalcium phosphate (Ca/P = 1.5) and tetracalcium phosphate (Ca/P = 2). Right: Phase diagram of the quasi-binary system CaO–P$_2$O$_5$–(H$_2$O) at a water partial pressure of 65.5 kPa (after Riboud, 1973). Note that incongruent melting of hydroxylapatite (HA) occurs beyond 1,570 °C under formation of α-C$_3$P (α-TCP) and C$_4$P (TTCP).

combinations of variables have the greatest impact on powder properties. Such studies can ultimately inform the process of creating robust standard operating procedures by statistical process control (SPC) measures.

Statistical design of experiments
Experiments are questions an experimenter poses to nature. By using SDE, Nature's inherent complexity can be reduced to a simplified "system," that is, to that part of the universe that is considered for a specific experiment or calculation. Nature's answers to these questions should be as straightforward, unequivocal and clear as possible, and amenable to predictions about the future behavior of the system outside the limited parameter space investigated. But as in human relationships, the quality of the answers depends very much on the clarity, shrewdness, subtlety, and cleverness of the question. However, to restrict the potentially infinite parameter space and hence, the possible number of experiments, only a limited subset of reasonable parameters (variables, factors) will be used and investigated. Statistical experimental designs vary all selected parameters simultaneously (as opposed to the classical experimental design that varies only one parameter at a time and keeps all others constant) by employing appropriate "factorial" matrices. This strategy is able to detect synergistic parameter interactions and thus, the "true" character of the "real world" outside the laboratory where a multitude of interacting factors tend to obscure the experimental response. Important tools of SDE are factorial designs, screening (Plackett–Burman) designs, response surface (Box–Behnken) designs, central composite (Box–Wilson) designs, and several others (Box et al., 2005; see also Heimann, 2008; Heimann, 2010b).

Statistical process control
SPC is defined as the application of statistical techniques to control a process or a production method. SPC tools and procedures are designed to monitor process behavior, discover procedural flaws in internal systems, and find solutions to avoid them. SPC is frequently used interchangeably with statistical quality control (SQC), although differences exist. A popular SPC tool is the *control chart* that assists in recording data and indicates when an unusual event occurs, for example, a very high or low observation compared with "typical" process performance. Control charts attempt to distinguish between two types of *process variations*: common cause variations (statistical variations) that are intrinsic to the process and will always be present, and special cause variations that originate from external sources and indicate that the process is out of statistical control. There are many tests available that assist in determining when an out-of-control event has occurred (Ramirez and Runger, 2006).

3.2.7.2 Powder consolidation and processing

Monolithic hydroxylapatite parts with controlled porosity can be produced by a plethora of techniques such as cold unidirectional and isostatic pressing, slip casting, gel casting, freeze casting, starch consolidation, gas foaming, EPD, and polymeric sponge techniques. In order to obtain ceramic monolithic parts with densities close to theoretical density, sintering at temperatures below the incongruent decomposition temperature of hydroxylapatite must be applied (Ramesh et al., 2013).

For deposition of coatings, a large variety of nonthermal and thermal techniques are available (Heimann and Lehmann, 2015). *Nonthermal deposition methods*

(Heimann, 2016b) are defined as techniques carried out at temperatures much below the incongruent melting point of hydroxylapatite, in particular at or near ambient temperature. However, frequently post-depositional heat treatment (Liu et al., 2019) must be applied to either crystallize amorphous calcium phosphate (ACP), transform (hydrated/hydroxylated) precursor phases such as octacalcium phosphate (OCP) to hydroxylapatite, and/or to remove organic compounds used in coating preparation. Typical nonthermal coating deposition techniques include biomimetic deposition, sol–gel deposition including dip and spin coating, electrochemical deposition (ECD) and EPD, hydrothermal deposition, as well as electron beam assisted deposition, IBAD, and RF magnetron sputtering. Occasionally, CGDS technology has been explored (Stoltenhoff et al., 2002; Vilardell et al., 2017, 2018) as well as aerosol deposition (AD) (Hahn et al., 2009; Lee et al., 2017).

Thermal deposition methods include foremost various plasma spraying techniques such as atmospheric plasma spraying (APS), low-pressure plasma spraying (LPPS) and vacuum plasma spraying (VPS), inductively coupled plasma spraying, microplasma spraying (MPS) and low-energy plasma spraying (LEPS), as well as suspension plasma spraying (SPS) and solution precursor plasma spraying (SPPS). Other thermal deposition methods include high velocity oxyfuel spraying, high-velocity suspension flame spraying (HVSFS), PEO/MAO, PLD, and CVD and PVD. Comprehensive accounts on these techniques and their influence on coating properties and biological performance can be obtained from Gadow et al. (2010) and Heimann and Lehmann (2015).

3.2.7.3 Plasma-sprayed hydroxylapatite coatings

The plasma spray process is essentially a rapid solidification technology during which material introduced into a high-temperature plasma jet will be melted and propelled by magneto-hydrodynamic forces with high velocity against a surface to be coated (Heimann, 2008). At the cool substrate surface, the melted droplets are quenched and solidify splat by splat as a more or less porous coating, frequently with a high proportion of amorphous material.

Figure 3.41 shows the working principle of a plasmatron (top), and the intrinsic and extrinsic parameters that influence development and properties of coatings (bottom).

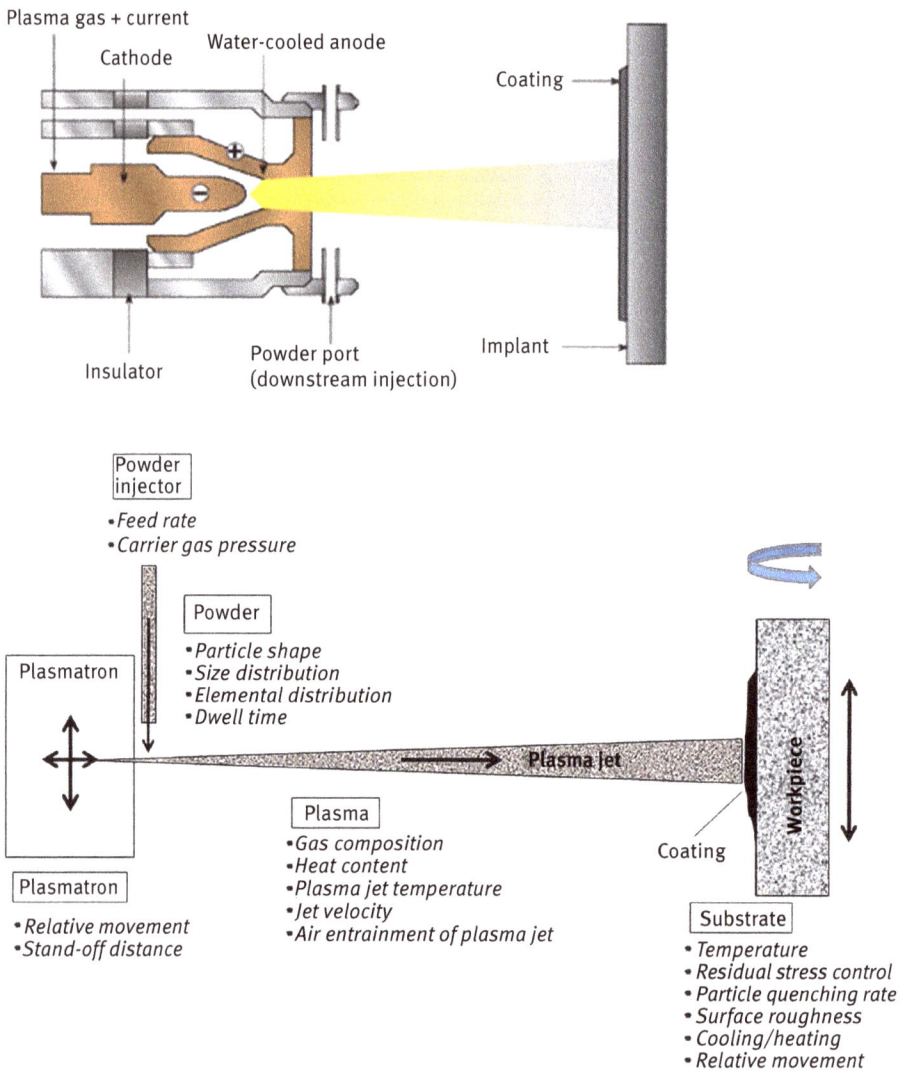

Plasma gas + current
Cathode
Water-cooled anode
Coating
Implant
Insulator
Powder port
(downstream injection)

Powder
injector
•Feed rate
•Carrier gas pressure

Plasmatron

Powder
•Particle shape
•Size distribution
•Elemental distribution
•Dwell time

Plasma jet

Workpiece

Plasma
•Gas composition
•Heat content
•Plasma jet temperature
•Jet velocity
•Air entrainment of plasma jet

Coating

Plasmatron
•Relative movement
•Stand-off distance

Substrate
•Temperature
•Residual stress control
•Particle quenching rate
•Surface roughness
•Cooling/heating
•Relative movement

Figure 3.41: Top: Schematic cross section of a typical plasmatron used for atmospheric plasma spraying. Bottom: Main plasma spray parameters influencing coating properties (Heimann, 2008).

Plasma spraying

The plasma originates from ionization in an electric potential field of a suitable gas, preferentially argon or nitrogen. Hence, a plasma by definition consists of positively charged ions and electrons but also neutral gas atoms and photons. Moving charges within the plasma column induce a magnetic field B perpendicular to the direction of the electric field characterized by the current j. The vector cross-product of the current and the magnetic field, $[j \times B]$ is the magnetohydrodynamic Lorentz force the vector of which is mutually perpendicular to j and B. Hence, an inward moving force is created that constricts the plasma jet by the so-called *magnetic* or *z-pinch*. In addition to the magnetic pinch, there is a *thermal pinch* that originates from the decreasing conductivity of the plasma gas at the cooled inner wall of the anode nozzle, leading to an increase in current density at the center of the jet. Consequently, the charged plasma tends to concentrate along the central axis of the plasmatron, thereby confining the jet. As the result of these two effects, the pressure in the plasma core increases drastically and the jet is blown out of the anode nozzle of the plasmatron with supersonic velocity (Figure 3.41, top).

Plasma spray technology is a versatile tool as any thermally reasonably stable metallic, ceramic, or even polymeric material with a well-defined congruent melting point can be coated onto nearly any surface. Frequently, some portion of the molten material rapidly quenched on contact with the relatively cool implant surface forms easily soluble amorphous phases such as ACP of variable composition. However, in practice many limitations persist related to coating porosity, thermal alteration, adhesion to the substrate, the presence of residual stresses, and line-of-sight restriction. For additional details see Heimann (2008).

Most commonly, hydroxylapatite coatings are being deposited on selected parts of implants by conventional APS although other deposition techniques exist such as RF-SPS to deposit nanostructured hydroxylapatite coatings with enhanced degree of crystallinity and limited contribution of thermal decomposition phases that appear to be able to control their bioactive behavior (Chambard et al., 2019).

Alternative deposition techniques include LPPS/VPS, SPPS, as well as MPS and LEPS. All these plasma-assisted deposition techniques offer a fast, well-controlled, economically advantageous, and, in its processing technology, mature way to coat almost any substrate with those materials that possess a defined congruent melting point. However, this requirement must be relaxed as hydroxylapatite melts incongruently, that is, decomposes on melting into tricalcium phosphates ($Ca_3(PO_4)_2$, α- and β-TCP) and tetracalcium phosphate ($Ca_4O(PO_4)_2$, TTCP), or even cytotoxic calcium oxide, CaO that can form following loss of P_2O_5 by evaporation or reduction to volatile phosphorus due to the presence of hydrogen as auxiliary plasma gas (Figure 3.46; see also Table 3.7). Moreover, a large portion of the molten powder material solidifies, adjacent to the implant surface, by rapid quenching to form ACP of various compositions.

Biomedical coatings deposited by thermal spray techniques have properties differing in chemical and phase composition, crystallinity, crystallite size, and defect density from natural bone-like apatite. Moreover, conventional APS of hydroxylapatite is unable to provide coatings with thickness below about 20 μm, a property that frequently does not meet medical requirements. To achieve thinner coating layers, recently suspension (SPS; see below) and SPPS (Candidato et al., 2016) techniques were developed. Furthermore, line-of-sight limitation makes coating of geometrically complex substrate shapes difficult, and undesirable local heating of the implant substrate may affect its metallic microstructure, as experienced, for example, by the forced α/β transition of alloyed titanium at high temperature. In addition, porosity control of the sprayed material is difficult, as plasma spraying frequently

Table 3.7: Plasma spray parameters, and resulting properties and compositions of hydroxylapatite coatings (Heimann et al., 2001; Heimann and Lehmann, 2015). Bold values refer to optimized coating properties as shown in Fig. 3.47 below.

Parameters	Coating type A	Coating type B	Coating type C
Plasma power (kW)	34	26	34
Argon flow rate (slpm)*	40	40	45
Hydrogen flow rate (slpm)*	4	7	7
Powder gas flow rate (slpm)*	10	5	5
Spray distance (mm)	85	115	85
Coating thickness (µm)	270 ± 65	159 ± 11	194 ± 76
Surface roughness (µm)	**20 ± 2**	9 ± 0.5	10 ± 0.5
Porosity (%)	4 ± 1	**8 ± 1**	5 ± 1
Adhesion strength (MPa)	45 ± 10	38 ± 9	**49 ± 9**
HAp content (mass%)	66 ± 3	51 ± 4	61 ± 3
β-TCP content (mass%)	1 ± 3	7 ± 3	4 ± 3
CaO content (mass%)	3 ± 0.5	1 ± 0.6	2 ± 0.7
ACP content (mass%); calculated	~ 30	~ 40	~ 33
Ca/P ratio	1.64	1.60	1.50

*slpm = standard liter per minute.

results in rather dense coating layers. Such coatings are unable to satisfy biomedical needs that call for pore sizes in excess of 75 µm required to guarantee unimpeded ingrowth of bone cells. Indeed, deposition of dense, stoichiometric, and well-crystallized hydroxylapatite coating layers is frequently ineffective. The reason is that those coatings tend to be bioinert because they have lost their osseoconductive property based on sufficient solubility. In general, the porosity of plasma-sprayed coatings is determined by both powder particle size and the degree of particle melting that, in turn, is controlled by plasma gas composition, plasma enthalpy, powder injection geometry, and spray distance. As an example, Figure 3.42 shows the effect of the fraction of molten particles (Khor et al., 2004) on porosity, degree of crystallinity, and bond strength of plasma-sprayed hydroxylapatite coatings (Li et al., 2019). With increasing fraction of molten particles, the coating porosity and degree of crystallinity decrease, whereas the adhesive bond strength increases.

In addition to thickness and porosity requirements, to attain bioactive, that is, osseoconductive functionality, hydroxylapatite ought to have some degree of

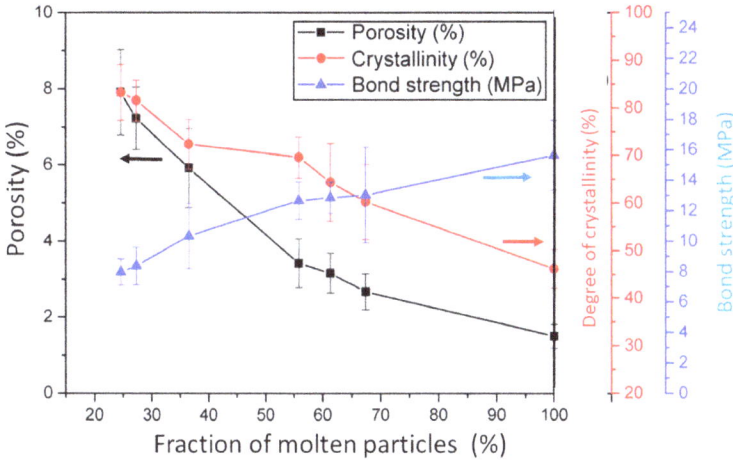

Figure 3.42: Dependence of porosity, crystallinity, and bond strength of plasma-sprayed hydroxylapatite coatings on the fraction of molten particles (Li et al., 2019). © With permission by Elsevier.

nonstoichiometry, expressed by both Ca deficiency caused by substitution of Ca cations by metabolic elements such as Mg, Sr, Na, and K, and substitution of carbonate ions for orthophosphate (type-B defect) or hydroxyl (type-A defect) anions. In particular, it has been found that in low-temperature apatites carbonate ions prefer to substitute for phosphate (type-B substitution) rather than for (monovalent) ions in channel sites (type-A substitution). Such nonstoichiometric, substituted, disordered, and sparingly soluble nanocrystalline carbonated hydroxylapatite compositions closely resemble so-called bone-like biological apatite with the approximate formula $Ca_{10-x}(HPO_4)_x$ $(PO_4)_{6-x}$ $(OH,O,Cl,F,CO_3,\Box)_{2-x} \cdot nH_2O$; $0 < x < 1$; $0 < n < 2.5$, the chemical variability of which aptly illustrates the complexity of the task at hand. Recently, Pasteris (2016) has pointed out similarities and differences between biological apatite and the calcium phosphate phases typically synthesized as biomaterials and stressed the fundamental difference between the natural mineral hydroxylapatite and the bone-like bioapatite.

To alleviate the apparent disadvantages of plasma-sprayed hydroxylapatite coatings, and, in particular to reduce the amount of nonapatitic calcium phosphate phases, the search is on to investigate, develop, and eventually clinically apply coatings deposited at temperatures much below the incongruent melting point of hydroxylapatite, ideally at or near ambient temperature (Heimann, 2016b). Postdepositional heat treatment must frequently still be applied as annealing may be required to crystallize ACP (Liu et al., 2019), to transform nonapatitic hydrated precursor phases such as OCP ($Ca_8(HPO_4)_2(PO_4)_4 \cdot 5H_2O$) or dicalcium phosphate dihydrate ($CaHPO_4 \cdot 2H_2O$, brushite) to hydroxylapatite, and/or to remove residuals of organic compounds used in coating preparation, for example, during sol–gel, dip coating, ECD, and EPD processes. However, lack of sufficient adhesive and cohesive strengths of low-temperature deposited coatings remains an issue that in some cases can be remedied by applying appropriate bond coats (e.g., Heimann, 1999) or by adding reinforcing polymeric materials (e.g., Martinez-Vázquez et al., 2013).

In response to the need for enhancing adhesion of coatings deposited at low temperature, attempts have been made to coat by AD titanium substrates with very dense (>98% theoretical density) nanostructured hydroxylapatite with adhesion strength exceeding 30 MPa (Hahn et al., 2009; Lee et al., 2017). The ceramic particles deform plastically on impact and are heated within a few picoseconds to temperatures slightly below their melting point. This mechanism may be paralleled with the plastic deformation of otherwise brittle particles occurring during CGDS by an adiabatic shear strain instability (Assadi et al., 2003). These effects, combined with high stress and deformation at high strain rates, cause minor penetration of the particle into the first atomic layers of the substrate surface and hence the physical fusing of the hydroxylapatite to the titanium substrate. After postdepositional heat treatment at 400 °C to crystallize ACP, these coatings have shown excellent biocompatibility in contact with preosteoblast MC3T3-E1 cells.

However, notwithstanding the acknowledged shortcomings of plasma-sprayed hydroxylapatite coatings that include thermal decomposition, line-of-sight limitation, and the inability to deposit coatings of less than about 20 μm thickness, thermal spraying is still the economically preferred method of choice to apply coatings to the metallic parts of commercially supplied hip and knee endoprostheses as well as endosseous dental root implants and dental abutments. Currently, plasma spraying of hydroxylapatite powder particles with diameters of tens to hundreds of micrometers onto the stems of hip endoprosthetic implants and dental roots is the most popular and the only Food and Drug Administration (FDA)-approved method to coat implant surfaces for clinical use (FDA, 2004; 2007). Indeed, hydroxylapatite-coated orthopedic biomedical devices are cleared by FDA through the 510(k) process (see box; Sun, 2018) or approved by the PMA (premarket approval application) process prior to commercial distribution in the USA and throughout the world (see Chapter 5.2).

What is the 510(k) process?
In the USA, section 510(k) of the Food, Drug and Cosmetic Act requires device manufacturers who must register, to notify FDA of their intent to market a medical device at least 90 days in advance. This is known as Premarket Notification – also called PMN or 510(k). This allows FDA to determine whether the device is equivalent to a device already placed into one of the three classification categories. Thus, "new" devices that have not been classified can be properly identified. Specifically, medical device manufacturers are required to submit a premarket notification if they intend to introduce a device into commercial distribution for the first time or reintroduce a device that will be significantly changed or modified to the extent that its safety or effectiveness could be affected.

In conclusion, even though deposition of hydroxylapatite coatings by atmospheric (air) plasma spraying comes across as a mature and well research-supported technique, there is an urgent need to address several shortcomings. Consequently, during the past decades, many attempts have been made to optimize essential properties of osseoconductive hydroxylapatite coatings deposited by conventional APS (Heimann, 2017a). These properties include coating cohesion and adhesion (Harun et al., 2018), phase composition, homogeneous phase distribution, crystallinity, porosity and surface roughness, nanostructured surface morphology, residual coating stress, and not in the least, coating thickness (Fazan and Marquis, 2000; Heimann, 2006, 2018; Heimann and Lehmann, 2015).

Suspension plasma spraying

Whereas coating development by APS can be described as the result of injecting the solid precursor powder by a stream of accelerated (inert) gas into the plasma jet, SPS uses an injected liquid as the powder carrier. The introduction of the powder material into the plasma jet in form of a solid–liquid suspension via an atomizer allows the utilization of very fine powders coupled with a substantial increase of the deposition efficiency and rate of coating deposition. In addition, the use of fine powders to deposit very thin coatings (<1 μm) provides advantages in terms of minimization of the residual coating stresses, increased adhesion, and improved biocompatibility (Hameed et al., 2019).

Based on these advantages, thermal spray coatings produced from liquid feedstock suspensions have received increasing interest due to the unique coating properties obtainable by these processes. Several research groups are working with different energy-supplying strategies such as plasmas (SPS) as well as combustion gases (HVSFS) to manufacture advanced nanostructured and nanophased materials (see, e.g., Killinger et al., 2006, 2011). These activities are centered on feedstock preparation, equipment and process design, modelling techniques, in-process diagnostics, coating characterization, and emerging applications.

The morphology of the powder formed by rapid evaporation of the liquid will be determined by the time constants of the vapor diffusion, the shrinking of the droplets, the diffusion of the suspended solids, as well as the heat conductivity in the surrounding gas, and the heat conduction within a droplet (Figure 3.43).

Beyond a critical supersaturation, solid material will be precipitated at the droplet surface that on further evaporation of the liquid moves toward the center of the ever-shrinking droplet. If the diffusion rate of the solid particles is below the evaporation rate of the solvent ($t_{PD} < t_E$), dense particles will be formed. In the opposite case ($t_{PD} > t_E$), hollow spheres are formed that will generate high coating porosity. However, if the solids content of the suspension is low, precipitation takes place within the volume of the droplet and dense spheres will be formed that on melting will behave akin to a solid powder particle during conventional APS.

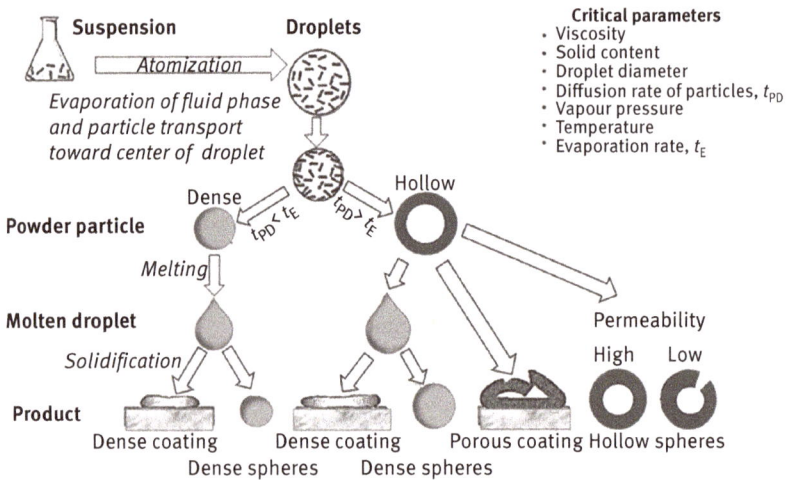

Figure 3.43: Formation and deposition of droplets during suspension plasma spraying (SPS) (Heimann and Lehmann, 2015). © With permission by Wiley-VCH.

Accordingly, the porosity of the coatings can be engineered throughout a wide range by selecting the appropriate solid/liquid ratio, the viscosity of the suspension, the suspension feed rate, injection point into the plasma jet, and ratio of feed rate to plate power, that is, plasma enthalpy (Müller, 2001; Xu et al., 2016). This possibility to control the coating porosity by adjusting SPS parameters is of particular importance for biomedical coatings as their biocompatibility, that is, the potential of ingrowth of bone cells into the fabric of the porous coating is of the utmost importance for implant performance. Figure 3.43 shows the development of different powder particle and coating morphologies dependent on the choice of critical deposition parameters, in particular the diffusion and evaporation rates.

As revealed in Figure 3.44, SPS coatings are much more porous than APS coatings (Gross and Saber-Samandari, 2009). Although the enhanced porosity leads to a decrease in hardness and elastic modulus of the bulk coating, site-specific indentations on dense areas of the SPS coatings demonstrated higher hardness values, possibly due to the finer grain size and crystal orientation. Thus, nanoindentation presents a valuable tool to assess the mechanical properties in solid areas of otherwise porous materials. The high porosity of hydroxylapatite coatings deposited by SPS is advantageous for enhanced ingrowth of bone cells and consequently, may impart improved biocompatibility if sufficient coating adhesion to the implant surface can be engineered. Wu et al. (2009) prepared porous hydroxyapatite coatings by the SPS process. They propose the use of a pore-forming agent into the hydroxyapatite suspension the solid content of which ranges between 16% and 45%. Pore formers are ammonium carbonate or ammonium bicarbonate added to the ethanol and demineralized water carriers.

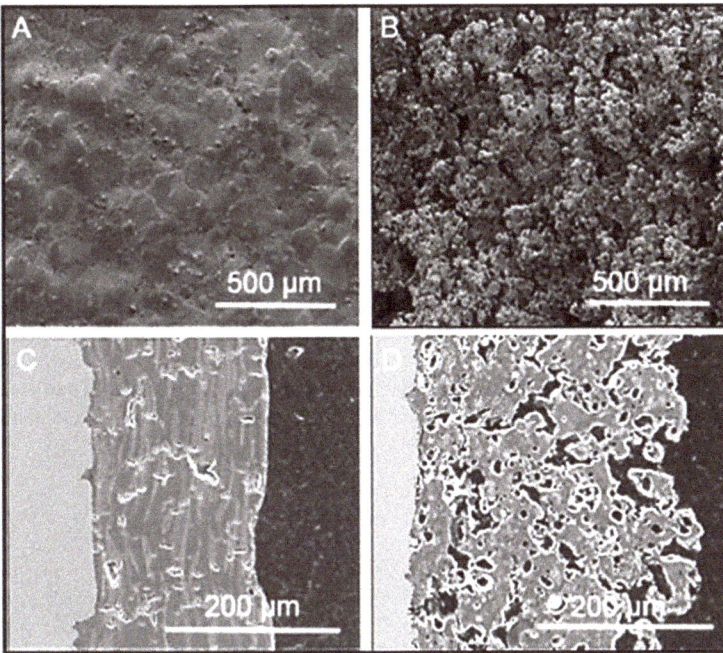

Figure 3.44: SEM images of top view (A, B) and cross sections (C, D) of conventional atmospheric plasma-sprayed (left column) and suspension plasma-sprayed (right column) hydroxylapatite coatings (Gross and Saber-Samandari, 2009). © With permission of Elsevier.

Thermal decomposition of hydroxylapatite

Hydroxylapatite powder injected into the extremely hot plasma jet undergoes first dehydroxylation into oxyhydroxylapatite and/or oxyapatite, and subsequent thermal decomposition to TCP and TTCP. Figure 3.45 shows a simple model of thermally induced phase transformation in a spherical hydroxylapatite powder particle.

During the short residence time of the hydroxylapatite particle in the plasma jet of only a few tens of microseconds, its core is still at a temperature well below 1,550 °C owing to its low thermal diffusivity, showing only hydroxylapatite (HAp) and oxyapatite (OA) as stable phases (steps 1 and 2 of Figure 3.46).

The second shell has been heated to a temperature above the incongruent melting point of hydroxylapatite (1,570 °C) and, thus, consists of a mixture of TCP and TTCPs (step 3 of Figure 3.46). The outermost spherical shell of the particle consists of CaO + melt (L) since evaporation of P_2O_5 shifts the composition along the liquidus toward CaO-richer phases (step 4 of Figure 3.46). The temperature increases to well beyond 1,800 °C, and the only unmelted composition is CaO. When the particle impinges at the metal implant surface to be coated, this clear phase separation will be lost, and an inhomogeneous calcium phosphate layer ensues in which hydroxylapatite,

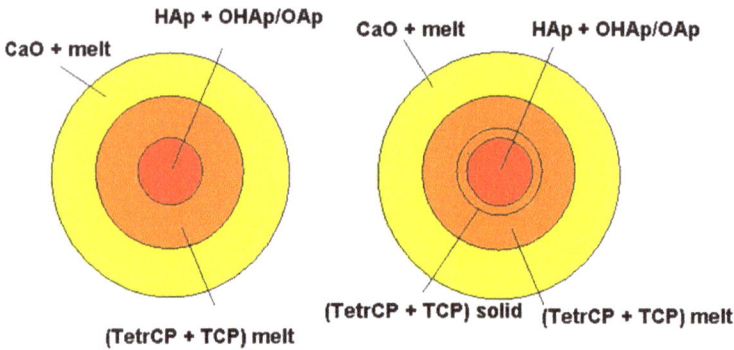

Figure 3.45: Schematic model of the thermal decomposition of a spherical hydroxylapatite particle subjected to high temperature in a plasma jet at a water partial pressure of 65.5 kPa (left) and at a water partial pressure of 1.3 kPa (right) (Heimann, 2006). © With permission by Elsevier.

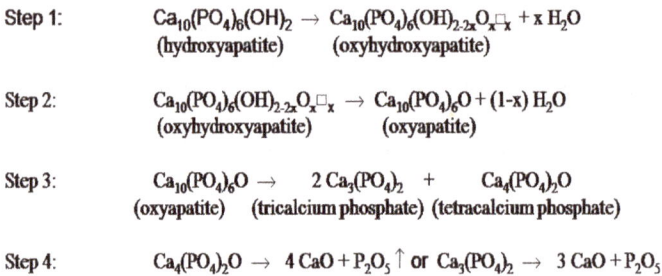

Step 1:　　$Ca_{10}(PO_4)_6(OH)_2 \rightarrow Ca_{10}(PO_4)_6(OH)_{2-2x}O_x\square_x + x\,H_2O$
　　　　　　(hydroxyapatite)　　　(oxyhydroxyapatite)

Step 2:　　$Ca_{10}(PO_4)_6(OH)_{2-2x}O_x\square_x \rightarrow Ca_{10}(PO_4)_6O + (1-x)\,H_2O$
　　　　　　(oxyhydroxyapatite)　　　(oxyapatite)

Step 3:　　$Ca_{10}(PO_4)_6O \rightarrow 2\,Ca_3(PO_4)_2 + Ca_4(PO_4)_2O$
　　　　　　(oxyapatite)　　(tricalcium phosphate)　(tetracalcium phosphate)

Step 4:　　$Ca_4(PO_4)_2O \rightarrow 4\,CaO + P_2O_5 \uparrow$ or $Ca_3(PO_4)_2 \rightarrow 3\,CaO + P_2O_5$

Figure 3.46: Thermal decomposition scheme of hydroxylapatite subjected to a hot plasma jet.

oxyapatite, TCP (α′-TCP, β-TCP), TTCP (hilgardite), CaO (oldhamite), and ACP are interspersed on a microcrystalline scale (Figure 3.48). At the immediate interface to the solid metallic substrate, a very thin layer of ACP forms by rapid quenching of the outermost melt layer with heat transfer rates beyond 10^6 K/s (Heimann and Wirth, 2006). This thin layer takes on a special significance as its enhanced solubility in the ECF may be one of the leading causes of coating delamination *in vivo*. Heat treatment of microplasma-sprayed hydroxylapatite coatings at 650 °C for several hours can be used to transform decomposition phases as well as ACP to crystalline hydroxylapatite (Liu et al., 2019).

The simple model shown in Figure 3.45 was modified to include the product of the solid-state dehydration transformation of hydroxylapatite into TCP and TTCP between 1,633 and 1,843 K that has been experimentally found to occur at very low water partial pressure (Figure 3.45, right). (Dyshlovenko et al., 2004).

Tailoring of coating properties

Coating thickness, surface roughness, porosity, adhesion strength, as well as phase composition can be tailored by appropriate adjusting of intrinsic and extrinsic plasma spray parameters (Heimann, 2008; Vahabzadeh et al., 2015). In survey experiments, coatings were engineered for maximum surface roughness (coating type A), porosity (coating type B), and tensile adhesion strength (coating type C) by varying five crucial intrinsic plasma spray parameters using an SDE methodology. The experimental design applied was of composite nature, comprising a total of 28 experimental runs with 16 design points 2^{5-1} + 10 star points + 2 center points. Figure 3.47 shows cross-sectional SEM micrographs of the three coating types. The plasma spray parameters, coating properties, and as-sprayed phase contents are shown in Table 3.7 (Heimann et al., 2001).

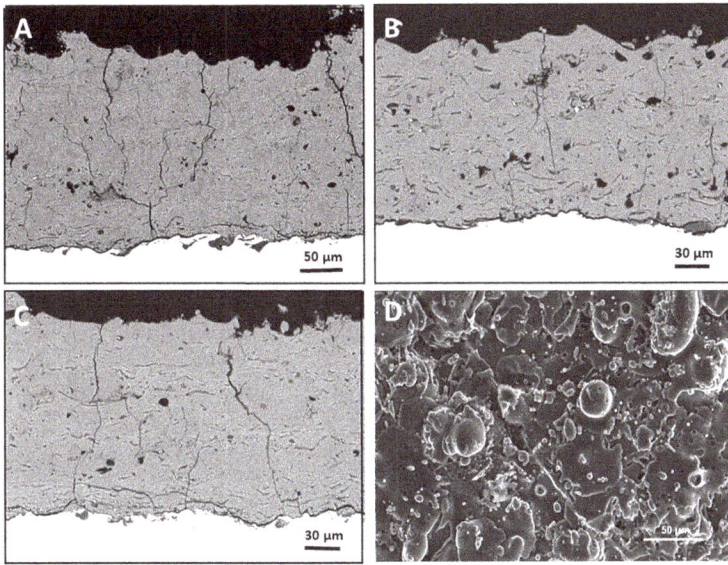

Figure 3.47: Cross-sectional SEM micrographs of plasma-sprayed hydroxylapatite coatings of types A, B, and C (Table 3.7), and the typical surface of a plasma-sprayed hydroxylapatite coating with well-developed stacked pancake-type splats and loosely adhering incompletely melted oversized spherical particles (D) (Heimann et al., 2001). © With permission by Wiley-VCH.

Recently, a related but more comprehensive study of the process–property–structure relationship of plasma-sprayed hydroxylapatite coatings was undertaken by a research group at Dublin City University, using a central composite design consisting of a 2^{5-1} fractional factorial design with 10 star points and 5 center points added (Levingston et al., 2017). The work was designed to investigate the influences and interaction effects of arc current, plasma gas flow rate, powder feed rate, spray distance, and carrier gas flow rate on the roughness, crystallinity, phase purity, porosity, and thickness of plasma-sprayed hydroxylapatite coatings. Coating roughness was found to be related to the particle velocity and particle melting and was highest at low gas flow rates. High crystallinity resulted at high current and low spray distance. Phase purity was highest at low carrier gas flow rate and high gas flow rate. Porosity was dependent on the degree of particle melting and was highest at low gas flow rate and powder feed rate, and at high current and spray distance. Coating thickness was highest at high current, low gas flow rate, high powder feed rate, and low spray distance. From this thorough analysis, predictive process equations were developed and used to optimize two distinct types of coating, that is, a "stable" coating and a "bioactive" coating, designed to form a layer adjacent to the substrate and a surface layer, constituting a functionally graded coating to provide enhanced osteogenesis, while maintaining long-term stability.

Investigation of the properties of plasma-sprayed hydroxylapatite coatings for biomedical applications using an L18-Taguchi experimental design (Cizek and Khor, 2012) confirmed that the particle in-flight characteristics are substantially influenced by the choice of the system parameters; the dominant control factor was found to be the spray distance. The microstructure of the deposited coatings is influenced by the in-flight temperature and velocity of the particles; contrary to other findings, high temperature particles ($\geq$2,500 K) formed individual splats with interconnected void chains, whereas low temperature particles ($\leq$2,470 K) gave rise to bulk-type, well-fused coatings.

Phase composition of coatings

The molten phase comprising the outermost shell of the heated hydroxylapatite particle (Figure 3.45) is quenched on impact with the comparatively cool substrate surface and thus, will solidify rapidly to produce ACP with various Ca/P ratios.

The force of impact of droplets accelerated to supersonic velocity triggers a series of events that profoundly affect the composition and the morphology of the resulting coating. On impact, the neatly ordered succession of spherical shells shown in Figure 3.45 will be destroyed and thus will form a mixture of ACP in close vicinity to hydroxylapatite and oxyhydroxylapatite. Patches of ACP are being scattered throughout the crystalline calcium phosphate phases (Figure 3.48). This patchy distribution of ACP is the result of the splashing of molten or semimolten droplets during impact at the substrate surface whereby the ACP will be folded into the crystalline constituents.

Scanning cathodoluminescence microscopic microanalyses of coatings conducted by Gross and Phillips (1998) confirmed that the darker regions in polished cross sections represent the amorphous phase that is shown in Figure 3.48 (right). The more energetic cathodoluminescence emission from the amorphous phase during electron beam irradiation compared with the lighter appearing crystalline phase can be used to distinguish between structurally ordered (crystalline) and disordered (amorphous) areas within the sample. By selecting the peak of the intrinsic electron emission at 450 nm, it is possible to scan the surface with the electron beam, thereby producing a map of ACP distribution. Hence, cathodoluminescence microscopy based on the different light emission from the amorphous phase and hydroxylapatite is a useful tool to identify and map the ACP constituent in plasma-sprayed coatings (Heimann and Lehmann, 2015).

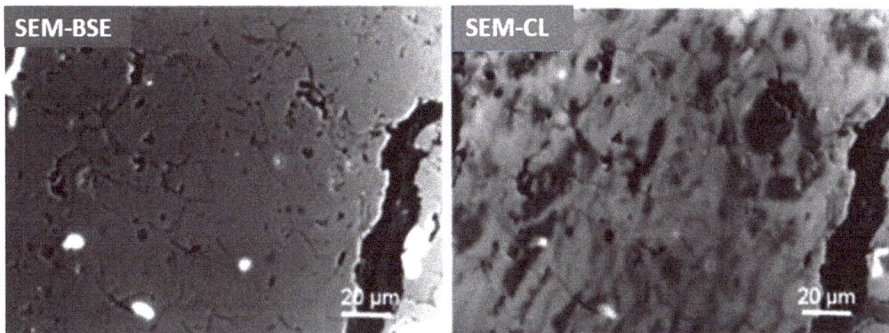

Figure 3.48: SEM images in backscattered electron (BSE) mode (left) and cathodoluminescence (CL) mode (right) of a plasma-sprayed hydroxylapatite coating immersed for 7 days in simulated body fluid (HBSS). The dark areas are predominantly composed of ACP (Götze et al., 2001; Heimann and Lehmann, 2015). © With permission by Wiley-VCH.

In addition to the compositional inhomogeneity caused by lateral splat spreading, the phase composition of plasma-sprayed hydroxylapatite coatings varies with coating depth. Depth-resolved diffraction studies with both conventional X-ray and synchrotron radiation energies revealed that the proportion of hydroxylapatite decreases linearly with coating depth from approximately 75 mass% at the free coating surface to 15 mass% at the coating-substrate interface (Figure 3.49, left). Moreover, at the free coating surface, the remainder of crystalline phases were determined to be 15 mass% TTCP, 4 mass% β-TCP, and 1.5 mass% CaO. These values decrease to 3 mass% TTCP, and 0.5 mass% each of β-TCP and CaO at the substrate-coating interface. The balance is thought to consist of ACP that reaches in excess of 80% at the substrate-coating interface owing to rapid quenching of the molten droplets arriving in sequence (Ntsoane et al., 2016). This ACP layer forms immediately adjacent to the metallic substrate and thus influences the adhesion of the entire coating. Figure 3.49 (right) shows that the crystallinity of the coating decreases exponentially with depth from about 93% at the free coating surface to approximately 50% at the coating-substrate interface, in accordance with the amount of ACP shown in Table 3.7.

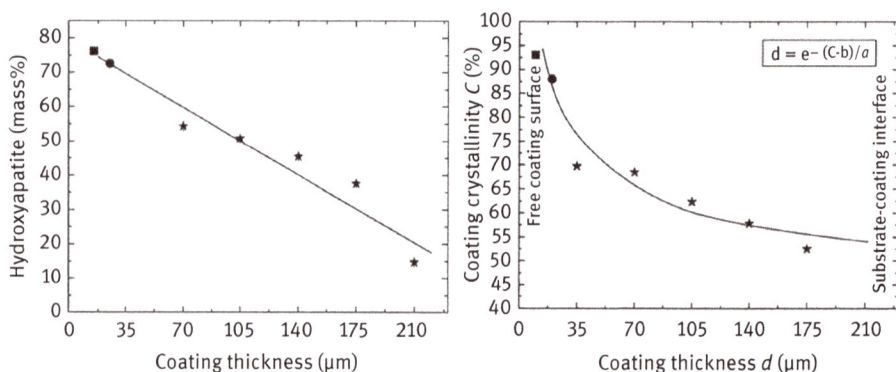

Figure 3.49: Left: Linear decrease of hydroxylapatite (HAp) content with coating thickness. Right: Exponential decrease of crystallinity of a plasma-sprayed HAp coating with coating thickness. The coefficients of the exponential decay equation were determined to be $a = 128.3$, and $b = 14.3$. The data were obtained by conventional X-ray diffraction (8 keV, square) and synchrotron radiation X-ray diffraction (11 keV, dot; 100 keV, stars) (Ntsoane et al., 2016) © With permission by Elsevier.

These findings are supported by the work of Hesse et al. (2008) who investigated the effect of the precursor powder grain size on the spatial distribution of calcium phosphate phases in air plasma-sprayed (APS) hydroxylapatite coatings incubated in revised r-SBF (Kim et al., 2001) for 8 weeks. The coatings were mechanically abraded under dry conditions in steps of 40 μm by abrasive carborundum (SiC) paper, and the newly created surfaces were analyzed by X-ray diffraction (XRD) with Rietveld refinement to record their quantitative phase composition. The results of this depth profiling are shown in Figure 3.50.

The coarse hydroxylapatite starting powder (Captal 90, Figure 3.50A) yields high HAp/OHAp/OAp contents around 80 mass % that vary little between the substrate/coating interface and the top surface of the coating. The amount of thermal decomposition products varies between 15 and 20 mass%, with TTCP dominating over α-TCP and β-TCP. At the as-sprayed surface, minor amounts of CaO (<1 mass %) are formed that quickly disappear during incubation. In contrast to this, plasma spraying of fine hydroxylapatite powder (XPT-D-701, Figure 3.50B) results in considerably lower HAp/OHAp/OAp contents, ranging from 30 mass% at the substrate/coating interface to about 50 mass% at the coating surface. The higher surface-to-volume ratio of the fine powder leads to substantially higher thermal decomposition compared to that encountered for coarser powders. Consequently, the CaO content at the surface of the as-sprayed coating is about 10–12 mass%. Incubation in r-SFB increased the hydroxylapatite content by dissolving predominately CaO, TTCP, and ACP (not shown) to about 30 mass% at the substrate/coating interface and 54 mass% at the

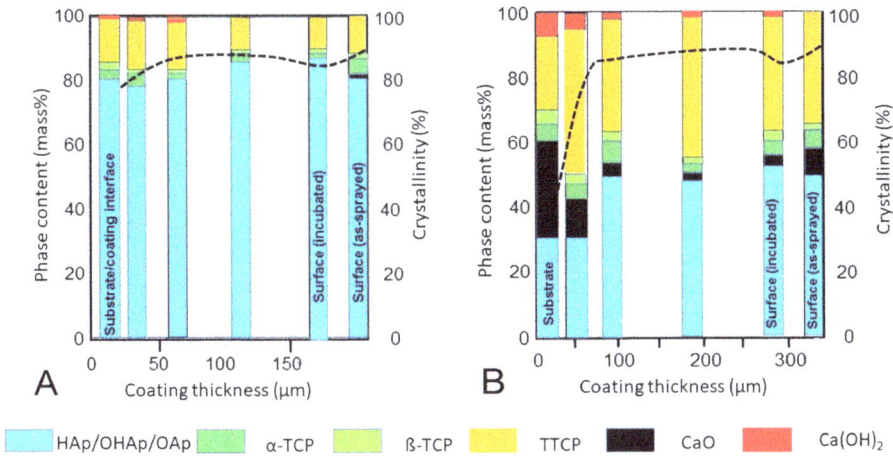

Figure 3.50: Spatial variation (depth profiling) of phase composition and crystallinity of air plasma-sprayed hydroxylapatite coatings incubated for 8 weeks in r-SBF as a function of the grain size of the precursor powder particles. (A) Coarse grain size (−140 + 100 µm; Captal 90, Plasma Coating Ltd., UK). (B) Fine grain size (<90 µm; XPT-D 701, Sulzer Metco). Plasma power: 30 kW; stand-off distance: 90 mm; traverse speed: 100 mm/s) (adapted from Hesse et al., 2008). © With permission by Springer Nature.

coating surface. In concurrence, the crystallinity increases steeply from a low 30% at the substrate/coating interface to about 90% at the coating surface. The low crystallinity at the interface is related to the formation of a thin layer of ACP produced by rapid quenching of the molten droplets arriving at the substrate surface during the first passes of the plasma torch. The coating porosity appears to be rather low as revealed by the retention of a considerable proportion of CaO of about 30 mass% at the very interface that diminishes monotonously toward the coating surface. Obviously, even after an incubation time of 8 weeks, the penetration of the r-SBF down to the coating/substrate interface is limited by the absence of an interconnected network of pores and cracks. Nevertheless, some of the CaO has been converted to portlandite, Ca$(OH)_2$ by the action of r-SBF. Since the occurrence and spatial distribution of this conversion phase can be correlated to the coating porosity, portlandite may serve as an indicator of the biodegradability of plasma-sprayed hydroxyapatite coatings (Hesse et al., 2008).

Origin and control of residual stresses

The occurrence of residual stresses (see box) at the interface hydroxylapatite coating-metallic implant as well as within the coating will result in reduced adhesion by delamination and crack formation, depending on the nature and sign of the stresses.

Residual coating stresses

The principal equation governing the generation of thermal coating stress, σ_c has been derived by the German glass chemist Adolf H. Dietzel and expressed by the equation

$$\sigma_c = \{E_c(\alpha_c - \alpha_s)\Delta T/(1 - v_c) + [(1 - v_s)/E_s]d_c/d_s\} \tag{3.23}$$

where E is the Young's modulus, α the coefficient of thermal expansion, T the temperature, v the Poisson's number, and d the thickness. The subscripts c and s refer to coating and substrate, respectively. Since at given values of v and E the thermal coating stress σ_c increases with increasing coating thickness d_c, the risk of spalling is much higher in thick coatings than in thin ones. Moreover, depending on the sign of $(\alpha_c - \alpha_s)$, the so-called thermal stress can either be tensile or compressive. Quenching and thermal stresses, combined with the complicated solidification process of the coating, are the two main contributors to the overall residual stress. Whereas during deposition the substrate is usually at some elevated temperature, postdepositional cooling to room temperature generates additional stress by thermal mismatch proportional to the differences in the thermal expansion coefficients of the coating and the substrate as well as the intrinsic elastic moduli (see eq. (3.23)).

Residual stresses originate from the large temperature gradients experienced during the plasma spraying process. When the molten particles strike the cold substrate, they will be rapidly quenched whereby their contraction is constrained by tight adherence to the rough and rigid substrate. This leads to accumulation of high levels of tensile stresses both within the coating and at the coating/substrate interface, commonly referred to as "quenching stresses." The first layer adjacent to the interface, found to consist of ACP (Heimann and Wirth, 2006), will crucially control the occurrence of residual stresses in terms of their magnitude as well as their signs. In addition, this thin ACP layer provides a low-energy fracture path that may lead to coating delamination in the presence of tensile or shear residual stresses.

The transformation of ACP to crystalline calcium phosphate phases during *in vitro* contact with SBF (Stammeier et al., 2018) and *in vivo* contact with ECF, respectively, will lead to stress relaxation as observed by Topić et al. (2006) and, thus, reduces the risk of coating failure by delamination. Figure 3.51 shows the lattice strain of hydroxylapatite coatings deposited by air plasma spraying on Ti6Al4V substrates measured by the $\sin^2\psi$ method, $\varepsilon = (d - d_0)/d_0 \cdot 10^{-3}$ as a function of $\sin^2\psi$ (see box), where ψ is the tilt angle toward the X-ray beam.

Stress analysis by X-ray diffraction

XRD can be applied to measure residual stress using the distance between crystallographic planes, that is, D-spacing, as a strain gage. When the material is in tension, the D-spacing increases and, when under compression the D-spacing decreases. Hence, stresses can be determined from the measured D-spacings. In other words, the determination of the stress state is based on the measurement of the lattice deformation of a polycrystalline materials subjected to

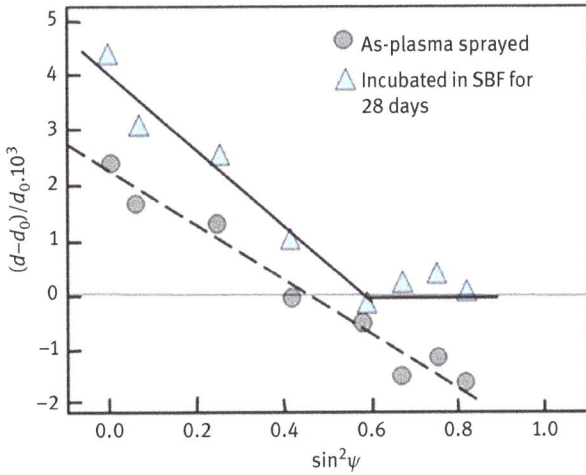

Figure 3.51: Development of residual stresses in as-sprayed hydroxylapatite coatings (dashed line) and coatings incubated in SBF for 28 days (solid line) (Heimann et al., 2000).

stresses. This is accomplished by measuring the change of the D-values of the interplanar spacing of selected lattice planes $\{hkl\}$ relative to the stress-free state, D_0:

$$dD = D - D_0 \tag{3.24}$$

Since the penetration of the radiation into the coating is rather limited (1–10 μm), only the stress state of the coating surface can be measured with accuracy. To obtain a stress distribution profile, the surfaces must be consecutively removed by polishing, sputtering or etching, and the measurement be repeated. Differentiation of the Bragg equation $n \cdot \lambda = 2D \cdot \sin\theta$ yields the (relative) lattice deformation

$$\varepsilon^L = dD/D_0 = -\cot\theta_0 \cdot d\theta \tag{3.25}$$

From eq. (3.25) it follows that the change of the Bragg angle, $d\theta$ is maximized for a given ε^L when θ_0 is large. Therefore, interplanar spacing with the largest possible Bragg angles must be chosen. Also, $d\theta$ increases with increasing stress σ and decreasing Young's modulus E. However, ceramic materials have generally large E values. Therefore, very small shifts of the interplanar spacing must be recorded with high accuracy. This requires highly sophisticated XRD hardware and appropriate software. Another problem exists related to the difficulty (impossibility?) to determine accurately the D-value of the stress-free state (D_0).

The as-sprayed coatings exhibit rather strong compressive surface stresses owing to thermal mismatch developed during cooling of the deposited layer to room temperature. This compressive stress will relax during incubation in SBF when high levels of ACP thought to be a main contributor to the residual stress crystallize to form calcium phosphate phases, most notably TTCP and hydroxylapatite. Hence, the layer of bone-like secondary apatite deposited at the outermost rim of the samples will be decoupled from the declining stress field, and thus, show close to zero stress as evident from Figure 3.51.

Figure 3.52 shows the residual stress state of as-sprayed hydroxylapatite coatings on a Ti6Al4V substrate as determined by the $\sin^2\Psi$-technique, using both conventional X-ray (8 keV) and synchrotron radiation (11 and 100 keV) diffraction (Heimann, 2016a). The principal Cauchy stress tensor components σ_{11} and σ_{33} are both tensile adjacent to the coating–substrate interface, but relax to zero within the first 80 μm of the coating (Figure 3.52A and B). With further accumulating coating thickness the component σ_{11} slightly increases to become tensile again with + 20 MPa at the free coating surface. In contrast, the tensile stress component σ_{33} at the coating–substrate interface decreases monotonously to compressive with approximately –30 MPa at the free coating surface. The shear stress tensor component σ_{13} shown in Figure 3.52C is likewise tensile near the coating–substrate interface

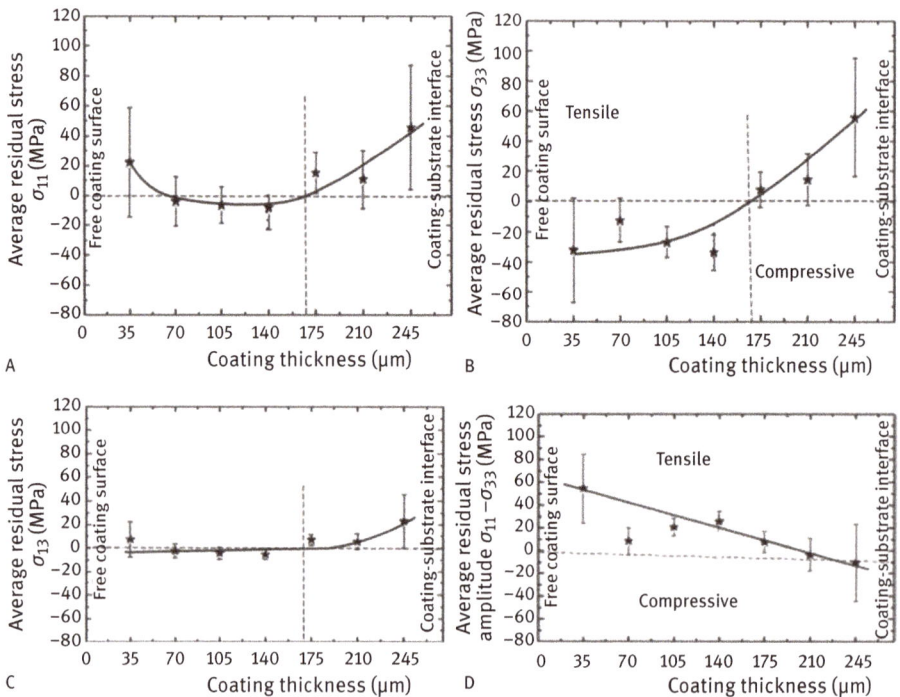

Figure 3.52: Distribution of the principal Cauchy stress tensor components σ_{11} (A) and σ_{33} (B) as well as the shear stress tensor component σ_{13} (C) and the residual stress amplitude $\sigma_{11}-\sigma_{33}$ (D) as functions of the thickness of an atmospheric plasma-sprayed hydroxylapatite coating deposited on a Ti6Al4V substrate. The data were averaged over three different measurement positions. The vertical dotted lines mark the position at which the residual stress changes sign (Ntsoane et al., 2016). © With permission by Elsevier.

with +20 MPa and relaxes to zero within the first 80 µm. After this, it remains neutral for the remainder of the coating thickness up to the free coating interface. In addition, as shown in Figure 3.52D the stress amplitude ($\sigma_{11}-\sigma_{33}$) follows a strictly linear trend, increasing from zero adjacent to the coating-substrate interface and remaining tensile up to the free coating surface, with maximum tensile stress of +60 MPa. In conclusion, the major stress component influencing residual stress formation in the coating is σ_{33}. As expected from the empirical Dietzel equation (eq. (3.23)), within the first 80 µm of the deposited coating, the residual stress was found to be tensile. With further increasing coating thickness, the stress state changes to become compressive. This is beneficial for coating integrity as the compressive stress counteracts crack formation and, hence, promotes coating cohesion. These findings are in general agreement with earlier work by Tsui et al. (1998) and Cofino et al. (2004).

As shown above, prevention and control of residual stresses are important to ensure the integrity of the coating–substrate system and in turn, its mechanical performance since high residual stresses can lead to cracking and delamination of the coating, shape changes of thin substrates, and in general can undermine the performance of the entire part. Tensile stresses exceeding the elastic limit of the coating cause cracking perpendicular to the direction of the tensile stress tensor. Whereas in general some degree of compressive stress is considered desirable as it closes cracks originating at the surface and thus improves fatigue properties, excessive compressive stress promotes adhesive failure as shown in Figure 3.53.

In biomedical service, the existing residual stresses superpose with the applied loading stress during movement of the patient and failure may occur or fatigue life be shortened if the residual stress is sufficiently high. Hence, stringent control of the occurrence of high residual stresses is of paramount importance in the quest for optimum mechanical coating performance.

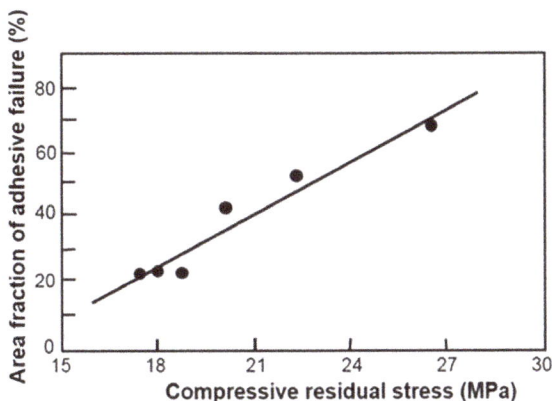

Figure 3.53: Relation between adhesive failure and compressive residual stresses of plasma-sprayed hydroxylapatite coatings on Ti6Al4V (Yang and Chang, 2001). © With permission by Elsevier.

As mentioned earlier, the adhesive bond strength of the coating is limited owing to a mismatch of the coefficients of thermal expansion between the metallic implant and the hydroxylapatite ceramic. To improve the adhesive strength, gradient Ti6Al4V/hydroxylapatite coatings have been deposited on Ti6Al4V by laser engineered net shaping (LENS™) followed by plasma spray deposition. The presence of an interfacial gradient layer enhanced the adhesive bond strength from 26 ± 2 MPa for common plasma-sprayed coatings to 39 ± 4 MPa for LENS™/plasma spray coatings (Ke et al., 2019). Details of this novel processing technology can be found in Chapter 1.3.7.

Additional details on deposition parameters, structure, mechanical, and chemical properties, residual stress, and biological function of plasma-sprayed hydroxylapatite coatings as well as other calcium phosphate-based coatings can be obtained from Heimann and Lehmann (2015) and Heimann (2006; 2016a; 2017a; 2017b; 2018). Recently, Cizek and Matejicek (2018) have reviewed patents granted between 2005 and 2018 that pertain to plasma-sprayed bioceramic materials.

3.2.8 Bioglasses

In the context of the history of development of biomaterials, the discovery of bonding of bone to specific compositions of glasses has led to a new generation of bioactive materials for tissue replacement. Subsequently, improved understanding of the mechanisms of gene activation of human progenitor cells by controlled release of dissolution products such as Ca^{2+} and SiO_4^{4-} ions from bioactive glasses has provided a sound basis for designing biomaterials that can be used for tissue regeneration. Bioactive glass science and technology continues to be at the forefront of providing innovative approaches to medicine (Hench, 2013). In particular, bioactive glass nanoparticles are being developed and strategies designed to control their size, shape, pore structure, and composition. Although these nanoparticles are particularly interesting biomaterials for bone-related applications, they also have potential for other biomedical applications, for example, in soft tissue regeneration and repair (Zheng and Boccaccini, 2017).

Bioglass compositions can be synthesized by several techniques including melt-quench synthesis as utilized in conventional glass making, ultrasonic- and microwave-assisted methods, flame pyrolysis and, most importantly, ultrastructure processing by sol–gel synthesis (Li et al., 1991; Zheng and Boccaccini, 2017) that today is the most promising route to synthesize and process bioglasses. By sol–gel technique, also known as chemical solution deposition, ceramic ultrastructures are generated at essentially ambient temperature by reactions of inorganic precursors or hydrolysis of organometallic compounds, usually alkoxides. From the precursor sol, a gel of hydrous oxides is produced that subsequently during sintering will attain a dense and homogeneous ceramic body (Brinker and Scherer, 1990). Sol–gel processing involves a wet-chemical technique used primarily for synthesis of materials starting from a chemical solution that acts as the precursor for an integrated network (or gel) of either discrete particles or network polymers (Innocenzi, 2019).

Figure 3.54 shows a schematic 2D rendering of the development of the ceramic microstructure, starting with very small colloidal particles, usually only about 3–4 nm in size that in sufficiently high concentration links up to form chains and, subsequently, 3D networks, filling the liquid phase as a gel (Figure 3.54A). On drying, these chains consolidate, and the gel hardens (Figure 3.54B) and shrinks (Figure 3.54C). During sintering, further shrinkage occurs by closure of pores (Figure 3.54D and E) until ideally a pore-free state is being reached (Figure 3.54F).

The advantages of the sol–gel method include the option to form intimate and uniform mixtures of different colloidal oxides on a molecular scale. Because of the lower fabrication temperatures used in this method, there is a greater level of control on the composition and homogeneity of the product. In addition, sol–gel bioglasses have much higher porosity, leading to a greater surface area and thus, a higher degree of integration in the body.

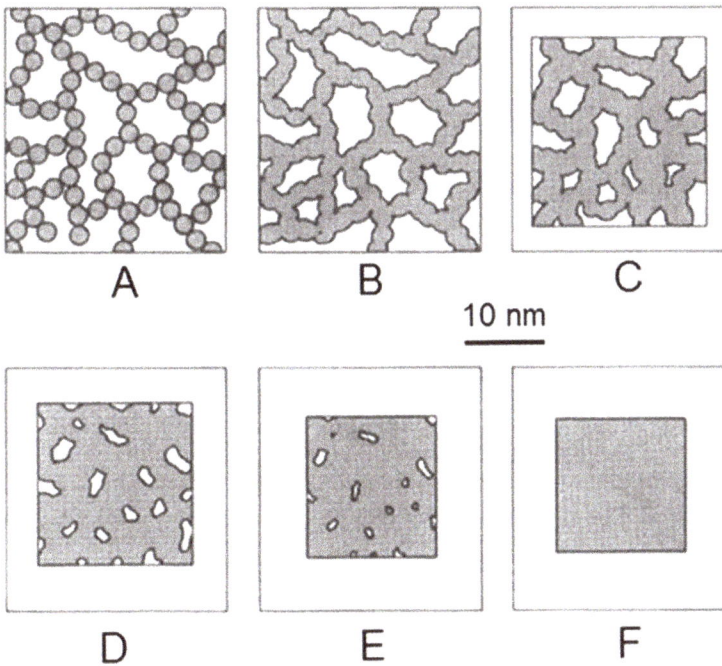

A   B   C

10 nm

D   E   F

Figure 3.54: Simplified 2D representation of the sol–gel process. (A) Colloidal particles with diameters around 4 nm form a gel network. (B) Coalescence of particles into chains. (C) Gel shrinkage during drying. (D)–(F) Further shrinkage during sintering accompanied by (complete) pore closure (adapted from Iler, 1986).

The resulting stiff gels can be molded into any desirable shape, the dimensions of which can be adjusted for shrinkage during drying and sintering. However, the process is marred by the generally high cost of the organometallic precursor materials and environmental concerns based on the frequently toxic solvents being used in forming the starting sols. In addition, the dried molded gel bodies tend to crack during sintering owing to the considerable shrinkage involved. Another setback to using the archetypal 45S5 Bioglass® is that it is difficult to process into porous 3D scaffolds. These porous scaffolds are usually prepared by sintering glass particles that are already formed into the 3D geometry and allowing them to bond to the particles into a strong glass phase made up of a network of pores. Since this particular type of bioglass cannot be fully sintered by viscous flow above its transition temperature T_g that, in addition, is close to the onset of crystallization, it is difficult to consolidate this material into a dense network (Rahaman, 2011). Improvements are being sought by tweaking the processing condition, for example, by altering the agent that catalyzes the hydrolysis reaction of alkoxides. Faure et al. (2015) used citric acid instead of nitric acid as a catalyst to obtain sol–gel powders with improved bioactivity, a result thought to be associated with an increase of porosity and specific surface area of the synthesized 45S5 Bioglass® powders.

A typical synthesis protocol for sol–gel-derived bioglass powder
A typical protocol describing the synthesis of sol–gel-derived bioglass powders uses precursors such as tetraethyl orthosilicate (TEOS), triethyl phosphate (TEP), calcium nitrate tetrahydrate (Ca $(NO_3)_2 \cdot 4H_2O$), potassium nitrate (KNO_3), magnesium nitrate hexahydrate ($Mg(NO_3)_2 \cdot 6H_2O$), and sodium nitrate ($NaNO_3$). To prepare the glass sols, 39.68 mL TEOS were added into 0.3 M HNO_3 aqueous solution at room temperature. The molar ratio of H_2O/TEOS was maintained at 15. The mixture was stirred for at least 60 min to initiate hydrolysis. Then, additional compounds, that is, 1.90 mL TEP, 3.34 g $NaNO_3$, 16.54 g $Ca(NO_3)_2 \cdot 4H_2O$, 6.32 g $Mg(NO_3)_2 \cdot 6H_2O$, and 5.23 g KNO_3 were added in this sequence but only after the previous solution had become clear. The final transparent solution was stirred overnight for homogenization and stored at room temperature for further processing. Gel formation occurred after 3 days at 25 ºC by hydrolysis of the TEOS accompanied by formation of silanol groups that interact by a condensation reaction to form Si-O-Si bonds (see Chapter 2.2.2.4). Subsequently, the gel was aged for 48 h at 60 ºC and then dried at 120 ºC for 24 h. During ageing, the aqueous gel continues to crosslink to form a glass with 3D-silicate network structure. The dried sample was ground and calcined at 625 ºC for 4 h to remove nitrate ions remaining in the structure. After the calcination step, the powder was ground for 5 min using a vibratory disk mill at 1,000 rpm (Deliormanh and Yildirim, 2016).

Increasingly, other techniques to synthesize and deposit bioglass-based coatings are being developed, for example, by APS (Cañas et al., 2019a), SPS (Cañas et al., 2017), or SPPS (Cañas et al., 2018; 2019b). It was found to be feasible to deposit bioactive glass coatings by SPPS with a dense microstructure, sufficient adhesion, and suitability to react with biological fluids (Cañas et al., 2019b). Bioactive 45S5 Bioglass® coatings were deposited under low vacuum conditions on a variety of metallic and ceramic substrates by AD (Eckstein et al., 2019). *In vitro* tests with human osteoblast-like cells on substrates with a 45S5 BG coating demonstrated high cell activity on the surfaces. Hence, utilization of the AD process for depositing amorphous bioglass coatings may be a promising technique for clinical biomedical application. However, more work is required aiming to further improvement of feedstock stability, coating adherence as well as to eliminate formation of undesirable crystalline phases. Moreover, doping elements will need to be introduced to improve the biological response of coatings.

3.3 Polymers (Matthias Schnabelrauch)

3.3.1 General synthesis principles for polymers

3.3.1.1 Biological polymers

Natural polymers a.k.a. biopolymers are synthesized by nature, primarily from plants, animals, and microorganisms during specific biochemical processes. Their syntheses involve multistep enzyme-catalyzed, chain-growth polymerization steps of activated monomers which are typically formed within animal or plant cells (intracellularly) by complex metabolic processes. In some cases, the biosynthesis of macromolecules can also take place extracellularly. In nature, nearly all types of molecular architecture can be found including homopolymers, alternating or random copolymers, block and graft copolymers, linear, branched, and cross-linked polymers. Among natural polymers, frequently used in medicine are various proteins, polysaccharides including glycosaminoglycans, and microbial polyesters, mainly poly(β-hydroxyalkanoic acid esters). All biological polymerization reactions are characterized by high specificity with regard to the selection of monomer (substrate) offered by cells, high regio- and stereoselectivities in terms of the chain growth and high polymerization rates at the physiological temperature region of living processes between 0 and 40 °C (Elias, 1977).

3.3.1.2 Synthetic polymers

Synthetic polymers are produced mainly by two basic polymerization mechanisms, *step-growth polymerization* and *chain polymerization*. Both types of polymerization reaction differ by the rate at which the size of polymer molecules increases, the time required for high monomer conversion, the reaction mechanism, the presence of an active center, and the termination reaction.

During *step-growth polymerization* reactions (Scheme 3.1), reactive monomers containing functional units such as carboxylic, hydroxylic, or amino groups bond together through an extension of chemical condensation reactions, thereby eliminating low-molecular-mass side products such as water or alcohols. Polymerization can occur between monomers containing at least two distinct functional groups able to react with each other. The step-growth process can take place between monomers, dimers, trimers, or oligomers in a multitude of distinct steps, resulting first in a slow increase of the molecular mass, and subsequently, a high turnover of reaction is required to achieve high molecular masses.

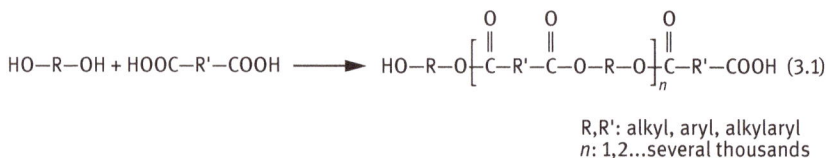

$$HO-R-OH + HOOC-R'-COOH \longrightarrow HO-R-O\left[\overset{\overset{O}{\|}}{C}-R'-\overset{\overset{O}{\|}}{C}-O-R-O\right]_n\overset{\overset{O}{\|}}{C}-R'-COOH \quad (3.1)$$

R,R': alkyl, aryl, alkylaryl
n: 1,2...several thousands

Scheme 3.1: Synthesis of a polyester as a typical example of a step-growth polymerization reaction.

In a special case of *step-growth polymerization,* no small molecule is eliminated but the reaction follows a similar mechanism as the latter. An example of this type of polymerization mechanism is the formation of polyurethane from diols and diisocyanates.

Figure 3.55 shows an overview of polymers typically prepared by step-growth polymerization.

In contrast, *chain polymerization* involves the bonding of unsaturated monomers by opening a double bond (Scheme 3.2). Due to the presence of active species initiating the polymerization step, three main mechanistic types of chain polymerization are known: free radical polymerization, ionic (cationic and anionic) polymerization, and coordination polymerization.

In general, chain polymerization reactions require an initiator step (3.2) able to generate the first active unit and to start the chain growth sequence. Subsequently, during the propagation reaction (3.3), monomers are added to the reactive sites of the growing polymeric chain consisting of an all-carbon backbone of single bonds. The last step is the termination reaction (3.4), in which the reactive end-site is deactivated. As a consequence of the *chain polymerization* mechanism, the molecular mass of the polymer chain increases rapidly at early stage and remains approximately constant throughout the entire polymerization process.

Well-known polymers, copiously used in medical applications, for example, polyethylene (PE) and poly(methylmethacrylate), are produced by a chain polymerization mechanism (see Figure 3.56 for an overview).

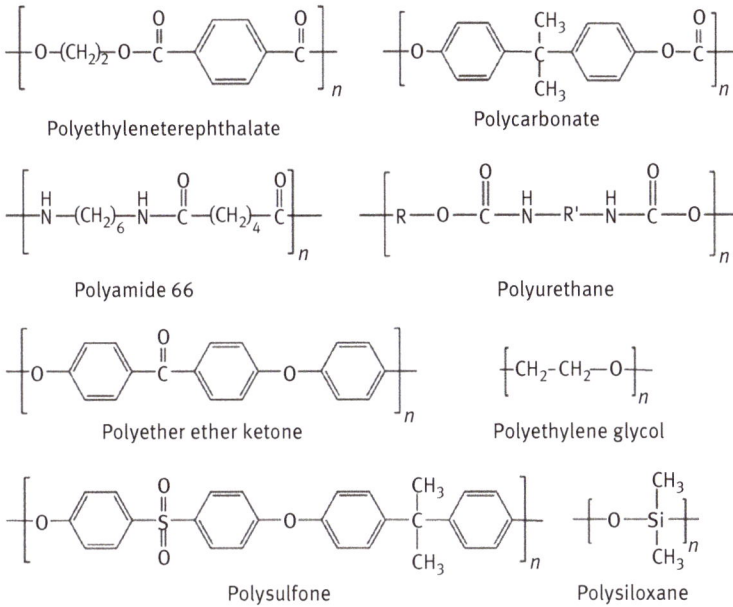

Figure 3.55: Polymers synthesized by step-growth polymerization reactions.

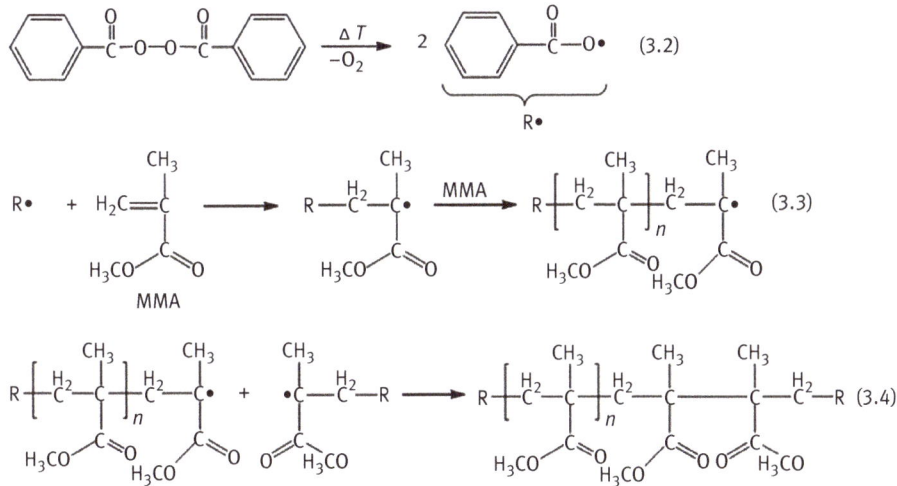

Scheme 3.2: Reaction mechanism of the radical polymerization of methylmethacrylate (MMA) as an example of chain polymerization.

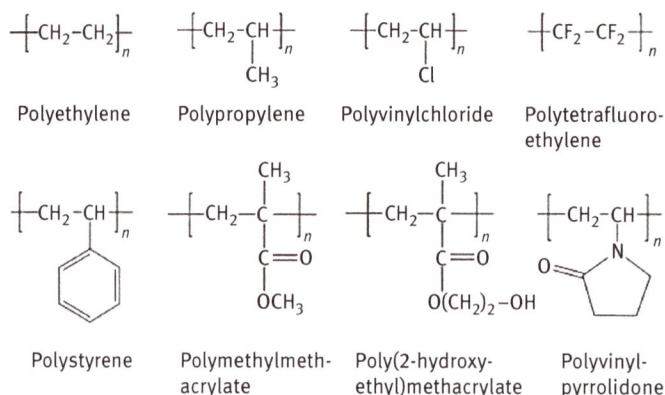

Figure 3.56: Polymers synthesized by chain polymerization mechanism.

3.3.1.3 Polymerization techniques

Polymerization reactions can be performed under homogeneous or heterogeneous conditions. In a homogeneous process, the pure or diluted monomers are directly added to one another and the reaction occurs within the media in which the reactants were mixed. In heterogeneous reactions, a phase boundary exists acting as an interphase at which the reaction takes place. Frequently used polymerizations techniques are described and characterized in Table 3.8.

Table 3.8: Polymerization techniques used on an industrial scale (Kariduraganavar et al., 2014).

Technique	Characteristics
Solution	– Both monomer and resulting polymer are soluble in the same solvent – Solvent has to be removed to receive the polymer – Often low molecular products due to chain transfer
Bulk (mass)	– Polymerization occurs in the monomer in liquid or gas phase – Process difficult to control due to exothermic polymerization – Resulting polymer pure but of nonuniform molecular mass distribution
Suspension	– Monomer-containing initiator, modifier, etc., are suspended in aqueous solvent by stirring – Suspension agent normally added to stabilize monomer droplets and avoid agglomeration – Very efficient heat transfer, reaction easily to control
Precipitation	– Polymerization starts as a homogeneous process whereby monomer is soluble but after initiation polymer becomes insoluble and precipitates – Polymer can be separated by filtration or centrifugation – High degree of polymerization (no problems in heat dissipation)

Table 3.8 (continued)

Technique	Characteristics
Emulsion	– Liquid monomer is dispersed in an insoluble liquid (most common type): oil-in-water emulsion wherein droplets of monomer are emulsified (with surfactants) in a continuous phase of water – Polymerization occurs in the latex particles (from individual polymer chains formed spontaneously in the first minutes of the process) – Normally high monomer to polymer conversion

3.3.1.4 Gel formation

Hydrogels are water-swollen, cross-linked polymeric structures containing either covalent bonds produced by simple reaction of one or more comonomers, physical cross-links from entanglements, association bonds such as hydrogen bonds or strong van der Waals interactions between macromolecular chains (Peppas, 1986, 2004) (Figure 3.57).

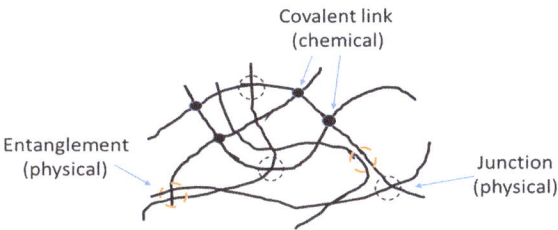

Figure 3.57: Network structure of a hydrogel with chemical links, junctions, and entanglements (adopted from Slaughter et al., 2009).

When hydrogels are dispersed in water, their molecules are tightly attached to water molecules resulting in the formation of a semisolid network. The water-retention capacity of a hydrogel is an intrinsic property depending on the chemical nature of the gel-forming polymer and on the nature of its chemical groups, whereby the shape and strength of the hydrogel strongly depends on the type and degree of cross-linking. When brought in contact with water, the macromolecular chains interact with water molecules, causing swelling of the gel (Agrawal, 2014).

Hydrogels can be classified in different way, for example, based on their source (natural/synthetic), their component of preparation (homopolymer, copolymer), their charge (anionic, cationic, zwitterionic, or neutral), their physical structure (amorphous, semicrystalline, hydrogen bonded), or their type of cross-linking (covalent (chemical) or physical (e.g., hydrogen bonds) cross-links) (Catoira et al. 2019).

Hydrogels are fabricated by swelling cross-linked structures in water or biological fluids. Methods to prepare the initial polymer networks include chemical cross-linking, photo-polymerization, or radiative cross-linking. In particular, chemical cross-linking reactions may occur by three different routes (Peppas, 2004):

- Reaction of a linear or branched polymer with at least one difunctional, small molecular-mass cross-linking agent;
- Copolymerization cross-linking reaction between one or more abundant monomers and one multifunctional monomer used in a relatively small amount;
- Using a combination of monomer and linear polymeric chains that are cross-linked by means of an interlinking agent (as in the preparation of polyurethanes).

Several of these reactions can be performed in the presence of UV or visible light and a suitable photoinitiator. Cross-linking by ionizing radiation utilizes electron beams or gamma rays to excite the polymer by initiating free-radical reactions.

The swelling behavior of a hydrogel can be described by the mass degree of swelling, $Q_m = m_s/m_d$, whereby m_s is the mass of the swollen gel and m_d is the mass of the dry gel sample. The degree of swelling influences the mechanical properties of a gel, the diffusion of solutes through the gel, and also the surface and optical properties of gels (Agrawal, 2014). Today, hydrogels are extensively used as soft contact lenses, wound dressings, and drug and gene delivery vehicles.

Cryogels are special types of gel that have gained growing interest and momentum as macroporous, 3D-support structures in biological separation columns (Mattiasson, 2014), as well as in cell culturing and tissue engineering to mimic the extracellular matrix (ECM) of mammalian cells (Henderson et al., 2013). Cryogels are formed by cross-linking reactions, similar to hydrogel formation at temperatures below the freezing point of the solvent (typically water) and thus, utilizing frozen solvent crystals as an interconnecting porogen. The porogen removal takes place by simply holding the temperature above the freezing point of the solvent.

Figure 3.58 illustrates the production route of porous gels by conventional versus cryogel formation processes. Cryogelling is a simple, cost-efficient method (Mattiasson, 2014), proper precursor candidates of which include monomers or macromers of both natural (Reichelt et al., 2014) and synthetic polymers.

3.3.2 Polymer processing

In most cases, polymer processing techniques are either thermal- or solvent-based. In thermal processes, the polymer is heated above its glass transition temperature T_g (see box) or its melting point T_m to make the material flow.

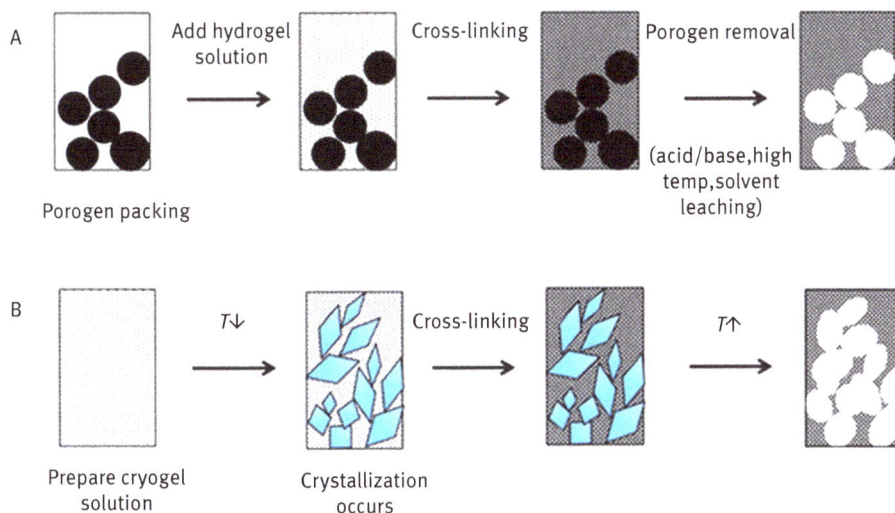

Figure 3.58: Illustration of conventional macroporous gel formation (A) versus cryogelling (B) Henderson et al., 2013. © With permission by The Royal Society of Chemistry.

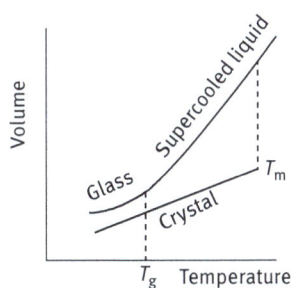

Glass transition temperature

The glass transition describes, in contrast to a fixed melting point T_m of crystalline materials, the gradual and reversible transition in amorphous materials from a hard, rigid and relatively brittle glass-like state into a viscous or rubbery state as the temperature is increased. The glass transition temperature T_g of amorphous materials including most polymers characterizes the range of temperatures over which this glass transition occurs. As shown in the figure, the glass transition temperature corresponds to the temperature of the intersection between the curve for the glassy state and that for the supercooled liquid.

The softened or molten polymer is then forced through a die or into a mold under pressure, and cooled rapidly to form the product of desired size and shape. In solvent-based processes, the polymer is dissolved in a solvent and a force is applied to eject the solution through a die or filled into a mold. After the forming process, the solvent is evaporated in vacuum and/or by heating. This technique is often used in laboratory-scale processes in small production volumes. A novel processing technique gaining increasing importance in biomaterial fabrication is AM, also known as rapid prototyping, rapid engineering, or solid free-form (SFF) fabrication. These various techniques combine related technologies including thermal- and solvent-based approaches used to fabricate physical objects directly from CAD data sources. AM enables to construct and fabricate anatomically exact, tailor-made implants based on medical imaging data (Damodaran et al., 2016).

In the following sections, the most important polymer processing techniques with regard to biomaterial fabrication will be presented more in detail.

3.3.2.1 Molding processes

The final polymer products leave the polymerization reactors in different forms and consistencies. Most of them are spherical particulates or powders but sometimes larger compact masses or even viscous liquids are formed. In most cases, these reaction products have to be converted to desired sizes and shapes (e.g., pellets, strands, films, plates) depending on the chosen processing method (Tadmor and Gogos, 2006). Furthermore, additives (stabilizers, antistatic agents, dyes, plasticizers, other polymers) have to be included homogeneously, for example, by extrusion (see below). At present, frequently fully compounded material mixtures, so-called master-batches, are used for polymer processing.

Traditionally, three polymer processing technologies exist: plastic, fiber, and elastomer or rubber processing. Although these methods have several elementary steps in common, individual processing techniques differ only in how forming of the final shape of the material is being attained (Billmeyer Jr., 1984).

The appropriate processing method for a polymeric material largely depends on its rheological behavior. It is a key consideration of polymer processing whether the polymer is of thermoplastic or thermosetting nature. Thermoplastic materials have the ability to flow at elevated temperatures for relatively long times. In thermosetting polymer materials, flow is rather quickly lost in favor of mechanical stability due to cross-linking reactions initiated at the temperatures at which flow is induced. Further considerations, important when choosing a suitable processing technology, are softening temperature, thermal stability, and the desired dimensional shape of the final product (Billmeyer Jr., 1984).

Extrusion and injection molding are the main technologies in plastic processing. Figure 3.59 shows these technologies with their two basic elements, the screw and the heat barrel system. Both single- and twin-screw systems are used for melting the resin, as well as for pumping the polymer melt through the extrusion die. Pumping is accomplished by generating the pressure required to push the melt through the die. The die in the extrusion system is used to shape the material into a continuous product, such as a film, a plate, a tube, a strand, or any desired profile. The twin-screw system is primarily used for mixing (compounding) and fabricates

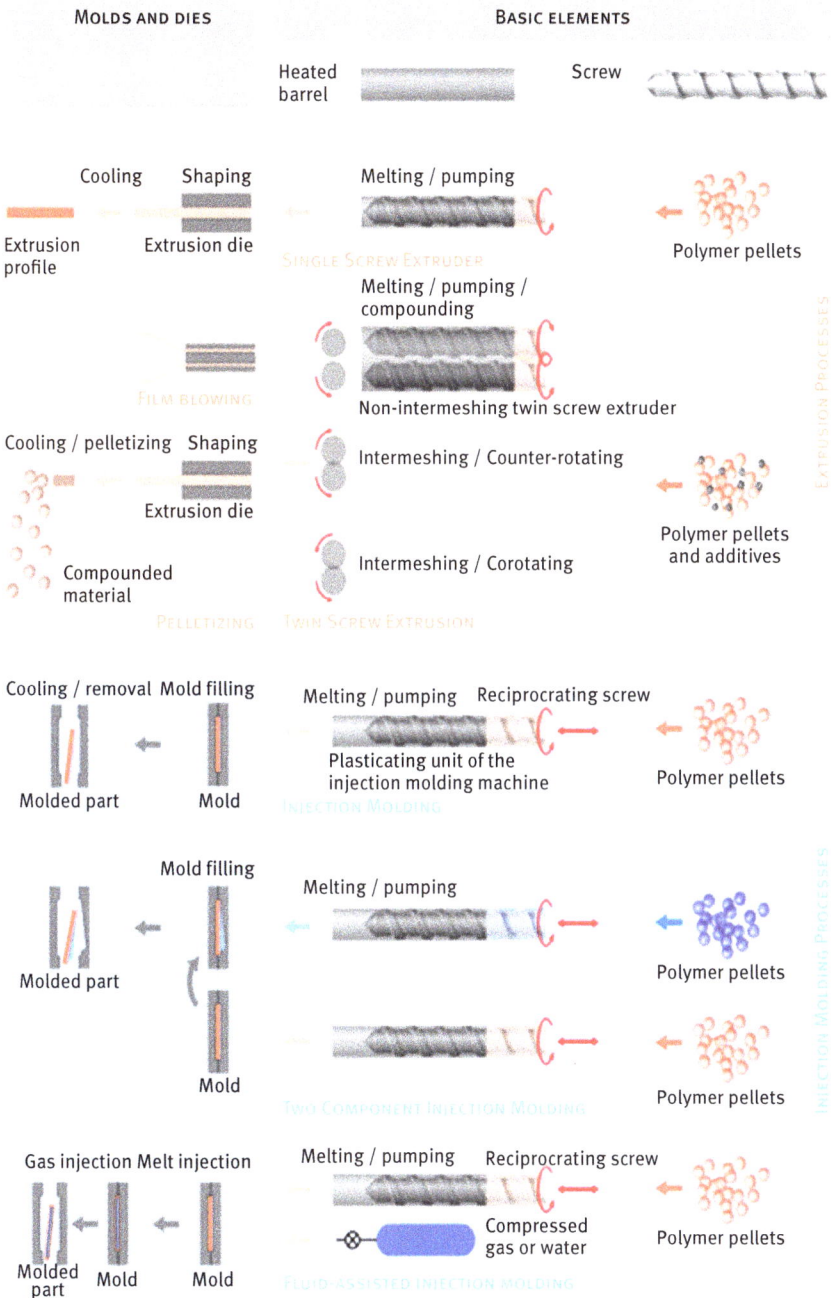

MOLDS AND DIES

BASIC ELEMENTS

Heated barrel

Screw

Cooling — Shaping

Melting / pumping

Extrusion profile — Extrusion die

SINGLE SCREW EXTRUDER

Polymer pellets

Melting / pumping / compounding

FILM BLOWING

Non-intermeshing twin screw extruder

Cooling / pelletizing — Shaping

Intermeshing / Counter-rotating

Extrusion die

Polymer pellets and additives

Compounded material

Intermeshing / Corotating

PELLETIZING — TWIN SCREW EXTRUSION

EXTRUSION PROCESSES

Cooling / removal — Mold filling

Melting / pumping — Reciprocating screw

Molded part — Mold

Plasticating unit of the injection molding machine

INJECTION MOLDING

Polymer pellets

Mold filling

Melting / pumping

Molded part

Polymer pellets

Mold

Polymer pellets

TWO COMPONENT INJECTION MOLDING

INJECTION MOLDING PROCESSES

Gas injection — Melt injection

Melting / pumping — Reciprocating screw

Molded part — Mold — Mold

Compressed gas or water

Polymer pellets

FLUID-ASSISTED INJECTION MOLDING

Figure 3.59: Extrusion and injection molding processes (Osswald, 2017). © With permission by Carl Hanser Verlag, München, Germany.

material strands that can be cut into pellets as required for subsequent extrusion or injection molding processes (Osswald, 2017).

Injection molding is one of the most widely applied and versatile processes to manufacture polymeric devices in the plastics industry, including the fabrication of devices made of polymeric biomaterials. Normally, an injection-molding machine (Figure 3.60) consists of a plasticizing/injection unit and a clamping unit in which the mold is fixed. For thermoplastic polymers, the temperature of the mold is usually controlled at about room temperature (for commodity polymers; higher for technical ones) by circulating water in order to solidify the polymer as quickly as possible. In contrast, for thermoset polymers or elastomers, the material is injected in a hot mold for cross-linking.

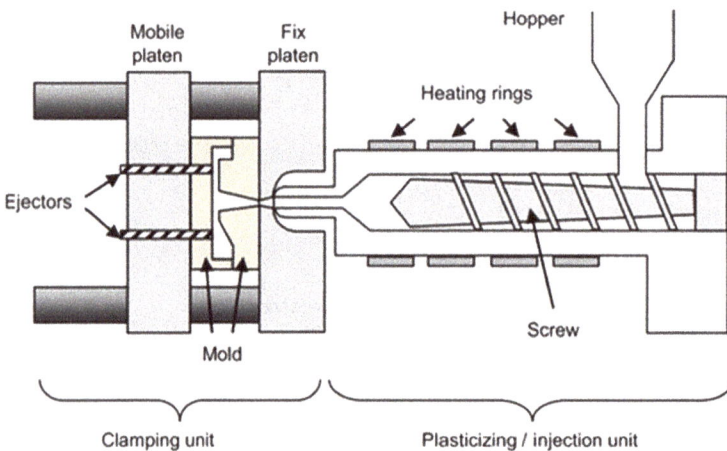

Figure 3.60: Injection-molding machine (Agassant et al., 2017). © With permission by Carl Hanser Verlag, München, Germany.

Compression molding is molding technique in which the preheated molding material is first placed in an open, heated mold cavity (Figure 3.61). After closing the

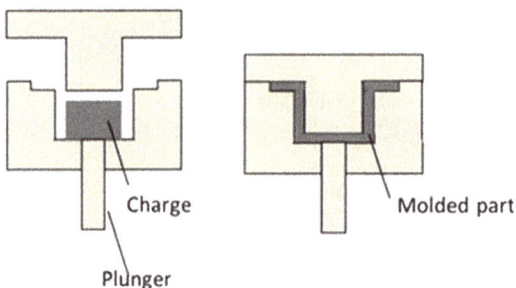

Figure 3.61: Compression-molding principle (Damodaran et al., 2016). © With permission by Springer Nature.

mold, pressure is applied to force the material into contact with all internal mold surfaces, whereby heat and pressure are maintained until the molding material has been cured. The process employs thermosetting resins in a partially cured stage, either in the form of granules, putty-like masses, or preforms. Advanced composite thermoplastics such as unsaturated polyesters reinforced with glass fibers or several grades of poly(etheretherketone) (PEEK) can also be processed by this technique (Osswald, 2017) (see 3.3.2.7).

Blow molding is widely used to fabricate bottles and similar hollow devices, either by an extruder (extrusion blow molding) or a reciprocating-screw injection machine (injection blow molding). The technique is similar to film blowing and explained more in detail in Chapter 3.3.2.2.

Rotational molding is characterized by the placement of the polymer in a divided mold, subsequent closure of the mold, followed by the simultaneous application of biaxial rotation and heat. The resin is deposited on the mold wall and after densification is completed, the mold is moved to a cooling station as rotation continues. Once demolding temperatures are reached, the finished part is removed from the mold. Advantages of rotational molding are lower residual stresses in molded parts, simpler machinery and molds, and the ability to produce single- or double-walled components with complex geometries and varied sizes (Ogila et al, 2017).

Reaction injection molding (RIM) is similar to injection molding and starts with pumping two unreacted components of a thermoset in predetermined amounts into a mixing head where they are thoroughly mixed and injected into a preheated mold under relatively low pressure. The polymerization reaction takes place within the mold. RIM is a process able to rapidly produce complex parts directly from monomers or oligomers. It has been commercialized for the fabrication of polyurethanes.

3.3.2.2 Film fabrication

The simplest way of film formation is *casting*, whereby a liquid is poured into a mold or on a smooth surface, and solidified by physical (e.g., cooling) or chemical (e.g., polymerization) means after which the solid film is removed from the mold. A squeegee can be used to adjust the film thickness.

Calendering is a technique to transform a molten polymer into films and sheets by squeezing it between pairs of corotating high-precision rollers. It is used for materials such as PVC or rubber.

In the *cast film extrusion* process, a thin film is extruded through a slit onto a chilled, turning roll, where it is quenched from one side. In this case, the film thickness is controlled by the speed of the roller. The film is then sent to a second roller, cooling the other side of the film. After passing through a system of rollers, the film is wound onto a roll.

In *film blowing,* the polymer is extruded through an annular die. Air is injected through a hole in the center of the die and the pressure causes the extruded melt to expand into a bubble. Constant pressure is maintained to ensure uniform thickness of the film. The bubble is pulled continually upwards from the die and cooled. After solidification, the film moves into a set of nip rollers which collapse the bubble and flatten it into two flat film layers. Finally, the guide rolls carry the film onto wind-up rollers.

The process is schematically depicted in Figure 3.62. The advantage of film blowing over casting is that the induced biaxial stretching renders possible stronger and less permeable films (Osswald, 2017).

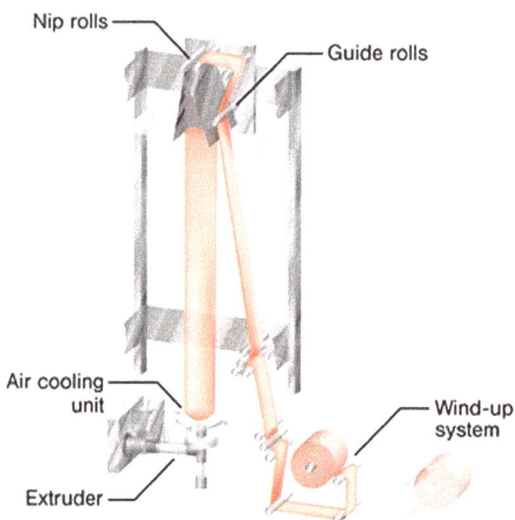

Figure 3.62: Film blowing system (Osswald, 2017). © With permission by Carl Hanser Verlag, München, Germany.

3.3.2.3 Coatings

Polymer coatings are deposited onto metallic, ceramic, polymer, and composite biomaterial substrates. Coatings have found growing interest in the biomaterial industry due to need to equip implantable devices with bioactive, cell adhesion-promoting layers, or layers designed to protect biomaterials against microbial contamination, and/or to control corrosion by aggressive biofluid. During coating processes, a liquid film is continuously deposited on a moving, flexible, or rigid substrate. The polymers can be used as melt, solution, paste, or particle suspension.

Dip coating is the simplest and oldest coating technique often used in biomaterial design. A substrate is continuously dipped into a fluid and withdrawn at constant speed with one or both sides coated with the fluid. Dip coating can also be used to coat individual objects that are dipped and withdrawn from the fluid. The fluid viscosity and density, and the speed and angle of the surface determine the coating thickness. In some cases, the dipping process is performed alternatively with a second dipping bath containing a coagulating bath or another polymer (formation of layer-by-layer coatings from oppositely charged polyelectrolytes). After several dips to build up the desired coating thickness, the device is stripped from the form and dried. There exist several similar coating techniques such as wire coating, knife coating, slide coating, curtain coating, and roll coating similar to calendering (Osswald, 2017).

Spray coating and further techniques derived thereof, including ultrasonic atomization, electrohydrodynamic jetting, and air-brush spray coating, are today the most commonly utilized cardiovascular stent coating techniques. These techniques use equipment that spray polymer and drug solutions onto a stent, enabling consistent deposit of a uniform drug release layer onto the stent surface (Livingston and Tan, 2016).

3.3.2.4 Fiber forming processes
The conversion of bulk polymers into fibers is accomplished by spinning. In most cases, spinning processes require polymer melts or solutions.

Melt spinning is the preferred fiber forming process if a polymer can be congruently melted under reasonable conditions. The process is very simple and fast. As a requirement, the polymer ought to be thermoplastic, and the viscosity of its melt must be controllable to achieve reasonable spinning rates and temperatures. A typical arrangement of melt spinning is shown in Figure 3.63. In this process, the molten polymer coming from an extruder is forced through a spinneret, a die with a

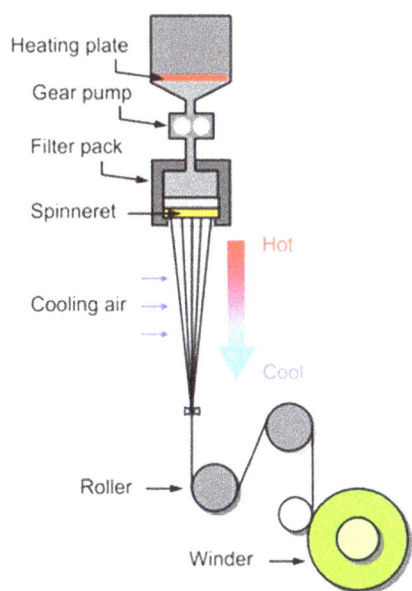

Figure 3.63: Scheme of a typical melt spinning process (Aranishi and Nishio, 2017). © With permission by Springer Nature.

small hole (or holes, to produce multiple filaments) with a diameter between 100 μm and 1 mm. When the melt leaves the spinneret, it is cooled to form a solid fiber. Filaments are normally drawn after extrusion to enhance the tensile properties of the fibers. Typical products are monofilaments used in surgical sutures, yarns, and ligaments. Fiber diameters are typically between 1 and 100 μm. Fibers used in biomedical textiles such as vascular grafts are often processed by knitting, braiding, or weaving. Many polymers with biomedical applications are processed by melt spinning, including nonbiodegradable polymers but also degradable polymers such as glycolide and lactide copolymers.

When melt spinning cannot be carried out, two options remain available, *dry spinning*, whereby the solvent is removed by evaporation into air or an inert gas atmosphere, and *wet spinning*.

Wet spinning is a classical spinning process, in which the solvent is leached out into another liquid which is miscible with the spinning solvent but by itself is not a solvent for the polymer (Figure 3.64). Although dry spinning has some advantages over wet spinning, all three processes have many features in common (Billmeyer Jr., 1984).

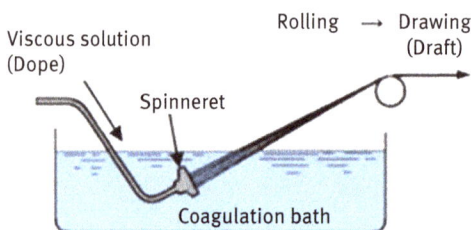

Figure 3.64: Scheme of a wet spinning process (Aranishi and Nishio, 2017). © With permission by Springer Nature.

Problems related to solution spinning include low spinning speed, that is, low productivity, limitation of the fiber dimension, and the use of specific, often harmful and toxic solvents or reagents.

Electrospinning is an old spinning technique, reinvented during the 1990s mainly for biomedical applications (Agarwal et al., 2008; Liu et al., 2012; Agarwal et al., 2016). The advantage of this process over other spinning techniques is that electrospinning is easily capable to generate fiber diameters in the nano- to microscale range, from several nanometers up to about 10 μm. Although electrospinning is currently most often used in solution, processing of polymer melts is also possible.

Electrospinning in solution is performed by injecting the polymer solution through a needle in the presence of a high electrical field (20–40 kV) as shown schematically in Figure 3.65. The charged solution is pushed out of a syringe pump, forming a liquid

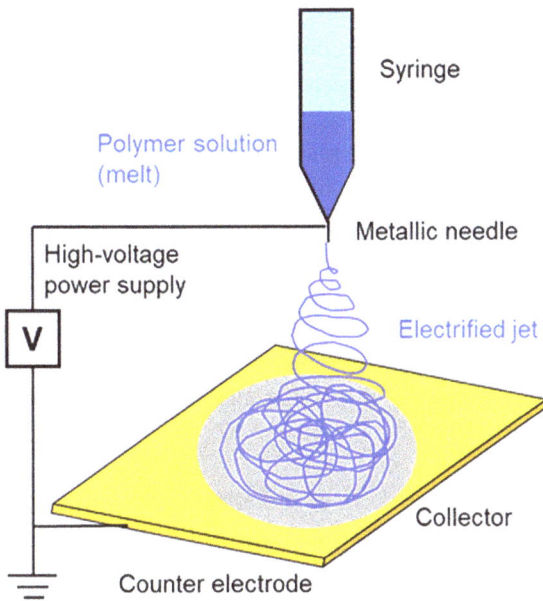

Figure 3.65: Scheme of an electrospinning process (adopted from Reise et al., 2012). © With permission by Elsevier.

jet directed to a grounded collector. The jet will experience both solvent evaporation and whipping instability before it reaches the collector. As result of stretching by electrostatic repulsion and whipping, the liquid jet will be continuously reduced in size until it has been solidified or deposited on the collector. By adjusting experimental parameters such as the concentration of polymer solution, the voltage, and the distance between spinneret and collector, fibers with uniform diameters can be routinely produced (Agarwal et al., 2008). Typically, nonwoven, randomly oriented 2D fiber mats or membranes are obtained that mimic the structure of the ECM surrounding living cells in their natural environment. Variation of the electrospinning setup and the experimental conditions allow the fabrication of mats with highly aligned fibers (Liu et al., 2012). Other modifications permit cospinning of two polymers, the spinning of tubular or 3D devices, and the production of core–shell structures and hollow fibers (coaxial spinning) (Teo et al., 2011).

A special technology of fiber manufacturing represents the production of hollow fibers with semipermeable walls, widely used in biology and medicine, especially in bioreactors or in hemodialysis equipment (Damodaran et al., 2016). Currently, hollow fiber membranes are produced by *phase inversion* in a solvent-phase separation process (Chung, 2008).

3.3.2.5 Fabrication of porous polymer materials

Foamed or porous polymers exhibit a cellular structure that has been generated by different processes. The cells formed can be completely enclosed (closed cell) or can be interconnected (open cell).

Direct foaming can be achieved by addition and reaction of physical or chemical blowing agents. In physical foaming, gases such as nitrogen or carbon dioxide are introduced into the polymer melt. Physical foaming can also be achieved by heating a melt containing a low-boiling liquid such as pentane that vaporizes. Foaming also occurs during volatilization of gases produced during polymerization, such as the production of carbon dioxide during the reaction of isocyanate with water (Prieto and Guelcher, 2014). Chemical foaming occurs when the chemical blowing agent thermally decomposes, releasing large amounts of gas. The most widely used agent for this purpose is azodicarbonamide (Casalini and Perale, 2012; Weber et al., 2016; Osswald, 2017).

Additional methods to fabricate porous polymer materials include porogen leaching, phase separation, freeze drying, cryogel formation (see Chapter 3.3.1.4), or AM (Sachlo, and Czernuszka, 2003; Liu and Ma, 2004; Wei and Ma, 2008; Janik and Marzed, 2015) (see Chapter 3.3.2.6).

3.3.2.6 Additive Manufacturing techniques

AM, also known under many other terms such as Rapid Prototyping, Rapid Engineering, or SFF Fabrication refers to a class of manufacturing processes based on buildup of solid objects from computer-aided, 3D model data by joining materials, usually layer-upon-layer addition. Three-dimensional model data for biomedical device development can be derived from medical imaging techniques used for diagnostic purposes, such as CT and magnetic resonance imaging (MRI), and are generally treated by CAD and CAM software. The typical work flow of an AM process is shown in Figure 3.66.

AM techniques are gaining great interest in producing medical devices, especially in tissue engineering for their suitability to achieve complex shapes and structures with a high degree of automation, good accuracy, and reproducibility. In addition, the ability to rapidly produce tissue-engineered constructs meeting specific requirements and expectations of patients in terms of tissue defect size and geometry as well as autologous biological features, makes AM a powerful tool to enhance clinical routine procedures (Hutmacher et al., 2004; Hutmacher and Woodruff, 2008; Hollister, 2005; Billiet et al., 2012).

Depending on the fabrication principle (Mota et al., 2015), the most extensively applied AM techniques for medical device fabrication are being conventionally classified into

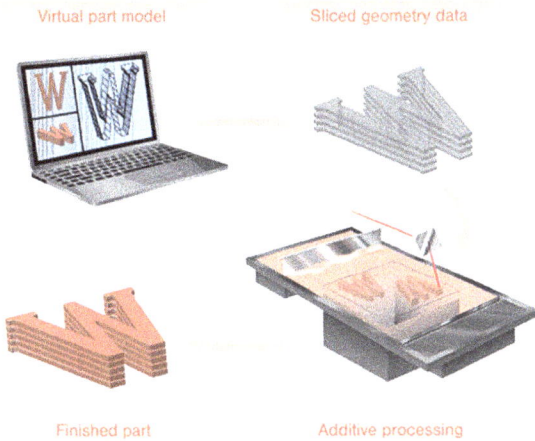

Figure 3.66: Typical workflow of an additive manufacturing process (Osswald, 2017). © With permission by Carl Hanser Verlag, München, Germany.

- powder-based techniques such as *SLS* (Eshraghi and Das, 2010), and *three-dimensional printing (3DP)* (Chia and Wu, 2015);
- photosensitivity-based techniques like *stereolithography (SLA)* (Melchels et al., 2010), *2-photon polymerization (2-PP)* (Ovsianikov et al., 2007), and *bioplotting* (Pfister et al., 2004); and
- melt extrusion-based techniques, for example, *fused deposition molding (FDM)* (Hutmacher et al., 2001).

Compared to classical subtractive fabrication methods (e.g., milling or turning), these techniques offer numerous advantages for biomedical device fabrication because objects with any geometric complexity or intricacy can be formed without the need for an elaborate machine setup or a final assembly. In addition to reducing the wastage of expensive raw material, AM systems reduce the construction of complex objects to a manageable, straightforward, and relatively fast process (Damodaran et al., 2016).

An overview of most commonly used AM techniques applied to fabricate medical devices including their advantages and current limitations are presented in Figure 3.67 and Table 3.9.

AM can be used to produce biodegradable and nondegradable biomaterials and also organic–inorganic hybrid materials. Under optimum conditions, various AM techniques have also shown the ability to successfully process polymers with living cells for 3D tissue engineering applications (Murphy and Atala, 2014).

A)

B)

C)

D)

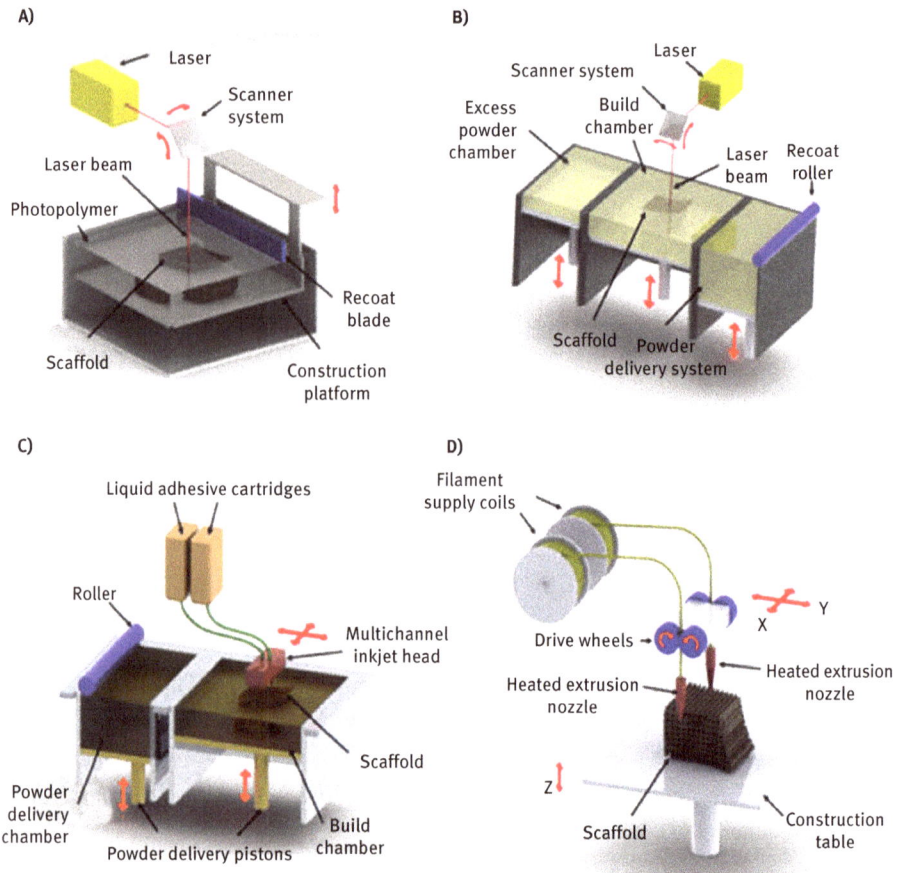

Figure 3.67: Schematic representation of the most common AM techniques employed in medical device fabrication (Mota et al., 2015). (A) Stereolithography; (B) selective laser sintering (SLS); (C) 3D printing (3DP); and (D) fused deposition modeling (FDM). © With permission by John Wiley and Sons.

3.3.2.7 Manufacture of multicomponent polymer materials

The term *multicomponent polymer systems* covering polymer blends, multilayer films and interpenetrating polymer networks was recently extended to composites comprising a polymeric matrix in which filler or reinforcing materials, often inorganic mineral components are dispersed. These material systems are also used to some extent in the field of biomaterials. Indeed, owing to the fact that most biological tissues can be regarded composite material systems, the fabrication of artificial composite materials mimicking or improving the properties of natural tissues has met with tremendous interest. Therefore, this chapter focuses on polymeric biomaterial composites.

Table 3.9: Comparison of most commonly applied AM techniques for medical device manufacture (Billiet et al., 2012; Mota et al., 2015; Damodaran et al., 2016).

Technique	Resolution (µm)	Advantages	Limitations
SLS	50–1000	Solvent free; no support material needed	High processing temperature; poor control of surface topography
3DP	100–300	No heating during fabrication; high production rate	(Toxic) solvents needed; rough surfaces
SLA	30–150	Supports relatively easy to remove	Past-curing needed; shrinkage; toxicity of photoinitiators; fabrication of biodegradable devices limited
2-PP	0-1-5	High accuracy	Choice of available monomers/photoinitiators limited; limited device size; time-consuming
Bioplotting	50–500	Wide range of materials usable (liquid melt, paste or gel)	Low mechanical strength, relatively low accuracy
FDM	100–500	Solvent free / Normally good compressive strength / No material trapping within small features	Support material for irregular structures needed; not thermally labile material required; anisotropy effects

Composite materials are those that contain two or more distinct constituent materials or phases, on a microscopic or macroscopic size scale. This composite definition encompasses fiber and particulate composite materials which are both of primary interest in biomaterials. Such composites consist of one or more discontinuous phases embedded within a continuous matrix phase. Since the discontinuous phase is normally mechanically harder and stronger than the continuous phase, it is considered as reinforcement whereas the continuous phase is called a matrix (Migliaresi and Alexander, 2004).

The properties of composites are strongly influenced by the properties of their constituent materials, their size, distribution and content, and the interaction among them. The composite properties may be controlled by the volume fraction sum of the constituents or the constituents may interact in a synergistic way due to a specific geometrical orientation. Most composites are designed and fabricated to provide improved mechanical properties such as strength, stiffness, toughness, and fatigue resistance. The strengthening mechanism strongly depends upon the geometry of the reinforcement. A common classification of composite materials is therefore based on the geometry of a representative unit of reinforcement (Figure 3.68).

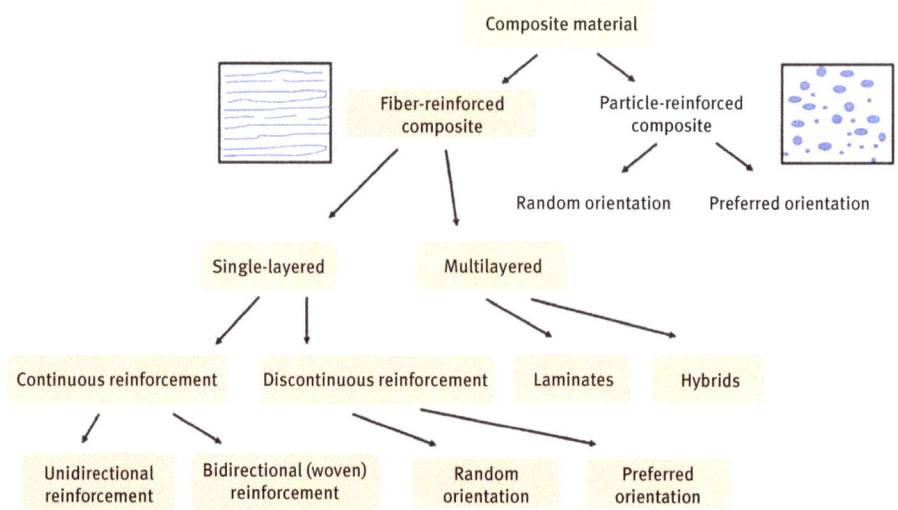

Figure 3.68: Classification of composite materials (modified after Agarwal et al., 2017). © With permission by John Wiley and Sons.

Usually, fibers are mechanically more effective than particles and hence, polymer fiber composites can attain stiffness and strength comparable to metals. Moreover, whereas particle-reinforced composites are isotropic, fiber-reinforced composites are in most cases anisotropic. Thus, properties in different directions can be properly designed (Migliaresi and Alexander, 2004).

Natural composites often exhibit a hierarchical structure at molecular and microstructural levels in which particulate, porous, and fibrous structural features are present on different length scales (Park and Lakes, 2007). As their properties are highly anisotropic, mimicking them is only possible by using composites. A perfect example of a composite material designed by Nature is bone in which nanosized calcium-deficient carbonate apatite crystals are embedded as reinforcing element in a collagen I fiber matrix forming a hierarchical structure (Fig. 2.54A). Other natural composites include dentin, cartilage, blood vessels, and skin.

Major reinforcing materials in biomedical polymer composites are carbon fibers, polymer fibers, ceramics, and glasses. Depending on the use, the reinforcements can be biologically inert or degradable.

The most common matrix materials are synthetic, nondegradable polymers such as poly(sulfone), PEEK, ultrahigh-molecular-weight PE, poly(tetrafluoroethylene), and poly(methylmethacrylate) (PMMA). To date, these matrices reinforced

with carbon or PE fibers, or ceramic particles have been used as stems for hip endoprostheses, osteosynthesis devices, artificial joint bearing surfaces, tooth roots, and
bone cements (Migliaresi and Alexander, 2004).

The production techniques of biomaterial composites depend on the selected
starting materials and the desired properties of the final composite. Decisive parameters include the processing temperature of the materials, the desired rate of control
over the distribution of the different materials, the required overall porosity, and
the required geometrical randomness or periodicity (Cardon et al, 2017).

In principle, most of the already described melt- and solvent-based polymer
processing techniques including AM methods can be applied to fabricate polymer-
based composites. For example, some biomedical composites are fabricated "*in
situ*" such as commercial bone cements. The most common manufacturing techni-
ques include
- Hand lay-up
- Bag molding
- Filament winding
- Pultrusion
- Resin transfer molding (RTM)
- Compression molding
- Injection molding

Hand lay-up is a common technique to make fiber-reinforced composites. Fibers
and resin are placed manually in an open mold consisting of a flat surface, a cavity,
or a convex shape. The resin mixture is then applied to the reinforcement and compacted by using appropriate rollers to remove entrapped air. The procedure can be
partially automated in a process called *spray-up* in which chopped fibers and resin
are deposited simultaneously onto the open mold by a spray gun (Migliaresi and
Pegoretti, 2002).

Vacuum bag molding is employed to manufacture high-performance composites, being an improvement over the hand lay-up and spray-up processes. Laid-up
molds are inserted into special bags consisting of flexible film of poly(vinyl alcohol)
(PVA), Nylon®, or PE which are then evacuated and/or placed in an autoclave for
curing under controlled heat and pressure conditions.

RTM is a technique during which a fibrous preform is placed in the cavity of a
closed mold and a certain amount of premixed resin is injected into the cavity by
a pressure injection device. Depending on the resin properties and the required
thickness of the final part, an appropriate precuring cycle is realized in the mold
to attain sufficient green strength required for device removal (demolding). Final
curing is normally performed in an electric oven. The technique is illustrated in
Figure 3.69. RTM can be employed to fabricate complex structures and large near-
net shapes at relatively low cost and with considerable time saving compared to
conventional lay-up techniques (Migliaresi and Pegoretti, 2002).

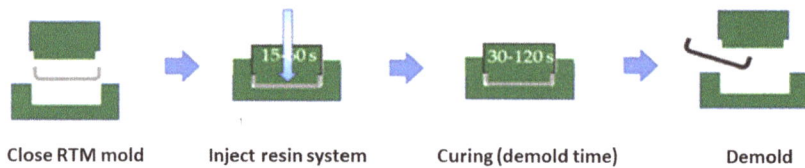

Close RTM mold   Inject resin system   Curing (demold time)   Demold

Figure 3.69: Flow chart of resin transfer molding (RTM) technique (Sun et al., 2019). © Permission granted under Creative Commons Attribution License.

Filament winding is a technique whereby continuous reinforcing fibers are wound on a mandrel. Fibers are passed through a resin bath before being placed on the mandrel with a certain winding angle, by a carriage device. The process is completed by a curing step usually performed in an oven, and the removal of the mandrel. The technique can be used to fabricate objects with curved surfaces such as tubes, pipes, pressure vessels, and tanks (Migliaresi and Pegoretti, 2002). This process can result in devices with extremely high tensile strength caused by a high degree of fiber orientation and high fiber loading. Biomedical applications of these products are intramedullary rods for fracture fixation, ligament prostheses, endoprosthetic hip stems, intervertebral disks, and vascular grafts.

Pultrusion is an extrusion-type process whereby a polymer resin is moved through a die or mold. The process can be used to manufacture continuous lengths of fiber-reinforced thermosetting resin. The reinforcing fibers or mats are continuously pulled through a resin bath impregnator that saturates each individual fiber (Billmeyer, Jr., 1984). Finally, the composite passes a heated die for curing. The process is schematically depicted in Figure 3.70.

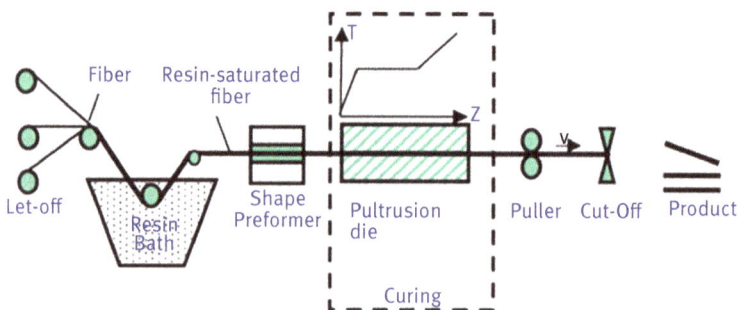

Figure 3.70: Illustration of pultrusion technique (Acquah et al., 2006). © With permission by Elsevier.

Currently, pultrusion is used to fabricate fiberglass-reinforced composites usable as root canal posts or burs designed to remove cement, stains, and colored coatings from the surface of dental enamel. A further application is the manufacture of carbon fiber reinforced (CFR) PEEK (e.g., Endolign™) (Green, 2019) that can be used for osteosynthesis implants as an attractive alternative to the metallic materials traditionally used in hard tissue implants.

3.4 Surface functionalization (Robert B. Heimann)

Since most biomaterials do not have all the ideal properties and desirable functions needed to fulfil completely their anticipated biological role, appropriate modification of their surfaces plays an important part in adapting surface properties of biomaterials. This is done to allowing better performance within their physiological surroundings, required to deliver the desired clinical results. Hence, it is imperative to control the interactions between biomaterials and living tissues to optimize their therapeutic effects and disease diagnostics ability (Treccani et al., 2013; Wu et al., 2015; Rana et al., 2017; Stewart et al., 2019).

The surfaces of biomaterials can be functionalized in several ways to alter their properties toward achieving improved biocompatibility and tissue-targeted performance (Hildebrand et al., 2005; Dorozhkin, 2019). Functionalization encompasses physical, chemical, biological, and radiative processes.

Physical functionalization processes do not modify the underlying substrate but instead add thin layers of material that interact with the biological environment in a positive way. Such processes include, for example, the formation of Langmuir–Blodgett (LB) films used as biological membranes of lipid molecules that can be used to study the modes of action of drugs as well as other reactions in biological systems, and the observation of immunological responses. *Chemical functionalization* includes oxidative modification of surfaces by strong acids or alkalis or ozone treatment as well as chemisorptive processes. Functionalization of polymer surfaces involving alkali hydrolysis forms, by cleaving ester bonds, carboxyl and hydroxyl moieties that can attach to proteins, thus binding them tightly to the surface of biopolymers. *Biological functionalization* involves the adsorption and fixation to the biomaterial of bone growth-supporting proteins such as recombinant human bone morphogenetic proteins (rhBMPs) as well as noncollagenous matrix proteins (NCPs) such as osteocalcin, osteonectin, osteopontin, and bone sialoprotein to assist osseoinductive bone reconstruction. Finally, *radiative functionalization* methods include treatment of biomaterial surfaces with glow and corona discharges as well as laser, infrared and gamma-ray irradiation, and plasma immersion.

As the processes applied to functionalize surfaces of biomaterials are vast and complex, a detailed description of their various ways to achieve improved performance of biomedical devices is beyond the scope of this synoptic treatise. Only a few typical methods will be addressed below.

3.4.1 Physical functionalization

To deposit ultrathin polymer films on the surface of biomaterials, two techniques can be applied: the self-assembly technique and the LB technique. For the self-assembly technique, special molecular systems such as α,ω-functionalized long alkyl chains are required to build up multilayer systems, whereas for the LB technique, a large range of different molecular structures are suitable including phospholipids and amphiphilic molecules consisting of a long alkyl chain with a hydrophilic head group that form monomolecular layers at the air/water interface and subsequently can be transferred onto the solid biomaterial substrates (Mathauer et al., 1989). LB films exhibit high order and uniformity and, thus, provide flexibility in incorporating a wide range of chemistries.

The stability of these films can be improved by cross-linking or internal polymerization of the surfactant molecules following film formation (Garcia, 2011). Two possibilities exist for carrying out polymerization: (i) polymerization of the monolayer floating at the air/water interface, followed by transfer of the then polymerized monolayer to a substrate; and (ii) building up multilayers from polymerizable amphiphiles via the LB technique, followed by polymerization in the form of LB assemblies. The latter may be considered a special case of topochemical polymerization. LB films, after fabrication, consist of low molecular mass units held together by noncovalent forces. However, these lipid-like subunits can be polymerized by free radical methods to form true polymeric structures exhibiting both the regularity of the LB structures and the desirable stability characteristics of high polymers. These films may find application as biocompatible coatings or as membranes to control specific biological processes (Ratner, 1989).

An interesting example of physical functionalization of biomaterials using LB films is the deposition of liposomes, that is, microspherical structures the walls of which are composed of lipid bilayer structures (Yager et al., 1988). The walls of liposomes can also be polymerized to obtain chemical, physical, and mechanical stabilization. Liposomes are subjects of active exploration with applications centering on drug and gene delivery systems and construction of artificial cells.

3.4.2 Chemical functionalization

Many ways exist to chemically modify surfaces, in particular polymeric surfaces, including alkali hydrolysis, covalent immobilization, wet chemical methods, or deposition of various coatings (Barillas et al., 2018a, 2018b). During alkali hydrolysis, protons diffuse between polymer chains and cause surface hydrolysis that cleaves ester bonds, resulting in the formation of carboxyl and hydroxyl moieties that can attach to proteins. In covalent immobilization, small fragments of proteins or short peptides are bonded to the surface to improve biocompatibility.

Wet chemical methods for protein immobilization comprise dissolution of chemical species in an organic solvent, able to reduce the hydrophobicity of the polymer.

Chemical covalent immobilization approaches allow for covalent attachment of proteins to a substrate through direct or linker-mediated immobilization. Several linker molecules have shown great success in immobilizing proteins to orthopedic implant surfaces (Stewart et al., 2019), including tresyl (2,2,2-*trifluoroethanesulfonyl*) chloride able to attach proteins by nucleophilic substitution to biomaterial surfaces thereby establishing a strong, noncovalent bond between protein and substrate.

Much work has been carried out on biomimetic formation of apatite at the surfaces of glass ceramic as well as titanium alloys pretreated in alkaline or in strongly acidic solutions followed by controlled heat treatment (Kokubo and Yamaguchi, 2015). Initial work included soaking apatite–wollastonite (A–W) glass ceramics in SBF (see box) the ion concentration, temperature, and pH of which were adjusted to equal those of human blood plasma (Abe et al., 1990).

Simulated body fluids

The composition of blood plasma is close to that of human extracellular fluid (hECF), and synthetic SBFs attempt to emulate this composition. SBFs are ubiquitously used to test biomaterials for their ability to behave osseoconductive, that is, induce bone formation by precipitating bone-like hydroxylapatite (Kokubo and Takadama, 2006). SBFs are also used as medium in potentiodynamic polarization and electric impedance spectroscopy utilized in corrosion testing of biometals, even though their buffering components may accelerate corrosion on their own (Mei et al., 2019).

An appropriate SBF should have compositional ranges as follows: *cations*: 142–145 mmol Na^+, 5–5.8 mmol K^+, 1.8–3.75 mmol Ca^{2+}, 0.8–1.5 mmol Mg^{2+}; *anions*: 103–133 mmol Cl^-, 22–27 mmol HCO_3^-, and 0.8–1.67 mmol HPO_4^{2-} (Tas, 2014). SBF within these compositional ranges are known to form nanospheres of ACP that on aging may be converted via OCP to bone-like calcium-deficient hydroxylapatite. SBF compositions most closely simulating the composition of hECF are SBF-T, SBF-H and SBF-L that use as hydrogen carbonate-complexing buffering agents TRIS (*tris*(hydroxymethyl) aminomethane), HEPES (2-*hydroxyethyl*)-1-*piperazine ethane sulfonic* acid, and L(+)-lactic acid, respectively (Kokubo et al., 1990; Kim et al., 2001).

This treatment resulted in the formation of a hydroxylapatite layer with strong bone-bonding ability. Subsequent work focused on chemical activation/functionalization of titanium surfaces with alkalis that lead to precipitation of amorphous sodium titanate, the Na^+ ions of which will be exchanged against Ca^{2+} ions in contact with SBF to form amorphous calcium titanate (Kokubo et al., 1996). In turn, this compound will be transformed into ACP acting as a template for subsequent precipitation of nanocrystalline hydroxylapatite that can be further functionalized by adsorption of relevant bone-growth promoting proteins. The complex chemical pathway of this type of chemical functionalization is thought to be as follows:

(i) $TiO_2 + OH^- \rightarrow HTiO_3^-$

(ii) $HTiO_3^- + 2Na^+ + OH^- \rightarrow Na_2TiO_{3(amorphous)} + H_2O$

(iii) $Na_2TiO_{3(amorphous)} + H_3O^+ \rightarrow TiO_2OH^- + 2Na^+ + H_2O$

(iv) $TiO_2OH^- + Ca^{2+} \rightarrow CaTiO_{3(amorphous)}^{\delta-} + H^+$

(v) $9CaTiO_{3(amorphous)} + 6PO_4^{3-} + 18H^+ \rightarrow \{Ca_3(PO_4)_2\}_3^* + 9TiO(OH)_{2(hydrated)}$

(vi) $\{Ca_3(PO_4)_2\}_3^* + 15Ca^{2+} + 12(PO_4)^{3-} + 6H^+ \rightarrow 3Ca_8(HPO_4)_2(PO_4)_{4(hydrated)}^{\delta+}$

(vii) $3Ca_8(HPO_4)_2(PO_4)_{4(hydrated)} + 6Ca^{2+} + 12OH^- \rightarrow$ **$3\ Ca_{10}(PO_4)_6(OH)_2$** $+\ 6\ H_2O$

*Posner's cluster

The conversion of amorphous calcium titanate into calcium phosphate (reactions v and vi) may be initiated by the formation of a thin layer of calcium phosphate with the conformation of the Posner's cluster, $\{Ca_3(PO_4)_2\}_3$ that has been shown to have the energetically most stable configuration (Onuma et al., 2000). With time, this amorphous layer takes up more Ca^{2+} and PO_4^{3-} ions from the SBF to form hydrated OCP (reaction vi) with Ca/P = 1.33 and, finally, near-stoichiometric hydroxylapatite (reaction vii) with Ca/P ~ 1.67. It should be emphasized that this reaction sequence is at odds with that postulated by Liu et al. (2001) who assumed early precipitation of OCP that in turn takes up more calcium ions to convert first to ACP with Ca/P = 1.5 and subsequently to hydroxylapatite with Ca/P = 1.67 via a Ca-deficient hydroxylapatite phase akin to biological apatite. However, since the activation energy required for this sequence appears to be very large, it is reasonable to assume that a transformation path via amorphous phases of sodium titanate and calcium titanate as described by the equations (i) to (vii) above may constitute an energetically more favorable situation.

3.4.3 Biological functionalization

On the one hand, extracellular NCPs attached to an implant surface serve to shield the implant from the innate immune system but, on the other hand, are able to elicit biological responses by providing chemical signals that are conducive to triggering the bone-forming cascade. These growth factors can be covalently bonded to biomaterials, in particular to polymers, to enhance the osseointegration ability of implants.

As far back as 1965, Marshall Urist (Urist, 1965), an orthopedic surgeon at the University of California, Los Angeles (UCLA), studied the ways how hydroxylapatite is being deposited on the collagen I-based matrix, a crucial process in bone formation. By implanting demineralized fragments of rabbit bone ectopically into rabbit muscle tissue he observed that new bone was created at the implantation site. Obviously, a trigger protein was responsible for this bone formation that coaxed muscle cells to kick-start production of new bone cells at places where normally no bone would grow. This new class of proteins, later called *bone morphogenetic proteins* (BMPs), provides chemical signals that cause them in natural bone matrix to attract stem cells from the bone marrow, and spur them to proliferate and differentiate to become bone-producing osteoblasts (Reddi, 2001). This is the hallmark of osseoinduction. Later, cloning of BMP-7 and BMP-2 opened the door to produce recombinant versions of the proteins that could be added to matrix implants. These signaling proteins proved that they were able to initiate the bone regeneration process even in cases when normal bone healing was impeded. This approach appears to provide a promising solution to several problems conventional biomaterials are plagued with. However, to date such therapeutic intervention is exceedingly expensive.

Proteins such as rhBMP can be incorporated into calcium phosphate layers when biomimetically coprecipitated with the inorganic components (Liu et al., 2004). For example, rhBMP-2 retains its osseoinductive potential while incorporated into the crystal lattice of hydroxylapatite in a dose-dependent manner as ascertained by protein blot staining and enzyme-linked immunosorbent assay (ELISA).

ELISA

ELISA is a sensitive method to quantitatively determine antibodies (Engvall and Perlmann, 1972). Antigens from the biological sample are attached to a surface. Then, a matching antibody is applied over the surface so it can bind to the antigen. This antibody is linked to an enzyme, and in the final step, a substance containing the enzyme's substrate is added. The subsequent reaction produces a detectable signal, most commonly a color change. The assay uses a solid-phase enzyme immunoassay to detect the presence of a ligand (commonly a protein) in a liquid sample using antibodies directed against the protein to be measured. ELISA is being used as a diagnostic tool in medicine, plant pathology, and biotechnology, as well as a quality control check in various industries (https://en.wikipedia.org/wiki/ELISA).

Apart from BMPs, other proteins were also found to interact with biomimetically precipitated calcium phosphate coatings. For example, bovine serum albumin incorporated into the crystal lattice of initially formed OCP was found to elicit transformation to carbonated hydroxylapatite structurally and functionally akin to natural bone mineral (Liu et al., 2003; Yu et al., 2009). Different strategies have been used to promote substrate-mediated guidance of osteogenic differentiation of immature osteoblasts, osteoprogenitor cells, and hMSCs through 2D- and 3D-chemically conjugated small moieties (Maia et al., 2013).

Hydroxylapatite-based osseoconductive implants were found to promote normal differentiation of cells in surrounding tissues by providing an enhanced environment for cell adhesion. Cytoskeletal microfilaments such as actin, myosin, actinin, and tropomyosin that control cell shape and migration (see Chapter 4.2) will be coupled through specialized cell membrane proteins (integrins) to extracellular adhesion molecules such as fibronectin, laminin, vitronectin, or thrombospondin (Nair and Laurencin, 2006). An interfacial layer of hydroxylapatite will adsorb these adhesion molecules in a favorable conformation thus generating focal adhesion centers. Particular growth factors may also be adsorbed at specific hydroxylapatite surfaces, further promoting osseointegration (Lobel and Hench, 1998). Growth-supporting cytokines such as transforming growth factor-β, insulin-like growth factor-1, tumor necrosis factor-α (TNF-α), or rhBMPs (Urist, 1965) provide a degree of osseostimulation that support the transformation of undifferentiated mesenchymal precursor cells into osteoprogenitor cells that precede endochondral ossification.

Recently, surgical and tissue engineering strategies have emerged that combine rhBMPs and bisphosphonates (BPs) to maximize anabolism, that is, bone growth by osteoblasts and, at the same time, minimize catabolism, that is, bone resorption by osteoclasts (Murphy et al., 2014). Collagen-based scaffolds, currently the surgical standard, can bind rhBMPs but fail to bind BPs. However, a biomimetic collagen–hydroxylapatite scaffold was found to bind both agents and hence was able to produce excellent *in vivo* bone defect healing capability. This study has demonstrated the relative advantages of codelivering anabolic and anticatabolic agents using a collagen–hydroxylapatite scaffold system.

However, BMPs also appear to be involved in triggering osteoclastic responses to wear particles produced by either degraded PMMA-derived bone cement or delaminated hydroxylapatite coatings. Numerous proinflammatory cytokines such as interleukin-1, interleukin-6, TNF-α, and prostaglandin E2 have pro-osteoclastogenic effects in response to implant-derived wear particles (see Chapter 4.3). Hence, there is the intriguing notion that both the immune system and skeletal homeostasis may be linked in the process of osteoclastogenesis and associated osteolysis (Holt et al., 2007).

Another important issue is the control of thrombogenicity. Cardiovascular devices such as stents or artificial vascular grafts are being designed to mimic properties of the specific tissue region the device is serving to replace. To reduce thrombogenicity, surfaces can be coated with fibronectin and RGD (Arg-Gly-Asp)-containing peptides that encourage attachment of endothelial cells. The peptides YIGSR (Tyr-Ile-Gly-Ser-Arg) and REDV (Arg-Glu-Asp-Val) have also been shown to enhance attachment and spreading of endothelial cells and ultimately reduce the thrombogenicity of vascular implants (Shin et al., 2003). For example, functionalizing CoCr stent surfaces with RGDS, REDV, YIGSR peptides, and their combinations were found to promote adhesion and proliferation of human umbilical vein endothelial cells. This finding may hold great potential to overcome clinical limitations of current metal stents by enhancing their capacity to support surface endothelialization (Castellanos et al., 2017).

3.4.4 Radiative functionalization

Plasma immersion and ion or electron beam techniques can tailor the surfaces of biomaterials to obtain the desirable biological functions (see Chapter 4.3.4.2). Cerium, hafnium, neodymium, tantalum, silver, and carbon are incorporated in biomaterials by ion implantation, magnetron sputtering, or plasma immersion. The metal oxide surface layers provide better anticorrosive and biocompatible properties for Mg-based implants, DLC coatings serve to enhance corrosion resistance and antibacterial properties of Mg-based alloys, and Nd-integrated polymeric delivery system may provide highly efficient anticancer therapies (Jin and Chu, 2018).

Plasma functionalization techniques are particularly useful as they can deposit nanometer-thin, well adherent, and highly conformal coating films (Morra and Cassinelli, 2006). A *glow discharge plasma* is created by ionizing gases such as argon, nitrogen or oxygen by microwaves or an electric current. The ionized plasma reacts with the surface to form a surface configuration appropriate for protein adhesion. Surface modification by glow discharge plasma technique is known to modify only the surface without altering the bulk properties of the (polymeric) material. For example, in PVA treated by a N_2/H_2 microwave plasma (Ino et al., 2013), X-ray photoelectron spectroscopy and Fourier transform infrared spectroscopy revealed the presence of carbonyl and nitrogen species, including amine and amide groups, while the main structure of PVA was retained. In addition, plasma modification induced an increase in surface wettability without significant change in surface roughness but allowing, in contrast to untreated PVA, successful culturing of mouse fibroblasts and human endothelial cells. These results evidenced that plasma amination of PVA promises to improve cell attachment and proliferation behavior on contact with synthetic hydrogels for tissue engineering.

In order to improve the performance of cardiovascular stents fashioned from L605 CoCr high-entropy alloy (Co20Cr15W10Ni3Fe), implant surfaces were subjected to PIII (Diaz-Rodriguez et al., 2019). The surface oxide layer formed was found to consist of a cobalt oxide outer layer followed by chromium oxide and tungsten oxide layers underneath as ascertained by ToF-SIMS (Figure 3.71). These oxide layers could be directly functionalized with primary amine groups, useful for further grafting of biomolecules. The plasma treatment decreased dramatically the water contact angle from about 90° for untreated alloy down to 15°, making the surface highly hydrophilic. Hence, a new generation of cardiovascular devices with reduced post-implantation complications could be envisaged.

Several plasma-based technologies have been developed to immobilize proteins depending on the final application of the resulting biomaterial (Cifuentes and

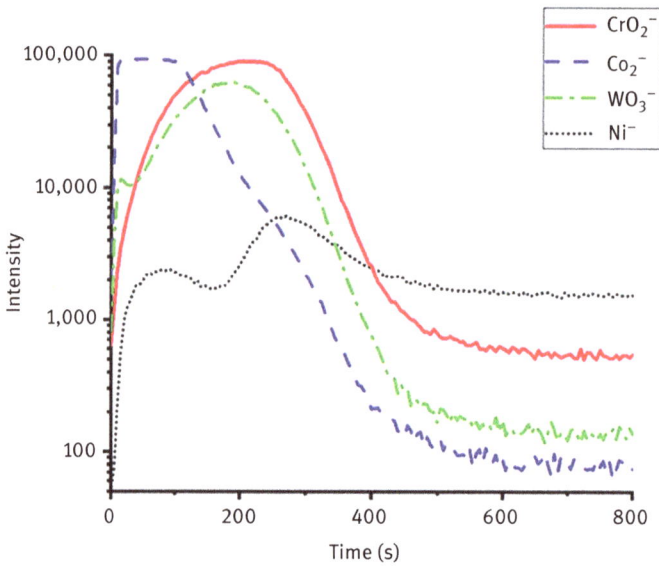

Figure 3.71: Time-of-Flight secondary ion mass spectroscopy (ToF-SIMS) results of plasma immersion ion implantation (PIII) of L605 CoCr alloy (Diaz-Rodriguez et al., 2019). © Permission granted under a Creative Commons Attribution 3.0 license.

Borros, 2013), making this technique a relatively fast approach to produce smart bioactive surfaces.

Cells, having membranes with fractal dimensions (Wang et al., 2010), are known to be very sensitive to the topographical features of their environment and able to respond to objects as small as 5 nm. Thus, well-defined *surface topographies* are known to guide selectively cell responses. While different technologies have been used including lithography, polymer demixing, plasma treatment, chemical etching, and others to generate suitable surface topographies, *femtosecond laser processing* of biomaterials has several advantages over other methods (Hao and Lawrence, 2005; Vilar, 2016). These advantages include low surface contamination, low mechanical damage, and controllable surface structuring of 3D components with complex geometries. The extremely short-time laser–material interaction confers upon femtosecond laser processing additional advantages over long-pulse laser processing, including higher quality and repeatability of machined structures, negligible bulk material damage, and a strongly reduced heat-affected zone (Fadeeva et al., 2014). The surface microtopography of the native Ti oxide layer on pure Ti has been modified toward improved cell compatibility by irradiation with a nanosecond-pulsed laser (Kurashina et al., 2019). As result of the laser interaction, pocket-like microtopographical structures formed along the laser trace that contained large amount of

OH^- groups, and thus, appeared to promote the growth of biomimetic hydroxylapatite when incubated in SBF.

Surface roughness and specific surface features including self-affine fractality (Heimann, 2011) introduced through surface structuring by radiative functionalization at nano- and micrometer scale, are able to affect cell orientation, morphology, proliferation, vitality, migration, and differentiation. Hence, adequate surface nano-topography was found to be an important prerequisite for optimum cell adhesion and proliferation. In this context, the origin and extent of surface topography has been described through the concept of fractality (see box) by Gentile et al. (2010). Fundamental experiments were conducted to study cell proliferation on electro-chemically etched silicon proxy surfaces with varying roughness but comparable surface free energies. The surface profile was found to be a self-affine fractal (see box), the average roughness R_a of which increased with etching time from ~2 to 100 nm, with fractal dimension ranging from 2 (an ideally smooth surface) to 2.6. A moderately rough surface with R_a between 10 and 45 nm yields a close to Brownian surface with $D \sim 2.5$. Gentile et al. (2010) concluded that the observed cell behavior could be satisfactorily explained by the theory of adhesion to randomly rough solids, and that a moderately rough surface with large fractal dimension (high roughness) is conducive to proliferation, suitable spreading, and vitality of bone cells. Gittens et al. (2014) have reviewed and interpreted the influence of surface topography including microroughness and nanostructures on the osseointegration of spinal implants.

Fractality

By definition, a geometric fractal is a rough or fragmented geometric shape that can be split into parts, each of which is (at least approximately) a reduced-size copy of the whole (Mandelbrot, 1983; Peitgen and Saupe, 1988; Taylor and Taylor, 1991). Fractal geometry is a natural description of disordered (non-Euclidean) objects ranging, in ascending size, from macro-molecules to self-similar vegetables (Broccoli Romanesco) to cloud shapes and topographical elements of the Earth's surface such as mountain ranges and coast lines. In the biological realm, the structure of the lung, the circulatory system, the digestive tract, and the heart and circadian rhythms are just a few examples of systems with fractal properties (Chang, 1993; Kumar et al., 2016). The common property of these dimensionally ill-defined objects is that they display "dilatation symmetry," meaning that they look geometrically self-similar under transformation of scale obtained by changing the magnification. Many surface structures can be characterized by a single parameter D, the fractal or Hausdorff-Besicovic dimension that is defined as the exponent that in the 2D case relates the surface area S of an object to its size R by the simple proportionality relation $S \propto R^{D_S}$, where D_S is the surface fractal dimension. An ideally smooth surface yields $D_S = 2$. However, fractally rough surfaces such as plasma-sprayed hydroxylapatite layers yield $2 < D_S < 3$, so that D_S becomes a measure of roughness. In plasma-sprayed coatings, the conditions in the direction perpendicular to the surface are very different from those parallel to the surface. Thus, one will deal essentially with self-affine geometries. *Self-affine fractals* are structures that can be rescaled by a transformation that involves a different change in length of scales in different directions. For example, whereas the scaling factor perpendicular to a plasma-sprayed surface may be on the order of a few micrometers, corresponding to the thickness of just one deposited lamella, the scaling factor along the coated surface will commonly be measured in millimeters or even centimeters, depending on the degree of structural homogeneity imposed, for example, by the relative movement of the plasmatron (Heimann, 2011).

Chapter 4
In vitro and *in vivo* performance of biomaterials and biomedical devices

Christine Knabe, Doaa Adel-Khattab, Christian Müller-Mai, Barbara Nebe, Michael Schlosser, Robert B. Heimann, Hans-Jürgen Kock

4.1 Biological performance of biomaterials and biomedical devices (Christine Knabe, Doaa Adel-Khattab, Christian Müller-Mai)

4.1.1 Principles of how to test new biomaterials and medical devices

A biomaterial is defined as a nondrug biological or synthetic substance that is implanted into the body as a medical device to augment or replace the function of bodily tissues or organs. The principles of how to test new biomaterials and medical devices prior to use in patients and introduction into the market have been well established since at least the early 1980s. For instance, these principles have already been published in 1984 in a series of books by leaders in the field of biomaterials and bioengineering (Ducheyne and Hasting, 1984). Biocompatibility has been defined 35 years ago at the 1986 Consensus Conference of the European Society for Biomaterials as "the ability of a material to perform with an appropriate host response in a specific application" (Williams, 2008).

The evaluation of biomaterials and medical devices typically involves five phases during a reasonable and effective design and design validation process:
1. *In vitro* evaluation of biocompatibility under simple, a-functionally loaded conditions
2. *In vitro* evaluation of acceptable performance under conditions mimicking functional *in vivo* conditions
3. *In vivo* evaluation of biocompatibility under simple, a-functionally loaded conditions
4. *In vivo* evaluation of satisfactory performance under functionally relevant conditions using appropriate clinically relevant animal models
5. Evaluation of materials and devices in well-conceived clinical trials

When a medical device is well designed for surgical use, one expects each of these evaluations to have been addressed and/or performed. In analyzing clinical performance of the product, one typically relies on experimental data following from studies wherein the product was used in a clinically relevant animal model. In such studies, the device is inserted into the tissue bed in such a manner that it approximates as closely as possible the actual function of the product in a human being. The company placing the product on the market for permanent implantation in

https://doi.org/10.1515/9783110619249-004

patients must have appropriate experimental evidence to demonstrate that there would not be adverse consequences, including adverse tissue responses, from the intended use of the device. A permanently implanted medical device is a foreign body. Further, the material properties must be such that the device does not promote infection, a chronic and perpetuated inflammatory tissue response, does not come apart or disintegrate over time after implantation, and does not elicit adverse immune responses. Thus, it is important to have experimental data to establish a relationship between the material properties of the medical device and what happens once the device is implanted in the human body. Consequently, it is the manufacturer's responsibility to conduct the appropriate animal and clinical studies supporting the appropriate use of the device in the general patient population, that is, a use whereby functional safety of the device is achieved when implanted in patients, before placing it on the market. If appropriately conducted animal studies demonstrate support for the intended use of the product, then the next tier of scientifically relevant studies is a controlled human study. Such studies frequently proceed in two phases: a first or "pilot" phase, and a second or "pivotal" phase. In the pilot phase, one verifies surgical procedures and, generally, absence of adverse events. In the pivotal phase, one documents safety and efficacy in a sufficiently large population. If results of the pilot study are reassuring, the larger controlled study ("pivotal" study) in humans is appropriate before the product is released into the market. In addition, after introduction to the market the manufacturer is obliged to conduct adequate post-market studies and to report adverse events. This should include clinical studies shedding light on the long-term performance of the respective device in patients.

As a result, translation of novel biomaterials or medical devices from the laboratory to the clinic entails detailed *in vitro* cell culture studies utilizing cell culture models that represent the respective tissue and analyze the cell and tissue responses at a molecular level, preclinical small and large clinically relevant animal models, as well as clinical studies. It furthermore involves the interdisciplinary efforts of clinicians, material scientists, chemists, bioengineers, biologists, medical doctors, dentists, and veterinarians. While the clinical problem and need are necessary to be formulated by the respective clinicians, developing the respective biomaterials and devices and taking them from basic research to the clinical arena require close collaboration and communication among these various disciplines. As a result, the path of biomaterials and medical device development is from bedside-to-bench and back from bench-to-bedside. Since medical practice in the twenty-first century needs to be evidence-based, also translation of novel biomaterials and medical devices needs to be evidence-based in a dual fashion; first evidence-based in the sense that we understand the cell and tissue responses to these devices at a molecular level, and secondly evidence-based in the sense that we need to provide long-term data of prospective clinical studies that demonstrate the superiority of these novel materials and devices over existing ones.

4.1.2 Tissue and host response and classification of biomaterials

When biomaterial science emerged as a discipline of its own in the 1970s and early 1980s, biomaterials were first categorized according to their bulk chemistry as metallic biomaterials (metals and alloys), ceramic biomaterials, polymer-based biomaterials, or composite biomaterials. Since then the focus has shifted to characterizing them according to the host and tissue responses which they elicit.

As outlined above, biocompatibility has been defined as "the ability of a material to perform with an appropriate host response in a specific application." Implantation of a biomaterial into the body induces reactions that affect the outcome of integration and the biological performance of the implant. The natural response of the body to a foreign body or material is trying to exterritorialize the foreign material, that is, isolating it from the body by encasing it in a fibrous capsule that in most cases also contains foreign body giant cells (FBGCs). Hence, the objective of developing biocompatible materials for biomedical applications is to suppress this foreign body response by creating materials that heal and are integrated into living tissue or that support and even stimulate tissue regeneration and are then replaced by the native tissue (Ratner, 2015, 2016, 2019). The interaction between the host tissues and the implant can be affected by degradation products released from the inserted implant. Implanted biomaterials, medical devices, and prostheses implanted in connective tissue induce an initial host response that can be graded according to (Hench and Best, 2004): If the material is (i) *toxic*, the surrounding tissue dies; (ii) *nontoxic and biologically inactive* (biotolerant, nearly inert), a fibrous tissue of variable thickness forms; (iii) *nontoxic and biologically active* (bioactive), an interfacial bond forms; and (iv) *nontoxic and dissolves in addition to stimulating tissue formation in certain cases (resorbable and bioactive materials)*, the surrounding tissue replaces it and thereby the defect heals.

4.1.2.1 Toxicity

A number of first-generation biomaterials elicited toxic tissue responses, such as lead-containing dental impression materials, some dental palladium copper alloys, which were susceptible to casting defects, disintegration, and corrosion, or polymeric biomaterials that leaked monomers or other chemicals such as softeners or others. Various formulations of early generation PMMA bone cements (Kalteis et al., 2004), which is an important polymer used in orthopedics for anchoring fracture fixation devices and joint replacement prostheses, displayed poor osseous integration, and tissue necrosis occurred due to heat production of up to 80°C during PMMA curing and due to toxic effects of the monomer (Revell et al., 1998). Also, some early generation resorbable biomaterials, for example, early generation tricalcium phosphate ceramic products displayed shortcomings with respect to biocompatibility due to the formation of particles smaller than 7 μm in size during their degradation, which led to macrophage activation and undesirable inflammatory responses (Peters and Reif, 2004; deGroot, 1988). The same was true for calcium phosphate bone grafting materials with insufficient phase purity. In total joint replacement, polyethylene and other wear particles have been shown to impede the performance of these prostheses by inducing adverse tissue responses involving macrophage activation and osteolysis, which can lead to aseptic loosening and failure of the device (McGee et al., 1997; Howlett et al., 2000).

4.1.2.2 Concept of biocompatibility, bioinert, and biotolerant biomaterials

With first-generation soft-tissue implants and metallic, ceramic, and polymeric biomaterials, occurrence of a mild foreign body response which entails formation of a thin fibrous capsule with a limited number of macrophages and foreign body giant cells (FBGCs) has been regarded as biocompatible host and tissue response (biotolerant and bioinert materials). This foreign body response may even persist as long as the implant is in place, i.e. for life in case of a permanent implant (Anderson et al., 2008; Bryers et al., 2012; Ratner, 2016). However, over the past 15 years, there has been a major paradigm shift regarding the concept of biocompatibility demanding that biocompatible biomaterials need to possess the capability to heal and regenerate the tissues they are supposed to augment or replace (Brown et al., 2012; Bryers et al., 2012; Ratner, 2015, 2016).

Over the past decade, there has been increasing controversy regarding the performance of some polymeric biomaterials, such as polypropylene, which were originally regarded as bioinert or biotolerant, when used in the form of surgical meshes in the abdomen for tissue reinforcement and long-lasting stabilization of fascial structures of the pelvic floor in women in treating pelvic organ prolapse, since an increasing number of complications were observed in patients over time, including pain, erosion, organ perforation, and infection. The adverse tissue reactions leading to these complications were found to involve a perpetuated chronic inflammatory response characterized as pro-inflammatory and tissue destruction-type response, which leads to fibrosis, enhanced production of degradative enzymes, weakening of neighboring tissues, pain, and erosion through mucosal tissues in some patients (Brown et al., 2015; Nolfi et al., 2016; Gigliobianco et al., 2015; Liang et al., 2016). This has led to the FDA (US Food and Drug Administration) issuing a report regarding complications in its "2011 Update on the Safety and Effectiveness of Transvaginal Placement of Urogynecologic Surgical Meshes for Pelvic Organ Prolapse (POP)." The scientific community agrees that there has clearly been a lack of adequate preclinical evaluation of these devices using clinically relevant large animal models before introducing these devices into the clinical arena for use in POP treatment, and that a better understanding of the forces within the pelvic floor, which the materials are subjected to after implantation needs to developed. In addition, the scientific literature points out that there is a considerable need to investigate the immune responses in patients, in whom these polymer mesh materials perform well over many years as compared with immune responses in patients in whom these meshes cause severe complications (Gigliobianco et al., 2015; Liang et al., 2016). In this context, it is noteworthy that the series of standard tests that are prescribed to determine the biological safety of bio-materials used for implantable medical devices in the context of regulatory approval and solely assess the effects that extractable substances have on cells *in vitro* and in simple short-term animal studies are not appropriate, that is, sufficient for determin-ing the suitability of materials for complex devices for tissue regeneration and tissue engineering (Williams, 2015).

4.1.2.3 Host response-related events

The host response to a biomaterial entails that within a few seconds after implant placement, blood is released from the damaged host blood vessels, which contains plasma components that include different proteins such as fibronectin, albumin, and fibrinogen. In addition to platelets and its components (growth factors as plate-lets derived growth factor), blood exudate also contains components that induce a coagulation cascade (Ekdahl et al., 2011). Platelets adhesion and recruitment of ad-ditional immune cells, proinflammatory cytokines, and chemokines contribute to provisional matrix formation. ECM (extracellular matrix) adhesion proteins include fibronectin and vitronectin that enhance cell adhesion. Cell adhesion to the pro-tein-coated biomaterials is mediated by integrin signals (Keselowsky et al., 2005). As outlined earlier, in some cases the development of a foreign body reaction (FBR) constitutes the method of healing rather than remodeling. This fibrotic encapsula-tion prevents further interaction between the biomaterial and the host tissues and, thereby, often interferes with the device being able to function properly (Ratner, 2016). Thus, implantation of a biomaterial induces a host response that has a great impact on the integration and biological performance of the implant. One important factor is that the degradation products released by devices can activate the immune system. The interplay between the host immune system and the biomaterial de-pends on the tissue surrounding the implant that will drive the tissue-specific in-nate defenses and the following induction of adaptive immune responses. In fact, it is becoming more apparent that macrophages resident in tissues or recruited from other sites play distinct roles in the healing process. In this context, macrophage polarization and macrophage phenotype shift from a M1 proinflammatory toward an M2 proremodeling phenotype have been recognized as important factors with re-spect to obtaining a tissue regeneration and healing rather than a foreign body en-capsulation response (Brown et al., 2012; Ratner, 2015, 2016, 2019). Furthermore, implantation of the same material into different sites elicits distinct responses (Anderson et al., 2008). In this context, relative movement of delicate soft tissue components and stiffening of biomaterials and medical devices due to fibrosis which may result from an undesirable perpetuated chronic inflammatory response need to be considered in various anatomical locations in order to avoid erosion and migration of devices into neighboring organs or through the skin or mucosal tissues (Giglibioanco et al., 2015; Liang et al., 2016). As a result, testing of medical devices using clinically representative large animal models in the intended anatomical location prior to use in patients is of paramount importance in order to protect patients from faulty devices (Olivarria et al., 2019).

The events at the implant surface-tissue interface are crucial in obtaining a biocompatible tissue response and the desired tissue integration and anchorage of the device *in vivo*. This is particularly true with respect to metallic implants. The biocompatibility of an implant and the response of the surrounding tissues depends on different biomaterial properties including both bulk chemical composition as well as surface chemistry and properties, such as hydrophobicity/hydrophilicity, wettability, surface charge, polarity, surface energy, mobility of surface molecules and smoothness, surface texture, and geometry (Boss et al., 1995).

4.1.2.4 Distance osteogenesis, contact osteogenesis, and osseointegration

With metallic implants, peri-implant bone healing can occur via distance osteogenesis or contact osteogenesis (Davies, 1996, 1998, 2003). In distance osteogenesis observed with bioinert surgical stainless steel, CoCrMo, tantalum, zirconia, and alumina implants, the bone grows from the old bone surface toward the implant surface in an appositional manner. In contact osteogenesis that occurs with implant materials and surfaces that osseointegrate, such as titanium and titanium alloys, *de novo* bone forms on the implant surface (Davies, 1996, 1998, 2003). In the absence of disrupting mechanical forces, contact osteogenesis leads to osseointegration (Brånemark et al., 1969; Davies, 1996, 1998, 2003), if the surface of the titanium implant displays the appropriate surface properties, that is, titanium oxide (rutile) layer (Steinemann and Straumann, 1984; Steinemann, 1996, 1998; Healy and Ducheyne, 1991, 1992a, 1992b) and surface texture/geometry (Cooper, 2000). The early stage of peri-implant bone healing is very important and involves protein adsorption, platelet activation, coagulation, formation of a stable fibrin clot that is a depot for growth factors and allows for migration of osteoprogenitor cells including pericytes to the implant surface that colonize the implant surface, secrete ECM within 24 h after implantation (Meyer et al., 2004), and differentiate into osteoblasts. Davies (1998) described the events occurring at the bone-titanium implant surface interface. First, a layer of calcified globular accretions is laid down on the implant surface by differentiating osteogenic cells. Second, this is followed by elaboration and deposition of collagen fibers, which subsequently (thirdly) mineralize. Collectively, these events lead to formation of a cement line at the bone-biomaterial interface, which has the biochemical composition and morphological structure found at natural interfaces in bone tissue (Davies, 1996; Müller-Mai et al., 1995) and thus equals the von Ebner cement lines present between osteons. As such titanium implants are integrated into the osseous tissue, such as the body's own bone structures rather than being encapsulated by fibrous capsule, which was the case with early stainless steel blade-form dental implants, such as early generation Linkow blade implants (Linkow, 1972; Linkow et al., 1995).

4.1.2.5 Bone-bonding bioactive glasses, glass ceramics, and calcium phosphates

Furthermore, it should be mentioned that with respect to endosseous implant materials there are two classes of implant material: non bone-bonding materials that osseointegrate such as titanium, and bone-bonding bioactive biomaterials such as bioactive glasses (Hench and Paschal, 1973; Hench, 1998, Ducheyne and Qui, 1999), glass ceramics, and calcium phosphates that bond to bone and are used as implant coatings or bone grafting materials, most commonly available in the form of granules (Knabe et al., 2011b, 2017b; Müller-Mai, 2003). The mechanisms underlying the bioactive behavior and bone-bonding properties of bioactive glasses and calcium phosphates have been outlined in detail in various review papers and recent book chapters (Ducheyne and Qui, 1999; Knabe et al., 2008, 2011b, 2017b). These events include surface-mediated as well as solution-mediated effects (Ducheyne and Qui, 1999). Upon contact of bioactive glasses and ceramics with body fluid, dissolution, reprecipitation, and ion-exchange phenomena occur in combination with protein adsorption leading to formation of a calcium phosphate transformation layer at the biomaterial surface. In this context, development at the biomaterial surface of calcium-deficient carbonated hydroxylapatite (HAp) resembling the natural mineral present in bone tissue is a key element of bioactive behavior of bone (Ducheyne and Qui, 1999). As a result, the material surface is recognized by the body as "own" rather than as a foreign body. This surface transformation layer affects the conformation of the adsorbed serum proteins, of which fibronectin is a key player (El-Ghannam et al., 1999), in such a manner that specific adhesive amino acid sequences are exposed to the osteogenic cells present at the wound healing site. The osteogenic cells adhere to these adhesive motif amino acid sequences via integrin receptors that mediate an outside-in signaling, which then activates intracellular signaling pathways and events that modulate cell differentiation and cell survival (see Chapter 4.2). By upregulation of cell differentiation and cell survival, enhanced osteoblast differentiation, ECM production, and mineralization, bone tissue formation is achieved, which underlies the stimulatory effect of bioactive bioceramic bone grafting materials on bone formation and regeneration (Ducheyne and Qui, 1999; Knabe et al., 2008, 2011b, 2017b). It is necessary to mention that in addition to the surface transformation layer, ionic dissolution products such as silicon and calcium ions have been shown to have a stimulatory effect on bone cell function by enhancing cell differentiation, bone matrix formation, and mineralization (Ducheyne and Qui, 1999; Xynos et al., 2000a,b; 2001; Yao et al., 2005; Knabe et al., 2008, 2011a, 2017b). Resorbable bioactive bone grafting materials, ideally, first stimulate bone tissue formation and regeneration in this manner and then gradually dissolve in the newly formed bone tissue resulting in substitution by functional bone tissue and restoration of the original osseous microanatomy and bone tissue, thereby constituting a *restitutio ad integrum*. It furthermore needs to be pointed out that the stimulatory effect of bioactive glasses and ceramics on bone formation is obtained solely based on these material-induced mechanisms outlined above without the addition of growth factors or biologicals. As a result, these materials are regulated as biomedical devices and require a premarket approval for medical devices, whereas obtaining the regulatory approval for clinical use of combination products, that is, biomaterials that contain and release biologicals or drugs for stimulating tissue formation, requires following the same pathway as for new drug approvals that need extensive financial resources and a long timeline involving phase I–III clinical trials (see Chapter 4.5).

4.1.3 Examples of polymer- and ceramic-based bone grafting materials

4.1.3.1 Polymer-based bone graft substitutes
Polymers are divided into natural polymers, synthetic polymers, degradable, and nondegradable types (see Chapters 2.3 and 3.3).

Natural polymers
Nature-derived polysaccharides such as alginate, chitosan, and collagen are widely used in bone tissue engineering applications. Natural polymers have advantages, as they are biocompatible, biodegradable, and nontoxic. However, some biodegradable polymers used to fabricate scaffolds for bone regeneration and tissue engineering, such as collagen-based sponges, poly-L-lactic acid, poly-L glycolic acid polymers, and gelatin sponges, have several drawbacks including immunogenicity and weak mechanical strength (Choi et al., 2001).

Alginate
Alginate hydrogel, a polysaccharide extracted from seaweeds, is mainly used for immobilization of enzymes or cells for bioreactors and also for tissue engineering. Alginate degradation is carried out by divalent calcium ions slowly diffusing away from the hydrogel allowing alginate molecules to be excreted by urine. Alginate has also been mixed with other materials, such as chitosan, to enhance its biological performance (Chang et al., 2001).

Chitosan
Chitosan has a structure similar to glycosaminoglycans (GAGs). It can be produced industrially by alkaline hydrolysis of chitin N-acetyl-glucosamine present in GAG of the ECM. Chitin is the most common polymer found in invertebrates as crustacean

shell or insect cuticles (Croisier and Jérôme, 2013). Chitosan can be easily processed into hydrogels, membranes, nanofibers, micro/nanoparticles, scaffolds, foams, and sponges for biomedical applications (drug delivery), wound healing, and tissue engineering. Chitosan can provide a nonprotein matrix for 3D cell in-growth, structurally similar to extracellular proteoglycans. Sponges that are used for tissue engineering are soft and flexible materials with interconnected microporous structure (Jayakumar et al., 2011). Chitosan is biodegradable, biocompatible, and osseoconductive; enhances osteogenesis and angiogenic activity; and inhibits fibrous tissue invasion that may prevent bone regeneration. The main drawback is its rather weak mechanical properties to maintain the desired shape until newly formed bone tissue matures (Domard and Domard, 2002; Lee et al., 2002).

Collagen

Collagen is the primary protein component of the ECM of bone. It is composed of different polypeptides containing mostly glycine, proline, hydroxyproline, and lysine. Collagen is enzymatically degradable and has weak antigenic property (Zhang et al., 2006). Collagen-based materials are considered favorable biomaterials for scaffolds for both cartilage and bone repair, as collagen type I is the major component in bone. Collagen is processed into a variety of forms as porous sponges, gels, and sheets. Chemical cross-linking of collagen using glutaraldehyde or diphenylphosphoryl azide can improve its physical properties. However, a disadvantage of collagen is its low physical strength, potential immunogenicity when derived from animal sources, and sometimes the related cost (Asti and Gioglio, 2014).

Synthetic polymers

PLA (poly-lactic acid), PGA (poly-glycolic acid), PLGA (poly-lactide-co-glycolide)

These are the most frequently used biodegradable synthetic polymers for scaffolds with different designs, such as meshes, fibers, sponges, and foams. Drawbacks of PLA and PLGA are their degradation characteristics, which involve fragmentation and an acidic pH that can elicit inflammatory and macrophage responses. Biodegradable synthetic polymers, such as PLA, PGA, and copolymers based on PLA and PGA, have been adapted for use as bone substitutes. There are two important drawbacks to PLA/PGA as the degradation products have the potential to acidify the local environment and cause inflammation. These drawbacks have led investigators to develop alternative polymers for use in bones as polypropylene fumarate, polycaprolactone, tyrosine-derived polymers, and polyanhydrides, which also have been increasingly investigated for producing scaffolds for bone tissue engineering either alone or in combination with CaPs (Boyan et al., 2011; Knabe et al., 2011a, 2017a).

Hybrid approach

The combination of degradable polymers and inorganic bioactive ceramic particles can yield the highest achievable mechanical and biological performances in hard tissue. Blending two polymers, natural and synthetic, has gathered growing interest in mimicking the ECM of natural tissues. These new hybrid structures can present a wide range of physicochemical properties and processing techniques of synthetic polymers as well as the biocompatibility and biological interactions of natural polymers (Asti and Gioglio, 2014).

4.1.3.2 Ceramic- and bioactive glass-based bone grafting substitutes

Owing to their ability to stimulate bone formation, bioactive calcium phosphate ceramics and bioactive glasses are excellent candidate grafting materials for bone augmentation and regeneration. Furthermore, alloplastic synthetic bone substitute materials are superior to freeze-dried human allografts due to their safety in terms of disease transmission and immunological aspects. Among the bioactive ceramics most commonly investigated for use in bone regeneration are β-tricalcium phosphate (β-TCP), hydroxylapatite (HAp) and bioactive glass 45S5 (BG45S5). All of these materials are biocompatible and osseoconductive. However, they differ considerably in the rate of resorption. HAp resorbs very slowly compared to β-TCP and bioactive glass, which have been shown to resorb within 1-2 years in humans. For more details regarding these various bone grafting materials, the interested reader is referred to the respective book chapters on bone grafting materials (Chapters 1.2.2.2, 2.2.2, 3.2.7 and 4.5 and Knabe et al., 2017a,b,c,d).

4.1.4 Translation from the laboratory to the clinic of rapidly resorbable calcium alkali orthophosphate bone grafting materials

4.1.4.1 Development of rapidly resorbable calcium alkali orthophosphate

Over the last 20 years, a number of clinical studies revealed that resorption of β-TCP and BG 45S5 bone grafting granules took 1-2 years in humans. This resulted in increasing efforts to develop rapidly resorbable bone substitute materials, which exhibit both excellent bone-bonding behavior by stimulating enhanced bone formation at their surface in combination with a degradation rate higher than that of the clinically established calcium phosphate-based bone grafting materials β-TCP and BG 45S5. This has led to the development of a series of novel, bioactive, rapidly resorbable glassy crystalline calcium alkali orthophosphate (CAOP) materials based on the main crystalline phase $Ca_2KNa(PO_4)_2$ sample code (GB) series by substituting alkali ions for Ca^{2+} in the crystal structure, which was synthesized in the mid-1990s (Berger et al., 1995; Knabe et al., 1998). By varying the composition of the amorphous phase, a variety of CAOP materials were produced with the most important being GB 14 that contains a small amount of amorphous magnesium-potassium phosphate, GB 9 that contains a small portion of amorphous silicophosphate, and GB9/25 that contains a small portion of calcium diphosphate ($Ca_2P_2O_7$) to further increase the solubility, and 352i that contains calcium diphosphate and a small amount of silica (Knabe et al., 1998).

4.1.4.2 Effect of CAOP on osteoblast differentiation *in vitro*
In vitro studies using human osteoblasts were performed to evaluate the effect of various CAOPs on osteoblastic differentiation in an effort to assess their osteogenic potential. Osteoblastic differentiation is defined by three principal biological periods: cellular proliferation, cellular maturation, and ECM mineralization. Proliferation is characterized by collagen type I (Coll) expression, the cellular maturation phase by alkaline phosphatase (ALP) expression, and the period of ECM mineralization by osteopontin (OP), osteonectin (ON), osteocalcin (OC), and bone sialoprotein (BSP) expression. The results showed that the silica-containing GB9 displayed the greatest stimulatory effect on osteoblastic proliferation and differentiation, suggesting that this material possesses the highest potency to enhance osteogenesis. GB14 and 352i supported osteoblast differentiation to a higher degree than β-TCP, whereas, similar to bioactive glass 45S5, GB9/25 also displayed an enhancement effect on osteoblastic phenotype expression (Knabe et al., 2008, 2017a).

4.1.4.3 Effect of CAOP on osteoblast differentiation and bone formation *in vivo*
The first animal study was performed to correlate quantitative expression of the osteogenic markers *in vitro* with the amount of bone formed after bioceramics implantation *in vivo* and with quantifying the expression of different osteogenic markers in histological sections obtained from *in vivo* experiments. Different CAOP materials with particle sizes ranging from 300 to 355 µm were implanted in sheep mandibles for 1, 4, 12, and 24 weeks with the goal to regenerate critical-size membrane-protected defects and to compare the behavior of the CAOP materials to that of β-TCP and BG 45S5. Autogenous bone chips and empty defects that were filled with collagen sponges served as controls. The study showed enhanced bone formation and bone-particle contact for GB9 followed by GB14. Furthermore, this was associated with enhanced expression of collagen type I, osteopontin, osteocalcin, osteonectin, and bone sialoprotein *in vivo*. Staining for tartrate-resistant acid phosphatase showed that biodegradation of the various CAOPs as well as β-TCP occurred by dissolution rather than by osteoclastic activity (Knabe et al., 2010, 2011a, 2017a).

Another animal study was performed to evaluate the effect of four injectable CAOP cements as compared to β-TCP particles on osteogenesis *in vivo* after implantation in critical size effects in rabbit femora for 1, 3, 6, and 12 months (Müller-Mai et al., 2010; Knabe et al., 2011a, 2017a). Of the various bone substitute cements studied, the GB9 cement showed the best bone-bonding behavior and had the greatest stimulatory effect on bone formation and expression of osteogenic markers, while exhibiting the highest biodegradability. After 12 months, the original trabecular structure of the rabbit femur was almost fully restored. The biodegradability of the GB9 cement, however, was lower than that of the β-TCP granules (Müller-Mai et al., 2010; Knabe et al., 2011a, 2017a).

Furthermore, moldable CAOP-based putty-like gel cements for restoring outer bony contours were implanted into critical size-contour defects in a sheep mandible (Knabe et al., 2011a, 2017a) for 4, 12, 24, and 48 weeks. GB9 and GB14 gel cements displayed excellent surgical handling properties and facilitated restoring the outer contour of the mandible in the critical size-defects as well as maintaining the augmented bone volume without any resorption over the 48 week observation period. Both materials induced woven bone formation at their surface as early as 1 week after implantation and stimulated bone formation at their surface inducing moderate to strong osteogenic marker expression of Col I, ALP, OC, and BSP in osteoblasts, osteocytes, and the bone matrix of the surrounding bone tissue at the various time points studied. Furthermore, they displayed excellent bone-bonding behavior and were gradually replaced by newly formed bone. Due to the results of this study, GB9 and GB14 gel cements can be regarded very promising moldable bone grafting materials for restoring contour defects and augmenting bony defects and voids resulting from orthognathic surgery (Knabe et al., 2011a, 2017a).

4.1.4.4 Effect of CAOP ceramics on cell adhesion and intracellular signaling mechanisms

To elucidate the underlying mechanisms of the stimulatory effect of CAOP on osteoblast differentiation and bone formation, *in vitro* and *in vivo* studies were performed investigating the involved cell adhesion and intracellular signaling mechanisms. Immersion of CAOP in physiologic solutions showed that these materials exhibited a high solubility in combination with adsorption of serum fibronectin proteins (Kim et al., 2010). GB9 surfaces displayed higher calcium and silicon ion release as well as calcium uptake and bioactive surface transformations after immersion in physiological solutions, as well as fibronectin and extensive collagen type I serum protein adsorption (Knabe et al., 2011b) when compared to β-TCP and BG45S5 and other CAOPs (GB14, GB9/25, GB352i). Cell adhesion to CAOP was mainly mediated through integrin $\alpha_5\beta_1$ and to a slightly lesser degree through $\alpha_2\beta_1$ and to a minor degree to β_3 that led to a synergistic effect of simultaneous enhanced activation of (i) the Ras/MAPkinase cell differentiation pathway, (ii) alternate p38 pathway (which is probably activated after cell adhesion to $\alpha_2\beta_1$), and (iii) of key signaling factors of the PI3K/Akt-cell survival pathway and antiapoptotic factors resulting in the greatest antiapoptotic effect. Thus, taken together GB9 displayed the greatest stimulatory effect on activation of all three signaling pathways (cell differentiation, cell survival, p38 pathway) as well as the greatest antiapoptotic effect (Knabe et al., 2011b, 2017b; Kim et al., 2010).

4.1.4.5 CAOP degradation

Degradation of CAOP occurred mainly by physicochemical dissolution rather than by osteoclastic activity. Dissolution of CAOP leads to a more alkaline pH. Resorption of HAp in contrast is associated with a lower acidic pH that favors osteoclastic activity (Knabe et al., 2010, 2011b).

4.1.4.6 Translation to the clinic – effect of silica-containing CAOP bone grafting material on bone formation and osteogenic marker expression after sinus floor augmentation in humans

Collectively, the data generated in the various *in vitro* cell culture and *in vivo* animal studies using a clinically relevant large animal model resulted in clearance of the silica-containing CAOP bone grafting material (Si-CAOP) by the FDA as well as in obtaining the CE mark (Figure 4.1A).

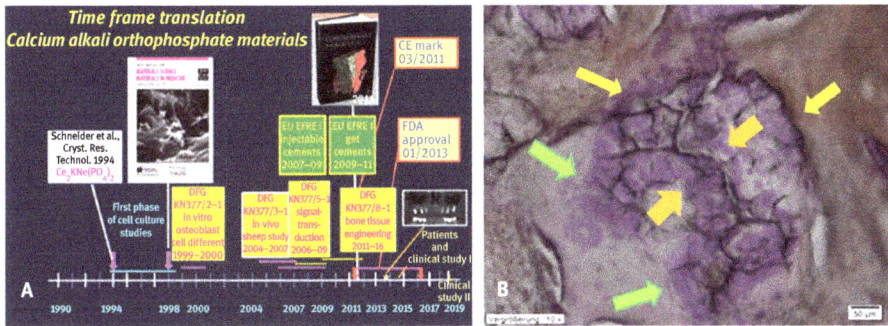

Figure 4.1: (A) Schematic timeframe required to develop a novel bioactive bone grafting material and its translation to the clinic. (B) Histomicrograph of resin-embedded human biopsy sampled 6 months after sinus floor augmentation stained immunohistochemically for osteocalcin after deacrylation. The image shows highly degraded Si-CAOP particle fragments that induce apically bone matrix mineralization (yellow arrows) and bone formation (green arrows) at their surface, that is, at great distance from the native bone of the sinus floor, while resorbing within the newly formed bone (orange arrowheads). Sawed section counterstained with hematoxylin, bar = 50 μm.

As a result, a first clinical study was performed to examine the effect of silica-containing CAOP ceramic particulate bone grafting material (tradename Osseolive; Knabe et al., 2019) and β-TCP on bone formation and osteogenic marker expression in human biopsies sampled 6 months after sinus floor augmentation (SFA) at implant placement. A total of 86 dental implants were inserted into the augmented maxillary sinus floors of 19 patients, in which Si-CAOP was used, and 19 patients, in which β-TCP was used. Intraoperatively, when preparing the implant bed, in sites, in which Si-CAOP was used, a homogenous tissue consistency was noted and excellent bone formation and vascularization was clinically observed in the area, in which the access window had been prepared. In this area, it was hardly possible to distinguish the augmented bone tissue from the native one. By contrast, when preparing the implant bed at sites in which β-TCP was used, the grafted area displayed a more heterogeneous tissue consistency with greater, more uneven drilling resistance. Furthermore, in the area of the access window it was easily possible to visually distinguish the grafted bone area from the native bone, when β-TCP was used. Cone beam computed tomography (CT) showed excellent bone formation in the sinus floor 6 months after SFA with Si-CAOP and that the bone grafting material had been widely replaced by newly formed bone, while at sites in which β-TCP was used, a higher amount of residual grafting material was visible. Biopsies sampled 6 months after SFA with Si-CAOP already macroscopically displayed a mature healthy bone structure. Histological analysis showed that both materials facilitated excellent bone formation and matrix mineralization, which were still actively progressing from the sinus floor in an apical direction 6 months after SFA. Biopsies of patients, in which Si-CAOP was used, however, displayed a significantly greater bone area fraction, and bone-biomaterial contact when compared to biopsies from patients, in which β-TCP was used. At the same time, 6 months after SFA, the particle area fraction was lower in biopsies from patients, in which Si-CAOP was used compared to when β-TCP were statistically significant. Furthermore, histologic examination revealed excellent cancellous bone formation in the sinus floor with few residual highly degraded Si-CAOP particle fragments embedded in the newly formed bone without signs of inflammatory tissue response or scar tissue formation in any of the grafted sites. These highly degraded Si-CAOP particle fragments displayed excellent bone-bonding behavior (Figure 4.1B) and ingrowth of bone tissue in the pores of the degrading granules (Knabe et al., 2017e).

4.1.4.7 3D-CAOP-based scaffolds for bone tissue engineering

Three-dimensional calcium alkali phosphate-based scaffolds were colonized with mesenchymal osteogenic stem cells cultured under perfusion bioreactor conditions for bone tissue engineering purposes with the aim to generate a tissue engineered synthetic bone graft with homogenously distributed osteoblasts and mineralizing bone matrix *in vitro*, thereby mimicking the advantageous properties of autogenous bone grafts and facilitating usage for reconstructing large segmental discontinuity defects *in vivo*.

3D scaffolds were developed from a silica-containing CAOP (code: GB9S14) utilizing two different fabrication processes, first a replica technique (SSM) using polyurethane foam templates, and second 3D printing (RP). The mechanical and physical properties of the scaffolds (porosity, compressive strength, solubility) and their potential to facilitate homogenous colonization by osteogenic cells and extracellular bone matrix formation throughout the porous scaffold architecture were examined prior to *in vivo* implantation. To this end, murine preosteoblastic cells (MT3T3-E1) were dynamically seeded and cultured for 7 days on both scaffold types under perfusion with two different concentrations of 1.5 and 3×10^6 cells/mL. The amount of cells and ECM formed and osteogenic marker expression were evaluated using hard tissue histology, immunohistochemical, and histomorphometric analyses. SSM scaffolds displayed an architecture resembling that of cancellous bone, and a significantly higher total porosity (87%) than RP scaffolds (RPS) (50%), while RPS had a geometric architecture with macrochannels and exhibited significantly more open micropores, greater compressive strength and higher silica release. RPS seeded with 3×10^6 cells/mL displayed maximum cell and ECM formation, mineralization, and osteocalcin expression. In summary, RPS displayed superior mechanical and biological properties and facilitated generating a tissue engineered synthetic bone graft *in vitro*, which mimics the advantageous properties of autogenous bone grafts, by containing homogenously distributed terminally differentiated osteoblasts and mineralizing bone matrix and therefore is suitable for subsequent *in vivo* implantation for regenerating segmental discontinuity bone defects (Adel-Khattab et al., 2018).

In vivo analysis of 3D calcium alkali phosphate-based bone grafts for reconstruction of segmental mandibular defects

Early establishment of angiogenesis is critical for bone tissue engineering. Recently, a technique was introduced based on the idea of using axial vascularization of the host tissues in engineered grafts, called "intrinsic angiogenesis chamber" technique that utilizes an artery and a vein to construct an arteriovenous bundle (AVB). An *in vivo* rat study was performed to evaluate the effect of varying scaffold architecture of CAOP scaffolds, resulting from two different fabrication procedures, namely 3D printing (RP) or a Schwartzwalder–Somers replica technique (SSM), on angiogenesis and bone formation *in vivo* when combining a microvascular technique with bioceramic scaffolds colonized with stem cells for bone tissue engineering. A total of 72 adult female Wistar rats, in which critical size segmental discontinuity defects 6 mm in length were created in the left femur, were divided into four groups. Group 1 received an RPS colonized with rat stem cells after 7 days of dynamic cell culture and an AVB, group 2 an SSM scaffold with rat stem cells after 7 days of dynamic cell culture and an AVB, group 3 an RP control scaffold (without cells and AVB), and group 4 an SSM control scaffold (without cells and AVB). After 3 and 6 months, the following analyses were performed: (i) angiomicro-CT after perfusion with a contrast agent, (ii) image reconstruction and analysis, (iii) histomorphometric analysis of the amount of bone formation, and (iv) immunohistochemical analysis of angiogenic and osteogenic markers utilizing antibodies to collagen IV, von Willebrand factor (vWF) and CD-31, and osteogenic markers. At 6 months, a statistically significant higher blood vessel volume%, blood vessel surface/volume, blood vessel thickness, blood vessel density, and blood vessel linear density were observed with RPSs with cells and AVB than with the other groups. At 6 months, RP with cells and AVB displayed the highest expression of collagen IV, CD31, and vWF, which is indicative of highly dense blood vessels. This was associated with significantly greater bone formation with RPSs, cells, and AVB after 3 and 6 months and higher expression of osteogenic markers, when compared to other scaffold types, as well as with greater scaffold resorption. Both angio-CT and immunohistochemical analysis demonstrated that AVB is an efficient technique to achieve scaffold vascularization in critical size segmental defects after 3 and 6 months of implantation and that the most mature vasculature and greatest bone formation and defect bridging was obtained when using the 3D printed scaffolds after precolonization with osteogenic cells in combination with an AVB. In this study, it was possible to create 3D-printed CAOP scaffolds that displayed superior quality when compared with SSM scaffolds with respect to physical material properties, such as scaffold architecture, porosity, and mechanical properties as well as biological properties such as their effect on cell and extracellular bone matrix, formation, differentiation, and maturation. RPSs displayed a precise reproducible architecture, micro- and macroporosity featuring a more homogenous pore size distribution. This was in addition to a more enhanced cell colonization, osteoblast differentiation, ECM production, and maturation after 7 days of dynamic perfusion cell culture *in vitro*. *In vivo*, when combined with a microvascular technique, that is, an AVB, 3D printed scaffolds, which contained homogenously distributed terminally differentiated osteoblasts and mineralizing bone matrix at implantation, showed superior bone formation, defect bridging as well as a high bone biomaterial contact, and high resorbability without eliciting any inflammatory reaction. As a result, it was possible to develop 3D-printed bioactive CAOP scaffolds possessing excellent physical and biological properties and allowing for a fabrication process that can produce synthetic bone grafts exactly matching the shape of the patient´s defect by utilizing the patient's CT data. These 3D-printed CAOP scaffolds meet the requirements for bone tissue engineering applications and producing synthetic bone grafts *in vitro* that contain terminally differentiated osteoblasts and mineralizing bone matrix homogenously distributed throughout the scaffold, thereby mimicking the advantageous properties of autogenous bone grafts. Thus, they are well suited for subsequent *in vivo* implantation to reconstruct segmental discontinuity defects both in craniomaxillofacial surgery as well as in orthopedics (Adel-Khattab et al., 2017, 2018).

4.1.5 Drug-releasing biomaterials and medical devices

Biomaterials and medical devices releasing drugs or biologicals constitute a category of their own including drug-eluting cardiac stents (Kandzari et al, 2017; Tzafriri et al., 2018), fracture fixation devices with antibiotics-releasing polymer- or silica-based coatings (Fuchs et al., 2011; Bhattacharyya et al., 2014; Qu et al., 2014, 2015), or growth factor-releasing biomaterials for bone generation (Kowalczewski and Saul, 2018) including polylactic and polyglycolic acid polymers which offer the possibility of integrating growth factors, drugs, and other compounds to create multiphase delivery systems (Choi et al., 2001). If the marketed product contains the respective drug, these devices are regulated like drugs, and the extensive regulatory pathway for approval of new drugs needs to be followed prior to clinical use. If the device is designed in such a fashion that the marketed product does not contain any drug or biological material, and that the respective drug is added to the device by the surgeon during implantation, the regulatory pathway for medical devices can be followed for obtaining approval for clinical use.

4.1.6 Significance of host factors and device registries

Owing to the significantly increased application of biomaterials and medical devices in modern medicine, many of these devices and implants functioning for extended periods of time and a globally aging population, there is great demand for increasing research effort that shed more light upon the role of host factors including age, gender, presence of diabetes, smoking, hormone levels (Knabe et al., 2017c), genetic factors (Freedman and Mooney, 2019) on the performance of biomaterials and medical devices. This provides us with statistically sound data generated in a sufficiently large patient population. The knowledge generated will then also facilitate identifying parameters useful for predicting the performance of biomaterials and medical devices in various patient populations including the elderly and patients with various comorbidities and can then be used to tailor individual treatment regimes for patients and advancing the concept of "individualized medicine." In addition, establishing and expanding national and international databases and implant and devices registries is of paramount importance to increase our knowledge base regarding the long-term performance of various medical devices and biomaterials (Olivarria et al., 2019).

4.2 *In vitro* studies and cell adhesion to biomaterials (Barbara Nebe)

One of the major challenges for materials science is to provide materials that enable optimum performance within the human organism by meeting specific requirements of biological systems without side effects such as chronic inflammation or the failure of an implant caused by bacterial infections.

Successfully responding to these challenges requires knowledge of the molecular context of the interaction of the biomaterial with the biological system, in particular with living tissues and cells. *In vitro* laboratory studies can provide both detailed insights into the cell-material interaction and a screening method for newly developed materials before they are tested in animal experiments (see Chapter 4.3). At the interface of biomaterials, adhesion of cells can be regulated by diverse physicochemical characteristics. One important feature is the topography, that is, the shape of the surface in the macro-, micro-, and nanometer range (Williams, 2014; Davidson et al., 2015; Boyan et al., 2003; Moerke et al, 2016) as well as the overall stiffness of the material (Discher et al., 2005). In addition, chemical factors are playing a major role not only by providing superior biocompatibility but also optimum cell physiology at the material interface. The chemistry can be triggered by the general material composition (alloys, e.g., Ti6Al4V for biomedical applications; Cohen et al, 2017), by micro- and nanocoating techniques (Peng et al., 2006; Chen et al, 2014; Connelly et al, 2011; Rau et al., 2019; Finke et al., 2014; Niepel et al., 2016, Moerke et al., 2017) or by implantation of ions into subsurface areas of biomaterials (Williams, 2014; Kosobrodova et al., 2018).

The development of biomaterials for implants has accelerated rapidly and the determination of cytotoxicity of the materials is by no means the sole tool to assess their biocompatibility any longer. The established testing methods are, however, still focused on the cytotoxic influence of the materials' properties. The cell behavior on a biomaterial surface is highly complex. Based on new insights into cell biology, the research focus has now shifted to primary cellular mechanisms in order to provide a deeper understanding of the effect of the material on the cellular processes taking place at its interface. This is the only reasonable way to address issues arising, for instance, in orthopedics, that are concerned with promotion of cell growth at the implant surface.

The process of adhesion and proliferation, that is, the occupation of the biomaterial surface by cells of the target tissue, is divided into stages. Figure 4.2 shows this sequence, beginning with the initial adhesion of the cells (1), their spreading (2), the cell migration along chemical gradients (3) and/or the topography of the biomaterial surface, and followed by cell proliferation (4) and differentiation of cell functions. Cell migration is playing an essential role in colonization of biomaterial scaffolds and

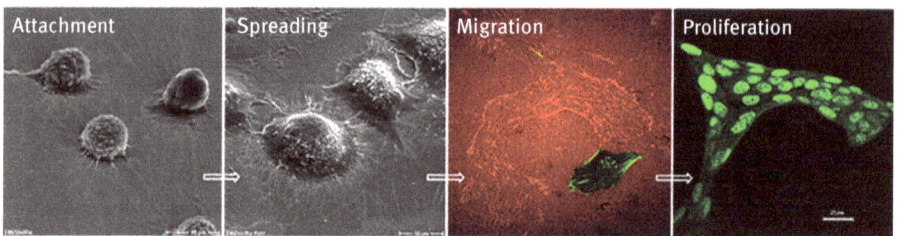

Figure 4.2: Stages of cell–biomaterial interaction *in vitro*. Main cellular processes are involved during the cell's occupation of biomaterial surfaces *in vitro*: cell attachment → cell spreading → cell migration → cell growth (proliferation). SEM and LSM images: Dept. of Cell Biology/Rostock University Medical Center: **1**. SEM, MG-63 osteoblasts, 30 min on polished Ti, 45°, scale bar 10 µm. Courtesy Henrike Rebl. **2**. SEM, MG-63 osteoblasts on Ti6Al4V modified with plasma polymer, 45°, scale bar 10 µm. Courtesy Henrike Rebl. **3**. LSM, MG-63 osteoblast, actin staining (green), on collagen-I (red), with footprints left in a semicircle (red) produced by the migrating cell. Courtesy Frank Lüthen. **4**. LSM, human primary breast cells, sub confluent, BrdU staining for S-phase cells (nuclei, green), scale bar 25 µm, BN.

is based on physically integrated molecular processes (Lauffenburger and Horwitz, 1996) that are controlled by the adhesion receptors of the cells – the integrins.

4.2.1 Cellular structures important for cell adhesion and signaling

4.2.1.1 Actin cytoskeleton

The actin cytoskeleton (Figure 4.3) is a multifunctional system within cells and supports diverse cellular activities.

Cytoskeleton

The cytoskeleton, an important part of the cell structure, provides support, shape, elasticity, and protection to the contents of the cell, much like the larger skeleton found in many living organisms. Its structure is made up of proteins that assemble themselves into actin microfilaments, intermediate filaments formed by, for example, vimentin and laminin, and microtubules formed by α- and β-tubulin. One of the key functions of the cytoskeleton is to act as cellular scaffolding, providing support for the contents of the cell, and anchoring the nucleus in place. Cytoskeletal microfilaments also called actin cytoskeleton that control cell shape and migration will be coupled through specialized transmembrane receptors (integrins) to ECM molecules, such as collagen, fibronectin, laminin, or vitronectin. The cytoskeleton forms specialized structures, such as flagellae, ciliae, lamellipodia, and podosomes, which can serve as template for the construction of cell walls, and are involved in cell division and muscle contraction (Cooper, 2018).

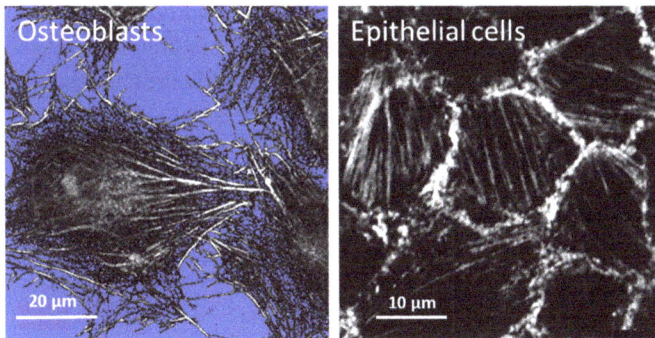

Figure 4.3: Cell-type specific formation of the actin cytoskeleton within the cell. Whereas in bone-forming cells (osteoblasts, left) the actin cytoskeleton is expressed in long filaments spanning the entire cell width, the actin fibers in epithelial cells (right) are additionally submembranously localized. Confocal microscopy, LSM 780, Carl Zeiss. Left: Human MG-63 osteoblastic cells (ATCC), phallacidine BODYPI, false color, BN. Right: Murine liver epithelial cells mHepR1 (Henning et al., 1994), phalloidin TRITC 546. Courtesy Anna-Christin Waldner, Rostock University Medical Center, Germany.

Actin filaments form by polymerization of globular, monomeric actin (G-actin) into twisted strands of filamentous actin (F-actin) (Mofrad and Kamm, 2006). The importance of the actin cytoskeleton network is reflected by the fact that up to 10% of all proteins in the cell consist of actin (Mofrad and Kamm, 2006). The molecular mechanisms underlying actin cytoskeleton activities are complex due to a huge number of accessory proteins (Zaidel-Bar et al., 2007; Kanchanawong et al., 2010). This subnet consists of the actin cytoskeleton, the actin modulators (e.g., α-actinin, profilin, ARP2/3), the adaptor proteins (e.g., paxillin, vinculin, talin, tensin, and focal adhesion kinase) and transmembrane molecules (e.g., integrin receptors, see Chapter 4.2.1.2) (Zaidel-Bar et al., 2007; Zamir and Geiger, 2001; Lauffenburger and Horwitz, 1996). The accessory proteins support formation of the actin filaments with different dynamics, spatial organizations, and interactions with other cellular structures (Svitkina et al., 2018).

On biomaterial surfaces with nano- and microscale roughness, the cells react sensitively to these physical features and reorganize their actin filaments accordingly. Hence, a clear cell architecture – cell function relationship on geometric and stochastic microstructures was found. Titanium surfaces with sharp edges produced, for example, by blasting with corundum grit or the presence of cubic pillars, were found to disturb the actin formation in osteoblasts and, thus, the stress filaments/fibers were distinctly shortened and locally arranged around the underlying structure (Moerke et al., 2016; Matschegewski et al., 2010). As a result, the synthesis of ECM proteins (Moerke et al., 2016) and the intracellular calcium ion (Ca^{2+}) mobilization capacity were inhibited (Staehlke et al., 2015).

Actin-based filopodia structures are slender cytoplasmic projections (microspikes) of migrating cells (Mattila and Lappalainen, 2008). The intracellular bundles of actin filaments provide the cells' filopodia with stability so that they are able to span the ridges of a structured biomaterial surface or the distances of scaffold pores horizontally and parallel to the gravitational force (Figure 4.4) (Kunz et al., 2010). When treated with Cytochalasin D, the agent disrupting the actin filaments, cells lose their ability to span the pores and the overall cell growth is hampered (Neumann and Klinkenberg, 2014). Therefore, the actin cytoskeleton of cells is an important prerequisite for the occupation of a biomaterial surface or a 3D scaffold.

4.2.1.2 Integrin receptors: Structure, functions, and dynamics

Integrin receptors belong to the group of transmembrane proteins and are involved in the adhesion of cells to one another and to the ECM thereby functioning as integrators between the extra- and intracellular spaces (Yamada et al., 2003). These receptors are heterodimeric and consist of noncovalently bonded polypeptide chains (α- and β-chains) that determine the specificity of binding to the ligands, in particular to ECM proteins (Hynes, 1992). At present, the existence of eight β-chains and 18 α-subunits has been established, resulting in a multitude of functional

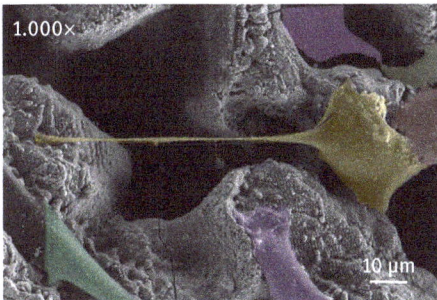

Figure 4.4: Cells are able to span horizontally the pores of a laser-structured titanium surface with their actin-based filopodia. SEM image Rostock University Medical Center: FE-SEM Merlin, Carl Zeiss, false colored. Cells: MG-63 osteoblasts. Dept. of Cell Biology. Courtesy Susanne Stählke. Material: Laser structuring provided by Schweißtechnische Lehr- und Versuchsanstalt Mecklenburg-Vorpommern GmbH, SLV, Rostock, Germany. Courtesy Rigo Peters.

heterodimers (about 24) (Calderwood, 2004). These are specific for binding to the matrix proteins such as, for instance, fibronectin as the prime target of $\alpha 5\beta 1$ (Hynes, 1999), collagen as the target of $\alpha 2\beta 1$ integrin or vitronectin as the binding partner of $\alpha v\beta 3$ (Ekblom and Timpl, 1996). However, since most integrins bind with more than one matrix protein, their binding specificity is reduced (Fässler et al., 1996). The connection of an integrin receptor to its extracellular ligands (e.g., fibronectin) leads to the accumulation and interaction of a hierarchy of cytoskeletal proteins and signal molecules in the local transmembrane complexes (Yamada and Geiger, 1997; Geiger et al., 2001) (Figure 4.5).

Since the cytoplasmic domains of the integrins are connected with the cell's actin cytoskeleton via submembrane adaptor proteins in focal adhesions, biochemical signals can be activated (Brakebusch and Fässler, 2003; Bershadsky et al., 2003; Giancotti and Ruoslahti, 1999) and, above all, mechanical forces can be transmitted into the interior of the cells (Ingber, 2003a). The adaptor proteins adapt either directly with the integrin's cytoplasmic domains such as talin and α-actinin (Cram and Schwarzbauer, 2004) or they are spatially associated, as in the case of vinculin, paxillin, and tensin (Dedhar and Hannigan, 1996).

Figure 4.6 illustrates schematically the cell–biomaterial interaction at the interface of a biomaterial, the material factors which directly regulate the cell physiology and the components required for the cell adhesion and signaling.

Integrin receptors are sensitive to the physicochemical surface characteristics of biomaterials such as titanium, and are therewith suitable cellular structures to regulate the cell response to the biomaterial and to detect how a material's surface affects primary adhesion mechanisms (Garcia, 2005; Olivares-Navarrete et al., 2015; Lüthen et. al, 2005; Lopes et al, 2019). For instance, Nebe et al. (2004) have shown that changing the structural organization as a result of a sharp-edged topography of a metal is associated with losses of various functions; in osteoblasts this is

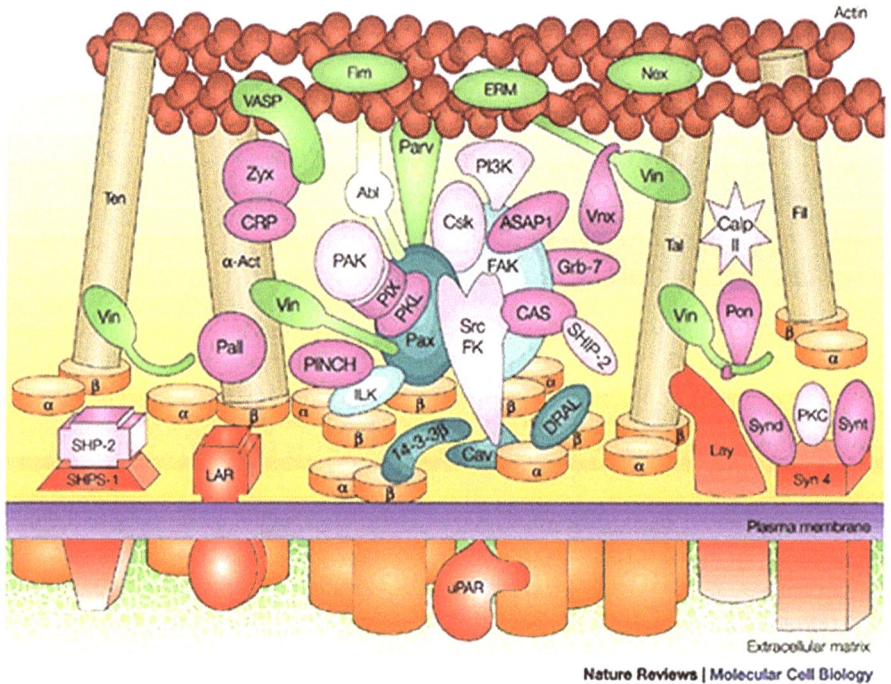

Nature Reviews | Molecular Cell Biology

Figure 4.5: Schematic depiction of the huge complexity of the main molecular domains of cell–matrix adhesions. The primary adhesion receptors are α- and β-integrins, shown as brown columns penetrating the plasma membrane. Proteins that interact with both integrin and actin include tensin (Ten), α-actinin (α-Act), talin (Tal) and filamin (Fil), shown as long buff columns. Integrin-associated molecules include: focal adhesion kinase (FAK), paxillin (Pax), integrin-linked kinase (ILK), 14-3-3β and caveolin (Cav). Actin-associated proteins include vasodilator-stimulated phosphoprotein (VASP), fimbrin (Fim), ezrin–radixin–moesin proteins (ERM), Abl kinase, nexillin (Nex), parvin/actopaxin (Parv) and vinculin (Vin). Several enzymes play a role in this complex, such as SH2-containing phosphatase-2 (SHP-2), SH2-containing inositol 5-phosphotase-2 (SHIP-2), p21-activated kinase (PAK), phosphatidyl inositol 3-kinase (PI3 K), Src-family kinases (Src FK), carboxy-terminal src kinase (Csk), the protease calpain II (Calp II) and protein kinase C (PKC). Further abbreviations: syndecan-4 (Syn4), layilin (Lay), phosphatase leukocyte common antigen-related receptor (LAR), SHP-2 substrate-1 (SHPS-1), urokinase plasminogen activator receptor (uPAR), rhabdomyosarcoma LIM-protein (DRAL), zyxin (Zyx), cysteine-rich protein (CRP), palladin (Pall), PINCH, paxillin kinase linker (PKL), PAK-interacting exchange factor (PIX), vinexin (Vnx), ponsin (Pon), Grb-7, ASAP1, syntenin (Synt), and syndesmos (Synd). Adopted from Geiger et al., 2001. © With permission by Springer Nature.

recognizable by curtailing of mineralization. In addition, cells attempt to phagocytize sharp-edged microstructures beneath the cells, which is accompanied by a stress response, for example, the increase of intracellular reactive oxygen species (ROS) (Moerke et al, 2016).

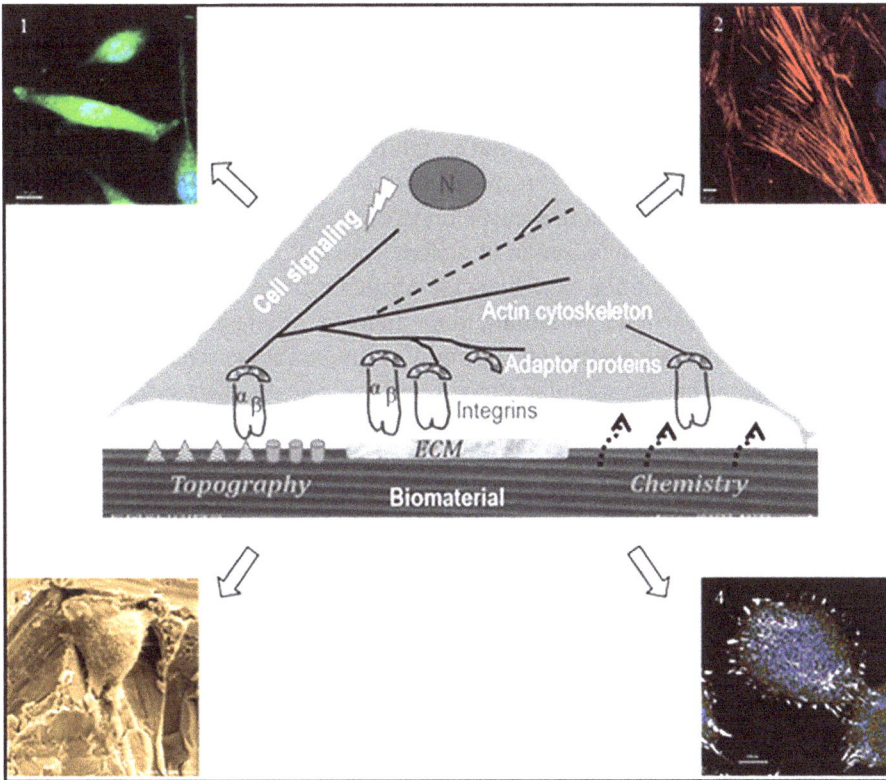

Figure 4.6: Schematic depiction of the cell–biomaterial interaction. The physicochemical characteristics of the material surface such as topography, chemistry, and ECM molecules influence decisively the cell adhesion processes and overall cell physiology. The adhesion of a cell to the biomaterials surface is generally mediated via transmembrane adhesion receptors (integrins) that are associated with the actin cytoskeleton through adaptor proteins (such as paxillin and vinculin) for subsequent transmission of intracellular signals to the nucleus. SEM and LSM images: Dept. of Cell Biology, Rostock University Medical Center: 1. Cytoplasmic Ca^{2+} ion increase in fluo-3-stained MG-63 cells on plasma polymer layer, scale bar 20 µm, LSM 780, Carl Zeiss. Courtesy Martina Grüning. 2. Actin cytoskeleton of human primary osteoblasts, scale bar 10 µm; LSM 780, Carl Zeiss. Courtesy Susanne Stählke. 3. Osteoblast adhering to corundum-blasted titanium, scale bar 8 µm, SEM DSM 960A, Carl Zeiss, BN. 4. Paxillin contacts in MG-63 cells on collagen I-coated Ti, scale bar 10 µm, LSM 410, Carl Zeiss, BN. Abbreviations: ECM, extracellular matrix; N, nucleus.

So far, the fundamental insight into the composition of adhesion complexes in which the integrin receptors are also accumulated is based on investigation conducted *in vitro* on cells in 2D-cultures. In contrast, when cells are cultivated on 3D matrices that more closely resemble the conditions in the tissues of an organism, it was proven that adaptor proteins interact with the integrins in the fibrillar adhesion complexes. An example is paxillin that otherwise can only be found in focal adhesion contact sites (Cukierman et al., 2001). In this context, the research group led by Marcus Textor discovered that dimensionality controls cytoskeleton assembly and cell metabolism and, in particular, that the reduction of actin filament assembly due to limited adhesive surface area in 2D can be overcome by "going 3D" (Ochsner et al., 2010). This means that in future *in vitro* studies, culturing conditions should be adopted that correspond more closely to *in situ* 3D tissue conditions. This could enable a more realistic understanding of the effect of stimulating agents used for biofunctionalization of materials on cell proliferation (e.g., growth factors) or cell differentiation (e.g., bone morphogenic proteins, TGF-β).

An important discovery pertaining to the function of integrins is that these receptors are not only responsible for the physical support of the cells of a tissue, but also for transmitting extracellular signals into the cell via their cytoplasmic domains, therefore effectively becoming signal receptors (Geiger et al., 2001).

By organizing the cytoskeletal "scaffold," that is, the framework for the signal components, an intracellular signal cascade is induced (Aplin et al., 1999) that triggers the mobilization of calcium ions (Sjaastad and Nelson, 1997). Calcium ions, as "second messengers," control cellular functions (Berridge et al., 2000). The calcium mobilization mediated by integrins is significant for the growth and differentiation of the cells, as well as for gene expression (Zhao et al., 1998).

In conclusion, the signals transmitted via integrins act to regulate cell functions such as migration, growth and differentiation, and are ultimately crucial for the fate of the cell – survival or apoptotic cell death.

4.2.2 Cell adhesion mechanisms

The cells are connected to the ECM via their integrin receptors through which cell growth and function are controlled. The ECM is a complex network of macromolecules such as proteins (e.g., collagens, fibronectin, laminin) and polysaccharides (glycosaminoglycans, GAGs) excreted by the cells and forming the intercellular substance (Figure 4.7). In tissues, the ECM serves as a structural element and influences the development and physiology of the cells (Alberts et al., 2017). The fibrous proteins in particular play a role in binding the integrin receptors of the cells to the matrix. These fibrous proteins are present in all multicellular organisms and make up about 25% of the protein in the organism. Determined by their function, fibrous proteins act as structural proteins (e.g., collagen, elastin) or adhesion proteins (e.g., fibronectin, laminin). The extracellular organization of the collagen in the fibrillar matrix appears to function under cellular control via the matrix protein fibronectin. It is therewith apparent that the cell physiology is controlled in a tight dovetailing of the cellular adhesion components.

As basic components of the ECM, the GAGs, in particular hyaluronic acid (HA; see Chapter 2.3), have become the focuses of scientific attention. Recent findings show that these macromolecules mediate initial mechanisms of cell adhesion (Zimmermann

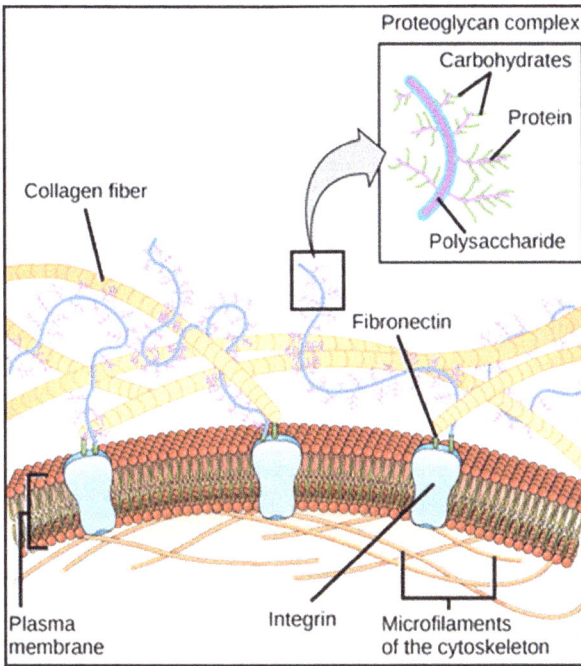

Figure 4.7: Schema of the interplay between the extracellular matrix (ECM), the transmembrane adhesion receptors and the intracellular adhesion structures building a "bridge" from outside the cell to its interior for cell signaling and the transmission of mechanical forces. The ECM molecules, proteins (e.g., collagen, fibronectin) and polysaccharides (proteoglycans) provide the cells with ligands for stable cell-matrix binding via integrins. (Molnar and Gair, 2015). © Creative Commons Attribution 4.0 International.

et al., 2002). Hyaluronic acid is a nonsulfated GAG consisting of D-glucuronic acid and *N*-acetylglucosamine that have a glycosidic link to form repetitive, repeating disaccharide units beta 1–4 (Alberts et al., 2017). These long, linear polysaccharides are negatively charged and form hydrated gels owing to their high capacity to bind water that provides the tissue with resistance to compressive forces. HA generally influences cell growth and migration, and serves as an important "filler" in the development of the embryo, that is, it is initially secreted by the cells in great amounts into the cell-free spaces into which new cells can then migrate.

The cells emit hyaluronic acid. Whereas around chondrocytes a hydrated, nearly 4.4 μm thick pericellular coating was observed, around epithelial cells this coating was only 2.2 μm thick. This membrane-bound hyaluronan mediates the initial adhesive interactions between cells (e.g., chondrocytes, epithelial cells, and osteoblasts) and their external surfaces (Cohen et al., 2006). Hyaluronan is negatively charged and therefore, the cells express a surface potential of –15.6 mV as determined by Zeta-potential measurements of suspended human osteoblastic cells (Rebl et al., 2016).

An enzymatic splitting of the hyaluronic acid by hyaluronidase shows clearly that the initial adhesion of the osteoblasts on, for example, titanium surfaces can be inhibited (Nebe and Lüthen, 2008). In addition, the time-dependent spreading of cells is suppressed when the cells lack their pericellular coat consisting of hyaluronic acid. This is accompanied by decelerated formation of the actin cytoskeleton in the cell (Finke et al., 2007).

In conclusion, besides the integrin–matrix interplay on the boundary layer of the biomaterial, hyaluronic acid secreted by cells is a further cellular adhesion agent.

Presently, in the quest to develop optimum implant surfaces, there is an international "race" among the proponents of the strategy of binding peptides to the surface as an ECM equivalent for the adhesion receptors, and those who advocate simply "helping" the cell adhesion with functionally active chemical groups (e.g., amino groups) on the material surface to provide positive charges for the initial hyaluronan-mediated cell adhesion. For the development of biomaterials, these findings harbor a highly explosive perspective, as elaborate immobilization of peptides and proteins on the implant's surface to promote integrin-mediated cell adhesion would not be necessary; instead, chemical functionalization of the surface to promote preadhesion mechanisms could prove to be sufficient. In addition, it appears that the hydrophilicity of a material surface (such as PCO®) alone does not facilitate fast cell adhesion and spreading processes, but instead, the positive surface charges mediated, for example, by plasma polymer nano-coatings (Rebl et al., 2016; Nebe et al, 2007; Nebe et al., 2019).

4.2.3 Cell-matrix binding and its influence on apoptosis

The interaction of cells with the ECM is essential for cell growth and the cell phenotype, and is generally decisive for the fate of cells by preventing apoptosis, the programmed cell death (Bissell and Nelson, 1999). This phenomenon is known in the literature as "anchorage dependence" (Boudreau et al., 1995), and is tantamount to the dependence of cells on their adherence to a set cell cluster.

When adhesion to the ECM is lost, or generally, when the cells are prevented from attaining adhesion contact as on hydrophobic material surfaces (see Figure 4.8), the cells undergo apoptosis, that is, they die off as a result of "homelessness" (Greek ανοικις, anoikis) (Frisch and Screaton, 2001).

Apoptosis can be ascertained by using several cellular examination techniques including flow cytometry (see box) that reveals the sub-G1 peak indicative of apoptotic cell behavior. This peak on the DNA histogram indicates the presence of apoptotic cells. The DNA strand is shortened due to cleaving off the oligonucleotides during the last phase of the apoptotic process. This causes less amount of propidium iodide dye to collect intercalatively in the DNA helix of the nucleus and, thus, the fluorescence intensity is reduced compared to vital cells.

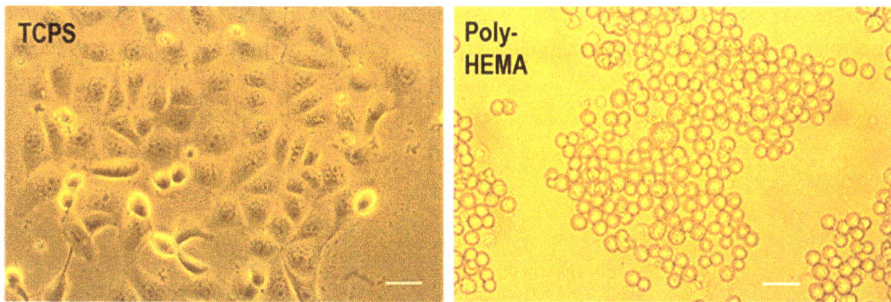

Figure 4.8: Epithelial cells (murine liver, mHepR1; Henning et al., 1994) normally grow in cell islands and are closely connected to each other via tight junctions in cell-cell contacts (left, on tissue culture polystyrene, TCPS). On hydrophobic surfaces (right), on poly-HEMA: poly(2-hydroxyethyl methacrylate), the cells lose their adhesion, become round, and undergo apoptosis (light microscopy, scale bars 50 µm).

Flow cytometry

Flow cytometry uses the Coulter principle whereby cells pass single file through the center of a laser beam with the intent to measure their optical properties that eventually allows cell sorting and quantification in near-real time. Precise positioning of the cells can be achieved by hydrodynamic or acoustic-assisted hydrodynamic focusing. Flow cytometry results in a high-throughput, automated quantification of selected optical parameters on a cell-by-cell basis. Light signals scattered off the individual cells pass through filters and are recorded by a set of photomultiplier tubes and analyzed, as shown in the figure (https://en.wikipedia.org/wiki/Flow_cytometry).

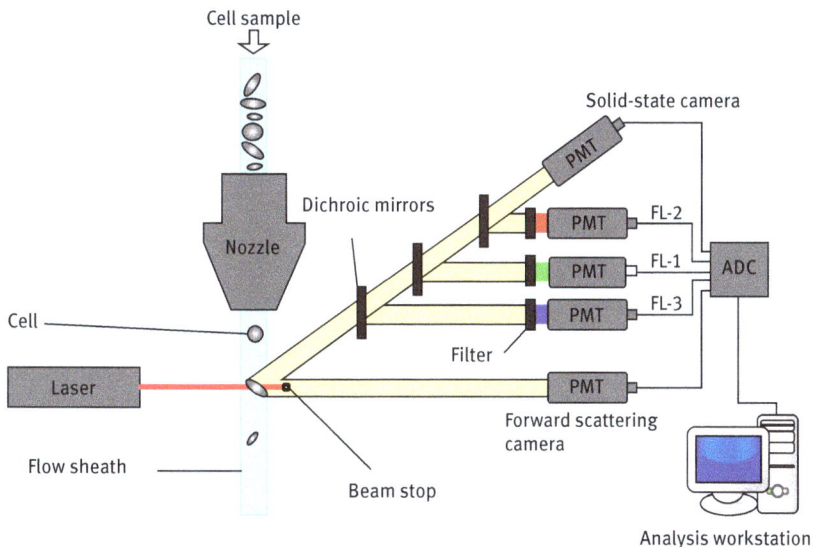

Changes in cell morphology during apoptosis can also be traced by transmission electron microscopy or confocal laser scanning microscopy suitable to verify apoptotically effective proteins such as the proapoptotic Bax or caspase, effective in both early and late stages. Knowledge on apoptotic processes at the boundary layer of a biomaterial is valuable information when assessing the biocompatibility of a potential biomaterial.

4.2.4 Implications

Understanding the complexity of cellular reactions is among the great challenges facing future research, for "The whole is truly greater than the sum of its parts" (Ingber, 2003b). Breaking down the molecular components of cell adhesion and the single proteins involved in linear signaling paths does not necessarily lead to knowledge of the overall cell responses to chemical, physical, or mechanical signals. Superordinate structures are, presumably, also crucial to the fate of cells and ultimately, the tissue. Information on this biocomplexity would be of clinical relevance to develop optimum biomaterial designs and in the area of tissue engineering, respectively.

Prospectively, systems biology will be able to answer questions on the superordinate regulatory mechanisms of a cell (Ideker, 2004). This field of research that brings together concepts from biology, information technology and systems science, is increasingly dedicated to mathematical modeling of intracellular signaling mechanisms and their interactions so that conclusions can be drawn about the functioning of the single components in a cell (Ideker and Lauffenburger, 2003). The vision here is to create a comprehensive simulation of the cell behavior *in silico*, that is, by computer simulation, and also the behavior of larger biological systems such as tissues and organisms (Palsson, 2000; Tomita, 2001). Applied in practice, these simulations are destined to reduce the number of cell and animal experiments in the process to develop novel pharmaceuticals or biomaterials (Kohn et al., 2007) and thus, will lead to time and cost savings, and avoidance of ethical issues connected with animal experimentation.

4.3 *In vivo* evaluation of inflammatory and immunological reactions (Michael Schlosser)

4.3.1 Medical implants and biocompatibility

This century and especially the last 50 years of the twentieth century have seen an explosion in the number and varieties of implantable medical devices used for treatment of various medical problems. These devices range from sutures at the one end of the spectrum to complicated electronic devices at the other, and include, for example, intraocular and contact lenses, vascular grafts and cardiovascular stents, heart valves and pace makers, catheters, hip and knee prostheses, orthopedic implants for bone reconstruction, breast and dental implants, dialysis membranes, implantable biosensors, and others (see also Chapter 1.3). These materials include synthetic and natural polymers/copolymers, several metals such as stainless steel or titanium and its alloys, ceramics, and combinations of these materials (see Chapter 2). The main characteristic of biomaterials is that they are used in contact with living tissue, resulting in an interface between living and nonliving substances (Szycher, 1992). Hence, it is important that they do not cause direct injury to living tissues, any adverse immunologic response, adverse systemic effects, or delayed adverse effects (Wallin, 1995a).

Although medical science has a long experience with synthetic biomaterials, very few have proved to be both safe and effective over a long time period. Materials that initially appeared to be appropriate and successful according to their physico-chemical characteristics or their excellent performance *in vitro* were later found to cause adverse effects *in vivo* after a long implantation time. Thus, withdrawal of specific silicones, polyurethanes, polyesters, fluorinated hydrocarbon polymers, and other polymers around the end of the twentieth century reflects the critical situation in the field of biomaterials at that time (Gould et al., 1993). In this context, it should be underlined that no biomaterial is currently available that is completely inert when implanted in the body (Williams, 1994a). So it appears that indeed the clinical application of several biomaterials has exceeded in quality and quantity our knowledge about the complex interaction between these biomaterials and the biological system.

However, the characteristics of biomaterials science have significantly changed during the last 50–70 years from a materials-dominated science to a multidisciplinary science. More than any other field of technology, biomaterials science connects researchers from diverse backgrounds to identify the need for a given biomaterial or device, to its manufacture process, biocompatibility testing, sale, and finally its clinical application. Current subjects which are integral to biomaterials science are studies on toxicology and especially biocompatibility issues which have to be more comprehensive than in the past, assuming that a material must be biocompatible as the recipient (patients or experimental animal) is still alive following implantation. Moreover, biomaterials science includes studies on healing, functional tissue structure and pathobiology, the distinctive characteristics of specific implantation sites, investigations of mechanical, physicochemical and performance requirements, and industrial involvement.

Furthermore, a wide range of ethical considerations as well as regulatory concerns impact biomaterials science (see Chapter 5). For example, the U.S. government has developed a complex regulatory system administered by the FDA. Most developed countries worldwide have similar medical device regulatory bodies and the International Standards Organization (ISO) has introduced international standards for the world community (see Appendix A). Finally, the continuing increase in multidisciplinarity of biomaterials research is amply illustrated by the number of relevant publications published in recent years, with only about 100 per year in the mid-1970s up to more than 7,500 in 2012.

The biocompatibility which currently concerns the complex interactions between a biomaterial and its host/recipient was traditionally only equated with a lack of toxicity and other parameters resulting from several *in vitro* tests. However, the difficulty with this concept has led to a definition termed as "the ability of a material to perform with an appropriate host response in a specific application" (Williams, 1989). Therefore, biocompatibility is concerned with all aspects of material-tissue reactions of the body but concentrates on the appropriate host response. It includes the following: (i) the energetics and kinetics of all physical, chemical, and biochemical processes at the interface between the biomaterial and the biological system; (ii) the direct reaction of the biological system induced by processes at the interface; (iii) all changes of physical and chemical properties of the material such as surface composition, corrosion, biodegradation, surface free energy, and other; and finally (iv) the sum of changes of the biological system beyond the interface, triggered or induced by the interaction (toxic or immune reactions, cancer, etc.).

To investigate the biocompatibility of a given implant/biomaterial, it is mandatory to note that biocompatibility refers to a specific situation as the performance or biocompatibility characteristics of a biomaterial or a medical device in one situation may be quite different to those in another. Hence, biocompatibility describes a phenomenon or various characteristics and it is not justified to use the term "biocompatible" only to describe a property of a material/device (Williams, 2014). In fact, it is necessary to define biocompatibility specifically in relation to those functions, but also to have a more profound overarching concept. Thus, for long-term implantable medical devices the following definition has been proposed: "The biocompatibility of a long term implantable medical device refers to the ability of the device to perform its intended function, with the desired degree of incorporation in the host, without eliciting any undesirable local or systemic effects in that host" (Williams, 2008). In contrast, in regenerative medicine using tissue-engineering scaffolds as examples, inert materials are not useful but more importantly, for biomaterials that have nonspecific or inappropriately directed activity the following definition is proposed: "The biocompatibility of a scaffold or matrix for a tissue engineering product refers to the ability to perform as a substrate that will support the appropriate cellular activity, including the facilitation of molecular and mechanical signaling systems, in order to optimize tissue regeneration, without eliciting any undesirable local or systemic responses in the eventual host" (Williams, 2008). Therefore, taking into account both aspects with long-term implantable biomaterials and biomaterials used in regenerative medicine, the current definition of biocompatibility has been proposed and redefined as follows: "Biocompatibility refers to the ability of a biomaterial to perform its desired function with respect to a medical therapy, without eliciting any undesirable local or systemic effects in the recipient or beneficiary of that therapy, but generating the most appropriate beneficial cellular or tissue response in that specific situation, and optimising the clinically relevant performance of that therapy" (Williams, 2008). However, in all these definitions, the two main variables significantly influencing the biocompatibility of an implantable biomaterial are the responses of the material and those of the recipient/host.

From the materials site, any heterogeneity of chemical nature in relation to bulk and surface chemistry and structure, electrical properties and resistance to corrosion and/or biodegradation, is an important factor which has to be taken into account, because the critical events in interfacial reactions may be particularly controlled at a micro-structural level by minor constituents of the biomaterial/medical device. In addition, the surface topography, the implant size and shape are important. Thus, significant differences in both material stability and cellular responses could be demonstrated between structures with ultrasmooth surfaces and porous structures. Depending on implantation site and duration, surface roughness will control many of the parameters of biocompatibility (Inoue et al., 1987). Also, the size of the implant influences the biological response in that larger devices cause a greater tissue injury which can lead to a pronounced inflammatory response and to formation of a thicker fibrous capsule around the implanted biomaterial/medical device. Otherwise, materials of small size such as micro- or nanoparticles may induce a phagocytic response, stimulating and activating macrophages and FBGCs. Regarding the shape of an implant, it has been shown that the local tissue response was pronounced around the corners, for example, shape contours of a device, suggesting that a certain degree of the tissue reaction around implants is presumably due to mechanical trauma produced by the presence of an implant (Wood et al., 1970). Thus, in recent investigations, material responses have been considered to be most relevant for biocompatibility. However, it is now becoming clear that the complexity of host reactions is of increasing importance and the main focus of current biomaterials research.

Among the host/recipient response, inflammation appears to play a central role (Williams, 1989; Anderson, 1993; Tang and Eaton, 1995). Many mediators of inflammation, each of them influenced by additional physiological and pharmaceutical variables, are acting within the local host response which represents a balance between inflammatory and repair processes. In addition, these processes are significantly influenced by the age, gender, health, immunological, and pharmaceutical status of the host/recipient, furthermore the implantation site and, of course, the individuality of the organism and the individuality of the response. For example, immunization experiments with defined immunogens performed in inbred animals have shown a broad variability in specific antibody responses including high-, low-, and nonresponders. However, other experimental animals and especially patients are very heterogeneous regarding their genetic background, which further makes the situation much more complicated. Moreover, when trying to assess biocompatibility characteristics of a given biomaterial/medical device from several animal experiments into the daily clinical routine, there will be inevitable species differences.

Inflammation, wound healing and the foreign body response are generally considered parts of the tissue or cellular host responses to the injury. From the view of implanted biomaterials, the placement in the *in vivo* environment usually involves injections, insertions, or surgical, orthopedic, or dental implantations. All these procedures cause an injury to the involved tissues or organs as well as blood–material contact. After implantation, complex mechanisms are activated resulting in wound healing as a response to the injury. The nature and degree of these mechanisms may determine and/or affect the biocompatibility and functional biostability of the implant. However, the complexity of the inflammatory response is difficult to understand in the classical view of the "nontoxic," "inert," and "nonimmunogenic" nature of biomaterials. Therefore, it is assumed that the surface of the implanted material alone may not be the proximate cause of inflammation.

Although it was convenient in the past to consider blood–material interactions separately from tissue-material interactions, one should realize that the involved cells and mechanisms occur closely related in both, blood and tissue (Cotran et al., 1989; Anderson, 1993). Regardless of the tissue or organ into which a biomaterial/ medical device is implanted, the initial inflammatory response is activated by the injury to vascularized connective tissue, and therefore blood and its components are primarily involved.

4.3.2 Interactions between implanted biomaterials and the host's inflammatory response

4.3.2.1 The initial event: blood–implant interaction

The aqueous soluble components in blood plasma can be classified into low molecular solutes (ca. 1%) such as ions, gases, vitamins, fatty acids, hormones, organic substances, and into the high molecular composites (ca. 7–8%) including enzymes and several other proteins. Most biological fluids have a near-neutral pH, but the concentration of low molecular components can vary up to 20%, resulting in altered ionic strength and therefore biological variability of buffer capacity. Furthermore, the surface deposition of low molecular solutes on metallic implants may also promote the processes of degradation/corrosion (Williams, 1994a).

The main components of solutes present in biological fluids are a variety of macromolecular proteins which occur in concentrations of ca. 80 g/L in plasma. Among the plasma proteins, albumin represents the main component (ca. 60%) beside the heterogeneous fractions of α1-, α2-, β-, and γ-globins. The adsorption of proteins onto implantable devices – and this is in fact the initial step of interaction – differs from these of low molecular weight components mainly that proteins are large and chemically heterogeneous: they contain hydrophobic, electrically positively and negatively charged, and polar regions. In addition, implant surfaces may also be heterogeneous and contain a similar variety of surface chemical domains so that a number of different protein–surface interactions are possible. In a very short time of less than one second proteins are already present on biomaterial surfaces. In seconds to minutes, a protein layer adsorbs to most implant surfaces, followed by a competitive exchange between different proteins, in particular albumin, immunoglobulins and fibrinogen which were found to be adsorbed at high levels on the biomaterial surface (Anderson et al., 1990; Tang et al., 1993a, b). Furthermore, high molecular weight kininogen (HMWK), Hageman factor (FXII), vWF (FVIII), fibronectin, complement components, and others have been detected and quantified on biomaterials (Ziats et al., 1990). As these processes occur well before any cell arrives, cells primarily encounter and recognize the protein layer rather than the real biomaterial surface. Since cells respond specifically to proteins, this protein film may be the event that controls subsequent bioreactions to implants. However, the adsorbed protein film might also be of concern for biologically active devices such as biosensors, but also for immunoassays, array diagnostics, marine fouling, and several other phenomena.

The relatively large size of proteins allows that adsorption of a single molecule may involve several types of binding sites resulting in a different strength of binding of high and low affinity proteins to different biomaterials. Many studies investigating the protein-biomaterial interface in blood or plasma *in vitro*, ex vivo, and *in vivo* have demonstrated that adsorbed proteins may be desorbed and exchanged over time with others and may be reoriented or denatured (Vroman, 1988; Anderson et al., 1990). Despite many unanswered questions, it is generally accepted that there is no "steady state" in protein adsorption after a biomaterial has been implanted. Since native soluble proteins have a well-defined but not symmetric conformation but are frequently of chiral nature, they can exist after adsorption in distinct orientations/conformations. These conformational changes have exemplarily been shown for two common human plasma proteins, albumin and fibrinogen, after adsorption to both hydrophobic and cross-linked hydrophilic polyurethanes (Barbucci and Magnani, 1994). Thus, it has to be taken into account that even "self" plasma proteins may represent "nonself" conformational epitopes by changing their conformation and, if adsorbed to implant surfaces, may induce and promote inflammatory/immune response of the host much more aggressively.

4.3.2.2 The central event: The inflammatory response to biomaterials

Following initial protein adsorption onto implanted biomaterials, cellular interactions occur by adhesion (see Chapter 4.2) and activation of cells as main feature of the inflammatory response. Inflammation is generally defined as the reaction of vascularized living tissue to local injury, aimed to destroying or diluting the injurious agent/implant and repairing the injured tissue. The degree of the host's reaction depends on factors such as the size, shape, and the physicochemical properties of the

implant and therefore the dimensions of the injury, all of which determine the duration of inflammatory and wound healing processes (Williams, 1989). Furthermore,
bacterial infection can amplify, prolong, or otherwise adversely affect these inflammatory processes and therefore appropriate and effective sterilization methods are
important.

4.3.2.3 The common wound healing process

Following damage of tissue and blood vessels by injury, well-defined mechanisms
characterized by two distinct and frequently overlapping phases aimed to wound
healing are initiated. In the context of this process, the acute inflammatory phase is
the initial reaction which involves changes of the micro-vascularity and cellular composition of the tissue. Subsequently, the tissue attempts to restore and repair the
structural and functional damage caused by the injury.

   Immediately after injury, the blood vessels dilate and changes in vascular flow,
caliber, and permeability occur allowing the diffusion of neutrophil granulocytes as
well as an extracellular exudate containing plasma proteins and other biochemical
mediators. The immediate response is bleeding, followed by thrombus formation involving the activation of the extrinsic and intrinsic coagulation systems, the complement system, the fibrinolytic system, the generation of kinine, and platelets.
Furthermore, phagocytic cells attend to the injury to remove cellular and other debris. Concomitantly with these reactions, the repair process is initiated by the vascular regeneration. During this response, new capillaries grow into the wound area
and fibroblasts generate collagen, the essential structural protein of connective tissue. In minor injuries, the formation of a layer of fibrous connective scar tissue may
differ only slightly from the original tissue. In cases of severe injury, inflammation
is more extensive, resulting in both a delayed repair phase and an increased formation of reparative tissue.

   If the injury is followed by implantation of a so called "inert" biomaterial, the initial event of the interaction between blood and the material is the adsorption of plasma
proteins. The first, acute inflammatory phase will be comparable to those of an injury
including dilated vessels, coagulation cascade and complement factors, occurrence of
neutrophil granulocytes, macrophages, and fibroblasts. However, regenerated blood
vessels and morphology of generated collagen are disrupted by the implant. Its presence will probably inhibit and therefore prolong the inflammatory and the repair process and the cellular infiltration will persist for a longer time compared to a normal
incision case. However, this description is too simple and ideal encapsulation is rarely
seen since absolute inertness of implanted biomaterials does not exist (Williams, 1989;
Black, 1995). Furthermore, implants which are ignored by the body might be not really
incorporated with less functional biostability (Williams, 1989, 1994b).

4.3.2.4 The inflammatory response following implantation of biomaterials and the central role of phagocytic cells

The predominant cell type in the different phases of inflammation varies following the injury. Whereas neutrophil granulocytes with a half-life of 6–9 h disappear within 1–2 days, monocytes which further differentiate into macrophages are very long-lived and their migration into the implant site may continue for days to weeks depending on the implanted biomaterial (Ward, 1994). The inflammatory response involves a series of complex reactions whereby different proteins are either adsorbed on biomaterial surface or be present in body fluids and different cells such as neutrophil granulocytes, monocytes/macrophages, platelets, fibroblasts, B and T lymphocytes, FBGCs, epithelial cells, and other. The sequence of local events following implantation of biomaterial may be subdivided into acute inflammation, chronic inflammation, FBR, and granulation of tissue and fibrosis (Anderson, 1993).

The first phase of acute inflammation following implantation of a biomaterial is of relatively short duration depending on the extent of injury, commonly from minutes to few days. The main characteristics of this early phase are fluid and plasma proteins which immediately adsorb onto the biomaterial surface or covalent-bonded complement proteins (e.g., C3b and iC3b) and further the migration and accumulation of leukocytes, consisting mainly of neutrophil granulocytes and monocytes (Colton, 1994; Ward, 1994). Furthermore, it has been shown that beside neutrophil granulocytes monocytes and macrophages are also frequently detectable in the surrounding tissue of implanted biomaterials during the acute inflammatory phase. In addition, both B and T lymphocytes have been found at increased levels at the implantation site (Marchant et al., 1983; Gong et al., 1993; Holgers et al., 1995), clearly indicating an involvement of immunological mechanisms.

Nearly all biomaterial surfaces can activate complement proteins to some degree via the alternative pathway by biomaterial-bound C3b (Chenoweth et al., 1988; Brett, 1992). Platelets and immune cells can adhere to proteins on the surface. For example, phagocytic cells, red cells, B cells and some T cells have CR1 receptors that bind to C3b, whereas particularly natural killer (NK) cells, monocytes and polymorphnuclear neutrophils with their CR3 receptors bind to iC3b which is produced due to cleavage of complement factor I and therefore acts as an *opsonin*. The complement fragment generation causes a C5a-induced increase in expression of adhesion receptors by phagocytes, also activating other cells including eosinophil granulocytes and basophil granulocytes and mast cells, and thus, complete the complement cascade by producing the final membrane attack complex leading to cell lysis and death and potentially to pathophysiological consequences. In addition to the alternative pathway, the classical pathway of complement activation is based on the reaction of circulating C1 with immunoglobulins which were initially adsorbed on biomaterial surfaces. The C3b can either bind to its receptor on different cells (granulocytes, monocytes, B lymphocytes) or adsorb to the biomaterial surface to complete the complement cascade as shown for the alternative pathway.

Besides the complement system, pro-inflammatory cytokines that may be relevant in inflammatory response to biomaterials include IL-1 and tumor necrosis factor (TNF) which have received the most attention because they are released by immune and inflammatory effector cells (Dinarello, 1990; Anderson, 1993). A variety of stimuli can activate the monocytes by inducing gene expression for one or more cytokines. These stimuli include the adhesion to foreign surfaces, a nonphysiological chemical environment caused by biomaterial leachables, complement fragments, or the secretion of cytokines by other cells during the immune response as has been shown, for example, for IL-1 synthesis due to sequential treatment with interferon followed by IL-2 (Ward, 1994).

Thus, measuring cytokines in body fluids such as serum might be a feasible approach to examine the inflammatory and immunological response of the body over an extended time period due to the diverse and distinct effects of cytokines within the immune system. For example, the T cell-derived pro-inflammatory cytokine IL-2 is essential for regulation of T cell function and proliferation especially for antigen-activated $CD8^+$ T lymphocytes (Smith, 2001). It also augments the production of other cytokines including $IFN\gamma$ typically released by various T cell populations. Furthermore, IL-2 is important for growth and differentiation of NK cells and has synergistic effects with a number of cytokines including IFN-y and IL-12 (Smith, 2001).

IFN-y, another important pro-inflammatory cytokine and mainly produced by NK cells as well as $CD4^+$ and $CD8^+$ T cells, is essential for regulation of antigen presentation to $CD4^+$ T cells and for stimulation of NK cells and macrophages, converting them from a resting to an activated state (Billiau and Vandenbroeck, 2001). In recent models of different macrophage activation pathways, $IFN\gamma$ is the key regulator of the classic pathway leading to the pro-inflammatory M1 phenotype (Martinez et al., 2009). Furthermore, IFN-y is a central link between the innate immune system and the T_H1-mediated cellular part of the adaptive immune system and therefore represents one of the most important immune regulators (Billiau and Matthys, 2009).

IL-6, a pleiotropic cytokine, that is, a cytokine with multiple phenotypic expression but with mainly pro-inflammatory function, is predominantly produced by macrophages, fibroblasts, and endothelial cells but also by T and B lymphocytes, mast cells and some other cell types (Matsuda and Hirano, 2001). IL-4, another pleiotropic cytokine mainly secreted by mast cells, stimulates proliferation of B cells, T cells, and NK cells as well as the differentiation of $CD4^+$ T cells into T_H2 cells (Keegan, 2001). It has mostly anti-inflammatory effects including antagonizing the action of $IFN\gamma$ by enhancing the humoral immune response and switching macrophages to the anti-inflammatory M2 phenotype via an alternative activation pathway (Martinez et al., 2009). Comparable to the role of IFN-y, it links the innate immune response with T_H2-mediated humoral reactions of the adaptive immune system.

Among the T_H2-associated anti-inflammatory cytokines, IL-10 is secreted by alternatively activated monocytes and macrophages but also by T and B lymphocytes

(Mosser and Zhang, 2008). IL-10 production is enhanced by other cytokines such as IL-4, IL-12, IFNα, and IL-10 itself and inhibits the expression of pro-inflammatory cytokines by directing the immune response toward humoral functions and attenuates cellular immune reactions. Another anti-inflammatory cytokine, the IL-13 is mainly produced by T_H2 cells, mast cells and NK cells and has inhibitory effects on monocytes and macrophages by downregulating a number of pro-inflammatory cytokines (McKenzie and Matthews, 2001).

At this point it is of interest to consider which component might be the initiator of the inflammatory response to a biomaterial. As mentioned above, the initial event is the adsorption of albumin, immunoglobulins, and fibrinogen. Whereas albumin is unlikely to mediate inflammatory reactions because of passivation of surfaces, immobilized IgG together with activated complement components might be powerful mediators of phagocytic activation and the inflammatory response. However, as demonstrated by Tang et al. (1993a), the acute inflammatory response to the biomaterial polyethylene terephthalate, measured by phagocyte accumulation, is normal in both severely hypogammaglobulinemic mice and complement-depleted mice, suggesting that neither surface-bound IgG nor complement activation are required to initiate the acute inflammatory response. However, it has been shown that materials precoated with fibrinogen or plasma (but not with serum or albumin) elicit large numbers of phagocytes onto implants (Tang et al., 1993b). Most interestingly, mice lacking fibrinogen fail to show an inflammatory response to implants whereas plasma- or fibrinogen-coated materials elicit a large number of phagocytes in these animals. Therefore, phagocytes probably together with fibroblasts may initiate the long-term inflammatory and fibrotic response which is seen around implants (Tang and Eaton, 1995). Thus, fibrinogen adsorption per se or conformational changes of fibrinogen structure (Barbucci and Magnani, 1994) might be of great importance in the inflammatory response to implanted biomaterials.

Although most unaffected biomaterials are unlikely to be phagocytized by neutrophils or macrophages because of their size, certain events in phagocytosis after biodegradation may occur. *First*, neutrophils and macrophages become highly activated in their attempt to digest and to degrade the biomaterial. The release of large quantities of lysosomal hydrolytic enzymes by direct extrusion or exocytosis from the cells may lead to biodegradation of biomaterials (Henson et al., 1971; Weiss et al., 1989). Furthermore, activated macrophages can release large quantities of oxidants such as H_2O_2, O_2, and HO radicals which may affect implanted biomaterials such as polymers which are susceptible to autoxidation, the autocatalytic reaction with oxygen (Stokes, 1993). *Second*, biomaterials are not homogenous. For example, polymers contain functional groups, a certain range of molecular weight distribution, hard and soft segments as well as leachables including solvents and microsized particles that can be phagocytized (Hill, 1995; Ruyter, 1995). *Third*, the biological host contains a large amount of water in intracellular and extracellular fluids with an average of 57% of the body weight. Water is, of course, the primary agent for hydrolytic degradation and especially ester and amide links in biopolymers such as cellulosic esters, aliphatic esters, Nylons®, and others are very susceptible to hydrolytic degradation (Stokes, 1993). *Fourth*, mechanical deterioration of polymers such as ultrahigh molecular weight polyethylene (UHWPE) has been observed in total hip replacement systems because of wear and loading. It is suggested that the average wear rate for UHMWPE of 0.1 mm per year generates 20 million particles per day, corresponding to 7 billion particles per year (Tang and Eaton, 1995).

Following this persistent inflammatory stimulus, chronic inflammation will be initiated which is characterized by the presence of phagocytes, FBGCs fused from monocytes/macrophages in attempt to phagocytose the material, monocytes, and lymphocytes (Anderson, 1993; Tang and Eaton, 1995). In parallel, the granulation of tissue initiates the healing process by action of monocytes/macrophages followed by proliferation of fibroblasts, which synthesize collagen and proteoglycans to form a fibrous capsule around the implant, and endothelial cells, which are involved in vascularization at the implant site (Williams, 1989; Anderson, 1993).

In this process, phagocytic cells, present in all body regions, are the most important cell population because of their interaction with different immune cells as well as their wide spectrum of biological active products including chemotactic factors, neutral proteases, arachidonic acid metabolites, reactive oxygen metabolites, complement components, coagulation factors, growth-promoting factors, and important cytokines (Anderson et al., 1990). Growth factors in particular, such as platelet-derived growth factor, fibroblast growth factor, transforming growth factor, and epithelial growth factor, are released by activated cells, and can stimulate a variety of cells. They are involved in wound healing and are important for the growth of fibroblasts and blood vessels and the generation of epithelial cells (Anderson, 1993).

In addition to their phagocytic function as a nonspecific innate reaction during the chronic inflammatory response, monocytes/macrophages have other important functions in the specific adaptive immune response. They express the human leukocyte antigen class II molecules which are essential for the presentation of foreign antigens to lymphoid immunocompetent cells (Unanue et al., 1987). In the immune response, lymphocytes and plasma cells are the key mediators of cellular immunity and antibody production consisting of a number of specialized cells. Laboratory and clinical evidence have shown that components of biomaterials such as leachables released by extraction in the biological system or micro-particles following biodegradation can migrate and be carried by macrophages and lymphocytes to distant areas. Consequently, Teflon®, silicone, polyurethane, and other polymers have been found in regional lymph nodes and in other tissues and can be presented as antigens to immunocompetent cells (Brunstedt et al., 1990; Glaser, 1993; Kaiser, 2009).

Immunological reactions to a foreign body/biomedical implant are thought to result from the attempts of the biological host to distinguish between self and nonself. Many complicated mechanisms exist by which the biological system recognizes and resists a foreign material. However, these self-recognition mechanisms are imperfect and can sometimes break down, possibly initiated by denaturation and/ or conformational changes of adsorbed proteins on biomaterials which may also lead to an autoimmune reaction. There are numerous reports in the past suggesting significant immunologic and related responses to certain implanted polymers. These adverse reactions including scleroderma, connective tissue disease, human adjuvant disease, lymphadenopathy, systemic sclerosis, rheumatic disease, granulomas, neoplastic transformations, sarcomas, carcinomas, and others (Kircher, 1980; Kossovsky et al., 1987; Endo et al., 1987; Snow and Kossovsky, 1989; Varga et al., 1989; Thomsen et al., 1990; Nakamura et al., 1992; Rowley et al., 1994; Tsuchiya et al., 1995; Bar-Meir et al., 1995; Smith, 1995).

A major concern regarding these problems is the lack of an early marker indicating ongoing or escalating development of bioincompatibility or latent complications. In addition, it should be kept in mind that the individual immune response to a biomaterial shows a broad variability and depends on several distinct factors of the recipient/host. It this context, humoral components induced by specific immune reactions, that is, the detection of specific antibodies directed to biomaterial components, but also the time course of pro- and anti-inflammatory serum cytokines might be an early marker for the degree of the individual local chronic inflammatory response (Goldblum et al., 1992; Shirakawa et al., 1992; Ziegler et al., 1994; Schlosser et al., 1994a; Yang et al., 1994). In addition, the quantification of humoral serum markers against implanted biomaterials might be an important tool and new parameter for evaluation of their biocompatibility.

There are numerous unanswered questions regarding the complex process of inflammatory/immunological response to biomaterials. However, prediction of future complications of individual implant recipients resulting in possible implant failure on the one hand and the search for the "right" biomaterial or surface modification for the intended application on the other hand will probably be major challenges of the future in terms of biomaterial/biocompatibility research.

4.3.3 The complex and interdisciplinary issue of biocompatibility testing

During the past decades, many different biocompatibility tests have been published and recommended. However, at present there is no universal test available for objective evaluation of the biological safety of a medical device. Although the main principles applied in testing implants/biomaterials have changed very little over the last 20 years, the standards, specific laboratory methods, regulatory views toward testing and interpretation of results have changed greatly in this period including the availability of new materials. However, basic problems such as a failure of standardization including "real" standards as well as a broad inter-laboratory variability still exist. The main unsolved problems in this context are different test results and interpretations of *in vitro*- and *in vivo*-based methods.

The European Commission has issued the Medical Device Directive (MDD) operational as of January 1, 1995, and the Active Implantable Medical Device Directive (AIMDD) for evaluation of biomaterials and medical devices. Furthermore, assistance in the design of many biomaterial tests is available through national and international standards and organizations. Thus, the American Society for Testing Materials (ASTM) and the ISO can often provide detailed protocols for widely accepted, carefully thought-out testing procedures. Other testing protocols are available through government agencies (e.g., the FDA) and through commercial testing laboratories (see Chapter 5 and Appendix A).

General principles of biological evaluation according to ISO 10993, "Biological Evaluation of Medical Devices, Part 1: Guidance on Selection of Tests," are that (i) the assays used must be performed on the final product or on representative samples of the biomaterial, and that (ii) the test selection must be performed according to the nature, duration, frequency, and exposure conditions of devices, to the toxicological data of the components of the device, to the physicochemical nature of the final product, and to the ratio between the surface of the device and the volume of the implantation site (Eloy, 1994).

Initial *in vitro* cytotoxicity tests including extract tests as well as direct and indirect contact tests are at least acceptable for material screening, but they are not sufficient for devices. The tests are performed using different cell lines such as mouse fibroblasts L-929, baby hamster kidney cells BHK-21, and others (Zhang et al., 1991). Cell lines present many advantages as they are easy to grow, can be subcultured indefinitely, and are well adapted to *in vitro* conditions. However, a serious concern could be raised on the possibility to extend the information obtained using such types of cells which are most likely to come in contact with the biomaterial or with substances released from them. If the cells are directly seeded on polymer discs, the effectiveness in promoting cell adhesion can be tested. Determination of cytotoxicity is generally performed by morphological assessment of cell damage, cell growth, and cellular metabolism.

More complex and closely connected are the examination of blood–material and/or tissue–material interactions involving red blood cell damage, activation of coagulation, of platelets, of the immune system, of the activation/inhibition of fibrinolysis, and after implantation in subcutaneous, muscular or bone tissue the evaluation of the tissue reaction around the implant. The respective test systems are performed *in vitro* either under static or dynamic conditions, ex vivo or *in vivo*.

Genotoxicity evaluation involves *in vitro* assays realized on biomaterial extracts including gene mutation, chromosomal observations, and DNA effects. However, to evaluate carcinogenicity no acceptable test is available which defines the clinical risk for carcinogenesis. It is well known that most biomaterials can cause tumors in laboratory animals especially in rodents, and components of many biomaterials are known as chemical carcinogens. However, in clinical use of implants/biomaterials, malignant tumors are rare events (Eloy, 1994; Wallin, 1995b).

Therefore, only cytotoxicity assayed by direct contact test, agar diffusion test or elution tests on cell lines such as mouse fibroblasts L-929 can be measured by means of established standardized protocols including established and recommended standards. However, other risks, involving more complex biological events, that is, blood–device and/or tissue–device interactions have to be evaluated according to test protocols of lower precision. In addition, prolonged clinical use or some specific situations may induce additional risks, based on the individuality of host response, that are not currently investigated by established biocompatibility testing protocols. Therefore, new testing models involving carcinogenesis, immunogenicity, thromboresistance, wound healing, blood compatibility, and chronic toxicity are urgently needed (Williams, 1991; Wallin, 1995b).

4.3.4 The *in vivo* evaluation of immunological/inflammatory response

The immune system plays a central role in the inflammatory process following implantation of a foreign "nonself" body which involves determination of complement activation (Brett, 1992), phagocytic cell function by quantification of the number of circulating phagocytes and their expression of adhesion receptors, and their products, the cytokines (Ward, 1994). Furthermore, hypersensitivity reactions are involved, since, for example, about two-thirds of patients with dialysis anaphylaxis on ethylene oxide (ETO)-sterilized dialyzer also have IgE against ETO–protein conjugates (Grammer and Patterson, 1987) as well as basophilic leukocytes and mast cells with IgE-specific antibodies and histamine release (Götze, 1994), and the production of cytokines and cytokine-inhibitory proteins were described (Pereira and Dinarello, 1994).

From my point of view, and in accordance with findings by others (Kossovsky et al., 1987; Goldblum et al., 1992; Yang et al., 1994), especially the immunogenicity, that is, cellular and humoral immune response to components of polymeric implants might be a new, important tool to evaluate biocompatibility after implantation. Since the cellular immunity is difficult to quantify and to standardize, humoral immune reactions by means of quantification of serological markers, such as specific antibodies or pro- and anti-inflammatory cytokines possibly reflect the degree of material–host interaction in general. In particular cases, and this might be the new quality for biocompatibility testing of newly developed biomaterials and/or their surface modifications such as coatings or functionalization, the individual host response/humoral immune response to an implant might therefore be an early sign of functional bioinstability and subsequent implant failure.

4.3.4.1 *In vivo* evaluation of systemic humoral immune response against implants

The rationale to this approach is the fact that large numbers of phagocytes are present in the local vicinity of biomaterials which may affect the biostability of the latter and, further, may initiate the process of biodegradation (Vince et al., 1991). Even micro-sized particles or monomers/oligomers which are "extracted" by the highly aggressive environment surrounding the implant may be phagocytized. This has in fact been demonstrated for many polymers including polyethylene terephthalate (PET), silicone, Teflon® (PTFE), L-lactic acid/glycolic acid homo- and copolymer, poly(L-lactide acid), and other (Glaser, 1993; Kaiser, 2009; Tang and Eaton, 1995). Since T and B lymphocytes are also present at increased levels at the implantation site (Gong et al., 1993; Holgers et al., 1995), phagocytes may act in their function as antigen-presenting cells of processed components of biomaterials to induce a specific immune response.

Ziegler et al. (1994) detected specific serum IgG antibodies against the outer cellulose acetate membrane without cross reactivity to the inner polyethylene membrane following subcutaneous implantation of glucose sensors into Lewis rats. Consequently, in further studies different polymeric membranes were subcutaneously implanted in rats to investigate the time course and prevalence of the polymer-specific antibody response in sera of animals (Schlosser et al., 1994b). Surprisingly, the highest antibody formation was obtained against Teflon® (PTFE), which is well known to be chemically inert. Whereas polycarbonate and Pellethane® (thermoplastic polyurethane) demonstrated only a moderate immunogenicity, polyimide did not induce specific antibody formation. Moreover, although inbred rats were used for this implantation study, the individual antibody formation of rats to the different polymers demonstrated a remarkably broad variability. It thus becomes possible to evaluate the individual biocompatibility, that is, the individual host response to a biomaterial which will much more vary in experimental animals of different genetic background or patients compared to inbred animals. Therefore, the detection and quantification of specific antibodies potentially enables assessing the degree of bio(in)compatibility/host immune response to an implanted biomaterial. As further demonstrated, repeated implantations result in an increased antibody formation resembling a boost effect as commonly known for immunization experiments. Moreover, the polymer-specific antibody response could be significantly enhanced by the polyclonal activation of the immune system using one application of complete Freund's adjuvant (CFA), containing heat-killed mycobacteria. Since CFA simulates a bacterial infection or an inflammatory response, these findings further demonstrate the importance of sterility of implantable biomaterials. In contrast to the antibody formation against the polymer membranes, no correlation could be observed with results obtained *in vitro* after direct and indirect cytotoxicity test or cell adhesion experiments on mouse fibroblast cell line L-929. This clearly indicates that *in vitro* biocompatibility tests frequently do not reflect the *in vivo* situation.

A further study was aimed to analyze the long-term antibody response following repeated intraperitoneal implantation of bovine collagen type I-impregnated PET (polyester; Hemashield®) prosthetic segments, clinically used in vascular surgery, into Lewis rats (Schlosser et al., 2002). It was observed that polymer IgG antibodies were significantly increased by both repeated implantation (boost effect) and by combined application of CFA (bacterial infection simulation) and that antibody positivity was subsequently followed up until experimental day 293. Furthermore, no collagen antibodies were detectable and the polyester-specific antibodies did not bind to an irrelevant polymer.

Following repeated intramuscular implantation and explantation of three polyester vascular prostheses coated with either collagen (Hemashield®), gelatin (Vascutek®), or human serum albumin (Bard®) into rats, the antibody response and prevalence was compared to that induced against a PTFE prosthesis (Gore®) and to sham-operated control animals (Wilhelm et al., 2007). In contrast to the PTFE prosthesis and the controls, all animals which received a polyester-based prosthesis demonstrated a high antipolyester antibody response with a broad individual variability graduated according to the prosthesis coating/impregnation as follows: gelatin > albumin > collagen. This was further significantly increased after the second implantation/first explantation and declined following the last explantation. Only animals with albumin-coated implants revealed specific antibodies to the coating/impregnation as well as the strongest overall immunological reaction against the prosthesis.

Beside the boost effect detected following repeated implantation/explantation, the specificity of antipolyester antibodies could be demonstrated as follows: *First*, and as also demonstrated later in pigs, the antipolyester IgG antibodies could be significantly competitively inhibited up to 56% after preincubation of individual positive rat sera with the polyester homogenate (Zippel et al., 2008). *Second*, no correlation exists between antipolyester Abs and serum IgG concentration of individual rat sera on any experimental day (Schlosser et al., 2002). *Third*, the blocking of polyester particles in the immunoassay results in a near complete reduction of nonspecific rat serum IgG binding, demonstrated for all experimental and control animals during the whole investigational period (Schlosser et al., 2002). Furthermore, none of the animals of the control group were found to be antipolyester antibody-positive on any experimental day. *Fourth*, the implantation of three segments of an uncoated polyester prosthesis (Microvel®) resulted in a further increase of antipolyester antibody concentration (Schlosser et al., 2002; Wilhelm et al., 2007).

Following studies in rodents, Zippel et al. (2001) analyzed the antipolymer and anticollagen antibody response after functional implantation of three different collagen-impregnated Dacron® vascular prostheses of low, medium, and high primary porosity in pigs, by replacing the infrarenal aorta with a segment of prosthesis. This study as well as a further study in sheep confirmed the antibody response against the polyester prosthesis matrix, and again the antibody formation demonstrated a broad individual variability (Walschus et al., 2008). In contrast to the results in rodents, specific anticollagen type I antibodies could also be detected, possibly indicating a higher coating/impregnation load compared to the relatively small segments used for implantation in rodents (Schlosser et al., 2002; Schlosser et al., 2005).

However, the nature of polymeric antigens is not yet known, and it is unlikely that unaffected synthetic biomaterials are immunogenic. Potential immunogens might rather be micro-sized particles, monomers, oligomers, and other leachables released after biodegradation. The presence of macrophages, FBGCs, T and B lymphocytes together with the important role of complement activation as well as the detection of specific antibodies to silicone elastomers (Goldblum et al., 1992) and to corrosion products (CoCr protein complexes; Yang et al., 1994) strongly suggests that beside nonspecific also specific immune reactions are involved in local phenomena of bio(in)compatibility during the process of inflammatory response to biomaterials.

Beside specific implant-associated antibodies, another potential humoral marker reflecting the *in vivo* behavior of implants/biomaterials might be the detection and quantification of pro- and anti-inflammatory cytokines in serum samples of recipients as mentioned in Chapter 4.3.2.4. Regarding this, the time course was investigated of the pro-inflammatory cytokines IL-2, IFN-*y* and IL-6 and the anti-inflammatory cytokines IL-4, IL-10, and IL-13 up to 56 days following intramuscular implantation of titanium-based implants. For this, the surfaces of titanium implants were modified by low-temperature plasma coatings resulting either in an amino group-rich positively charged surface from allylamine or a carboxyl group-rich negatively charged surface created by acrylic acid. Overall, the results demonstrate a distinct pattern of cytokine levels and associations depending on the surface treatment which is in agreement with data previously reported investigating the cytokine response in exudate from steel implant cages containing different polymer samples (Schutte et al., 2009). Furthermore, especially during the acute phase of the inflammatory response, surface-specific changes for IL-4 and IL-10 as well as for IL-4 in the chronic inflammatory phase could be observed. Thus, the findings indicate that the systemic cytokine profile could be an additional parameter for *in vivo* examination of new or modified biomaterials as well as for *in vivo* monitoring or outcome prediction after implantation as recently demonstrated for patients with left ventricular assist devices (Caruso et al., 2010). This could also help to understand the relationship between material properties, the systemic and local material-related inflammatory and/or immune reactions and the implant function, thereby elucidating the mediating role of cytokines within these interactions.

On the basis of the results presented here, it is suggested to generally include the examination of the humoral immune response by means of detection of biomaterial-specific antibodies as well as by differentiation of pro- and anti-inflammatory cytokines as standardizable quantitative parameters in biocompatibility testing.

A disadvantage of the serological approach of *in vivo* biocompatibility testing is that only one biomaterial either modified or unmodified can be implanted into one recipient, since otherwise a mixed host response against different implants would be induced. Furthermore, and taking the individual variability into account, at least eight experimental animals for one new developed biomaterial and/or its modification are needed to reach statistical significance and to enable comparison with another same-sized control group (Figure 4.9A).

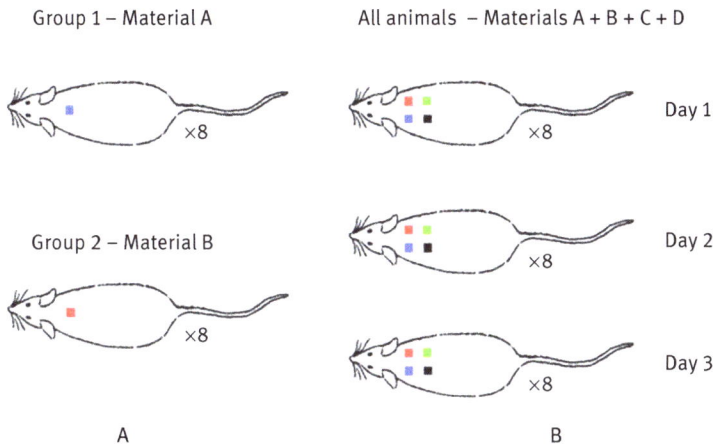

Figure 4.9: Schematic representation of two different study designs for *in vivo* evaluation of biocompatibility. (A) Single implantation to examine serological markers such as antibodies or pro- and anti-inflammatory cytokines. (B) Simultaneous implantation of different material variations aimed to comparative evaluation of acute and chronic local tissue response.

Thus, a relatively large number of experimental animals are needed to perform *in vivo* testing by means of serological markers only. However, the main advantage of this serological approach is that by simple blood collections over the whole experimental time course, distinct serological markers such as specific antibodies or cytokines can be analyzed and compared for all experimental groups.

Another approach, which is the subject of the following passage, is proposed as the first *in vivo* examination step, being the simultaneous implantation of differently modified biomaterials in comparison to the unmodified biomaterial and followed by subsequent investigation of the local inflammatory reaction at different experimental time points (Figure 4.9B). In order to cover distinct phases of local inflammatory response it is suggested, for example, for a rat animal model to examine day 7 postoperative to analyze the acute inflammatory phase, followed by day 14 as the transition phase between acute and chronic inflammation and finally day 56 for the chronic inflammatory phase. The main advantage of this simultaneous implantation approach is that fewer experimental animals are required, for example, for each experimental day a group of eight experimental animals receive four different implants per animal, and the local inflammatory reaction against a given modified implant can be pairwise analyzed and compared to another differently modified or to an unmodified control implant.

4.3.4.2 Morphometric evaluation of local immunological/inflammatory response against modified biomaterials:

An important problem of histological studies is the subjectivity in examining the tissue sections by manual microscopic evaluation. Even a quantitative approach to manual examination is limited by the low speed of manual counting and the limitation of the counting process to a small sector of the total sample. Furthermore, it is difficult or even impossible to manually quantify markers that are diffusely expressed in the tissue due to their presentation by many different cell types. By contrast to qualitative and semi-quantitative approaches often used in this field, a quantitative technique would facilitate a more accurate determination and better comparability of different studies. As a possible solution, quantification of cells or cell markers in histological specimens using digital image analysis procedures can be used to obtain more precise and reproducible results, thereby increasing the objectivity of the evaluation (Hunt et al., 1997; Al-Kofahi et al., 2011).

For morphometric evaluation of local tissue response following implantation of vascular prostheses in pigs, the commercially available digital image analysis system KS400 was used following development of user-defined software macros. Using determination of the proliferation rate by detection of Ki-67 positive cells as a corpuscular marker and beta-1-integrin expression as a diffusely stained marker (Garcia, 2005) as well as the *neointima* formation, the KS400 system was found to be well suited to analyze and quantify the local host–implant interactions with high reproducibility in the specific model chosen with the unique strength of possible individual adaptation of the system for specific problems (Zippel et al., 2006). Moreover, the simultaneous availability of macros and reference images combined with standardized investigation protocols was found to be an important step toward an objective and thus comparative evaluation of tissue responses.

Furthermore, the KS400 system was used to study *intima* hyperplasia, resulting from ECM secretion by analyzing the fibronectin levels following functional implantation of vascular prostheses with different porosity. Throughout the study, fibronectin levels were highest at the proximal anastomosis region (Patrzyk et al., 2010). The results further indicate a relationship between fibronectin and *neointima* formation with the prosthesis porosity, demonstrating the importance of the textile design for tissue reactions following implantation.

In a further study, the applicability of the free image-analysis software ImageJ was demonstrated for fast, easy and reproducible quantification of the tissue response following implantation of titanium samples in rats with subsequent immunohistochemical examination of peri-implant tissue samples for monocytes and macrophages (ED1) and MHC-class-II-positive antigen-presenting cells (OX6). For this, the quantification of positively stained cells in the vicinity of the implant pockets was based on a grid-supported manual count carried out using the two ImageJ plugins CellCounter and Grid (Walschus et al., 2011). The manual procedure for counting stained cells was found to be a reproducible, flexible, and versatile method for the quantitative examination of the local inflammatory response. However, analyzing either samples with very low cell numbers could require additional efforts and otherwise, the method has its limits for samples with very high cell numbers for which a near-automatic approach based on color thresholding might be more suitable to save time.

Calcium phosphate preparations are established coatings for titanium-based medical implants used for bone reconstruction (see Chapter 3.2.7). However, biodegradation of the coating might result in micro-sized particles, subsequently causing inflammatory reactions such as the so-called "particle disease." Therefore, Walschus et al. (2009) investigated the short-term inflammatory response following simultaneous implantation of titanium-brushite and titanium-HAp implants in comparison to sol-gel derived silica/calcium phosphate in rats using the ImageJ-based approach described above. Whereas the number of tissue macrophages, T lymphocytes, MHC-class-II-positive cells and proliferating cells remained constant, a distinct time course of monocytes/macrophages was found for both titanium implants with a lower inflammatory response against the HAp coating. In contrast, the silica/calcium phosphate implants demonstrated a high degree of degradation and scattering into smaller particles with an increase in all cell populations except the proliferating cells.

Another approach to potentially improve the *in vivo* biocompatibility and functionality of implants are surface modifications by biomembrane-derived phospholipids due to their biomimetic properties. This could be demonstrated by the beneficial effects of phospholipid coatings regarding interactions with cells and bacteria *in vitro* (Willumeit et al., 2007a; Willumeit et al., 2007b). Thus, an *in vivo* study was performed which aimed to examining the time course of local inflammatory reactions after simultaneous implantation of titanium plates with phospholipids, either attached noncovalently or covalently via octadecylphosphonic acid, in comparison to uncoated controls (Kochanowski et al., 2011). Overall, both phospholipid-coated implants were comparable to the titanium controls, demonstrating the potential of phospholipids for implant surface modification.

In orthopedic surgery, implants are mostly used for treatment of fractures, bone defects and degenerative changes in joints. Depending on their application, the implants remain either temporarily or permanently in the body. In the case of permanent implants such as artificial joint replacement, a stable long-term integration in the bone is required with the adhesion and in-growth of bone cells on the implant surface, for example, by applying positively charged amino group-rich polymer coatings (Nebe et al., 2019). In contrast, for temporary devices, cell-material contact should be restricted for easy removal after bone healing which could be achieved, for example, by surface modification with antiadhesive fluorocarbon compounds.

Recently, low temperature plasmas, a.k.a. cold plasmas are intensively investigated as a tool to modify biomaterial surfaces (see Chapter 3.4.4). They are characterized by a low degree of ionization at low or atmospheric pressure and thereby avoid structural or chemical deterioration of biomaterials (Roth, 1995; Roth, 2001; Hippler et al., 2008). These processes can be combined with an appropriate functionalization of the surface by applying positively charged, nanometer-thin polymer coatings of plasma-polymerized allylamine (PPAAm) or ethylenediamine onto the implant surface and accelerating the in-growth of bone cells or, otherwise, by plasma-polymerized fluorocarbon compounds resulting in antiadhesive coatings.

To investigate low-temperature plasma coatings *in vivo*, the effect of a cell-adhesive, PPAAm layer on titanium implants prepared at three different plasma conditions on the local inflammation following simultaneous implantation was studied in a rat animal model (Hoene et al., 2010). In particular, in the chronic inflammatory phase on day 56 following implantation, two modified titanium samples, both prepared at higher duty cycle (ratio between the time plasma on to plasma off) had significantly lower numbers of inflammatory cells compared to the third modified sample (prepared at lowest duty cycle) and to titanium controls. Physicochemical analysis of the three differently prepared PPAAm implants revealed that the implant with increased local chronic inflammatory response had the lowest water absorption, the highest nitrogen loss, and the lowest oxygen uptake. Overall, the results demonstrate that the PPAAm samples and the controls were comparable regarding local inflammation, but that different plasma conditions also lead to variations in the material properties which further influence the tissue reaction *in vivo* (Hoene et al., 2010).

Antiadhesive plasma-generated fluorocarbon polymer films were synthesized using micro-wave discharge plasmas with two different low-pressure plasma sources consisting of an octafluoropropane/hydrogen mix and a hexafluorobenzene/hydrogen mix, or in a radiofrequency discharge plasma with an octafluoropropane/hydrogen mixture (Finke et al., 2014). Since cell adhesion and spreading of human osteoblasts were clearly reduced on these antiadhesive surfaces *in vitro*, first *in vivo* data on the biocompatibility of films deposited in the radiofrequency discharge plasma demonstrate that the local inflammatory tissue response was comparable to that of controls, while an antiadhesive coating deposited in a micro-wave discharge plasma induced stronger inflammatory tissue reactions (Finke et al., 2016).

Another general problem with implants, not only with those used in orthopedic surgery, relates to external bacteria or protracted endogenous bacteria which can colonize implants and can lead to the formation of biofilms on the implant, finally resulting in infection of the surrounding tissue. The usual outcome is that the implant has to be removed due to functional limitations or simply because it is a chronic infection site. Therefore, in order to avoid bacterial colonization of the implant, coating of the implant with antibacterial agents such as antibiotics, silver or copper is currently being investigated (see Chapter 3.4.3). In addition to the adhesive or antiadhesive surfaces, this can be achieved by low-temperature plasma coating processes in which copper in varying quantities can be applied to the implant surface in an organized, uniform and temporary fashion. Thus, these coatings can effectively prevent the colonization of several biomaterial surfaces by dangerous bacterial infectious agents and potentially avoid implant failure and therefore clinical complications (Nebe et al., 2013).

To examine the balance of antibacterial versus adverse tissue effects, a preliminary study was undertaken aimed to evaluate *in vivo* copper release and the time course of early acute local inflammatory reactions following implantation in rats of titanium plates that received a plasma electrolytic oxidation only or in combination with galvanic copper deposition (Hoene et al., 2013a). In general, the inflammatory tissue reactions increased beyond copper release, indicating effects of either surface-bound copper or more likely the implants themselves. Altogether, copper-treated samples possessed antibacterial effectiveness *in vitro*, released measurable copper amounts *in vivo*, and caused a moderately decreased local inflammatory response, thereby demonstrating the anti-infective potential of copper coatings.

Another approach to create anti-infective implants could be, for example, the incorporation of copper into the titanium implant surface using plasma immersion ion implantation. In this preparation, the known cytotoxicity of copper might be circumvented by an additional cell-adhesive positively charged amino group-rich film of PPAAm. Thus, *in vivo* local inflammatory reactions of titanium implants treated with copper alone or with an additional cell-adhesive amino group-rich film were compared to untreated titanium samples by simultaneous implantation into the neck muscles of rats (Walschus et al., 2017). The results demonstrated a significantly stronger chronic inflammatory reaction of titanium implants treated with copper alone compared to titanium controls for tissue macrophages, antigen-presenting cells, T lymphocytes, and to titanium implants with copper together with an amino group-rich surface for tissue macrophages, T lymphocytes, and mast cells. Thus, the additional PPAAm surface significantly reduced the inflammatory reactions caused by copper and therefore the combination of both plasma processes could be useful to create antibacterial as well as tissue compatible, cell adhesive titanium-based implants.

A further study was performed aimed to *in vitro* and *in vivo* evaluation of low-temperature plasma treatment of titanium with a cell-adhesive polymerized ethylene diamine film alone, copper by means of magnetron sputtering alone or a combination of both processes. As demonstrated *in vitro*, both titanium–copper implants were comparable in terms of copper release and antibacterial effectiveness.

Following simultaneous implantation in a rat animal model, it was noted that the differences obtained for the number of NK cells are in line with the results for the tissue macrophages, the T lymphocytes and the mast cells, in particular observable for both implants with the cell-adhesive polymerized ethylene diamine surfaces (Hoene et al., 2013b). This indicates that the NK cells might play a regulatory role through reciprocal interactions with macrophages and T lymphocytes as described previously for other situations (Vivier et al., 2008). The results of this study clearly demonstrate that the combination of both plasma processes results in creation of implants that possess antibacterial properties without eliciting an increased inflammatory response *in vivo*.

Biological scaffold materials derived from ECM of mammals, extensively studied regarding their structural properties, mechanical behavior, and degradation characteristics *in vitro* and *in vivo*, have been successfully used in several medical applications (Badylak et al., 2009). Among these, collagen as a major component of the ECM is known to act as a matrix which mediates the migration and adhesion of cells as well as subsequent vascularization and formation of connective tissue (Badylak, 2007). Regarding their use for medical purposes, these materials should cause only a low inflammatory response and have no or negligible antigenicity and a good stability *in vivo*.

After implantation of two cross-linked acellular porcine dermal collagen matrices into rats, an inflammatory reaction with different, material-dependent characteristics was observed which changed from an acute to a chronic phase demonstrated by a decrease of all investigated immune cell populations (Lucke et al., 2015). In addition, the change from acute to chronic inflammation was associated with a shift from pro-inflammatory M1-like to anti-inflammatory M2-like macrophages. Moreover, indications for humoral immune reactions such as the generation of specific antibodies against porcine collagen as described earlier in this chapter were seen, since macrophages, antigen presenting cells, and T lymphocytes were found to be involved in both the acute and chronic phases of inflammation.

Electrospinning is a well-known polymer processing technique that has recently received remarkable interest for biomedical applications such as tissue engineering or drug delivery (Sill and von Recum, 2008; Agarwal et al., 2009; see Chapter 3.3.2.4). The technique is applicable to both polymer solutions and melts (Huang et al., 2003; Reneker et al., 2007) and enables the fabrication of tissue engineering matrices resembling major structural features of the native ECM (Szentivanyi et al., 2011). Those artificial matrices are of great interest for tissue reconstruction acting as support for cell adhesion, proliferation, and differentiation. Natural polymers such as collagen and synthetic polymers such as the well-known biodegradable polylactones (Piskin et al., 2007) and other polymers are well-suited for processing by this technique. As it is commonly known, implanted biomaterials can cause chronic local inflammation called FBR with formation of FBGC by monocyte/macrophage fusion depending on various implant-specific characteristics. Thus, studies were conducted aimed to examining in a rat animal model the FBGC and inflammatory cells after simultaneous implantation of poly(L-lactide-*co*-D/L-lactide) samples differing in their structure either as solid membranes and or electro-spun fiber meshes, with the latter either uncoated or coated with a ultra-thin positively charged plasma-polymer allylamine layer (Schnabelrauch et al., 2014; Lucke et al., 2018). FBGC occurrence was primarily found to be determined by material morphology, as their numbers were significantly increased for both coated and uncoated PLA meshes during the acute and chronic inflammatory response compared to the PLA membranes. This is exemplary shown in Figure 4.10 for a PET vascular implant. FBGC were predominantly CD68$^+$ (M1-like) and CD163$^-$ (M2-like).

Furthermore, the nestin-stained tissue area as a marker for tissue regeneration was negatively correlated with the number of CD68$^+$ monocytes/macrophages but positively correlated with CD163$^+$ macrophages, highlighting the differing roles in

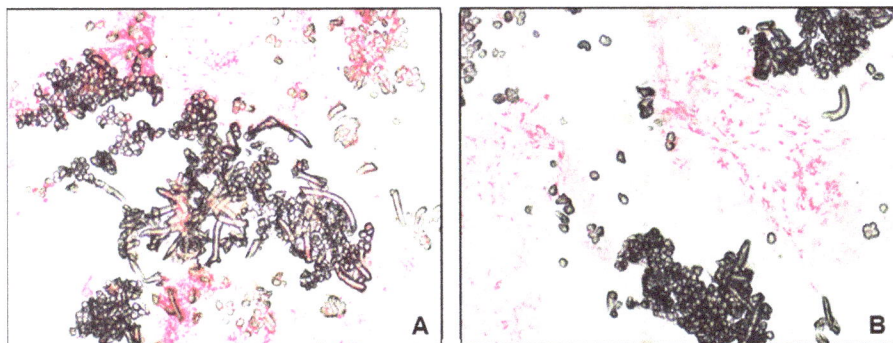

Figure 4.10: Proinflammatory CD68+ (M1-like) monocytes and macrophages (A) and anti-inflammatory CD163+ (M2-like) macrophages (B) in the peri-implant tissue of a human serum albumin-coated polyethylene terephthalate (PET)-based vascular prosthesis segment following implantation for 23 days in rats, demonstrating intense inflammation and signs of material biodegradation.

FBGC formation and tissue regeneration (Lucke et al., 2018). However, currently neither the exact morphological differentiation nor the formation and function of the FBGC are fully clarified (Boyce et al., 2000). Thus, further studies are required to confirm whether FBGC also possess different functional characteristics comparable to macrophages and if so, how they participate in the removal of foreign materials and in anti-inflammatory and/or regenerative processes.

4.3.4.3 Association of systemic and local inflammatory response

The use of synthetic biomaterials such as textile prostheses in vascular procedures has been subjected to extensive investigations in the past. As with most implants, vascular prostheses have not yet achieved the goal of perfectly imitating the biological characteristics or structures for which they are used as replacement. Research is still ongoing not only to optimize their functional properties but also to improve their biocompatibility. However, studies regarding the relationship between the local inflammatory tissue response and systemic humoral response such as immunogenicity of vascular prostheses and biomaterials in general have been given less priority. As described above, specific antibodies against polymers were detected in several animal models and quantitative methods were established to differentiate the local inflammatory reactions against various implants and their modifications. Thus, this experimental study was intended to examine possible associations of the local cellular inflammatory reactions with the humoral immune response by detection of specific antibodies against the prosthesis matrix and/or the coating/impregnation. For this, three bovine collagen type I coated/impregnated vascular prostheses made from the same polyester matrix but differing in their primary porosity were functionally implanted in pigs and examined for 116 days postoperative (Zippel et al., 2008). For morphometric examination of the local tissue reactions, the *neointima* formation and the expression of β-1-integrins were analyzed that are important transmembrane receptors and regulators of interactions between cells as well as between cells and the ECM. In the first three weeks, the prosthesis of medium porosity caused the highest tissue reactions and antipolyester antibody response, but the lowest anticollagen antibodies, whereas the prosthesis with highest porosity induced the highest anticollagen antibody prevalence but the lowest tissue reaction and antipolyester antibody response in the early phase after three weeks. In contrast, on day 116, the prosthesis with lowest porosity induced the highest tissue reactions and antipolyester antibody prevalence (Zippel et al., 2008). Therefore, the results obtained from this experimental study indicate a possible association between local inflammatory reactions and humoral immune responses, influenced by the structure and properties of vascular grafts.

Despite long-term success of titanium-based implants, sometimes complications including infections, septic loosening or peri-implantitis occur, often associated with inflammatory reactions that consist of an early acute phase and is followed by a chronic phase throughout the implantation period (Esposito et al., 1998; Campoccia et al., 2006; Baumann et al., 2007). In this process, macrophages are of central interest and are accompanied by other immune/inflammatory cells such as T lymphocytes or NK cells (Davis et al., 2003; Huss et al., 2010).

However, repeated postoperative evaluation of the implant site would be a severe burden on patients. Alternatively, examining blood components with analysis of inflammatory serum markers could potentially be useful to reflect the local cellular response for detection and/or prediction of inflammation-related complications since clinical blood examination is common standard. Therefore, following intramuscular implantation of different surface-modified titanium implants in rats, Hoene et al. (2015) aimed to examining possible associations between the postimplantation time course of the pro-inflammatory cytokines INF-*y* and IL-2 and the anti-inflammatory cytokines IL-4 and IL-10 in serum samples obtained weekly, and the local peri-implant tissue response in the chronic inflammatory phase at postoperative day 56. The parameters chosen for morphometric analysis are the number of pro-inflammatory CD68$^+$ (M1-like) monocytes/macrophages, anti-inflammatory CD163$^+$ (M2-like) macrophages, MHC class II-positive cells, activated NK cells, and mast cells. Using multivariate correlation analysis, a significant interaction between the serum IFN-*y* concentration and the number of pro-inflammatory (M1-like) CD68$^+$ monocytes/macrophages was obtained, while no interactions were found for the other investigated cytokines and cell types. Furthermore, a correlation analysis of the serum IFN-*y* concentrations on each experimental day and the CD68$^+$ monocytes/macrophages response on day 56 demonstrated a consistently positive correlation which was strongest during the first three postoperative weeks (Hoene et al., 2015). Altogether, these results indicate that in this implantation study an early and increased pro-inflammatory IFN*y* serum concentration was associated with a pronounced chronic inflammatory response demonstrated by high number of pro-inflammatory CD68$^+$ (M1-like) monocytes/macrophages in the peri-implant tissue and vice versa low serum IFN-*y* concentrations in the early acute inflammatory phase were found to be associated with low number of CD68$^+$ monocytes/macrophages in the postoperative chronic inflammatory phase.

4.3.5 Conclusion

In general, implantation of biomaterials results in an injury of tissue and inevitably initiates an inflammatory response. This inflammatory response with its main phases of acute and chronic inflammation involves a variety of different cell-biomaterial and cell-cell interactions which includes both blood and tissue and is leading to a microenvironment which might also affect the implant by initiation of biodegradation. The whole complexity of innate nonspecific and adaptive specific host defense mechanisms including coagulation system, complement cascade, phagocytosis, and the immune response might be involved. Although the diversity, complexity and individuality of interactions between the host's biological system and the foreign body implant/ biomaterial is largely unknown yet, activated phagocytic cells appear to play a central and most important role in the response to an implanted biomaterial. Owing to their various cellular and humoral interactions, their secretion products, but also their ability to fuse into FBGCs, they create an aggressive environment surrounding the implant/biomaterial, thereby potentially affecting the biofunctionality and biostability of implanted biomaterials, also by inducing biodegradative processes.

Thus, the aim of future attempts to develop implants/biomaterials and their modifications must be, depending on their purpose and application, to adequately perform regarding biostability and biofunctionality but also to avoid chronic inflammatory processes as the host's response against the foreign body. For this, characterization and standardization of implants/biomaterials and their modifications according to international regulatory bodies, for example, ISO criteria are essential (see Chapter 4.5; Appendix A).

Furthermore, well-established methods to evaluate biocompatibility according to ISO criteria but most importantly, new *in vivo* biocompatibility tests including the network of immune and inflammatory responses must be established, applied, standardized, and interpreted in a concerted and multidisciplinary action. However, it should be pointed out that presently and also in the immediate future, there will neither exist a completely biocompatible biomaterial nor a universal biocompatibility test, since the uniqueness and individuality of the host response is the most significant variable. Therefore, the individual humoral immune response is highlighted, for example, by simple detection of implant/biomaterial-specific antibodies or quantification and differentiation of pro- and anti-inflammatory cytokines in serum samples of implant/biomaterial recipients. On the one hand, this should enable to estimate both the degree of local individual tissue responses to a selected biomaterial in order to avoid or reduce complications. On the other hand, it will allow selecting from a variety of biomaterials, modified or unmodified, the "implant/biomaterial of choice" for an intended application, based on their mechanical and physicochemical characteristics.

4.4 Interaction of biomaterials with living tissue: osseoinduction (Robert B. Heimann)

4.4.1 Introduction

In this chapter, salient aspects of the interaction of inorganic biomaterials with living tissue will be briefly discussed, and important biological pathways of implant integration highlighted. Since hydroxylapatite (HAp) plays an important role in this complex process, the discussion will center on the functionalization of implant surfaces with HAp, most often provided by thin coatings. As important as the elucidation of the subtle interaction between implant material and tissue is, the present author is very conscious about the skimpiness of this account and hence, the perceived need to expand on this important subject. He feels, however, that within the general context of this treatise biochemical exactitude has to take second place to an account on technological developments of bioceramic coatings that are designed to prepare a nurturing bed for bone apposition and cell ingrowth controlled by the principles of biomineralization. To somewhat remedy this deficiency, the interested reader is referred to an excellent and necessarily much more detailed overview of biomineralization by Weiner and Dove (2003).

4.4.2 Types of tissue

To understand the various ways through which a biomaterial interacts with living tissue, the nature of the tissue at the interface to an implant has to be thoroughly understood. Any organ in the human body is made up of four types of tissue combined in varying proportions: epithelial, muscle, nervous, and connective tissues. *Epithelial tissue* is a membranous tissue composed of one or more layers of cells separated by small amounts of intercellular substance, forming the covering of most internal and external surfaces of the body and its organs. *Muscle tissue* is composed of fibers capable of contracting to effect bodily movement, and controlled either voluntarily as in the skeleton or involuntary, that is, biochemically as in the cardiovascular, digestive, and respiratory systems. *Nervous tissue* is made up of different types of nerve cells, all of which have an axon, the long stem-like part of the cell that sends action potential signals to neighboring cells. Functions of the nervous system are sensory input, integration, control of muscles and glands, homeostasis, and mental activity. Finally, *connective tissue*, arising chiefly from the embryonic mesoderm, is characterized by a highly vascular matrix and includes collagenous, elastic, and reticular fibers, adipose tissue, cartilage, and bone. Since connective tissue forms the supporting and connecting structures of the body, it is of particular importance in the context of those biomaterials it is interacting with.

Indeed, biomaterials including implant coatings interact chiefly with connective tissue and as such, bone. If a biomaterial is introduced into the body, the always present tissue response is that of an inflammation (see Chapter 4.3). Damaged tissue will be recognized and eventually be repaired by leukocytes, the white blood cells that are of two types, granular leukocytes (microphages) and nongranular leukocytes (lymphocytes, monocytes), that on contact with foreign invading material or cells develop into phagocytes (macrophages). Macrophages activated by the presence of foreign bodies secrete enzymes that interact with fibroblasts to create the collagen of the fibrous tissue capsule separating an (uncoated) implant.

However, no foreign material placed in contact with living tissue is completely biocompatible (Williams, 2014). The only substances that conform completely are those manufactured by the body itself (autogenous). Any other material (allogenous) recognized by the body as foreign initiates some more or less benign type of host-tissue response. The four general types of responses allowing different means of achieving attachment of implants to the muscular–skeletal system are shown in Fig. 1.2.

4.4.3 Osseoconduction and osseoinduction

Returning to HAp, its bone growth-stimulating function can best be expressed by two biomedically derived properties, termed osseoconductivity and osseoinductivity. On the one hand, *osseoconductivity* is the ability of a biomaterial to support the in-growth of bone cells, blood capillaries, and perivascular tissue into the operation-induced gap between implant body and existing (cortical) bone bed. Interconnected coating pores of 100–300 μm size are known to foster the process of osseogenesis and osseointegration, thus underscoring the need to enhance and control pore sizes. Indeed, development of such pore networks in HAp coatings is of the utmost importance since pore-free coatings that are too dense lose their bioactivity, thus acting like bioinert materials. Hence, their eventual substitution by bone tissue is not guaranteed. On the other hand, the term *osseoinductivity* refers to the ability to transform undifferentiated mesenchymal precursor stem cells into osseoprogenitor cells that precede endochondral ossification. This process relies crucially on the osseoinductive action of noncollagenous proteins as will be discussed below.

Biocompatibility, osseoconductivity, and osseoinductivity of calcium phosphate-based bioceramic materials are generally attributed to properties that include (i) a chemical composition resembling that of the inorganic component of natural bone; (ii) a nano-structured surface topography; (iii) an appropriate macro- and microporosity; (iv) bioadhesion capability; and (v) favorable dissolution kinetics. However, as convincingly argued by Wopenka and Pasteris (2005), bone-like apatite present in biological tissue should not be called "hydroxylapatite" per se since its composition, characterized by the absence of hydroxyl ions, defect density, and stoichiometry deviate strongly from "geological" apatite.

According to current views on calcium phosphate osseogenicity, Ca^{2+} and $[PO_4]^{3-}$ ions released from dissolving HAp during interaction with body fluid profoundly affect the migration, proliferation, and differentiation of osteoblasts during bone formation. However, the detailed molecular mechanisms guiding *de novo* bone formation are still under debate (Chai et al. 2012).

The timeline of basic steps occurring at the bone-implant interface during osseointegration of an implant in the presence of osseoconductive calcium phosphate coatings can be summarized as follows:
– Formation of a thrombus after approximately two days
– Tissue reorganization by formation of a callus after approximately six days
– Phenotypical expression of undifferentiated cells toward osteocytes
– Bone remodeling

Interface reactions of newly formed osteoblasts are dependent on intrinsic coating properties such as
– Phase composition of the coating
– Degree of crystallinity
– Surface structure including roughness, fractality, and porosity
– Presence of metabolically important trace elements and bone growth–supporting proteins

4.4.4 Levels of interaction

4.4.4.1 Molecular level

At a molecular level, apatite nucleation is thought to be initiated by nucleation of calcium phosphate protonuclei (embryos) from the extracellular body fluid that is supersaturated with respect to HAp. *Ab initio* calculations suggest that the first products of precipitation involve the so-called Posner's cluster, $\{Ca_3(PO_4)_2\}_3$ that has the energetically most favored and hence, most stable configuration (Onuma et al., 2000). Nevertheless, at present it is not clear how the transition occurs from these clusters with $Ca/P = 1.5$ to HAp with $Ca/P = 1.67$, and how the presence of organic species will affect this conversion mechanism in detail. Experimental evidence obtained by *in situ* synchrotron small-angle X-ray scattering suggests that carboxylate ligands such as citrate and oxalate anions delay the onset of HAp nucleation, whereas noncollagenous proteins such as osteocalcin, osteonectin, and proteoglycans lead to enhanced nucleation (Schofield et al., 2005). In particular, osteocalcin shows high affinity for

HAp and appears to play a significant role in cell signaling for bone formation (Hoang et al., 2003). In addition, octacalcium phosphate ($Ca_8(HPO_4)_2(PO_4)_4 \cdot 5H_2O$, OCP) is a known precursor of HAp nucleation and appears to support significantly enhanced appositional bone formation (Shiwaku et al., 2012).

However, not only the precipitation kinetics but also the morphology of HAp micro- or nanocrystals will be modified by structure-mediated (epitaxial) adsorption of organic constituents such as poly(amino acids) at prominent lattice planes of HAp. For example, adsorption of poly(L-lysine) on {00.1} planes triggers the formation of polycrystalline HAp nanocrystals whereas adsorption of poly(L-glutamic acid) supports the precipitation of large flat micron-sized single crystals (Stupp and Braun, 1997). Adsorption experiments involving recombinant human-like collagen, and citrate and cetyl(trimethylammonium) bromide showed comparable relations. However, in this context it should be emphasized that aqueously precipitated carbonated, hydrated, calcium-, and hydroxyl-deficient HAp has size and shape similar to bone-like HAp crystallites, in the total absence of any biological agent. This has led to the assumption that the carbonate incorporation into the crystal lattice of HAp is sufficient to control both the size and shape of the apatite.

4.4.4.2 Cellular level

On a cellular level, bone mineralization is thought to originate in cell-derived microstructures called matrix vesicles by major influx of calcium and phosphate ions into the cells (Anderson, 1969). Within the plasma membrane of the vesicles, phosphatidylserine–calcium phosphate complexes are being produced, mediated by proteins such as annexins, integrins, and the hydrolase enzyme ALP. These enzymes cleave phosphate groups off phosphatidylserine and thus act as foci of calcium phosphate deposition. Hence, both Ca^{2+} cations bound to phospholipids and PO_4^{3-} anions released from the dissolving calcium phosphate biomaterial combine to nucleate amorphous calcium phosphate, which is converted to nanocrystalline HAp (n-HA) at the vesicle membrane (Luo et al., 2015). Eventually, matrix vesicles bud from the plasma membrane at sites of interaction with the ECM, and in this way provide calcium and phosphate ions, lipids, and proteins that act to nucleate biological apatite, apparently controlled by the speciation of the phosphate carrier. Indeed, pyro $[P_2O_7]^{4-}$ and polyphosphate $[P_nO_{3n+1}]^{(n+2)-}$ ions were found to inhibit mineralization, whereas hydrogen phosphate $[HPO_4]^{2-}$ ions appear to stimulate nucleation outside the matrix vesicles in the interstitial space (ECM) as well as directly on the triple-helical collagen I strands.

Moreover, HAp-based bioactive implant coatings promote normal differentiation in surrounding tissues by providing a fertile environment for enhanced cell adhesion (Hynes, 1992) and biocompatibility, including reduction of bacterial adhesion in dental implantology (Mandracci et al., 2016). Cytoskeletal microfilaments (see Chapter 4.2) such as actin, myosin, actinin, and tropomyosin known to control cell shape and migration will be coupled through specialized cell membrane proteins, so-called integrins to extracellular adhesion molecules such as fibronectin, laminin, vitronectin, or thrombospondin (Nair and Laurencin, 2006). An interfacial layer of HAp will adsorb these adhesion molecules in a favorable crystallographic conformation and, thus, promote the formation of focal adhesion centers. Particular growth factors, so-called cytokines adsorbed at specific HAp surfaces sites further promote osseointegration (Lobel and Hench, 1998). Growth-supporting cytokines include transforming growth factor-β (TGF-β), insulin-like growth factor-1, TNF-α or recombinant human bone morphogenetic proteins. These cytokines provide a degree of osseostimulation that supports the transformation of undifferentiated mesenchymal precursor cells into osteoprogenitor cells and eventually osteoblasts. This is at the very heart of the mechanism of osseoinduction.

Figure 4.11 shows a much simplified schematic representation of osseoinduction induced by adsorption and incorporation of cell membrane proteins into the

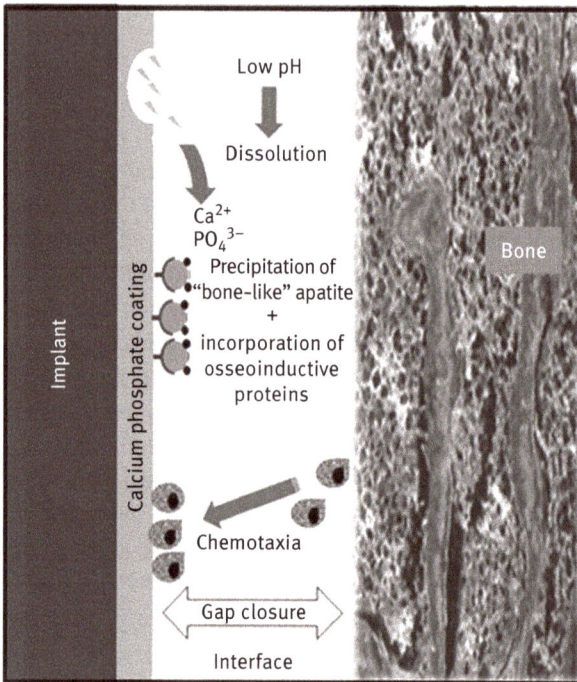

Figure 4.11: Schematic representation of cell-implant interaction mediated by a thin calcium phosphate coating layer (adapted and redrawn from Rahbek et al., 2001). A local decrease of pH results in partial dissolution of the coatings, triggering the release of chemotaxia from bone. Addition of Ca^{2+} and PO_4^{3-} ions leads to increased supersaturation of the extracellular fluid (ECF) with respect to hydroxylapatite, precipitating "bone-like" apatite, and promoting subsequent incorporation of osseoinductive proteins such as osteocalcin, osteonectin as well as annexins and integrins.

bone-like HAp layer precipitated onto the dissolving calcium phosphate coating. The increased concentration of Ca^{2+} and PO_4^{3-} ions appears to stimulate the chemotactical release of chemokines from the cortical bone bed (Rahbek et al., 2001). Immediately following the implantation, a space filled with biofluid exists adjacent to the HAp-coated implant surface (left) (see also Fig. 2.64). With time, bone growth-supporting proteins will be adsorbed at the calcium phosphate coating surface that will give rise to osseointegration by proliferation of stem cells and their differentiation toward osteoblastic bone cells, revascularization, and eventual gap closing. The direction of closing is a two-way affair as new bone matter grows into the gap from the cortical bone site as well as from the coated implant surface.

4.5 Long-term performance of surgical implants (Hans-Jürgen Kock)

4.5.1 Introduction

Worldwide, medical devices have been used routinely in the medical practice for more than 50 years and routine hip arthroplasty is considered among the most successful surgical procedures of the twentieth century.

Since 1995, it has become a standard for registration in the European Community (EC) to classify all medical devices according to the MDD into four classes according to their inherent risk for patients. Long-term implants such as hip and knee prostheses, heart valves, vascular stents, breast implants, and many others are to be registered into class III (MDR, 2017). Whereas class I devices present little or no risk to their users, the class IIa and IIb devices present some limited risks and, thus, are subject to performance standards. In contrast, long-term implants are to be registered into class III (MDR, 2017). They involve some unreasonable risks and consequently, require premarket approval prior to their widespread distribution, and extensive safety and effectiveness testing prior to commercial sale.

Since the early 1960s, hip implants need to address some special considerations related to long-term results that involve the study of the consequences of the long-term survival of manufactured biomaterials in implants. Historically, improved biomaterials and devices emerged from short-term animal studies and clinical observations that have eliminated undesirable material- and/or design-related performances on a case-by-case or small group study basis (trial-and-error principle). The result is that now only a small number of biomaterials are highly successful when used in a broad variety of designs, and that their use for many medical indications has become routine and standard.

In order to reliably assess the long-term performance of class III implants such as hip endoprostheses, one has to properly follow-up on these implants in their hosts for more than 10 years to gain insight into their long-term performance in the human body under daily life conditions. Although it is quite clear that the economic and medical impact of medical devices such as hip, knee and other implants is very large, one underscores the importance of continued research and development to improve the quality of health care by selecting the best long-term implants available worldwide and refrain from using those that do not perform well over time.

However, long-term success has produced a new set of problems that can be summarized as follows:

- Large-scale, routine clinical use of implants, even with low failure rates, produces significant numbers of patients whose implants fail to meet expectations in the long-term.
- Surgeons' clinical confidence in implants results in pragmatic extension of the indications for their use, especially to more difficult medical problems and to earlier intervention in disease processes. Routine use and earlier intervention produce an increasing mismatch between typically short development and evaluation cycles and longer intended (and actual) service periods.

Together, these factors have generated the need for certain types of data concerning biomaterials, implant designs, and other factors for all medical implants:

- Long-term effects of *in vivo* environments on biomaterial and implant properties
- Chronic (including systemic and remote site) effects of manufactured biomaterials and implants on human physiological processes
- Comparative service experience of different biomaterials and implants in similar or different device designs used for the same clinical application/indication

Because of the size and importance of this field of modern medicine, it is impossible to cover all implant applications in detail in this textbook. Therefore, only the classic hip endoprosthetic implant in its well examined long-term usage has been selected to be covered below. The purpose of this reduction toward the most successful surgical implant of the last century allows showing the strengths and weaknesses of the current state-of-the-art in medicine and engineering to obtain long-term insight into the overall performance of hip implants in modern societies.

4.5.2 Preclinical evaluation in the long-term perspective

This aspect has been explained and discussed for preclinical testing in Chapters 4.1, 5.1, and 5.2. The following deals mainly with the long-term consequences of the standard testing procedures.

4.5.2.1 Medical device design

Normally, only a few test specifications are evaluated so that time is not wasted conducting extensive testing if the material does not meet the most important design criteria. If initial evaluation or screening studies show positive results, a complete evaluation of the specifications is warranted and undertaken. Once a complete specification evaluation is achieved, the design is modified based on analysis of the test results. The design process is cyclic since the results of each step may lead to new hypotheses that, in turn, lead to continued testing until the prototype performs in a similar manner as the tissue to be replaced.

Many medical device designs have been studied in animals for decades with limited progress because designers or scientists are either unwilling to interpret experimental results carefully or are infatuated with "his/her" own designs. Although the success of any design is dependent on the choice of the design specifications and final end-use testing, before end-use testing is warranted, a series of laboratory studies must be completed that are directed at evaluating the safety and efficacy of the prototype.

4.5.2.2 Safety and efficacy testing

The general requirement for all preclinical tests is acute screening based upon *in vitro* and tissue culture techniques as outlined in Chapters 4.1 and 4.2 (see also Chapter 5). The tests conducted on a biomaterial intended for use in a medical device must address safety and effectiveness criteria as outlined in detail in several classic textbooks (Ciarkowski, 1986; Black, 1988). Although the specific tests required vary with the type of device and application, some general testing is usually recommended.

Normally, animal testing is conducted to demonstrate that a medical device is safe and that, when implanted in humans, the device will reduce, alleviate or eliminate the possibility of adverse medical reactions or conditions. Biological reactions that are detrimental to the success of a material in one device application may not be applicable in a different end-use. Nevertheless, for practical and economic reasons animal studies are normally carried out only over a period of several months up to 2 years and hence, are no particularly good indicators for long-term results in humans.

To estimate the biological safety of a biomaterial or a biomedical device, a plethora of tests is available that include (DIN e.V., 2017).

- *Cell culture cytotoxicity* studies evaluate the *in vitro* toxicity of substrate materials toward cultured cells. Cell death is indicated by the inability of cells to incorporate vital dyes. Nevertheless, no long-term results in the human body can be predicted by this test.
- *Skin irritation assay* involves applying a patch of the material to be evaluated to an area of rabbit skin that has been shaved or in some cases, abraded. After 24 h of contact, the patch is removed and the skin graded for redness and swelling. This test provides a first indication of acute skin irritation, but no long-term guarantee to avoid delayed skin-irritation or the potential for allergies.

- *Short-term intramuscular implantation* is designed to evaluate the reaction of tissue to an implant for periods of 7 and 30 days. At the conclusion of the test period, the samples are graded both visually and histologically. Although this test conveys an idea of the effect to be expected in the human body, it does not consider any effects of material debris and other long-term release of substances from the implant material. The metal-on-metal (MoM) experience with hip implants in recent years are an example how chromium and cobalt ions released from these implants can cause severe health problems in patients who had received these implants.
- *Blood compatibility* is normally assessed by determining the clotting times and degree of platelet aggregation initiated by a test surface or by blood exposed to a surface. Tests are conducted *in vitro*, *in vivo*, and *ex vivo*. Hemolysis is determined by placing powder, rods, or extracts of the material in human or animal plasma for about 90 min at 37 °C. The amount of hemoglobin released into solution is determined by measuring the absorbance at a characteristic wavelength after red blood cells have been removed by centrifugation. Hemolysis of less than 5% of the red blood cells is generally considered acceptable. Hemolysis is also measured *in vivo* by determining the red cell's half-life after implantation of a device and hence, provide some degree of safety for the recipients of implants made from these materials.
- *Carcinogenicity* testing involves long-term implantation to evaluate the potential for cell transformation and tumor formation. Nevertheless, there is long-standing discussion about long-term sequelae-like cancer caused be implant debris that can only be answered by watching all patients carefully over long-term periods.
- *Long-term implant tests* are governed by ASTM and DIN ISO specifications for muscle and bone, respectively. Implant materials are being placed in a muscle serving as a soft tissue model and in bone as a model of hard tissue. The implantation site is being evaluated globally and histologically for inflammation triggered by giant cell formation, and signs of implant movement and tissue necrosis.
- *Mucous membrane irritation* is evaluated by placing the material in close proximity to a mucous membrane. The amount of irritation/inflammation is determined from global and histological measurements. The hamster cheek pouch is a model that is becoming more frequently used for this test procedure.
- *Systemic injection (acute toxicity)* is designed to determine the biological response to a single intravenous or intraperitoneal injection of extract (50 mL/kg) of a material over a 72-h time period. Extracts are prepared in saline or other solutions that simulate body fluids. Animals are then monitored for immediate signs of toxicity and at specific time intervals, giving a general indication for

acute toxicity but not for long-term and organ-specific damage after decades of implantation periods in the human body.

- *Intracutaneous injection* involves the reaction of an animal to a single dose of a saline or vegetable extract of a material. Rabbits are commonly used and studied for signs of redness and swelling at the injection site for periods of 72 h.
- *Sensitization assay* involves mixing the material or an extract with CFA and injecting the material into subcutaneous tissue of animals during a 2-week induction period. After 2 weeks, the animal, commonly a guinea pig, is challenged with the material or extract by placing it in contact with the skin near the injection site for 24 h. The skin is graded for allergic reactions.
- *Mutagenicity* is evaluated using the Ames test (Ames et al., 1973, OECD (2017)). This test employs genetically altered bacteria that are placed in contact with an extract of the material to be tested. The bacteria have altered nutritional needs. Mutations that cause reversion back to the "wild-type" phenotype generate bacteria that will grow only under the original nutritional conditions but not under conditions under which the mutant grows. This test is used to screen materials for their carcinogenic potential.
- *Pyrogenicity* is used for fever-producing substances that are either a component of gram-negative bacterial cell membranes (endotoxins) or are materials of chemical origin. Endotoxins are determined by injecting an extract of material into the rabbit circulatory system and measuring the resulting elevation in body temperature. Another method involves contacting the material with cells that are lysed specifically by endotoxins. Chemical pyrogens are determined only by the rabbit test.

As a minimum requirement for all preclinical testing, the following parameters must be considered:
- Use of multiple species
- Identical control (reference) materials must be used in all tests
- Group sizes and sacrifice schedules must be of sufficient size
- Operative sites must be similar to those of the intended human application

If as result of these types of tests a new material considered for implants demonstrates that it is in its performance equal or superior to materials in present use, and if no extraordinary hazards specifically associated with it arise, the planning and execution of phases I and II of clinical trials appear warranted.

The requirements for such clinical trials are probably more stringent than those needed to demonstrate performance of new device designs. This is the case due to the subtlety of many materials' problems and the general "endorsement" that a new material may achieve inferentially after successfully completing its first clinical trial series.

It can be concluded that, although quite a variety of cell culture and animal tests are standard safety procedures required by health authorities in the USA and in the European Union to avoid acute and chronic damages of implant materials in humans, there is a remaining amount of uncertainty as to which long-term results are to be inspected in routine clinical use of novel materials and implants. This fact makes clinical testing in humans inevitable before any new biomaterial or any class III implants can allowed to be used clinically on a larger scale.

4.5.2.3 Need for clinical testing/evaluation

After material selection, device design, *in vitro* testing, and implantation in animals, the material under scrutiny must be tested in humans. Some aspects of designing and conducting of clinical trials will be briefly outlined below.

What has happened is that workers in the field of biomaterials have been blinded by success. The techniques used in the 1960s of the last century to qualify materials (limited *in vitro* studies, 12- to 104-week animal studies, 2-year human clinical studies) are still current practice in the early 2000s. This is despite the widespread use of bio-materials in long-term clinical applications that, in at least one device type (total hip replacement; Malchau et al., 2002), exceed 25 years in individual and group patient experience. It now appears that two critical aspects of the study of biomaterials have been neglected: the epidemiology and human physiology of biomaterials.

Each component of an implant has three elements: a functional element, a connecting element, and a structural element. This means that each component has a desired function related to its site of implantation. Now suppose that a certain material has been in use as an articulating (functional) element in the hip, but using it as an articulating element in the shoulder is desired. The temptation is to revert to the initial question of biological performance. However, this is not logical because there is (or should be) information on the actual, rather than conjectural, performance of the material in the biological environment of a human joint, being in this case only remotely similar to that of the new application (hip vs. shoulder). Then the appropriate initial question becomes: "Which data and clinical experience support the assertion that the material in question will prove safe and effective in the proposed (new) clinical application?"

A study toward answering this question would have the following challenge now: Considering the material in question (a ceramic used for femoral heads in the hip), in the proposed design in the proposed application (prosthetic replacement of the humeral head), do find and analyze the laboratory and clinical predicates that support the assertion that its use will meet appropriate regulatory standards for safety and effectiveness.

The analysis then proceeds to the following secondary questions: For each component of the proposed design fabricated in whole or part from the selected material, the question is "What are the laboratory and clinical predicates for friction and wear, structural integrity, and fixation (including possible adverse local and systemic reactions to wear debris and other degradation products) that would guarantee good long-term results in both indications (e.g., hip and shoulder arthroplasty)?"

"Safe" and "effective" are fundamental descriptors in medical device regulation. However, in case of biocompatibility, they cannot be defined or determined on an absolute basis. Therefore, the usual test provides satisfaction of locally prevailing regulatory definitions and standards, rather than any intrinsic de novo considerations. Note, however, that in U.S. experience, no legal connection exists between a regulatory decision that a device is safe and effective for a given set of indications and the actual observed clinical performance. Whereas this issue has been extensively litigated, pro and con, discussion of this complex topic is beyond the scope of this work.

For interfaces between different components of the proposed design, the question arises: "What are the predicates to support the assertion that the use of the material in question will not produce clinical outcomes inferior to those experienced with materials now in general clinical use in the proposed market?" It should be clear that the ability to answer this progression of questions depends acutely on the quality and quantity of data that exist concerning the actual material in question, its host responses in patients, and the overall clinical outcome as a function of time.

To decide which tests are necessary and sufficient before phase III clinical trials of new materials are warranted is extremely controversial. It would be very inviting to develop a consensus viewpoint or general matrix into which all new materials, material combinations, and material applications could be classified, thus settling the generic problem once and for all. At present, a number of working groups within various US and EU health authorities and international standards organizations, are examining this question. Progress over the last decades has been slow, and the desired answer is clearly a long space of time away.

4.5.2.4 Design of clinical trials

Before considering a clinical trial, it is wise to consider the exact goal of such an evaluation. Unless an implant is designed for acute use, trials cannot extend beyond a short fraction of intended device life. Furthermore, testing a new biomaterial in a particular device cannot result in qualification of the material itself as long as other components are part of the implant.

Therefore, clinical trials must be regarded as serving primarily as detectors of bad news. Considering the long-term sequelae of new implants, a clinical trial is only capable of a limited, controlled, well-observed introduction of a new material and/or design and does normally not indicate long-term safety.

The use of the term "introduction" is deliberate because, unlike in the case of animal trials, the experimental subjects will continue to be exposed to the device and its material components even after the period of observation. The longer the time elapsed and the greater the number of patients studied during such limited introduction, the greater the confidence in acceptable biological performance following general introduction without detecting adverse results.

In general, there are four types or sequential phases of clinical trials known to both researchers and medical device industry and to health authorities as well (Burdette and Gehan, 1970):

- Phase I (early trial): selecting a new treatment from among several options for further study
- Phase IIA (preliminary trial): if the new treatment is not effective in the early trial, this phase examines whether further studies should be performed or the treatment abandoned
- Phase IIB (follow-up trial): estimating the effectiveness of a new treatment that appears promising based upon phase I or phase IIA trials
- Phase III: comparison of the effectiveness of the new treatment with a standard method of management or some other treatment

During human testing of materials of devices, phases I, IIA, and IIB are rarely planned in a formal sense. Their function is usually fulfilled by the use of individual custom devices for selected patients under the direct care or supervision of the surgeon member of the research group. Only when the new material/device (usually in comparison to other material/device arrangements) is perceived to have a relative benefit does formal clinical testing begin with a phase III trial. This phase in implant development may be further subdivided into two subphases:

- Phase IIIA: examination of clinical outcome of a defined new treatment for a group of patients with defined indications
- Phase IIIB: following success in a subphase IIIA trial, examination of the clinical outcome for a defined (refined) new treatment for a group of patients with defined (refined) indications, usually involving multiple investigators and institutions.

It is standard practice in new drug trials to employ the double-blind method, that is, a drug and a harmless inactive material (placebo) are used in a treatment plan for a defined group of patients with a common set of symptoms. Which patients receive the drug and which the placebo is randomly pre–determined. Neither the patients (single blind) nor the treating physician (double blind) know who is receiving the active drug. When the experimental trial is complete, the identifying code assigned to the drug and placebo doses is deciphered and an analysis of effectiveness of the treatment is made. A further sophistication is employed in some designs that include a "crossed-over" treatment of the two groups or an interchange half–way through the trial. Thus, each test subject has a near equal chance to benefit from the drug or to suffer from adverse effects of the drug or the placebo because administering the placebo prevents the use of other (previous) drugs known to have beneficial effects on the patient's condition.

Once an implant is surgically inserted, whatever the phase of the trial, it is not possible to pair the implanted patient with a placebo-treated patient. The case of no insertion is clearly not blind to patient or doctor, and would be ethically or practically inacceptable in longer term trials. A study comparing identical devices made of different materials is at best blind to the patient. Differences in appearance, weight, shape, and so on between devices made of different materials usually render them easily distinguishable to the physician and the patient. Therefore, clinical trials of implant materials must be based upon different experimental designs.

Comparisons may be made as follows:
- Between the condition of the patient before and after implant surgery. This is useful to detect acute changes that may take place in an individual with underlying disease (for which the implant is indicated) that may be associated with response to the implant material.
- Among patients with similar implants made of different materials. When possible, this is useful to investigate acute and chronic differences in material and host responses.
- Between patients with implants and nondiseased (control) individuals of the same age and sex, and with a similar home/workplace environment. This may be useful to detect subtle chronic effects of materials (host response). Spousal or partner controls are used in many such studies.

A number of efforts have been made to set standards for selecting and treating patients in clinical trials. Some general rules that have emerged are:
- Medical care must be under the direction of a medical professional.
- The patients must give informed consent to any experimental procedure. This consent can only be obtained after a full explanation of possible benefits and risks of the proposed procedure in comparison to alternatives, including no treatment.
- The identity of the patients must be protected and the confidentiality of their medical records must be preserved.

Perhaps the most important point is that, for the trial to be justified, whichever the phase, there must be a reasonable possibility of specific benefit to the patients in the trial combined with reasonable assurance of the absence of unusual risk. The governing ethical considerations of which these rules are a part of are the 12 basic principles of the Declaration of Helsinki II, revised and extended by the 29th World Medical Assembly (Silverman, 1985).

Finally, the time will arrive when confidence in a new material has risen sufficiently high, so that clinical trials can be started with caution. All clinical trials must be performed under the control of a defined (written) prospective protocol that includes the following provisions:
- Description of the implant device (note that the implant site must be that of the proposed application)

- Outline of indications for the surgical procedure
- Outline of the uniform surgical procedure used
- Outline of postoperative treatment
- Outline of follow-up schedule and postoperative evaluation techniques

The following information are needed for adequate consideration and evaluation of clinical testing results (with respect to biological performance of materials):
- Protocol as listed previously
- Certification of 200 patients, by code number, constituting consecutive individuals seen by the treating physician and meeting item 2 of the protocol
- Results of follow-up of these patients for a minimum of 5 years and average of 7–10 years for those not lost to follow-up at an earlier date (minimum of 100)
- Summary of all adverse results (note that a statistical summary is desirable; individual results with code should not be reported)

The last point should be dwelt upon. Presumably, at the time at which the trial protocol was being developed, consideration was given to each of the questions to be asked and the statistical measures to be used in answering them. At the end of the trial, it is thus appropriate to suggest that statistical measures be employed. Therefore, reports of clinical trials should take care of the following requirements:
- Discuss accuracy and precision of all measurements, where possible
- Define a minimum confidence level for all statistical measures of the data, usually $p < 0.05$
- Report confidence intervals or other measures of significance associated with all derived parameters
- Indicate the significance of any conclusion arrived at by analysis of the trial

Furthermore, to detect long-term problems with implants, a Phase IV study has been implemented by health authorities in the USA and in the European Union. This requires postmarket studies of newly introduced surgical implants over realistic long-term periods to gain insight into the real long-term results in broad clinical use in different hospitals. To fulfil this new and high safety standard, the manufacturers of implants will be highly interested to have their implants used in countries with well-functioning and well-established implant registers as in Sweden (shpr.registercentrum.se) and Norway (http://nrlweb.ihelse.net).

In conclusion, it is clear that clinical trial protocols for both drugs and devices are difficult to design and implement. As an aid in such efforts, virtually all medical research and treatment facilities involved with patients maintain Human Subjects Committees (also known as Internal Review Boards, IRBs) or Ethics Committees. These committees are available to help preparing protocols and generally must review the procedures and safeguards in any experimental program involving human subjects before a clinical trial is started. In addition, many of the Device Classification Panels of the

Center for Medical Devices and Radiological Health of the FDA and European Health Authorities directly involved in the MDD have developed various guidelines for design of clinical trials and for statistical treatment, as well as forms to report results of new materials in novel implants. Since 1995, in the European Union the MDD regulated all prerequisites to be fulfilled before an implant is CE-marked by the authorities and allowed to be used in patients on a regular base. In May 2020, MDD is replaced by the Medical Device Regulation (MDR, 2017).

4.5.3 Clinical evaluation

4.5.3.1 Long-term clinical performance of implants

Biomaterials researchers, busy with studying the success and failure of implanted materials in experiments with animals numbering in the tens and twenties, have largely overlooked the vast clinical "experiments" under way during which thousands, in many cases tens or hundreds of thousands, of human patients receive virtually identical biomaterials as chronic implants.

With the exception of the occasional report on clinical failures and studies of the materials aspects of retrieved devices, almost nothing is known about biomaterials' performance in these implants in the human clinical environment. Dependable incidence and prevalence data on the devices are hard to obtain as such data on the materials from which they are made, including their exact (not merely specified) composition and processing, are still essentially nonexistent.

Today, making the same mistake as certain penologists who, wishing to know about crime, study only failed criminals (that very small nonrandom proportion of the criminal population actually apprehended, convicted, and incarcerated), one persists in studying only random device failures. Even these limited studies, based upon clinical or postmortem retrievals, are frequently incomplete, focusing on the clinical features of the failure or upon the physical attributes of the failed device, depending upon the background and interests of the principal investigator, but rarely dealing with both aspects in a balanced way.

Recently increased interest in studying the outcomes of surgical procedures including change in patient lifestyle, satisfaction level, and relative cost, rather than merely in the success of the procedure (rated as excellent, good, etc., on largely subjective bases) may improve matters, but only if accurate information on the implant and its materials of construction is made part of the permanent cycle of quality measurement and improvement of newly developed medical devices (Figure 4.12).

This is in stark contrast to many other countries in which device registries and resultant up-to-date statistics are now available for periods exceeding two decades of clinical use and clinical record. Suggestions have been repeatedly made concerning the need for registration systems for implants in the USA so that this information may follow patients as they move from place to place. Countries with national health

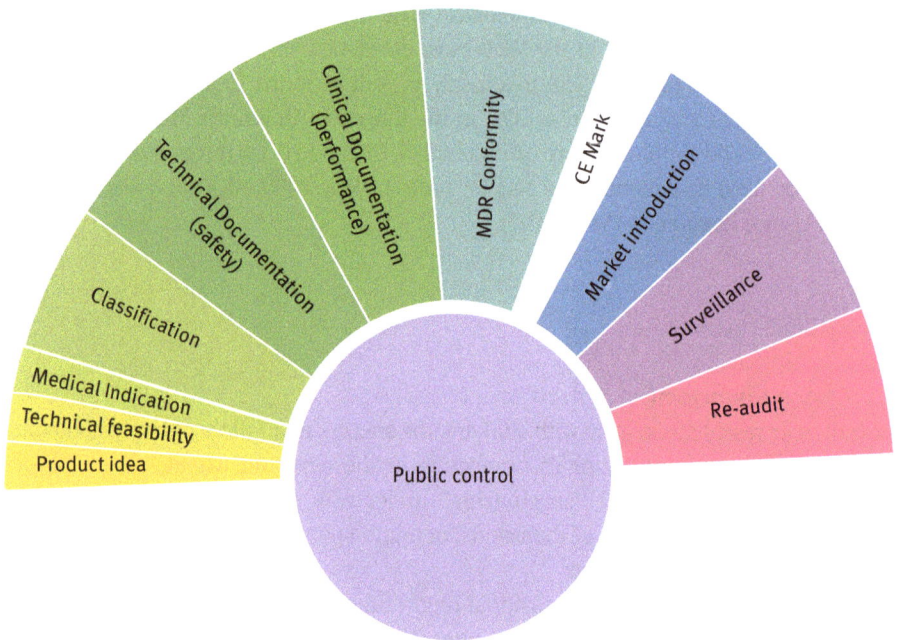

Figure 4.12: Development of medical products according to the **M**edical **D**evice **R**egulation.

services, such as the UK, have had some success in developing such national registries. National systems exist for registration of certain classes of implants such as hip replacements, for example, in Sweden (from 1979 onward) and Norway (from 1987 onward).

However, countries with predominantly private health care systems such as the USA have encountered difficulties in establishing such systems. Numerous proposals have been made for a national system for all permanent implants in the USA (Black, 1996).

4.5.3.2 Clinical long-term studies

As pointed out earlier, no clinical trial can approach the numbers and period of exposure that will be experienced when a material enters into general use. Thus, in a sense, it is imperative that clinical testing never ends. The treating physician and his or her consultants should distinguish between new and old materials, and should continue to be sensitive to possible biological performance problems associated with the use of either kind.

A well-designed and -conducted clinical trial does not serve to qualify a material. In addition to design studies, physical measurements, and acute and animal tests, it provides data that are required for a decision to release a product for general use.

The decisions made along the path to release should all be based on appropriate statistical tests at a minimum confidence level of 95%. Such release, when it occurs with the approval of the appropriate regulating agencies, is necessarily a risk. Hence, the benefits attendant to the use of the device incorporating the material in question that are felt at the moment of decision must outweigh the potential risks involved.

Throughout this book, aspects of biological performance, including material and host responses, have been considered. The latter sections of this work have begun to examine the details of test methods for examining biological performance leading to clinical use. Two factors are common to these methods:

- Large investments of time and money are required to produce results with reasonable levels of reliability and statistical significance due the variability of the biological systems involved.
- Large quantities of inductive reasoning must be used to apply the results of *in vitro* and animal testing to the prediction of clinical performance.

These two general conclusions overshadow all considerations of clinical testing and introduction into routine clinical use of new materials or old materials in new designs or applications.

4.5.3.3 General problems of clinical long-term follow-up studies

Prospective long-term follow-up studies are rarely found in the medical literature because hardly any patient with a severe health problem would like to serve the community as a "rabbit-model for new implants" deliberately as long as there are well-examined implants available to help with the patient´s health problem. Therefore, prospective research studies using well-designed, in advance planned, and modern data collection-supported methods such as stratified randomization into different treatment groups of study patients according to gender and weight, and individual risk factors such as diabetes mellitus or smoking are very difficult to uphold over decades, as both researchers and patients will not be physically at the same place over long periods for personal reasons involving changing houses, following career moves and many other reasons to be not available for clinical follow-ups.

As a consequence, on the one hand most research in the surgical field is done after design of the study in a retrospective *post factum* research way using existing data of implants already implanted many years ago and trying to find and follow-up as many survivors as possible with an acceptable low "drop-out rate" of patients lost to follow-up due to reasons such as relocation, death, revision operation in other hospitals, or just being unwilling to participate in a study. In this quite common type of research, "the study is planned after data collection," which gives it far less validity and insight into the true long-term performance of the implant under investigation. For most "long-term" implants, it is desired by both surgeon and patient that the implant functions for the entire remaining life of the patient. Thus, "long-term" analyses have to be done within a time frame of decades. For example, between 6 and 21% of patients undergoing primary total hip replacement are under 55 years old (data from the USA, UK, Australia, and Canada). Taking into account todays' life expectancy, long-term follow-up studies within or above 2–3 decades would be desirable (Kurtz et al., 2009; Canadian Institute for Health Information, 2017; National Joint Registry, 2015; Australian Orthopaedic Association National Joint Replacement Registry, 2014).

Prospective research on the other hand has also its own strengths and weaknesses and therefore has some limitations as a "golden standard" of clinical research (Boyko, 2013). Hence, a general limitation of observational research (prospective and retrospective) is that causal inference can be difficult or even impossible: correlation does not imply causation. Unknown confounding factors or reverse causation may be possible. The only possibility to exclude any known or unknown confounding factor is to assign the treatment under investigation randomly to patients. If treatment was chosen randomly, any relationship with confounders is also random and thus, causal inference is possible. Therefore, randomized controlled trials (RCTs) represent the highest level of evidence for therapeutic studies.

Alas, prospective long-term observational research is not attractive for researchers. Young researchers need publications to advance their career. Thus, planning research that is going to provide results only after decades is not attractive. Furthermore, young researchers usually have no access to sufficient resources. Although senior researchers do have access to resources, it is foreseeable that they will not profit from a new study during their career but most likely will retire before the study will be finished.

Prospective long-term observational research requires sufficient resources for maintenance of the data collection. Data providers have to be motivated to contribute and possibly sign contracts that even bind their successors for decades. Infrastructure (data storage, personnel) has to be planned and supported for decades in advance. Regular updates will be required to ensure comparability as used software may be outdated and no longer supported or data formats may change.

Furthermore, this disadvantage is not only limited to data storage systems and the format of databases but also to the data itself. A good example of these limitations of long-term follow-up studies is the commonly used so-called International Classification of Diseases (ICD) coding that are updated regularly (Table 4.1) and may not be easily comparable anymore to each other after several decades. Providers need to be very aware of "high-risk" codes. These are ICD-9 codes that map to multiple different ICD-10 variations. In fact, there are > 3600 instances in which ICD-10-CM codes can map to multiple different ICD-9-CM codes. Conversely, and more unusual, there are > 500 codes that are more specific in the ICD-9-CM than in ICD-10-CM (Hirsch et al., 2016).

In contrast, retrospective observational research is easier to handle regarding data collections as it could make use of already existing data. However, retrospective research has several limitations too (Boyko, 2013) that include:

Table 4.1: International Classification of Diseases (ICD) coding.

Year	Version	New content
1957	ICD-7	Coding relevant for categorizing needs for hospitals
1968	ICD-8	Focus on morbidity and mortality reporting
1977	ICD-9	Increased granularity (from 3 digits to 4–5 digits), diagnostic coding
1992*	ICD-10	Increased granularity (up to 7 digits, ~ 9fold increase)

*mandatory in the US since 2015.

- Patient selection may not be representative for the overall population because patients are selected based on their availability and disease status for case–control studies that could result in selection bias.
- Retrospective research is limited by data availability. If some data was not recorded, for example, during medical routine, it is also not available for analysis. This is particularly important for new questions that could depend on data that were of no relevance to caregivers decades ago or impossible to collect (e.g., because a lab test was unknown/not available that time)
- Lack of control groups is a major problem of case–control studies and, thus, cannot be assessed if an adverse outcome occurred rarely or often because no comparable reference level is available. Comparability to data from other studies is usually limited as patient populations or environments (e.g., due to advancement of medical knowledge in historical controls) may not be comparable. Nevertheless, case studies may be useful for safety research (adverse event reporting).
- Long-term follow-up data may not be relevant anymore when they become available because of advancements in medical science, or because products under investigation may be outdated and/or no longer available.

Health-care insurances may have long-term follow-up data on their records. However, the use of these data may be limited as they were collected for reimbursement. Thus, medical coding may be of insufficient detail (e.g., minor diagnoses irrelevant for reimbursement are not reported; Wolf et al., 2012; Kaafarani et al., 2009) and data items relevant for research might not be collected due to irrelevance for administrative purposes (Wolf et al., 2012; Kaafarani et al., 2009; Romano et al., 2002; Silverman and Skinner, 2004). Economically driven coding may result in false coding from a medical point of view, and changes of reimbursement could affect reporting frequency (Silverman and Skinner, 2004).

There are several other limitations of studies that are common in analyzing large claims databases. In general, there is risk of bias resulting from underreporting of outcomes like complications or patient co-morbidities. Underreporting of complications could inflate risk-adjusted outcomes. Furthermore, coding of co-morbidities and outcomes in claims databases requires a clinical suspicion and confirmation of that co-morbidity or outcome (usually with a test), and finally, documentation of the result (Evans et al., 2019).

Reviews of medical records have shown that risk estimates from claims data are good surrogates for the actual occurrence of major procedures and complications. However, studies have also shown that the completeness of such data could be limited for minor complications or cost-irrelevant outcomes such as surgical site skin infections or deep venous thromboses (Evans et al., 2019).

In conclusion, as a consequence, large prospective long-term follow-up research is mainly done by national registries. Retrospective studies are usually small and of lower quality. Nevertheless, there are a number of outstanding long-term follow-up publications from quite a number of implant- developing surgeons that may give additional insight into some data not available from larger data collection of the same implants in large national implant registries.

To evaluate the strengths and weaknesses of an implant over decades, it appears advisable to follow it up ideally on all 4 evidence levels (Table 4.2).

Table 4.2: Evidence levels (Greenhalgh, 2014).

Level	Evidence
1	Systematic reviews of randomized controlled trials (RCTs)
2	Randomized controlled trials (RCTs)
3	Observational studies (case–control and cohort)
4	Case studies, anecdote, expert opinion

4.5.3.4 Implant retrieval

Implant retrieval and evaluation offer the unique opportunity to investigate and study the intended use of biomaterials. Implant retrieval and evaluation programs, in general, are designed to determine the efficacy and safety, or biocompatibility of biomaterials, prostheses, and medical devices. Appropriate programs include assessment of safety and/or biocompatibility. The goal of safety testing is to determine if a biomaterial presents potential harm to the patient; it evaluates the interaction of the biomaterial with the *in vivo* environment and determines the effect of the host on the implant (see Chapter 4.3). Biocompatibility assessment is the determination of the ability of a biomaterial to perform with an appropriate host response in a specific application. In the *in vivo* environment, biocompatibility assessment is considered a measure of the degree (magnitude) and extent (duration) of adverse alteration(s) in homeostatic mechanism(s). The terms "safety assessment" and "biocompatibility assessment" may or may not be considered to be synonymous.

The response to these needs has long since been an effort, led primarily by bioengineers, to study implants and explants (retrieved devices) in a field that has come to be termed device retrieval and analysis (DRA). Early DRA efforts tended to focus on the disease state and viewed the device generically (medical or clinical pathology model) or to study the device closely, with little attention given to the generic disease or to individual patient conditions and use (engineering failure analysis model).

However, despite early and continuing perceptions that the data and the knowledge that can be gained from study of retrieved devices are of vital importance, most of these efforts continue to focus on the implant and on the implications of its physical condition (engineering failure analysis model). The seven major parties to DRA (patient, physician, manufacturer, treating institution, insurer, regulatory agency, and society at large) recognize a shared common interest in DRA and its outcomes. Nevertheless, individual benefit/risk calculations by each concerned party are in constant conflict (Black and Fielder, 1992) and have stood in the way of emergence of a unified analytical model or of comprehensive and/or multiinstitutional DRA programs.

While many implant failures can be characterized as either implant- or material-dependent, or clinical- or biological-dependent, many modes and mechanisms of failure are interdependent, that is, they are dependent on both implant nature and biological factors. To appropriately appreciate the dynamics and temporal variations of tissue–material and blood–material interactions of implants, a fundamental understanding of these interactions is important. Hence, an implant retrieval and evaluation program should provide for the appreciation of materials, design, and biological factors and result in design criteria for future development. Finally, implant retrieval and evaluation programs should offer the opportunity to determine the adequacy and appropriateness of animal models used in preclinical testing of implants and biomaterials. Thus, the strengths and weaknesses as well as the advantages and disadvantages of animal models can be evaluated.

Simple evaluation of the implant without attention to the tissue will produce an incomplete evaluation without understanding the host response. In general, analytical protocols and techniques for assessing host and implant responses can be divided into nondestructive and destructive testing procedures.

Only after appropriate accessing, cataloguing and identification, and after a complete review of the patient's medical history and radiography available, the analytical protocols and techniques for implant evaluation can be sufficiently specified. It should be noted that the techniques for implant evaluation are frequently destructive, that is, the implant or portions of the implant must be destroyed or altered to obtain the desired information on implant or material properties. The availability of the implant and tissue specimens will dictate the choice of technique.

DRA is a general term that covers a wide variety of activities, each of which has specific goals such as:

- Premarket approval clinical evaluation: goals may include determining normal and abnormal device and/or material performance, meeting regulatory requirements for reporting adverse outcomes, and providing feedback for device and/or surgical technique modification.
- Routine clinical use of a device: goals may include monitoring of appropriateness of device/ patient matching, determining specific device-related "failure" rates as part of survivorship calculations, establishing mechanisms underlying survivorship estimates, and planning future treatment for individual patients.
- Autopsy retrieval: goals may include investigation of local and systemic host response, long-term material property changes, and distribution and storage of implant degradation products.

In any of these or other activities, there should be formal written statements of and agreement to goals because such goals strongly affect the design and implementation of DRA studies.

DRA today is, by necessity, a "team sport" brought about by the complexities of medical practice and the social and legal setting of the early twenty-first century. Before undertaking such activities, it is necessary to establish a supporting organization and, for the parties involved, to agree on a number of assumptions. The issues to be dealt with include goals, responsibility, methodology, and reporting as detailed below.

Any DRA study should be conducted under the direction or a group of interested, appropriately trained, and committed individuals. The issues raised by DRA are such that careless or unplanned activities can produce extremely adverse outcomes for many of the parties involved. There should be a written, agreed upon set of procedures and methods. One individual, preferably a Ph.D.-trained person with experience in DRA, should have final responsibility for the program and should be designated as the custodian for the devices between their surgical recovery and their discharge from DRA study.

Methodology

In its most general form, DRA is an example of discovery science. Thus, it is inappropriate, except in very small, tightly focused studies, for all recovered devices to undergo a fixed set of procedures. Such an approach, especially in the usual setting of routine clinical practice, would produce insupportable costs without returning commensurably valuable information. Hence, it is good practice to classify devices prospectively before study.

Reporting

Timely reporting of results is the key to sustained and useful DRA studies and should take the following forms:

1. Rapid reports should be made to the treating physicians, on a time schedule parallel to that in the treating institution for clinical pathology studies. In addition to a personal report to the primary physician, a note should be placed in the hospital chart, over the signature of the DRA supervising professional. Although it is unusual for such individual studies to have an impact on the further treatment of the patient in question, such reporting is simply good practice and helps to maintain the professional relationships needed for DRA studies.
2. In the case of class I and II devices, reports should be made directly to the manufacturer. This is particularly required if more than one set of similar findings occur during a study. Sensitivity should be shown to studies of class I devices: because of the possibility of litigation in such cases, reports should focus on physical findings and descriptions but omit theories of causation, unless they are supportable by peer-reviewed, published studies.
3. Depending upon the nature of the findings and, in the case of a class 1 device, the degree of patient involvement, timely reports may need to be filed with a regulatory agency, such as the FDA in the U.S.
4. Because this is a legal requirement, the treating institution should have already established an approved procedure for submitting such reports as a part of its device recovery system. DRA personnel should be aware of such a system and provide rapid access to their findings for the individuals responsible for making the report.
5. Finally, there is a responsibility to make results of studies of groups of retrieved devices available to the scientific and engineering community. Such reports might include publication, presentation at professional meetings, and participation in workshops and instructional courses. It cannot too strongly be emphasized that, except in reports to physicians within the treating institution, great care must be taken to preserve the privacy and identity of the patients involved. This is an ethical and legal requirement (Black and Fielder, 1992).

4.5.3.5 Common concerns about device retrieval and analysis

Implant retrieval and evaluation must also take the risk of bias into account. The findings from individual groups or hospitals on specific implants may not necessarily represent the norm regarding the implant-host interactions for that implant. In particular, the surgeon's experience, technique, and preference for a specific type of implant must be considered. In addition, the clinical failure of an implant versus failure of the implant itself must also be considered. Implant retrieval and evaluation should be viewed as a continuous process that provides numerous clinical benefits.

It is a truism that one can only really learn about the clinical performance of biomaterials by actually examining that clinical long-term performance. Knowing and accepting this situation means that a vast human experiment is underway; significant numbers of patients now have had chronic devices *in situ* for more than 20 or even 30 years. The time when important new discoveries about the biological performance of biomaterials can be made in the laboratory or in limited animal studies without primary reference to this clinical experience has probably passed. Device retrieval and analysis, metastudies, national data systems, and autopsy retrieval programs will come to play important roles in obtaining data and insight to benefit future generations of patients.

In discussions of clinical evidence for materials incompatibility, that is, failure to exhibit adequate levels of biological performance, time and again, incidence rates of complications are small. For instance, the "variant poppet" problem in early heart valve designs probably did not affect more than 3% of patients receiving the prosthesis. The more adverse experience with an early design of anterior cruciate ligament replacement prosthesis (Chen and Black, 1980) resulted in failure in a larger percentage of cases, but still probably less than 20%. In many high-use applications, such as total hip and knee joint replacement, device-related failures of all causes, including failure of biological performance, are now appreciably less than 1% per year after surgery.

Why then is concern expressed here for examining biological performance? A manufacturer could fairly argue that modern implant materials have proven beneficial for 95 to 99% of patients receiving them. Similarly, a study conducted by the Carnegie-Mellon Institute on costs and benefits associated with research to improve the then current orthopedic prosthetic devices concluded that failure/complication rates, even in the 1970s, were acceptably small. Complication rate data assumed that the investment required to reduce these rates significantly would be costly, out of proportion to the resulting benefits (Piehler and Crichlow, 1978).

This remains clearly the case today, with an additional problem that in some cases, such as total hip replacement, the present technology is so successful that, even if they were actually superior to current choices, newer developments, such as changes in the materials of the articulating wear pair, cannot reasonably be expected to demonstrate statistical superior clinical outcomes due to practical limitations on size and duration of clinical trials.

In the US, this humane view on individuals rather than on average statistics has resulted in a dramatic rise in medical malpractice and damage suits associated with device failure (Kiplinger, 1991). The test of adequate performance laid out by law is that the device be "safe and effective" and expose the patient to no "unreasonable" risk or hazard.

Thus, decisions on what will be an acceptable failure rate remain subjective and depend on the continued interaction of public opinion, expert advice, and administrative action. Individuals are torn between two views. The heart says that no failure is acceptable, especially if it were to happen to oneself or to one dear to oneself. The mind says that such a goal may be unattainable and approachable only at prohibitive cost.

In conclusion, DRA today is, by necessity, a "team sport" brought about by the complexities of medical practice and of the social and legal setting of the early twenty-first century. Before undertaking such activities, it is necessary to establish a supporting organization and for the parties involved to agree on a number of assumptions. The issues to be dealt with include goals, responsibility, methodology, and reporting. Only recently the AE (Arbeitsgemeinschaft Endoprothetik, www.AE.de) has released practical guidelines for device retrieval analysis that are easily accessible on the AE homepage, giving a certain state-of-the-art account on this field.

4.5.4 Total hip replacement – An example of long-term result evaluation

Even in the absence of data on clinical epidemiology and the development of clinical tests of biomaterials' performance, it is still possible to gain some knowledge from actual clinical experience. A clinical internship and, if possible, continuous contact with clinic personnel can be extremely valuable for biomaterials scientists or engineers, in particular if they are observant and analytical in approach.

Further information about state of the art in hip arthroplasty can be found in Evans et al. (2019), Claes et al. (2009), Dreinhöfer et al. (2009), Knahr (2013), and Walenkamp et al. (2001).

4.5.4.1 Role of registries for arthroplasty

Patients are exposed to medical devices such as implants in hip replacement surgery for several years, possibly even decades. Therefore, not only acute complications and adverse effects are relevant for evaluation of the device's performance but also long-term outcomes in particular. This causes several problems.

First, a long-term follow-up of real-life performance is a challenging task. It is not trivial to trace patients for evaluation of their outcome over long periods of time, in particular if they do not experience complications with the device. Then, patients might stop follow-up visits with the surgeon who performed the implantation. Patients might also move on to another health care provider who then would possibly have follow-up data but cannot connect them to the original data at the time of the implantation (Malchau et al., 2018).

Second, it may be challenging to actually measure the long-term performance of a medical device. Several patient-related outcomes like quality of life and hip scores (e.g., Harris Hip Score, Oxford Hip Score) are important for patients. However, quality of life scores may not only depend on the implant itself but also on several independent factors. In addition, the measurement of such outcomes could depend on the person performing the measurement and thus, might be less objective and reproducible. Whereas such outcomes are considered as "weak," outcomes referred to as "hard" are objective and unambiguous such as death or implant failure (Malchau et al., 2018; Lübbeke et al., 2017). An additional limitation specific to implants is that their performance may also or (in the worst case) even predominantly depend on the skills and experience of the implanting surgeon, but not on the quality of the device itself.

A further common problem with medical devices is the lack of RCTs. Because such studies may not be required for market entry of medical devices, data about efficacy and adverse effects might not be available. For example, this has been claimed for about 24% of medical devices in the UK (Lübbeke et al., 2018).

Finally, the long-term use of implants can result in a limited comparability of data over time. During the long time periods that are typical for the use of implants, major advancements of medical science causing changes of standards of care and different availability of diagnostic procedures, as well as product updates could frequently occur. As a result, the outcome of devices might be hardly comparable over time, for example, it might no longer be relevant when product features have changed substantially.

Several problems described above can be at least partially solved by clinical registries in hip replacement surgery. Registries collect institutional, regional, or national data in order to analyze them and to identify associations between patient outcome, surgical technique, and implant associated risk factors (Malchau et al., 2018)

Joint replacement registries are common in Europe, especially in Scandinavian countries, and international initiatives have been developed to harmonize data collection (e.g., International Society of Arthroplasty Registries, Network of Orthopaedic Registries of Europe, and Nordic Arthroplasty Register Association) (Havelin et al., 2009; Sedrakyan et al., 2011). These registries are mostly run by academic researchers and/or national authorities.

National registries such as those established in Scandinavian countries could help resolving the problem of structured follow-up. If patient data are collected in a central database, correct assignment of follow-up to implantation data is possible based on patient names, demographic data and nationally used identifiers, for example, in residents registration systems or the healthcare system. Furthermore, a "hard" endpoint is available in hip replacement surgery for measuring performance: the need for revision surgery. It is an important indicator of implant quality and reproducibility, and allows comparing between different patient groups and surgeons (Lübbeke et al., 2017).

The dependence of implant performance on experience and skills of surgeons is still a problem in outcome evaluation. However, since registries usually collect data of a large number of patients from different clinics, this could become less relevant or can be addressed by statistical analysis. Still, registries cannot replace RCTs as they are only observational studies, resulting in the limitations mentioned above (see Chapter 4.5.3.4).

Changes in medical practice or product updates can also not be resolved by registries. Nevertheless, registries may enhance comparability of data over time by promoting standardized data collection and by maintaining the required infrastructure to allow a long-term perspective.

Today, registries about outcomes associated with medical devices are particularly important owing to the new regulatory environment established in the EU in 2017 (European Parliament and Council (2017)). These regulations include a tightening of postmarket surveillance requirements for manufacturers. Thus, registries for postmarketing surveillance will be key tools for product evaluation in this new regulatory environment (European Parliament and Council, 2017; Lübbeke et al., 2017).

A list of well-established national registries can be found in Table 4.3.

Table 4.3: Examples of national joint (hip) replacement registries with long-term follow-up.

Country	Registry	URL
Australia	Australian Orthopaedic Association; National Joint Replacement Registry	https://aoanjrr.sahmri.com
Denmark	Danish Hip Arthroplasty Register (DHR)	http://danskhoftealloplastikregister.dk
Finland	Finnish Arthroplasty Register	https://www.thl.fi
New Zealand	NZOA Joint Registry	https://nzoa.org.nz/nzoa-joint-registry
Norway	The Norwegian Arthroplasty Register	http://nrlweb.ihelse.net/eng/
Sweden	Swedish Hip Arthroplasty Register	https://shpr.registercentrum.se
The Netherlands	Dutch Arthroplasty Register	http://www.lroi-rapportage.nl/
United Kingdom	National Joint Registry for England, Wales, Northern Ireland and Isle of Man	http://www.njrcentre.org.uk

4.5.5 Summary and conclusion

It is a truism that one can only really learn about the clinical performance of a biomaterial by actually examining its clinical performance. Therefore, a vast human experiment is underway since the early sixties of the last century and significant numbers of patients now have had chronic devices *in situ* for more than 20 years. The time when important new discoveries about the biological performance of biomaterials can be made in the laboratory or in limited animal studies without primary reference to this clinical experience has probably passed and most recent failure in broad clinical use of "innovative" material combinations like "MoM" hip arthroplasty dramatically shows the limits of our knowledge, memory, and safety assessments in introducing new materials and design into clinical use.

Device retrieval and analysis studies, national data systems like arthroplasty registries, and autopsy retrieval programs will continue to play important roles in obtaining data and insight to benefit future generations of patients. In other words, better understanding of biological performance of devices and their materials of construction in actual clinical settings is and will remain important and necessary. It should be clear that the ability to answer the progression of questions raised in this chapter depends acutely on the quality and quantity of short and long-term data concerning the actual physical material in question, its material and host responses in patients, and the overall clinical long-term outcomes as a function of time.

The second frame of reference that should be briefly examined with respect to cost/risk/benefit aspects of material introduction is that of the future. How can one judge when a new material or device is ready to enter use and perhaps supplant present, apparently less effective products?

The answer is rather simple. If the ideas of safe failure modes in design, redundancy in testing, and, in particular, continuing evolutionary performance improvement are adopted, a real distinction between old and new materials will no longer be relevant. No one would knowingly substitute an inferior new product for a current well-functioning product unless driven by inhumane motives.

With this in mind, there is hope that progressive attitudes on the parts of researchers, manufacturers, surgeons, and regulatory authorities will lead to a continuing upgrading of the performance of current materials and the gradual introduction of new materials when subjective levels of safety and efficacy, most probably defined by then-current experience, are reached.

Critics would suggest the need for some absolute (minimum) level of safety that must be obtained before introduction of a new material. With respect to devices, it has been proposed that the following definition is used:

"A device is safe enough to use when it is no worse than others in use and present no greater hazard than the condition it is to be used to treat."

This appears clear enough in instances in which large improvements in devices (and materials) can be demonstrated and the conditions treated are life-threatening.

In situations in which a new material represents an evolutionary change in composition and/or processing and in subsequent behavior and the goal is to improve the quality of life of the patient by alleviating a condition of low mortality and morbidity, such a statement is a rather poor guide.

One can pass judgment, however, on the increasing trend toward market-driven rather than technology-driven introduction of new devices and materials. Clinical experience with many existing materials now exceeds three decades.

Consequently, it is very difficult to argue that short-term (2–5 years) testing of new materials might be capable of revealing their subtle or long-term defects or, more directly, of providing the information needed to determine whether the new material is equal to or exceeds the performance of the older material that it may replace simply on a novelty basis.

However, one can expect to avoid future problems associated with inadequate biological performance of materials in patients only through the bioengineer's endorsement of the Hippocratic injunction to do, in the first place, no harm.

Chapter 5
Quality management and regulatory affairs

Robert B. Heimann

5.1 Quality management of biomaterials and biomedical devices

Manufacturers of biomaterials are required to operate an extensive quality management program (see box) during which materials are thoroughly evaluated and tested before being released to the general practitioners to assure conformity with standard specifications (see Chapters 5.1.1 and 5.2; Appendix A). Before a biomaterial or biomedical device is allowed to be commercialized, several interdependent quality assurance (QA) and control steps must be adhered to. To achieve this, a quality management system (QMS) needs to be established according to ISO 9001:2015 that "specifies requirements . . . whereby an organization (i) needs to demonstrate its ability to consistently provide product or service that meets customer and applicable statutory and regulatory requirements, and (ii) aims to enhance customer satisfaction through the effective application of the system, including processes for continual improvement of the system and the assurance of conformity to customer and applicable statutory and regulatory requirements. All requirements of this document are generic and are intended to be applicable to all organizations, regardless of type, size and product provided" (ISO 9001, 2015).

Quality management

Quality has been defined as fitness for use, conformance to requirements, and the pursuit of excellence, summarized under the tenet "doing the right things right, the first time, every time." Even though the concept of quality has existed from early times, the study and definition of quality have been given prominence only in the last century (https://asq.org/quality-resources/quality-assurance-vs-control). QMS contain two interrelated aspects: quality assurance (QA) and quality control (QC) as defined by ISO 9000:2015: Quality management systems. Fundamentals and vocabulary. On the one hand, QA focuses on providing confidence that quality requirements will be fulfilled. The confidence provided by QA operates on two levels: internally to management and externally to customers, government agencies, regulators, certifiers, and third parties. On the other hand, QC can be defined as that part of quality management that focuses on actually fulfilling quality requirements. In other words, while QA relates to how a process is performed or how a product is made or service provided, QC addresses the inspection aspect of quality management. Inspection involves measuring, examining, and testing to gauge characteristics of a product or service and the comparison of these measured or tested characteristics with specified requirements to determine conformity (standards specifications; see below). Inspection protocols are designed to ensure that the object coming off a production line, or the service being provided, are correct and meet predetermined specifications. Inspection differs from auditing that is only involved with the evaluation of the process and the

https://doi.org/10.1515/9783110619249-005

controls covering the production and verification activities but is not required to carrying out any verification activities per se that lead to the actual acceptance or rejection of a product or service. Hence, audit and inspection are not interchangeable (Mill, 1988).

5.1.1 Standard specifications

Today, many standard specifications issued by both national (e.g., American Society for Testing and Materials (ASTM), Deutsches Institut für Normung (DIN) Comité Européen de Normalisation (CEN), and Japanese Standard Association (JSA)) and International Standards Organizations (ISO) are available as guide to how to effectively maintain quality levels of biomaterials, biomedical devices, and their production processes (see Appendix A). For example, German DIN standards are the results of work at national, European, and/or international level, whereby anyone can submit proposals for new standards. Once the relevance of this demand has been accepted, the standards project is carried forward according to set rules of procedure by the relevant DIN Standards Committee, the relevant technical committee of the European standards organization CEN or the relevant committee at the ISO.

Definition of standard
A standard is a rule, principle, or measure established as a model or an example by government authority, established custom, or general consent. Standards generally are in the form of baseline specifications according to which manufacturers can develop products with the assurance that they will interconnect and interoperate with those of other manufacturers, at least at a fundamental level.

Ideally, all stakeholders are involved in this procedure: materials and device manufacturers, consumers, research institutions, public authorities, government bodies, and testing organizations. These stakeholders delegate experts to represent their, frequently antagonistic, interests within DIN's working bodies that are overseen by some 70 standards committees, each of which is responsible for a specific subject area. For work at European and international levels, the DIN standards committees appoint acknowledged experts to represent German interests within CEN and ISO, respectively. DIN staff members coordinate the standardization process and are responsible for overall project management, ensuring the uniformity and consistency of the German standards collection. Standards are developed with full consensus, meaning that they are developed by experts aiming to arrive at a common understanding. All DIN standards are reviewed at least every 5 years. If a standard no longer reflects the current state of technology, it is either revised or withdrawn (https://www.din.de/en/about-standards/din-standards).

Specifications (see Appendix A) normally provide thorough details of (i) the testing of certain materials and products, (ii) the method of calculating the results including statistical deviations, and (iii) the minimum permissible result that is acceptable.

5.1.2 Laboratory evaluation

5.1.2.1 *In vitro* testing

In vitro laboratory tests are being used to indicate the suitability of certain biomaterials for certain application in biomedical devices and the field of medicine in general. It is of crucial importance that methods used to evaluate materials in the laboratory provide results that can be correlated with clinical experience. Foremost, *in vitro* incubation tests of metallic, ceramic and polymeric biomaterials in simulated body fluid (SBF) provide information on biocompatibility, chemical stability, and, to a limited extent, bioactivity. It must be emphasized that precipitation of bone-like apatite *in vitro*, frequently taken as hallmark of bioactive behavior, is only a necessary but not a sufficient indication of a positive outcome *in vivo*. To establish bioactivity, cell proliferation, vitality, and spreading test must be conducted as well as measurement of markers of bone growth including upregulation of proteins that signal such behavior. There are many SBFs available the compositions of which attempt to emulate those of blood plasma and human extracellular fluid (hECF), respectively. Aqueous solutions simulating the composition and the overall ionic strength of hECF ought to have compositional ranges as follows (in mmol): 142–145 Na^+, 1.8–3.75 Ca^{2+}, 0.8–1.5 Mg^{2+}, 5–5.8 K^+, 103–133 Cl^-, 22–27 HCO_3^-, and 0.8–1.67 HPO_4^{2-} (Tas, 2014). SBF within these compositional ranges are known to form nanospheres of amorphous calcium phosphate that on aging may be converted via octacalcium phosphate to bonelike, that is, calcium-deficient hydroxylapatite. This *in vitro* apatite-forming ability measured by the degree of interaction of biomaterials with SBF is a popular predictor of *in vivo* bioactivity. However, whereas researchers in the field have argued both in favor and against this *bona fide* assumption, the question of its validity is still open as actual experimental confirmation of this contention has not fully been obtained yet. Certain guidelines have been devised by Zadpoor (2014) for designing SBF immersion test protocols and interpretation of the test results. These guidelines could assist researchers in designing improved SBF test protocols that have better chances of predicting correctly the bioactivity of biomaterials for potential application in clinical settings.

To ensure that a designated laboratory produces accurate and precise results, proficiency testing is an indispensable requirement (see box).

Proficiency testing

Proficiency tests, as known as round robin tests or ring trials, intend to check the performance of laboratories. It is a key element in the laboratory accreditation process, alongside reference materials, enabling laboratories to monitor the quality of their analytical results as stipulated in DIN EN ISO/IEC 17025:2018-03. It is often referred to as external quality assessment, especially in the medical/clinical arena. For the purpose of proficiency testing, identical samples are sent out to several laboratories, who then analyze the samples under specified conditions. The test results are then collected and processed statistically to evaluate the accuracy of measurement carried out in the laboratory under surveillance. According to ISO standard, accredited testing laboratories must participate in proficiency tests regularly to ensure the quality of their analysis.

An example of *in vitro* testing of the adhesion of cells to biomaterials, mediated by integrins, and the influence of integrin–matrix binding on cell apoptosis has been provided in Chapter 4.2.

5.1.2.2 Cell proliferation and viability tests

To confirm bioactivity, cell proliferation, vitality, and spreading, as well as the absence of cytotoxicity, tests are being conducted that involve measuring certain markers of bone growth including monitoring alkaline phosphatase (ALP) activity and upregulation of other noncollagenous matrix proteins. In addition, AlamarBlue® and MTT assays are popular methods to establish cell viability and proliferation. *In vitro* staining with acridine orange (AO) and ethidium bromide (EthB) is used to color individual cells or cell structures that have been removed from their biological context (see below). Combined with specific protocols for fixation, sample preparation and investigative methods such as fluorescence and confocal laser scanning microscopy, these standard staining techniques are used as consistent, repeatable diagnostic tools.

ALP is a hydrolase enzyme that is responsible for cleaving off phosphate groups from many types of biomolecules, including nucleotides and proteins, a process called dephosphorylation. On a cellular level, bone mineralization is known to originate in cell-derived microstructures called matrix vesicles formed by a major influx of calcium and phosphate ions into the cells. Within the plasma membrane of the vesicles, phosphatidylserine–calcium phosphate complexes are being produced that will be mediated by several noncollagenous proteins (NCPs) including ALP that by cleaving off phosphate groups act as foci of calcium phosphate deposition. More precisely, phosphate is actively produced by a combination of secretion of phosphate-containing compounds, including ATP, and by ALP that cleaves off phosphate to create a high concentration of phosphate at the mineralization front. Hence, ALP is a characteristic marker of osteoblast development since its concentration increases strongly during osseogenesis. The action of hydroxylapatite and BMP-2 increases cell adhesion, ALP activity, and gene expression of osteogenic markers, which are potentially useful in the development of advanced medical devices. For details see Chapter 4.1.

Monitoring the expression of other NCPs (see box Chapter 2.2.2.1) is an additional important test to establish biocompatibility of biomaterials. The bone matrix is a highly hierarchical organized and complex tissue, consisting of inorganic hydroxylapatite and an organic component comprised of proteins and proteoglycans. In addition to the collagen I framework, NCPs are present, including osteocalcin, the phosphoprotein osteopontin, osteonectin, and bone sialoprotein (BSP) II.

Osteocalcin is exclusively secreted by osteoblasts and thought to play a role in the body's metabolic regulation and bone-building ability. It is also implicated in bone mineralization and calcium ion homeostasis. As osteocalcin is produced by osteoblasts, it is frequently used as a marker for the bone formation process when upregulated. Many research studies have indicated that higher osteocalcin levels are relatively well correlated with increase in bone mineral density during treatment with anabolic bone formation drugs indicated to treat osteoporosis. Hence, determination of osteocalcin levels is used as a preliminary biomarker on the effectiveness of a given drug on bone formation as well as the osseoconductive action of biomaterials. The function of osteocalcin was found to depend on the presence of a negatively charged protein surface that places five calcium ions in positions complementary to those in hydroxylapatite. Using this topotactic recognition mechanism, osteocalcin is thought to modulate the morphology as well as the growth rate of hydroxylapatite crystallites and to promote the adhesion of osteoblasts and osteoclasts during bone replacement (Hoang et al., 2003).

Osteopontin (formerly known as BSP I) binds tightly to hydroxylapatite in bone and thus forms an integral part of the mineralized matrix. It has been postulated to be a ligand for the vitronectin receptor (see Chapter 4.2.1), and this suggests a possible role of osteopontin in osteoclast attachment and function. Hence, it functions as a potent marker of ossification and consequently its expression serves to quantify the degree of mineralization of bone tissue in contact with biomaterials.

Both osteocalcin and osteopontin appear to be closely linked in bone repair mechanisms. Higher osteocalcin levels were found on bone fractures, thus creating tiny holes in healthy bone tissue. These holes, called dilatational bands, function as a defense mechanism that helps to prevent further damage to the surrounding bone structure. However, if the force of impact exceeds the yield strength of the bone or if the bone lacks either osteopontin, osteocalcin, or both, then the bone will be prone to fracture. The implications of this discovery may presage new strategies and therapeutic treatment routes to prevent osteoporosis and lower the risk of bone fracture in the elderly (Poundarik et al., 2012).

Osteonectin is a calcium-binding bone glycoprotein that is secreted by osteoblasts during bone formation, thus initiating mineralization and promoting Ca-deficient hydroxylapatite crystal nucleation. High levels of osteonectin are found in active osteoblasts and marrow progenitor cells, odontoblasts, periodontal ligament and gingival cells, and some chondrocytes. Osteonectin has also been detected in the osteoid, bone matrix proper, and dentin.

The search for a protein-based promoter of hydroxylapatite crystal nucleation has been a focus for the isolation and characterization of the major NCPs in bone. Of the proteins characterized to date, BSP *II* has emerged as the only *bona fide* candidate for hydroxylapatite nucleation. BSP is a highly glycosylated and sulfated phosphoprotein found almost exclusively in mineralized connective tissue and has been suggested to constitute approximately 8% of all NCPs found in the bone.

Analyses of proliferation of cells spread on biomaterials and evaluation of the potential cytotoxicity of the biomaterial are vital steps in establishing cellular health. *AlamarBlue*® is an established indicator for cell viability that uses the reducing power of living cells to convert blue resazurin to red fluorescing resorufin (Scheme 5.1), thereby producing very bright red fluorescence and thus, generating a quantitative measure of cell viability and cytotoxicity. Previous studies have shown that the AlamarBlue® test results in improved sensitivity and performance compared to MTT assays (Hamid et al., 2004).

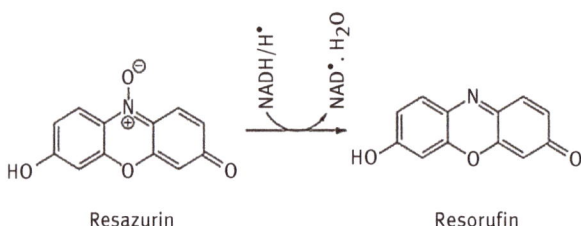

Resazurin   Resorufin

Scheme 5.1: Structures of resazurin and resorufin. The transformation from blue resazurin to red fluorescing resorufin by reduction is catalyzed by the redox reaction of nicotinamide adenine dinucleotide (NAD +, coenzyme 1) to NAD dehydrogenase (NADH).

The *MTT assay* is another well-known colorimetric method to assess cell viability. Under defined conditions, cellular oxidoreductase enzymes will reflect the number of viable cells present. The action of enzymes consists of reducing the yellow tetrazolium dye MTT, 3-(4,5-dimethylthiazol-2-yl)-2,5-diphenyltetrazolium bromide to its insoluble purple-colored formazan (Scheme 5.2). Solubilizing solutions, for example, dimethyl sulfoxide or sodium dodecyl sulfate dissolved in diluted hydrochloric acid are added to dissolve the insoluble purple formazan product into a colored solution. Their absorbance is monitored by a spectrophotometer and used to quantify cell proliferation and viability.

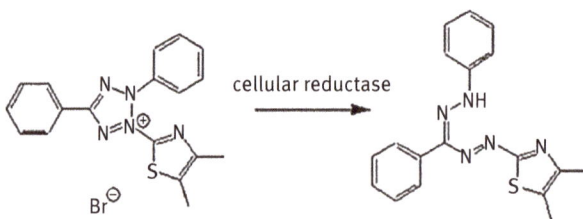

Scheme 5.2: Structure of MTT (left) and formazan (right). NAD(P)H-dependent cellular oxidoreductase enzymes reduce the tetrazolium dye MTT 3-(4,5-dimethylthiazol-2-yl)-2,5-diphenyltetrazolium bromide to its insoluble purple-colored formazan compound.

5.1.2.3 Detection of live and dead cells

AO is a nucleic acid-selective fluorescent cationic dye useful for cell cycle determination. Owing to its ability to permeate cells, it interacts with DNA and RNA by intercalation or electrostatic attraction. When bound to DNA, it is spectrally very similar to fluorescein, with similar excitation maximum at 502 nm and emission maximum at 525 nm (green). When AO associates with RNA, the excitation maximum shifts to 460 nm (blue) and the emission maximum shifts to 650 nm (red). AO will also enter acidic compartments such as lysosomes, and become protonated and sequestered. Under these low pH conditions, the dye will emit orange light when excited by blue light. Thus, AO can be used to identify engulfed apoptotic cells, because it will fluoresce upon engulfment. The dye is often used in epifluorescence microscopy. The cell nucleus emits yellowish-green fluorescence in normal condition and deep-green fluorescence when the RNA synthesis is inhibited by inhibitors such as chloroquine. AO can be used in conjunction with EthB to differentiate between viable, apoptotic, and necrotic cells. Scheme 5.3 shows the structure of AO and EthB.

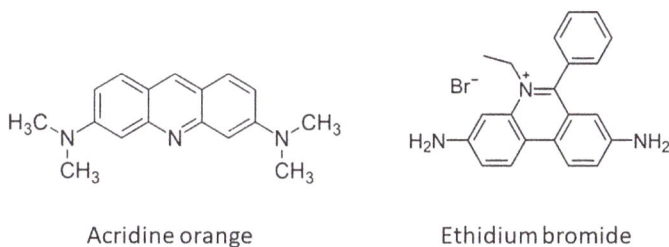

Acridine orange   Ethidium bromide

Scheme 5.3: Structures of acridine orange and ethidium bromide.

Ethidium bromide (EthB-1) intercalates and stains both DNA and RNA, providing a fluorescent red-orange stain. Although it will not stain healthy cells, it can be used to identify cells that are in the final stages of apoptosis since such cells have much more permeable membranes. It is also frequently used together with AO (see above) to differentiate between viable, apoptotic, and necrotic cells. The EthB/AO combined stain causes live cells to fluoresce green whereas apoptotic cells retain their distinctive red-orange fluorescence.

Differentiation between living and dead (apoptotic) cells can also be achieved by fluorescent staining with calcein-*acetoxymethyl* (AM) combined with EthB (Figure 5.1).

Calcein-AM is a nonfluorescent compound that easily permeates live cells. The hydrolysis of calcein-AM by intracellular esterases produces calcein by removing the AM group, a strongly green fluorescing compound that is well-retained in the cell cytoplasm and serves as an indicator of healthy cells, that is, intact cells since

Calcein-AM	EthB-1	Merge

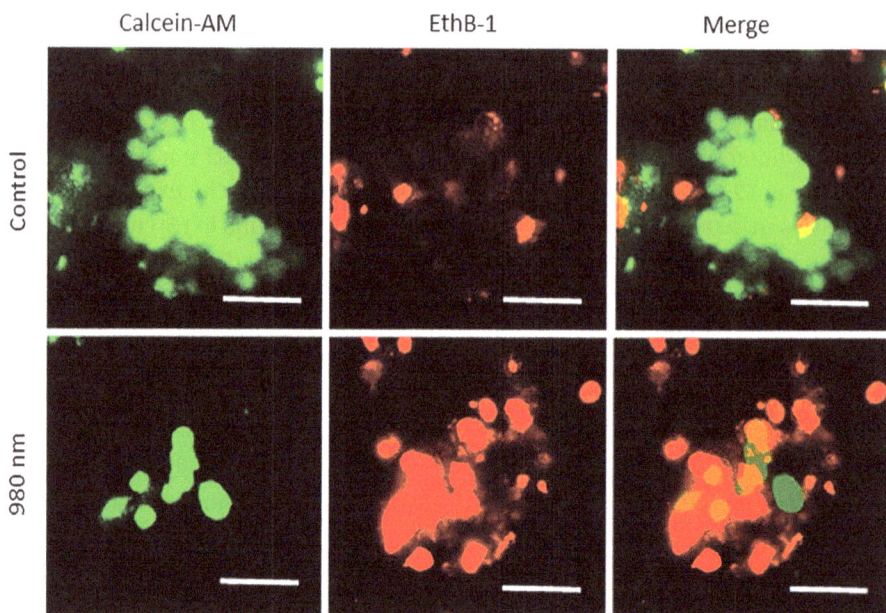

Figure 5.1: Fluorescence (980 nm) staining images for live and dead human breast adenocarcinoma MCF-7 cells in a tumorsphere in the presence of hybrid nanoparticles. Control: live (calcein-AM, green) and dead (EthB-1, red) cells in the absence of nanoparticles and light exposure. The scale bar is 50 µm (Wang et al., 2017). © Permission granted under Creative Commons Attribution (CC BY) license.

dead cells lack the esterases required to convert calcein-AM to fluorescing calcein (Scheme 5.4).

An additional, frequently used QC check involves the *enzyme-linked immunosorbent assay* (ELISA). This assay uses a solid-phase enzyme immunoassay to detect the presence of a ligand (commonly a protein) in a liquid sample using antibodies directed against the protein to be measured. ELISA is being used as a diagnostic tool in medicine, plant pathology, and biotechnology (see box, Chapter 3.4.3).

5.1.2.2 *In vivo* testing

Laboratory evaluation of biomaterials frequently involves animal tests, a thorny issue that has provoked much antagonistic activity in the past (Rollin, 1990; Andre and Velasques, 2014). To test the functional performance of biomaterials, in particular to prove that no adverse tissue reactions will occur, many animal studies are conducted worldwide. An animal model is a living, nonhuman entity used during the research and investigation of human disease, for the purpose of better understanding the disease process without risking to harm human beings. The animal

Calcein-AM (A)

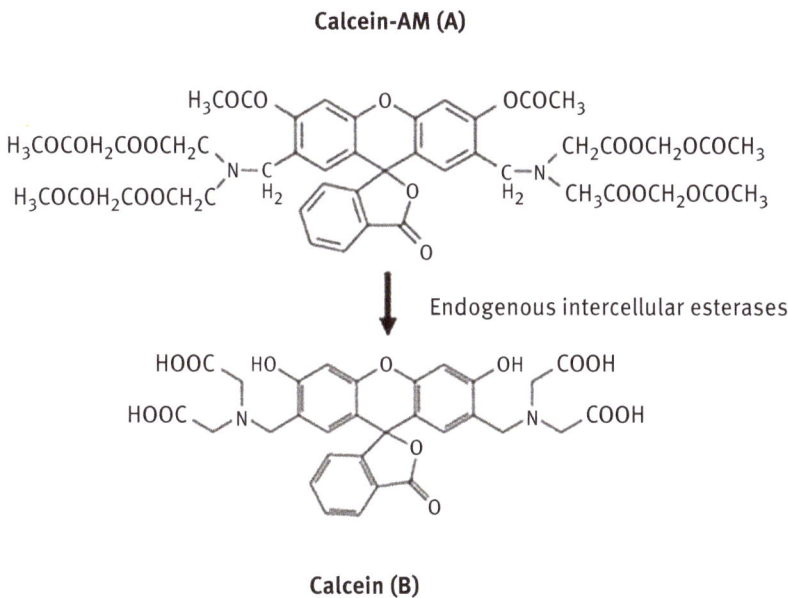

Endogenous intercellular esterases

Calcein (B)

Scheme 5.4: Hydrolysis of nonfluorescent calcein-AM (A) to green fluorescing calcein (B) by the action of endogenous intracellular esterases that cleave off the acetoxymethyl (AM) groups of calcein-AM.

types chosen range from prokaryotes such as *Escherichia coli*, unicellular eukaryotes such as the baker's yeast *Saccharomyces cerevisiae* and multicellular eukaryotes such as the nematode species *Caenorhabditis elegans*, invertebrates such as the fruit fly *Drosophila melanogaster* to vertebrates such as rats, mice, rabbits, and certain pigs (Marchant-Forde and Herskin, 2018). They will bear, to a certain degree, a taxonomic equivalency to humans in their capacity to react to disease or its treatment in ways that resembles human physiology. Animal models representing specific taxonomic groups in the research and study of developmental processes are also referred to as model organisms (Kari et al. 2007; see box).

Model organisms
There are three main types of model organisms: homologous, isomorphic, and predictive. Homologous (animal) model organisms have the same causes, symptoms, and treatment options as would humans with the same disease. Isomorphic model organisms share only the same symptoms and treatments. Predictive models are similar to a particular human disease in only a few aspects. However, these are useful in isolating and making predictions about mechanisms of a set of disease features. One of the first model systems for molecular biology was the prokaryote bacterium *E. coli*. In eukaryotes, several yeasts, particularly *S. cerevisiae* have been widely used in genetics and cell biology since the cell cycle in a simple yeast is very similar to the cell cycle in humans and is regulated by homologous proteins. The invertebrate fruit fly

D. melanogaster is easy to grow, has various visible congenital traits and has a giant chromosome in its salivary glands that can be examined easily under a light microscope. The roundworm *Caenorhabditis elegans* is studied because it has very defined development patterns involving fixed numbers of cells, and it can be rapidly assayed for abnormalities. Among vertebrates, guinea pigs, mice, rats, and zebrafish are useful as toxicological and neurological models and source of primary cell cultures. Eggs and embryos from African clawed frogs are used in developmental biology, cell biology, toxicology, and neuroscience. Other important model organisms include hydras, rabbits, cats, dogs, and nonhuman primates such as rhesus macaques and chimpanzees.

The interaction between implanted biomaterials and the host's inflammatory response is of overriding importance to consider when designing biomedical devices. Inflammation is generally defined as the reaction of vascularized living tissue to local injury, aimed to destroying or diluting the injurious agent/implant and repairing the injured tissue. The degree of the host's reaction depends on many factors such as the size, shape, and the physicochemical properties of the implant and therefore the dimensions of the injury, all of which determine the duration of inflammatory and wound healing processes. The detailed description of inflammatory and immunological responses and their biochemical foundations has been dealt with in Chapter 4.3.

In addition to proving the potential of animal tests for the benefit of human health, alleviation of pain and mental distress, and timely and humane euthanasia, experimenters are required to justify their research protocols based on the concept of replacement, reduction, and refinement ("RRR") strategy.

- *Replacement* refers to using alternatives to animals. This may include computer models, nonliving tissues, and standardized cell lines, as well as replacement of "higher order" animals (primates and other mammals) with "lower order" animals (cold-blooded animals and invertebrates) or even bacteria.
- *Reduction* refers to attempts to minimize the number of animals used during an experiment, as well as the prevention of unnecessary replication of confirmatory experiments. To satisfy this requirement, statistical experimental design methodology and other mathematical techniques of statistical power are employed to determine the minimum number of animals that should be used to obtain a statistically significant experimental result.
- *Refinement* refers to the quest to design experiments that are as painless and efficient as possible to reduce suffering of animal subjects. One of these refinement routes consist of noninvasive resonance frequency analysis to assess the osseointegration of implants (Rozé et al., 2014).

Important ethical questions to be addressed include proof whether (i) the animal model chosen is relevant to human physiology, (ii) the experiments performed are well designed, and (iii) the outcome is sufficiently important so that the data obtained will justify the suffering and sacrifice of the life of a living and feeling creature (Ratner et al., 2013).

Standardization is urgently required to streamline and unify future studies on osseoinduction and osseoconduction using animal models. Owing to biological differences in the osseoinduction pathways of various animal species, an international consensus needs to be developed regarding the type of an animal model that all studies should use to simplify interstudy comparisons and data analyses. The type of animal model used should consider differences between human and animal biology and clarify whether the results obtained in animal models on osseoinduction can be translated for use in humans (Zhang et al., 2014).

Attempts are underway to simulate comprehensive cell behavior *in silico*, that is, by computer simulation and also the behavior of larger biological systems such as tissues and organisms These simulations are likely to reduce the number of animal experiments in the process of development of pharmaceuticals or biomaterials and thus, will lead to time and cost savings and avoidance of ethical issues connected with animal experimentation (see Chapter 4.3). It ought to be mentioned that artificial intelligence will likely play a dominating role in this endeavor in the near future.

5.1.3 Clinical trials

Although laboratory tests including studying the molecular biological aspects of adhesion of cells to biomaterial surfaces as discussed in much detail in Chapter 4.2 can provide many important and useful data on biomaterials, the ultimate test is the controlled clinical trial and evaluation by clinical practitioners after a period of use (see Chapter 4.5). Many biomaterials produce suitable and promising laboratory results *in vitro*, only to be found deficient when subjected to clinical application *in vivo*. In this context, Gross (2014) has cautioned in a survey on calcium phosphate research in Australia that tests of *in vitro* mineralization behavior of calcium phosphate coatings in SBF as a one-on-one confirmation of their suitable *in vivo* performance must be seriously questioned. Examples exist showing that *in vitro* mineralization was observed on calcium phosphate-coated electrospun poly(caprolactone) fibers but that the same coatings did not show any bone apposition *in vivo* (Vaquete et al., 2013). Obviously, positive recognition of bioceramic coatings by the human body as a germane template for bone attachment appears to depend as much on surface topography including microroughness and porosity, thickness and hydrophilicity as on phase composition, and the presence of metabolically important trace elements in the coating material. Consequently, much further research is still needed to design bioceramic coatings for a variety of applications that will be able to fulfil the numerous requirements for optimum *in vivo* performance (Zadpoor, 2014; Heimann and Lehmann, 2015). As a first step, the majority of manufacturers carry out extensive clinical trials of new bio-materials, normally in cooperation with a university or hospital department, prior to seeking acceptance by government bodies and releasing a product for use by general practitioners. Such clinical trials required for regulating future application and deployment to the medical market rely on endpoint strategies involving testing for variables such as mechanical, thermal, morphological, biological, and other factors that may impact the safety and/or efficacy of the final device (Amato, 2015c).

In 2014, the European Union adopted Regulation EU No. 536/2014 of the European Parliament and of the Council on clinical trials on medicinal products for human use. This regulation aims to simplify current rules, streamline the application procedure for new clinical trials, provide more transparency in general, and harmonize the process of performing clinical trials throughout the member states of the European Union. The changes are supposed to make the European Union more attractive for future clinical trials and turn around the decreasing number in new clinical trials observed in recent years (Geysels et al., 2017). More details on the role of clinical tests in research and development of biomaterials and biomedical devices can be found in Chapter 4.5.

5.1.4 Reviewing

In the USA, the Food and Drug Administration (FDA) recommends and encourages manufacturers to initiate early discussions with the appropriate review divisions in the Office of Device Evaluation of the Center for Devices and Radiological Health (CDRH) prior to beginning testing of novel biomaterials applications or new intended uses of a device. Consensus standards recognized by the FDA have facilitated the review of biocompatibility data for medical devices (Kammula and Morris, 2001). Although these standards are frequently used to support regulatory decisions for market clearance or clinical investigations, they generally do not provide all of the information needed to make such decisions. In particular, standards for biocompatibility evaluation (Appendix A) may require additional guidance for their proper implementation and to address the issues of a particular intended use. Therefore, it is always recommended to interact early with CDRH premarket review staff regarding specialized testing that may be required for a particular device in a specific application. As an example, Sun (2018) recently described the FDA's review process of thermally sprayed coatings for orthopedic devices and thus, alerted potential applicants to problems and pitfalls that might arise in this review process. More information can be found in Chapter 4.5.3.4.

5.1.5 Other considerations and ethical guidelines

As amply discussed in this volume, significant basic research effort is under way to understand how biomaterials function *in vitro* and which measures have to be taken to optimize their performance *in vivo*. At the same time, companies are producing implants for use in humans and, appropriate to the mission of the company, earning profits on the sale of medical devices. Industry normally deals well with optimizing technologies such as packaging, sterilization, storage, distribution, QC, and materials and product analysis. In contrast, many of these technological considerations are often ignored by academic researchers. Hence, a large portion of information gained at university laboratories will never make it up to the commercialization level of a biomedical product.

Furthermore, a wide range of ethical considerations impact biomaterials and their applications as rapid growth in the field of applied biomaterials poses new ethical problems in the design, manufacture, deployment, evaluation, and standardization and regulation of medical devices and implants (Saha and Saha, 1987; Merryman, 2008). As typical for most ethical questions, absolute and binding answers are difficult to obtain. Emerging technologies such as nanotechnology, stem cell research, and tissue engineering have given rise to numerous ethical challenges including biological safety, long-term viability, and their utility for human implantation. In addition, fair distribution of and equitable access to emerging biomedical technologies among interest groups and nations have not been adequately addressed yet and hence, warrant further discussion on an international level including needed action by the World Health Organization (WHO). At present, there are no clearly formulated ethical guidelines available to assist researchers in the field of biomedical engineering to deal with real-world crisis situations associated with biomedical research (Monzon, 1999; Mishra, 2011; Rajab et al., 2013; Kashi and Saha, 2017).

5.2 Regulatory affairs for biomaterials and biomedical devices

The standardization of biomaterials (see Appendix A) and the regulations for commercialization of biomedical products do not differ substantially from the regulation of other medical devices and biologic products from a general framework perspective, but from their unique application perspective they certainly do. Hence, understanding of the crucial regulatory processes involved in biomedical device commercialization is a vital step to developing an understanding of the particulars of the processes for their design, development, manufacturing, testing, validation, and marketing. In Appendix A, key pieces of international standardization, both ASTM and ISO, are listed that impact biomaterials product regulation (Amato, 2015a). As a reference document, the volume "Regulatory Affairs for Biomaterials and Medical Devices," published by Woodhead Publ. Series in Biomaterials is recommended that addresses global regulations and standards applicable to biomaterials and medical devices and, in particular, focuses on procedures and policies including risk management, intellectual property protection, marketing authorization (MA), university patent licensing, and general good manufacturing practices (GMP; see box) of biomaterials and biomedical devices (Amato and Ezzell, 2015a).

Good manufacturing and good distribution practices
GMP for medical devices is the minimum standard that manufacturers must meet in their production processes. According to the rules and regulations of GMP, biomedical device products must be of consistently high quality, be appropriate to their intended use, and meet the requirements of the MA or product specification. In addition, good distribution practice requires that medicines are obtained from the licensed supply chain and are consistently stored, transported, and handled under suitable conditions, as required by the MA or product specification (Guidance, 2019).

Major components of a typical GMP quality systems established in the United States and the European Union with particular emphasis on design controls have been discussed by Tarabah (2015), focusing on quality system regulation issues such as ISO 13485:2016 harmonized with ISO 9001:2015, and considering the integration of risk management principles into an extended QMS. Regulations for biomedical devices made from polymers were described by Sastri (2014).

Today, companies involved in producing biomaterials and biomedical devices as well as regulatory professionals can draw on a highly specialized and continuously growing range of guidance documents issued by the US FDA, other national regulatory agencies around the world, and nongovernmental organizations such as the WHO and the Pharmaceutical Inspection Cooperative Scheme. Recently, this situation has led to attempts to secure international regulatory convergence for standards for regulatory approval, risk management, and compliance with quality management requirements. To this end, a Global Harmonization Task Force has been established consisting of the USA, Canada, the European Union, Japan, and Australia with the objective to streamline and harmonize regulatory standards. In October 2011, representatives from the medical device regulatory authorities of Australia, Brazil, Canada, China, EU, Japan, and the USA, as well as the WHO, congregated in Ottawa to found a new organization, the International Medical Device Regulatory Forum aimed to accelerate international medical device regulatory harmonization and convergence as a powerful tool to further the development of innovative, life-improving, and life-saving medical technologies (ITA, 2016).

To guarantee producer and customer satisfaction as stipulated by ISO 9001:2015, supply chain controls are designed that aim to minimize risk to a finished product. As an example, Amato (2015b) has reviewed key steps in supply chain control and the ways in which FDA encourages organizations through guidance documents and regulatory harmonization activities designed to improve supply chain controls.

However, at the end of the day serious questions remain: do the regulations and standards developed by national and international government bodies truly address all relevant safety issues? Do these regulatory agencies have sufficient information to define appropriate tests for biomaterials and biomedical devices, to properly regulate their deployment and clinical application, and to enforce these regulations for all stakeholders in industry, governments, and academia? Answering these salient questions will depend on a concerted effort of suppliers of biomaterials, researchers in laboratories and clinics, scientists in testing organizations, personnel of regulatory bodies, and not in the least, politicians for enforcement.

In May 2017, the new Medical Device Regulation (MDR) became effective and on May 26, 2020 its validity was supposed to become binding throughout the European Union member states. However, at the time of this writing this was postponed to 2021, on account of the disruptive impact of the Sars-CoV-2 pandemic. The MDR will replace the EU's Medical Device Directive of 1993 (93/42/EEC) (MDD) and the EU's directive on active implantable medical devices (90/385/EEC). The MDR is supposed to be an improved version of the MDD. The major difference is that as a regulation, the MDR has general validity and immediate effectiveness in all EU member states. Thus, the MDR provides an optimized and standardized regulation for placing biomedical devices on the market, with stronger emphasis on product quality and safety as those expressed in precursor legislation. Guidance on transitioning from MDD to MDR includes transition timeline, standard operating procedures, reporting of adverse events, requirement of new forms, and new structuring of technical documentation (FDA News, 2019).

5.3 The route of biomaterials from inception to market

The route of a novel biomaterial or biomedical device from inception to the market is tortuous, extremely expensive, and time-consuming, and consists of a myriad of multidisciplinary development steps with a multitude of actors involved. Figure 5.2 shows the most important steps and their facilitators, and the progression in the development of a biomaterial or a biomedical device from identifying a need to a clinical product that has been tested, regulated, manufactured, sold, and implanted (Ratner et al., 2013). In particular, it identifies the important role of regulatory activities (see Chapter 5.2) that encompasses premarket approval, limited clinical studies, and clinical trials as well as long-term follow-up to ensure that a biomaterial or biomedical device is safe, reliable, and functions in the expected way (see Chapter 4.5).

Additional useful information on the commercialization process can be obtained from Zenios et al. (2010). This book tells prospective start-up entrepreneurs how to identify, invent, and implement biomedical devices and move them smoothly to markets. Another important document to consult pertains to the development of thermally sprayed coatings on orthopedic devices and the review process required by the FDA (Sun, 2018; FDA, 2019).

Action   Who is involved?

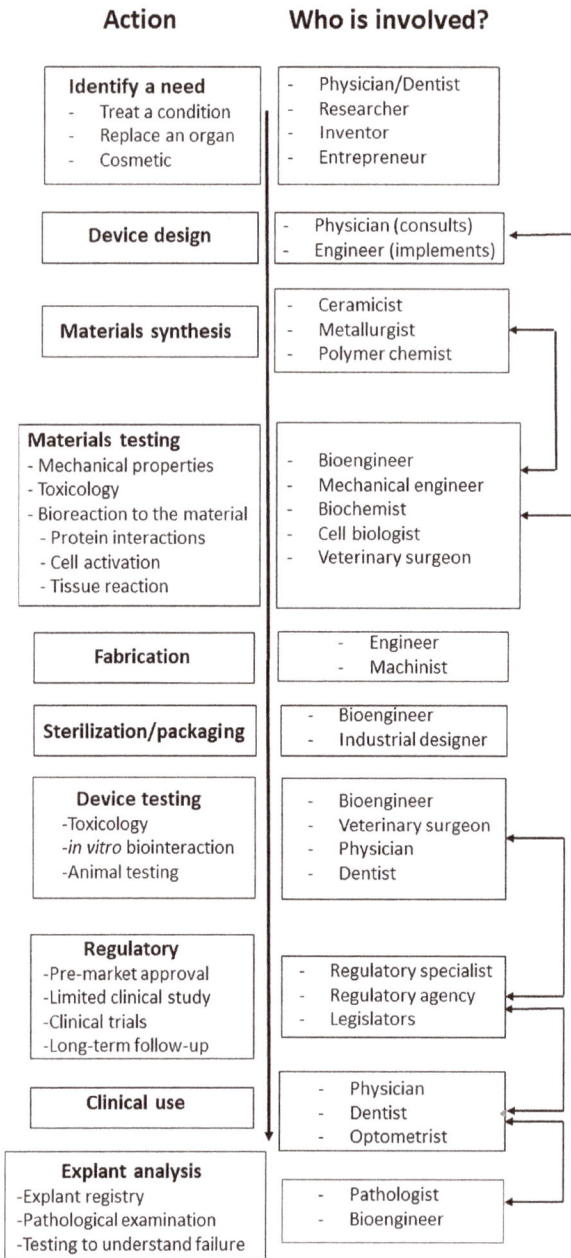

Figure 5.2: Key steps, their facilitators involved, and their interaction in the complex developmental way from inception to fabrication to clinic of a biomaterial or biomedical device (Ratner et al., 2013). © With permission by Elsevier.

Chapter 6
Future challenges, anticipated developments, and outlook

Robert B. Heimann

6.1 Introduction

"Prediction is very difficult, especially if it's about the future" quipped reportedly famous physicist Niels Bohr, paraphrasing an ancient Danish proverb. Alas, this witticism is intrinsically true. However, it will not deter the present authors from inquisitively peering into the crystal ball that ever so faintly indicates future trends and associated required research and development in the area of biomaterials. As it goes, such predictions will be entirely subjective as they are garnered from studying the content of scientific journals and the available patent literature, sources that will necessarily neglect the largely unpublished undercurrent of research carried out in many biomedical research laboratories around the world. Paradigmatic shifts within future biomaterial-based technologies might be disruptive, and might be able to alter profoundly experience-generated tenets and ways of doing things, in both industry and academia, in the laboratory, and in clinical setting and the operational theater. Nevertheless, those technologies are able to provide significant and decisive benefits to patients and society as a whole, and to economics in a comparatively narrow time span. Moreover, although many novel materials and technologies are in a very early stage of development and being studied by only few isolated research groups, they may generate excitement among experts and being considered worthy of further developmental follow-up by health-related industry, eventually attracting sustained investment from venture capital, and a continuous flow of governmental research subsidies and grants. But there is a caveat: large-scale production and return on investment (ROI) of biomedical industrial production expenditures have been deemed close to impossible for deployment of truly novel biomaterials. This means that, in many cases, conventional "entrenched" biomaterials are being merely further purified, microstructurally refined, or applied in a different biomedical context. For example, biological functionalization of conventional biomaterials is being achieved by adsorption of tissue growth–supporting proteins, including bone morphogenetic proteins (BMPs), and cyto- and chemokines to elicit osseoinductive mechanisms as a crucial step for successful osseointegration. In contrast, to go beyond economically motivated limitations, a paradigmatic shift toward completely new materials and technologies is required, including fourth-generation biomaterials based on integrating electronic systems with the human body or the application of artificial intelligence (AI) to develop and deploy entirely new materials for biomedical purposes.

https://doi.org/10.1515/9783110619249-006

However, at present, there are substantial intellectual and economic hurdles to overcome to score in these advanced fields.

6.2 Biomimetics

Progress in the realm of design, development, validation, and deployment of novel or improved biomaterials and biomedical devices is uniquely connected and intrinsically tied to man's innate ability to learn from nature and to manipulate cleverly her most successful process routes and age-tested structural designs. In this quest, humanity has always tried to adapt nature's efficient and effectual ways by developing routes toward solving problems that occur in day-to-day life, including attempts to improve on existing materials, systems, and technological processes. This "learning from nature" by tailoring materials and processes to address actual human needs is indeed the hallmark of biomimetics. Its range of application is vast, and examples of successful emulation of the principles of biomimicry can be found in various science and engineering fields, including modern medicine and bioengineering. Although biology is applied in technology at all levels, including materials, structures, mechanisms, apparatus, monitoring, and control devices, a substantial gap still remains to be bridged between biology and its technological realization in the real world. Hence, the novel science of biomimetics challenges the current technological paradigms by suggesting more sustainable and environmentally friendly routes to manipulate the world that surrounds us and, thus, to better the lives for all of us. This is indeed a worthwhile endeavor to be part of.

Exciting future developments using biomimetic design routes pertain to the rapidly evolving field of *drug and gene carrier* materials. For example, monomeric silica can be covalently coupled to lipids, forming self-assembled vesicles in water akin to natural liposomes. Polymerization of such silica–lipid structures results in spherical constructs, the so-called cerasomes that may find application in drug or gene delivery, development of primary cells responsible for skeletal homeostasis, and expression of osteoblasts and osteoclasts. Likewise, drug and gene delivery vehicles are being developed based on dendrimers, biodegradable poly(ethylene glycol) (PEG)–poly(L-lactide)-grafted copolymers as well as PEG–y-poly(L-lactide)–DNA condensates. This may include highly sophisticated DNA "origami" structures of synthesized nanocontainers that are able to enclose molecular cargoes to be used as drug or gene delivery ferries without eliciting the body's immune defense system. Along this line, novel approaches such as spherical nucleic acids are being developed that, unlike linear oligonucleotides, are able to penetrate cancerous cells and bind to disease-causing mRNAs, thus interrupting their messenger role.

Nanoparticles with soft magnetic properties ("magnetic" liposomes) will further be developed for *in vivo* drug targeting, for separation of leukemia cell from blood plasma, for hypothermia application to fight cancer cells, and as a contrast agent in magnetic resonance imaging (MRI).

Compared with complex biomolecules presently attached to the surface of biomaterials to induce osseogenicity, the use of simple compounds such as chemical moieties, low-molecular-weight peptides, and other small molecules can be advantageous in the quest to design less expensive and easily tunable biomaterial formulations. Such molecules may include cyclic adenosine monophosphate, and plant-based forskolin and resveratrol that are thought to activate the protein kinase A signaling pathway and to act as cancer chemopreventive agent, respectively. They likewise activate the Wnt/β-catenin (canonical) pathway that regulates stem cell pluripotency and cell fate decisions during their development cascade, and promote BMP action and upregulation of the RUNX2 gene expression. Here, promising ways to tailor tissue regeneration appear to open up in future developments. Finally, genetically engineered bacteriophage viruses may allow the formation of nontoxic biocomposite quantum dot (QD) structures to be applied as antibacterial films and coatings.

6.3 Arthroplasty and scaffolds for regenerative medicine

Within the wide range of future applications of biomaterials, the development of novel and improvement of existing biomaterials for *arthroplastic* uses take on a particularly significant role as bone stands as the primary organ requiring surgical intervention in excess of 1.2 million cases annually worldwide. Hence, the implications linked to the development of efficient bone substitutes and scaffolds are of dramatic socioeconomic importance and act as a powerful driver for the emergence of new, more efficient, more reliable, and less expensive materials as well as improved manufacturing technologies for biomedical devices. These technologies include, for example, solid free form and laser-engineered net shaping (LENS®) strategies for 3D-printed scaffolds for bone tissue engineering (see below).

Consequently, the quest for novel biomaterials with improved mechanical strength, reduced modulus, prolonged lifetime, increased reliability, and better osseoconductive functionality is high up on the agenda of worldwide research and development. Such developments include testing and deployment of new β-type Zr- and Nb-containing Ti alloys with comparatively low elastic moduli such as Ti13Nb13Zr or the medium-entropy alloys Ti29Nb13Ta4.6Zr and Ti12Mo6Zr2Fe that are designed to reduce the effect of stress shielding of endoprosthetic implants and, thus, to mitigate the risk of bone tissue resorption by osteolysis and impaired bone remodeling. Indeed, the quest for so-called isoelastic implants has been an ongoing challenge that has not been satisfyingly solved to this day. Ideally, isoelastic polymer composites are able to reduce or even eliminate the elasticity mismatch between implant and bone by providing a two-prong approach: enhanced biocompatibility and enhanced mechanical, that is, elastic compatibility with the host bone. Among several candidate materials, polymer composites, including composites fabricated using poly(ether etherketone), have achieved a reasonable degree of isoelasticity in experimental orthopedic prosthetic designs, with clinical evidence of acceptably sustained fixation and reduced stress shielding when compared with traditional metallic endoprosthetic stems. However, even though isoelastic hip implants are a clinical reality today, it must be emphasized that they still require decades of development and, thus, remain in the intermediate stages of clinical evaluation.

Novel so-called *high-entropy alloys* including equal or relatively high proportions of metals such as TiZrNbTaFe or CoCrNiFeMn that appear as metallic solid solutions without forming ordered intermetallic phases may provide implants with exceptional mechanical properties, including high hardness, yield strength, fracture toughness, fatigue life, ductility, oxidation and corrosion resistance, and endurance limit, but display comparatively low moduli of elasticity as desired to mitigate stress shielding.

Since early corrosion of some surgical stainless steels prior to delayed passivation tends to release potentially cytotoxic metal ions into the tissue surrounding an implant, novel nitrogen-alloyed nickel-free austenitic stainless steel grades such as Fe25Mn20Cr3Mo1.3N are currently entering the market, which will likely further secure the important role stainless steels are playing in modern implantology.

To counteract the always occurring wear and tear of artificial joint components, several strategies are being pursued today. Since the risk of osteolysis depends on the bulk material as well as the nature and amount of wear debris, development, testing, and finally deployment to the market of novel materials are important aspects of the global research agenda. Acquisition of comprehensive clinical experience with the combined wear couple of alumina as femoral head material and ultrahigh-molecular-weight poly(ethylene) (UHMW-PE) as acetabular cup lining has triggered research in two directions. First, reduction of wear debris by utilizing other, low-wear articulation partners, and second, replacement of UHMW-PE by other materials including zirconia-toughened alumina or silicon nitride (Si_3N_4) as well as hard tribological coatings of titanium nitride (TiN), titanium carbonitride, tantalum nitride, and diamond-like carbon (DLC). Another approach researches the use of so-called hindered amine light stabilizers that can be used in total hip, knee, shoulder, and ankle implants to stabilize UHMW-PE that has been damaged during cross-linking by irradiation with y-rays or otherwise altered by sterilization with ionizing radiation prior to implantation. Also, chemical diffusion-controlled modification of the bulk implant metal or high-conformity cushion bearings made from polyurethane or advanced hydrogels is being developed and preclinically tested *in vitro* and *in vivo*. This paradigmatic shift involves the move from UHMW-PE to polyurethane acetabular cup lining materials that can be processed by cost-effective injection molding techniques.

In this context, the *bearing surfaces* of joint implants require improvement to counteract enhanced friction caused by the absence of lubricating hyaluronic acid-based synovial fluid. Ways to do this rely on deposition of ultra-hard thin films of DLC, TiN, tantalum nitride, or Si_3N_4 and/or silicon carbide (SiC) on implant bearing surfaces. Important developmental goals include improved film adhesion to the metallic substrate, crack-free film morphology, enhanced tribological behavior, and limitation of release of ultra-hard particulates that may be distributed by the lymphatic system throughout the body with possibly metabolically deleterious effects. In addition, novel additive manufacturing processes such as extremely high-velocity laser material deposition (EHLA), similar to conventional laser cladding, allow joining of dissimilar metals with velocities exceeding 3 m/s or, on rotating workpieces as high as 150 m/s in a fast, economical, and ecofriendly way. In contrast to laser cladding, the heat-affected zone is orders of magnitude lower. Rapid quenching yields dense layers with thickness as low as 25 μm, frequently as metallic glasses with wear and corrosion resistances vastly higher than those of conventional metallic biomaterials. The EHLA process can easily be adapted to additive manufacturing of near net-shaped 3D objects such as implants with complex geometry.

6.4 Biodegradable metals

In the future, special magnesium-based alloys may be developed and used for arthroplastic applications. However, many still unanswered questions are waiting to be tackled, and presumably there is a long way to go before Mg alloy-based implants will become a routine implant option in the clinical operation theater. An ultimate future challenge pertains to the application of Mg alloys for load-bearing *endoprosthetic hip and knee implants*. Their favorable mechanical properties in terms of a high stiffness-to-mass ratio and, in particular, an elastic modulus close to that of cortical bone suggest, in the long run, the emergence of novel biomaterials able to mitigate or even alleviate stress shielding. This would enormously benefit the outcome of total hip replacement operations that are frequently plagued by diverting the load-imposed forces away from low-modulus cortical bone bed toward the high-modulus metal implant, a process that eventually results in osteoclastic bone resorption, and hence, atrophy. Today, truly isoelastic endoprosthetic implants possibly based on magnesium might be considered the holy grail of implantology. However, to reach this lofty goal, there are still very substantial thermodynamic barriers to overcome. First among future challenges is the need to improve the corrosion resistance of Mg-based alloys by orders of magnitude, either by discovering novel biocompatible alloys with drastically improved electrochemical corrosion resistance, or by developing appropriate well-adhering biocompatible protective surface coatings that synergistically will guarantee long-term *in vivo* stability of future Mg-based load-bearing implants.

The paradigmatic shift from environmentally very stable corrosion-resisting metals for *osseosynthetic musculoskeletal fixation devices* and *cardiovascular stents* such as stainless steels, CoCr alloys and titanium alloys to biodegradable metals such as magnesium, zinc or even iron alloys requires stringent control of the enhanced corrosion behavior of the latter. For example, as shown earlier, low-modulus magnesium-based implants reduce dramatically the stiffness gradient between implant and bone, and thus decrease considerably the risk of deleterious stress shielding. In addition, their biodegradable behavior makes obsolete a second surgical intervention to remove the device when healing has taken place. Recently developed Mg- and Fe-based biomaterials as well as other metallic biomaterials such as Zn and its alloys have interesting properties in terms of microstructure, mechanical, and degradation behavior, and *in vitro* and *in vivo* performances as emphasized by preclinical and clinical trials. Currently, approaches are being developed to control biodegradation rates of magnesium by applying a host of biocompatible nanocomposite coatings to match the healing rates of the host tissues with various surface modification techniques. The directions of future research and development will likely revolve around the challenge of making the transition from raw material to semiproducts to final medical devices as well as steering biodegradable metals and their corrosion control coatings toward biomaterials with multifunctional capabilities in a controllable manner to benefit local tissue reconstruction. In conclusion, future research on biodegradable metals will presumably concentrate on a two-pronged approach: (i) design and development of novel magnesium alloy compositions such as special Mg–Nd–Zn alloys that show pronounced layer-by-layer corrosive material removal instead of localized pitting or crevice corrosion modes, and (ii) application of advanced protective coatings for Mg-based devices such as cardiovascular stents that limit hydrogen evolution and the built-up of deleterious corrosion products, allowing improved prediction of meaningful residence time within the human body.

6.5 Bioceramics

As far as *bioinert ceramic materials* are concerned, in the past decades high-purity alumina and partially stabilized zirconia used for femoral balls of hip endoprosthetic implants have undergone impressive performance improvement, largely by deploying highly purified precursor materials and consequently reducing grain size down to the low nanometer range. However, further grain size reduction appears unlikely or even undesirable as the relative grain boundary area increases dramatically with decreasing grain size. This means that the overall material behavior will be dominated and, thus, controlled by grain boundary phenomena such as induction of deleterious stresses at grain boundaries on cooling with an increased tendency for solute segregation. Nevertheless, development of nanostructured alumina and zirconia ceramics and composites as well as nanostructured calcium phosphate ceramics and porous bioactive glasses, possibly as hybrid composites with organic constituents, will provide desired properties for bone substitution and tissue engineering for a long time to come.

Further development of somewhat neglected *coralline bone graft scaffolds* with enhanced biocompatibility and resorption kinetics may shift attention from entirely inorganic to organic-based implant materials. However, although corals display excellent biocompatibility, a well-interconnected 3D porous structure, and appropriate mechanical properties, their high dissolution rate *in vivo* has limited their clinical applications, especially when a high load-bearing capacity of bone grafts is required. Future application of composite technology to strengthen coralline materials, for example, by infusion with resorbable polymers or partial transformation to hydroxylapatite by hydrothermal conversion routes may improve the mechanical as well as chemical properties of these biologically derived ceramics.

Silica–collagen and silica–chitin nanocomposites found naturally in siliceous tissue of the skeleton of marine sponges (Porifera) may be promising candidates for modern bone replacement applications since bioinspired, silicified collagen scaffolds are able to enhance recruitment of progenitor cells and, thus, promote osteogenesis and angiogenesis by immunomodulation of monocytes.

Advanced Si_3N_4 coatings are currently under investigation as bearing surfaces for joint implants, owing to their low wear rate and good biocompatibility of both the coating and its potential wear debris. Hence, the propitious mechanical properties of Si_3N_4, associated with high biocompatibility, high chemical stability, and promising tribological features including a low friction coefficient, high fracture toughness, and high wear resistance render this ceramic material a promising future candidate for replacing joint components. Since lifetime predictions revealed that Si_3N_4 implants are mechanically very reliable, this indicates that high-strength nonoxide ceramics such as Si_3N_4 and possibly SiC may become future biomaterial candidates, in particular for highly loaded, thin-walled implants such as ceramic hip resurfacing prostheses.

Recently, several studies have demonstrated that bioceramics with specific chemical composition and surface topography have the potential to regulate the development of stem cells and the interaction with their microenvironment, resulting in enhanced hard and soft tissue regeneration. These findings may open up a new direction toward exploring the possibility to design novel biomaterials for tissue engineering and regenerative medicine based on the specific biological function of the chemical and topographical characteristics of these bioceramic materials. Radiative surface functionalization using femtosecond laser irradiation is being researched for micropatterning of metallic and polymeric biomaterials to guide cell attachment and enhance cell proliferation, vitality, and spreading.

6.6 Carbon materials

Among novel carbon materials waiting in the wings to be introduced to the biomaterials market are *graphene* and *graphene oxide* that promise to provide a vast potential for application in various fields of biomedical devices including biosensors for detection of glucose, cholesterol, cytochrome c, coenzyme 1 (nicotinamide adenine dinucleotide), hemoglobin, uric acid, as well as the neurotransmitters dopamine and serotonin. Graphene-based devices could be employed for optical, two-photon, Raman, positron emission tomography, single photon emission computed tomography (CT), MRI, photoacoustic tomography, and CT molecular imaging. Furthermore, graphene could presumably be utilized as drug delivery vehicle.

Likewise, work is being carried on to further develop *pyrolytic graphitic carbon* with enhanced hemocompatibility and anticlotting capability for a new generation of heart valves. The exceptional physical, mechanical, and electronic properties of *carbon nanotubes* (CNTs) allow them to be used in biosensors, electronic bioprobes, bioelectric actuators, nanoelectronic devices, and drug and gene delivery systems. The increasing interest shown by the nanotechnology research community in this field suggests that a plethora of applications of CNTs will be explored in the near future. *Carbon quantum dots* (CQDs) likewise to be applied in biosensing devices, drug and gene delivery matrices, as well as medical diagnostics and theragnostics, and as photodynamic therapy agents need to be further developed with the intention to replace existing, potentially toxic semiconductor-based QDs.

6.7 Osseoconductive coatings

Osseoconductive hydroxylapatite coatings of, for example, the stems of hip endoprostheses and the roots of dental implants are commonly applied by thermal, most often plasma spray technology. However, the limited stability of hydroxylapatite and its thermal decomposition products in the aggressive body environment calls for only sparingly soluble materials or hydroxylapatite-based composites with increased *in vivo* resorption resistance. Such materials with higher stability against *in vivo* resorption include, for example, novel sodium (Na) super-ionic conductor (NaSiCon)-type compositions within the quaternary system $CaO\text{-}P_2O_5\text{-}TiO_2\text{-}ZrO_2$. In this system, stoichiometric compounds exist with much improved mechanical and chemical stability compared to calcium phosphates but with comparable osseoconductive potential, and, when doped with small alkali ions, solid-state ionic conductivity that suggests the use in novel bioelectric systems including bone growth–stimulating devices for treatment of pseudoarthrosis.

Whereas hydroxylapatite has a wide range of applications both in bone grafts and for coating of metallic implants, to more accurately mirror the chemistry of biological hydroxylapatite, various substitutions, both cationic (substituting for calcium ions) and anionic (substituting for phosphate or hydroxyl groups), are being synthesized and studied. Increasingly, precursor powder compositions are being tailored toward the real composition of biological, that is, bone-like hydroxylapatite through substituting in its structure Ca^{2+} ions by Na^+, Mg^{2+}, Sr^{2+}, and other important metabolic ions such as Si^{4+}, as well as replacing PO_4^{3-} and OH^- by CO_3^{2-}. Even small substitutions were found to have very significant effects on thermal stability, solubility, osteoclastic and osteoblastic response *in vitro*, and degradation and bone regeneration *in vivo*. An important aspect to be considered in the future is the design and widespread application of antimicrobial Ag-doped hydroxylapatite coatings that prevent biofilm formation and, thus, promise to combat nosocomial infection with methicillin-resistant or multiresistant *Staphylococcus aureus* (MRSA) colonies and others.

It has been amply confirmed that the surface microstructure of a biomaterial plays a crucial role in the osteogenic response of calcium phosphate ceramics such as hydroxylapatite. However, the presence of surface irregularities and random distribution of microstructural surface features in present-day calcium phosphate coatings renders it difficult to assess the exact effect surface microstructures have on bone formation. Laser ablation patterning can be applied to generate custom-designed ordered surface microstructures to minimize randomness and to introduce a suitable surface fractality, a property that has been found to influence the adhesion capacity of cells. In the future, the development of laser-induced controlled surface microstructures of hydroxylapatite may be helpful in the quest to further optimize calcium phosphate bulk ceramics as well as coatings for bone regeneration. Suitable microstructures could be characterized and tailored toward maximized biological performance by determining and standardizing their fractal properties that were found to influence adhesion of bone growth–supporting proteins.

In addition to well-established phosphate-based coatings for endoprosthetic implants and bone-substituting scaffolds, *silicate-based coatings* are being increasingly researched, including calcium silicates (wollastonite, larnite), calcium magnesium silicates (diopside, merwinite, monticellite, åkermanite), calcium zirconium silicate (baghdadite), and calcium zinc silicate (hardystonite). These materials are thought to be future promising sources for enhanced bone regeneration with properties useful to adjust the coated implant performance in terms of dissolution rate, apatite formation ability, and mechanical stability.

6.8 Bioinspired hybrid nanocomposites

A different route toward new promising biomaterials is based on *inorganic/organic nanocomposites* that are being designed to mimic the mechanical and biological performance of natural bone, for example, composites of hydroxylapatite with bioresorbable poly(lactide), poly(ε-caprolactone), collagen, and polysaccharides such as cellulose, hyaluronic acid, chitosan, alginate, as well as silk fibroin. Such human-friendly novel materials have been designated "intelligent" since they are able to change their characteristic properties in response to surrounding conditions, for example, responding to varying stress fields by enhanced or delayed dissolution. Biodegradable ferro- and piezoelectric polymers such as poly(vinylidene fluoride), glycine, diphenylalanine peptide nanotubes, or poly(L-lactic acid) fibers may provide future opportunities when integrated into biological systems to be used for miniaturized bioelectronics and biomechanical devices.

6.9 Surface-active bioglasses

Bioactive glass science and technology continues to be at the forefront of providing innovative approaches to medicine. The wide range of application of bioglasses will be further augmented by tailoring bioglasses for tissue scaffolds with the intent to mimic the porous structure of bone, in particular trabecular bone. The challenge is to prepare bioresorbable scaffolds of suitable geometry and osseoconductive potential to support the growth of artificially seeded tissues that can be adjusted to fit specific bone defects. Research is currently being carried out toward fine-tuning the architecture and resorption characteristics of sol–gel-derived bioactive glasses. Novel foaming agents and surfactants are being incorporated into the sol–gel reaction mixture to introduce 3D-interconnected porosity that is supposed to emulate the porous structure of trabecular bone. In addition, novel porogens and processing technologies must be developed to introduce graded porosity to mimic more closely the structure of natural bone: a porous "cancellous" core surrounded by a dense "cortical" rim. Furthermore, bioglass compositions are being researched, seeded with the patient's own cells to design rejection-free prosthetic implants.

Mesoporous bioactive glasses need to be developed for antibacterial strategies, primarily because of their capability of acting as potent carriers for the local release of antimicrobial agents. The incorporation of antibacterial metallic ions including silver (Ag^+), zinc (Zn^{2+}), copper (Cu^+ and Cu^{2+}), cerium (Ce^{3+} and Ce^{4+}), and gallium (Ga^{3+}) ions into the mesoporous glass structure and their controlled release is thought today among the most attractive ways to inhibit bacterial growth and reproduction. Furthermore, in the future therapeutic inorganic ions with pronounced angiogenic and osteogenic effects ought to be developed that must be delivered into the body environment in a controlled and sustained manner from bone tissue-engineering drug delivery scaffolds composed of mesoporous bioactive glasses.

6.10 Hydrogels

Supramolecular hydrogels rely on low-molecular-mass compounds likely to be widely used in various biomedical applications, including uses to target specific cells, to concurrent drug delivery, and to provide a template of scaffolds for promotion of precipitation of bone mineral and, thus, bone growth. Attempts are being made to design modern synthetic biomaterials intended to integrate bioactive ligands within hydrogel scaffolds for cells to respond and assimilate within the matrix. These advanced biomaterials are now only beginning to be used to simulate the complex spatiotemporal control of the natural regenerative microenvironment. In future developments, the increasing understanding of the role of protein-based growth factors and cytokines, and their interaction with components of the extra-cellular matrix (ECM) will possibly lead to the design of novel hydrogels that more closely mimic the natural healing environments of tissues, resulting in increased efficacy in applications of tissue repair and regeneration. Numerous novel designs of "smart" hydrogels may involve attempts to control the morphology of self-assembling peptide fibers, artificial glycoproteins for controlling cell responses, building material for microchemotactic devices, and enhanced use of DNA recognition motifs. An outstanding example of the potential of "smart," that is, stimuli-sensitive hydrogels is the design of optical systems with a tunable liquid lens that permits autonomous focusing based on a temperature-sensitive hydrogel integrated into a microfluidic system.

Presently, problems associated with repair of damaged *cartilage* have not been satisfyingly solved. Options include the use of biological approaches that try to re-grow cells or using synthetic materials emulating the mechanical and biological properties of cartilage. A promising technology for cartilage repair may rely on *in situ*-formed hybrid hydrogels based on fibrinogen and PEG. These hydrogels are de-signed to match the biomechanical properties of cartilage tissue. Preclinical studies have demonstrated that, over time, the hydrogel is replaced by cartilage tissue rather than nonfunctional fibrous connective tissue.

Biomaterials such as fibrin or hydrogel scaffolds must be developed for stem cell transplantation to treat patients with spinal cord injury. Research is needed to investigate the optimal chemical structure, porosity, and morphology of hydrogel-based biomaterials used for transplantation of stem cells.

6.11 Polymers

Major future research considerations will include the study of drug–polymer inter-actions, drug transformation, and diffusion properties of drugs through bulk poly-mer matrices. This requires developing a more complete understanding of matrix degradation in the case of biodegradable polymers and developing new engineered polymers designed for specific purposes such as vaccination. For example, research is being conducted on new PEG-based acid-stable macromonomers with methacry-late units that enable 3D cross-linking. This type of macromonomer is of interest for the formulation of drug delivery systems to transport therapeutic proteins, as it combines the excellent biological properties of PEG, including biocompatibility, water solubility, and low immunogenicity with stimuli-responsive units that enable triggered drug release.

Polymersomes are a class of artificial vesicles, forming tiny hollow spheres that enclose a solution and are made using amphiphilic synthetic block copolymers to form the vesicle membrane. Multifunctional polymersomes loaded with ferrimagnetic maghemite (γ-Fe_2O_3) nanoparticles and grafted with an antibody that is directed against human endothelial receptor 2 are developed as novel MRI contrast agents for bone metastasis imaging. Magnetic resonance images show targeting and enhanced retention of antibody-labeled polymersomes at the tumor site.

In addition, cell–polymer interactions need to be clarified, including the fate of inert polymers, the use of polymers as templates for tissue regeneration, and the study of polymers aiding cell transplantation.

6.12 Hemocompatible coatings

In the field of *cardiovascular medicine*, DLC and TiO_2 coatings are being researched to provide stability, abrasion resistance, and enhanced hemocompatibility and anti-clotting capability to pyrolytic graphitic carbon used to design and construct artificial heart valves. When biomedical devices contact blood, blood will respond aggressively because most current device surfaces are thrombogenic. This will over time result in clotting that impairs device performance and can harm the patient. Hemocompatible coatings are an option to improve the blood compatibility of medical devices. Biodegradable Zn- and Fe-based alloys are being tested for application in cardiovascular stents. In addition, novel protective coatings for biodegradable Mg-based stents with enhanced hemocompatible property promise improved ability to tailor the survival time of these devices *in vivo*.

6.13 Dentistry

In *dentistry*, future perspectives of bioceramic materials including coatings for dental restoration will likely continue to focus on the development of tough and esthetically pleasing glass ceramics, and bioinert ceramics such as zirconia or titania with suitable optical properties such as tooth-matching color and shine. Such ceramics should not only mimic the appearance of natural teeth but also provide long-term durability. Also, control of *in vivo* degradation of bone and existing dental cements by applying novel formulations based on stable magnesium ammonium phosphate (struvite) will likely prevent formation of sparingly soluble hydroxylapatite. In addition, advanced long-lived dental porcelain compositions are needed with enhanced fracture toughness and resistance against tensile and shear stresses. Several studies showed that the future role of restorative dental materials and treatments is conceptually based on three areas of highest importance perceived: the preservation of existing enamel and dentin tissues, the prevention of secondary caries, and the maintenance of pulp vitality.

6.14 Three-dimensional-printed scaffolds

One of the key challenges in regenerative medicine is the engineering of 3D microenvironments that can simulate the complex features of physiological tissues, in particular bone. In this context, biomaterials are often used in regenerative medicine because they can mimic the native ECM and provide structural support to the engineered tissue constructs. During the past decade, 3D bioprinting techniques have been used to engineer structures with precise control over the arrangement of biomaterials and cells within the engineered biomimetic tissue constructs. Such effort is evident in emerging future technologies such as 3D printing of bone growth scaffolds. Although the high hierarchical complexity of natural bone limits the creation of optimum materials and fabrication techniques for ideal custom-designed scaffolds required for guided bone tissue engineering, attempts abound to solve this problem by 3D printing, using a combination of different calcium phosphate powders selectively solidified with various binders. The basic technological concept focuses on the chemical interaction between precursor calcium phosphate powders, for example, tricalcium phosphate (β-TCP) and a printing fluid (ink) such as orthophosphoric acid. β-TCP is known to react with orthophosphoric acid to form bone cement, consisting of bioresorbable monetite ($CaHPO_4$) or brushite ($CaHPO_4 \cdot 2H_2O$), and dicalcium pyrophosphate ($Ca_2P_2O_7$). The setting and solidification processes during 3D printing rely on two types of interaction, either acid–base reactions with the formation of a neutral compound or a hydrolysis reaction of the metastable phosphate resulting in an adhesive effect among particles. The main final phase of the 3D product is bone-like calcium-depleted hydroxylapatite. The chemical and phase composition of the 3D-printed scaffold can be adjusted toward controlling its rate of biodegradation and its specific ion release/absorption kinetics. This control can be achieved by soaking the scaffold in simulated body fluids (SBFs) such as Dulbecco's modified Eagle's medium with controlled pH and temperature for a predetermined time. Moreover, this final procedure can also improve both mechanical integrity and osseointegrative potential.

In addition, there is a need to design multilayer composite scaffolds that contain collagen, bone, and calcified layers intended to simulate full-thickness bone-cartilage structures. Addition of collagen or some polysaccharides was shown to significantly improve the flexural strength of the scaffold as well as cell proliferation, viability, and spreading.

However, there is a serious caveat. The lack of vascularity in scaffolds and tissue-engineered constructs is a major challenge, and improving the vascularization strategy is considered one of the areas requiring the most extensive research in the field of tissue engineering. In addition, innovation is required in the areas of tissue engineering to solve the socioeconomic demands of a rapidly growing and aging population worldwide. Such innovative tasks conceptually include

- further development of advanced third-generation biomaterials that activate germane genetic repair mechanisms,
- tissue engineering by molecular scaffolding,
- stem cell engineering including marrow stem cell therapy,
- novel scaffolds as delivery systems for therapeutic genes, and
- rapid, inexpensive, and highly predictive *in vitro* testing techniques for biomaterial cell response evaluation.

Evidently, stem cell engineering is a highly contentious issue that at present is hotly debated among various segments of the population and by politicians as well as ethically concerned self-appointed activists. Rapid and predictive *in vitro* testing methods for bioceramic cell responses could largely alleviate the use of costly and ethically compromised animal models. These are challenging aspects of future bioceramics research and development, even though, at present, there are no clearly formulated ethical guidelines available to assist researchers in the field of biomedical engineering. However, implementation of RRR (replacement, reduction, refinement) strategies and the future use of AI (see below) will likely obliterate ethical concerns associated with animal models.

6.15 Third-generation biomaterials

In spite of impressive advances made in biomaterials and biomedical device research, today there is consensus among experts that we have reached a limit to the current medical practices that emphasize replacement of tissue by predominantly materials science–based approaches, and many of which have been discussed in the preceding chapters.

Historically, research and development of *first-generation biomaterials* have centered on essentially bioinert or biotolerant materials that were incorporated into the body to replace missing parts or functions, or to fill bone cavities with dental amalgam, gypsum, ivory, or even wood. Hence, these replacement parts acted entirely as mechanical and space-filling parts of the body. This early development was followed by the invention of *second-generation biomaterials*, including osseoconductive and also bioresorbable bulk metals and ceramics, as well as coatings designed to elicit biological responses to induce growth of bone and other tissues.

Predictably, today and in the future there will be a necessity to apply a much more biological-based approach that will concentrate on repair and regeneration of tissue through *third-generation biomaterials*. This includes functionalization of bioceramic and biopolymer surfaces by adsorption of bone growth–promoting proteins such as recombinant human BMPs, osseoinductive noncollagenous proteins such as osteocalcin, osteopontin, and osteonectin as well as other extracellular proteins. Thus, current and potential future applications of biological coatings for orthopedic implants are being engineered toward facilitating osseointegration and preventing potential adverse tissue responses including foreign body reaction (FBR) and implant infection with bacteria, viruses, or prions.

While many of these coatings are still in the preclinical testing stage, bioengineers, material scientists, and surgeons alike continue to explore surface coatings as a means of improving clinical outcome of patients undergoing orthopedic reconstructive surgery. Despite a plethora of research, many aspects of the *in vivo* behavior of bone graft substitutes coated with hydroxylapatite and functionalized with bone growth–promoting proteins are still poorly understood. What is the exact mechanism behind their osteoinductivity? What are the optimum chemistry and architecture of hydroxylapatite-coated implants? Are identical surgical instruments being used for different implant insertions? Answering these questions will lead to major advancements in the properties and performance of implants functionalized with calcium phosphate coatings that are deposited by different techniques.

6.16 Fourth-generation biomaterials

Recently, the idea of *fourth-generation biomaterials* was suggested, based on future electronic systems integrated with the human body to provide powerful diagnostic as well as therapeutic tools for basic research and clinical use. The functionalities of such biomaterial systems include monitoring and manipulating cellular bioelectric responses for tissue regeneration with the aim to communicate with host tissues via bioelectric signals. The molecular biology revolution and advances in genomics and proteomics have significantly promoted the development of present-day biomaterials. In future developments, fourth-generation biomaterials will likely involve the molecular tailoring of microenvironments to achieve specific bioelectric cellular responses that could be monitored and appropriately manipulated by influencing ion channel activity. Study of such fourth-generation biomaterials is important to understand both intra- and intercellular signaling pathways and may elucidate the way how cells communicate across large networks. Cellular electrical monitoring may be able to reveal the fundamental behavior of cells and their response to external environmental stimuli.

In the future, the combination of fourth-generation biomaterials with biosensors, novel drug and gene delivery systems, and tissue scaffolds may trigger the development of advanced electronic medical devices. Along these lines, so-called *electroceuticals* are being developed, that is, electronic devices that are designed to affect biological systems by alleviating or mitigating symptomatological and/or pathological processes in the body and the brain. They utilize specific electric waveforms to stimulate electronically the nervous system and, thus, modify bodily functions including easing of cluster headaches and migraine-related syndrome.

Furthermore, there is anticipation that fourth-generation bioceramic materials based on ion-conductive alkali metal–doped calcium (Ti, Zr) hexaorthophosphates with NASICON structure will play a role in the development and clinical testing of novel devices including bone growth stimulators based on their elevated solid-state ionic conductivity, allowing them to transmit bioelectric signals adapted to this task. Consequently, future research should focus on development of a composite coating system for implants with the equivalent circuit of a capacitor that, by appropriate poling, could store negative electrical charges close to the interface with the growing bone, thus enhancing bone apposition rate as well as bone density, important figures of merit in reconstructive medicine.

6.17 Biosensors

Biosensors based on the *surface plasmon resonance* (SPR) effect will find increasing use. They are based on the interaction of an electromagnetic field, provided by laser light, with free electrons at the surface of metallic nanoparticles such as gold. In photothermal cancer therapy, plasmonic light-activated gold nanoparticles (GNPs) infused into the blood of a patient tend to concentrate inside a tumor. When an external laser light beam of the same frequency as that of the surface plasmon is shone into the tumor mass, the GNPs will heat up by resonance, thereby selectively killing cancer cells without affecting the surrounding healthy tissue. Although controversial at the present time, novel information gained in the future may be of great potential in curing certain cancers.

Interesting applications emerge for related *SPR microscopy* (SPRM), being a promising optical label-free technique in which polarized light is incident on a glass prism covered by thin gold layer on its backside. Free electrons at the surface of the gold show SPR by absorbing incident photons and create plasmon polaritons, that is, electromagnetic surface waves coupled to oscillating free electrons of the metallic surface that propagate along a metal/dielectric (air) interface. Biomolecules can be immobilized on the gold surface, and target molecules are allowed to interact with these molecules that cause a change in the index of refraction, captured by a CCD camera. This technique is being used for recording the kinetic behavior and surface interaction of biomolecules as well as detection of biominerals. For example, the kinetics of biomineralization can be quantitatively followed by SPRM. Poly(amido-amine) dendrimers are being immobilized via thiol groups on GNP surfaces and exposed, for example, to Ca^{2+}- and PO_4^{3-} ion-bearing solutions such as SBFs. SPRM is able to detect changes of the angle of refracted light and the refractive index, respectively, in response to mineral precipitation.

6.18 Biological materials

For the sake of completeness, future challenges should be mentioned in the realm of *biological materials* including their structure, function, and property design. Biological materials as opposed to biomaterials can be defined as materials derived from biological molecular building blocks such as proteins, carbohydrates, lipids, and nucleic acids including DNA and RNA. As for biomaterials discussed earlier, biological materials span a comparable tripartite evolutionary trend: *classic* biological materials based on advanced biomolecules with well-studied structure–function–property relationships, *emerging* biological materials that have been developed within the past decade or so, and *future* biological materials that are only experimentally studied at present or synthesized *in silico*. Future biological materials will likely play an important role, for example, by using bacteriophages and viruses as building blocks and potent gene ferries, and by exploring and eventually applying both a rational design of protein structure–function as well as the application of directed evolution for molecular discovery toward biomaterials development.

Bioinspired computers operating via artificial neural networks with nucleic acid-based integrated circuits may revolutionize parallel computing in the future. This approach uses the fact that the organization and complexity of all living beings are based on a coding system that functions with only four key components of the DNA molecule: the nucleobases cytosine, guanine, adenosine, and thymine. Multiple methods exist for constructing future computing devices based on DNA. Most of these methods are able to build the basic Boolean logic gates (AND, OR, NOT) associated with digital logic from a DNA basis. In this, nature's high-fidelity mechanism to translate the genetic information into the synthesis of biochemical molecules such as proteins is analogous to silicon-based computing technology still used today to store information and translate them into specific outputs. Hence, the DNA appears to be well suited as a medium for massive parallel computing architectures, with an enormous theoretical maximum of data storage capacity, processing speed, and unsurpassed energy efficiency. Although such computers may be many decades away from realization, the potential exists to move away from storing data in clunky silicon-based memory devices, the data storage capacity of which is suspected to deteriorate over time.

DNA bar codes to test and track nanoparticles designed for efficient drug delivery are already being used to screen large numbers of nanoparticles simultaneously to determine which of them gain entry to which organ tissues. This capability will allow establishing rapidly which nanoparticle design will be best suited for different drug delivery goals and strategies while reducing negative side effects to other organs.

However, for a stable self-assembled product, a relatively high cation concentration is required to prevent denaturation of DNA. Physiological and cell-culture conditions do not provide the required concentrations. In addition, the presence of nucleases may cause a serious threat to the integrity of DNA-based materials. Hence, for the translation of the promising DNA technology toward bioengineering challenges, stability needs to be guaranteed. Over the past years, various methods addressing the stability-related weaknesses of DNA origami have been developed.

6.19 Antibacterial medical textiles

The worldwide rising incidence of antibiotic drug-resistant infections has been listed by the WHO and in the World Economic Forum's Global Risks Report among the major threats to global health and, hence, constitutes a growing global emergency. This present and future threat arises clearly from the overuse of antibacterial drugs for even minor ailments. For example, aggressive and long-lived bacteria such as MRSA have emerged rapidly and spread globally in recent years. Other multidrug-resistant bacteria include *Escherichia coli*, which may cause hemorrhagic diarrhea, kidney failure, or pneumonia. Hence, among urgent research needs is the development of surface treatment routes toward long-lasting, passively acting antibacterial medical textiles that are clinically and economically effective in disrupting bacterial development and transmission cycles. Silver, copper, or graphene oxide nanoparticles embedded in medical textiles as well as incorporation of quaternary ammonium salts may provide novel routes to combat microbial resistance.

6.20 Cell–biomaterial interactions

In June 2012, the National Science Foundation had convened a workshop in Arlington, VA, USA, to assess the status of the field of biomaterials science and engineering and to identify particularly promising directions for the future of biomaterials research. As far as cell–biomaterial interactions are concerned, future general R&D requirements revolve around the improvement of biocompatibility of implanted biomaterials since FBRs may compromise their biological performance. To achieve this development goal, biosensors, electrodes, drug delivery devices, and vascular grafts must be enlisted and appropriately designed and applied. This also requires engineering of responsive and multifunctional biomaterials for cellular control, bidirectional signaling, and dynamic adaptation. Additional research tasks may include harnessing of developmental and regenerative biology, stem cell renewal and differentiation, patterning, generation of tissues and organs, cellular dedifferentiation as well as temporal regulation of signaling.

So-called bio-orthogonal (cell-friendly) chemistries are needed to probe and direct cell behavior by analysis of cellular- and molecular-level responses to biomaterials. Needs have been expressed to gain deeper knowledge of cell-induced remodeling of materials, protein adsorption, secretion, and degradation, and changes in mechanical properties, for example, by using *in situ* strain gauges. This also includes real-time *in situ* 3D cell monitoring, receptor presentation, transcriptional and epigenetic events, and secretion of cytokines.

6.21 Artificial intelligence

In future developments, AI will allow clinical trials without the involvement of human beings, thus avoiding ethical issues, the need for approval of dangerous drugs and devices, and the inevitably ensuing class-action lawsuits and litigation cases. AI is poised to simplify the lives of patients, doctors, and hospital administrators by performing tasks that are typically carried out by humans, but in less time, at a fraction of the ensuing cost, and with higher accuracy and, hence, reliability. As one of the world's highest growth industries, the AI sector was valued at about US$ 600 million in 2014 and is projected to reach US$ 150 billion by 2026. Among many other targets, AI is used to find new links between genetic codes or to drive surgery-assisting robots and, thus, is reinventing and reinvigorating modern healthcare through intelligent machines that can predict, comprehend, learn, and act. The WHO is working on a repository of used cases of AI in healthcare to identify data formats as well as interoperability mechanisms required to amplify their impact. The NIH is actively exploring ways to incorporate AI into biomedical research and bringing in outside talent to help it better embrace the emerging technology.

AI could also be instrumental in thorough understanding of the complexity of molecular biological aspects of cell adhesion by bringing together concepts and experience from biology, information technology, and systems science. Mathematical modeling of intracellular signaling pathways could allow comprehensive simulation not only of cellular processes but also of larger biological systems such as tissues and organs. With this tool, the number and extent of cell- and animal-based experiments could be substantially reduced or even alleviated, leading to time and cost savings and the avoidance of ethically compromised issues related to the use of animal models.

Presently established high-throughput screening platforms have necessitated the requirement for massive data analysis as part of biomaterials research and engineering. Further advancement in AI, miniaturization of materials fabrication, and application of robotics will eventually lead to the emergence of autonomous, intelligent systems able to perform biomaterials research independently.

However, as concerned experts admonish, there are two forms of AI: "narrow or primitive AI" that does mundane, repetitive tasks very well and can learn to do them better over time, and the fear-inducing "general or deep-learning AI" that could respond dynamically to any conceivable situation but presumably will not exist for a long time. Nevertheless, AI is not enthusiastically welcomed by everybody. The late theoretical physicist Stephen Hawking acknowledged that primitive forms of AI developed so far have already proved to be very useful, but he expressed grave concern of the consequences of creating something new that can match or even surpass humans: "It would take off on its own, and re-design itself at an ever increasing rate and humans, who are limited by slow biological evolution, couldn't compete, and would be superseded." Is a "brave new world" in the offing?

In a European context, the Council of Europe's Committee on Bioethics is preparing a Strategic Action Plan on technologies and human rights in the field of biomedicine 2020–2025, which includes deployment of AI and focus on the protection of particularly vulnerable population groups such as children and the elderly. A guide is also under preparation on public engagement on fundamental questions raised by developments in biology and medicine with particular focus on new technologies including AI.

6.22 Far-future materials and bioengineering tools

Research needs and development strategies into biomaterials and biomedical devices discussed hitherto are confined to rather well-known ways how to do things in university and industry laboratories concerned with biomaterial R&D, and thus, truly paradigmatic shifts in perception, application, and implication of this research are not to be expected soon. However, there may be conceivable ways and novel techniques to overcome such limitations in the far future. We should be clear about the fact that "far future" may be just 20 or 50 years away given the exponential growth of penetrating knowledge as well as the deployment of efficient and effective research tools at the disposal of the research community worldwide.

In their 2010 book on *Advanced Ceramics and Future Materials*, the authors Fritz Aldinger and Volker A. Weberruß have presented a sophisticated algorithmic formalism aimed to model biomimetic materials in a self-organizing, synergetic manner based on a nonlinear extension of conventional quantum mechanics. With this approach, completely novel materials would be first simulated *in silico* by advanced modeling, then synthesized, developed, tested, and deployed in preclinical settings. Today, the properties and applications of such novel bioinspired materials are still unimaginable.

A more philosophically oriented look much farther into the distant future has recently been taken by the Israeli historian and author Yuval Noah Hariri in his controversial 2016 book *Homo Deus: A Brief History of Tomorrow*. The author suggests that in the (not too far?) future, contemporary humans could be upgraded by either biological engineering, cyborg engineering, or engineering of nonorganic entities acting like humans. By these measures, *Homo sapiens* may become an amortal "god-like" super-being with a bioengineered body, and physical and mental abilities vastly superior to those that limit present-day humans. Presumably, yet unknown biomaterials, biomedical devices, bioengineering technologies, advanced algorithms, and deep-learning AI need to be enlisted to achieve this epic task, the nature, extension, and societal consequences of which cannot even dimly be fathomed today. Indeed, human genome editing may become a reality, using, for example, CRISPR/Cas software platforms and other bioinformational tools built to facilitate the design of appropriate guide RNAs to recruit and direct the nuclease activity to the region of interest. Even though at present there exist very powerful ethical barriers against using these tools to engineer the human genome, which way humanity's decision using such concepts is tilting in the far future is a matter of conjecture only.

6.23 Conclusion and implications

At present, many exciting avenues are opening up toward design, development, and implementation of future biomaterials and biomedical devices. Synergistic interaction of scientists, engineers, and clinical professionals in academia and industry promises great new breakthroughs in the immediate future. Major drivers for the global biomaterials market include the increase of aging population, availability and global spreading of advanced materials and technologies, enhanced benefits to patients, generally increased acceptance of novel devices and technologies, slowly rising ROI, and heightened awareness for biomaterial products and their benefits. Further research and development efforts in biomaterials and biomedical devices are expected to create increasing opportunities for the worldwide biomaterials market in the near future.

However, sophisticated materials science still needs to be developed with the aim to match details of the biological complexity at the molecular level. The task of tailoring the surfaces of biomaterials for at least some of the different purposes of implant integration and tissue regeneration appears as a feasible challenge in the future, achieved by the synergistic interdisciplinary work of materials scientist, engineers, biologists, chemists, physicists, and medical doctors.

As shown earlier, today, biomaterials research is focusing preferentially on the development of new materials and scaffolds for regenerative medicine, stem cell therapy, drug and gene delivery, and biosensing applications. Biomaterials engineering includes synthesis and characterization of novel materials such as biodegradable metals, bioceramics, bioglasses, polymers, and hybrid nanocomposites. In addition, a new area into which the biomaterial research activity is moving comprises developing and tailoring of inorganic nanoparticles that include QDs with bioactive peptides attached to semiconducting substrates, enabling to act as monitors to detect the enzyme activity. Research is being performed to develop CQDs, a new generation of optically active nanoparticles potentially capable of replacing semiconductor QDs, but with the advantage of much lower cytotoxicity.

Recent fertile research areas include cardiovascular biomaterials such as pyrolytic graphitic carbon and DLC, and analysis of their hemo- and biocompatibilities. Moreover, future research activities will likely include functionalized polymeric surfaces and coatings for biosensors and actuators, as well as research into nanoscale orthopedic biomaterials, and hard or soft biomechanics.

There is also a politicoeconomic aspect. Within the European Union, the fragmented nature of research activities requires an integrated approach to overcome the new technological challenges that lie ahead, and thus, to take advantage of the various possibilities for development, growth, and expansion of the biomaterials market. The many national stakeholders involved in activities ranging from basic research through innovation and eventual deployment in clinical settings (bench-to-bedside strategy) will need to be focused and amalgamated, allowing for increased upscaling of production and financial risk sharing, and the simultaneous optimization of the use of available knowledge and resources within the 27 countries of the EU. Furthermore, the ability of novel and improved biomaterials to adapt to their biological environments in a beneficial manner will determine the success of those therapeutic interventions with which biomaterials are associated. This particular aspect of interactive or smart biomaterials is expected to be a major area of growth for research and development activities in the EU's Horizon 2020 Framework Programme and can thus assist in moving academic research on biomaterials into commercially and socioeconomically viable European and global solutions. This effort is being augmented by the new Medical Device Regulation (MDR) that was supposed to become binding for the European Union member states in May 2020. However, as a result of the Sars-CoV-2 pandemic this has been postponed. In addition to the already existing regulations, the new MDR is poised to provide an optimized and standardized regulation for placement of medical devices on the market, with a strong emphasis on product quality and safety.

Appendix A
Standard specifications for biomaterials and biomedical devices

Robert B. Heimann

Below is a list of recognized standard specifications for biomaterials and biomedical devices (Government of Canada, 2018) that shows the standards most pertinent to the content of this textbook (see also Sun, 2018). Since these standard specifications are subject to constant revision, care should be taken when applying them to novel formulations and devices without checking their relevance and actuality.

A.1 Standard specifications for metals involved in manufacturing biomedical devices

Stainless steels

ASTM F138-08
    Standard specification for wrought 18chromium–14nickel–2.5molybdenum stainless steel bar and wire for surgical implants (UNS S31673)
ASTM F139-08
    Standard specification for wrought 18chromium–14nickel–2.5molybdenum stainless steel sheet and strip for surgical implants (UNS S31673)
ASTM F621-12
    Standard specification for stainless steel forgings for surgical implants
ASTM F899-12
    Standard specification for wrought stainless steel for surgical instruments
ASTM F1314-07
    Standard specification for wrought nitrogen strengthened 22chromium–13nickel–5manganese–2.5molybdenum stainless steel alloy bar and wire for surgical implants (UNS S20910)
ASTM F1350-08
    Standard specification for wrought 18chromium–14nickel–2.5molybdenum stainless steel surgical fixation wire (UNS S31673)
ASTM F1586-08
    Standard specification for wrought nitrogen strengthened 21chromium–10nickel–3manganese–2.5molybdenum stainless steel bar for surgical implants (UNS S31675)

https://doi.org/10.1515/9783110619249-007

ASTM F2229-12

Standard specification for wrought, nitrogen strengthened 23manganese–21chromium–1molybdenum low-nickel stainless steel alloy bar and wire for surgical implants (UNS S29108)

ASTM F2257-09

Standard specification for wrought seamless or welded and drawn 18 chromium–14nickel–2.5molybdenum stainless steel small diameter tubing for surgical implants (UNS S31673)

ISO 5832-1:2008

Implants for surgery – Metallic materials – Part 1: Wrought stainless steel

ISO 5832-9:2007

Implants for surgery – Metallic materials – Part 9: Wrought high nitrogen stainless steel

ISO 7153-1:1991

Surgical instruments – Metallic materials – Part 1: Stainless steel

CoCr alloys

ASTM F75-12

Standard specification for cobalt–28chromium–6molybdenum alloy castings and casting alloy for surgical implants (UNS R30075)

ASTM F90-09

Standard specification for wrought cobalt–20chromium–15tungsten–10nickel alloy for surgical implant applications (UNS R30605)

ASTM F562-07

Standard specification for wrought 35cobalt–35nickel–20chromium–10molybdenum alloy for surgical implant applications (UNS R30035)

ASTM F688-05

Standard specification for wrought cobalt–35nickel–20chromium–10molybdenum alloy plate, sheet, and foil for surgical implants (UNS R30035)

ASTM F799-11

Standard specification for cobalt–28chromium–6molybdenum alloy forgings for surgical implants (UNS R31537, R31538, R31539)

ASTM F961-08

Standard specification for 35cobalt–35nickel–20chromium–10molybdenum alloy forgings for surgical implants (UNS R30035)

ASTM F1091-08

Standard specification for wrought cobalt–20chromium–15tungsten–10nickel alloy surgical fixation wire (UNS R30605)

ASTM F1537-08
  Standard specification for wrought cobalt–28chromium–6molybdenum alloy
  for surgical implants (UNS R31537, UNS R31538, and UNS R31539)
ISO 5832-4:1996
  Implants for surgery – Metallic materials – Part 4: Cobalt–chromium–molybdenum
  casting alloy
ISO 5832-5:2005
  Implants for surgery – Metallic materials – Part 5: Wrought cobalt–chromium–
  tungsten–nickel alloy
ISO 5832-6:1997
  Implants for surgery – Metallic materials – Part 6: Wrought cobalt–nickel–
  chromium–molybdenum alloy
ISO 5832-12:2007
  Implants for surgery – Metallic materials – Part 12: Wrought cobalt–chromium–
  molybdenum alloy
ISO 5832-12:2007/Cor.1:2008

Titanium and titanium alloys

ASTM F67-06
  Standard specification for unalloyed titanium for surgical implant applications
  (UNS R50250, UNS R50400, UNS R50550, UNS R50700)
ASTM F136-12
  Standard specification for wrought titanium–6aluminum–4vanadium ELI (extra
  low interstitial) alloy for surgical implant applications (UNS R56401)
ASTM F620-11
  Standard specification for titanium alloy forgings for surgical implants in the
  alpha plus beta condition
ASTM F1108-04 (R2009)
  Standard specification for titanium–6aluminum–4vanadium alloy castings for
  surgical implants (UNS R56406)
ASTM F1295-05
  Standard specification for wrought titanium–6 aluminum–7 niobium alloy for
  surgical implant applications (UNS R56700)
ASTM F1472-08
  Standard specification for wrought titanium–6aluminum–4vanadium alloy for
  surgical implant applications (UNS R56400)
ASTM F1580-12
  Standard specification for titanium and titanium–6aluminum–4vanadium alloy
  powders for coatings of surgical implants

ASTM F1713-08
  Standard specification for wrought titanium–13niobium–13zirconium alloy for surgical implant applications (UNS R58130)
ASTM F1854-15
  Standard test method for stereological evaluation of porous coatings on medical implants
ASTM F1978-12
  Standard testing method for measuring abrasion resistance of metallic thermal spray coatings by using Taber Abraser
ASTM F2066-13e1
  Standard specification for wrought titanium–15molybdenum alloy for surgical implant applications (UNS R58150)
ASTM F2146-13
  Standard specification for wrought titanium–3aluminum–2.5vanadium alloy seamless tubing for surgical implant applications (UNS R56320)
ISO 4287:1997
  Geometrical Product Specifications (GPS) – Surface texture: Profile method – Terms, definitions, and surface texture parameters
ISO 5832-2:1999
  Implants for surgery – Metallic materials – Part 2: Unalloyed titanium
ISO 5832-3:1996
  Implants for surgery – Metallic materials – Part 3: Wrought titanium 6–aluminium 4–vanadium alloy
ISO 5832-11:1994
  Implants for surgery – Metallic materials – Part 11: Wrought titanium 6–aluminium 7–niobium alloy
ISO 13179-1: 2014
  Implants for surgery –Plasma-sprayed unalloyed titanium coatings on metallic surgical implants – Part 1: General requirements

Magnesium alloys

ASTM 02.02-18
  Aluminum and magnesium alloys
ASTM B80-15
  Standard specification for magnesium-alloy sand castings
ASTM B91-12
  Standard specification for magnesium-alloy forgings
ASTM B94-18
  Standard specification for magnesium-alloy die castings
ASTM B199-17
  Standard specification for magnesium-alloy permanent mold castings

ASTM B403-12

Standard specification for magnesium-alloy investment castings

ASTM B879-17

Standard practice for applying non-electrolytic conversion coatings on magnesium and magnesium alloys

ASTM F3268-18a

Standard guide for *in vitro* degradation testing of absorbable metals

ISO 16220

Magnesium and magnesium alloys – Magnesium alloy ingots and castings

ISO DIS 3116

Magnesium and magnesium alloys – Wrought magnesium and magnesium alloys

Other metals

ASTM F560-08

Standard specification for unalloyed tantalum for surgical implant applications (UNS R05200, UNS R05400)

ASTM F2063-12

Standard specification for wrought nickel-titanium shape memory alloys for medical devices and surgical implants

ASTM F2384-10

Standard specification for wrought zirconium-2.5niobium alloy for surgical implant applications (UNS R60901)

ISO 13402:1995

Surgical and dental hand instruments – Determination of resistance against autoclaving, corrosion and thermal exposure

ISO 13782:1996

Implants for surgery – Metallic materials – Unalloyed tantalum for surgical implant applications

A.2 Standard specifications related to corrosion tests for metallic biomaterials

ASTM F746

Standard test method for pitting or crevice corrosion of metallic surgical implant materials

ASTM F897

Standard test method for measuring fretting corrosion of osteosynthesis plates and screws

ASTM F1089
   Standard test method for corrosion of surgical instruments
ASTM F1801
   Standard practice for corrosion fatigue testing of metallic implant materials
ASTM F2129
   Standard test method for conducting cyclic potentiodynamic polarization measurements to determine the corrosion susceptibility of small implant devices

A.3 Standard specifications for bioceramics involved in manufacturing biomedical devices

Given the plethora of standards pertaining to composition, structure, and processing of metallic biomaterials, at a first glance it is surprising that only relatively few standards and regulations applicable to bioinert and bioactive ceramics exist. Techniques used to evaluate the mechanical strength, fracture toughness, and fatigue properties of bioceramics are derived from the evaluation methods used for industrial ceramics, but with added consideration of the biological environment in which bioceramics are supposed to function (Ishikawa et al., 2003). Since most methods to evaluate bioceramics in terms of their clinical requirements have not been standardized, researchers need to be clear about the specifics of the biological environment. For example, unlike metals that behave, to a large extent, bioinert, and thus, do little interact with human metabolism, for bioactive ceramics that elicit biological responses there is no standard on how to evaluate their dissolution characteristics. One reason for the lack of standards is the difficulty to simulate the chemical dissolution of bioceramics and subsequent precipitation of bone-like apatite *in vitro* as is it easier to dissolve materials *in vivo* than in distilled water or SBF. Many researchers in the field take precipitation of bone-like apatite on bioceramics incubated in simulated body fluid (SBF) as sure-fire confirmation of their "bioactivity." However, this phenomenon is only a necessary but not a sufficient requirement. To establish bioactivity unambiguously, a set of biological tests must be employed that include cell proliferation, spreading and vitality of established cell lines, measurement of alkaline phosphatase levels, as well as observation of upregulation of bone growth-specific proteins such as osteocalcin, osteonectin, osteopontin, and silylated glycoproteins. Another problem exists when micron-sized particles formed by wear or flaking of bioceramic coatings that elicit strong cytotoxicity are phagocytized by macrophages. Standard specifications of bioceramic materials include

ASTM F1044-05(2017)
   Standard test method for shear testing of calcium phosphate coatings and metallic coatings

ASTM F1147-05(2011)
Standard test method for tension testing of calcium phosphate and metallic coatings
ASTM F1160 (2014)
Standard test method for shear and bending fatigue testing of calcium phosphate and metallic medical and composite calcium phosphate/metallic coatings
ASTM F1088-04a (R2010)
Standard specification for beta-tricalcium phosphate for surgical implantations
ASTM F1185-03(2014)
Standard specification for composition of hydroxylapatite for surgical implants
ASTM F1538-03(2017)
Standard specification for glass and glass ceramic biomaterials for implantation
ASTM F1609-08(2014)
Standard specification for calcium phosphate coatings for implantable materials
ASTM F1854-15
Standard test method for stereological evaluation of porous coatings on medical implants or an alternative recognized validated method
ASTM F1926-14/F1926M-10
Standard test method for evaluation of the environmental stability of calcium phosphate granules, fabricated forms, and coatings
ASTM F2024-10 (2016)
Standard practice for X-ray diffraction determination of phase content of plasma-sprayed hydroxyapatite coatings
ASTM F2224-09(2014)
Standard specification for high-purity calcium sulfate hemihydrate or dihydrate for surgical implants
ASTM F2393-12
Standard specification for high-purity dense magnesia partially stabilized zirconia (Mg-PSZ) for surgical implant applications
ISO 4287:1997
Geometrical product specifications (GPS) – Surface texture: Profile method – Terms, definitions, and surface texture parameters
ISO 6474-1:2010
Implants for surgery – Ceramic materials – Part 1: Ceramic materials based on high purity alumina
ISO 6474-2:2012
Implants for surgery – Ceramic materials – Part 2: Composite materials based on a high purity alumina matrix with zirconia reinforcement.
ISO 6872:2008
Dentistry – Ceramic materials

ISO 13356:2016-02.
    Implants for Surgery – Ceramic materials based on yttria-stabilized tetragonal zirconia (Y-TZP)
ISO 13779-2:2018
    Implants for Surgery – Hydroxyapatite – Part2: Coatings of hydroxyapatite
ISO 13779-3:2008 Part 3: Chemical analysis and characterization of crystallinity and phase purity
ISO 13779-4:2002 Part 4: Determination of coating adhesion strength
ISO 13779-6:2015 Part 6: Powders

A.4 Standards specifications for polymers involved in manufacturing biomedical devices

ASTM F451-08
    Standard specification for acrylic bone cement
ASTM F639-09
    Standard specification for polyethylene plastics for medical applications
ASTM F648-14
    Standard specification for ultra-high-molecular-weight polyethylene powder and fabricated form for surgical implants
ASTM F665-09
    Standard classification for vinyl chloride plastics used in biomedical application
ASTM F702-10
    Standard specification for polysulfone resin for medical applications
ASTM F732-00(2011)
    Standard test method for wear testing of polymeric materials used in total joint prostheses
ASTM F754-08
    Standard specification for implantable polytetrafluoroethylene (PTFE) sheet, tube, and rod shapes fabricated from granular molding powders
ASTM F755-99(2011)
    Standard specification for selection of porous polyethylene for use in surgical implants
ASTM F997-10
    Standard specification for polycarbonate resin for medical applications
ASTM F1839-08(2012)
    Standard specification for rigid polyurethane foam for use as a standard material for testing orthopaedic devices and instruments

ASTM F1855-00(2011)
Standard specification for polyoxymethylene (acetal) for medical applications
ASTM F1925-09
Standard specification for semicrystalline poly(lactide) polymer and copolymer resins for surgical implants
ASTM F1983-99(2008)
Standard practice for assessment of compatibility of absorbable/resorbable biomaterials for implant applications
ASTM F2003-02(2008)
Standard practice for accelerated aging of ultra-high molecular weight polyethylene after gamma irradiation in air
ASTM F2026-14
Standard specification for polyetheretherketone (PEEK) polymers for surgical implant applications
ASTM F2038-00(2011)
Standard guide for silicone elastomers, gels, and foams used in medical applications – Part I – Formulations and uncured materials
ASTM F2042-00(2011)
Standard guide for silicone elastomers, gels, and foams used in medical applications – Part II – Crosslinking and fabrication
ASTM F2214-02(2008)
Standard test method for *in situ* determination of network parameters of crosslinked ultra-high molecular weight polyethylene (UHMWP)
ASTM F2313-10
Standard specification for poly(glycolide) and poly(glycolide-co-lactide) resins for surgical implants with mole fractions greater than or equal to 70% glycolide
ASTM F2565-06
Standard guide for extensively irradiation-crosslinked ultra-high molecular weight polyethylene fabricated forms for surgical implant applications
ISO 5834:2011
Implants for surgery – ultra-high molecular weight polyethylene
ISO 5834-1:2005 Part 1: Powder form
ISO 5834-2:2011 Part 2: Moulded forms
ISO 5834-3:2005 Part 3: Accelerated ageing methods
ISO 5834-4:2005 Part 4: Oxidation index measurement method
ISO 5834-5:2005 Part 5: Morphology assessment method
ISO 3826-1:2003
Plastic collapsible containers for human blood and blood components – Part 1: Conventional containers
ISO 5833:2002
Implants for surgery – Acrylic resin cements

ISO 13781:2017

Implants for surgery – Homopolymers, copolymers and blends on poly(lactide) – *In vitro* degradation testing

A.5 Standard specifications related to biocompatibility

ASTM F981-04

Standard practice for assessment of compatibility of biomaterials for surgical implants with respect to effect of materials on muscle and bone

ISO 7405:2008

Dentistry – Evaluation of biocompatibility of medical devices used in dentistry

ISO 10993-1:2010

Biological evaluation of medical devices – Part 1: Evaluation and testing within a risk management process

ISO 10993-2:2006

Biological evaluation of medical devices – Part 2: Animal welfare requirements

ISO 10993-3:2003

Biological evaluation of medical devices – Part 3: Tests for genotoxicity, carcinogenicity and reproductive toxicity

ISO 10993-4:2006

Biological evaluation of medical devices – Part 4: Selection of tests for interactions with blood

ISO 10993-5:2009

Biological evaluation of medical devices – Part 5: Tests for *in vitro* cytotoxicity

ISO 10993-6:2007

Biological evaluation of medical devices – Part 6: Tests for local effects after implantation

ISO 10993-7:2009

Biological evaluation of medical devices – Part 7: Ethylene oxide sterilization residuals

ISO 10993-9:2009

Biological evaluation of medical devices – Part 9: Framework for identification and quantification of potential degradation products

ISO 10993-10:2010

Biological evaluation of medical devices – Part 10: Tests for irritation and skin sensitization

ISO 10993-11:2006

Biological evaluation of medical devices – Part 11: Tests for systemic toxicity

ISO 10993-12:2007
Biological evaluation of medical devices – Part 12: Sample preparation and reference materials

ISO 10993-13:2010
Biological evaluation of medical devices – Part 13: Identification and quantification of degradation products from polymeric medical devices

ISO 10993-14:2001
Biological evaluation of medical devices – Part 14: Identification and quantification of degradation products from ceramics

ISO 10993-15:2000
Biological evaluation of medical devices – Part 15: Identification and quantification of degradation products from metals and alloys

ISO 10993-16:2010
Biological evaluation of medical devices – Part 16: Toxicokinetic study design for degradation products and leachables

ISO 10993-17:2002
Biological evaluation of medical devices – Part 17: Establishment of allowable limits for leachable substances

ISO 10993-18:2005
Biological evaluation of medical devices – Part 18: Chemical characterization of materials

A.6 Standard specifications for aseptic processing of biomedical devices and cleanroom requirements

ISO 13408-1:2008
Aseptic processing of health care products – Part 1: General requirements

ISO 13408-2:2003
Aseptic processing of health care products – Part 2: Filtration

ISO 13408-3:2006
Aseptic processing of health care products – Part 3: Lyophilization

ISO 13408-4:2005
Aseptic processing of health care products – Part 4: Clean-in-place technologies

ISO 13408-5:2006
Aseptic processing of health care products – Part 5: Sterilization in place

ISO 13408-6:2005
Aseptic processing of health care products – Part 6: Isolator systems

ISO 13408-7:2012
Aseptic processing of health care products – Part 7: Alternative processes for medical devices and combination products

ISO 14644-1:1999
Cleanrooms and associated controlled environments – Part 1: Classification of air cleanliness
ISO 14644-2:2000
Cleanrooms and associated controlled environments – Part 2: Specifications for testing and monitoring to prove continued compliance with ISO 14644-1
ISO 14644-3:2005
Cleanrooms and associated controlled environments – Part 3: Test methods
ISO 14644-4:2001
Cleanrooms and associated controlled environments – Part 4: Design, Construction and Start Up
ISO 14644-5:2004
Cleanrooms and associated controlled environments – Part 5: Operations
ISO 14644-6:2007
Cleanrooms and associated controlled environments – Part 6: Vocabulary
ISO 14644-7:2004
Cleanrooms and associated controlled environments – Part 7: Separative devices (clean air hoods, glove boxes, isolators and mini-environments)
ISO 14644-8:2012
Cleanrooms and associated controlled environments – Part 8: Classification of air cleanliness by chemical concentration (ACC)
ISO 14644-9:2012
Cleanrooms and associated controlled environments – Part 9: Classification of surface cleanliness by particle concentration
ISO 14644-10:2013
Cleanrooms and associated controlled environments – Part 10: Classification of surface cleanliness by chemical concentration
ISO 14698-1:2003
Cleanrooms and associated controlled environments – Biocontamination control – Part 1: General principles and methods
ISO 14698-2:2003
Cleanrooms and associated controlled environments – Biocontamination control – Part 2: Evaluation and interpretation of biocontamination data.

A.7 Risk management of medical devices

ISO 14971:2012
Medical devices – Application of risk management to medical devices (ISO 14971:2007, corrected version 2007-10-01); German version EN ISO 14971:2012.
ISO 14971:2019
Under revision

References

Government of Canada (2018). List of recognized standards for medical devices – draft guidance document. Issued Sept 11, 2018. https://www.canada.ca/en/health-canada/services/drugs-health-products/medical-devices/standards/list-recognized-standards-medical-devices-draft (accessed Feb 24, 2019).

Ishikawa, K., Matsuya, S., Miyamoto, Y. and Kawate, K. (2003). 9.05 Bioceramics. In: Reference Module in Materials Science and Materials Engineering. Comprehensive Structural Integrity, vol. 9. p. 169–214. Elsevier.

Sun, L. (2018). Thermal spray coatings on orthopedic devices: When and how the FDA reviews your coatings. J. Thermal Spray Technol., 27, p. 1280–1290.

Appendix B
Current world market situation of biomaterials and biomedical devices

Robert B. Heimann

Major drivers for the global biomaterials market include the increase of aging population, availability of advanced technology, enhanced benefits to patients, slowly increasing return on investment (ROI), and rising awareness for the breadth and capabilities of biomaterial products. Further research in medical applications is expected to create strongly increasing opportunities for the biomaterials market in the near future.

Today, biomaterials research and development is focusing, among many other applications, on the development of new scaffolds for regenerative medicine, stem cell therapy, drug and gene delivery, and biosensing devices. Biomaterials engineering include processing, synthesis and characterization of novel materials such as bioceramics, bioglasses, polymers, bone cement composites, and hybrid compositions. In addition, a new area into which the biomaterial activity is moving comprises tailoring of inorganic nanoparticles that include *quantum dots* with bioactive peptides (Hauser and Zhang, 2010) and gold, which enables them to act as reporters to detect enzyme activity. Recent research on biomaterial applications includes cardiovascular biomaterials and analysis of their hemo- and biocompatibilities. Moreover, other active research areas include polymeric surface coatings for sensors and implants as well as research in nanoscale orthopedic biomaterials, and hard or soft biomechanics (Persistence Market Research, 2018).

B.1 Biomaterials market

The global market for biomaterials has been estimated to be US $70 billion in 2016 and is expected to more than double to US $150 billion by 2021 with an impressive average annual growth rate (AAGR) of 16% (Mordor Intelligence, 2019). According to a different estimate, the global biomaterials market is projected to reach US $207 billion by 2024 from US $105 billion in 2019, at a AAGR of 14.5% between 2019 and 2024 (MarketsandMarkets, 2019).

The medical disciplines using biomaterials include cardiovascular medicine, orthopedics, dental medicine, ophthalmology, immunology, and infection. The launch of new products in areas such as plastic surgery, neurology, and wound healing is expected to fuel the growth of this market. The cardiovascular medicine field dominates the global biomaterials market in terms of share due to high prevalence rates of cardiovascular diseases (CVDs), with the orthopedic segment, that is, hip and knee

https://doi.org/10.1515/9783110619249-008

arthroplasties being the second largest market. Important economic drivers of this market are technological advancements, rise in the number of aging population, increased funds and grants by government bodies worldwide, the growing implantable devices market, rising number of hip and knee replacement operations, increasing awareness of the benefits of tissue engineering, and growing incidence of chronic diseases, many of them are individual lifestyle-imposed. In addition, high growth rates are expected for plastic surgery and wound healing applications, strong drivers for the growth of the biomaterials market.

On the downside, stringent safety and standardization requirements, validation regulations, increasing number of litigation cases, and compatibility issues are important factors counteracting the growth of the biomaterials market. This also includes the notion that large-scale production and ROI of industrial production expenditures are close to impossible for deployment of truly novel biomaterials. For the European market, in particular, the fragmented nature of European research activities requires an integrated approach at the level of the European Union to overcome the new technical challenges that lie ahead and thus, to take advantage of the various possibilities for market development, growth, and expansion. The many stakeholders involved in activities ranging from research through innovation and eventual market placement in clinical settings ("bench-to-bedside") will need to be amalgamated to allowing for increased upscaling of production and financial risk sharing, and the simultaneous optimization of the use of available knowledge and resources.

The global biomaterials market is segmented on the basis of type of material, application, and geographical region as follows:

Type of material

Based on the type of material, the biomaterials market is divided into metallic, ceramic, polymeric, composite, and biological materials. The metallic segment is categorized into surgical austenitic stainless steel, titanium alloys, CoCrMo alloys, silver, gold, and magnesium. The ceramic segment is further subdivided into alumina, zirconia, calcium phosphate, calcium sulfate, carbon, and bioglasses. In particular, the high-purity alumina market is estimated to be worth US $5.3 billion by 2023 (Nathwani, 2017) of which biomedical alumina is only a part.

The polymer segment is divided into poly(methylmethacrylate); polyethylene; polyester; poly(vinyl chloride); silicone; polyamide; poly(etheretherketone); poly (lactic acid) (PLA); poly(glycolic acid) (PGA), and many other polymeric biomaterials. The natural biomaterials, that is, biological materials market is further segmented into hyaluronic acid, collagen, gelatin, fibrin, cellulose, chitin, alginate, and silk. In 2016, the metallic segment was estimated to command the largest share of the market, by type of material. The polymeric biomaterials segment is projected to grow at the highest AAGR during the forecast period, owing to the increasing use of polymers in soft tissue applications such as plastic surgeries. In particular, the global polymers market is expected to reach over US $19 billion by 2022, with an estimated AAGR of 13%. Among the medically relevant polymers, bioresorbable plastics such as PLA and PGA are thought to grow at an AAGR of 17% during the forecast period (Thompson, 2016).

Applications

On the basis of application, the market is segmented into cardiovascular, orthopedic, dental, plastic surgery, wound healing, tissue engineering, ophthalmology, neurological/central nervous system, and other applications which include drug delivery systems, gastrointestinal applications, bariatric surgery, and urinary applications. In 2016, the cardiovascular segment accounted for the largest share of the biomaterials market. This was attributed to rising incidence of CVDs worldwide. However, the plastic surgery segment is projected to grow at the highest AAGR from 2016 to 2021. Preference for minimally invasive cosmetic surgeries and positive public perception for these surgeries are significant factors that are expected to drive market growth.

Geographical region

Based on geography, the market is segmented into North America, Europe, Asia-Pacific, and rest of the world. North America is still the largest market for biomaterials owing to the sophisticated healthcare infrastructure, closely followed by the European Union. The biomaterials market is expected to witness the highest growth rate in the Asia-Pacific region. Emerging countries such as India and China are strongly contributing to the Asia-Pacific biomaterials market because of rising patient awareness coupled with rising healthcare expenditure, and increasing gross domestic product. For Europe, biomaterials will become a major focus of European research effort in the coming years as part of the Horizon 2020 Framework Programme for Research and Innovation. In particular, biomaterials will find applications as integral parts of advanced therapy medicinal products (ATMPs) or as complete or parts of medical devices (see Section B.3).

Major players

Major players on the biomaterials market are BASF SE, Bayer AG, Berkeley Advanced Biomaterials Inc., Biomed Inc., Corbion N.V., Invibio Ltd., Koninklijke DSM N.V., and Noble Biomaterials Inc. (Mordor Intelligence, 2019).

However, there is a caveat. All forecasts and predictions discussed above are wide open to educated criticism and healthy skepticism, and as such are subject to the uncertain overall economic growth in the geographical regions investigated. As we all are keenly aware, the world as we know it is in turmoil, and known and not yet known obstacles to global economic growth such as the establishment of new trade barriers, protective taxes, and unforeseen environmental and political upheavals including wars may severely curb the optimistic growth scenarios for biomaterials and medical devices alluded to above.

B.2 Medical device market

A special impediment of current market development of biomaterials and medical devices is associated with the fact that in litigious countries such as the USA, smaller specialized suppliers tend to pull out of the market owing to the excessive cost of defense against legal action in case devices truly or allegedly fail in the hostile body environment (Galletti, 1998). Notwithstanding these impediments, the USA is still leading the world both in production and in the export trade of medical devices. In 2015, the combined value of US industry shipments for medical devices was worth US $43 billion, with an annual growth of about 1.5 percent. However, the US biomedical industry faces increasingly stiff competition from Germany (Siemens®, Braun®), Japan (Hitachi®, Toshiba®), and the Netherlands (Philips Electronics®) in high-technology products. In terms of geographic segmentation, the American market for medical devices is forecast to grow from US $167 billion in 2016 to US $209 billion in 2020, the Asia-Pacific market from US $69 billion (2016) to US $89 billion (2020), and the European market from US $80 billion (2016) to US $106 billion (2020). The German medical device market is the largest in Europe and the third largest in the world. It was valued at US $26 billion in 2014 and continues to grow despite the present still sluggish European economy.

In 2019, the fastest growing segments of the medical device markets were brain monitoring devices (CAGR of 9%), cardiac monitoring and therapy devices (CAGR of 8%), and orthopedic monitoring devices (CAGR of 5%). The market growth of brain monitoring devices is expected to be mainly driven by the rising population with brain and neurological diseases, such as epilepsy, Alzheimer's disease, and brain tumors. The growth in market size of cardiac monitoring and therapy devices is being attributed to the huge demand driven by the large number of individuals across the globe who are increasingly suffering from CVDs such as angina, myocardial infarction, stroke, heart failure, venous thromboembolism, and heart arrhythmia. Finally, the advent of 3D imaging has enhanced orthopedic care by allowing better bone and implant position examination compared with other imaging modalities such as conventional fluoroscopy or X-ray imaging. The emergence of 3D holographic imaging has revolutionized medical imaging by providing more flexible and accurate images in orthopedic and sports medicine at a faster rate by converting the two-dimensional computed tomography and magnetic resonance imaging data into interactive virtual reality images (TechnavioBlog, 2019).

Table B.1 shows the major market forecast figures for biomedical devices divided into geographical regions based on a 2015 forecast for the period between 2016 and 2020. A more recent 2016 forecast predicted a total worldwide sale of biomedical device products of US $530 billion by 2022. However, the huge negative economical impact of the Sars-CoV-2 pandemic will render this optimistic forecast obsolete.

Table B.1: Medical devices market forecast for growth, in US$ billions (rounded).

Region	2016	2017	2018	2019	2020	2022
Americas	167	177	187	198	209	
Asia-Pacific	69	73	78	83	89	
Middle East/Africa	10	11	12	13	13	
Eastern Europe	15	16	17	18	19	
Western Europe	80	85	93	101	106	
Total	341	362	387	413	436	530

Source: Worldwide Medical Device Forecast to 2020; 2015; Worldwide Medical Device Forecast to 2022; 2016. ITA Medical Devices Top Markets Report 2016.

To quote an example, according to a report by Zion Market Research (2017), the global heart valve device market was valued at approximately US $5.6 billion in 2016 and is expected to generate revenue of around US $10.3 billion by 2022, growing at an AAGR of around 13% between 2017 and 2022.

A 2011 census of the 11 most implanted medical devices in America showed pseudophakia (artificial eye lenses) on top of the list with 2.6 million procedures per year, followed by tympanostomy tubes (715,000 procedures), coronary stents (560,000), artificial knees (543,000), traumatic fracture repair devices (453,000), intrauterine devices (IUDs) (425,000), spine screws, rods and disks (413,000), breast implants (366,000), heart pacemakers (235,000), artificial hips (230,000), and cardioverter defibrillators (133,200). In terms of total annual expenditures, artificial knees were first with US $12 billion, followed by pseudophakia (US $8–10 billion), and spine repair devices (US $10 billion). The total annual expenditure of the 11 listed procedures was US $61.3 billion in 2011 (Allen, 2011).

B.3 European perspective

Important development trends of biomaterials and biomedical devices in a EU context were recognized and analyzed by the Directorate-General for Research and Innovation of the European Commission in Brussels (Donnelly, 2015), According to this document, the major quantum leap for the short to medium-term future is a move away from merely replacing a natural function towards promotion of the body's capacity for self-healing, thereby facilitating the regeneration and renewal of the body's own functions. This will involve modulating an immune response that already exists in an organism to allow other actions to follow, such as the infiltration of a diseased organ with an ATMP or a medical device so as to repair damaged tissue. Collectively, the ability of biomaterials to adapt to their environments in a beneficial manner will determine the success of the therapeutic interventions with which the biomaterials are associated. This particular aspect of interactive or smart biomaterials is expected to be a major area of growth for research and development activities in the EU's Horizon 2020 Framework Programme and can thus assist to move academic research into commercially and socioeconomically viable solutions.

Next to regulating affairs, ATMPs, medical devices, and wound healing by tissue engineering, emphasis is being put on education and training in the biomaterials realm. Owing to the multifarious and diverse nature of the organization of the European Union, the following actions are urgently required (Donnelly, 2015):

EU recommendation for education and training in biomaterials research
- Designing curricula with appropriate opportunities for interdisciplinarity as well as the possibility of research as a long-term career option.
- Modernizing resources and strategies to reflect new developments and challenges related to healthcare systems, thereby rendering to be more attractive for medical researchers.
- Facilitating the mutual recognition of degrees and diplomas, thereby fostering better international collaboration while maintaining the highest possible clinical and quality standards
- Improving cooperation between academia and the industrial and clinical sectors to facilitate practical hands-on training

For medical devices in particular, in 2017 the European Parliament and the Council have issued a Medical Device Regulation (MDR) that intents to provide general safety and performance requirements, EU conformity, implantation rules, certification, verification, and quality assurance of medical devices, as well as rules for clinical investigation, evaluation and application, and post-market clinical follow-up of devices (EU, 2017). Details on international standards and regulations have been listed in Appendix A (see also Chapter 4.5).

References

Allen, B. (2011). The eleven most implanted medical devices in America. 24/7 Wall Str. News, https://www.businessinsider.com/au/the-11-most-implanted-medical-devices-in-america-2011 (accessed August 29, 2019).

Donnelly, F. (2015). European perspectives on biomaterials for health. EWMA Journal, 15(1), p. 54–58.

EU (2017). Regulation 2017/745 of the European Parliament and Council on medical devices. Official J. Europ. Union, L 117, p. 1–175. May 5, 2017.

Galetti, P.M. (1998) Biomaterials: facts and fiction. In: Akutsu T. and Koyanagi H. (eds). *Heart Replacement*. Springer, Tokyo. p. 103–109.

Hauser, C.A.E. and Zhang, S.G. (2010). Peptides as biological semiconductors. Nature, 468(7323), p. 516–517.

Markets and Markets (2019). Biomaterials market by type of materials (metallic, ceramic, polymers, natural) & by application (cardiovascular, orthopedic, dental, plastic surgery, wound healing, neurological disorders, tissue engineering, ophthalmology) – Global Forecast to 2024.

Mordor Intelligence (2019). Biomaterials Market-Growth, Trends, and Forecast (2019–2014). Mordor Intelligence, Hyderabad, India. info@mordorintelligence.com

Nathwani, S. (2017). High purity alumina market by type (4N, 5N, and 6N), application, and technology – Global opportunity analysis and industry forecast, 2017–2023. Allied Market Research. https://www.alliedmarketresearch.com/high-purity-alumina-market (accessed March 8, 2019).

Persistence Market Research (2018). Biomaterials market: global industry analysis and forecast 2016–2024. Persistence Market Research, report code PMRREP2781, Dec. 2018.

TechnavioBlog (2019). Medical device market research for 2019: Top 3 fast growing segments to watch. https://blog.technavio.com/blog/medical-device-market-research-top-fast-grwoing-segments (accessed February 6, 2020).

Thompson, H. (2016). Global medical polymers market expected to hit $19 billion by 2022. Medical Design & Outsourcing. July 5, 2018.

Worldwide Medical Device Forecast to 2020. EvaluateMedTech®, 4th edition, October 2015.

Worldwide Medical Device Forecast to 2022. EvaluateMedTech®, 5th edition, October 2016.

Zion Market Research (2017). *Heart valve devices market by procedure*. Heart valve replacement devices (mechanical valve, bioprosthetic valve, and TAVR) and position (mitral valve, aortic valve, other); heart valve repair devices (surgical valve, balloon valvuloplasty devices, TMVR)), by end user (hospitals and ambulatory surgical centers), and by region: global industry perspective, comprehensive analysis and forecast, 2014–2022.

Appendix C
Socioeconomic role of biomaterials: the example of hip endoprostheses

Robert B. Heimann

The important (and frequently underestimated) role biomaterials are playing in the current economy of developed countries can best be illustrated by the example of hip endoprosthetic implants used to repair the compromised ambulatory knee–hip kinematic. The increasing demand for endoprostheses is the result of the wear and tear the hip and knee joints are subjected to during a human lifetime. An average person walks about 1 million steps per year with a frequency of about 1 Hz. Using a conservative step length of 0.8 m, this amounts in an average lifespan of 75 years to $15 \cdot 10^7$ load changes during walking a distance of roughly 60,000 km or 1.5 times the circumference of the Earth. The load on the joints is equivalent to the body mass during rest, 2 to 3 times the body mass during normal walking, up to 5 times while jogging, and up to 8 times during jumping. Beyond this threshold, the risk of irreversible damage to the joints increases drastically. Since people generally live longer and get increasingly overweight due to overeating fatty and starchy food and lack of exercise, the load acting on the joints constantly increases over time. Eventually, the protective cartilage tissue linings of the femoral head and the acetabular cup wear away, friction during movement increases, and inflammation, pain, and finally immobilization will inevitably result. At this point, a total hip replacement (THR) operation is the only reasonable option to maintain mobility, freedom from pain, and hence, a rewarding life in older age. However, with increasing frequency, younger people require such an operation because their lifestyle includes damaging sports activities that tend to promote premature wear on their joints. Here a problem arises since younger patient will generally outlive the average lifetime of even the most advanced contemporary implants so that in later life a remediation operation is required with substantial additional cost to the healthcare system. Moreover, the early onset of pain and associated mental stress on the patient adds an ethical dimension to the problem of the incompatibility of the lifespan of both the patient and the implant.

It is not surprising then that the number of patients requiring and receiving large-joint reconstructive implants of the hip and the knee is constantly on the rise. Currently, in the United States and in the European Union, in excess of 1,200,000 hip and knee arthroplasties are being performed annually, and this number is expected to double by the year 2025. The global hip replacement implants market was valued at approximately US\$ 7.0 billion in 2017 and is anticipated to expand at a CAGR of 4.2% from 2018 to 2026 to reach a value of approximately US\$ 9.0 billion by 2026. This market is driven by a rise in demand for hip replacement procedures,

https://doi.org/10.1515/9783110619249-009

increase in government expenditure for the advancement of health care, rise in the incidence of trauma/accidental injuries, and increase in the prevalence of hip-related disorders. Furthermore, the global demand for hip replacement is increasing primarily due to a rise in clinical education among patients, increase in demand for hip replacement implants from the growing geriatric population, and rise in the prevalence of obesity and osteoarthritis (Transparency Market Research, 2018). Indeed, according to a census by the United Nations, the number of people aged 60 and over is estimated to swell up to 2.1 billion by 2050, more than doubling from 962 million as of 2017.

On a more general note, a rough estimate of the number of metallic, ceramic, and polymeric implants of all kinds delivered worldwide to patients is in the range of 10 million annually. Consequently, at present, the global number of orthopedic surgeries increases by 10–12% per year. These trends are indeed a very strong incentive to embark on research and development aiming to providing novel and improving existing biomaterials for all kinds of implant (see Chapter 1.3). Major players in the hip endoprosthetic market are Zimmer Biomet Holdings Inc., Smith & Nephew plc, Stryker Corporation, OMNI, DePuy Synthes, Autocam Medical, B. Braun Melsungen AG, MicroPort Scientific Corporation, Exactech Inc. and Corin Connected Orthopaedic Insight.

According to many observers in the field, in Germany the growing numbers of THR operations and the frequency of such surgical interventions are exceptionally high in comparison to other countries. This fact is currently the subject of critical discussion among physicians, hospital administrators, health insurance officials, and the concerned public.

As shown in Table C.1, among the OECD countries, Germany occupied the first place with 309 THR operations per 100,000 population in 2017. This ought to be compared to the situation in the USA with 205/100,000 in 2017. At the low end of the scale is India with only 5.7 THR/100,000 population in 2017. As one might expect, there is a strong correlation between the gross domestic product and healthcare expenditure per capita with utilization rates.

Between 2011 and 2015, the average of THR/100,000 population of all 34 OECD countries (OECD, 2017) increased slightly by about 4% from 160 to 166/100,000. Interestingly, in Poland, the number of THR increased strongly by 47% between 2011 and 2015, from 76/100,000 to 112/100,000 in 2015, and further increasing to 160/100,00 in 2017; this is presumably owing to the noticeably increased standard of living after the demise of communism in 1989. However, this figure is topped by South Korea with an unprecedented increase by 300%, up from 17/100,000 in 2011 to 56/100,000 in 2017. For India, a strong annual average growth rate of 137% was recorded between 2011 (2.4/100,000) and 2017 (5.7/100,000), although the actual numbers of THR remain very low in comparison. Contrariwise, in Greece the number of THR decreased by 23% from 167/100,000 in 2011 down to 132/100,000 in 2015, likely the result of severe financial constraints experienced by the Greek health care system during this period of time. More data on the growth of the biomedical device market are shown in Appendix B.

Table C.1: Number of total hip replacement (THR) operations in selected OECD and some Asian countries per 100,000 population in 2011, 2015, and 2017.

Country	2011	2015	2017
Switzerland	306	308	307
Germany	286	299	309
USA	149		205
OECD-34	**160**	**166**	
Greece	163	132	
Estonia	134	156	156
Poland	76	112	160
South Korea	17	53	56
Mexico	8	8	9
India	2.4		5.7

AAGR = average annual growth rate (Health at a glance 2017: OECD indicators. https://read.oecd-ilibrary.org).

In terms of economic impact, there is another factor to consider. The majority of materials utilized today for biomedical implants are standard commercial formulations that were originally developed for other industrial purposes, including Ti alloys such as Ti6Al4V and Mg alloys such as AZ61 both originally designed for aerospace application. This also holds for alumina and zirconia and many engineered polymers such as poly(methylmethacrylate) and ultrahigh-molecular-weight polyethylene. Those materials that were found to be appropriate for a specific medical application or device are subject to established industrial norms, specifications, regulations, and quality control measures that are only weekly supported by continuing feedback from the clinical experience. Notwithstanding these facts, a major obstacle to develop and implement custom-designed biomaterials is the fallout of a remarkable market dichotomy. According to several analysts, the medical device market is still very small in terms of the mass of material used per implant device but large in numbers of implants and their cost, a fact that has a decisive impact on governmental healthcare budgets. In addition, this means that large-scale production and return on investment of industrial production expenditures are close to impossible for deployment of truly novel biomaterials. Instead, conventional "entrenched" biomaterials are being further purified and microstructurally refined. This includes, in some cases, biological functionalization by adsorption of tissue-growth supporting proteins, including bone morphogenetic proteins, and cyto- and chemokines to elicit osseoinductive mechanisms as a crucial step for successful osseointegration.

Table C.2 shows the staggering mechanical property improvement of biomedical alumina as an example of this trend. Flexural strengths and fracture toughness are maximized by reduction of alumina grain size. This can be implemented by grain boundary engineering during which suppression of grain growth at high sintering temperatures is achieved by addition of minor amounts of magnesium oxide in the range of a few tenths of a percent. Accumulation of magnesium oxide along the grain boundaries of alumina will result in a thin surface layer consisting of spinel ($MgAl_2O_4$) that acts as a barrier toward grain boundary movement normally associated with the process of recrystallization. The strengthening mechanism of BIOLOX®delta involves crack-arresting particles of strontium hexaluminate distributed among the tiny alumina grains. In general, polycrystalline alumina of the BIOLOX® family of CeramTec GmbH exceeds the ISO 6474 norm requirements.

Table C.2: Mechanical property improvement of biomedical alumina.

Property	High alumina	ISO 6474–1:2010	ISO 6474–2:2012	BIOLOX® forte	BIOLOX® delta
Average grain size (µm)	>10	<7	<4.5	<2	<0.6
Flexural strength (MPa)	280–420	400–500	>450	>500	~1,400
Fracture toughness (MN·Vm)	3–4	4–6	–	4	6.5
Modulus of elasticity (GPa)	350–400	380–420	–	380	360

According to the ISO 6474-1:2010 norm, the grain size of alumina applied for femoral head balls was specified to be <7 µm, resulting in flexural strength of 400–500 MPa. The ISO 6474-2:2012 norm established in 1994 specified a much lower average grain size of <4.5 µm with a concurrent increase in the flexural strength to beyond 450 MPa.

References

OECD (2017). Hip replacement surgery trends, 2000 to 2015 (or nearest year). In: *Health Care Activities*. Paris: OECD Publishing. https://doi.org/10.1787/health_glance-2017-graph167-en.
Transparency Market Research (2018). Global Industry Analysis, Size, Share, Growth, Trends and Forecast 2018–2026. https://transparencymarketresearch.com/hip-replacement-implant-market.html (accessed January 17, 2020).

References

A selection of recent textbooks related to biomaterials

Agrawal, C.M., Ong, J.L., Appleford, M.R. and Mani, G. (2013). Introduction to Biomaterials: Basic Theory with Engineering Applications. Cambridge University Press: Cambridge. ISBN 978-0-5211-1690-9.

Balakrishnan, P., Sreekala, M.S. and Thomas, S. (eds) (2018). Fundamental Biomaterials: Metals. Woodhead Publ. Series in Biomaterials. Woodhead Publ.: Sawston, UK. ISBN 978-0-0810-2205-4.

Ben-Nissan, B. (ed.) (2014). Advances in Calcium Phosphate Biomaterials. Springer Series in Biomaterials Science and Engineering. Springer: Berlin, Heidelberg. ISBN 978-3-642-53979-4.

Bergmann, C.P. and Stumpf, A. (2012). Dental Ceramics. Microstructure, Properties and Degradation. Springer: Berlin, Heidelberg. ISBN 978-3-642-38223-9.

Bezerra, U.T., Ferreira, H.S. and Barbosa, N.P. (eds) (2019). Bio-Inspired Materials. Frontiers in Biomaterials, vol.6. Bentham Sci. Publ. Pte. Ltd.: Singapore. ISBN 978-981-14-0688-1.

Black, J. and Hastings, G. (eds) (1998). Handbook of Biomaterial Properties. Springer US. ISBN 978-0-412-60330-3.

Chaughule, R. (ed.) (2018). Dental Applications of Nanotechnology. Springer Intern. Publ. ISBN 978-3-319-97633-4.

Chen, Q.Z. and Thouas, G. (2014). Biomaterials: A Basic Introduction. CRC Press: Boca Raton. ISBN 978-1-1387-4966-5.

Choi, A.H. and Ben-Nissan, B. (eds) (2019). Marine-Derived Biomaterials for Tissue Engineering Applications. Springer Series in Biomaterials Science and Engineering. Springer: Singapore. ISBN 978-9-8113-8854-5.

Chun, H.J., Park, K., Kim, C.H. and Kang, G. (eds) (2018). Novel Biomaterials for Regenerative Medicine. Springer: Singapore. ISBN 978-9-8113-0947-2.

Davis, J.R. (ed.) (2003). Handbook of Materials for Medical Devices. ASM International, Materials Park, OH. ISBN 0-87170-790-X.

Dos Santos, V., Brandalise, R.N. and Savaris, M. (2017). Engineering of Biomaterials. Springer Intern. Publ. ISBN 978-3-319-58606-9.

Ducheyne, P., Grainger, D.W., Healy, K.E., Hutmacher, D.W. and Kirkpatrick, C.J. (eds) (2017). Comprehensive Biomaterials II. Oxford: Elsevier. ISBN 978-0-08-055294-1.

Guelcher, S.A. and Hollinger, J.O. (eds) (2005). An Introduction to Biomaterials. CRC Press: Boca Raton. ISBN 978-0-8493-32282-2.

Hasirci, V. and Hasirci, N. (2018). Fundamentals of Biomaterials. Springer: New York. ISBN 978-1-4939-8854-9.

Heimann, R.B. (ed.) (2012). Calcium Phosphate. Structures, Synthesis, Properties, and Applications. Biochemistry Research Trends. Nova Science Publishers, Inc.: New York. ISBN 978-1-62257-299-1.

Heimann, R.B. and Lehmann, H.D. (2015). Bioceramic Coatings for Medical Implants: Trends and Techniques. Wiley-VCH: Weinheim. ISBN 978-3-527-33743-9.

Helsen, J.A. and Missirlis, Y. (eds) (2010). Biomaterials: Implants, Materials and Tissues. Springer: Berlin. ISBN 978-3-6421-2531-7.

Hench, L.L. (ed.) (2013). An Introduction to Bioceramics. 2nd ed., Imperial College Press: London. ISBN 978-1-908977-15-1.

Ishikawa, K., El-Ghannam, A. and Kokubo, T. (eds) (2020). Bioceramics of Their Clinical Application. 2nd ed. Woodhead Publ. Series in Biomaterials. Woodhead Publ.: Sawston, UK. ISBN 978-0-0810-3001-1.

https://doi.org/10.1515/9783110619249-010

Kirkland, N.T. and Birbilis, N. (2013). Magnesium Biomaterials: Design, Testing, and Best Practice. Springer. ISBN 978-3-3190-2122-5.

Li, B.Y. and Webster, T. (eds) (2017). Orthopedic Biomaterials. Springer: Cham. ISBN 978-3-319-73664-8.

Li, J., Osada, Y. and Cooper-White, J. (eds) (2018). Functional Hydrogels as Biomaterials. Springer Series in Biomaterials Science and Engineering. Springer: Berlin, Heidelberg. ISBN 978-3-662-57509-3.

Li, Q. and Mai, Y.W. (eds) (2017). Biomaterials for Implants and Scaffolds. Springer Series in Biomaterials Science and Engineering. Springer: Berlin, Heidelberg. ISBN 978-3-662-53572-1.

Love, B. (2017). Biomaterials. A Systems Approach to Engineering Concepts. Elsevier. ISBN 978-0-12-809478-5.

Murphy, W., Black, J. and Hastings, G. (eds) (2016). Handbook of Biomaterial Properties. 2nd ed. Springer: New York. ISBN 978-1-4939-3303-5.

Niinomi, M., Narushima, T. and Nakai, M. (eds) (2015). Advances in Metallic Biomaterials. Tissues, Materials and Biological Reactions. Springer: Berlin, Heidelberg. ISBN 978-3-662-46835-7.

Niinomi, M., Narushima, T. and Nakai, M. (eds) (2015). Advances in Metallic Biomaterials. Processing and Applications. Springer: Berlin, Heidelberg. ISBN 978-3-662-46841-8.

Ong, K.L., Lovald, S. and Black, J. (2014). Orthopaedic Biomaterials in Research and Practice. 2nd ed. CRC Press: Boca Raton. ISBN 978-1-4665-0350-3.

Park, J. and Lakes, R.S. (2007). Biomaterials. Springer: New York. ISBN 978-0-387-37879-4.

Piskin, E. and Hoffman, A.S. (eds) (1986). Polymeric Biomaterials. Springer: Dordrecht. ISBN 978-9-4010-8452-9

Ratner, B.D., Hoffman, A.S., Schoen, F.J. and Lemons, J.E. (eds) (2013). Biomaterials Science. An Introduction to Materials in Medicine. Elsevier, ISBN 978-0-12-374626-9.

Rezaie, H.R., Bakhtiari, L. and Öchsner, A. (2015). Biomaterials and their Applications. Springer: Cham. ISBN 978-3-319-17846-2.

Saltzman, W.M. (2015). Biomedical Engineering: Bridging Medicine and Technology. Cambridge University Press: Cambridge. ISBN 978-1-1070-3719-9.

Schmalz, G. and Arenholdt-Bindsley, D. (2008). Biocompatibility of Dental Materials. Springer. Berlin: ISBN 978-3-5407-7782-3.

Shi, D. (ed.) (2004). Biomaterials and Tissue Engineering. Springer: Berlin, Heidelberg. ISBN 978-3-642-06067-0.

Taubert, A., Mano, J.F. and Rodriguez-Cabello, J.C. (eds) (2013). Biomaterials Surface Science. Wiley-VCH: Weinheim. ISBN 978-8-3527-64962-4.

Temenoff, J.S. and Mikos, A.G. (2008). Biomaterials: The Intersection of Biology and Materials Science. Pearson Intern. Ed., 1634. ISBN 978-0-1300-9710-1.

Tripathi, A. and Melo, J.S. (eds) (2017). Advances in Biomaterials for Biomedical Applications. Springer: Singapore. ISBN 978-9-8110-3327-8.

Vincent, J. (2012). Structural Biomaterials. 3rd ed. Princeton University Press: Princeton. ISBN 978-0-6911-5400-8.

Williams, D.F. (2014). Essential Biomaterials Science. Cambridge Texts in Biomedical Engineering. Cambridge University Press: Cambridge. ISBN 978-0-52-189908-6.

Wong, J.Y., Bronzino, J.D. and Peterson, D.R. (eds) (2012). Biomaterials: Principles and Practices. CRC Press: Boca Raton. ISBN 978-1-4398-7251-2.

Yock, P.G., Zenios, S. Makower, J. and 6 additional authors (2015). Biodesign: The Process of Innovating Medical Technologies. 2nd ed. Cambridge University Press: Cambridge. ISBN 978-1-1070-8735-4.

Zenios, S., Makower, J., Yock, P., Kumar, U.N., Denend, L. and Krummel, T.M. (2010). Biodesign: The Process of Innovating Medical Technologies. Cambridge: Cambridge University Press. ISBN 978-0-5115-1742-3.

Zhang, M., Naik, R.R. and Dai, L.M. (eds) (2016). Carbon Nanomaterials for Biomedical Applications. Springer Series in Biomaterials Science and Engineering. Springer: Cham. ISBN 978-3-319-22860-0.

Zhang, X.D. and Williams, D.F. (2019). Definitions of Biomaterials for the Twenty-First Century. Elsevier. ISBN 978-0-1281-8291-8.

Chapter 1

Abd El-Ghany, O.S. and Sherief, A.H. (2016). Zirconia-based ceramics, some clinical and biological aspects: review. Future Dent. J., 2(2), p. 55–64.

Abraham, G.E. and Himmel, P.B. (1997). Management of rheumatoid arthritis: Rationale for the use of colloid metallic gold. J. Nutr. Env. Med., 7(4), p. 295–305.

Adleman, L.M. (1998). Computing with DNA. Sci. Amer., 279(2), p. 54–61.

Aisenberg, S. and Chabot, R. (1971). Ion-beam deposition of thin films of diamond-like carbon. J. Appl. Phys., 42(7), p. 2953–2957.

Albrektsson, T., Brånemark, P.I., Hansson. H.A. and Lindström, J. (1981). Osseointegrated titanium implants. Requirements for ensuring a long-lasting, direct bone anchorage in man. Acta Orthop. Scand., 52, p. 155–170.

Albrektsson, T., Chrcanovic, B., Jacobsson, M. and Wennerberg, A. (2017). Osseointegration of implants – a biological and clinical overview. JSM Dent. Surg., 2 (3),p. 1022 (6 pages).

Anderson, P.W. (1995). Through the glass lightly. Science, 267(5204), p. 1615–1616.

Antunes, L.H.M. and de Lima, C.R.P. (2018). Cobalt-chromium alloys – properties and applications. Reference Module Materials Science and Materials Engineering. Elsevier. https;//doi.org/10.1016/B978-0-12-803581-8.04089-3.

Aubriot, J.H., Deburge, A. and Genet, J.P. (2014). GUEPAR hinge knee prosthesis. Orthop. Traumatol. Surg. Res., 100(1), p. 27–32.

Bache, F. (1819). A System of Chemistry for the Use of Students of Medicine. Printed and published for the author. Philadelphia: William Fry. 624 pp.

Baino, F. (2018). Bioactive glasses-when glass science and technology meet regenerative medicine. Ceram. Intern. 44(13), p. 14953–14966.

Becker, R.O. and Selden, G. (1985). The Body Electric: Electromagnetism and the Foundation of Life. New York: William Morrow and Company. ISBN 978-0-6880-0123-0.

Benenson, Y., Gil, B., Ben-Dor, U., Adar, R. and Shapiro, E. (2004). An autonomous molecular computer for logical control of gene expression. Nature, 429 (6990), p. 423–429.

Bezerra, U.T., Ferreira, H.S. and Barbosa, N.P. (eds) (2019). Bio-Inspired Materials. Frontiers in Biomaterials, vol.6. Singapore: Bentham Sci. Publ. Pte. Ltd. ISBN 978-981-14-0688-1.

Bila, H., Kurisinkal, E.E. and Bastings, M.M.C. (2019). Engineering a stable future for DNA-origami as a biomaterial. Biomater. Sci., 7, p. 532–541.

Billmeyer, Jr., F.W. (1984). Textbook of Polymer Science. John Wiley & Sons: New York. ISBN: 0-471-82834-3. p. 25–81.

Biomaterials Market by Type of Materials (Metallic, Ceramic, Polymers, Natural) & Application (Cardiovascular, Orthopedic, Dental, Plastic Surgery, Wound Healing, Neurology, Tissue Engineering, Ophthalmology) – Global Forecast to 2021. Markets and Markets, Magarpatta city, Hadapsar, Pune, Maharashtra 411013, India. 2017.

Bohner, M. and Miron, R.J. (2019). A proposed mechanism for material-induced heterotopic ossification. MaterialsToday, 22, p. 132–141.

Bonnet, J., Yin, P., Ortiz, M.E., Subsoontorn, P. and Drew, E. (2006). Amplifying genetic logic gates. Science, 340, p. 599–603.

Bose, S., Roy, M. and Bandyopadhyay, A. (2012). Recent advances in bone tissue engineering scaffolds. Trends Biotechnol., 30(10), p. 546–554.

Bothe, R.T., Beaton, L.E. and Davenport, H.A. (1940). Reaction of bone to multiple metallic implants. Surg. Gynecol. Obstet., 71, p. 598–602.

Boutin, P. (1972). L'arthroplastie total de la hanche par prothèse en alumine frittée. Rev. Chir. Orthop., 58, p. 229–246.

Bowen, P.K., Drelich, J. and Goldman, J. (2013). Zinc exhibits ideal physiological corrosion behavior for bioabsorbable stents. Adv. Mater., 25(18), p. 2577–2582.

Brand, R.A., Mont, M.A. and Manring, M.M. (2011). Biographical sketch: Themistocles Gluck (1853-1942). Clin. Orthop. Rel. Res., 469(6), p. 1525–1527.

Brånemark, P.I. (2016). Promotional brochure Associated Brånemark Osseointegration Centres. www.branemark.se/osseointegration.

Buchhorn, G.H. and Willert, H.G. (1994). Technical Principles, Design and Safety of Joint Implants. Seattle, Toronto, Bern, Göttingen: Hogrefe & Huber Publishers, 431 pp. ISBN 3-456-82161-1.

Burlacov, I., Jirkovský, J., Kavan, L., Ballhorn, R. and Heimann, R.B. (2007). Cold gas dynamic spraying (CGDS) of TiO_2 (anatase) powders onto poly(sulfone) substrates: Microstructural characterisation and photocatalytic efficiency. J. Photochem. Photobiol. Chem., 187(2-3), p. 285–292.

Carbone, E.J., Rajpura, K., Allen, B.N., Cheng, E., Ulery, B.D. and Lo, K.W.H. (2017). Osteotropic nanoscale drug delivery systems based on small molecule bone-targeting moieties. Nanomed., 13(1), p. 37–47.

Chader, G.J., Weiland, J. and Humayun, M.S. (2009). Artificial vision: Needs, functioning, and testing of a retinal electronic prostheses. Prog. Brain Res., 175, p. 317–332.

Chang, H.C. and Yeo, L. (2009). Electrokinetically Driven Microfluidics and Nanofluidics. Cambridge University Press: Cambridge, UK. ISBN 978-0-521-86025-3.

Chatters, J.C. (2000). The recovery and first analysis of an early Holocene human. Am. Antiquity, 65(2), p. 291–316.

Chen, Q.Z. and Thouas, G.A. (2015). Metallic implant biomaterials. Mater. Sci. Eng. R, 87, p. 1–57.

Chillag, K.J. (2016). Giants of orthopedic surgery: Austin T. Moore MD. Clin. Orthop. Rel. Res., 474, p. 2606–2610.

Chouard, C.H. (2014). The French surgical and electrophysiological researches towards development of the multichannel cochlear implant. <recorlsa.online.fr/implant cochleaire/ historicfrancaisenanglais.html> (accessed July 29, 2018).

Chung, S.J. (2016). Fundamental study of the design and development of iron-based alloys for biodegradable implant device application. Ph.D. Thesis, Swanson School of Engineering, University of Pittsburgh, PA. 209 pp.

Church, G.M., Gao, Y. and Kosuri, S. (2012). Next-generation digital information storage in DNA. Science, 337 (6102), p. 1628–1629.

Clark, A.E., Hench, L.L. and Paschall, H.A. (1976). The influence of surface chemistry on implant interface bonding: a theoretical basis for implant material selection. J. Biomed. Mater. Res., 10, p. 161–177.

Clarke, I.C. and Willmann, G. (1994). Structural ceramics in orthopedics. In: Cameron, H.U. (ed.). Bone Implant Interface, p. 203–252. St. Louis, Baltimore, Boston, Chicago, London, Madrid, Philadelphia, Sydney, Toronto: Mosby. ISBN 0-8016-6483-7. 390 pp.

Clatworthy, M. (2015). An early outcome study of the ATTUNE® Knee System vs. the SIGMA® CR150 Knee System. DePuy Synthes Companies White Paper. DSUS/JRC/ 0814/0418(1).

Cohen, D.J., Nelson, W.J. and Maharbiz, M.M. (2014). Galvanotactic control of collective cell migration in epithelial monolayers. Nat. Mater., 13, p. 409–417.

Cohen, R. (2002). A porous tantalum trabecular metal: basic science. Am. J. Orthop., 31(4), p. 216–217.

Colas, A. and Curtis, J. (2004). Silicone biomaterials: history and chemistry. In: Ratner, B.D. et al. (eds). Biomaterials Science. An Introduction to Materials in Medicine. Elsevier, ISBN 978-0-12-374626-9. p. 80–86.

Dahlman, J.E. (2019). All the world's data could fit in an egg. Sci. Amer., 320(6),63–67.

Damaraju, S.M., Shen, Y.Y., Elele, E., Khusit, B., Eshghinejad, A., Li, J.Y., Jaffe, M. and Livingston Arinzeh, T. (2017). Three-dimensional piezoelectric fibrous scaffolds selectively promote mesenchymal stem cell differentiation. Biomaterials, 149, p. 51–62.

Daschner, R., Rothermel, A., Rudorf, R., Rudorf, S. and Stett, A. (2018). Functionality and performance of the subretinal implant chip alpha AMS. Sens. Mater., 30(2), p. 179–192.

De Jong, W.F. (1926). La substance minerale dans les os. Recl. Trav. Chim. Pays-Bas Belg., 45, p. 445–448.

De Leeuw, N., Bowe, J.R. and Rabone, J.A.L. (2007). A computational investigation of stoichiometric and calcium-deficient oxy- and hydroxyapatite. Faraday Discuss., 134, p. 195–214.

D'Este, M. and Alini, M. (2018). Lessons to be learned and future directions for intervertebral disc biomaterials. Acta Biomater., 78, p. 13–22.

Diebold, U. (2003). The surface science of titanium dioxide. Surf. Sci. Rep., 48, p. 53–229.

Djourno A. et Eyries, C. (1957). Prothése auditive par excitation électrique a distance du nerf sensorial a l'aide d'un bobinage inclus a demeure. Presse Med., 65 (63), p. 1417.

Donath, K., Laas, M. and Günzl, H. (1991). The histopathology of different foreign body reactions in oral soft tissue and bone tissue. Virchows Arch. Pathol. Anat. Histopathol., 420, p. 131–137.

Dorozhkin, S.V. (2012). Calcium orthophosphates and Man: a historical perspective from the 1770s to 1940. In: Heimann R.B. (ed.) Calcium Phosphate. Structure, Synthesis, Properties, and Applications. New York: Nova Science Publishers Inc., p. 1–40.

Dorozhkin, S.V. (2013). A detailed history of calcium orthophosphates from 1770s till 1950. Mater. Sci. Eng. C, 33(6), p. 3085–3110.

Dorozhkin, S.V. (2018). Biocomposites and hybrid biomaterials of calcium orthophosphates with polymers. CRC Press. ISBN 978-1-1383-4310-8. 119 pp.

Ducheyne, P., El-Ghannam, A. and Shapiro, I. (1994). Effect of bioactive glass templates on osteoblast proliferation and in vitro synthesis of bonel-like tissue. J. Cell Biochem., 56, p. 162–167.

Ducheyne, P. and Qiu, Q. (1999). Bioactive ceramics: The effect of surface reactivity on bone formation and bone cell function. Biomaterials, 20, p. 2287–2303.

EIRD (2019). Entwurf eines Gesetzes zur Errichtung des Implantatregisters in Deutschland und zu weiteren Änderungen des Fünften Buches Sozialgesetzbuch (EIRD) [Draft of legislation to establish an implant register and to further amendments of the fifth part social code of law]. Deutscher Bundestag, Drucksache 19/10523.

Enke, K., Dimigen, H. and Hübsch, H. (1980). Frictional properties of diamond-like carbon layers. Appl. Phys. Lett., 36(4), p. 291–292.

Fernandez-Yague, M.A., Akogwu Abbah, S., McNamara, L., Zeugolis, D.I., Pandit, A. and Biggs, M.J. (2015). Biomimetic approaches in bone tissue engineering: integrating biological and physico-mechanical strategies. Adv. Drug Delivery Rev., 84, p. 1–29.

Fousová, M., Vojtech, D., Jablonska, E., Fojt, J. and Lipov, J. (2017). Novel approach in the use of plasma spray: Preparation of bulk titanium for bone augmentation. Materials, 10(9): 987.

Fu, H.L., Cheng, S.X., Zhang, X.Z. and Zhuo, R.X. (2008), Dendrimer/DNA complexes encapsulated functional biodegradable polymers for substrate-mediated gene delivery. J. Gene Med., 10(12), p. 1334–1342.

Fujishima, A., Zhang, X. and Tryk, D.A. (2008). TiO_2 photocatalysis and related surface phenomena. Surf. Sci. Rep., 63, p. 515–582.

Fujishima, A. and Honda, K. (1972). Electrochemical photolysis of water at a semiconductor electrode. Nature, 238 (5358), p. 37–38.

Galkowski, V., Petrisor, B., Drew, B. and Dick, D. (2009). Bone stimulation for fracture healing: what's all the fuss? Indian J. Orthop., 43(2), p. 117–120.

Garvie, R.C. and Nicholson, P.S. (1972). Structure and thermodynamical properties of partially stabilized zirconia in the $CaO\text{-}ZrO_2$ system, J. Am. Ceram. Soc., 55, p. 152–157.

Gee, E.C.A., Jordan, R., Hunt, J.A. and Saithna, A. (2016). Current evidence and future directions for research into the use of tantalum in soft tissue re-attachment surgery. J. Mater. Chem. B, 4, p. 1020–1034.

Gluck, T. (1891). Referat über die durch das modern chirurgische Experiment gewonnenen positiven Resultate, betreffend die Naht und den Ersatz von Defekten höherer Gewebe, sowie über die Verwendung resorbirbarer und lebendiger Tampons in der Chirurgie [Report on the positive results obtained by modern surgical experimentation related to sutures and the replacement of defects of higher tissues, as well as the use of resorbable and living tampons in surgery]. Langenbecks Arch. Klin. Chir., 41, p. 187–239.

Gombay-McGill, K. (2016). Hip implant mimics real bone to last longer. Futurity, https://www.futurity.org/hip-replacement-implant-1288372 (accessed August 5. 2019).

Good, J.M., Olinthus, G. and Newton, B. (1813). Pantologia: A new encyclopædia, comprehending a complete series of Essays, Treatises, and Systems, alphabetically arranged; with a general dictionary of arts, sciences, and words: the whole presenting a distinct survey of Human Genius, Learning, and Industry. Vol. IX. P – PYX. London.

Graham, T. (1861). Über die Diffusion von Flüssigkeiten und ihre Anwendung zur Analyse [On the diffusion of liquids and its application to analysis]. Compt. Rend., 162, p. 223–227.

Griffin, M. and Bayat, A. (2011). Electrical stimulation in bone healing: critical analysis by evaluating levels of evidence. Open Access Journal of Plastic Surgery, 11, e34 (online July 26, 2011).

Gupta, U. and Perumal, O. (2014). Dendrimers and its biomedical applications. In: Kumbar, S.G., Laurencin, C.T., Deng. M. (eds) Natural and Synthetic Biomedical Polymers. Chapter 15, p. 243–257. Elsevier. ISBN 978-0-12-396983-5.

Ha, S.W., Viggeswarapu, M., Habib, M. and Beck Jr., G.P. (2018). Bioactive effects of silica nanoparticles on bone cells are size, surface, and composition dependent. Acta Biomater., 82, p. 184–196.

Hagen, T. (2012). Knee endoprosthesis. US patent 8246688 B2.

Han, H.S., Loffredo, S., Jun, I.D., Edwards, J., Kim, Y.C., Seok, H.K., Witte, F., Mantovani, D. and Jones, S.G. (2019). Current status and outlook on the clinical translation of biodegradable metals. MaterialsToday, 23, p. 57–71.

Heimann, R.B. (2008). Plasma Spray Coating. Principles and Applications. 2nd ed. Weinheim: Wiley-VCH. 427 pp. ISBN 978-3-527-32050-9.

Heimann, R.B. (2010). Classic and Advanced Ceramics. Weinheim: Wiley-VCH. 553 pp. ISBN 978-3-527-32517-7.

Heimann, R.B. (ed.) (2012). Calcium Phosphate: Structure, Synthesis, Properties and Applications. Biochemistry Research Trends Series. New York: Nova Science Publishers. 498 pp. ISBN 978-1-62257-299-1.

Heimann, R.B. (2015). Tracking the thermal decomposition of plasma-sprayed hydroxylapatite. Amer. Mineral., 100, p. 2419–2425.

Heimann, R.B. (2017). Calcium (Ti, Zr) hexaorthophosphate bioceramics for electrically stimulated biomedical implant devices: a position paper. Amer. Mineral., 102, p. 2170–2171.

Heimann, R.B. (2018). Plasma-sprayed hydroxylapatite coatings as biocompatible intermediaries between inorganic implant surfaces and living tissue. J. Thermal Spray Technol., 27(8), p. 1212–1237.

Heimann, R.B. and Lehmann, H.D. (2015). Bioceramic Coatings for Medical Implants. Trends and Techniques. Weinheim: Wiley-VCH. 467 pp. ISBN 978-3-527-33743-9.

Heimann, R.B. and Lehmann, H.D. (2017). Recent research and patents on controlling corrosion of bioresorbable Mg alloy implants: Towards next-generation biomaterials. Recent Patents on Mater. Sci., 10, p. 2–19.

Heimann, R.B., Graßmann, O., Zumbrink, T. and Jennissen, H.P. (2001). Biomimetic processes during in vitro leaching of plasma-sprayed hydroxyapatite coatings for endoprosthetic applications. Materialwiss. Werkstofftech., 32, p. 913–921.

Helmer, J.D. and Driskell, T.D. (1969). Research on bioceramics. Symposium on Use of Ceramics as Surgical Implants, Clemson Univ., Clemson, S.C.

Hench, L.L. (1971). Mechanisms of interfacial bonding between ceramics and bone. J. Biomed. Mater. Res., 2, p. 485–497.

Hench, L.L. (2006). The story of Bioglass®. J. Mater. Sci. Mater. Med., 17(11), p. 967–978.

Hench, L.L. (2008). Genetic design of bioactive glass. J. Eur. Ceram. Soc., 29, p. 1257–1265.

Hench, L.L. and Polak, J.M. (2002). Third-generation biomedical materials. Science, 295(5557), p. 1014–1017.

Hench, L.L., Splinter, R.J., Allen, W.C. and Greenlee Jr., T.K. (1972). Bonding mechanisms at the interface of ceramic prosthetic materials. Biomed. Mater. Res. Symp., 2(1), p. 116–141.

Hench, L.L. and Jones, J.R. (2015). Bioactive glasses: Frontiers and challenges. Front. Bioeng. Biotechnol., 3: 194.

Hendley, M.A., Murphy, K.P., Isely, C., Struckman, H.L., Annamalai, P. and Gower, R.M. (2019). The host response to poly(lactide-co-glycolide) scaffolds protects mice from diet induced obesity and glucose intolerance. Biomaterials, 217, 119281.

Hendricks, S.B., Hill, W.A., Jakobs, K.D. and Jefferson, M.E. (1931). Structural characteristics of apatite-like substances and composition of phosphate rock and bone as determined from microscopical and X-ray examinations. Ind. Eng. Chem., 23(12), p. 1413–1418.

Heness, G. and Ben-Nissan, B. (2004). Innovative bioceramics. Mater. Forum, 27, p. 104–114.

Hermanaw, H., Dubé, D. and Mantovani, D. (2010). Developments in metallic biodegradable stents. Acta Biomater., 6, p. 1693–1697.

Hofmann-Axhelm, W. (1985). Die Geschichte der Zahnheilkunde. Die Quintessenz, Berlin. ISBN 3-87652-160-2

Holland, L. and Ojha, S.M. (1976). Deposition of hard and insulating carbonaceous films on an r.f. target in a butane plasma. Thin Solid Films, 38(2), p. L17-L19.

Hutmacher, D.W., Sittinger, M. and Risbud, M.V. (2004). Scaffold-based tissue engineering: rationale for computer-aided design and solid free-form fabrication systems. Trends Biotechnol., 22(7), p. 354–362.

ISO 14971:2012. Medical devices – Application of risk management to medical devices (ISO 14971:2007, Corrected version 2007-10-01); German version EN ISO 14971: 2012.

Ivanova, E.P., Bazaka, K. and Crawford, R.J. (2014a): Advanced synthetic and hybrid polymer biomaterials derived from inorganic and mixed organic-inorganic sources. In: Ivanova, E.P., Bazaka, K. and Crawford, R.J. (eds). New Functional Biomaterials for Medicine and Healthcare. p. 100–120. Woodhead Publishing Series in Biomaterials. Woodhead Publishing Limited.

Ivanova, E.P., Bazaka, K. and Crawford, R.J. (2014b): Metallic biomaterials: types and advanced applications-new functional biomaterials for medicine and healthcare. In: Ivanova, E.P., Bazaka, K. and Crawford, R.J. (eds). New Functional Biomaterials for Medicine and Healthcare. p. 121–147. Woodhead Publishing Series in Biomaterials. Woodhead Publishing Limited.

Ivanova, E.P., Bazaka, K. and Crawford, R.J. (2014c): Bioinert ceramic biomaterials: advanced applications. In: Ivanova, E.P., Bazaka, K. and Crawford, R.J. (eds). New Functional Biomaterials for Medicine and Healthcare. p. 173–186. Woodhead Publishing Series in Biomaterials. Woodhead Publishing Limited.

Ivanova, E.P., Bazaka, K. and Crawford, R.J. (2014d): Advanced bioactive and biodegradable ceramic biomaterials. In: Ivanova, E.P., Bazaka, K. and Crawford, R.J. (eds). New Functional Biomaterials for Medicine and Healthcare. p. 187–219. Woodhead Publishing Series in Biomaterials. Woodhead Publishing Limited.

Ivanova, E.P., Bazaka, K. and Crawford, R.J. (2014e): Natural polymer biomaterials: advanced applications. In: Ivanova, E.P., Bazaka, K. and Crawford, R.J. (eds). New Functional Biomaterials for Medicine and Healthcare. p. 32–70. Woodhead Publishing Series in Biomaterials. Woodhead Publishing Limited.

Jarcho, M. (1981). Calcium phosphate ceramics as hard tissue prosthetics. Clin. Orthop. Rel. Res., 157, p. 259–278.

Jarcho, M., Bolen, C.H., Thomas, M.B., Bobick, J., Kay, J.F. and Doremus, R.H. (1976). Hydroxyapatite synthesis and characterization in dense polycrystalline form. J. Mater. Sci., 11, p. 2027–2035.

Jelinek, R. (2013). Biomimetics. A molecular perspective. Berlin, Boston: De Gruyter. 252 pp. ISBN 978-3-11-028117-0.

Justino, C.I.L., Gomes, A.R., Freitas, A.C., Duarte, A.C. and Rocha-Santos, T.A.P. (2017). Graphene based sensors and biosensors. TrAC Trends Anal. Chem., 91, p. 53–66.

Kafri, A., Ovadia, S., Yosafovich-Doitch, G. and Aghion, E. (2018). In vivo performances of pure Zn and Zn-Fe alloy as biodegradable implants. J. Mater. Sci. Mater. Med., 29(7), p. 94.

Kasemo, B. and Lausmaa, J. (1991). The biomaterials-tissue interface and its analogoues in surface science and technology. In: Davies, J.E. and Albrektsson, T. (eds). The Bone-Biomaterials Interface, p. 19–32. University of Toronto Press: Toronto. ISBN 978-1-4426-7150-8. 352 pp.

Kingery, W.D., Bowen, H.K. and Uhlmann, D.R. (1976). Introduction to Ceramics. 2nd ed. Wiley Series on the Science and Technology of Materials. New York, Chichester, Brisbane, Toronto, Singapore: John Wiley & Sons. 1032 pp. ISBN 978-0-471-47860-1.

Kirsch, D.L. and Marksberry, J.A. (2015). The evolution of cranial electrotherapy stimulation for anxiety, insomnia, depression, and pain and its potential for other indications. In: Rosch, P. (ed.) Bioelectromagnetic and Subtle Energy Medicine. 2nd ed. Boca Raton, FL: CRC Press. p. 189–211.

Knabe, C., Adel-Khattab, D. and Ducheyne, P. (2017). Bioactivity: Mechanisms. In: Ducheyne, P., Grainger, D.W., Healy, K.E., Hutmacher, D.W. and Kirkpatrick, C.J. (eds). Comprehensive Biomaterials II. p. 291–310. Oxford: Elsevier. ISBN 978-0-08-055294-1.

Knoevenagel, E. (1900). Thiele's Theorie der Partialvalenzen im Lichte der Stereochemie [Thiele's theory of partial valences in the light of stereochemistry]. Justus Liebigs Ann. Chem., 311/312, p. 241–255.

Kohn, W. and Sham, L.J. (1965). Self-consistent equations including exchange and correlation effects. Phys. Rev., 140(4A), p. 1133–1138.

Kokubo, T. (1993). Bioactivity of glasses and glass ceramics. In: Ducheyne, P., Kokubo, T. and van Blitterswijk, C. A., (eds). Bone-Bioactive Biomaterials. Leidersdorp: Reed Healthcare Communications. p. 31–46.

Koupaei, N. and Karkhaneh, A. (2016). Porous crosslinked polycaprolactone hydroxylapatite networks for bone tissue engineering. Tissue Eng. Regen. Med., 13(3), p. 251–260.

Kraus, T., Fischerauer, S., Treichler, S., Martinelli, E., Eichler, J., Myrissa, A., Zötsch, S., Uggowitzer, P., Löffler, J.F. and Weinberg, A.M. (2018). The influence of biodegradable magnesium implants on the growth plate. Acta Biomater., 66, p. 109–117.

Kuzum, D., Takano, H., Shim, E. and 10 additional authors (2014). Transparent and flexible low noise graphene electrodes for simultaneous electrophysiology and neuroimaging. Nat. Commun., 5, 5259.

Lambotte, A. (1909). Technique et indication des prothèses dans le traitement des fractures. Presse Med., 17, p. 321.

Lambotte, A. (1932). L'utilisation du magnesium comme materiel perdu dans l'osteosynthèse. Bull. Mémoires Soc. National Chirugie, 28, p. 1325–1334.

Launey, M.E., Munch, E., Alsem, D.H. and 4 additional authors (2009). Designing highly toughened hybrid composites through Nature-inspired hierarchical complexity. Acta Mater., 57, p. 2919–2932.

Leibundgut, G. (2018). A novel, radiopaque, bioresorbable tyrosine-derived polymer for cardiovascular scaffolds. Cardiac Interv. Today Europe, 2(2), p. 26–30.

Leng, Y.X., Wang, F., Yang, P., Chen, J.Y. and Huang, N. (2019). The adhesion and clinical application of titanium oxide film on a 316L vascular stent, Surf. Coat. Technol., 363, p. 430–435.

Levy, G.K., Goldman, J. and Aghion, E. (2017). The prospects of zinc as a structural material for biodegradable implants – a review paper. Metals, 7, 402. 18 pages.

Li, Z., Jia, S., Xiong, Z., Long, Q., Yan, S., Hao, F., Liu, J. and Yuan. Z. (2018). 3D-printed scaffolds with calcified layer for osteochondral tissue engineering. J. Biosci. Bioeng., 126(3), p. 389–396.

Li, F.X., Guo, P.S., Han, S.B. and 8 additional authors (2019). A novel magnesium alloy with enhanced mechanical property, degradation behavior and cytocompatibility. Mater. Letters, 244, p. 70–73.

Lin, S., Ran, X.L., Yan, X.H., Wang, Q.L., Zhou, J.G., Hu, T.Z. and Wang, G.X. (2019). Systematic evolution on a Zn-Mg alloy potentially developed for biodegradable cardiovascular stents. J. Mater. Sci.: Mat. Med., 30:122.

Ling, G. and Lathan, C.E. (2018). Electroceuticals. Sci. Am., 319(6), p. 24.

Love, B. (2017). Orthopedic biomaterials and strategies. In: Love, B. (ed.) Biomaterials. A Systems Approach to Engineering Concepts. Chapter 11. New York: Academic Press. ISBN 978-0-1280-9478-5.

Love, C.A., Cook, R.B., Harvey, T.J., Dearnley, P.A. and Wood, R.J.K. (2013). Diamond-like carbon coatings for potential application in biological implants – a review. Tribolog. Inter., 63, p. 141–150.

Lu, D.Z., Huang, Y.L., Jiang, Q.T., Zheng, M., Duan, J.H. and Hou, B.R. (2019). An approach to fabricating protective coatings on a magnesium alloy utilizing alumina. Surf. Coat. Technol., 367, p. 336–340.

Luk, Y.Y. and Abbott, N.L. (2002). Application of functional surfactants. Curr. Opin. Coll. Interface Sci., 7(5-6), p. 267–275.

Maitz, M.F. (2015). Application of synthetic polymers in clinical medicine. Biosurf. Biotribol., 1(3), p. 161–176.

Mani, G. (2016). Metallic biomaterials: cobalt-chromium alloys. In: Murphy, W., Black, J. and Hastings, G. (eds). Handbook of Biomaterial Properties. New York: Springer. ISBN 978-1-4939-3303-7. p. 159–166.

Mann, S. (1996). Biomineralization and biomimetic materials chemistry. In: Mann, S. (ed.), Biomimetic Materials Chemistry, Weinheim: VCH, p. 1–40.

Mann, S. (2001). Biomineralization. Principles and Concepts in Bioinorganic Materials Chemistry. Oxford: University Press. ISBN 978-0-1985-0882-3.

Mao, L., Yuan, G. and Wang, S.A. (2012). Novel biodegradable Mg–Nd–Zn–Zr alloy with uniform corrosion behavior in artificial plasma. Mater. Letters, 88, p. 1–4.

Meißner, R., Bertol, L., Rehman, M.A.U., Loureiro dos Santos, L.A. and Boccaccini, A. (2019). Bioprinted 3D calcium phosphate scaffolds with gentamicin releasing capability. Ceram. Inter., 45 (6), p. 7090–7094.

Moravej, M. and Mantovani, D. (2011). Biodegradable metals for cardiovascular stent applications: interests and new opportunities. Int. J. Mo. Sci., 12(7), p. 4250–4270.

Moriguchi, H., Ohara, H. and Tsujioka, M. (2016). History and applications of diamond-like carbon manufacturing processes. SEI Techn. Review, 82, p. 52–58.

Mostaed, E., Sikora-Jasinska, M., Drehlich, J.W. and Vedani, M. (2018). Zinc-based alloys for degradable vascular stent applications. Acta Biomater., 71, p. 1–23.

Moussaoui, K., Mousseigne, M., Senatore, J., Lameste, P. and Chiaragatti, R. (2015). Influence of milling on the fatigue lifetime of a Ti6Al4V titanium alloy. Metals, 5(3), p. 1148–1162.

Murray, D.G. (1991). History of total knee replacement. In: Laskin, R.S. (ed.). Total Knee Replacement. London: Springer. p. 3–15. ISBN 978-1-4471-1827-5.

Niinomi, M. (ed.) (2010). Metals for Biomedical Devices. Woodhead Publ. Series in Biomaterials. 432 pp. ISBN 978-1-84569-434-0.

Niinomi, M., Liu, Y., Nakai, M., Liu, H.H. and Li, H. (2016). Biomedical titanium alloys with Young's moduli close to that of cortical bone. Regen. Biomater., 3(3), p. 173–185.

Ning, C.Y., Zhou, L. and Tan, G.X. (2016). Fourth-generation biomedical materials. Materials Today, 19(1), p. 2–3.

Noll, W. (1968). Chemistry and Technology of Silicones. New York: Academic Press. 716 pp. ISBN 978-0-124-12227-7.

Nordenstrom, B.E.W. (1998). Exploring BCEC-Systems (Biologically Closed Electric Circuits). Stockholm: Nordic Medical Publications.

O'Brien, F.J. (2011). Biomaterials and scaffolds for tissue engineering. MaterialsToday, 14(3), p. 88–95.

Ogunleye, A., Bhat, A., Irorere, V.U., Hill, D., Williams, C. and Radecka, I. (2015). Poly-γ-glutamic acid: production, properties and applications. Microbiol., 116(1), p. 1–7.

Palmero, P., De Barra, E. and Cambier, F. (eds) (2017). Advances in Ceramic Biomaterials: Materials, Devices and Challenges. Woodhead Publ., 500 pp. ISBN 978-0-081-00882-9.

Pasqualini, U. and Pasqualini, M.E. (2009). The history of implantology. In: Pasqualini, U. and Pasqualini, M.E. (eds). Treatise on Implant Dentistry: The Italian Tribute to Modern Implantology. Carimate, Milano: Ariesdue.

Pasteris, J.D. (2016). A mineralogical view of apatitic biomaterials. Am. Mineral., 101, p. 2594–2610.

Perale, G. and Hilborn, J. (2017). Bioresorbable Polymers for Biomedical Applications. Elsevier. ISBN 978-0-08-100262-9. 628 pp.

Pert, C.B. (1999). Molecules of Emotion: The Science between Mind-Body Medicine. Scribner. ISBN 0-684-84634-9. 368 pp.

Pert, C.B. and Snyder, S.H. (1973). Opiate receptor: demonstration in nervous tissue. Science, 179(4077), p. 1011–1014.

Phillips, H. and Taylor, J.G. (1975). The Walldius hinge arthroplasty. J. Bone Joint Surg. Br, 57(1), p. 59–62.

Poinern, G.E.J., Brundavanam, S. and Fawcett, D. (2012). Biomedical magnesium alloys: A review of materials properties, surface modification and potential as a biodegradable orthopedic implant. Am. J. Biomed. Eng., 2(6), p. 218–240.

Prado da Silva, M.H. (2016). Osteoconductive biomaterials. Reference Module Materials Science and Materials Engineering. Elsevier. https://doi.org/10.1016/B978-0-12-803581-8.04089-3.

Quinn, G. and Brachmann, S. (2014). The evaluation of hip replacement: A patent history. IP News, https://www.IPWatchdog.com. April 21, 2014.

Raabe, D. (2019). Introduction to titanium alloys. www.dierk-raabe.com/titanium-alloys/ (accessed September 07, 2019).

Raabe, D., Sander, B., Friák, M., Ma, D. and Neugebauer, J. (2007). Theory-guided bottom-up design of ß-titanium alloys as biomaterials based on first principle calculations: Theory and experiments. Acta Mater., 55, p. 4475–4487.

Radin, S. R. and Ducheyne, P. (1993). The effect of calcium phosphate ceramic composition and structure on in vitro behavior. II. Precipitation. J. Biomed. Mater. Res., 27, p. 35–45.

Radin, S., Reilly, G., Bhargave, G., Leboy, P. S. and Ducheyne, P. (2005). Osteogenic effects of bioactive glass on bone marrow stromal cells. J. Biomed. Mater. Res. A, 73, p. 21–29.

Rajabi, A.H., Jaffe, M. and Livingston Arinzeh, T. (2015). Piezoelectric materials for tissue regeneration: A review. Acta Biomater., 24, p. 12–23.

Ratner, B.D., Hoffman, A.S., Schoen, F.J. and Lemons, J.E. (2013) (eds). Biomaterials Science. An Introduction to Materials in Medicine. Elsevier, ISBN 978-0-12-374626-9.

Rey, C., Combes, C., Drouet, C. and Glimcher, M.J. (2010). Bone mineral: update on chemical composition and structure. Osteopor. Int., 20(6), p. 1013–1021.

Rieger, W. (2001). Ceramics in orthopedics – 30 years of evolution and experience. In: Rieker, C., Oberholzer, S. and Wyss, S. (eds), World Tribology Forum in Arthroplasty. Bern: Hans Huber. p. 309–318.

Rock, M. (1933). Künstliche Ersatzteile für das Innere und Äussere des menschlichen und tierischen Körpers [Artificial replacement parts for internal and external locations of the human and animal body]. Deutsches Reichspatent DRP 583 589, 24.08.1933.

Roseti, L., Parisi, V., Petretta, M., Cavallo, C., Desandro, G., Bartolotti, I. and Grigolo, B. (2017). Scaffolds for bone tissue engineering: state of the art and new perspectives. Mater. Sci. Eng. C., 78, p. 1246–1262.

Rotman, S.G., Grijpma, D.W., Richards, R.G., Moriarty, T.F., Eglin, D. and Guillaume, O. (2018). Drug delivery systems functionalized with bone mineral seeking agents for bone targeted therapeutics. J. Control. Release, 269, p. 88–99.zt6

Ruiz-Hitzky, E., Darder, M., Avanda, P. and Ariga, K. (2010). Advances in biomimetic and nanostructured biohybrid materials. Adv. Mater., 22(3), p. 323–336.

Saini, M., Singh, Y., Arora, P., Arora, V. and Jain, K. (2015). Implant biomaterials: a comprehensive review. World J. Clin. Cases, 3(1), p. 52–57.

Saltzman, W.M. and Torchilin, V.P. (2018). Drug Delivery Systems. AccessScience, McGraw-Hill Corp. https://doi.org/10.1036/1097-8542-757275 (accessed July 30, 2018).

Sampathkumar, S.G. and Yarema, K.J. (2005). Targeting cancer cells with dendrimers. Cell Chem. Biol., 12(1), p. 5–6.

Sanchez, A.H.M., Luthringer, B., Feyerabend, F. and Willumeit, R. (2014). Mg and Mg alloys: How comparable are in vitro and in vivo corrosion rates? A review. Acta Biomater., 13, p. 16–31.

Sandhaus, S. (1966). Bone implants, and drills and tapes for bone surgery. European patent EP 1083769 (A).

Savolainen, V., Cowan, R.S., Vogler, A.P., Roderick, G.K. and Lane, R. (2005). Towards writing the encyclopaedia of life: an introduction to DNA barcoding. Phil. Trans. Royal Soc. B: Biolog. Sci., 360 (1462), p. 1805–1811.

Schepers, E., Declercq, M., Ducheyne, P. and Kempeneers, R. (1991). Bioactive glass particulate material as a filler for bone lesions. J. Oral Rehab., 18, p. 439–452.

Scott, R.D. (2015). Total Knee Arthroplasty. 2nd ed., Elsevier. ISBN 978-0-3232-8663-3.

Semlitsch, M. (1984). Metallische Werkstoffe für Implantate [Metallic materials for implants]. In: Rettig, H. (ed.). Biomaterials und Nahtmaterials. Jahrestagung Deutsche Gesellschaft für Plastische und Wiederherstellungschirugie, vol. 21, Berlin, Heidelberg: Springer.

Sezer, N., Evis, Z., Kayhan, S.M., Tahmasebifar, A. and Koç, M. (2018). Review of magnesium-based biomaterials and their applications. J. Magn. Alloys, 6(1), p. 23–43.

Sherman, W.O. (1912). Vanadium steel bone plates and screws. Surg. Gynecol. Obstet., 14, p. 629–634.

Spriano, S., Yamaguchi, S., Baino, F. and Ferrari, S. (2018). A critical review of multifunctional titanium surfaces: new frontiers for improving osseointegration and host responses, avoiding bacterial contamination. Acta Biomater., 79, p. 1–22.

Strohbach, A. and Busch, R. (2015). Polymers for cardiovascular stent coatings. Int. J. Polymer Sci., article ID 782653, 11 pp.

Sun, R., Tai, C.W., Strømme, M. and Cheung, O. (2019). The effects of additives on the porosity and stability of amorphous calcium carbonate. Micropor. Mesopor. Mater., 292:109736.

Surmeneva, M.A., Ivanova, A.A., Tian, Q.M., Pittman, R., Jiang, W.S., Lin, J.J., Liu, H.N. and Surmenev, R.A. (2019). Bone marrow derived mesenchymal stem cell response to the RF magnetron sputter deposited hydroxylapatite coating on AZ91 magnesium alloy. Mater. Chem. Phys., 221, p. 89–98.

Suzuki, M. and Ikada, Y. (2004). Biodegradable polymers in medicine. In: Reis, R. L., San Román, J. (eds). Biodegradable Systems in Tissue Engineering and Regenerative Medicine. Boca Raton: CRC Press. ISBN: 978-0-8493-1936-5.

Tandon, B., Blaker, J.J. and Cartmell, S.H. (2018). Piezoelectric materials as stimulatory biomaterials and scaffolds for bone repair. Acta Biomater., 73, p. 1–20.

Thompson, L.A., Law, F.C., Rushton, N. and Franks, J. (1991). Biocompatibility of diamond-like carbon coating. Biomaterials, 12(1), p. 37–40.

Vince, K.G. and Insall, J.N. (1991). The total condylar knee prosthesis. In: Laskin, R.S. (ed.). Total Knee Replacement. London: Springer. p. 85–111. ISBN 978-1-4471-1827-5.

Waksman, R. (2006). Biodegradable stents: they do their job and disappear. J. Invasive Cardiol., 18 (2). February 2006.

Wang, L., Shi, C.Y., Wang, X., Guo, D.D., Duncan, T.M. and Luo, J.T. (2019a). Zwitterionic Janus dendrimer with distinct functional disparity for enhanced protein delivery. Biomaterials, 215, article ID 119233.

Wang, X., Shao, X.X., Dai, T.Q. and 6 additional authors (2019b). In vivo study of the efficacy, biosafety, and degradation of a zinc alloy osteosynthetic system. Acta Biomater., 92, p. 351–361.

Watson, J.R., Wood, H. and Hill, R.C.J. (1976). The Shiers arthroplasty of the knee. The Bone & Joint J., 58(3), p. 300–304.

Werner, A. (1893). Beitrag zur Konstitution anorganischer Verbindungen [Contribution to the coordination of inorganic compounds]. Z. Anorg. Chem., 3, p. 267–330.

Wessinghage, D. (1995). Themistocles Gluck: Von der Organexstirpation zum Gelenkersatz [Themistocles Gluck: from organ exstirpation to joint replacement]. Dtsch. Ärzteblatt, 92(33), p. A2180-A2184.

Williams, D.F. (1987). Definitions in Biomaterials: Proceedings of a consensus conference of the European Society for Biomaterials, Chester, England, March 3-5,1986. Amsterdam, New York: Elsevier. ISBN 978-0-44-442858-5.

Williams, D.F. (2009). On the nature of biomaterials. Biomaterials, 264-268, p. 2051–2054.

Willmann, G. (1996). Development in medical grade alumina during the past two decades. J. Mater. Proc. Technol., 56, p. 168–176.

Witte, F. (2010). The history of biodegradable magnesium implants: A review. Acta Biomater., 6(5), p. 1680–1692.

World Health Organization (WHO). www.who.int/medical_devices/full_definition/en/ (accessed July 16, 2018).

Wubneh, A., Tsekura, E.K., Ayranci, C. and Uludag, H. (2018). Current state of fabrication technologies and materials for bone tissue engineering. Acta Biomater., 80, p. 1–30.

Xynos, I. D., Edgar, A. J., Buttery, L. D., Hench, L. L. and Polak, J. M. (2000). Ionic products of bioactive glass dissolution increase proliferation of human osteoblasts and induce insulin-like growth factor II mRNA expression and protein synthesis. Biochem. Biophys. Res. Commun., 276, p. 461–465.

Xynos, I. D., Edgar, A. J., Buttery, L. D. K., Hench, L. L. and Polak, J. M. (2001). Gene-expression profiling of human osteoblasts following treatment with the ionic products of bioglass® 45S5 dissolution. J. Biomed. Mater. Res., 55, p. 151–157.

Yang, Y., Asiri, A.M., Tang, Z.W., Du, D. and Lin, Y.H. (2013). Graphene based materials for biomedical applications. MaterialsToday, 16(10), p. 365–373.

Yao, J., Radin, S., Leboy, P. S. and Ducheyne, P. (2005). Solution-mediated effect of bioactive glass in poly (lactic-co-glycolic acid)-bioactive glass composites on osteogenesis of marrow stromal cells. J. Biomed. Mater. Res. A, 75, p. 794–801.

Zhang, B.G.X., Myers, D.E., Wallace, G.G., Brandt, M. and Choong, P.F.M. (2014). Bioactive coatings for orthopedic implants-Recent trends in development of implant coatings. Int. J. Mol. Sci., 15, p. 11878–11921.

Zhang, Z.Z., Chen, Y.R., Wang, S.J. and 11 additional authors (2019). Orchestrated biomechanical, structural, and biochemical stimuli for engineering anisotropic meniscus. Science Transl. Med., 11 (487), eaao0750. DOI: 10.1126/scitranslmed.aao0750.

Zhou, C.Y., Geng, H.M. and Guo, C.L. (2018). Design of DNA-based innovative computing system of digital comparison. Acta Biomater., 80, p. 58–65.

Zivic, F., Grujovic, N., Pellicer, E., Sort, J., Mitrovic, S., Adamovic, D. and Vulovic, M. (2018). Biodegradable metals as biomaterials for clinical practice: iron-based materials. In: Zivic, F., Affatato, S., Trajanovic, M., Schnabelrauch, M., Grujovic, N. and Choy, K.L. (eds). Biomaterials in Clinical Practice: Advances in Clinical Research and Medical Devices. p. 225–280. Springer Intern. Publ. AB. ISBN 978-3-319-68024-8.

Chapter 2

Abendroth, M. (2017). FEM analysis of small punch tests. Key Eng. Mater., 734, p. 23–36.

Agarwal, S., Curtin, J., Duffy, B. and Jaiswal, S. (2016), Biodegradable magnesium alloys for orthopaedic applications: A review on corrosion, biocompatibility and surface modifications. Mater. Sci. Eng. C, 68, p. 948–963.

Agrawal, C.M., Ong, J.L., Appleford, M.R. and Mani, G. (2014). Introduction to biomaterials. Basic theory with engineering application. Cambridge: Cambridge University Press. ISBN: 978-1-316-61130-2. p. 198–232.

Ahmed, T., Long, M., Silvestri, J., Ruiz, C. and Rack, H. J. (1996). A new low modulus, biocompatible titanium alloy. In: Blenkinsop, P. A., Evans, W. J. and Flower, H. M. (eds). Titanium '95, Science and Technology. Institute of Metals, London, UK, Vol. II, p. 1760–67.

Alamo, J. (1993). Chemistry and properties of solids with the [NZP] skeleton. Solid State Ionics, 63/65, p. 547–561.

Aldosari. A., Sukumaran, A., Al Amri, M.D., Bart Van Oirschot. A. and Jansen, J.A. (2018). A comparative study of the bone contact to zirconium and titanium implants after 8 weeks of implantation in rabbit femoral condyles. Odontology, 106, p. 37–44.

Al-Qtaitat, A. and Aldalaen, S. (2014). A review of non-collagenous proteins; their role in bone. Am. J. Life Sci., 2(6), p. 351–355.

Alley, W.M. and Alley, R. (2013). Too Hot to Touch. The Problem of High-Level Nuclear Waste. Cambridge: Cambridge University Press. ISBN 978-1-107-03011-4. 370 pp. p. 52.

Anderson, M.C. and Olsen, R. (2010). Bone ingrowth into porous silicon nitride. J. Biomed. Mater. Res. A, 92(4), p. 1598–1605.

Anitha, A., Sowmya, S., Sudheesh Kumar, P.T., Deepthi, S., Chennazhi, K.P., Ehrlich, H., Tsurkan, M. and Jayakumar, R. (2014). Chitin and chitosan in selected biomedical applications. Progr. Polym. Sci., 39, p. 1644–1667.

Ardakani, M.H., Moztarzadeh, F., Rabiee, M. and Talebi, A.R. (2011). Synthesis and characterization of nanocrystalline merwinite ($Ca_3Mg(SiO_4)_2$) via sol-gel method. Ceram. Inter., 37(1), p. 175–180.

Ashida, S., Kyogaku, H. and Hosoda, H. (2012). Fabrication of Ti-Sn-Cr shape memory alloy by PM and its properties. Mater. Sci. Forum, 706-709, p. 1943–1947

ASTM F799-11 (2011). Standard specification for cobalt-28chromium-6molybdenum alloy forgings for surgical implants (UNS R31537, R31538, R31539), ASTM International, West Conshohocken, PA.

ASTM F75-12 (2012). Standard specification for cobalt-28chromium-6molybdenum alloy castings and casting alloy for surgical implants (UNS R30075), ASTM International, West Conshohocken, PA.

ASTM F562-13 (2013). Standard specification for wrought 35cobalt-35nickel-20chromium-10molybdenum alloy for surgical implant applications (UNS R30035), ASTM International, West Conshohocken, PA.

ASTM F90-14 (2014). Standard specifcation for wrought cobalt-20chromium-15tungsten-10nickel alloy for surgical implant applications (UNS R30605). ASTM International, West Conshohocken, PA.

ASTM F1713. Standard specification for wrought titanium-13niobium-13zirconium alloy for surgical implant applications (UNS R58130). ASTM International, West Conshohocken, PA, 19428-2959, USA.

ASTM F1813. Standard specification for wrought titanium–12 molybdenum–6 zirconium–2 iron alloy for surgical implant (UNS R58120), ASTM International, West Conshohocken, PA, 19428-2959, USA.

ASTM F2066. Standard specification for wrought titanium-15 molybdenum alloy for surgical implant applications (UNS R58150), ASTM International, West Conshohocken, PA, 19428–2959, USA.2959 USA.

ASTM designation draft #3. Standard specification for wrought taitanium-35Niobium-7zirconium-5tantalum alloy for surgical implant applications (UNS R58350), ASTM, Philadephia. PA, U.S.A.

Babu, J-P., Alla, R.K., Alluri, V.R., Datla, S.P. and Konakanchi, A. (2015). Dental ceramics: Part I – An overview of composition, structure and properties. Am J. Mater. Eng. Technol., 3(1), p. 13–18.

Balas, F., Pérez-Pariente, J. and Vallet-Regí, M. (2003). In vitro bioactivity of silicon-substituted hydroxyapatites. J. Biomed. Mater. Res., 66, p. 364–375.

Balasubramanian, P., Büttner, T., Pacheco, V.M. and Boccaccini, A.R. (2018). Boron-containing bioactive glasses in bone and soft tissue engineering. J. Eur. Ceram. Soc., 38(3), p. 855–869.

Barrias, C.C., Ribeiro, C.C., Lamghari, M., Miranda, C.S., and Barbosa, M.A. (2005). Proliferation, activity, and osteogenic differentiation of bone marrow stromal cells cultured on calcium titanium phosphate microspheres. J. Biomed. Mater. Res. A, 72(1), p. 57–66.

Baur, W.H. (1974). The geometry of polyhedral distortions. Predictive relationships for the phosphate group. Acta Cryst. B, 30(5), p. 1195–1215.

Ben-Nissan, B. (ed.) (2014). Advances in Calcium Phosphate Biomaterials. Springer Series in Biomaterials Science and Engineering. ISBN 978-3-642-53979-4. 547 pp.

Ben-Nissan, B., Choi, A.H. and Cordingley, R. (2008). Alumina ceramics. In: Bioceramics and their Clinical Applications. Woodhead Publ. Ser. Biomater., p. 223–242.

Ben-Nissan B., Choi A.H. and Green D.W. (2019). Marine derived biomaterials for bone regeneration and tissue engineering: Learning from Nature. In: Choi A., Ben-Nissan B. (eds) Marine-Derived Biomaterials for Tissue Engineering Applications. Springer Series in Biomaterials Science and Engineering, 14, p. 51–78. Singapore: Springer.

Beraha, E. and Sphigler, B. (1977). Color Metallography. Metals Park OH: Am. Soc. Metals. 160 pp.

Bergamo, E.T.P., Bordin, D., Ramalho, I.S. and 7 additional authors (2019). Zirconia-reinforced lithium disilicate crowns: Effect of thickness on survival and failure mode. Dental Mater., https://doi.org/10.106/j.dental. 2019.04.007.

Berger, G., Gildenhaar, R. and Ploska, U. (1995a). Rapid resorbable, glassy crystalline materials on the basis of calcium alkali orthophosphates. Biomaterials, 16, p. 1241–1248.

Berger, G., Gildenhaar, R. and Ploska, U. (1995b). Rapidly resorbable materials based on a new phase: $Ca_2KNa(PO_4)_2$. In: Wilson, J., Hench, L.L., Greenspan, D.C. (eds.). Bioceramics 8. Oxford: Butterworth-Heinemann. p. 453–456.

Berglund, I.S., Dir, E.W., Ramaswamy, V., Allen, J.B., Allen, K.D. and Manuel, M,V. (2018). The effect of Mg-Ca-Sr alloy degradation products on human mesenchymal stem cells. J. Biomed. Mater. Res., B: Applied Biomater., 106B, p. 697–704.

Bewilogua, K. and Hofmann, D. (2014). History of diamond-like carbon films – from first experiments to worldwide applications. Surf. Coat. Technol., 242, p. 214–225.

Bezwada, R.S., Jamiolkowski, D.D. and Cooper, K. (1997). Poly(p-dioxanone) and its copolymers. In: Domb, A.J., Kost, J. and Wiseman, D.M. (eds). Handbook of Biodegradable Polymers. Amsterdam: Harwood Academic Publishers, ISBN: 90-5702-153-6. p. 29–61.

Bhattacharjee, S., Zhao, Y.H., Hill, J.M., Percy, M.E. and Lukiw, W.J. (2014). Aluminum and its potential contribution to Alzheimer's disease (AD). Front. Aging Neurosci., 6:62.

Bhattacherjee, A., Gupta, A., Verma, M. and 5 additional authors (2019). Site-specific antibacterial efficacy and cyto/hemo-compatibility of zinc substituted hydroxyapatite. Ceram. Intern., 45(9), p. 12225–12233.

Biasetto, L., Elsayed, H., Bonollo, F. and Colombo, P. (2016). Polymer-derived sphene biocoatings on cp-Ti substrates for orthopedic and dental implants. Surf. Coat. Technol., 301, p. 140–147.

Bigi, A., Boanini, E., Capuccini, C. and Gazzano, M. (2007). Strontium-substituted hydroxylapatite nanocrystals. Inorg. Chim. Acta, 360(3), p. 1009–1016.

Biran, R. and Pond, D. (2017). Heparin coatings for improving blood compatibility of medical devices. Adv. Drug Deliv. Rev., 112, p. 12–23.

Birkby, I. and Stevens, R. (1996). Applications of zirconia ceramics. Key Eng. Mater., 122-124, p. 527–552.

Black, J. (1999). Biological Performance of Materials. Fundamentals of Biocompatibility, 3rd edition. New York: Marcel Dekker.

Boccardi, E., Philippart, A., Beltrán, A.M., Schmidt, J., Liverani, L., Peukert, W. and Boccaccini, A.R. (2018). Biodegradability of spherical mesoporous silica particles (MCM-41) in simulated body fluid (SBF). Am. Mineral., 103, p. 350–354.

Boehm, R.D., Jin, S. and Narayan, R.J. (2017). Carbon and diamond. In: Reference Module in Materials Science and Materials Engineering. Comprehensive Biomaterials II. Vol. 1, p. 145–164. Elsevier. ISBN 978-0-08-100692-4.

Boesel, L.F., and Reis, R.L. (2005). Injectable biodegradable systems. In: Reis, R. L., San Roman, J. (eds). Biodegradable systems in tissue engineering and regenerative medicine. Boca Raton: CRC Press. ISBN: 0-8493-1936-6, p.13–27.

Bohner, M. (2010). Resorbable biomaterials as bone graft substitutes, Materials Today, 13(1/2), p. 24–30.

Bohner, M. and Miron, R.J. (2019). A proposed mechanism for material-induced heterotopic ossification. MaterialsToday, 22, p. 132–141.

Borenfreund, E. and Puerner, J. A. (1985). Toxicity determined in vitro by morphological alternations and neutral red absorption. Toxicol. Lett., 24, p. 119–124.

Borie, E., Rosas, E., Kuramochi, G., Etcheberry, S., Olate, S. and Weber, B. (2019). Oral applications of cyanoacrylate adhesives: A literature review. BioMed Res. Int., ID 8217602.

Bouler, J.M., Pilet, P., Gauthier, O. and Verron, E. (2017). Biphasic calcium phosphate ceramic for bone reconstruction: a review of biological response. Acta Biomater., 53, p. 1–12.

Boussif, O., Lezoualcht, F., Zanta, M. Mergny, M.D., Scherman, D., Demeneix, B. and Behr, J.-P. (1995). A versatile vector for gene and oligonucleotide transfer into cells in culture and in vivo: polyethylenimine. Proc. Natl. Acad. Sci. USA, 92, p. 7297–7301.

Boutin, P. (1972). L'arthroplastie total de la hanche par prothèse en alumine frittée. Revue. Chir. Orthop., 58, p. 229–246.

Brodkin, D., Panzera, C., Panzera, P., Pruden, J.N., Kaiser, L. and Brightly, R. (2000). Low firing dental porcelain containing fine-grained leucite. US Patent 6120511 A. Sept 19, 2000.

Brophy, G.P. and Nash, T. (1968). Compositional, infrared, and x-ray analysis of fossil bone. Am. Min., 53(3-4), p. 445–454.

Brow, K-H and Kramer, K-H. (1985). On the properties of a new titanium alloy (TiAl5Fe2.5) as implant material. In: Lütjering, G., Zwicker, U. and Bunk, W. (eds). Titanium Science and Technology, 2, p. 1381–1386. Deutsche Gesellschaft für Metallkunde e. V.

Brunetto, G., Elsayed, H. and Biasetto, L. (2019). Bioactive glass and silicate-based ceramic coatings on metallic implants: Open challenge or outdated topic? Materials, 12: 2929. doi:10.3390/ma1282929.

Buehler, M.J. (2006) Nature designs tough collagen: explaining the nanostructure of collagen fibrils. PNAS, 103(33), p. 12285–12290.

Buga, C., Hunyadi, M., Gácsi, Z., Hegedüs, C., Hakl, J., Schmidt, U., Ding, S.J. and Csik, A. (2019). Calcium silicate layer on titanium fabricated by electrospray deposition. Mater. Sci. Eng. C, 98, p. 401–408.

Bussola Tovani, C., Gloter, A., Zais, T., Selmani, M., Ramus, A.P. and Nassif, N. (2019). Formation of stable strontium-rich amorphous calcium phosphate: Possible effects on bone mineral. Acta Biomater., 92, p. 315–324.

Calvo, C. and Gopal, R. (1975). The crystal structure of whitlockite from the Palermo Quarry. Am. Min., 60, p. 120–133.

Cama, G., Mogosanu, D.E., Houben, A. and Dubruel, P. (2017). Synthetic biodegradable medical polyesters: poly-ε-caprolactone. In: Zhang, X. (ed.) Science and principles of biodegradable medical polymers. Duxford: Woodhead Publishing. ISBN: 978-0-08-100372-5. p. 79–105.

Canillas, M., Moreno, B., Carballo-Vila, M., Jurado, J.R. and Chinarro, E. (2019). Bulk Ti nitride prepared from rutile TiO_2 for its application as stimulation electrode in neuroscience. Mater. Sci. Eng. C, 96, p. 295–301.

Cantor, B., Chang, I.T.H., Knight, P. and Vincent, A.J.B. (2004). Microstructural development in equiatomic multicomponent alloys. Mater. Sci. Eng. A, 375-377, p. 213–218.

Cappi, B., Neuss, S., Salber, J., Telle, R., Knüchel, R. and Fischer, H. (2010). Cytocompatibility of high strength non-oxide ceramics. J. Biomed. Mater. Res. A, 93(1), p. 67–76.

Carlisle, E.M. (1981). Silicon: a requirement in bone formation independent of vitamin D1. Calcif. Tissue Intern., 33, p. 27–34.

Carlisle, E.M. (1982). The nutritional essentiality of silicon. Nutrit. Rev., 40, p. 193–198.

Chandra Ray, S. and Ranjan Jana, N. (2017). Application of carbon-based nanomaterials as drug and gene delivery carriers. In: Chandra Ray, S and Ranjan Jana, N. Carbon Nanomaterials for Biological and Medical Applications. Elsevier. ISBN 978-0-323-47906-6. 250 pp. p. 163-203.

Chang, Y.Y., Huang, H.L., Chen, H.J., Lai, C.H. and Wen, C.Y. (2014). Antibacterial properties and cytocompatibility of tantalum oxide coatings. Surf. Coat. Technol, 259, p. 193–198.

Chaya, A., Yoshizawa, S., Verdelis, K. and 7 additional authors (2015). In vivo study of magnesium plate and screw degradation and bone fracture healing. Acta Biomater., 18, p. 262–269.

Chen, R.Z. and Tuan, W.H. (2000). Thermal etching of alumina. Am. Ceram. Soc. Bull., 79(10), p. 83–86.

Chen, S.S., Tan, L., Zhang, B.C., Xia, Y.H., Xu, K, and Yang, K. (2017) In vivo study on degradation behavior and histologic response of pure magnesium in muscles. J. Mater. Sci. Technol., 33(5), p. 469–474.

Chen, J.X., Tan, L. and Yang, K. (2016). Recent advances on the development of biodegradable magnesium alloys: a review. Mater. Technol., 31(12), p. 681–688.

Chen, X.C., Ou, J., Wei, Y., Huang, Z.B., Kang, Y.Q. and Yin, G.F. (2010). Effect of MgO contents on the mechanical properties and biological performances of bioceramics in the $MgO–CaO–SiO_2$ system. J. Mater. Sci. Mater. Med., 21, p. 1463–1471.

Cheng, S. and Lu, X. (2012). Connecting piece for human bone substitute for use as biological medical material provided with a porous titanium surface layer coated with a titanite layer. Patent CN202982311-U.

Chevalier, J. (2006). What future for zirconia as a biomaterial? Biomaterials, 27(4), p. 535–543.

Chevalier, J. and Gremillard, L. (2017). Zirconia as a biomaterial. In: Reference Module in Materials Science and Materials Engineering. Comprehensive Biomaterials II. Vol. 1, p. 122–144. Elsevier. ISBN 978-0-08-100692-4.

Chou, J., Hao, J., Hatoyama, H., Ben-Nissan, B., Milthorpe, B. and Otsuka, M. (2014). Effect of biomimetic zinc-containing tricalcium phosphate (Zn-TCP) on the growth and osteogenic differentiation of mesenchymal stem cells. J. Tissue Eng. Regen. Med., doi:10.1002/term.

Chu, C. C. (2003). Biodegradable polymeric biomaterials: An updated overview. In: Park, J.B., and Bronzino, J.D. (eds). Biomaterials: Principles and Applications.Boca Raton: CRC Press. ISBN: 0-8493-1491-7. p. 95–115.

Cicco, S.R., Vona, D., Gristina, R., Sardella, E., Ragni, R., Lo Presti, M. and Farinola, G.M. (2016). Biosilica from living diatoms: Investigations on biocompatibility of bare and chemically modified Thalassiosira weissflogii silica shells. Bioengineering, 3(4), 35. https://doi.ord/10.3390/bioengineering3040035.

Ciobanu, G. and Harja, M. (2019). Cerium-doped hydroxyapatite/collagen coatings on titanium for bone implants. Ceram. Intern., B 45(2), p. 2852–2857.

Cizek, J., Brozek, V., Chraska, T. and 11 additional authors (2018). Silver-doped hydroxyapatite coatings deposited by suspension plasma spraying. J. Thermal Spray Technol., 27 (8), p. 1333–1243.

Clarke, I.C., Green, D., Williams, P., Donaldson, T. and Pezzotti G. (2006). U.S. perspective on hip simulator wear testing of BIOLOX® delta in' severe' test modes. In: Benazzo, F., Falez, F., Dietrich, M. (eds), Bioceramics and Alternative Bearings in Joint Arthroplasty. Ceramics in Orthopaedics. Darmstadt: Steinkopff. ISBN 978-3-7985-1634-2.

Clarke, S.A., Walsh, P., Maggs, C.A. and Buchanan, F. (2011). Designs from the deep: Marine organisms for bone tissue engineering. Biotech. Adv., 29(6), p. 610–617.

Colas, A. and Curtis, J. (2004). Silicone biomaterials. In: Ratner, B.D., Hoffman, A.S., Schoen, F.J., and Lemons, J.E. (eds). Biomaterials Science: An Introduction to Materials in Medicine. San Diego: Elsevier. ISBN: 0-12-582463-7. p. 80–86.

Colen, T.P., Paridaens, D.A., Lemij, H.G., Mourits, M.P. and Van den Bosch, W.A. (2000). Comparison of artificial eye amplitudes with acrylic and hydroxyapatite spherical enucleation implants. Ophthalmology, 107(10), p. 1889–1894.

Combes, C. and Rey, C. (2010). Amorphous calcium phosphates: Synthesis, properties and uses in biomaterials. Acta Biomater., 6(9), p. 3362–3378.

Comodi, P., Liu, Y., Stoppa, F. and Woolley, A.R. (1999). A multi-method analysis of Si-, S- and REE-rich apatite from a new find of kalsilite-bearing leucitite (Abruzzi, Italy). Min. Mag., 63(5), p. 661–672.

Corrêa, J.M., Mori, M., Sanches, H.L., Dibo da Cruz, A., Poiate, E. and Venturini Pola Poiate, I.A. (2015). Silver nanoparticles in dental biomaterials. Intern. J. Biomater., article ID 485275, 9 pages.

Cowley, A. and Woodward, B. (2011). A healthy future: Platinum in medical applications. Platinum Met. Rev., 55(2), p. 98–107.

Curtis, J. and Colas, A. (2004). Medical applications of silicones. In: Ratner, B.D., Hoffman, A.S., Schoen, F.J., and Lemons, J.E. (eds). Biomaterials Science: An Introduction to Materials in Medicine. San Diego: Elsevier. ISBN: 0-12-582463-7. p. 697–707.

Daculsi, G., LeGeros, R. Z., Nery, E., Lynch, K. and Kerebel, B. (1989). Transformation of biphasic calcium phosphate ceramics in vivo: Ultrastructural and physicochemical characterization. J. Biomed. Mat. Res., 23, p. 883–894.

Dalibón, E.L., Escalada, L., Simison, S., Forsich, C., Heim, D. and Brühl, S.P. (2017). Mechanical and corrosion behavior of thick and soft DLC coatings. Surf. Coat. Technol., 312, p. 101–109.

D'Amora, M. and Giordani, S. (2018). Carbon nanomaterials for nanomedicine. In: Ciofani, G. (ed.). Smart Nanoparticles for Biomedicine. Elsevier. ISBN 978-0-12-814156-4. 268 pp. p. 103–113.

D'Arros, C., Rouillon, T., Veziers, J., Malard, O., Borget, P. and Daculsi, G. (2020). Bioactivity of biphasic calcium phosphate granules, the control of a needle-like apatite layer formation for further medical device development. Front. Bioeng. Biotechnol., https://doi.org/10.3389/fbioe.2019.00462 (accessed February 6, 2020).

Das, R., Bandyapadhyay, R. and Pramanik, P. (2018). Carbon quantum dots from natural resources: A review. MaterialsToday, Chemistry, 8, p. 96–109.

Dash, M., Chiellini, F., Ottenbrite, R.M. and Chiellini, E. (2011). Chitosan-A versatile semi-synthetic polymer in biomedical applications. Progr. Polym. Sci., 36, p. 981–1014.

Davies, R.L. and Etris, S.F. (1997). The development and function of silver in water purification and disease control. Catal. Today, 36, p. 107–114.

De Bruijn, J. D., Davies, J. E., Klein, C. P. A. T., de Groot, K. and van Blitterswijk, C. A. (1993). Biological responses to calcium phosphate ceramics. In: Ducheyne, P., Kokubo, T. and van Blitterswijk, C. A. (eds). Bone-Bonding Biomaterials. Reed Healthcare Communications: Leiderdorp, p. 57–72.

De Bruijn, J. D., Bovell, Y. P. and van Blitterswijk, C. A. (1994). Structural arrangements at the interface between plasma sprayed calcium phosphates and bone. Biomaterials, 15, p. 543–550.

De Groot, K., Klein, C.P.A.T., Wolke, J.G.C. and de Blieck-Hogervorst, J. (1990). Plasma-spraying of calcium phosphate. In: Yamamuro, T., Hench, L.L. and Wilson-Hench, J. (eds). Handbook of Bioactive Ceramics. Vol 2: Calcium phosphate and hydroxylapatite ceramics, Boca Raton: CRC Press.

Deliormanli, A.M. (2013). Size-dependent degradation and bioactivity of borate bioactive glass. Ceram. Intern., 39(7), p. 8087–8095.

De Melo Alencar, C., Freitas de Paulo, B.L., Guanipa Ortiz, M.I., Magno, M.B., Silva, C.M. and Maia, L.C. (2019). Clinical efficacy of nano-hydroxyapatite in dentin hypersensitivity: a systematic review and meta-analysis. J. Dent., https://doi.org/10.1016/j.jdent.2018.12.014.

Demers, C., Hamdy, C.R, Corsi, K., Chellat, F., Tabrizian, M. and Yahia, L. (2002). Natural coral exoskeleton as a bone graft substitute: a review. Biomed. Mater. Eng., 12(1), p. 15–35.

Denry, I. and Holloway, J.A. (2010). Ceramics for dental applications: a review. Materials, 3(1), p. 351–368.

Derry, I. and Kuhn, L.T. (2016). Design and characterization of calcium phosphate ceramic scaffolds for bone tissue engineering. Dental Mater., 32(1), p. 43–53.

Deville, S., Gremillard, L., Chevalier, J. and Fantozzi, G. (2017). A critical comparison of methods for the determination of the aging sensitivity in biomedical grade yttria-stabilized zirconia. https://arxiv.org/pdf/1710.04449 (accessed October 4, 2018).

Deymier, A.C., Nair, A.K., Depalle, B. and 8 additional authors (2017). Protein-free formation of bone-like apatite: New insights into the key role of carbonation. Biomaterials, 127, p. 75–88.

Diba M., Goudouri O.M., Tapia F. and Boccaccini A.R. (2014). Magnesium-containing bioactive polycrystalline silicate-based ceramics and glass-ceramics for biomedical applications. Curr. Opin. Solid State Mater. Sci., 18, p. 147–167.

Dicker, K.T., Gurski, L.A., Pradhan-Bhatt, S., Witt, R.L., Farach-Carson, M.C. and Jia, X. (2014). Hyaluronan: A simple polysaccharide with diverse biological functions. Acta Biomater., 10, p. 1558–1570.

Dobrzynski, P., Kasperczyk, J. and Li, S. (2017). Synthetic biodegradable medical polyesters: poly (trimethylene carbonate). In: Zhang, X. (ed.). Science and Principles of Biodegradable Medical Polymers. Duxford: Woodhead Publishing. ISBN: 978-0-08-100372-5. p. 107–152.

Doi, Y. (1997). Polyhydroxyalcanoates. In: Domb, A.J., Kost, J. and Wiseman, D.M. (eds). Handbook of Biodegradable Polymers. Amsterdam: Harwood Academic Publishers, ISBN: 90-5702-153-6. p. 79–86.

Domb, A., Jain, J.P. and Kumar, N. (2011). Polyanhydrides. In: Lendlein, A. and Sisson, A. (eds). A Handbook of Biodegradable Polymers. Weinheim: Wiley-VCH. ISBN: 978-3-527-32441-5. p. 45–75.

Dorozhkin, S.V. (2010). Amorphous calcium (ortho) phosphates. Acta Biomater., 6(12), p. 4457–4475.

Dräger, G., Krause, A., Möller, L. and Dumitriu, S. (2011). Carbohydrates. In: Lendlein, A. and Sisson, A. (eds). A Handbook of Biodegradable Polymers. Weinheim: Wiley-VCH. ISBN: 978-3-527-32441-5. p. 155–193.

Drouet, C. (2013). Apatite formation: Why it may not work as planned, and how to conclusively identify apatite compounds. Biomed. Res. Int., 2013: 4909946.

Duarte, A.P., Coelho, J.F., Bordado, J.C., Cidade, M.T. and Gil, M.H. (2012). Surgical adhesives: Systematic review of the main types and development forecast. Progr. Polym. Sci., 37, p. 1031–1050.

Ducheyne, P., Kim, C. S. and Pollack, S. R. (1992). The effect of phase differences on the time-dependent variation of the Zeta potential of hydroxyapatite. J. Biomed. Mater. Res., 26, p. 147–168.

Ducheyne, P., El-Ghannam, A. and Shapiro, I. (1994). Effect of bioactive glass templates on osteoblast proliferation and in vitro synthesis of bonel-like tissue. J. Cell Biochem., 56, p. 162–167.

Ducheyne, P. and Qui, Q. (1999). Bioactive ceramics: The effect of surface reactivity on bone formation and bone cell function. Biomaterials, 20, p. 2287–2303.

Ebrahimi, M., Botelho, M.G. and Dorozhkin, S.V. (2017). Biphasic calcium phosphates bioceramics (HA/TCP): concept, physicochemical properties and the impact of standardization of study protocols in biomaterials research. Mater. Sci. Eng. C, 71, p. 1293–1312.

Edamatsu, H., Kawai, T., Matsui, K., Nakai, M., Takahashi, T., Echigo, S., Niinomi, M. and S. Kamakura, S. (2015). High bone bonding ability and high bone affinity of new low rigidity β- type Ti-29Nb-13Ta-4.6Zr alloy as a dental implant. J. Dental Oral Health. Open Access, 8 pages.

Edén, M. (2011). The split network analysis for exploring composition-structure correlation in multicomponent glasses: I. Rationalizing bioactivity-composition trends of bioglasses. J. Non-Cryst. Sol., 357, p. 1595–1602.

Ehrlich, H. (2015). Biological Materials of Marine Origin. Biologically-Inspired Systems. Vol. 4. Dordrecht: Springer. ISBN 978-9-4007-5729-5.

Eisenbarth, E., Velten, D., Müller, M., Thull, R. and Breme, J. (2006). Nanostructured niobium oxide coatings influence osteoblast adhesion, J. Biomed. Mater. Res. A, 79(1), p. 166–175.

El-Ghannam, A., Ducheyne, P. and Shapiro, I. M. (1995). Bioactive material template for in vitro synthesis of bone. J. Biomed. Mater. Res., 29, p. 359–370.

El-Ghannam, A., Ducheyne, P. and Shapiro, I. M. (1997). Formation of surface reaction products on bioactive glass and their effects on the expression of the osteoblastic phenotype and the deposition of mineralized extracellular matrix. Biomaterials, 18, p. 295–303.

El-Ghannam, A., Ducheyne, P. and Shapiro, I. M. (1999). Effect of serum protein adsorption on osteoblast adhesion to bioactive glass and hydroxyapatite. J. Orthop. Res., 17, p. 340–345.

Elliot, J.C., Mackie, P.E. and Young, R.A. (1973). Monoclinic hydroxylapatite. Science, 180, p. 1055–1057.

Elsayed, H. and 8 additional authors (2018). Bioactive sphene-based ceramic coatings on cpTi substrates for dental implants: An in vitro study. Materials, 11 (11), 2234.

Fambri, L. and Migliaresi, C. (2010). Crystallization and thermal properties. In: Auras, R., Lim, L.T., Selke, S.E., Tsuji, H. (eds). Poly(lactic acid): Synthesis, structures, properties, processing, and applications. Hoboken: Wiley. ISBN: 978-0-470-29366-9, p. 113–124.

Ferreira de Sousa, R.M., Rosa da Silva, T.A., do Couto Almeida, J. and Guerra, W. (2013). Tântalo: Breve histórico, propiedades e aplicações [Tantalum: Brief history, properties and applications]. Educ. Quim., 24 (3), p. 343–346.

Fielding, G.A., Roy, M., Bandyopadhyay, A. and Bose, S. (2012). Antibacterial and biological characteristics of silver-containing and strontium-doped plasma sprayed hydroxyapatite coatings. Acta Biomater., 8(8), p. 3144–3152.

Filho, L., Schmidt, S., Leifer, K., Engqvist, H., Högberg, H. and Persson, C. (2019). Towards functional silicon nitride coatings for joint replacement. Coatings, 9(2),73. https://doi.org/10.3390/coatings9020073.

Forestier, J. (1929). L'aurothérapie dans les rhumatismes chroniques. Bull. Mem. Soc. Med. Hop. Paris, 45, p. 323–327.

Frasnelli, M., Cristofaro, F., Sglavo, V.M. and 6 additional authors (2017). Synthesis and characterization of strontium-substituted hydroxyapatite nanoparticles for bone regeneration Mater. Sci. Eng. C, 71, p. 653–662.

Friederichs, R.J., Bonfield, W. and Best, S.M. (2013). Silicon substituted hydroxyapatite. In: Hench, L.L. (ed.). An Introduction to Bioceramics. 2nd ed., p. 279–286. London: Imperial College Press. ISBN 978-1-908977-15-1.

Gao, J., Gu, H. and Xu, H.B. (2009). Multifunctional magnetic nanoparticles: design, synthesis, and biomedical applications Acc. Chem. Res., 42(8), p. 1097–1107.

Garcia, E., Miranzo, P., and Sainz, M.A. (2018). Thermally sprayed wollastonite and wollastonite-diopside compositions as new modulated bioactive coatings for metal implants, Ceram. Int., 44(11), p. 12896–12904.

Gelli, R., Ridi, F. and Baglioni, P. (2019). The importance of being amorphous: Calcium and magnesium phosphates in the human body. Adv. Coll. Interf. Sci., 269, p. 219–235.

George, N. and Nair, A.B. (2018). Porous tantalum: A new biomaterial in orthopedic surgery. In: Preetha, B., Sreekala, M.S. and Sabu, T. (eds). Fundamental Biomaterials: Metals. Woodhead Publ. Ser. Biomater., p. 243–268. ISBN 978-0-08-102205-4. 439 pp.

Ghani, Y., Coathup, M.J., Hing, K.A. and Blunn, G.W. (2011). Development of a hydroxylapatite coating containing silver for the prevention of peri-prosthetic infection. J. Orthop. Res., 30(3), p. 356–363.

Gheisari Dehsheikh, H. and Karamian, E. (2016). Characterization and synthesis of hardystonite (HT) as a novel nanobioceramic powder. Nanomed. J., 3(2), p. 143–146.

Ghiban, B., Popescu, G., Dumitrescu, D. and Soare, V. (2017). New high entropy alloy for biomedical applications. Key Eng. Mater., 750, p. 180–183.

Ghosh, S., Sanghavi, S. and Sancheti, P. (2018). Metallic biomaterials for bone support and replacement. In: Preetha, B., Sreekala, M.S. and Sabu, T. (eds). Fundamental Biomaterials: Metals. Woodhead Publ. Ser. Biomater., p. 139–165. ISBN 978-0-08-102205-4. 439 pp.

Gibson, I.R., Huang, J., Best, S.M. and Bonfield, W. (1999). Enhanced in vitro cell activity and surface apatite layer formation on novel silicon-substituted hydroxyapatites. In: Ohgushi, H., Hasting, G.W. and Yoshikawa, T. (eds). Proc.12th Intern. Symp. Ceramics in Medicine, Nara, Japan. p. 191–194.

Gilbert, R.D., and Kadla, J.F. (1998). Polysaccharides – Cellulose. In: Kaplan, D.L. (ed.). Biopolymers from Renewable Resources. Berlin: Springer. ISBN: 3-540-63567-X. p. 47–95.

Gotfredsen, K. (2015). Implant coatings and its application in clinical reality. In: Wennerberg, A., Albrektsson, T. and Jimbo, R. (eds). Implant Surfaces and their Biological and Clinical Impact. Berlin, Heidelberg: Springer. p. 147–155. ISBN 978-3-662-45378-0.

Göpferich, A. (1997). Mechanisms of polymer degradation and elimination. In: Domb, A.J., Kost, J. and Wiseman, D.M. (eds). Handbook of Biodegradable Polymers. Amsterdam: Harwood Academic Publishers, ISBN: 90-5702-153-6. p. 451–471.

Goto, T. and Katsui, H. (2015). Chemical vapor deposition of Ca-P-O film coating. In: Sasaki, K., Suzuki, O. and Takahashi, N. (eds). Interface Oral Health Science. Tokyo: Springer. ISBN 978-4-432-55125-6.

Gotzmann, G., Beckmann, J., Wetzel, C., Scholz, B., Herrmann, U. and Neunzehn, J. (2017). Electron-beam modification of DLC coatings for biomedical applications. Surf Coat Technol., 311, p. 248–256.

Graziani, G., Boi, M. and Bianchi. M. (2018). A review on ionic substitution in hydroxyapatite thin films. Coatings, 8(8), p. 269, doi:10.3390/coatings8080269

Gross, U., Müller-Mai, C., Berger, G. and Ploska, U. (2003). The tissue response to a novel calcium zirconium phosphate ceramics. Key Eng. Mater.,Trans Tech Publ.: Zürich, 240-242, p. 629–632.

Gross, U., Müller-Mai, C., Berger, G. and Ploska, U. (2004). Do calcium zirconium phosphate ceramics inhibit mineralization? Key Eng. Mater.,Trans Tech Publ.: Zürich, 254-256, p. 635–638.

Guarino, V., Marica Marrese, M. and Ambrosio, L. (2015). Chemical and physical properties of polymers for biomedical use. In: Puoci, F. (ed.). Advanced Polymers in Medicine. Cham: Springer. ISBN: 978-3-319-12477-3. p. 67–90

Guedes e Silva, C.C., König Jr, B., Carbonari. M.J., Yoshimoto, M., Allegrini Jr, S. and Bressiani, J.C. (2008). Bone growth around silicon nitride implants – an evaluation by scanning electron microscopy. Mater. Character., 59(9), p. 1339–1341.

Guedes e Silva, C.C. (2012). Silicon nitride as biomaterial. In: Hierra, E.J. and Salazar, J.A. (eds). Silicon Nitride: Synthesis, Properties and Applications. p. 149–156. New York: Nova Science Publ. Inc. ISBN 1-619-428-652.

Guelcher S.A. (2008). Biodegradable polyurethanes: Synthesis and applications in regenerative medicine. Tissue Eng. Pt. B., 14, p. 3–17.

Gunawarman, B., Niinomi, M., Akahori, T., Takeda, J. and Tida, H. (2005). Mechanical properties of Ti-4.5Al-3V-2Mo-2Fe and possibility for healthcare applications. Mater. Sci. Eng. C. 25 (3), p. 296–303.

Guo, S., Ng, C., Lu, J. and Liu, C.T. (2011). Effect of valence electron concentration on stability of fcc or bcc phase in high-entropy alloys. J. Appl. Phys., 109(10), p. 103505.

Guo, Z.J., Pang, X.L., Yan, Y., Gao, K.W., Volinsky, A.A. and Zhang, T.Y. (2015). CoCrMo alloy for orthopedic implant application enhanced corrosion and tribocorrosion properties by nitrogen ion implantation. Appl. Surf. Sci., 347, p. 23–34.

Gupta, K.P. (2005). The Co-Cr-Mo (cobalt-chromium-molybdenum) system. J. Phase Equ. Diff., 26(1), p. 87–92.

Ha, S.-W., Wintermantel, E. and Maier, G. (2009). Biokompatible Polymere. In: Wintermantel, E., Ha, S.-W. (eds). Medizintechnik. Life Science. Springer-Verlag: Berlin. ISBN: 978-3-540-939356-8. p. 219–276.

Habraken, W., Habibovic, P., Epple, M. and Bohner, M. (2016). Calcium phosphates in biomedical application: Materials for the future? Materials Today, 19(2), p. 69–87.

Hagiwara, M. (2009). Basic Characters. In: Niinomi, M. (ed.) Basic Materials Science, Manufacturing and Newly Advanced Technologies of Titanium and its Alloys. Tokyo: CMC. ISBN978-4-7813-0107-5.

Hammer, B. and Norskov, J.K. (1995). Why gold is the noblest of all the noble metals. Nature, 376, p. 238–240.

Hanawa, T. (2019). Overview of metals and applications. In: Niinomi, M. (ed.) Metals for Biomedical Devices, 2nd ed. p. 3–25. Duxford: Elsevier.

Harada, K., Nagashima, K., Nakao, K. and Kato, A. (1971). Hydroxyellestadite, a new apatite from Chichibu Mine, Saitama Prefecture, Japan. Amer. Mineral, 56(9), p. 1507–1518.

Hatanaka, I. S., Ueda, M., Ikeda, M. and Niinomi, M. (2010). Isothermal aging behavior in Ti-10Cr-Al alloys for medical applications. Adv. Mater. Res. 89-91, p. 232–2337

Hartmann, M.H. (1998). High molecular weight polylactic acid polymer. In: Kaplan, D.L. (ed.). Biopolymers from Renewable Resources. Berlin: Springer. ISBN: 3-540-63567-X. p. 367–411.

Hauert. R. and 7 additional authors (2012). Retrospective lifetime estimation of failed and explanted diamond-like carbon coated hip joint balls. Acta Biomater., 8 (8), p. 3170–3176.

Hauert, R., Thorwarth, K. and Thorwarth, G. (2013). An overview on diamond-like carbon coatings in medical applications. Surf. Coat. Technol., 233, p. 119–130.

He, F.P., Zhang, J., Yang, F.W., Zhu, J.X., Tian, X.M. and Chen, X.M. (2015). In vitro degradation and cell response of calcium carbonate composite ceramic in comparison with other synthetic bone substitute materials. Mater. Sci. Eng. C, 50, p. 257–265.

He, L., Li, H.Y., Chen, X.Y. and 9 additional authors (2019). Selenium-substituted hydroxyapatite particles with regulated microstructure for osteogenic differentiation and anti-tumor effects. Ceram. Intern., 45(11), p. 13787–13798.

Hee, A.C., Zhao, Y., Jamali, S.S., Bendavid, A., Martin, P.J. and Guo, H.B. (2019). Characterization of tantalum and tantalum nitride films on Ti6Al4V substrate prepared by filtered cathodic vacuum arc deposition for biomedical applications. Surf. Coat. Technol., 365, p. 24–32.

Heimann, R. B. (1975). Auflösung von Kristallen [Dissolution of Crystals]. Appl. Mineral., vol. 8, p. 70–78. Wien, New York: Springer. ISBN 3-211-81278-4. 270 pp.

Heimann, R.B. (2008). Plasma Spray Coating. Principles and Applications. 2nd ed., Weinheim: Wiley-VCH. Chapter 10. ISBN 978-3-527-32050-9. 449 pp.

Heimann, R.B. (2010). Classic and Advanced Ceramics. From Fundamentals to Applications. Weinheim: Wiley-VCH. ISBN 978-3-527-32517-7. 553 pp.

Heimann, R.B. (2012). Transition metal-substituted calcium orthophosphates with NaSiCON structure: A novel type of bioceramics. In: Heimann, R.B. (ed.). Calcium Phosphate: Structure, Synthesis, Properties, and Applications. New York: Nova Science Publ. p. 363–379. ISBN 978-1-62257-299-1. 498 pp.

Heimann, R.B. (2015). Tracking the thermal decomposition of plasma-sprayed hydroxylapatite. Amer. Mineral., 100, p. 2419–2425.

Heimann, R.B. (2017). Calcium (Ti, Zr) hexaorthophosphate bioceramics for electrically stimulated biomedical implant devices: A position paper. Amer. Mineral., 102, p. 2170–2179.

Heimann, R.B. (2018a). Plasma-sprayed hydroxylapatite coatings as biocompatible intermediaries between inorganic implant surfaces and living tissue. J. Thermal Spray Technol., 27(8), p. 1212–1237.

Heimann, R.B. (2018b). Weathering of ancient and medieval glasses-potential proxy for nuclear fuel waste glasses. A perennial challenge revisited. Int. J. Appl. Glass Sci., 9(1), p. 29–41.

Heimann, R.B. (2019). Plasma-sprayed bioactive ceramic coatings with high resorption resistance based on transition metal-substituted calcium hexaorthophosphates. Materials, 12, 2059; doi:10.3390/ma12132059. 14 pp.

Heimann, R.B. and Lehmann, H.D. (2015). Bioceramic Coatings for Medical Implants. Trends and Techniques. Weinheim: Wiley-VCH. ISBN 978-3-527-33743-9. 467 pp.

Heimann, R.B. and Lehmann, H.D. (2017). Recent research and patents on controlling corrosion of bioresorbable Mg alloy implants: towards next-generation biomaterials. Recent Patents on Mater. Sci., 10, p. 2–19.

Heimann, R.B. and Wirth, R. (2006). Formation and transformation of amorphous calcium phosphates on titanium alloy surfaces during atmospheric plasma spraying and their subsequent in vitro performance. Biomaterials, 27, p. 823–831.

Heimann, R.B., Schürmann, N. and Müller, R.T. (2004). In vitro and in vivo performance of Ti6Al4V implants with plasma-sprayed osteoconductive hydroxyapatite-bioinert titania bond coat 'duplex' systems: An experimental study in sheep. J. Mater. Sci. Mater. Med., 15, p. 1945–1952.

Heinze, T., Liebert, T., Heublein, B. and Hornig, S. (2006). Functional polymers based on dextran. Adv. Polym. Sci., 205, p. 199–291.

Helgen, J.A. and Breme, H.J. (eds) (1998). Metals as Biomaterials. Biomaterials Science and Engineering Series. John Wiley and Sons. ISBN 978-0-471-96935-8. 522 pp.

Heller, J., Barr, J., Ng, S.Y., Abdellauoi, K.S., and Gurny, R. (2002). Poly(orthoesters): synthesis, characterization, properties and uses. Adv. Drug Del. Rev., 54, p. 1015–1039.

Hench, L.L. (1991). Bioceramics: From concept to clinic. J. Am. Ceram. Soc., 74(7), p. 1487–1510.

Hench, L.L. (1998). Bioceramics. J. Am. Ceram. Soc., 81(3), p. 1705–1728.

Hench, L.L. (2008). Genetic design of bioactive glass. J. Eur. Ceram. Soc., 29, p. 1257–1265.

Hench, L.L. (2013a). Chronology of bioactive glass development and clinical applications. New J. Glass Ceram., 3(2), Article ID: 30885,7 pages.

Hench, L.L. (2013b) (ed.). An Introduction to Bioceramics, Chapters 6-12 and 31. 2nd ed., London: Imperial College Press. ISBN 978-1-908977-15.

Hench, L.L., Splinter, R., Greenlee, T. and Allen, W. (1971). Bonding mechanisms at the interface of ceramic prosthetic materials. J. Biomed. Eng., 2, p. 117–141.

Hench, L.L. and Polak, J.M. (2002). Thrid-generation biomedical materials. Science, 295, p. 1014–1017.

Herring, C. (1951). Some theorems on the free energy of crystal surfaces. Phys. Rev., 82, p. 87–93.

Hesse, K.F., Küppers, H. and Suess, E. (1983). Refinement of the structure of ikaite, $CaCO_3 \cdot 6H_2O$. Z. Krist., 163, p. 227–231.

Hill, R. (1996). An alternative view of the degradation of bioglass. J. Mater. Sci. Lett., 15, p. 1122–1125.

Hill, R.G. and Brauer, D.S. (2011). Predicting the bioactivity of glasses using the network connectivity or split network models. J. Non-Cryst. Solids, 357(24), p. 3884–3887.

Hillert, M. (2007). Phase Equilibria, Phase Diagrams and Phase Transformations: Their Thermodynamic Basis. Cambridge: Cambridge University Press. ISBN 978-0-521-85351-4. 526 pp.

Hirano, T., Murakami, T., Taira, M., Narushima, T. and Ouchi, C. (2007). Alloy design and properties of new (α+β)- titanium alloy with excellent cold workability, superplasticity and cytocompatibility. ISIJ International. 47(5). p. 745–752.

Höhling, H.J., Kreilos, R., Neubauer, G. and Boyde, A. (1971). Electron microscopy and electron microscopical measurements of collagen mineralization in hard tissue. Cell Tissue Res., 122(1), p. 36–52.

Holliday, L. (1977). On the connectivity of modified glass networks. J. Polymer Sci., 15, p. 675–682.

Hoppe A., Güldal N.S. and Boccaccini A.R. (2011). A review of the biological response to ionic dissolution products from bioactive glasses and glass-ceramics. Biomaterials, 32, p. 2757–2774.

Hornberger, H., Virtanen, S. and Boccaccini, A. (2012). Biomedical coatings on magnesium alloys – a review. Acta Biomater., 8, p. 2442–2455.

Hornbogen, E., Eggeler, G. und Werner, E. (2019). Werkstoffkunde. Aufbau und Eigenschaften. 12th ed., Berlin, Heidelberg, New York: Springer. ISBN 978-3-662-58846-8.

Hospital, A., Goñi, J.R., Orozco, M. and Gelpi, J.L. (2015). Molecular dynamic simulations: Advances and applications. Adv. Appl. Bioinform. Chem., 8, p. 37–47.

Hotokebuchi, T. and Noda, I. (2013). Antimicrobial product useful for biological implants, e.g. artificial dental roots, comprising a thermally sprayed brookite-type titanium oxide film formed by high-speed flame spraying on metal, ceramic or plastic substrates. Japanese patent JP5308754-B2,

Huang, L., Ji, H., Liang, Y., Xie, Y. and Zheng, X. (2011). Bone replacing material of baghdadite coated-titanium alloy, useful for inducing formation of bone-like apatite in simulated body fluid, comprises titanium and its alloy as matrix and coating is deposited on matrix by plasma spray coating. Chinese patent CN102049065-A.

Huang, L., Zhou, B., Wu, H.Y, Zheng, L. and Zhao, J.M. (2017). Effect of apatite formation of biphasic calcium phosphate ceramic (BCP) on osteoblastogenesis using simulated body fluid (SBF) with and without bovine serum albumin (BSA). Mater. Sci. Eng. C, 70, p. 955–961.

Huang, L.F., Grabowski, B., Zhang, J., Lai, M.J., Tasan, C.C., Sandlöbes, S., Raabe, D. and Neugebauer, J. (2016). From electronic structure to phase diagrams: A bottom-up approach to understand the stability of titanium-transition metal alloys. Acta Mater., 113, p. 311–319.

Huang, X.Y., Ackland, G.J. and Rabe, K.M. (2003). Crystal structure and shape-memory behavior of NiTi. Nature Mater., 2, p. 307–311.

Hulbert, S.F. (2013). The use of alumina and zirconia in surgical implants. In: Hench, L.L. (ed.). An Introduction to Bioceramics, p. 27–48. London: Imperial College Press. ISBN 978-1-908977-15-1.

Hulbert, S.F., Morrison, S.J. and Klawitter, J.J. (1972). Tissue reaction to three ceramics of porous and non-porous structures. J. Biomed. Mater. Res., 6(5), p. 347–374.

Hulmes, D.J.S. (2008). Collagen diversity, synthesis and assembly. In: Fratzl, P. (ed.). Collagen: Structure and Mechanics. New York: Springer. ISBN: 978-0-387-73905-2. p. 15–47.

Husain, M.S.B., Gupta, A., Alashwal, B.Y. and Sharma, S. (2018). Synthesis of PVA/PVP based hydrogel for biomedical applications: a review. Energy Sources, Pt. A, 40, p. 2388–2393.

Ichinose, N. (ed.) (1987). Introduction to Fine Ceramics. Applications in Engineering. Chichester, New York, Brisbane, Toronto, Singapore: Wiley. ISBN 0 471 91445 2. 160 pp.

ICRP (1991). 1990 recommendations of the International Commission on Radiological Protection. ICRP Publication 60. Ann. ICRP, 21, p. 1–3. Pergamon Press.

Ikeda, M. and Sugano, M. (2005). The effect of aluminum content on phase constitution and heat treatment behavior of Ti-Cr-Al alloys for healthcare applications. Mater. Sci. Eng. C. 25. p. 377–381.

Ikeda, M., Ueda, M., Matsunaga, R., Ogawa, M. and Niinomi, M. (2009). Isothermal aging behavior of beta titanium-manganese alloys, Mater. Trans. 50, p. 2737–2743.

Ikeda, M., Ueda, M., Matsunaga, R. and Niinomi, M. (2010). Phase constitution and heat treatment behavior of Ti-7 mass%Mn-Al alloys. Mater. Sci. Forum. 654-656, p. 855–858.

Ikeda, M., Ueda, M., Kinoshita, T., Ogawa, M. and Niinomi, M. (2012). Influence of Fe content of Ti-Mn-Fe alloys on phase constitution and heat treatment behavior. Mater. Sci. Forum. 706-709, p. 1893–1898.

Ishida, K. and Nishikawa, T. (1990). The Co-Cr (cobalt-chromium) system. Bull. Alloy Phase Diagrams, 11(4), p. 356–370.

ISO 5892-1:2007. Implants for Surgery-Metallic Materials-Part 1: Wrought stainless steel.

ISO 5892-9:2007. Implants for Surgery-Metallic Materials-Part 9: Wrought high nitrogen stainless steel.

ISO 6474-1:2010. Implants for Surgery-Ceramic Materials-Part 1: Ceramic materials based on high purity alumina.

ISO 6474-2:2012. Implants for Surgery-Ceramic Materials-Part 2: Ceramic materials based on a high purity alumina matrix with zirconia reinforcement.

ISO 7153-1:2016. Surgical Instruments-Metallic Materials-Part 1: Stainless steel.

ISO 13356:2016-02. Implants for Surgery – Ceramic materials based on yttria-stabilized tetragonal zirconia (Y-TZP).

Jackson, M.J. and Ahmed, W. (eds) (2007). Titanium dioxide coatings in medical device applications. In: Surface Engineered Surgical Tools and Medical Devices. Boston: Springer. p. 49–63. ISBN 978-0-387-27026-5.

Jäger, C., Maltsev, S. and Karrasch, A. (2006a). Progress of structural investigation of amorphous calcium phosphate (ACP) and hydroxyapatite: Disorder and surfaces seen by solid state NMR. Key Eng. Mater., 309/311, p. 69–72.

Jäger, C., Welzel, T., Meyer-Zaika, W. and Epple, M. (2006b). A solid-state NMR investigation of the structure of nanocrystalline hydroxyapatite. Magn. Res. Chem., 44, p. 573–580.

Jahanmir, S., Ozmen, Y. and Ives, L.K. (2004). Water lubrication in silicon nitride in sliding. Tribol. Lett., 17(3), p. 409–417.

Jain, J.P., Modi, S., Domb, A.J., Kumar, N. (2005). Role of polyanhydrides as localized drug carriers. J. Controlled Rel., 103, p. 541–563.

Jain, J.P., Ayen, W.Y., Domb, A.J., and Kumar, N. (2011). Biodegradable polymers in drug delivery. In: Domb, A.J., and Kumar, N. and Ezra, A. (eds). Biodegradable Polymers in Clinical Use and Clinical Development. Hoboken: John Wiley& Sons. ISBN: 978-0-470-42475-9. p. 3–58.

Jain, S., Hirst, D.G. and O'Sullivan, J.M. (2012). Gold nanoparticles as novel agents for cancer therapy. Br. J. Radiol., 85(1010), p. 101–113.

Jakobsen, S.S. (2008). CoCrMo Alloy. In vitro and in vivo Studies. Ph.D. Thesis, Dept. of Endocrinology and Metabolism, University Hospital Aarhus, Aarhus, Denmark. 51 pp.

Jasim, K.M., Rawlings, R.D. and West, D.R.F. (1992). Stability of ZrO_2-Y_2O_3 t'-phase formed during laser sealing. Mater. Sci. Technol., 8(1), p. 83–91.

Jelinek, R. (2013). Biomimetics. A Molecular Perspective. Berlin, Boston: De Gruyter. 252 pp. ISBN 978-3-11-028117-0.

Jinawath, S., Hengst, M. and Heimann, R.B. (2004). Plasma-sprayed $DCPD$/$CaCO_3$ coatings on Ti6Al4V substrates. J. Sci. Res. Chulalongkorn Univ., 29(1), p. 33–43.

Johnstone, T.C., Park, G.Y. and Lippard, S.J. (2014). Understanding and improving platinum anticancer drugs-phenantriplatin. Anticancer Res., 34(1),p.471–476.

Jones, J.R., Gentleman, E. and Polak, J. (2007). Bioactive glass scaffolds for bone regeneration. Elements, 3, p. 393–399.

Jones, J.R. (2013). Review of bioactive glass: From Hench to hybrid. Acta Biomater., 9(1), p. 4457–4486.

Kalantari, E. and Naghib, S.M. (2019). A comparative study on biological properties of novel nanostructured monticellite-based composites with hydroxyapatite bioceramic. Mater. Sci. Eng. C, 98, p. 1087–1096.

Kamrani, S. and Fleck, C. (2019). Biodegradable magnesium alloys as temporary orthopaedic implants: A review. Biometals, 32(2), p. 185–191.

Kareva, M., Wang, Y., Kriegel, M.J., Peng, J. and Kuznetsov, V. (2017). Co-Cr-Mo ternary phase diagram evaluation.https://www.researchgate.net/publication/243187726_The_Co-Cr-Mo_Cobalt-Chromium-Molybdenum_System.

Kargozar, S., Montazerian, M., Hamzehlou, S., Kim, H.W. and Baino, F. (2018). Mesoporous bioactive glasses: Promising platforms for antibacterial strategies. Acta Biomater., 81, p. 1–19.

Kargozar, S., Mozafari, M., Hamzehlou, S., Kim, H.W. and Baino, F. (2019). Mesoporous bioactive glasses (MBGs) in cancer therapy: Full of hope and promise. Mater. Letters, https://doi.org/10.1016/ j.matlet.2019.05.019.

Kasano, Y., Inamura, T., Kanetaka, H., Miyazaki, S. and Hosoda, H. (2010). Phase constitution and mechanical properties of Ti-(Cr, Mn)-Sn biomedical alloys. Mater. Sci. Forum. 654-656, p. 2118–2121.

Kasemo, B. and Lausmaa, J. (1991). The biomaterials-tissue interface and its analogues in surface science and technology. In: Davies J.E. (ed.) The Bone-Biomaterials Interface. Toronto: University of Toronto Press. p. 19–32. ISBN 978-1-4426-7150-8.

Kasper, C., Bloh, J.Z., Wagner, S., Bahnemann, D.W. und Scheper, T. (2010). Untersuchungen zur Zytotoxizität von photokatalytisch aktiven Titandioxid-Nanopartikeln [Investigation of cytotoxicity of photocatalytically active titanium dioxide nanoparticles]. Chem. Ing. Technik, 82(3), p. 335–341.

Kaufmann, E. A., Ducheyne, P., Radin, S., Bonnell, D. A. and Composto, R. (2000). Initial events at the bioactive glass surface in contact with protein containing solutions. J. Biomed. Mater. Res., 52, p. 825–830.

Kawahara, H., Ochi, S., Tanetani, K., Kato, K., Isogai, M., Mizuno, Y., Yamamoto, H. and Yamaguchi, A. (1963). Biological test of dental materials. Effect of pure metals upon the mouse subcutaneous fibroblast. Starin L cell in tissue culture. J. Jpn. Soc. Dent. Apparat. and Mater., 4, p. 65–75.

Kaya, S., Cresswell, M. and Boccaccini, A. (2018). Mesoporous silica-based bioactive glasses for antibiotic-free antibacterial applications. Mater. Sci. Eng. C, 83, p. 99–107.

Kazemi, A., Abdellahi, M., Khajeh-Sharafabadi, A., Khandan, A. and Ozada, N. (2017). Study of in vitro bioactivity and mechanical properties of diopside nano-bioceramic synthesized by a facile method using eggshell as raw material. Mater. Sci. Eng. C, 71, p. 604–610.

Kelland, L.R. and Farrell, N.P. (eds) (2000). Platinum-Based Drugs in Cancer Therapy (Cancer Drug Discovery and Development). New York: Humana Press. ISBN 978-0-8960-3599-7.

Kemnitzer, J., and Kohn, J. (1997). Degradable polymers derived from the amino acid L-tyrosine. In: Domb, A.J., Kost, J. and Wiseman, D.M. (eds). Handbook of Biodegradable Polymers. Amsterdam: Harwood Academic Publishers. ISBN: 90-5702-153-6. p. 251–272.

Keshk, S.M.A.S. and El-Kott, A. F. (2017). Natural bacterial biodegradable medical polymers: bacterial cellulose. In: Zhang, X. (ed.). Science and Principles of Biodegradable Medical Polymers. Duxford: Woodhead Publishing. ISBN: 978-0-08-100372-5. p. 295–319.

Kim, H.M., Miyazaki, T., Kokubo, T. and Nakamura, T. (2001). Revised simulated body fluid. Bioceramics, 13, p. 47–50.

Kim, H., Camata, R.P., Chowdhury, S. and Vohra, Y.K. (2010). In vitro dissolution and mechanical behavior of c-axis preferentially oriented hydroxyapatite thin films fabricated by pulsed laser deposition, Acta Biomater., 6(8), p. 3234–3241,

Kingery, W.D., Bowen, H.K. and Uhlmann, D.R. (1976). Introduction to Ceramics. 2nd ed. Wiley Series on the Science and Technology of Materials. Chichester, Brisbane, Toronto, Singapore: Wiley. 1032 pp. ISBN 978-0-471-47860-1.

Klemm, D., Schumann, D., Kramer, F., Hessler, N., Hornung, M., Schmauder, H.-P. and Marsch, S. (2006). Nanocellulose as innovative polymer in research and application. Adv. Polym. Sci., 205, p. 49–96.

Klueh, U, Wagner, V., Kelly, S., Johnson, A. and Bryers, J.D. (2000). Effect of silver-coated fabric to prevent bacterial colonization and subsequent device-based biofilm formation. J. Biomed. Mater. Res. B: Appl. Biomater., 53, p. 621–631.

Knabe, C., Ostapowicz, W., Radlanski, R.J., Gildenhaar, R., Berger, G., Fitzner, R. and Gross, U. (1998). In vitro investigation of novel calcium phosphates using osteogenic cultures. J. Mater. Sci. Mater. Med., 9, p. 337–345.

Knabe, C., Berger, G., Gildenhaar, R., Klar, F. and Zreiqat, H. (2004). The modulation of osteogenesis in vitro by calcium titanium phosphate. Biomaterials, 25(20), p. 4911–4919.

Knabe, C., Adel-Khattab, D. and Ducheyne, P. (2017). Bioactivity: Mechanisms. In: Ducheyne, P., Grainger, D.W., Healy, K.E., Hutmacher, D.W. and Kirkpatrick, C.J. (eds). Comprehensive Biomaterials II. p. 291–310. Oxford: Elsevier. ISBN 978-0-08-055294-1.

Knabe, C., Stiller, M., Adel-Khattab, D. and Ducheyne, P. (2017a). Dental graft materials. In: Ducheyne, P., Healy, K., Hutmacher, D., Grainger, D. and Kirkpatrick, J.P. (eds). Comprehensive Biomaterials II. Oxford: Elsevier, 7, p. 378–405.

Koga, Y., Fujieda, H., Meguro, H., Ueno, Y., Aoki, T., Miwa, K. and Kainoh, M. (2018). Biocompatibility of polysulfone hemodialysis membranes and its mechanisms: Involvement of fibrinogen and its integrin receptors in activation of platelets and neutrophils. Artific. Organs, 42, p. E246-E258.

Kokubo, T. (1993). Bioactivity of glasses and glass ceramics. In: Ducheyne, P., Kokubo, T. and van Blitterswijk, C. A., (eds). Bone-Bioactive Biomaterials. Leidersdorp: Reed Healthcare Communications. p. 31–46.

Kokubo, T. (2008). Bioceramics and Their Clinical Application. Abingdon: Woodhead Publ. Ltd. ISBN 978-184-569-204-9.

Kokubo, T., Kushitani, H., Ohtsuki, C., Sakka, S. and Yamamuro, T. (1992). Chemical reaction of bioactive glass and glass-ceramics with a simulated body-fluid. J. Mater. Sci. Mater.Med. Sci., 3, p. 79–83.

Kolmas, J., Groszyk, E. and Kwiatkowska-Rázycka, D. (2014). Substituted hydroxyapatites with antibacterial properties. Biomed. Res. Int., 2014: 178123.

Koronfel, M.A., Goode, A.E., Weker, J.N. and 11 additional authors (2018). Understanding the reactivity of CoCrMo-implant wear particles. npj Materials Degradation, 2 (8), 5 pages. doi:10.1038/s41529-018-0029-2.

Kudoh, Y., Tokonami, M., Miyazaki, S. and Otsuka, K. (1985). Crystal structure of the martensite in Ti-49.2 at% Ni alloy analyzed by the single crystal X-ray diffraction method. Acta Metall., 33 (11), p. 2049–2056.

Kühn, K.D. (2014). PMMA cements. Springer: Heidelberg. ISBN: 978-3-642-41535-7. 291 pp.

Kuntz, M. (2015). Einfluss einer Chromoxid-Dotierung auf die Härte von BIOLOX® delta. Whitepaper. CeramTec GmbH.

Kurita, K. (2001). Controlled functionalization of the polysaccharide chitin. Progr. Polym. Sci., 26, p. 1921–1971.

Kuroda, D., Niinomi, M., Morinaga, M., Kato, Y. and T. Yashiro, (1998). Design and mechanical properties of new beta type titanium alloys for implant materials. Mater. Sci. Eng. A. A243, p. 244–249.

Kurtz, S.M., and Devine J.N. (2007). PEEK Biomaterials in trauma, orthopedic, and spinal implants. Biomaterials, 28, p. 4845–4869.

Kutikov, A.B. and Song, J. (2015). Biodegradable PEG-based amphiphilic block copolymers for tissue engineering Applications. ACS Biomater. Sci. Eng., 1, p. 463–480.

Lamari, F.N., and Karamanos, N.K. (2006). Structure of chondroitin sulfate. In: Volpi, N. (ed.). Chondroitin Sulfate: Structure, Role and Pharmacological Activity. San Diego: Academic Press. ISBN: 13-978-012-032955-7. p. 33–48.

Laskus, A. and Kolmas, J. (2017). Ionic substitution in non-apatitic calcium phosphates. Int. J. Mol. Sci., 18(12), Article ID 2542.

Layrolle, P. (2017). Calcium phosphate coatings. In: Reference Module in Materials Science and Materials Engineering. Comprehensive Biomaterials II. Vol. 1, p. 360–167. Elsevier. ISBN 978-0-08-100692-4.

Lazarinis S., Maekelae, K.T., Eskelinen, A., Havelin, L., Hallan, G., Overgaard, S., Pedersen, A.B., Kaerrholm, J. and Hailer, N.P. (2017). Does hydroxyapatite coating of uncemented cups improve long-term survival? An analysis of 28605 primary total hip arthroplasty procedures from the Nordic Arthroplasty Register Association (NARA). Osteoarthritis Cartilage, 25, p. 1980–1987.

Lee, H.B., Khang, G. and Lee, J.H. (2003). Polymeric biomaterials. In: Park, J.B., and Bronzini, J.D. (eds). Biomaterials: Principles and Applications. Boca Raton: CRC Press. ISBN: 0-8493-1491-7. p. 55–77.

Lee, J.W., Han, H.S., Han, K.J. and 17 additional authors (2016). Long-term clinical study and multiscale analysis of in vivo biodegradation mechanism of Mg alloy. PNAS, 113 (6), p. 716–721.

Lee, S., Nomura, N. and Chiba, A. (2008). Significant improvement in mechanical properties of biomedical Co-Cr-Mo alloys with combination of N addition and Cr enrichment. Mater. Trans., 2, p. 260–264.

LeGeros, R.Z. and LeGeros, J.P. (1984). Phosphate minerals in human tissues. In: Nriagu, J.O. and Moore, P.B (eds). Phosphate Minerals. Springer: Berlin, Heidelberg, pp. 351–385. ISBN 978-3-642-61738-6.

LeGeros, R. Z., Daculsi, G., Orly, I. et al. (1993). Formation of carbonate apatite on calcium phosphate materials: Dissolution/precipitation processes. In: Ducheyne, P., Kokubo, T. and van Blitterswijk, C. A. (eds). Bone-Bonding Biomaterials. Leidersdorp: Reed Healthcare Communications, p. 201–212.

LeGeros, R.Z., Lin, S., Rohanizadeh, R., Mijares, D. and LeGeros, J.P. (2003). Biphasic calcium phosphate bioceramics: preparation, properties and applications. J. Mater. Sci: Mater. Med., 14(3), p. 201–209.

Lei, Z.F., Liu, X.J., Wang, H., Wu, Y., Jiang, S.H. and Lu, Z.P. (2019). Development of advanced materials via entropy engineering. Scripta Mater., 165, p. 164–169.

Leng, Y.X., Wang, F., Yang, P., Chen, J.Y. and Huang, N. (2019). The adhesion and clinical application of titanium oxide film on a 316L vascular stent, Surf. Coat. Technol., 363, p. 430–435.

Li, G.D., Zhang, K.L., Pei, Z.J. and 6 additional authors (2019a). A novel method to enhance magnetic property of bioactive glass-ceramics for hypothermia. Ceram. Intern., 45 (4), p. 4945–4956.

Li, J., Li, Z.L., Tu, J.P., Jin, G., Li, L.L., Wang, K.T. and Wang, H.R. (2019b). In vitro and in vivo investigations of a-C/a-C:Ti nanomultilayer coated Ti6Al4V alloy as artificial femoral head. Mater. Sci. Eng. C, 99, p. 816–826.

Li, P.H., Zheng, W.J., Ma, W.Y., Li, X., Li, S.Q., Zhao, Y.C., Wang, J. and Huang, N. (2018). In-situ preparation of amino-terminated dendrimers on TiO_2 films by generational growth for potential and efficient surface functionalization. Appl. Surf. Sci., 459, p. 438–445.

Liao, Y., Pourzal, R., Stemmer, P., Wimmer, M.A, Jacobs, J.J., Fischer, A. and Marks, L.D. (2012). New insights into hard phases of CoCrMo metal-on-metal hip replacements. J. Mech. Behav. Biomed. Mater., 12, p. 39–49.

Liao, Y.F., Hoffman, E., Wimmer, M., Fischer, A., Jacobs, J. and Marks, L. (2013). CoCrMo metal-on-metal hip replacements. Phys. Chem. Chem. Phys., 15(3), p. 746–756.

Liu, C., Huang, Y., Shen, W. and Cui, J. (2001). Kinetics of hydroxyapatite precipitation at pH 10 and 11. Biomaterials, 22, p. 301–306.

Liu, X., Huang, A., Ding, C. and Chu, P.K. (2006). Bioactivity and cytocompatibility of zirconia (ZrO_2) films fabricated by cathodic arc deposition. Biomaterials, 27(21), p. 3904–3911.

Liu, Y. and van Humbeeck, J. (1997). On the damping behaviour of NiTi shape memory alloy. J. Physique IV Colloque, 7(5), p. 519–524.

Lobo, S.E. and Arinzeh, T.L. (2010). Biphasic calcium phosphate ceramics for bone regeneration and tissue engineering applications. Materials, 3, p. 815–826.

Love, B. (2017). Metallic biomaterials. In: Biomaterials. A Systems Approach to Engineering Concepts. Chapter 7. New York: Academic Press. ISBN 978-0-1280-9478-5. 410 pp.

Love, B. (2017). Ceramic biomaterials. In: Biomaterials. A Systems Approach to Engineering Concepts. Chapter 8. p. 185–204. New York: Academic Press. ISBN 978-0-1280-9478-5. 410 pp.

Lucas-Girot, A., Langlois, P., Sangleboeuf, J.C., Ouammou, A., Rouxel, T. and Gaude, J. (2002). A synthetic aragonite-based bioceramic: influence of process parameters on porosity and compressive strength. Biomaterials, 23(2), p. 503–510.

Lütjering, G. and Williams, J. C. (2003). Titanium. Berlin, Heidelberg, New York: Springer. ISBN 3-540-42990-5.

Luginina, M., Orrú, R., Cao, G., Grossin, D., Brouillet, F., Chevallier, G., Thouron, C. and Drouet, C. (2019). First successful stabilization of consolidated amorphous calcium phosphate (ACP) by cold sintering: Toward highly-resorbable reactive bioceramics. J. Mater. Chem. B, doi: 10.1039/C9TB02121C.

Lukyanova, E., Anisimova, N., Martynenko, N., Kiselevsky, M., Dobatkin, S. and Estrin, Y. (2018). Features of in vitro and in vivo behaviour of magnesium alloy WE43. Mater. Letters, 215, p. 308–311.

Mackert, J.R., Butts, M.B. and Fairhurst, C.W. (1986). The effect of the leucite transformation on dental porcelain expansion. Dental Mater., 2(1), p. 32–36.

Malament, K.A. (2015). Study report on IPS e.max CAD/press. Tufts University School of Dental Medicine, Boston, MA. In: IPS e.max® Scientific Report, 3, p. 11.

Mao, L., Yuan, G. Wang, S. (2012). A novel biodegradable Mg–Nd–Zn–Zr alloy with uniform corrosion behavior in artificial plasma. Mater. Letters, 88, p. 1–4.

Marukawa, E., Masato, T., Takahashi, Y., Hatakeyama, I. and Sata, M. (2016). Comparison of magnesium alloys and poly-l-lactide screws as degradable implants in canine fracture model. J. Biomed. Mater. Res., B: Applied Biomater., 104B, p. 1282–1289.

Matejicek, J. and Sampath, S. (2003). In situ measurement of residual stresses and elastic moduli in thermal sprayed coatings. Part 1: Apparatus and analysis. Acta Mater., 51(3), p. 863–872.

Matos, M., Terra, J. and Ellis, D.E. (2010). Mechanism of Zn stabilization in hydroxyapatite and hydrated (001) surfaces of hydroxyapatite. J. Phys.: Cond. Matter, 22(14), 145502.

Matsuda, T. and Davies, J.E. (1987). The in vitro response of osteoblasts to bioactive glass. Biomaterials, 8, p. 275–284.

Mazumder, J., Abu, M., Sheardown, H. and Al-Ahmed, A. (2019). Functional Biopolymers. In: Kar, K.K. (ed). Polymers and Polymeric Composites: A Reference Series. Springer Intern. Publ. ISSN 2510-3458.

Mello, D.C.R., de Oliveira, J.R., Cairo, C.A.A. and 6 additional authors (2019). Titanium alloys: In vitro biological analyzes on biofilm formation, biocompatibility, cell differentiation to induce bone formation, and immunological response. J. Mater. Sci. Mater. Med., 30:108.

Meyer, L.W., Krüger, L., Hockauf, K., Halle, T. and Hockauf, M. (2008). Dynamic strength and failure behavior of titanium alloy Ti6Al4V for a variation of heat treatments. Mechan. Time-Depend. Mater., 12(3), p. 237–247.

Miola, M., Pakzad, Y., Banijamali, S. and 7 additional authors (2019a). Glass ceramics for cancer treatment: So close, or yet so far? Acta Biomater., 83, p. 55–70.

Miola, M., Bellare, A., Laviano, F., Gerbaldo, R. and Verné, E. (2019b). Bioactive superparamagnetic nanoparticles for multifunctional composite bone cements. Ceram. Intern., 45(12), p. 14533–14545.

Miranda, M., Fernandez, A., Diaz, M., Esteban-Tejeda, L., Lopez-Esteban, S., Malpartida, F., Torrecillas, R. and Moya. J.S. (2010). Silver-hydroxyapatite nanocomposites as bactericidal and fungicidal materials. Int. J. Mater. Res., 101(1), p. 122–127.

Mishra, A. K., Davidson, J. A., Poggie, R. A., Kovacs, P. and Fitzgerald, J. (1996). Mechanical and tribological properties and biocompatibility of diffusion hardened Ti-13Nb-13Zr – A new titanium alloy for surgical implants. In: Brown, S. A. and Lemons, J. E. (eds) Medical Applications of Titanium and its Alloy. ASTM STP 1272, ASTM, West Conshohocken, PA, USA, p. 96–116.

Mold, M., Umar, D., King, A. and Exley, C. (2018). Aluminium in brain tissue in autism. J. Trace Elem. Med. Biol., 46, p. 76–82.

Mouriño, V., Vidotto, R., Cattalini, J.P. and Boccaccini, A. (2019). Enhancing biological activity of bioactive glass scaffolds by inorganic ion delivery for bone tissue engineering. Curr. Opin. Biomed. Eng., 10, p. 23–34.

Mozafari, M., Banijamali, S., Baino, F., Kargozar, S. and Hill, R.G. (2019). Calcium carbonate: adored and ignored in bioactivity assessment. Acta Biomater., 91, p. 35–47.

Mulloy, B. (2012). Structure and physicochemical characterization of heparin. In: Lever, R., Mulloy, B. and Page, C.P. (eds). Heparin-a Century of Progress. Heidelberg: Springer. ISBN: 978-3-642-23055-4. p. 77–98.

Murayama, Y. and Sasaki, S. (2009). Mechanical properties of Ti-Cr-Sn-Zr alloys. Univ. Res. J. Niigata Inst. Tech., 14, p. 1–8.

Nabiyouni, M., Brückner, T., Zhou, H., Gburek, U. and Bhaduri, S.B. (2018). Magnesium-based bioceramics in orthopedic applications. Acta Biomater., 66, p. 23–43.

Nakai, M., Niinomi, M., Akahori, T., Ohtsu, N., Nishimura, H., Toda, H., Fukui, H. and Ogawa, M. (2008). Surface hardening of biomedical Ti-29Nb-13Ta-4.6Zr and Ti-6Al-4V ELI by gas nitriding. Mater. Sci. Eng. A, 486, p. 193–201.

Nakai, M., Niinomi, M., Zhao, X. F. and X. L. Zhao,. (2011). Self-adjustment of Young's modulus in biomedical titanium alloy during orthopaedic operation. Mater. Letters, 65, p. 688–690.

Nakai, M., Niinomi, M. and Oneda, T. (2012). Improvement in fatigue strength of biomedical β-type Ti−Nb−Ta−Zr alloy while maintaining low Young's modulus through optimizing ω-phase precipitation. Metal. Mater. Trans. A, 43(1), p. 294–302.

Narushima, N. and Niinomi, M. (2010), Titanium and its alloys. In: Hanawa, T. (eds). Metals for Medicine. p. 92–106. JIM. Sendai, Japan. ISBN 978-4-88903-075-4.

Narushima, T. (2010). New-generation metallic biomaterials. In: Niinomi, M. (ed.). Metals for Biomedical Devices. p. 355–378. Woodhead Publ. Ser. Biomater., ISBN 978-1-0456-9434-0. 432 pp.

Narushima T., Ueda K. and Alfirano (2015). Co-Cr alloys as effective metallic biomaterials. In: Niinomi M., Narushima T., Nakai M. (eds). Advances in Metallic Biomaterials. Springer Series in Biomaterials Science and Engineering, 3, p. 157–178. Berlin, Heidelberg: Springer. ISBN 978-3-662-46835-7.

Navarro, M., Michiardi, A., Castaño, O. and Planell, J.A. (2008). Biomaterials in orthopaedics. J. R. Soc. Interface, 5(23), p. 1137–1158.

Nene, S.S., Frank, M., Liu, K., Sinha, S., Mishra, R.S., McWilliams, B.A. and Cho, K.C. (2019). Corrosion-resistant high entropy alloy with high strength and ductility. Scripta Mater., 166, p. 168–172.

Niedhart, C. und Niethard, F.U. (1998). Klinische Anforderungen an Knochenersatzstoffe [Clinical demands on bone replacement materials]. Proc. 3rd BIOLOX® Symp., 2-1, p. 46–50.

Niinomi, M. (1998). Mechanical properties of biomedical titanium alloys. Mater. Sci. Eng. A, 243, p. 231–236.

Niinomi, M. (2000). Development of high biocompatible titanium alloys. Function & Materials, 20, p.36–44.

Niinomi, M., Fukui, H., Hattori, T., Kyo, K. and Suzuki, A. (2002a). Development of high biocompatible titanium alloy. Materials Jpn., 41. p. 221–223.

Niinomi, M., Akahori, T., Nakamura, S., Fukui,. H. and Suzuki, A. (2002b). Characteristics of surface oxidation treated new biomedical β-type titanium alloy in simulated body environment. Tetsu-to-Hagane, 88(9), p. 567–574 (in Japanese).

Niinomi, M. (2002c). Recent metallic materials for biomedical applications. Met. Mat. Trans. A, 33A(3), p. 477–486.

Niinomi, M, Hattori, T., Morikawa, K., Kasuga, T., Suzuki, A., Fukui, H. and Niwa, S. (2002d). Development of low rigidity β-type titanium alloy for biomedical applications. Mater. Trans., 43(12), p. 2970–2977.

Niinomi, M. (2003). Fatigue performance and cytotoxicity of low rigidity titanium alloy, Ti-29Nb-13Ta-4,6Zr. Biomaterials, 24(16), p. 2673–2683.

Niinomi, M., Nakai, M. and Akahori, T. (2007). Frictional wear characteristics of biomedical Ti-29Nb-13Ta-4.6Zr alloy with various microstructures in air and simulated body fluid. Biomed. Mater., 2, p. S167-S174.

Niinomi, M. (2008). Metallic biomaterials. J. Artif. Organs, 11:15. https://doi.org/10.1007/s10047-008-0422-7.

Niinomi, M., Nakai, M. and Hieda, J. (2012a). Development of new metallic alloys for biomedical applications, Acta Biomater., 8, p. 3888–3903.

Niinomi, M. (2012b). Shape memory, super elastic and low Young's modulus alloys. In: Ambrosio, L. and Tanner, E. (eds). Biomaterials for Spinal Surgery. Woodhead Publishing Ltd. Cambridge, UK. p. 462–490.

Niinomi, M. (2013). Development of titanium alloys with high mechanical biocompatibility with focusing on controlling elastic modulus. Materials Jpn., 52(5), p. 219–228.

Niinomi, M. and Boehlert, C. (2015). Titanium alloys for biomedical applications. In: Niinomi, M., Narushima, T., and Nakai, M. (eds). Tissues, Materials and Biological Reactions, Advances in Metallic Biomaterials, Part I. Springer-Series in Biomaterials Science and Engineering, 3, p. 179–213. Berlin, Heidelberg, New York: Springer. ISBN 978-3-4662-46835-7.

Niinomi, M. (2018). Recent progress in research and development of metallic structural biomaterials with mainly focusing on mechanical biocompatibility. Mater. Trans., 59(1), p. 1–13.

Niinomi, M. (2019). Development of strengthening and toughening of ß-type titanium alloys. Materials Jpn., 58(4), p. 193–200.

Ning, C.Y., Zhou, L. and Tan, G.X. (2015). Fourth generation biomedical materials. MaterialsToday, 19(1), p. 2–3.

Ning, C.Y., Wang, Y.J., Chen, X.F., Zhao, N.R., Ye, J.D. and Wu, G. (2005). Mechanical performance and microstructural characteristics of plasma-sprayed biofunctionally gradient HA-ZrO_2-Ti coatings. Surf. Coat. Technol., 200(7), p. 2403–2408.

Nishihara, H., Adanez, M.H. and Att, W. (2019). Current status of zirconia implants in dentistry: preclinical tests. J. Prosthod. Res., 63(1), p. 1–14.

Niu, L., Jiao, K., Ryou, H., Yiu, C.K.Y., Chen, J., Breschi, L., Arola, D.D., Pashley, D.H. and Tay, F.R. (2013). Multiphase intrafibrillar mineralization of collagen. Angew. Chem. (Intern. Ed.), 52, p. 5762–5766.

Nivison-Smith, L. and Weiss, A. (2011). Elastin based constructs. In: Eberli, D. (ed.). Regenerative Medicine and Tissue Engineering – Cells and Biomaterials. Rijeka: InTech. ISBN: 978-953-307-663-8, p. 323–340.

Noori, A., Ashrafi, S. J., Vaez-Ghaemi, R., Hatamian-Zaremi, A. and Webster, T.J. (2017). A review of fibrin and fibrin composites for bone tissue engineering. Intern. J. Nanomed., 12, p. 4937–4961.

O'Brien, B.J., Stinson, J.S., Larsen, S.R., Eppihimer, M. and Carroll, W.M. (2010). A platinum-chromium steel for cardiovascular stents. Biomaterials, 31(4), p. 3755–3761.

O'Brien, B. and Carroll, W. (2009). The evolution of cardiovascular stent materials and surfaces in response to clinical drivers: A review. Acta Biomater., 5, p. 945–958.

Okazaki, Y., Ito, Y., Tateishi, T. and Ito, A. (1995). Effect of heat treatment on microstructure and mechanical properties of new titanium alloys for surgical implantation. J. Japan Inst. Metals, 59, p. 108–15.

Okazaki, Y., Rao, Y., Asao, S. and Tateishi, T. (1998). Effects of metallic concentrations other than Ti, Al and V on cell viability. Mater. Trans., 39, p. 1070–1079.

Olofsson, J., Pettersson, M., Teuscher, N., Heilmann, A., Larsson, K., Grandfield, K., Persson, C., Jacobson, S. and Engqvist, H. (2012). Fabrication and evaluation of Si_xN_y coatings for total joint replacements. J. Mater. Sci. Mater. Med., 23, p. 1879–1889.

Ottoboni, T., Gelder, M.S. and O'Boyle, E. (2014). Biochronomer™ technology and the development of aPF530, a sustained release formulation of granisetron. J. Exp. Pharmacology, 6, p. 15–21.

Pace, N.R. (2001). The universal nature of biochemistry. PNAS, 98(3), p. 805–808.

Paddock, C. (2013). Bone grafts may be better with new sea coral material. Medical News Today. November 2013. Retrieved from http://www.medicalnewstoday.com/articles/269512.php.

Papadimitriou-Olivgeri, I., Brown, J.M., Kilpatrick, A.F.R., Gill, H.S. and Athanasou, N.A. (2019). Solochrome cyanine: A histological stain for cobalt-chromium particles in metal-on-metal periprosthetic tissues. J. Mater. Sci. Mater. Med., 30:103.

Paganias, C.G., Tsakotos, G.A., Koutsostathis, S.D. and Macheras, G.A. (2012). Osseous integration of porous tantalum implants. Indian J. Orthop., 46(5), p. 505–513.

Parisi, O.I., Manuela Curcio, M., and Puoci, F. (2015). Polymer chemistry and synthetic polymers. In: Puoci, F. (ed.). Advanced Polymers in Medicine. Cham: Springer. ISBN: 978-3-319-12477-3. p. 1–31.

Parveen, S., Misra, R. and Sahoo, S.K. (2012). Nanoparticles: a boon to drug delivery, therapeutics, diagnostics and imaging, nanomedicine: nanotechnology. Biol. Med., 8(2), p. 147–166.

Pasteris, J.D. (2016). A mineralogical view of apatitic biomaterials. Am. Mineral., 101, p. 2594–2610.

Pasteris, J.D., Wopenka, B., Freeman, J.J., Rogers, K., Valsami-Jones, E., van der Houwen, J.A.M. and Silva, M.J. (2004). Lack of OH in nanocrystalline apatite as a function of degree of atomic order: implications for bone and biomaterials. Biomaterials, 25(2), p. 229–238.

Pasteris, J.D., Wopenka, B. and Valsami-Jones, E. (2008). Bone and tooth mineralization: why apatite? Elements, 4, p. 97–104.

Patat, G.G., Fournier, J. and Chetail, M. (1987). The use of coral as a bone graft substitute. J. Biomed. Mater. Res., 21(5), p. 557–567.

Patil N. and Goodman S.B. (2015). The use of porous tantalum for reconstructing bone loss in orthopedic surgery. In: Niinomi M., Narushima T., Nakai M. (eds) Advances in Metallic Biomaterials. Springer Series in Biomaterials Science and Engineering, 3, p.223–243. Berlin, Heidelberg: Springer.

Pauline, S.A., Rajendran, N. (2014). Biomimetic novel nanoporous niobium oxide coating for orthopaedic applications, Appl. Surf. Sci., 290, p. 448–457.

Pei, D.D., Meng, Y.C., Li, Y.C., Liu, J. and Lu, Y. (2019). Influence of nano-hydroxyapatite containing desensitizing toothpastes on the sealing ability of dentinal tubules and bonding performance of self-etch adhesives. J. Mech. Behav. Biomed. Mater., 91, p. 38–44.

Peitl, O., Zanotto, E. D. and Hench, L. L. (2001). Highly bioactive P_2O_5–Na_2O–CaO–SiO_2 glass-ceramics. J. Non-Cryst. Solids, 292, p. 115–126.

Peppas, N.A. (2004). Hydrogels. In: Ratner, B.D., Hoffman, A.S., Schoen, F.J., and Lemons, J.E. (eds). Biomaterials Science: An Introduction to Materials in Medicine. San Diego: Elsevier. ISBN: 0-12-582463-7. p. 100–107.

Perner, A., Haase, N., Guttormsen, A.B., Tenhunen, J., Klemenzson, G. et al. (2012). Hydroxyethyl starch 130/0.42 versus Ringer's acetate in severe sepsis. New Engl J. Med., 367, p. 124–134.

Peroos, S., Du, Z. and de Leeuw, N. (2006). A computer modelling study of the uptake, structure and distribution of carbonate defects in hydroxyapatite. Biomaterials, 27(9), p. 2150–2161.

Persaud-Sharma, D. and McGoron, A. (2012). Biodegradable magnesium alloys: a review of material development and applications. J. Biomim. Biomater. Tissue Eng., 12, p. 25–39.

Perrin, D.E., and English, J.P. (1997). Polyglycolide and polylactide. In: Domb, A.J., Kost, J. and Wiseman, D.M. (eds). Handbook of Biodegradable Polymers. Amsterdam: Harwood Academic Publishers, ISBN: 90-5702-153-6. p. 3–27.

Petrini, L. and Migliavacca, F. (2011). Biomedical applications of shape memory alloys. J. Metall., Article ID 501483, 15 pages.

Pfaff, H.G. and Willmann, G. (1998). Stability of Y-TZP zirconia. Proc. 3rd BIOLOX® Symp., 1-6, p. 29–31.

Piconi, C. (2017). Alumina. In: Reference Module in Materials Science and Materials Engineering. Comprehensive Biomaterials II. Vol. 1, p. 92–121. Elsevier. ISBN 978-0-08-100692-4.

Pieger, S., Salman, A. and Bidra, A.S. (2014). Clinical outcomes of lithium disilicate single crowns and partial fixed dental prostheses: A systematic review. J. Prosth. Dent., 112(1), p. 22–30.

Platt, J.A. (2016). Dental materials. In: Dean, J.A. (ed.). McDonald and Avery's Dentistry for the Child and the Adolescent, p. 206–220. 10th ed., Maryland Heights: Mosby. ISBN 978-0-323-28745-6.

Popescu, G., Ghiban, B., Popescu, C.A. and Rosu, L. (2018). New TiZrNbTaFe high entropy alloy used for medical applications. IOP Conf. Ser. Mater. Sci. Eng., 400 (2),p. 022049. ISSN 1757-899X.

Poinern, G.E.J., Brundavanam, S. and Fawcett. D. (2012). Biomedical magnesium alloys: A review of material properties, surface modifications and potential as biodegradable orthopedic implant. Am. J. Biomed. Eng., 2(6), p. 218–240.

Porath, J. and Flodin, P. (1959). Gel filtration: A method for desalting and group separation. Nature, 183, p. 1657–1659.

Posner, A.S., Perloff, A. and Diorio, A.F. (1958). Refinement of the hydroxyapatite structure. Acta Cryst., 11, p. 308–309.

Pountos, I. and Giannoudis, P.V. (2016). Is there a role of coral bone substitutes in bone repair? Injury, 47(12), p. 2606–2613.

Pouroutzidou, G.K., Theodorou, G.S., Kontonasaki, E. and 7 additional authors (2019). Effect of ethanol/TEOS ratios and amount of ammonia on the properties of copper-doped calcium silicate nanoceramics. J. Mater. Sci. Mater. Med., 30:98.

Porter, A.E., Patel, N., Skepper, J.N., Best, S.M. and Bonfield, W. (2004). Effect of sintered silicate-substituted hydroxyapatite on remodeling processes at the bone-implant interface. Biomaterials, 25, p. 3303–3314.

Posset, U., Löcklin, E., Thull, R. and Kiefer, W. (1998). Vibration spectroscopic study of tetracalcium phosphate in pure polycrystalline form and as a constituent of a self-setting bone cement. J. Biomed. Mater. Res., 40, p. 640–645.

Prabaharan, M. and Tiwari, A. (2011). Chemical modification of chitosan intended for biomedical applications. In: Chitin, Chitosan, Oligosaccharides and their Derivatives: Biological Activities and Applications. Boca Raton: CRC Press. ISBN: 978-1-4398-1603-5. p. 173–184.

Pradhan, D., Wren, A.W., Misture, S.T. and Mellott, N.P. (2016). Investigating the structure and biocompatibility of niobium and titanium oxides as coatings for orthopedic metallic implants, Mater. Sci. Eng. C, 58, p. 918–926.

Predoi, D., Iconaru, S.L., Predoi, M.V., Stan, G.E. and Buton, N. (2019). Synthesis, characterization, and antimicrobial activity of magnesium-doped hydroxyapatite suspensions. Nanomaterials, 9(9): 1295.

Prentice, T.C., Pickford, M.E.L., Lewis, D.R. and Turner, A.D. (2013). Implant for contacting with a bone during e.g. prosthetic surgery, comprises metal structure having a surface with a ceramic coating containing hydroxyapatite and silver ions that gradually leach out in body fluids after implantation, European patent EP2316499-B1.

Prestwich, G.D. and Kuo, J. (2008). Chemically-modified HA for therapy and regenerative medicine. Curr. Pharmaceut. Biotechnol., 9, p. 242–245

Prieto, E.M. and Guelcher, S.A. (2014). Tailoring properties of polymeric biomedical foams. In: Netti, P.A. (ed.). Biomedical Foams for Tissue Engineering Applications. Cambridge: Woodhead Publishing. ISBN: 978-0-85709-696-8, p. 129–162.

Quinn, G.D. and Quinn, J.D. (2010). A practical and systematic review of Weibull statistics for reporting strength of dental materials. Dental Mater., 26(3), p. 135–147.

Radin, S. R. and Ducheyne, P. (1993). The effect of calcium phosphate ceramic composition and structure on in vitro behavior. II. Precipitation. J. Biomed. Mater. Res., 27, p. 35–45.

Radin, S. and Ducheyne, P. (1996). Effect of serum proteins on solution-induced surface transformations of bioactive ceramics. J. Biomed. Mater. Res., 30, p. 273–279.

Radin, S., Ducheyne, P., Berthold, P. and Decker, S. (1997a). Effect of serum proteins and osteoblasts on the surface transformation of a calcium phosphate coating: A physicochemical and ultrastructural study. J. Biomed. Mater. Res., 39, p. 234–243.

Radin, S., Ducheyne, P., Rothman, B. and Conti, A. (1997b). The effect of in vitro modeling conditions on the surface reactions on bioactive glass. J. Biomed. Mater. Res., 37, p. 363–375.

Radin, S., Reilly, G., Bhargave, G., Leboy, P. S. and Ducheyne, P. (2005). Osteogenic effects of bioactive glass on bone marrow stromal cells. J. Biomed. Mater. Res. A, 73, p. 21–29.

Rahmati, B., Sarhan, A.A.D., Zalnezhad, E., Kamiab, Z., Dabbagh, A., Choudhury, D. and Abas, W.A.B.W. (2016). Development of tantalum oxide (Ta-O) thin film coating on biomedical Ti-6Al-4V alloy to enhance mechanical properties and biocompatibility. Ceram. Inter., 42(1A), p. 466–480.

Rahmani, H. and Salahinejad, E. (2019). Incorporation of monovalent cations into diopside to improve biomineralization and cytocompatibility. Ceram. Inter., 44(16), p. 19200–19206.

Rahmitasari, F., Ishida, Y., Kurahashi, K., Matsuda, T., Watanabe, M. and Ichikawa, T. (2017). PEEK with reinforced materials and modifications for dental implant applications. Dentistry J., 5, Article #35.

Raj, T., Dimitrijevic, N.M., Bissonnette, M., Koritarov, T. and Konda, V. (2014). Titanium dioxide in the service of the biomedical revolution. Chem. Rev., 114(19), p. 10177–10216.

Rajendran, A., Sugunapriyadharshini, S., Mishra, D. and Pattanayak, D.K. (2019). Role of calcium ions in defining the bioactivity of surface modified Ti metal. Mater. Sci. Eng. C, 98, p. 197–204.

Ramaswamy, Y., Wu, C., Dunstan, C.R., Hewson, B., Eindorf, T., Anderson, G.I. and Zreiqat, H. (2009). Sphene ceramics for orthopedic coating applications: an in vitro and in vivo study. Acta Biomater., 5, p. 3192–3204.

Ramesh, S., Sara Lee, K.Y. and Tan, C.Y. (2018). A review on the hydrothermal ageing behaviour of Y-TZP ceramics. Ceram- Intern., 44(17), p. 20620–20634.

Raphey, V.R., Henna, T.K., Nivitha, K.P., Mufeedha, P., Sabu, C. and Pramad, K. (2019). Advanced biomedical applications of carbon nanotube. Mater. Sci. Eng., 100, p. 616–630.

Ratner, B.D., Hoffman, A.S., Schoen, F.J. and Lemons, J.E. (2013). Biomaterials Science. An Introduction to Materials in Medicine. Elsevier, ISBN 978-0-12-374626-9.

Razawi, M., Fathi, M., Savabi, O., Beni, B.H., Vashaee, D. and Tayebi, L. (2014). Nanostructured merwinite bioceramic coating on Mg alloy deposited by electrophoretic deposition. Ceram. Intern. A, 40(7), p. 9473–9484.

Reddy, T.S., Privér, S.H., Rao, V.V., Mirzadeh, N. and Bhargava, S.K. (2018). Gold(I) and gold(III) phosphine complexes: Synthesis, anticancer activities towards 2D and 3D cancer models, and apoptosis inducing properties. Dalton Trans., 47, p. 15312–15323.

Reisel, G. (1996). Entwicklung von HVOF- und APS-gespritzten biokeramischen Schichten für die Endoprothetik [Development of HVOF- and APS-sprayed bioceramic coatings for endoprosthetic uses]. Master Thesis. Rheinisch-Westfälische Technische Hochschule (RWTH) Aachen, Aachen, Germany. September 1996.

Rey, C., Combes, C., Drouet, C. and Glimcher, M.J. (2009). Bone mineral: Update on chemical composition and structure. Osteoporos. Int., 20, p. 1013–1021.

Rifai, A., Pirogova, E. and Fox, K. (2019). Diamond, carbon nanotubes and graphene for biomedical applications. In: Narayan, R. (ed.). Encyclopedia of Biomedical Engineering. Elsevier, Inc. ISBN 978-0-12-805144-5. 2054 pp.

Rising, A. (2014). Controlled assembly: A prerequisite for the use of recombinant spider silk in regenerative medicine? Acta Biomater. 10, 1627–1631.

Ritz, J.P. and Scott, C. (2015). Making bone implant used for repair of the ends of bones at orthopedic joints involves creating substrate of structurally strong isotropic graphite of the shape desired for bone implant, and coating with microporous isotropic pyrocarbon. US patent US8932663-B2.

Roach, M.D., Williamson, R.S. and Thomas, J.M. (2012). Noble and precious metal applications in biomaterials with emphasis on dentistry. In: Narayan, R.J. (ed) ASM Handbook Materials for Medical Devices. Vol. 23, p. 251–264. ISBN 978-1-62708-198-6.

Rosen, A. (2012). Effects of the Fukushima nuclear meltdown on environment and health. www. ippnw.de/commonFiles/pdfs/Atomenergie/FukushimaBackgroundPaper.pdf.

Rosenberg, B., VanCamp, L., Trosko, J.E. and Mansour, V.H. (1969). Platinum compounds: A new class of potent antitumour agents. Nature, 222(5191), p. 385–386.

Sagbas, S. and Sahiner, N. (2019). Carbon dots: preparation, properties, and application. In: Khna, A., Jawaid, M. and Asiri, A.M. (eds). Nanocarbon and its Composites. Preparation, Properties and Applications. Woodhead Publ. Ser. Comp. Sci. Eng., ISBN 978-0-0-8-102509-3. 872 pp. p. 651–676.

Sahai, N. and Anseau, M. (2005). Cyclic silicate active site and stereochemical match for apatite nucleation on pseudowollastonite bioceramics-bone interfaces. Biomaterials, 26, p. 5763–5770.

Sahasrabudha, H., Bose, S. and Bandhyapadhyay, A. (2018). Laser processed calcium phosphate reinforced CoCrMo for load-bearing applications: processing and wear induced damage evaluation. Acta Biomater., 66, p. 118–128.

Sahin, O., Uzunoglu, S. and Sahin, E. (2015). Mechanical characterization of CoCrMo alloys consisting of different palladium ratios produced by investment casting method. Acta Phys. Polon. A, 128(2B), p. 149–151.

Sainz, M.A., Pena, P., Serena, P. and Caballero, A. (2010). Influence of design on bioactivity of novel $CaSiO_3$-$CaMg(SiO_3)_2$ bioceramics: In vitro simulated body fluid test and thermodynamic simulation, Acta Biomater., 6, p. 2797–2807.

Sakaguchi, R.L. and Powers, J.M. (eds) (2012). Restorative materials-ceramics. In: Craig's Restorative Dental Materials. Chapter 11, p. 253–275. Mosby: Maryland Heights. ISBN 978-0-323-08108-5.

Sakiyama-Elbert, S.E. (2014). Incorporation of heparin into biomaterials. Acta Biomater. 10, p. 1581–1587.

Sakona, A.N., McDonald, D.W., Sharma, P., Medel, F. and Kurtz, S.M. (2010). Contemporary alternatives to zirconia: retrieval analyses of Oxinium® and BIOLOX® delta femoral heads. www.abstracts.biomaterials.org/data/papers/2010/107.pdf (accessed September 26, 2018).

Salahinejad, E. and Vahedifard, R. (2017). Deposition of nanodiopside coatings on metallic biomaterials to stimulate apatite-forming ability. Mater. Design, 123, p. 120–127.

Santos, P. F., Niinomi, M., Liu, H., Nakai, M., Cho, K., Narushima, T., Ueda, K., Ohtsu, N. and Hirano, M. (2017). Development and performance of low cost beta-type Ti-based alloys for biomedical applications using Mn additions. In: Sasaki, K, Suzuki, O. and Takahashi, N. (eds). Interface Oral Health Science 2016: Innovative Research on Biosis-Abiosis Interface. Berlin, Heidelberg, New York: Springer. p. 229–245.

Sasaki, Y., Doi, K. and Matsushita, T. (1996). Titanium: new titanium alloys for artificial hip joints. Kinzoku, 66, p. 812–817 (in Japanese).

Schaller, B., Saulacic, N., Imwinkelried, T., Beck, S., Liu, E.W., Gralla, J., Nakahara, K., Hofstetter, W. and Iizuka, T. (2016). In vivo degradation of magnesium plate/screw osteosynthesis implant systems: Soft and hard tissue response in a calvarial model in miniature pigs. J. Craniomaxillofac. Surg., 44(3), p. 309–317.

Schepers, E., Declercq, M., Ducheyne, P. and Kempeneers, R. (1991). Bioactive glass particulate material as a filler for bone lesions. J. Oral Rehab., 18, p. 439–452.

Scherrer, S.S., Mekki, M., Crottaz, C. and 5 additional authors (2019). Translational research on clinically failed zirconia implants. Dental Mater., 35(2), p. 368–388.

Schiller, J., Becher, J., Möller, S., Nimptsch, K.1, Riemer, T., and Schnabelrauch M. (2010). Synthesis and characterization of chemically modified hyaluronan and chondroitin sulfate. Mini-Rev. Organic Chem., 7, p. 290–299.

Schnabelrauch, M. (2018). Chemical bulk properties of biomaterials. In: Zivic, F., Affatato, S., Trajanovic, M., Schnabelrauch, M., Grujovic, N., and Choy, K.L. (eds). Biomaterials in Clinical Practice. Cham: Springer International. ISBN: 978-3-319-68024-8. p. 431–459.

Schneider K., Heimann R.B. and Berger G. (2001). Plasma-sprayed coatings in the system CaO-TiO_2-ZrO_2-P_2O_5 for long-term stable endoprostheses. Mat.-wiss. Werkstofftechn., 32, p. 166–171.

Schoeppler, V., Reich, E., Vacelet, J., Rosenthal, M., Pacureanu, A., Rack, A., Zaslansky, P., Zolotoyabko, E. and Zlotnikov, I. (2017). Shaping highly regular glass architectures: A lesson from nature. Science Advances, 3, eaao2047.

Schrieber, R., and Gareis, H. (2007). Gelatin Handbook: Theory and Industrial Practice. Weinheim: Wiley-VCH. ISBN: 978-3-527-31548-2. Chapter 2, p. 45–117.

Schwartzwalder, K. and Somers, A.V. (1963). Method of making porous ceramic articles. US Patent 3090094 (May 21, 1963).

Scott, J.E. (2000). Secondary and tertiary structures in solutions of hyaluronan and related "shape module" anionic glycosaminoglycans. In: Abatangelo, G., Weigel, P.H. (eds). New Frontiers in Medical Sciences: Redefining Hyaluronan. Amsterdam: Elsevier. ISBN: 0-444-50357-9. p. 11–19.

Seth, D., Choudhury, S.R., Pradhan, S. et al. (2011). Nature-inspired novel drug design paradigm using nanosilver: Efficacy on multi-drug-resistant clinical isolates of tuberculosis. Current Microbiol., 62 (3), p. 715–726.

Semlitsch, M. (1984). Metallische Werkstoffe für Implantate [Metallic materials for implants]. In: Rettig, H. (ed.). Biomaterials und Nahtmaterials. Jahrestagung Deutsche Gesellschaft für Plastische und Wiederherstellungschirugie, 21, Berlin, Heidelberg: Springer.

Semlitsch, M. (1986). Classic and new titanium alloys for production of artificial hip joints. Titan, 2, p. 721–740.

Senbhagaraman, S., Guru Row, T.N. and Umarji, A.M. (1993). Structural refinement using high-resolution powder X-ray diffraction data of $Ca_{0.5}Ti_2P_3O_{12}$, a low-thermal-expansion material. J. Mater. Chem., 3(3), p. 309–314.

Šesták, P., Černý, M. and Pokluda, J. (2010). Elastic constants of austenitic and martensitic phases of NiTi shape memory alloy. In: Brezina, T. and Jablonski, R. (eds). Recent Advances in Mechatronics. p. 1–6. Berlin, Heidelberg: Springer. ISBN 978-3-642-05021-3.

Shabalovskaya, S. and van Humbeeck, J. (2009). Biocompatibility for biomedical applications. In: Yoneyama, T. and Miyazaki, S. (eds). Shape Memory Alloys for Biomedical Applications. Woodhead Publ. CRC. 2009. Chapter 9. ISBN 978-1-84569-344-2.

Shadjou, N. and Hasanzadeha, M. (2015). Bone tissue engineering using silica-based mesoporous nanobiomaterials: Recent progress. Mater. Sci. Eng. C, 55, p. 401–409.

Shalukho, N.M. and Kuz'menkov, M.I. (2011). Polymorphic transformation of leucite in a crystal-glass material for bioceramics. Glass and Ceramics (Steklo i Keramika), 68(7-8), p. 231–233.

Shamray, V.F., Sirotinkin, V.P., Smirnov, I.V., Kalita, V.I., Fedotov, A.Yu., Barinov, S.M. and Komlev, V.S. (2017). Structure of the hydroxyapatite plasma-sprayed coatings deposited on pre-heated titanium substrates. Ceram. Int., 43, p. 9105–9109.

Shamray, V.F., Sirotinkin, V.P., Kalita, V.I., Komlev, V.S., Barinov, S.M., Fedotov, A.Yu. and Gordeev, A.S. (2019). Study of the crystal structure of hydroxyapatite in plasma coating. Surf. Coat. Technol., 372, p. 201–208.

Shelby, J.E. (2005). Introduction to Glass Science and Technology. 2nd ed. Royal Chem. Soc., p. 72–110. ISBN 978-0-85404-639-3.

Shepherd, J.H., Shepherd, D.V. and Best, S.M. (2012). Substituted hydroxyapatites for bone repair. J. Mater. Sci. Mater. Med., 23(10), p. 2335–2347.

Shogren, R.L. (1998). Starch: Properties, and material applications. In: Kaplan, D.I. (ed.). Biopolymers from Renewable Resources. Berlin: Springer. ISBN: 3-540-63567-X. p. 30–46.

Smirnov, V.V., Goldberg, M.A., Shvorneva, L.I., Fadeeva, I.V., Shibaeva, T.V. and Barinov, M. (2010). Synthesis of composite biomaterials in the hydroxyapatite-calcite system. Dokl. Chem., 432, p. 151–154.

Smith, W.F. (1996). Principles of Materials Science and Engineering. 3rd ed. New York: McGraw-Hill. ISBN 0.07–059241-1.

Sommerdijk, N.A.J.M., van Leeuwen, E.N.M., Vos, M.R.J. and Jansen, J.A. (2007). Calcium carbonate thin films as biomaterial coatings using DNA as crystallization inhibitor. Cryst. Eng. Comm., 9, p. 1209–1214.

Song, G., Atrens, A., John, D., Nairn. J. and Li, Y. (1997). The electrochemical corrosion of pure magnesium in 1N NaCl. Corr. Sci., 39(5), p. 855.

Song, G.L. (2011). Corrosion of Magnesium Alloys. Woodhead Publishing Series in Metal and Surface Engineering. Elsevier.Song, X., Niinomi, M., Tsutsumi, H., Nakai, M. and Wang,

L. (2011). Effects of TiB on mechanical properties of Ti-29Nb-13Ta-4.6Zr for use in biomedical applications. Mater. Sci. Eng. A, 528(16-17), p. 5600–5609.

Sorrell, C., Hardcastle, P.H., Druitt, R.K., Howlett, C.R. and McCartney, E.R. (2004). Results of 15-years clinical study of reaction bonded silicon nitride intervertebral spacers. In: Proc. 7th World Biomaterial Congress, May 17- 21, 2004, Sydney. Sydney: Australian Society for Biomaterials, 2004, p. 1872.

Spiess, K., Lammel, A. and Scheibel, T. (2010). Recombinant spider silk proteins for applications in biomaterials. Macromol. Biosci., 10, p. 998–1007.

Spotnitz, W.D. (2014). Fibrin Sealant: The Only Approved Hemostat, Sealant, and Adhesive—a Laboratory and Clinical Perspective. ISRN Surgery, ID 203943

Srichana, T. and Domb, A.J. (2009). Polymeric biomaterials. In: Narayan, R. (ed.). Biomedical Materials. New York: Springer. ISBN: 978-0-387-84871-6. p. 83–119.

Srivastav, A., Chandanshive, B., Dandekar, P., Khushalani, D. and Jain, R. (2019). Biomimetic hydroxyapatite: A potential universal nanocarrier for cellular internalization and drug delivery. Pharm. Res., 36: 60.

Steegmüller, R., Wagner, C., Fleckenstein, T. and Schuessler, A. (2002). Gold coating on Nitinol devices for medical applications. Mater. Sci. Forum, 394/395, p. 161–164.

Steinemann, S. G. (1980). Corrosion of surgical implants – in vivo and in vitro tests. In: Winter, G. D., Leray, J. L and de Groot, K. (eds). Evaluation of Biomaterials. p. 1–34. John Wiley & Sons Ltd.

Steinemenan, S. (1972). Implants of titanium or a titanium alloy for the surgical treatment of bones. US Patent 3643658 A.

Stemmer, P., Pourzal, R., Liao, Y., Marks, L.D., Morlock, M.M., Jacobs, J.J., Wimmer, M.A. and Fischer, A. (2013). Microstructure of retrievals made from standard cast HC-CoCrMo alloys. STP1560. ASTM International, West Conshohocken, PA.

Stevels J.M. (1957). The electrical properties of glass. In: Elektrische Leitungsphänomene II. Handbuch der Physik, 20(4), p. 350–391. Springer: Berlin, Heidelberg.

Stilwell, R.L., Marks, M.G., Saferstein, L. and Wiseman, D.M. (1997). Oxidized cellulose: Chemistry, processing and medical applications. In: Domb, A.J., Kost, J. and Wiseman, D.M. (eds). Handbook of Biodegradable Polymers. Amsterdam: Harwood Academic Publishers, ISBN: 90-5702-153-6. p. 291–306.

Subramanian, B., Muraleedharan, C.V., Ananthakumar, R. and Jayachandran, M. (2011). A comparative study of titanium nitride (TiN), titanium oxy nitride (TiON) and titanium aluminum nitride (TiAlN), as surface coatings for bioimplants. Surf. Coat. Technol., 205 (21-22), p. 5014–5020.

Suchanek, W., Yashima, M., Kakihana, M. and Yoshimura, M. (1998). ß-rhenanite (ß-NaCaPO$_4$) as weak interphase for hydroxyapatite ceramics. J. Eur. Ceram. Soc., 18, p. 1923–1929.

Sugiura, Y., Ishikawa, K., Onuma, K. and Makita, Y. (2019). PO$_4$ adsorption on the calcite surfaces modulates calcite formation and crystal size. Am. Mineral., 104, p. 1381–1388.

Sun, L., Berndt, C.C., Gross, K.A. and A. Kucuk (2001). Material fundamentals and clinical performance of plasma-sprayed hydroxyapatite coatings: A review. J. Biomed. Mater. Res., 58(5), p. 570–592.

Surmenev, R.A., Surmeneva, M.A. and Ivanova, A.A. (2014). Significance of calcium phosphate coatings for the enhancement of new bone osteogenesis: A review. Acta Biomater., 10(2), p 557–579.

Surmenev, R.A. and Surmeneva, M.A. (2019). A critical review of decades of research on calcium phosphate-based coatings: how far are we from their widespread clinical application? Curr. Opinion Biomed. Eng., 10, p. 35–44.

Suzuki, M. and Ikada, Y. (2005). Biodegradable polymers in medicine. In: Reis, R. L., San Roman, J. (eds). Biodegradable Systems in Tissue Engineering and Regenerative Medicine.Raton: Boca CRC Press. ISBN: 0-8493-1936-6, p. 3–12.

Szmukler-Moncler, S., Daculsi, G., Delécrin, J., Passuti, N., and Deudon, C. (1992). Calcium-metallic phosphates: a new coating biomaterial? Advanc. Biomater., 10, p. 377–383.

Tanaka, K., Tamura, J., Kawanabe, K., Nawa, M., Oka, M. and Uchida, M. (2002). Ce-TZP/Al$_2$O$_3$ nanocomposites as a bearing material in total joint replacement. J. Biomed. Mater. Res, 63, p. 262–270.

Thian, E.S., Konishi, T., Kawanabe, Y., Lim, P.N., Choong, C., Ho, B. and Aizawa, M. (2013). Zinc-substituted hydroxyapatite: a biomaterial with enhanced bioactivity and antibacterial properties. J. Mater. Sci. Mater. Med., 24(2), p. 437–445.

Thomas, S., Balakrishnan, P. and Sreekala, M.S. (2018). Fundamental Biomaterials: Metals. Woodhead Publ. Inc. ISBN 978-0-081-02205-4. 438 pp.

Thomaz, T.R., Weber, C.R., Pelegrini Jr., T., Dick, L.F.P. and Knörnschild, G. (2010). The negative difference effect of magnesium and of the AZ91 alloy in chloride- and stannate-containg solutions. Corros. Sci., 52 (7), p. 2235–2243.

Thorwarth, G., Falub, C.V., Müller, U., Weisse, B., Voisard, C., Tobler, M. and Hauert, R. (2010). Tribological behavior of DLC-coated articulating joint implants. Acta Biomater., 6(6), p. 2335–2341.

Tilocca, A. and Cormack, A. N. (2009). Modeling the water–bioglass interface by ab-initio molecular dynamics simulations. ACS Appl. Mater. Interfaces, 1, p. 1324–1333.

Tilocca, A. and Cormack, A.N. (2011). The initial stages of bioglass dissolution: a Car-Parrinello molecular-dynamic study of the glass-water interface. Proc. Royal Soc. A, 467(2132). Article ID 20100519.

Todai, M., Nagase, T., Hori, T., Matsugaki, A., Sekita, A. and Nakano, T. (2017). Novel TiNbTaZrMo high-entropy alloy for metallic biomaterials. Scripta Mater., 129, p. 65–68.

Torroni, A., Xiang, C.C., Witek, L., Rodriguez, E.D., Coelho, P.G. and Gupta, N. (2017). Biocompatibility and degradation properties of WE43 Mg alloys with and without heat treatment: in vivo evaluation and comparison in a cranial bone sheep model. J. Cranio-Maxill. Surg., 45(12), p. 2075–2083.

Touchet, T.J. and Cosgriff-Hernandez, E.M. (2016). Hierarchal structure–property relationships of segmented polyurethanes. In: Cooper, S.L. and Guan, J. (eds). Advances in Polyurethane Biomaterials. Cambridge: Woodhead Publishing. ISBN: 978-0-08-100614-6. p. 3–22.

Toyama, T., Kameda, S. and Nishimiyo, N. (2013). Synthesis of sulfate-ion-substituted hydroxyapatite from amorphous calcium phosphate. Bioceram. Dev. Appl., Conf. Proc. ISACB-6. doi: 10.4172/2090-5025 S1-011.

Tran, L.K., Stepien, K.R., Bollmeyer, M.M. and Yoder, C.H. (2017). Substitution of sulfate in apatites. Amer. Mineral., 102, p. 1971–1976.

Ueno, M., Miyamoto, H., Tsukamoto, M., Eto, S., Noda, I., Shobuike, T., Kobatake, T., Sonohata, M. and Mawatari. M. (2016). Silver-containing hydroxyapatite coating reduces biofilm formation by methicillin-resistant staphylococcus aureus in vitro and in vivo. BioMed. Res. Int., 2016, p. 1–7.

Vander Voort, G.F. (2004). Color metallography. In: Vander Voort, G.F. (ed.). ASM Handbook, vol. 9: Metallography and Microstructures, p. 493–512.

Vander Voort, G.F. (2005). Color metallography. Microscopy Today, November 2005, p. 22–27.

Vander Voort, G.F. (2009). Specimen preparation for metallographic examination of medical devices. Adv. Mater. & Proc. April 2009, p.51.

Vander Voort, G. (2011). Microstructure of ferrous alloys. VacAero Intern. Inc., March 2, 2011.

Vander Voort, G. (2011a). Color metallography. VacAero Intern. Inc., May 4, 2011.

Vander Voort, G. (2014). Metallographic preparation of titanium and its alloys. VacAero Intern. Inc., July 9, 2014.

van Howe, R.P., Sierevelt, I.N., van Royen, B.J. and Nolte, P.A. (2015). Titanium-nitride coating of orthopaedic implants: a review of the literature. Biomed. Res. Intern., 2015: 485975.

van Landuyt, K.L., Snauwaert, J., De Munck, J., Pneumans, M., Yoshida, Y., Poitevin, A., Coutinho, E., Suzuki, K., Lambrechts, P. and van Meerbeek, B. (2007). Systematic review of the chemical composition of contemporary dental adhesives. Biomaterials, 28, 3757–3785.

Vassal, M.F., Nuñes-Pereira, J., Miguel, S.P., Correira, I.J. and Silva, A.P. (2019). Microstructural, mechanical and biological properties of hydroxyapatite-$CaZrO_3$ biocomposites. Ceram. Intern., 45(7A), p. 8195–8203.

Vecstaudza, J., Gasik, M. and Locs, J. (2019). Amorphous calcium phosphate materials: Formation, structure and thermal behavior. J. Europ. Ceram. Soc., 39(4), p. 1642–1649.

Vecstaudza, J. and Locs, J. (2017). Novel preparation route of stable amorphous calcium phosphate nanoparticles with high specific surface area. J. Alloys Comp., 700, p. 215–222.

Veronese, F.M., and Pasut, G. (2005). PEGylation, successful approach to drug delivery. Drug Discovery Today, 10, p. 1451–1458.

Von Euw, S., Wang, Y., Laurent, G., Drouet, C., Barbonneau, F., Nassif, N. and Azais, T. (2019). Bone mineral: New insights into ist chemical composition.

Walsh, W.R., Chapman-Sheath, P.J., Cain, S., Debes, J., Bruce, W.J.M., Svehla, M.J. and Gillies, R.M. (2006). A resorbable porous ceramic composite bone graft substitute in a rabbit metaphyseal defect model. J. Orthop. Res., 21(4), p. 653–661.

Wang, H.I., Tang, X.G., Li, W.K., Chen, J.Y., Li, H., Yan, J.Y., Yuan, X.F., Wu, H. and Liu, C.X. (2019). Enhanced osteogenesis of bone marrow stem cells cultured on hydroxyapatite/collagen I scaffold in the presence of low-frequency magnetic field. J. Mater. Sci. Mater. Med., 30:89. https://doi.org/10.1007/s10856-019-6289-8.

Wang, I. C., Lin, J.H.C. and Ju, C.P. (2005). Transmission electron microscopic study of tetracalcium phosphate surface-treated with diammonium hydrogen phosphate solution. Mater. Trans., 46(4), p. 885–890.

Wang, J., Ma, X.Y., Feng, Y.F. and 6 additional authors (2017). Magnesium ions promote the biological behaviour of rat calvarial osteoblasts by activating the PI3K/Akt signaling pathway, Biol. Trace. Elem. Res., 179, p. 284–293.

Wang, K. K., Gustavson, L. J. and Dumbleton, J. H. (1996). Microstructure and properties of a new beta titanium alloy, Ti-12Mo-6Zr-2Fe, developed for surgical implants. In: Brown, S. A. and Lemons, J. E. (eds). Medical Applications of Titanium and its Alloy. ASTM STP 1272. ASTM, West Conshohocken, PA, USA, p. 76–87.

Wang, K. K. (1996). The use of titanium for medical applications in the USA. Mater. Sci. A, 213, p. 134–137.

Wang, S.Q. (2013). Atomic structure modelling of multi-principal-element alloys by the principle of maximum entropy. Entropy, 15(12), p. 5536–5541.

Wang, W.H. and Yeung, K.W.K. (2017). Bone grafts and biomaterials substitutes for bone defect repair: a review. Bioactive Mater., 2(4), p. 224–247.

Wang, Y., Fan, T., Zhou, Z. and He, D. (2016). Hydroxyapatite coating with strong (002) crystallographic texture deposited by micro-plasma spraying. Mater. Letters, 185, p. 484–487.

Watanabe, T. (2011). Grain boundary engineering: Historical perspective and future prospects. J. Mater. Sci., 46(12), p. 4095–4115.

Wauthle, R., van der Stok, J., Yavari, S.A. and 6 additional authors (2015). Additively manufactured porous tantalum implants. Acta Biomater., 14, p. 217–225.

Webster, T.J., Ergun, C., Doremus, R.H. and Lanford, W.A. (2003). Increased osteoblast adhesion on titanium-coated hydroxyapatite that form $CaTiO_3$. J. Biomed. Mater. Res. Part A, 67A(3), p. 976–980.

Wei, D., Zhou, Y., Jia, D. and Wang, Y. (2007). Structure of calcium titanate/titania bioceramic composite coatings on titanium alloy and apatite deposition on their surfaces in a simulated body fluid. Surf. Coat. Technol., 201, p. 8715–9722.

Weinans, H. and Zadpoor, A.A. (eds) (2016). Special issue "Metallic Biomaterials". Metals, 6(3-9). ISSN 2075–4701.

Weinberger, C.R., Boyce, B.L. and Battaile, C.C. (2013). Slip planes in bcc transition metals. Intern. Mater. Rev., 58(5), p. 296–314.

Weiner, S., Traub, W. and Wagner, H. (1999). Lamellar bone structure-function relation. J. Struct. Biol., 126, p. 241–255.

Wendler, M., Beiil, R. and Lohbauer, W. (2019). Factors influencing development of residual stresses during crystallization firing in a novel lithium silicate glass-ceramic. Dental Mater., 35(6), p. 871–882.

WHO/IAEA/UNDP (2005). Chernobyl: the true scale of the accident. 20 years later a UN Report provides definitive answers and ways to repair lives. https://www.who.int/mediacentre/news/releases/2005/pr38/en (accessed March 14, 2019).

Williams, D. (2014). Essential Biomaterials Science. Cambridge: Cambridge University Press. ISBN: 978-0-521-89908-6. p. 511–583.

Willumeit-Römer, R. and Müller, W.D. (eds) (2017). Degradable biomaterials based on magnesium alloys. Special issue. Materials, 10 (1), ISSN 1996–1944.

Wise, S.G., Yeo, G.C., Hiob, M.A., Rnjak-Kovacina, J., Kaplan, D.L. Ng, M.K.C. and Weis, A.S. (2014). Tropoelastin: A versatile, bioactive assembly module. Acta Biomater., 10, p. 1532–1541.

Wolf, M.P., Salieb-Beugelaar, G.B. and Hunziker, P. (2018). PDMS with designer functionalities – properties, modifications strategies, and applications. Progr. Polym. Sci., 83, p. 97–134.

Wolf, S.E., Schüßler, M., Böhm, C.F. and Demmert, B. (2019). Frontiers in bio-inspired mineralization: Addressing mimesis of four-dimensional, hierarchical, and nonclassical growth characteristics of biominerals. In: Bezarra, U.T., Ferreira, H.S. and Barbosa, N.P. (eds). Bio-Inspired Materials. Frontiers in Biomaterials, 6, p. 160–176. Singapore: Bentham Science Publishers Pte. Ltd. ISBN 978-981-14-0688-1.

Wopenka, B. and Pasteris, J.D. (2005). A mineralogical perspective on the apatite in bone. Mater. Sci. Eng. C, 25, p. 131–143.

Wu, C.T., Chang, J. and Zhai, W.Y. (2005). A novel hardystonite bioceramic: preparation and characteristics. Ceram. Inter., 31(1), p. 27–31.

Wu, G.S., Ibrahim, J.M. and Chu, P.K. (2013). Surface design of biodegradable magnesium alloys – a review. Surf. Coat. Technol., 233, p. 2–12.

Wysokowski, M., Jesionowski, T. and Ehrlich, H. (2018). Biosilica as a source for inspiration in biological materials science. Am. Mineral., 103, 665–691.

Xue, L. and Greisler, H.P. (2003). Biomaterials in the development and future of vascular grafts. J. Vasc. Surg., 37, p. 472–480.

Xynos, I. D., Edgar, A. J., Buttery, L. D., Hench, L. L. and Polak, J. M. (2000). Ionic products of bioactive glass dissolution increase proliferation of human osteoblasts and induce insulin-like growth factor II mRNA expression and protein synthesis. Biochem. Biophys. Res. Commun., 276, p. 461–465.

Xynos, I. D., Edgar, A. J., Buttery, L. D. K., Hench, L. L. and Polak, J. M. (2001). Gene-expression profiling of human osteoblasts following treatment with the ionic products of bioglass® 45S5 dissolution. J. Biomed. Mater. Res., 55, p. 151–157.

Yamamoto, A., Honma, R. and Sumita, M. (1998). Cytotoxicity evaluation of 43 metal salts using murine fibroblasts and osteoblastic cells. J. Biomed. Mater. Res., 39, p.331–340.

Yamashita, K., Oikawa, N. and Umegaki, T. (1996). Acceleration and deceleration of bone-like crystal growth on ceramic hydroxyapatite by electric poling. Chem. Mater., 8, p. 2697–2700.

Yang, Y.X., Wang, Q.C., Li, J., Tan, L. and Yang, K. (2019). Enhancing general corrosion resistance of biomedical high nitrogen nickel-free stainless steel by water treatment. Mater. Letters, 251, p. 196–200.

Yannas, I.V. (2004). Natural materials. In: Ratner, B.D., Hoffman, A.S., Schoen, F.J., and Lemons, J.E. (eds). Biomaterials Science: An Introduction to Materials in Medicine. San Diego: Elsevier. ISBN: 0-12-582463-7. p. 127–137.

Yao, J., Radin, S., Leboy, P. S. and Ducheyne, P. (2005). Solution-mediated effect of bioactive glass in poly (lactic-co-glycolic acid)-bioactive glass composites on osteogenesis of marrow stromal cells. J. Biomed. Mater. Res. A, 75, p. 794–801.

Yashima, M., Sakai, A., Kamiyama, T. and Hoshikawa, A. (2003). Crystal structure analysis of ß-tricalcium phosphate $Ca_3(PO_4)_2$ by neutron diffraction. J. Solid State Chem., 175(2), p. 272–277.

Yashima, M., Kawaike, Y. and Tanaka, M. (2007). Determination of precise unit cell parameters of the α-tricalcium phosphate $Ca_3(PO_4)_2$ through high-resolution synchrotron powder diffraction. J. Am. Ceram. Soc., 90(1), p. 272–274.

Yilmazer, Y., Niinomi, M., Akahori, T., Nakai, M. and Tsutsumi, H. (2009). Effects of severe plastic deformation and thermo-mechanical treatments on microstructures and mechanical properties of β-type titanium alloys for biomedical applications In: Proc. PFAMXIII. (2009). p. 1401–1410.

Yin, L. and Stoll, R. (2014). Ceramics in restorative dentistry. In: Low, I.M. (ed.) Advanced Ceramic Matrix Composites, p. 624–655. Abingdon: Woodhead Publ. Ltd. ISBN 978-0-85709-120-8. 734 pp.

Yoneyama, T. and Miyazaki, S. (eds) (2009). Shape Memory Alloys for Biomedical Applications. Woodhead Publ. CRC. ISBN 978-1-84569-344-2. 352 pp.

Yuan, H., Barbieri, D., Luo, X., van Blitterwijk, C.A. and De Bruijn, J.D. (2017). Calcium phosphates and bone induction. In: Comprehensive Biomaterials II, 1, p. 333–349.

Zardiackas, D., Mitchell, D. W. and Disegi, J. A. (1996). Characterization of Ti-15Mo beta titanium alloy for orthopedic implant. In: Medical applications of titanium and its alloy. In: Brown, S. A. and Lemons, J. E. (eds). Medical Applications of Titanium and its Alloy. ASTM STP 1272. ASTM, West Conshohocken, PA, USA. P. 60–75.

Zarone, F., Russo, S. and Sorrentino, R. (2011). From porcelain-fused-to-metal to zirconia: Clinical and experimental considerations. Dental Mater., 27, p. 83–96.

Zarone, F., Ferrari, M., Mangano, F.G., Leone, R. and Sorrentino, R. (2016). „Digitally Oriented Materials": Focus on lithium disilicate ceramics. Int. J. Dent., 2016: 9840594.

Zdrahala, R.J. and Zdrahala, I.J. (1999). Biomedical applications of polyurethanes: A review of past promises, present realities, and a vibrant future. J. Biomater. Appl., 14, p. 67–90.

Zeng, Y.Z., Hoque, J. and Varghese, S. (2019). Biomaterial-assisted local and systemic delivery of bioactive agents for bone repair. Acta Biomater., 93, p. 152–168.

Zhang, J., Tasan, C.C., Lai, M.J., Dippel, A.C. and Raabe, D. (2017). Complexion-mediated martensitic phase transformation in titanium. Nat. Commun., 8, 14210.

Zhang, J.W., Luo, X.M., Barbieri, D., Barradas, A.M.C., De Bruijn, J.D., Van Blitterswijk, C.A. and Yuan, H.P. (2014). The size of surface microstructures as an osteogenic factor in calcium phosphate ceramics. Acta Biomater., 10, p. 3254–3263.

Zhang, T.F., Deng, Q.Y., Liu, B., Wu, B.J., Jing, F.J., Leng, Y.X. and Huang, N. (2015a). Wear and corrosion properties of diamond like carbon (DLC) coatings on stainless steel, CoCrMo and Ti6Al4V substrates. Surf. Coat. Technol., 273, p. 12–19.

Zhang, W., Titze, M., Cappi, B., Wirtz, D., Telle, R. and Fischer, H. (2010). Improved mechanical long-term reliability of hip resurfacing prostheses by using silicon nitride. J. Mater. Sci. Mater. Med., 21(11), p. 3049–3057.

Zhang, Y., Reed, B.W., Chung, F.R. and Koski, K.J. (2015c). Mesoscale elastic properties of marine sponge spicules. J. Struct. Biol., 193, p. 67–74.

Zhao, D.L., Wang, T.T., Hoagland, W. and 8 additional authors. (2016b). Visual H_2 sensor for monitoring biodegradation of magnesium implants in vivo. Acta Biomater., 45, p. 399–409.

Zhao, D.L., Wang, T.T., Nahan, K. and 8 additional authors. (2017). In vivo characterization of magnesium alloy biodegradation using electrochemical H_2 monitoring, ICP-MS, and XPS. Acta Biomater., 50, p. 556–565.

Zhao, D.W., Witte, F., Lu, F.Q., Wang, J.L., Li, J.L. and Qin, L. (2016a). Current status on clinical application of magnesium-based orthopedic implants: a review from clinical translational perspective. Biomater., 112, p. 287–302.

Zhao, S.F., Jiang, Q.H., Peel, S., Wang, X.X. and He, F.M. (2013). Effects of magnesium-substituted nanohydroxyapatite coating on implant osseointegration, Clin. Oral. Implan. Res., 24 Suppl. A100, p. 34–41.

Zhao, X.F., Niinomi, M., Nakai, M. and Hieda, J. (2012). Beta-type Ti-Mo alloys with changeable Young's modulus for spinal fixation applications. Acta Biomater., 8, p. 1990–1997.

Zhao, X. L., Niinomi, M. and Nakai, M. (2011a). Relationship between various deformation-induced products and mechanical properties in metastable Ti-30Zr-7Mo alloys for biomedical applications. J. Mech. Behav. Biomed. Mater., 4, p. 2009–2016.

Zhao, X. L., Niinomi, M., Nakai, M., Miyamoto, G. and Furuhara, T. (2011b). Microstructures and mechanical properties of metastable Ti-30Zr-(Cr,Mo) alloys with changeable Young's modulus for spinal fixation applications. Acta Biomater., 7, p. 3230–3236.

Zimmermann, H. (2013). Thermal sprayed surface layer made of titanium on a non-metallic substrate of an orthopedic implant, comprises an X-ray-sensitive mixture made of biocompatible indicator metal in relation to titanium. European patent EP2199423-B1.

Chapter 3

Abe, Y., Kokubo, T. and Yamamuro, T. (1990). Apatite coatings on ceramics, metals and polymers utilizing a biological process. J. Mater. Sci.: Mater. Med., 1, p. 233–238.

Acquah, C., Datskov, I., Mawardi, A., Zhang, F., Achenie, L. E. K., Pitchumani, R. and Santos, E. (2006). Optimization under uncertainty of a composite fabrication process using a deterministic one-stage approach. Computer Chem. Eng., 30, p. 947–960.

Adams, J.W., Duz, V.A., Moxson, V.S. and Roy, W.R. (2007). Low cost blended elemental titanium powder metallurgy. https://cdn.ymaws.com/titanium.org/resource/resmgr/2005_2009_pa pers/Moxson-Adams_Final_2007.pdf (accessed December 6, 2019).

Agarwal, B. D., Broutman, L. J. and Chandrashekhara, K. (4th Ed.). (2017). Analysis and Performance of Fiber Composites. John Wiley & Sons: Hoboken. ISBN: 978-1-119-38998-9.

Agarwal, S., Wendorff, J. H. and Greiner, A. (2008). Use of electrospinning technique for biomedical applications. Polymer, 49, p.5603–5621.

Agarwal, S., Burgard, M., Greiner, A., and Wendorff, J. H. (2016). Electrospinning – A Practical Guide to Nanofibers. Berlin: De Gruyter. ISBN: 978-3-11-033180-6.

Agassant, J.-F., Avenas, P., Carreau P. J., Vergnes, B. and Vincent, M. (2017). Polymer Processing, Principles and Modeling. München: Carl-Hanser-Verlag. ISBN 978-1-56990-605-7.

Agrawal, C. M., Ong, J. L., Appleford, M. R. and Mani, G. (2014). Introduction to Biomaterials. Basic Theory with Engineering Application. Cambridge: Cambridge University Press. ISBN: 978-1-316-61130-2. p. 155–161.

Agrawal, K., Singh, G., Puri, D. and Prakash, S. (2011). Synthesis and characterization of hydroxyapatite powder by sol-gel method for biomedical application. J. Min. Mater. Charact. Eng., 10(8), p. 727–734.

Akahori, T., Niinomi, M., Isohama, R. and Suzuki, A. (2000). Effect of thermomechanical treatment on mechanical properties of cast Ti-6Al-7Nb alloy for dental applications. J. Jpn. Inst. Metal. Mater., 64, p. 895–902.

Akahori, T., Niinomi, M., Harada, M., Takeda, J., Katsura, S., Takeuchi, T., and Fukui, H. (2005). Mechanical properties and cyto-toxicity of newly designed ß- type Ti alloys for dental applications. J. Jpn. Metal. Mater., 69, p. 96–102.

Aldinger, F. and Weberruß, V.A. (2010). Advanced Ceramics and Future Materials. An Introduction to Structures, Properties, Technologies, Methods. Weinheim: Wiley-VCH. ISBN 978-3-527-32157-5.

Araci, K., Mangabhai, D. and Akhtar, K. (2015). Production of titanium by the Armstrong Process. In: Ma, Q. and Froes, F.H. (eds). Titanium Powder Metallurgy. p. 149–162. Elsevier. ISBN 978-0-12-800054-0. 648 pp.

Aranishi, Y. and Nishio, Y. (2017). Cellulosic fiber produced by melt spinning. In: Nishio, Y., Teramoto, Y., Kusumi, R., Sugimura, K. and Aranishi, Y. (eds). Blends and Graft Copolymers of Cellulosics – Toward the Design and Development of Advanced Films and Fibers. Cham: Springer. ISBN 978-3-319-55321-4. p. 109–125.

Assadi, H., Gärtner, F., Stoltenhoff, T. and Kreye, H. (2003). Bonding mechanism in cold gas spraying. Acta Mater., 51(15), p. 4379–4394.

Barillas, L., Cubero-Sesin, J.M. and Vargas-Blanco, I. (2018a). Hydroxyapatite coatings of polymers using a custom low-energy plasma spray system. IEEE Trans. Plasma Sci., 46(7), p. 2420–2424.

Barillas, L., Testrich, H., Cubero-Sesin, J.M., Quade, A., Vargas, V.I., Polak, M. and Fricke, K. (2018b). Bioactive plasma sprayed coatings on polymer substrates suitable for orthopedic applications: a study with PEEK. IEEE Trans. Rad. Plasma Med. Sci., 2(5), p. 520–525.

Bastan, F.E., Erdogan, G., Moskalewicz, T. and Ustel, F. (2017). Spray drying of hydroxyapatite powders: The effect of spray drying parameters and heat treatment on the particle size and morphology. J. Alloys Comp., 724, p. 586–596.

Bewilogua, K. and Hofmann, D. (2014). History of diamond-like carbon films – from first experiments to worldwide applications. Surf. Coat. Technol., 242, p. 214–225.

Billiet, T., Vandenhaute, M., Schelfhout, J., Van Vlierberghe, S. and Dubruel, P. (2012). A review of trends and limitations in hydrogel-rapid prototyping for tissue engineering. Biomaterials, 33, p. 6020–6041.

Billmeyer, Jr., F.W. (1984). Textbook of Polymer Science. New York: Wiley. ISBN: 0-471-82834-3. p. 25–81.

Black, J.T. and Kohser, R.A. (2012). DeGarmo's Materials and Processes in Manufacturing, 11th ed. Wiley. ISBN 978-1-118-16373-3.

Bocanegra-Bernal, M.H., Dominguez-Rios, C., Garcia-Reyes, A., Aguilar-Elguezabal, A., Echeberria, J. and Nevarez-Rascon, A. (2009). Hot isostatic pressing (HIP) of α-Al$_2$O$_3$ submicron ceramics pressureless sintered at different temperatures: Improvement in mechanical properties for use in total hip arthroplasty (THA). Intern. J. Refract. Metals Hard Mater., 27(5), p. 900–906.

Bose, S. and Saha, S.K. (2003). Synthesis and characterization of hydroxyapatite nanopowders by emulsion technique. Chem. Mater., 15(23), p. 4464–4469.

Box, G.E.P., Hunter, J.S. and Hunter, W.G. (2005). Statistics for Experimenters: Design, Innovation, and Discovery. 2nd ed. Wiley-Interscience. ISBN 978-0-4717-1813.0. 633 pp.

Brinker, C.J. and Scherer, G.W. (1984). Relation between sol-to-gel and gel-to-glass conversions. In: Hench. L.L. and Ulrich, D.R. (eds). Ultrastructure Processing of Ceramics, Glasses and Composites. p. 43–59. New York: Wiley-Interscience. ISBN 978-0-4718-9669-2.

Brinker, C.J. and Scherer, G.W. (1990). Sol-Gel Science: The Physics and Chemistry of Sol-Gel Processing. Gulf Professional Publ., ISBN 978-0-1213-4970-7. 908 pp.

Cañas, E., Vicent, M., Orts, M.J. and Sánchez, E. (2017). Bioactive glass coatings by suspension plasma spraying from glycolether-based solvent feedstock. Surf. Coat. Technol., 318, p. 190–197.

Cañas, E., Orts, M.J., Boccaccini, A.R. and Sánchez, E. (2018). Solution precursor plasma spraying (SPPS): A novel and simple process to obtain bioactive glass coatings. Mater. Lett., 223, p. 198–202.

Cañas, E., Vicent, M., Bannier, E., Carpio, F., Orts, M.J. and Sánchez, E. (2019a). Effect of particle size on processing of bioactive glass powder for atmospheric plasma spraying, J. Eur. Ceram. Soc., 36, p. 837–845.

Cañas, E., Orts, M.J., Boccaccini, A.R. and Sánchez, E. (2019b). Microstructural and in vitro characterization of 45S5 bioactive glass coatings deposited by solution precursor plasma spraying (SPPS). Surf. Coat. Technol., 371, p. 151–160.

Candidato, R.T., Sokołowski, P., Pawłowski, L., Lecompte-Nana, G., Constantinescu, C. and Denoirjean, A. (2016). Development of hydroxyapatite coatings by solution precursor plasma spraying process and their microstructural characterization. Surf. Coat. Technol., 318, p. 39–49.

Cardon, L. K., Ragaert, K. J., De Santis, R. and Gloria, A. (2017). Design and fabrication methods for biocomposites. In: Ambrosio, L. (ed.). Biomedical Composites. Duxford: Woodhead Publishing. ISBN: 978-0-08-100752-5. p. 17–36.

Carter, G.A., Hart, R.D., Rowles, M., Ogden, M.I., Buckley, C.E. (2009). Industrial precipitation of yttrium chloride and zirconyl chloride: effect of pH on ceramic properties for yttria partially stabilised zirconia. J. Alloy Comp., 480(2), p. 639–644.

Casalini, T. and Perale, G. (2012). Processing of bioresorbable and other polymers for medical applications. In: Jenkins. M. and Stamboulis, A. (eds). Durability and Reliability of Medical Polymers. Cambridge: Woodhead Publishing. ISBN 978-1-84569-929-1. p. 49–76.

Castellanos, M.I., Mas-Moruno, C., Grau, A. and 8 additional authors (2017). Functionalization of CoCr surfaces with cell adhesive peptides to promote HUVECs adhesion and proliferation. Appl. Surf. Sci., 393, p. 82–92.

Catoira, M.C., Fusaro, L., Di Francesco, D., Ramella, M. and Boccafoschi, F. (2019). Overview of natural hydrogels for regenerative medicine applications. J. Mater. Sci.: Mater. Med., 30, 115.

Cengiz, B., Gokce, Y., Yildiz, N., Aktas, Z. and Calimli, A. (2008). Synthesis and characterization of hydroxyapatite nanoparticles. Coll. Surf. A, 322(1-3), p. 29–33.

Chambard, M., Marsan, O., Charvillat, C., Grossin, D., Fort, P., Rey, C., Gitzhofer, F. and Bertrand, G. (2019). Effect of the deposition route on the microstructure of plasma-sprayed hydroxyapatite coatings. Surf. Coat. Technol., 371, p. 68–77.

Chandrasekaran, M. (2019). Forging of metals and alloys for biomedical applications. In: Metals for Biomedical Devices, 2nd. Woodhead Publ. Ser. Biomater. p. 293–310. Duxford: Elsevier.

Chang, A. (1993). Fractals in biological systems. https://poignance.coiraweb.com/math/Fractals/FractBio/FractBio.html (accessed February 10, 2020).

Chen, G.Z., Gordo, E. and Fray, D.J. (2004). Direct electrolytic preparation of chromium powder. Metall. Mater. Trans. B, 35(2), p. 223–233.

Chen, Q.Y., Zou, Y.L., Chen, X. and 5 additional authors (2019). Morphological, structural and mechanical characterization of cold sprayed hydroxyapatite coating. Surf. Coat. Technol., 357, p. 910–923.

Chia, H. N. and Wu, B. M. (2015). Recent advances in 3D printing of biomaterials. J. Biol. Eng., 9:4.

Chou, D.T. (2016). Fundamental study of the development and evaluation of biodegradable Mg-Y-Ca-Zr based alloys as novel implant materials. Ph.D. Thesis, University of Pittsburgh; Publication Number: AAT 10645879. ISBN: 978-0-3551-9133-2. Dissertation Abstracts Intern., 79-01(E), Section: B.; 248 pp.

Chung, T.S.N. (2008). Fabrication of hollow-fiber membranes by phase inversion. In: Li, N.N., Fane, A.G., Winston Ho, W.S. and Matsuura, T. (eds). Advanced Membrane Technology and Applications. Hoboken: John Wiley & Sons. ISBN 9780471731672. p. 821–839.

Cifuentes, A. and Borros, S. (2013). Comparison of two different plasma surface-modification techniques for the covalent immobilization of protein monolayers. Langmuir, 29(22), p. 6645–6651.

Cinca, N., Vilardell, A.M., Dosta, S. and 6 additional authors (2016). A new alternative for obtaining nanocrystalline bioactive coatings: Study of hydroxyapatite deposition mechanisms by cold gas spraying. J. Am. Ceram. Soc., 99, p. 1420–1428.

Cizek, J. and Khor, K.A. (2012). Role of in-flight temperature and velocity of powder particles on plasma-sprayed hydroxyapatite coating characteristics. Surf. Coat. Technol., 206, p. 2181–2191.

Cizek, J. and Matejicek, J. (2018). Medicine meets thermal spray technology: a review of patents. J. Thermal Spray Technol., 27(8), p. 1251–1279.

Clarke, I.C. and Willmann, G. (1994). Structural ceramics in orthopedics. In: Cameron, H.U. (ed.). Bone Implant Interface. St. Louis, Baltimore, Boston, Chicago, London, Madrid, Philadelphia, Sydney, Toronto: Mosby. p. 203–252. ISBN 0-8016-6483-7.

Cofino, B., Fogarassy, Millet, P. and Lodini, A. (2004). Thermal residual stresses near the interface between plasma-sprayed hydroxyapatite coating and titanium substrate: Finite Element Analysis and synchrotron radiation measurements, J. Biomed. Mater. Res. A, 70, p. 20–27.

Cox, S.C., Jamshidi, P., Grover, L.M. and Mallick, K.K. (2014). Low temperature aqueous precipitation of needle-like nanophase hydroxyapatite. J. Mater. Sci.: Mater. Med., 25(1), p. 37–46.

Dabkowska, H.A. and Dabkowski, A.B. (2010). Crystal growth of oxides by optical floating zone technique. In: Dhanara, J G., Byrappa, K., Prasad, V. and Dudley, M. (eds). Springer Handbook of Crystal Growth. Berlin, Heidelberg; Springer, ISBN 978-3-540-74182-4. p. 367–391.

Damodaran, V. B., Bhatnagar, D. and Murthy N. S. (2016). Biomedical Polymers – Synthesis and Processing. Heidelberg: Springer. ISBN 978-3-319-32053-3. p. 54–71

Deliormanh, A.M. and Yildirim, M. (2016). Sol-gel synthesis of 13-93 bioactive glass powders containing therapeutic agents. J. Austral. Ceram. Soc., 52(2), p. 9–19.

Demukai, N. (2003). Development of a new type titanium casting technology/LEVICAST process. Curr. Adv. Mater. Proc., Iron Steel Inst. Jpn., 16, p. 1206.

Demukai, N. and Tshuishima, K. -I. (2005). Development of a new type titanium casting technology LEVICAST process. J. Adv. Mater. -Covina. 37, p. 12–16.

Diaz-Rodriguez, S., Chevallier, P., Paternoster, C., Montaño-Machado, V., Noël, C., Houssiau, L. and Mantovani, D. (2019). Surface modification and direct plasma amination of L605 CoCr alloys: on the optimization of the oxide layer for application in cardiovascular implants. RSC Advances, 9, p. 2292–2301.

Dickson, J.S. (2015). EuroChem reveals phosphate expansion plant at Kovdor and beyond. http://www.eurochemgroup.com/wp-content/uploads/2016/07/EuroChem-reveals-phosphate-expansion-plant-at-Kovdor-and-beyond-Indust.pdf (accessed March 8, 2019).

Dietrich, A., Heimann, R.B. and Willmann, G. (1996). The colour of medical-grade zirconia (Y-TZP). J. Mater. Sci. Mater. Med., 7, p. 559–565.

Doi, H., Yoneyama, T., Kobayashi, E. and Hamanaka, H. (1998). Mechanical properties and corrosion resistance of Ti-5Al-13Ta alloy casting. Jpn. Soc. Dental Mater. Device,. 17, p. 47–52.

Dorozhkin, S.V. (2019). Functionalized calcium orthophosphates (CaPO₄) and their biomedical applications. J. Mater. Chem. B, 7, p. 7471–7489.

Dubbo Zirconia Project (2013). Environmental impact statement. Executive summary. Report No. 545/04. R.W. Corkery & Co. Pty Limited: Sydney, Orange, Brisbane.

Dyshlovenko, S., Pateyron, B., Pawlowski, L., Murano, D. (2004). Numerical simulation of hydroxyapatite powder behaviour in plasma jet. Surf. Coat. Technol., 179, p. 110–117.

Earl, J.S., Wood, D.J. and Milne, S.J. (2006). Hydrothermal synthesis of hydroxyapatite. J. Phys. Conf. Ser., 261, p. 268.

Eckstein, U.R., Detsch, R., Khansur, N.H. and 5 additional authors (2019). Bioactive glass coating using aerosol deposition. Ceram. Intern., https://doi.org/10.1016/j.ceramint.2019.04.197.

El-Barawy, K.A., El-Tawil, S.Z. and Francis, A.A. (1999). Production of zirconia from zircon by thermal reaction with calcium oxide. J. Ceram. Soc. Japan, 107(2), p. 97–102.

Elias, H.-G. (1977). Macromolecules 2, Synthesis and Materials. New York: Springer Science+Business Media. ISBN: 978-1-4615-7366.

Ellingham, H. J. T. (1944). Reducibility of oxides and sulphides in metallurgical processes. J. Soc. Chem. Ind. (London), 69(5), p. 125.

Engvall, E. and Perlmann, P. (1972). Enzyme-linked immunosorbent assay, Elisa. J. Immunol., 109 (1), p. 129–135.

Eshraghi, S and Das, S. (2010). Mechanical and microstructural properties of polycaprolactone scaffolds with 1-D, 2-D, and 3-D orthogonally oriented porous architectures produced by selective laser sintering. Acta Biomater., 6, p. 2467–2476.

Evans, K.A. (1996). The manufacture of alumina and its use in ceramics and related applications. Key Eng. Mater., 122-124, p. 489–526.

Fadeeva, E., Deiwick, A., Chichkov, B. and Schlie-Wolter, S. (2014). Impact of laser-structured biomaterial interfaces on guided cell responses. Interface Focus, 4(1), 20130048.

Fang, Y., Agrawal, D., Skandan, G. and Jain, M. (2004). Fabrication of translucent MgO ceramics using nanopowders. Mater. Letters, 58, p. 551–554.

Fang, Z. Z., Paramore, J. D., Sun, P., Chandran, K. S. R., Zhang, Y., Xia, Y., Cao, F., Koopman, M. and Free, M. (2018). Powder metallurgy of titanium-past, present, and future. Int. Mater. Review, 63, p. 407–459.

Faure, J., Drevet, R., Lemelle, A., Ben Jaber, N., Tara, A., El Btaouri, H., and Benhayoune, H. (2015). A new sol–gel synthesis of 45S5 bioactive glass using an organic acid as catalyst. Mater. Sci. Eng. C, 47, p. 407–412.

Fazan, F. and Marquis, P.M. (2000). Dissolution behavior of plasma-sprayed hydroxyapatite coatings. J. Mater. Sci.: Mater. Med., 11, p.787–792.

FDA, Guidance for Industry and FDA Staff – Class II Special Controls Guidance Document: Root-form Endosseous Dental Implants and Endosseous Dental Abutments. Silver Springs: U.S. Dept. of Health and Human Services. 2004.

FDA, Guidance for Industry and FDA Staff – Non-clinical Information for Femoral Stem Prostheses. Silver Springs: U.S. Dept. of Health and Human Services. 2007.

Fedel, M. (2013). Blood compatibility of diamond-like carbon (DLC) coatings. In: Narayan, R. (ed.). Diamond-Based Materials for Biomedical Applications. Woodhead Publ. Series in Biomaterials. ISBN 978-0-85709-340-0. 296 pp.

Fray, D. J., Chen, G. Z. and Farthing, T. W. (2000). Direct electrochemical reduction of titanium dioxide to titanium in molten calcium chloride. Nature, 407(6802), p. 361–364.

Froes, F.H., Eylon, D. and Suryanarayana, C. (1990). Thermochemical processing of titanium alloys. J. Min. Met. Mater. Soc., 42(3), p. 26–29.

Gadow, R., Killinger, A. and Stiegler, A. (2010). Hydroxyapatite coatings for biomedical applications deposited by different thermal spray techniques. Surf. Coat. Technol., 205(4), p. 1157–1164.

Garcia, A.J. (2011). Surface modification of biomaterials. In: Atala, A., Lanza, R., Thomson, J.A. and Nerem, R. (eds). Principles of Regenerative Medicine. Academic Press. ISBN 978-0-12-381422-7. 1202 pp.

Gentile, F., Tirinato, L., Battista, E., Causa, F., Liberale, C., di Fabrizio, E.M. and Decuzzi, P. (2010). Cells preferentially grow on rough substrates. Biomaterials, 31, p. 7205–7212.

Gittens, R.A., Olivares-Navarrete, R., Schwartz, Z. and Boyan, B.D. (2014). Implant osseointegration and the role of microroughness and nanostructures: Lessons for spine implants. Acta Biomater., 10, p. 3363–3371.

Götze, J., Hildebrandt, H. and Heimann, R.B. (2001). Charakterisierung des in vitro-Resorptionsverhaltens von plasmagespritzten Hydroxylapatit-Schichten [Characterization of the in vitro-resorption behavior of plasma-sprayed hydroxyapatite coatings]. BIOmaterialien, 2(1), p. 54–60.

Goldschmidt, H. (1898). Über ein neues Verfahren zur Darstellung von Metallen und Legierungen mittelst Aluminiums. Liebig's Ann. Chem., 301(1), p. 19–28.

Green, S. (2019). Compounds and composite materials (2nd Ed) In: Kurtz, S.M. (ed.). PEEK Biomaterial Handbook. Oxford: William Andrew. ISBN978-0-12-812524-3. : p. 17–51.

Gross, K.A. and Phillips, M.R. (1998). Identification and mapping of the amorphous phase in plasma-sprayed hydroxyapatite coatings using scanning cathodoluminescence microscopy, J. Mater. Sci. Mater. Med., 9(12), p. 797–802.

Gross, K.A. and Saber-Samandari, S. (2009). Revealing mechanical properties of a suspension plasma sprayed coating with nanoindentation. Surf. Coat. Technol., 203, p. 2995–2999.

Habashi, F. (2005). A short history of hydrometallurgy. Hydrometall., 79(1-2), p. 15–22.

Hao, L. and Lawrence, J. (2005). Laser Surface Treatment of Bio-Implant Materials. John Wiley & Sons. Ltd. ISBN 978-0-4700-1687-9.

Hagiwara, K., Tachibana, T., Sasaki, K., Yoshida, Y., Shirakawa, N., Nagasawa, T., Narushima, T. and Nakano, T. (2009). Oxygen distribution in titanium single crystal fabricated by optical floating-zone method under extremely low oxygen partial pressur. eMater. Trans., 50, p. 2709–2715.

Hahn, B.D., Park, D.S., Choi, J.J., Ryu, J., Yoon, W.H., Kim, K.H. and Kim, H.E. (2009). Dense nanostructured hydroxyapatite coating on titanium by aerosol deposition. J. Am. Ceram. Soc., 92(3), p. 683–687.

Hameed, P., Gopal, V., Bjorklund, S., Ganvir, A., Sen, D., Markocsan, N. and Manivasagam, G. (2019). Axial suspension plasma spraying: An ultimate technique to tailor Ti6Al4V surface with HAp for orthopaedic applications. Coll. Surf. B: Biointerfaces, 173, p. 806–815.

Harun, W.S.W. and 6 additional authors (2018). A comprehensive review of hydroxyapatite-based coatings adhesion on metallic biomaterials. Ceram. Intern., 44 (2), p. 1250–1268.

Hassan-Pour, S., Vonderstein, C., Achimovičová, M., Voigt, V., Gock, E. and Friedrich, B. (2015). Aluminothermic production of titanium alloys (part 2): Impact of activated rutile on process sustainability. Metall. Mater. Eng., 21(2), p. 101–114.

Hassler, M. (2012). Other commonly used biomedical coatings: pyrolytic carbon coatings. In: Driver. M. (ed.). Coatings for Biomedical Applications. Woodhead Publ. Series in Biomaterials. ISBN 978-1-84569-568-2. 376 pp.

Heimann, R.B. (1999). Design of novel plasma-sprayed hydroxyapatite-bond coat bioceramic systems. J. Thermal Spray Technol., 8(4), p. 597–604.

Heimann, R.B. (2006). Thermal spraying of biomaterials. Surf. Coat. Technol., 201(5), p. 2012–2019.

Heimann, R.B. (2008). Plasma Spray Coating. Principles and Applications. 2nd ed. Weinheim: Wiley-VCH. ISBN 978-3-527-32050-9. 427 pp.

Heimann, R.B. (2010a). Classic and Advanced Ceramics. From Fundamentals to Applications. Weinheim: Wiley-VCH. ISBN 978-3-527-32517-7. 553 pp.

Heimann, R.B. (2010b). Better quality control: Stochastic approaches to optimize properties and performance of plasma-sprayed coatings. J. Thermal Spray Technol., 19(4), p. 765–778.

Heimann, R.B. (2011). On the self-affine fractal geometry of plasma-sprayed surfaces. J. Thermal Spray Technol., 20(4), p. 898–908.

Heimann, R.B. (2016a). Plasma-sprayed hydroxyapatite-based coatings: Chemical, mechanical, microstructural, and biomedical properties. J. Thermal Spray Technol., 25(5), p. 827–850.

Heimann, R.B. (2016b). The challenge and promise of low-temperature bioceramic coatings: An editorial. Surf. Coat. Technol., 301, p. 1–5.

Heimann, R.B. (2017a). Hydroxyapatite coatings: Applied mineralogy research in the bioceramics field. In: Heuss-Aßbichler, S., Amthauer, G. and John, M. (eds). Highlights in Applied Mineralogy. Berlin: De Gruyter. p. 301–316. ISBN 978-3-11-04912-2-7.

Heimann, R.B. (2017b). Osseoconductive and corrosion-inhibiting plasma-sprayed calcium phosphate coatings for metallic medical implants. Metals, 7(11): 468.

Heimann, R.B. (2018). Plasma-sprayed hydroxylapatite coatings as biocompatible intermediaries between inorganic implant surfaces and living tissue. J. Thermal Spray Technol., 27(8), p. 1212–1237.

Heimann, R.B. and Lehmann, H.D. (2015). Bioceramic Coatings for Medical Implants. Trends and Techniques. Weinheim: Wiley-VCH. ISBN 978-3-527-33743-9. 467 pp.

Heimann. R.B. and Willmann, G. (1998). Irradiation-induced colour changes of Y-TZP ceramics, Brit. Ceram. Trans., 97, p. 185–188.

Heimann. R.B. and Wirth, R. (2006). Formation and transformation of amorphous calcium phosphates on titanium alloy surfaces during atmospheric plasma spraying and their subsequent in vitro performance. Biomaterials, 27, p. 823–831.

Heimann, R.B., Graßmann, O., Hempel, M., Bucher, R. and Härting, M. (2000). Phase content, resorption resistance and residual stresses of bioceramic coatings. In: Applied Mineralogy in Research, Economy, Technology, Ecology and Culture. Proc. 6th Intern. Congress on Applied Mineralogy, ICAM2000, Göttingen, Germany, July 17-19, pp. 155–158.

Heimann, R.B., Graßmann, O., Zumbrink, T. and Jennissen, H.P. (2001). Biomimetic processes during in vitro leaching of plasma-sprayed hydroxyapatite coatings for endoprosthetic applications. Mater.-wiss. Werkstofftechn., 32, p. 913–921.

Hench, L.L. (2013). Chronology of bioactive glass development and clinical applications. New J. Glass Ceram., 3 (2), Article ID: 30885, 7 pages.

Henderson, T. M. A., Ladewig, K., Haylock, D. N., McLean, K. M. and O'Connor, A. J. (2013). Cryogels for biomedical applications. J. Mater. Chem. B, 1, p. 2682–2695.

Hesse, C., Hengst, M., Kleeberg, R. and Götze, J. (2008). Influence of experimental parameters on spatial phase distribution in as-sprayed and incubated hydroxyapatite coatings. J. Mater. Sci. Mater. Med., 19(10), p. 3235–3241.

Hildebrand, H.F., Blanchemain, N., Mayer, G., Zhang, Y.M., Melnyk, O., Morcellet, M. and Martel, B. (2005). Functionalization of biomaterials. Key Eng. Mater., 288/289, p. 47–50.

Hollister, S. J. (2005). Porous scaffold design for tissue engineering. Nature Mater., 4, p. 518–524.

Hollitt, M.J., Mcclelland, R.A., Liddy, M.J., Grey, I.E. and Fleming, C.A. (1997). Removal of radioactivity from zircon. Patent EP0582598 B1.

Holt, G., Murnaghan, C., Reilly, J. and Meek, R.M, (2007). The biology of aseptic osteolysis. Clin. Orthop. Relat. Res., 460, p. 240–252.

Huang, L. G., Kong, F. T., Chen, Y. Y., Xiao, S. L. and Xu, L. J. (2013). Effects of trace TiB_2 on microstructure in cast titanium alloys. Int. J. Cast Metals Review, 25(6),p.358–363.

Hutmacher, D. W., Schantz, T., Zein, I., Ng, K. W., Teoh, S. H. and Tan, K. C. (2001). Mechanical properties and cell cultural response of polycaprolactone scaffolds designed and fabricated via fused deposition modeling. J. Biomed. Mater. Res., 55, p. 203–216.

Hutmacher, D. W., Sittinger, M. and Risbud, M. V. (2004). Scaffold-based tissue engineering: rationale for computer-aided design and solid free-form fabrication systems. Trends Biotechnol., 22, p. 354–362.

Hutmacher, D. W., and Woodruff, M. A. (2008). Design, fabrication, and characterization of scaffolds via Solid Free-Form fabrication techniques. In: Chu, P. K. and Liu, X. (eds). Biomaterials Fabrication and Processing Handbook. Boca Raton: CRC Press. ISBN 978-0-8493-7973-4. p. 45–67.

Hyslop, D. J. S., Abdelkader, A. M., Cox, A. and Fray, D. J. (2010). Electrochemical synthesis of a biomedically important Co-Cr alloy. Acta Mater., 58, p. 3124–3130.

Iizuka, A., Ouchi, T. and Okabe, T.H. (2019). Ultimate deoxidation method of titanium utilizing Y/YOCl/YCl$_3$ equilibrium. Metal. Mater. Trans. B, https://doi.org/10.1007/s11663-019-01742-6 (accessed Jan 14, 2020).

Iler, R. K. (1986). Inorganic colloids for forming ultrastructures, In: Hench, L.L. and Ulrich, D.R. (eds). Science of Ceramic Chemical Processing, pp. 3–20, New York, Chichester, Brisbane, Toronto, Singapore: Wiley. ISBN 0-471-82645-6.

Inagaki, I. (2008). Forging. In: Niinomi, M. (ed.) Fundamentals of Titanium and Its Working. p. 107–120. Tokyo: Korona Publishing Co., Ltd.

Innocenzi, P. (2019). The Sol-to-Gel Transition. 2nd ed. Springer Briefs in Materials. ISSN 2192-1091.

Ino, J.M., Chevallier, P., Letourneur, D., Mantovani, D. and Le Visage, C. (2013). Plasma functionalization of poly(vinyl alcohol) hydrogel for cell adhesion enhancement. Biomaterials, 3(4), e25414.

IPAI (1995). International Primary Aluminium Institute, London.

Ivanyuk, G.Yu., Kalashnikov, A.O., Pakhomovsky, Ya.A. and 6 additional authors (2016). Economic minerals of the Kovdor baddeleyite-apatite-magnetite deposit, Russia: Mineralogy, spatial distribution and ore processing optimization. Ore Geol. Rev., 77, p. 279–311.

Janik, H. and Marzec, M. (2015). A review: Fabrication of porous polyurethane scaffolds. Mater.Sci. Eng., C48, p. 586–591.

Jiang, D., Zhou, H., Wan, S., Cai, G.Y. and Dong, Z.H. (2018). Fabrication of superhydrophobic coating on magnesium alloy with improved corrosion resistance by combining micro-arc oxidation and cyclic assembly. Surf. Coat. Technol., 339, p. 155–166.

Jin, W.H. and Chu, P.K. (2018). Surface functionalization of biomaterials by plasma and ion beam. Surf. Coat. Technol., 336, p. 2–8.

Kar, A., Raja, K.S. and Misra, M. (2006). Electrodeposition of hydroxyapatite onto nanotubular TiO$_2$ for implant applications. Surf. Coat. Technol., 201(6), p. 3723–3731.

Kariduraganavar, M. Y. Kittur, A. A. and Kamble, R. R. (2014). Polymer synthesis and processing. In: Kumbar, S., Laurencin, C., Deng, M. (eds). Natural and Synthetic Biomedical Polymers. Burlington: Elsevier. ISBN: 9780123969835. p. 1–6.

Ke, D.X., Vu, A.A., Bandhyopadhyay, A. and Bose, S. (2019). Compositionally graded doped hydroxyapatite coating on titanium using laser and plasma spray deposition for bone implants. Acta Biomater., 84, p. 414–423.

Khor, K.A., Li, H. and Cheang, P. (2004). Significance of melt-fraction in HVOF-sprayed hydroxyapatite particles, splats and coatings. Biomaterials, 25(7), p. 1177–1186.

Killinger, A., Kühn, M. and Gadow, R. (2006). High-velocity suspension flame spraying (HVSFS), a new approach for spraying nanoparticles with hypersonic speed. Surf. Coat. Technol., 201(5), p. 1922–1929.

Killinger, A., Gadow, R., Mauer, G., Guignard, A., Vaßen, R. and Stöver, D. (2011). Review of new developments in suspension and solution precursor thermal spray processes. J. Thermal Spray Technol., 20(4), p. 677–695.

Kim, H.M., Miyazaki, T., Kokubo, T. and Nakamura, T. (2001). Revised simulated body fluid. Bioceramics, 13, p. 47–50.

Kingery, W.D., Bowen, H.K. and Uhlmann, D.R. (1976). Introduction to Ceramics. 2nd ed. Wiley Series on the Science and Technology of Materials. p. 469–501. New York, Chichester, Brisbane, Toronto, Singapore: Wiley. 1032 pp. ISBN 978-0-471-47860-1.

Kljajević, Lj., Matović, B., Radoslaljević-Mihajlović, A., Rosić, M., Bosković, S. and Devečerski, A. (2011). Preparation of ZrO_2 and ZrO_2/SiC powders by carbothermal reduction of $ZrSiO_4$. J. Alloys Comp., 509(5), p. 2203–2215.

Kokubo, T., Kushitani, H., Sakka, S., Kitsugi, T. and Yamamuro, T. (1990). Solutions able to reproduce in vivo surface-structure changes in bioactive glass-ceramic A-W. J. Biomed. Mater. Res., 24, p. 721–734.

Kokubo, T., Miyaji, F., Kim, H.M. and Nakamura, T. (1996). Spontaneous formation of bonelike apatite layer on chemically treated titanium metal. J. Am. Ceram. Soc., 79(4), p. 1127–1129.

Kokubo, T. and Takadama, H. (2006). How useful is SBF in predicting in vivo bone bioactivity? Biomaterials, 27, p. 2907–2915.

Kokubo, T. and Yamaguchi, S. (2015). Bioactive titanate layers formed on Ti metal and its alloys by simple chemical and heat treatments. Open Biomed. Eng. J., 9, p. 29–41.

Kou, H.C., Zhang, H.L., Chu, Y.D., Huang, D., Nan, H. and Li, J.S. (2015). Microstructure evolution and phase transformation of Ti-6.5Al-2Zr-Mo-V alloy during thermohydrogen treatment. Acta Metall. Sin. (Engl. Lett.), 28(4), p. 505–513.

Kreidler, E.R. and Hummel, F.A. (1967). Phase relations in the system $SrO-P_2O_5$ and the influence of water vapor on the formation of $Sr_4P_2O_9$. Inorg. Chem., 6(5), p. 884–891.

Kurashima, Y., Ezura, A., Murakami, R., Mizutani, M., Komotori, J. (2019). Effect of hydroxyl groups and microtopography generated by a nanosecond-pulsed laser on pure Ti surfaces. J. Mater. Sci.: Mater. Med., 30 (5), Article # 57.

Kumar, D., Arjunan, S.P. and Aliahmad, B. (2016). Fractals: Applications in Biological Signalling and Image Processing. Taylor & Francis Inc. ISBN 978-1-4987-4421-8.

LCA Zircon (2017). Life cycle assessment of zircon sand production applied to ceramic tiles. ISO 14040/44 standard LCA. Zircon Industry Association (ZIA).

Lebedev, V.N., Lokshin, E.P., Mel'nik, N.A., Shchur, T.E. and Popova, A. (2004). Possibility of integrated processing of the baddeleyite concentrate. Russ. J. Appl. Chem., 77(5), p. 701–710.

Lee, D.W., Shin, M.C., Kim, Y.N. and Oh, J.M. (2017). Brushite ceramic coatings for dental brace brackets fabricated via aerosol deposition, Ceram. Intern., 43(1), p. 1044–1051.

Levingston, T.J., Barron, N., Ardhaoui, M., Benyounis, K., Looney, L. and Stokes, J. (2017). Application of response surface methodology in the design of functionally graded plasma sprayed hydroxyapatite coatings. Surf. Coat. Technol., 313, p. 307–318.

Li, A., Zhang, S.Y., Reznik, B., Lichtenberg, S. and Deutschmann, O. (2010). Synthesis of pyrolytic carbon composites using ethanol as precursor. Ind. Eng. Chem. Res., 49(21), p. 10421–10427.

Li, H., Ma, Y.L., Zhao, Z.C. and Tian, Y.L. (2019). Fatigue behavior of plasma sprayed structural-grade hydroxyapatite coating under simulated body fluid. Surf. Coat. Technol., 368, p. 110–118.

Li, R., Clark, A.E. and Hench, L.L. (1991). An investigation of bioactive glass powders by sol-gel processing. J. Appl. Biomater., 2(4), p. 231–239.

Li, Y.H., Yang, C., Wang, F. and 5 additional authors (2015). Biomedical TiNbZrTaSi alloys designed by d-electron alloy design theory. Mater. Design, 85, p. 7–13.

Liu, C., Huang, Y., Shen, W. and Cui, J. (2001). Kinetics of hydroxyapatite precipitation at pH 10 and 11. Biomaterials, 22, p. 301–306.

Liu, W., Thomopoulos, S. and Xia, Y. (2012). Electrospun nanofibers for regenerative medicine. Adv. Healthcare Mater., 1, p. 10–25.

Liu, X. and Ma, P. X. (2004). Polymeric scaffolds for bone tissue engineering. Ann. Biomed. Eng., 32, p. 477–486.

Liu, X.M., He, D.Y., Zhou, Z., Wang, G.H., Wang, Z.J. and Guo, X.Y. (2019). Effect of post-heating treatment on the microstructure of micro-plasma sprayed hydroxyapatite coatings. Surf. Coat. Technol., 367, p. 225–230.

Liu, Y., Hunziker, E.B., Randall, N.X., de Groot, K. and Layrolle, P. (2003). Proteins incorporated into biomimetically prepared calcium phosphate coatings modulate their mechanical strength and dissolution rate. Biomaterials, 24(1), p. 65–70.

Liu, Y., Hunziker, E.B., Layrolle, P., de Bruijn, J.D. and de Groot, K. (2004). Bone morphogenetic protein 2 incorporated into biomimetic coatings retains its biological activity. Tissue Eng., 10(1-2), p. 101–108.

Livingston, M. and Tan, A. (2016). Coating techniques and release kinetics of drug-eluting stents. J. Med. Device, 10, 010801.

Lobel, K.D. and Hench, L.L. (1998). In vitro adsorption and activity of enzymes on reaction layers of bioactive glass substrates. J. Biomed. Mater. Res., 39(4), p. 575–579.

Lütjering, G. and Williams, J. C. (2003). Titanium. Berlin, Heidelberg, New York: Springer. ISBN 3-540-42990-5.

Maia, F.R., Bidarra, S.J., Granja, P.L. and Barrias, C.C. (2013). Functionalization of biomaterials with small osteoinductive moieties. Acta Biomater., 9, p. 8773–8789.

Mandelbrot, B.B. (1983). The Fractal Geometry of Nature. New York: W.H. Freeman. ISBN 978-0-7167-1186-5.

Manieh, A.A. and Spink, D.R. (1973). Chlorination of zircon sand. Can. Metal. Quarterly, 12(3), p. 331–340.

Martinez-Vázquez, F.J., Miranda, P., Guiberteau, F. and Pajares, A. (2013). Reinforcing bioceramic scaffolds with in situ synthesized ε-polycaprolactone coatings. J. Biomed. Mater. Res. A, 101(12), p. 3551–3559.

Mathauer, K., Embs, F. and Wegner, G. (1989). Structure, properties and applications of polymeric Langmuir-Blodgett films. In: Allen, G. and Bevingon, J.C. (eds). Comprehensive Polymer Science and Supplements. Vol 7 (Specialty Polymers and Polymer Processing), p. 449–470. Elsevier. ISBN 978-0-08-096701-1.

Mattiasson, B. (2014). Cryogels for biotechnological applications. In: Okay, O. (ed.). Polymeric Cryogels. Cham: Springer. ISBN: 978-3-319-05845-0. p. 245–281.

Melchels, F.P.W., Feijen, J. and Grijpma, D. W. (2010). A review on stereolithography and its applications in biomedical engineering. Biomaterials 31, p. 6121–6130.

Mei, D., Lamaka, S.V., Gonzalez, J., Feyerabend, F., Willumeit-Römer, R. and Zheludkevich, M.L. (2019). The role of individual components of simulated body fluid on the corrosion behavior of commercially pure Mg. Corr. Sci., 147, p. 81–93.

Migliaresi, C. and Pegoretti, A. (2002). Fundamentals of polymeric composite materials. In: Barbucci, R. (ed.). Integrated Biomaterials Science. New York: Kluwer Academic/Plenum Publishers. ISBN: 0-306-46678-3. p. 69–117.

Migliaresi, C. and Alexander, H. (2004). Composites. In: Ratner, B.D., Hoffman, A.S., Schoen, F.J., and Lemons, J.E. (eds). Biomaterials Science: An Introduction to Materials in Medicine. San Diego: Elsevier. ISBN: 0-12-582463-7. p. 181–197.

Miura. H. (2008). In: Niinomi, M. (ed.) Fundamentals of Titanium and Its Working. Tokyo: Korona Publishing Co., Ltd, p.150–159.

Miyamoto, M., Watanabe, M., Narushima, T., and Iguchi, Y. (2008). Deoxidation of NiTi alloy melt using metallic barium. Mater. Trans. 49, p. 289–293.

Moign, A., Vardelle, A., Themelis, N.J. and Legoux, J.G. (2010). Life cycle assessment of using powder and liquid precursors in plasma spraying: The case of yttria-stabilized zirconia. Surf. Coat. Technol., 205(2), p. 668–673.

Morinaga, M. (2016). Alloy design based on molecular orbital method. Mater. Trans., 57(3), p. 213–226.

Morinaga, M., Yukawa, N., Maya, T., Sone, K. and Adachi, H. (1988). Theoretical design of titanium alloys. Proc. 6th World Conf. on Titanium, Cannes, p. 1601–1606.

Morra, M. and Cassinelli, C. (2006). Biomaterials surface characterization and modification. Inter. J. Artificial. Organs, 29 (9), p. 824–833.

Mota, C., Puppi, D., Chiellini, F. and Chiellini, E. (2015). Additive manufacturing techniques for the production of tissue engineering constructs. J. Tissue Eng. Regen. Med., 9, p. 174–190.

Müller, M. (2001). Entwicklung elektrokatalytischer Oxidschichten mit kontrollierter Struktur und Dotierung aus flüssigen Precursoren mittels thermischer Hochfrequenzplasmen [Development of electrocatalytic oxide coatings with controlled structure and doping from liquid precursors by high-frequency plasma spraying]. Ph.D. thesis, Dept. of Mineralogy, TU Bergakademie Freiberg, Germany.

Murphy, C.M., Schindeler, A., Gleeson, J.P., Yu, N.Y.C., Cantrill, L.C., Mikulec, K., Peacock, L., O'Brien, F.J. and Little, D.G. (2014). A collagen-hydroxyapatite scaffold allows for binding and co-delivery of recombinant bone morphogenetic proteins and bisphosphonates. Acta Biomater., 10(5), p. 2250–2258.

Murphy, S. V. and Atala, A. (2014). 3D bioprinting of tissues and organs. Nature Biotechnol., 32, p. 773–785.

Murr, L. E., Esquivela, E. V., Quinones, S. A., Gaytan, S. M., Lopez, Martinez, E. Y., Medina, F., Hernandez, D. H., Martinez, E., Martinez J. L., Stafford, S. W., Brown, D. K., Hoppe, T., Meyers, W., Lindhe, U. and Wicker, R. B. (2009). Microstructures and mechanical properties of electron beam-rapid manufactured Ti–6Al–4V biomedical prototypes compared to wrought Ti–6Al–4V. Mater. Charact., 60, p. 96–105.

Nair, L.S. and Laurencin, C.T. (2006). Polymeric applications as biomaterials in the areas of tissue engineering and controlled drug delivery. Adv. Biochem. Engng/Biotechnol.: Special Issue, Tissue Engineering, 102, p. 47–90.

Nakahigashi, J. and Yoshimura, H. (2002). Superplasticity and its application of ultra-fine grained Ti-6Al-4V alloy obtained through protium treatment. Mater. Trans., 43 (11), p. 2768–2772.

Nayak, A.K. (2010). Hydroxyapatite synthesis methodologies: An overview. Int. J. ChemTech Res., 2(2), p. 903–907.

Niinomi, M. (2001). Casting of titanium and its alloys. J. Jpn. Foundry Eng. Soc., 73, p. 784–790.

Niinomi, M., Fukui, H., Takeuchi, T. and Katsura, S. (2001). Dental precision casting of titanium and its alloys. J. Jpn. Foundry Eng. Soc., 73, p. 798–804.

Niinomi, M. (2012). Recent trend in titanium research and development. In: Proc. Ti-2011. I, p. 30–37.

Niinomi, M. (2014). Trend of applications in medical field. In: Additive Manufacturing Technology of Metallic Powders. Japan Society of Powder and Powder Metallurgy. Tokyo, Japan, p. 20–32.

Niinomi, M. and Kagami, K. (2016). Recent topics of titanium research and development in Japan. Proceed. Ti-2015, 1, p. 27–40.

Niinomi, M. and Nakai, M. (2017). Melting and casting of titanium and its alloys. J. Jap. Inst. Light Metals, 67(7), p. 307–314.

Niinomi, M. (2019). Casting. In: Niinomi, M. (ed.). Materials for Biomedical Devices, p. 311–330. Duxford: Elsevier.

Ntsoane, T.P., Topić, M., Härting, M., Heimann, R.B. and Theron, C. (2016). Spatial and depth-resolved studies of air plasma-sprayed hydroxyapatite coatings by means of diffraction techniques: Part I, Surf. Coat. Technol., 294, p 153–163.

Ogila, K. O. Shao, M., Yang, W. and Tan, J. (2017). Rotational molding: A review of the models and materials. eXPRESS Polym. Letters, 11, p. 778–798.

Okabe, T. (2005). New smelting process of titanium. J. Jap. Inst. Light Metals, 55(11), p. 537–543.

Onuma, K., Oyane, A., Kokubo, T., Treboux, G., Kanzaki, N. and Ito, A. (2000). Precipitation kinetics of hydroxyapatite revealed by the continuous-angle laser light-scattering technique. J. Phys. Chem., B104, p. 11950–11956.

Osaki, S., Sakai, H. and Suzuki, R. O. (2009). Direct production of Ti-29Nb-13Ta-4.6Zr biomedical alloy from oxide mixture in molten CaCl₂. In: Niinomi, M., Morinaga, M., Nakai, M., Bhatnagar, N., and Srivatsan, T. S. (eds). Proc. Processing and Fabrication of Advanced Materials XVIII., 2, p. 815–824.

Osswald, T. A. (2017). Understanding Polymer Processing – Processes and Governing Equations. (2nd ed.). München: Carl Hanser. ISBN 978-1-56990-647-7.

Ovsianikov, A., Schlie, S., Ngezahayo, A., Haverich, A. and Chichkov, B. N. (2007). Two-photon polymerization technique for microfabrication of CAD-designed 3D scaffolds from commercially available photosensitive materials. J. Tissue Eng. Regen. Med., 1, p. 443–449.

Pal, U.B. and Powell, A.C. (2007). The use of solid oxide membrane technology for electrometallurgy. JOM, 59(5), p. 44–49.

Park, J. and Lakes, R. S. (2007). Biomaterials – An Introduction. 3rd ed., New York: Springer. ISBN 978-0-387-37879-4.

Pasteris, J.D., (2016). A mineralogical view of apatite biomaterials. Am. Mineral., 101(2), p. 2594–2610.

Pederson, H. (1927). A process for producing aluminum hydroxide. US Patent 1,618,105.

Peitgen, H.O. and Saupe, D. (eds) (1988). The Science of Fractal Images. New York: Springer. ISBN 978-1-4612-8349-2.

Peppas, N. A. (1986). Hydrogels in medicine and pharmacy. Vol 1 Fundamentals. Boca Raton: CRC Press Inc. ISBN 0-8493-5546-X.

Peppas, N. A. (2004). Hydrogels. In: Ratner, B.D., Hoffman, A.S., Schoen, F.J., and Lemons, J.E. (eds). Biomaterials Science: An Introduction to Materials in Medicine. San Diego: Elsevier. ISBN: 0-12-582463-7. p. 100–107.

Pfister, A., Landers, R., Laib, A., Hübner, U., Schmelzeisen, R. and Mülhaupt, R. (2004). Biofunctional rapid prototyping for tissue-engineering applications: 3D bioplotting versus 3D printing. J. Polym. Sci. Part A: Polym. Chem., 42, p. 624–638.

Pramanik, S., Kumar Agarwal, A., Rai, K.N. and Garg, A. (2007). Development of high strength hydroxyapatite by solid-state-sintering process. Ceram. Intern., 33(3), p. 419–426.

Prieto, E. M. and Guelcher, S. A. (2014). Tailoring properties of polymeric biomedical foams. In: Biomedical Foams for Tissue Engineering Application. Cambridge: Woodhead Publishing. ISBN: 978-0-85709-696-8. p. 129–162.

Rahaman, M. (2011). Bioactive glass in tissue engineering. Acta Biomater., 7, p. 2355–2373.

Ramesh, S., Aw, K.L., Tolouei, R. and 6 additional authors (2013). Sintering properties of hydroxyapatite powders prepared using different methods. Ceram. Intern., 39 (1), p. 111–119.

Ramirez, B. and Runger, G. (2006). Quantitative techniques to evaluate process stability. Quality Engineering, 18 (1), p. 53–68.

Rana, D., Ramasamy, K., Leena, M. Pasricha, R., Manivasagam, G. and Ramalingam, M. (2017). Surface functionalization of biomaterials. In: Vishwakarma, A. and Karp, J.M. (eds). Biology and Engineering of Stem Cell Niches. Chapter 21, p. 331–343. Academic Press. ISBM 978-0-12-802734-9. 642 pp.

Rao, R.R., Roopa, H.N. and Mariappan, L. (2016). Development of microporous scaffolds through slip casting of solution combustion derived nanohydroxyapatite. Ceram, Intern. Ceram. Review, 65(3), p. 100–105.

Ratner, B.D. (1989). Biomedical applications of synthetic polymers. In: Allen, G. and Bevington, J.C. Comprehensive Polymer Science and Supplements. vol.7 (Specialty Polymers and Polymer Processing), p. 201–247. Elsevier. ISBN 978-0-08-096701-1.

Ratner, B.D., Hoffman, A.S., Schoen, F.J. and Lemons, J.E. (2013). Biomaterials Science. An Introduction to Materials in Medicine. Elsevier, ISBN 978-0-12-374626-9.

Reclaru, L. and Ardelan, L.C. (2019). Alternative processing techniques for CoCr dental alloys. In: Narayan, R. (ed.). Encyclopedia of Biomedical Engineering, 1, p. 1–5. Elsevier. ISBN 978-0-1280-4829-0.

Reddi, A.H. (2001). Bone morphogenetic proteins: from basic science to clinical applications. J. Bone Joint Surg. Am., 83A, Suppl. 1 (pt. 1) p. 1–6.

Reichelt, S., Becher, J., Weisser, J., Prager, A., Decker, U., Möller, S., Berg, A. and Schnabelrauch, M. (2014). Biocompatible polysaccharide-based cryogels. Mater. Sci. Eng., C35, p.164–170.

Reise, M., Wyrwa, R., Müller, U., Zylinski, M., Völpel, A., Schnabelrauch, M., Berg, A., Jandt, K. D., Watts, D. C. and Sigusch, B.W. (2012). Release of metronidazole from electrospun poly(l-lactide-co-d/l-lactide) fibers for local periodontitis treatment. Dental Mater., 28, p. 179–188.

Rendtorff, N.M., Suárez, G., Conconi, M.S., Singh, S.K. and Aglietti, E.F. (2012). Plasma dissociated zircon (PDZ) processing: influence of the Zr: Si ratio in the composition, microstructure and thermal recrystallization. Proc. Mater. Sci., 1, p. 337–342.

Riaz, U., Shabib, I. and Haider, W. (2018). The current trends of Mg alloys in biomedical apllications-A review. J. Biomed. Mater. Res. B, Appl. Biomater., 107(6), p. 1970–1996.

Riboud, P.V. (1973). Composition et stabilité des phases a structure d'apatite dans le systeme $CaO-P_2O_5$-oxide de Fer-H_2O a haute temperature. Ann. Chim., 8, p. 381–390.

Righi, S., Simonetto, C., Bruzzi, L., Andretta, M. and Serro, R. (2002). Environmental impact and risk assessment of zircon mineral plant emissions. In: Brebbia, C.A. (ed.). Proc. Risk Analysis III. p. 457–466. WIT Press: Ashurst Lodge, Southampton, UK. ISBN 1-85312-915-1.

Robertson, J. (2002). Diamond-like amorphous carbon. Mater. Sci. Eng. R: Reports, 37(4-6), p. 129–281.

Sachlos, E. and Czernuszka, J. T. (2003). Making tissue engineering scaffolds work. Review on the application of solid freeform fabrication technology to the production of tissue engineering scaffolds. Eur. Cells & Mater., 5, 29–40.

Sadik, C., Moudden, O., El Bouari, A. and El Amrani, I.E. (2016). Review on the elaboration and characterization of ceramics refractories based on magnesite and dolomite. J. Asian Ceram. Soc., 4(3), p. 219–233.

Safarian, J. and Kolbeinsen, L. (2016). Sustainability in alumina production from bauxite. In: Kongdi, F. et al. (eds). 2016 Sustainable Industr. Proc. Summit and Exhib. 5, p. 75–81.

Satyendra, K.S. (2014). Stainless-steel manufacturing processes. http://ispatguru.com/stainless-steel-manufacturing-processes (accessed February 18, 2019).

Schlebrowski, T., Beucher, L., Bazzi, H., Hahn, B., Wehner, S. and Fischer, C.B. (2019). Prediction of a-C:H layer failure on industrial relevant biopolymer polylactide acide (PLA) foils based on the sp^2/sp^3 ratio. Surf. Coat. Technol., 368, p. 79–87.

Shin, H., Jo, S. and Mikos, A.G. (2003). Biomimetic materials for tissue engineering. Biomaterials, 24(24), p. 4353–4364.

Slaughter, B. V., Khurshid, S. S., Fisher, O. Z., Khademhosseini, A. and Peppas, N. A. (2009). Hydrogels in regenerative medicine. Adv. Mater., 21, p. 3307–3329.

Stammeier, J.A., Purgstaller, B., Hippler, D., Mavromatis, V. and Dietzel, M. (2018). In-situ Raman spectroscopy of amorphous calcium phosphate to crystalline hydroxyapatite transformation. MethodsX, 5, p. 1241–1250.

Stewart, C., Akhavan, B., Wise, S.G. and Bilek, M.M.M. (2019). A review of biomimetic surface functionalization for bone-integrating orthopedic implants: Mechanisms, current approaches, and future directions. Progr. Mater. Sci., 106: 100588.

Stoltenhoff, T., Kreye, H. and and Richter, H. (2002). An analysis of the cold spray process and its coatings. J. Thermal Spray Technol., 11(4), p. 542–550.

Suchanek, K., Bartkowiak, A., Perzanowski, M., Marszałek, M., Sowa, M. and Simka, W. (2019). Electrochemical properties and bioactivity of hydroxyapatite coatings prepared by MEA/EDTA double-regulated hydrothermal synthesis. Electrochim. Acta, 298, p. 685–693.

Sun, L.M. (2018). Thermal spray coatings on orthopedic devices: When and how the FDA reviews your coatings. J. Thermal Spray Technol., 27, p. 1280–1290.

Sun, Z., Xiao, J., Tao, L., Wei, Y., Wang, S., Zhang, H., Zhu, S., and Muhuo, Y. (2019). Preparation of high-performance carbon fiber-reinforced epoxy composites by compression resin transfer molding. Materials, 12, 13.

Suzuki, R. O., Nogachi, H., Hada, H., Natsui, S. and Kikuchi, T. (2017). Reduction of $CaTiO_3$ in molten $CaCl_2$ – as basic understanding of electrolysis. Mater. Trans., 58, p. 341–349.

Suzuki, T. and Moriguchi, Y. (1995). Story of Titanium, Japan Standards Assoc., Tokyo, Japan, JSA.

Szcześ, A., Holysz, L. and Chibowski, E. (2017). Synthesis of hydroxyapatite for biomedical applications. Adv. Coll. Interface Sci., 249, p. 321–330.

Tadmor, Z. and Gogos, C. G. (2006). Principles of Polymer Processing. (2nd ed.) Hoboken: John Wiley & Sons. ISBN: 0-471-38770-3. p. 144–177.

Takeda, O. and Okabe, T. (2017). Smelting and refining off titanium. J. Jap. Inst. Light Metals, 67, p. 257–263.

Tas, A.C. (2000). Combustion synthesis of calcium phosphate bioceramic powders. J. Europ. Ceram. Soc., 20, p. 2389–2394.

Tas, A.C. (2014). The use of physiological solutions or media in calcium phosphate synthesis and processing, Acta Biomater., 10(5), p. 1771–1792.

Taylor, C.C. and Taylor, S.J. (1991). Estimating the dimension of a fractal. J. R. Stat. Soc. Ser. B, 53, p. 353–364.

Teo, W.-E., Inai, R. and Ramakrishna, S. (2011). Technological advances in electrospinning of nanofibers. Sci. Technol. Adv. Mater., 12, 013002.

Topić, M., Ntsoane, T., Hüttel, T. and Heimann, R.B. (2006). Microstructural characterisation and stress determination in as-plasma sprayed and incubated bioconductive hydroxyapatite coatings, Surf. Coat. Technol., 201(6), p. 3633–3641.

Treccani, L., Klein, T., Meder, F., Pardun, K. and Rezwan, K. (2013). Functionalised ceramics for biomedical, biotechnological and environmental applications. Acta Biomater., 9(7), p. 7115–7150.

Tsui, Y.C., Doyle, C. and Clyne, T.W. (1998). Plasma-sprayed hydroxyapatite coatings on titanium substrate. Part 1. Mechanical properties and residual stress levels, Biomaterials, 19, p. 2013–2029.

Urist, M.R. (1965). Bone formation by autoinduction. Science, 150 (698), p. 893–899.

Vahabzadeh, S., Roy, M., Bandhyapadhyay, A. and Bose, S. (2015). Phase stability and biological property evaluation of plasma sprayed hydroxyapatite coatings for orthopedic and dental applications. Acta Biomater., 17, p. 47–55.

Vilar, R. (ed.) (2016). Laser Surface Modification of Biomaterials. Techniques and Applications. Woodhead Publ., ISBN 978-0-0810-0942-0. 350 pp.

Vilardell, A.M., Cinca, N., Cano, I.G. and 6 additional authors (2017). Dense nanostructured calcium phosphate coating on titanium by cold spray. J. Eur. Ceram. Soc., 37 (4), p. 1747–1755.

Vilardell, A.M., Cinca, N., Garcia-Giralt, N., Dosta, S., Cano, I.G., Nugués, X. and Guilemany, J.M. (2018). Functionalized coatings by cold spray: An in vitro study of micro- and nanocrystalline hydroxyapatite compared to porous titanium. Mater. Sci. Eng. C, 87, p. 41–49.

Wakabayashi, N. (1992). Mechanical property and adaptability of superplastic titanium alloy denture base. J. Stomatol. Soc. Jpn., 59, p. 48–67.

Wang, X., Becker, F.F. and Gascoyne, P.R. (2010). The fractal dimension of cell membranes correleates with its capacitance: A new fractal single-shell model. Chaos, 20(4); 043133.

Weber, H., De Grave, I., Röhrl, E. and Altstädt, V. (2016). Foamed plastics. In: Ullmann's Encyclopedia of Industrial Chemistry, p. 1–54. Wiley-VCH: Weinheim. ISBN 978-3-5273-0673-2.

Wei, G. and Ma, P. X. (2008). Nanostructured biomaterials for regeneration. Adv. Funct. Mater., 18, p. 3566–3582.

Willmann, G. (1995). Ceramic hip joint heads made from zirconia and alumina – a comparison. In: G. Fishman, A. Clare and L. Hench (eds), Bioceramics. Materials and Application, Ceramic Transactions 48, p. 83, The American Ceramic Soc.: Westerville. ISBN 978-1-5749-8006-6. 329 pp.

Willmann, G. (1998). Überlebensrate und Sicherheit von keramischen Kugelköpfen für Hüftendoprothesen [Survival rate and safety of ceramic spherical heads for hip endoprostheses]. Mat.-wiss. u. Werkstofftech., 29, p. 595–604.

Wolff, M., Mesterknecht, T., Bals, A., Ebel, T. and Willumeit-Römer, R. (2019). FFF of Mg-alloys for biomedical application. In: Joshi, V. et al. (eds). Magnesium Technology 2019. Cham: Springer. p. 43–49.

Wu, F., Huang, Y., Song, L., Liu, X., Xiao, Y., Feng, J. and Chen, J. (2009). Method for preparing porous hydroxyapatite coatings by suspension plasma spraying. US Patent 8877283 B2.

Wu, G.S., Li, P.H., Feng, N.Q., Zhang, X.M. and Chu, P.K. (2015). Engineering and functionalization of biomaterials via surface modification. J. Mater. Chem. B, 3, p. 2024–2042.

Wubneh, A., Tsekoura, E.K., Ayranci, C. and Uludag, H. (2018). Current state of fabrication technologies and materials for bone tissue engineering. Acta Biomater., 80, p. 1–30.

Xu, H.F., Geng, X., Liu, G.X. and 5 additional authors (2016). Deposition, nanostructure and phase composition of suspension plasma sprayed hydroxyapatite coatings. Ceram. Inter., 42 (7), p. 8684–8690.

Yager, P., Price, R.R., Schnur, J.M., Schoen, P.E., Singh, A. and Rhodes, D.G. (1988). The mechanism of formation of lipid tubules from liposomes. Chem. Phys. Lipids, 46(3), p. 171–179.

Yang, Y.C. and Chang, E. (2001). Influence of residual stress on bonding strength and fracture of plasma-sprayed hydroxyapatite coatings on Ti–6Al–4V substrate. Biomaterials, 22(13), p. 1827–1836.

Yelten, A. and Yilmaz, S. (2016). Various parameters affecting the synthesis of the hydroxyapatite powders by the wet chemical precipitation technique Materials Today, Proceedings, 3(9), Part A, p. 2869–2876.

Yelten-Yilmaz, A. and Yilmaz, S. (2018). Wet chemical precipitation synthesis of hydroxyapatite (HA) powders. Ceram. Intern., 44(8), p. 9703–9710.

Yu, X., Qu, H., Knecht, D.A. and Wei, M. (2009). Incorporation of bovine serum albumin into biomimetic coatings on titanium with high loading efficacy and its release behavior. J. Mater. Sci.: Mater. Med., 20(1), p. 287–294.

Zenios, S., Markower, J., Yock, P., Kumar, U.N., Denend, L. and Krummel, T.M. (2010). Biodesign: The Process of Innovating Medical Technologies. Cambridge: Cambridge University Press. ISBN 978-0-5115-1742-3.

Zhang, J.D., Wang, L.J. and Jiang, D.M. (2012). Decomposition process of zircon sand concentrate with CaO-NaOH. Rare Metals, 31(4), p. 410–414.

Zheng, K. and Boccaccini, A. (2017). Sol-gel processing of bioactive glass nanoparticles: A review. Adv. Coll. Interface Sci., 249, p. 363–373.

Chapter 4

Adel-Khattab, D., Peleska, B., Kampschulte, M., Stiller, M,. Gildenhaar, R., Berger, G., Gomes, C., Linow, U., Hardt, M., Günster, J., Houshmand, A., Ghaffar, K.A., Gamal, A., EL-Mofty, M. and Knabe, C. (2017). An intrinsic angiogenesis approach and varying bioceramic scaffold architecture affect blood vessel formation in bone tissue engineering in vivo. Key Eng. Mater., 720, p. 58–64.

Adel-Khattab, D., Giacomini, F., Peleska, B., Gildenhaar, R., Berger. G., Gomes, C., Linow, U., Hardt, M., Günster, J., Stiller, M,. Houshmand, A., Ghaffar, K.A., Gamal, A., EL-Mofty, M. and Knabe, C. (2018). Development of a synthetic tissue engineered 3D printed bioceramic-based bone graft with homogenously distributed osteoblasts and mineralizing bone matrix in vitro. J. Tissue Eng. Regen. Med., 12(1), p. 44–58.

Agarwal, S., Wendorff, J.H. and Greiner, A. (2009). Progress in the field of electrospinning for tissue engineering applications, Adv. Mater., 21, p. 3343–3351.

Alberts, B., Johnson, A.D., Lewis, J., Morgan, D., Raff, M., Roberts, K. and Walter, P. (2017). Molekularbiologie der Zelle [Molecular biology of the cell]. Weinheim: Wiley-VCH. 6. Auflage. ISBN 978-3-527-34072-9. 1676 pp.

Al-Kofahi, Y., Lassoued, W., Grama, K. and 6 additional authors (2011). Cell-based quantification of molecular biomarkers in histopathology specimens. Histopathol., 59(1), p. 40–54.

Ames, B.N., Lee, F.D. and Durston, W.E. (1973) An improved bacterial test system for the detection and classification of mutagens and carcinogens. Proc Natl Acad Sci USA, 70(3), p. 782–786.

Ames, B.C., McCann, J. and Yamasaki, E. (1975). Methods for detecting carcinogens and mutagens with the salmonella/mammalian-microsome mutagenicity test. Mutation Res., 31 (6), p. 347–364.

Anderson, H.C. (1969).Vesicles associated with calcification in the matrix of epiphyseal cartilage. J. Cell Biol., 41, p. 59–72.

Anderson, J.M. (1993). Mechanisms of inflammation and infection with implanted devices, Cardiovasc. Pathol., 2 (Suppl.), p. 33–41.

Anderson, J.M., Bonfield, T.L., and Ziats, N.P. (1990). Protein adsorption and cellular adhesion and activation on biomedical polymers, Int. J. Artif. Organs, 13, p. 375–82.

Anderson, J.M., Rodriguez, A. and Chang, D.T. (2008). Foreign body reaction to biomaterials. Semin. Immunol. 2008, 20.

Aplin, A.E., Howe, A.K. and Juliano, R.L. (1999). Cell adhesion molecules, signal transduction and cell growth. Curr. Opin. Cell Biol., 11, p. 737–744.

Asti, A. and Gioglio, L. (2014). Natural and synthetic biodegradable polymers: different scaffolds for cell expansion and tissue formation. Int. J, Artif. Organs, 37, p. 187–205.

Australian Orthopaedic Association National Joint Replacement Registry (ed.) (2014): Hip, knee and shoulder arthroplasty: annual report 2014. Adelaide (Australia).

Badylak, S.F. (2007). The extracellular matrix as a biologic scaffold material, Biomaterials, 28, p. 3587–3593.

Badylak, S.F., Freytes, D.O. and Gilbert, T.W. (2009). Extracellular matrix as a biological scaffold material: structure and function, Acta Biomater., 5, p. 1–13.

Bar-Meir, E., Teuber, S.S., Lin, H.C., Alosacie, I., Goddard, G., Terybery, J., Barka, N., Shen, B., Peter, J.B., Blank, M., Gershwin, M.E. and Shoenfeld, Y. (1995). Multiple autoantibodies in patients with silicone breast implants, J. Autoimm., 8, p. 267–277.

Barbucci, R. and Magnani, A. (1994). Conformation of human plasma proteins at polymer surfaces: the effectiveness of surface heparinization, Biomaterials, 15, p. 955–962.

Baumann, B., Hendrich, C., Barthel, T., Bockholt, M., Walther, M., Eulert, J. and Rader, C.P. (2007) 9- to 11-year results of cemented titanium mueller straight stem in total hip arthroplasty, Orthopedics, 30, p. 551–557.

Berger, G., Gildenhaar, R. and Ploska, U. (1995). Rapid resorbable, glassy crystalline materials on the basis of calcium alkali orthophosphates. Biomaterials, 16, p. 1241–1248.

Berridge, M.J., Lipp, P. and Bootman, M.D. (2000). The versatility and universality of calcium signaling. Nat. Rev. Mol. Cell Biol., 1, p. 11–21.

Bershadsky, A.D., Balaban, N.Q. and Geiger, B. (2003). Adhesion-dependent cell mechanosensitivity. Annu. Rev. Cell Dev. Biol., 219, p. 677–695.

Bhattacharyya, S., Agrawal, A., Knabe, C. and Ducheyne, P. (2014). Sol-gel silica controlled release thin films for the inhibition of methicillin-resistant Staphylococcus aureus. Biomaterials, 35(1), p. 509–517.

Billiau, A. and Vandenbroeck, K. (2001). IFNγ. In: Oppenheim, J.J, Feldmann, M. (eds). Cytokine reference: compendium of cytokines and other mediators of host defense. San Diego: Academic Press; p. 641–688.

Billiau, A. and Matthys P. (2009). Interferon-γ: a historical perspective. Cytokine Growth Factor Rev., 20, p. 97–113

Bissel, M. and Nelson, W.J. (1999). Cell-to-cell contact and extracellular matrix. Integration of form and function: the central role of adhesion molecules. Curr. Opin. Cell Biol., 11, p. 537–539.

Black, J. (1988). Orthopaedic Biomaterials in Research and Practice. New York: Churchill-Livingstone.

Black, J. and Fielder, J. H. (1992). Ethical aspects in device retrieval. Proc. Implant Retrieval Symposium 9/17-20/92, 14-1. St. Charles, LA: Society for Biomaterials.

Black, J. (1995). "Safe" biomaterials, J. Biomed. Mater. Res., 29, p. 791–792.

Black, J. (1996). Implant retrieval: An overview of goals and perspectives. Int. J. Risk Med., 8(1), p. 99–104.

Boss, J.H., Shajrawi, I., Aunullah, J. and Mendes, D.G. (1995). The relativity of biocompatibility A critique of the concept of biocompatibility. Isr. J. Med. Sci., 31(4), p. 203–209.

Boudreau, N., Sympson, C.J., Werb, Z. and Bissell, M. (1995). Suppression of ICE and apoptosis in mammary epithelial cells by extracellular matrix. Science, 267, p. 891–893.

Boyan, B.D., Lossdörfer, S., Wang, L., Zhao, G., Lohmann, C.H., Cochran, D.L. and Schwartz, Z. (2003). Osteoblasts generate an osteogenic microenvironment when grown on surfaces with rough microtopographies. Eur. Cells Mater., 6, p. 22–27.

Boyan, B.D., Baker, M.I., Lee, C.S.D., Raines, A. L., Greenwald, A.S., Olivares-Navarrete, R. and Schwartz, Z. (2011). Bone tissue grafting and tissue engineering concepts. In: Ducheyne, P. (ed.). Comprehensive Biomaterials., Elsevier, 6, p. 237–255.

Boyce, D.E., Jones, W.D., Ruge, F., Harding, K.G. and Moore, K. (2000). The role of lymphocytes in human dermal wound healing, Brit. J. Dermatol., 143, p. 59–65.

Boyko, E. J. (2013). Observational research – opportunities and limitations. J. Diab. Compl., 27(6), p. 642–648.

Brakebusch, C. and Fässler, R. (2003). The integrin-actin connection, an eternal love affair. EMBO J., 22, p. 2324–2333.

Brånemark, P.I., Adell, R., Breine, U., Hansson, B.O. Lindström, J. and Ohlsson, A. (1969). Intra-osseous anchorage of dental prostheses. I. Experimental studies. Scand. J. Plast. Reconstr. Surg., 3(2). p. 81–100.

Brett, T. (1992). The laboratory assessment of biocompatibility: The role of complement activation testing, Med. Dev. Technol., 3, p. 26–30.

Brodano, G.B., Cappuccio, M., Gasbarrini, A., et al. (2007). Vertebroplasty in the treatment of vertebral metastases: clinical cases and review of the literature. Eur. Rev. Med. Pharmacol. Sci., 11(2), p. 91–100.

Brown, B.N., Ratner, B.D., Goodman, S.B., Amar, S. and Badylak, S.F. (2012). Macrophage polarization: an opportunity for improved outcomes in biomaterials and regenerative medicine. Biomaterials, 33(15), p. 3792–3802.

Brown, B.N., Mani, D., Nolfi, A.L., Liang, R., Abramowitch, S.D. and Moalli, P.A. (2015). Characterization of the host inflammatory response following implantation of prolapse mesh in rhesus macaque. Am. J. Obstet. Gynecol., 213(5), p. 668.e1-10.

Brunstedt, M.R., Anderson, J.M., Spilizewski, K.L., Marchant, R. and Hiltner, A. (1990). In vivo leukocyte interactions on pellethane surfaces, Biomaterials, 11, p. 370–378.

Bryers, J.D., Giachelli, C.M. and Ratner, B.D. (2012). Engineering biomaterials to integrate and heal: the biocompatibility paradigm shifts. Biotechnol. Bioeng., 109(8), p. 1898–1911.

Burdette, W. J. and Gehan, E. A. (1970). Planning and Analysis of Clinical Studies. Springfield IL: Charles C Thomas.

Calderwood, D.A. (2004). Integrin activation. J. Cell. Sci., 117, p. 657–666.

Campoccia, D., Montanaro, L. and Arciola, C.R. (2006). The significance of infection related to orthopedic devices and issues of antibiotic resistance, Biomaterials, 27, p. 2331–2339.

Canadian Institute for Health Information (ed.) (2017). Hip and knee replacements in Canada, 2014–2015: Canadian Joint Replacement Registry Annual report. Ottawa.

Caruso, R., Trunfio, S., Milazzo, F., Campolo, J., De Maria, R., Colombo, T., Parolini, M., Cannata, A., Russo, C., Paino, R., Frigerio, M., Martinelli, L. and Parodi, O. (2010). Early expression of pro- and anti-inflammatory cytokines in left ventricular assist device recipients with multiple organ failure syndrome, ASAIO J., 56, p. 313–318.

Chai,Y.C., Carlier, A., Bolander, J., Roberts, S.J., Geris, L., Schrooten, J., Van Oosterwyck, H., Luyten, F.P. (2012). Current views on calcium phosphate osteogenicity and the translation into effective bone regeneration strategies. Acta Biomater., 8(11), p. 3876–3887.

Chang, S.C., Rowley, J.A. and Tobias, G. (2001). Injection molding of chondrocyte/alginate constructs in the shape of facial implants. J. Biomed. Mater. Res., 55, p. 503–511.

Chen, E. H. and Black, J. (1980). Materials design analysis of the prosthetic anterior cruciate ligament. J. Biomed. Mater. Res., 14 (5), p. 567–586.

Chen, Q., Cabanas-Polo, S., Goudouri, O.M. and Boccaccini, A.R. (2014). Electrophoretic co-deposition of polyvinyl alcohol (PVA) reinforced alginate-Bioglass® composite coating on stainless steel: mechanical properties and in-vitro bioactivity assessment. Mater. Sci. Eng. C, Mater. Biol. Appl., 40, p. 55–64.

Chenoweth, D.E. (1988). Complement activation produced by biomaterials, Artif. Organs, 12, p. 502–504.

Choi, Y.S., Lee, S.B., Hong, S.R., Lee, Y.M., Song K.W. and Park, M.H. (2001). Studies on gelatin-based sponges. Part III: A comparative study of cross-linked gelatin/alginate, gelatin/ hyaluronate and chitosan/hyaluronate sponges and their application as a wound dressing in full-thickness skin defect of rat. J. Mater. Sci. Mater. Med., 12, p. 67–73.

Chu, C.L., Reenstra, W.R., Orlow, D.L. and Hartford Svoboda, K.K. (2000). Erk and PI-3 kinase are necessary for collagen binding and actin reorganization in corneal epithelia. Invest Ophthalmol. Vis. Sci., 41, p. 3374–3382.

Ciarkowski, A. A. (1986). Preclinical testing evaluation of biomaterials: in vitro and in vivo. In: Von Recum, A.F. (ed.) Handbook of Biomaterials Evaluation: Scientific, Technical, and Clinical Testing of Implant Materials. New York: Macmillan Publishing Co.

Claes, L., Kirschner, P., Perka, C. and Rudert, M. (eds) (2009). Health Technology Assessment of Hip Arthroplasty in Europe. Heidelberg: Springer Verlag.

Cohen, D.J., Cheng, A., Sahingur, K., Clohessy, R.M., Hopkins, L.B., Boyan, B.D. and Schwartz, Z. (2017). Performance of laser sintered Ti-6Al-4V implants with bone-inspired porosity and micro/nanoscale surface roughness in the rabbit femur. Biomed. Mater., 12(2), 025021.

Cohen, M., Kam, Z., Addadi, L. and Geiger, B. (2006). Dynamic study of the transition from hyaluronan to integrin-mediated adhesion in chondrocytes. The EMBO Journal, 25, p. 302–311.

Colton, C.K., Ward, R.A. and Shaldon, S. (1994). Scientific basis for assessment of biocompatibility in extracorporeal blood treatment, Nephrol. Dial. Transplant., 9 (Suppl. 2), p. 11–17.

Connelly, J.T., Petrie, T.A., García, A.J. and Levenston, M.E. (2011). Fibronectin- and collagen-mimetic ligands regulate bone marrow stromal cell chondrogenesis in three-dimensional hydrogels. Eur Cell Mater., 22, p. 168–177.

Cooper, G.M. (2018). Actin, myosin, and cell movement, In: The Cell. A Molecular Approach. 8th ed. Sunderland: Sinauer Associates. ISBN 978-1-6053-5707-2. 816 pp.

Cooper, L.F. (2000). A role for surface topography in creating and maintaining bone at titanium endosseous implants. J. Prosthet. Dent., 84(5), p. 522–534.

Cotran, R.Z., Kumar, V. and Robbins, S.L. (1989). Inflammation and repair, Pathologic Basis of Disease, 4th ed. Philadelphia: W.B. Saunders Co., p. 33–86.

Cram, E.J. and Schwarzbauer, A. (2004). The talin wags the dog: new insights into integrin activation. Trends in Cell Biol., 14, p. 55–57.

Croisier, F.and Jérôme, C. (2013). Chitosan-based biomaterials for tissue engineering. Eur. Polym. J., 49, p. 780–792.

Cukierma, E., Pankov, R., Stevens, D.R. and Yamada, K.M. (2001). Taking cell-matrix adhesion to the third dimension. Science, 294, p. 1708–1712.

Davidson, P.M., Bigerelle, M., Reiter, G. and Anselme, K. (2015). Different surface sensing of the cell body and nucleus in healthy primary cells and in a cancerous cell line on nanogrooves. Biointerphases, 10, 031004.

Davis, M.J., Tsang, T.M. and Qiu, Y. (2013). Macrophage M1/M2 polarization dynamically adapts to changes in cytokine microenvironments in Cryptococcus neoformans infection. mBio, 4.

Davies, J.E. (1996). In vitro modelling of the bone/implant interface. Anat. Rec., 245, p. 426–445.

Davies, .J.E. (1998). Mechanisms of endosseous integration. Int. J. Prosthodont, 11, p. 391–401.

Davies, J.E. (2003). Understanding peri-implant endosseous healing. J. Den.t Educ., 67, p.932–949.

Dedhar, S. and Hannigan, G.E. (1996). Integrin cytoplasmic interactions and bidirectional transmembrane signalling. Curr. Opin. Cell Biol., 8, p. 657–669.

DeGroot, K. (1988). Effect of porosity and physiocochemical properties on the stability, resorption, and strength of calcium phosphate ceramics. Ann. N.Y. Acad. Sci., 523, p. 227–233.

DIN e.V. (2017). DIN EN ISO 10993-1 to DIN EN ISO 10993-23, Berlin: Beuth-Verlag.

Dinarello, C.A. (1990). Cytokines and biocompatibility, Blood Purif., 8, p. 208–213.

Discher, D.E., Janmey, P. and Wang, Y.L. (2005). Tissue cells feel and respond to the stiffness of their substrate. Science, 310, p. 1139–1143.

Domard, A. and Domard, M. (2002). Chitosan: Structure-properties relationship and biomedical applications. In: Dumitriu S. (ed.). Polymeric Biomaterials. Marcel Dekker., New York, USA, p. 187–212.

Dreinhöfer, K., Dieppe, P., Günther, K.P. and Puhl, W. (eds) (2009). EUROHIP – Health Technology Assessment of Hip Arthroplasty in Europe. Heidelberg: Springer Verlag.

Ducheyne, P. and Hastings, G.W. (eds) (1984). Structure-Property Relationship in Biomaterials, CRC Press, Inc., Boca Raton, Florida.

Ducheyne, P. and Qui, Q. (1999). Bioactive ceramics: The effect of surface reactivity on bone formation and bone cell function. Biomaterials, 20, p. 2287–2303.

Ekblom, P. and Timpl, R. (1996). Cell-to-cell contact and extracellular matrix. A multifaceted approach emerging. Curr. Opin. Cell Biol., 8, p. 599–601.

Ekdahl, K.N., Lambris, J.D., Elwing, H. and 8 additonal authors (1999). Effect of serum proteins on osteoblast adhesion to surface-modified bioactive glass and hydroxyapatite. J. Orthop. Res., 17, p. 340–345.

Ekdahl, K.N., Lambris, J.D., Elwing, H., Ricklin, D., Nilsson, P.H., Teramura, Y., Nicholls, I.A, and Nilsson, B. (2011). Innate immunity activation on biomaterial surfaces: A mechanistic model and coping strategies. Adv. Dru.g Deliv. Rev., 63(12), p.1042–1050.

El-Ghannam, A., Ducheyne, P. and Shapiro, I. M. (1999). Effect of serum protein adsorption on osteoblast adhesion to bioactive glass and hydroxyapatite. J. Orthop. Res., 17, p. 340–345.

Eloy, R. (1994). Overview of test procedures. In: llerton, A. (ed.) The Interaction between biocompatibility and device performance, Sem. Proceed. p. 6–31, Medical Device Technology, Advanstar Commun.

Endo, L.P., Edwards, N.L., Longley, S., Cormann, L.C. and Panush, R.S. (1987). Silicone and rheumatic disease, Semin. Arthr. Rheum., 17, p. 112–118.

Esposito, M. Hirsch, J.M., Lekholm, U. and Thomsen, P. (1998). Biological factors contributing to failures of osseointegrated oral implants. (I). Success criteria and epidemiology, Eur. J. Oral Sci., 106, p. 527–551.

European Parliament and Council (2017). Regulation (EU) 2017/745 of the European Parliament and of the Council on medical devices. Available online at https://eur-lex.europa.eu/legal-content /EN/ALL/?uri=CELEX:32017R0745, checked on 10/ 31/2019.

Evans, J.T., Evans, J.P., Walker, R.W., Blom, A.W., Whitehouse, M.R. and Sayers, A. (2019). How long does a hip replacement last? A systematic review and meta-analysis of case series and national registry reports with more than 15 years of follow-up. Lancet, 393, p. 647–54

Fässler, R., Georges-Labouesse, E. and Hirsch, E. (1996). Genetic analyses of integrin function in mice. Curr. Opin. Cell Biol., 8, p. 641–646.

Finke, B., Lüthen, F., Schröder, K., Müller, P.D., Bergemann, C., Frant, M., Ohl, A. and Nebe, B.J. (2007). The effect of positively charged plasma polymerization on initial osteoblastic focal adhesion on titanium surfaces. Biomaterials, 28, p. 4521–4534.

Finke, B., Rebl, H., Hempel, F., Schäfer, J., Liefeith, K., Weltman, K.D. and Nebe, B.J. (2014). Aging of plasma-polymerized allylamine nanofilms and the maintenance of their cell adhesion capacity. Langmuir 2014, 30, p. 13914–13924.

Finke, B., Testrich, H., Rebl, H., Nebe, B., Bader, R., Walschus, U., Schlosser, M., Weltmann, K., and Meichsner, J. (2014). Anti-adhesive finishing of temporary implant surfaces by a plasma-fluorocarbon-polymer, Mater. Sci. Forum, 783, p. 1238–1243.

Finke, B., Testrich, H., Rebl, H., Walschus, U., Schlosser, M., Zietz, C., Staehlke, S., Nebe, J.B., Weltmann, K.D., Meichsner, J. and Polak, M. (2016). Plasma-deposited fluorocarbon polymer films on titanium for preventing cell adhesion: a surface finishing for temporarily used orthopaedic implants, J. Phys. D Appl. Phys., 49, 234002 (14pp).

Florencio-Silva, R. G., Sasso, R.S., Sasso-Cerri, E., Simões, M.J. and Cerri, P.S. (2015). Biology of bone tissue: Structure, function, and factors that influence bone cells. Biomed. Res. Int., 2015: 421746.

Freedman, B.R. and Mooney, D.J. (2019). Biomaterials to mimic and heal connective tissues. Adv. Mater., 31(19):e1806695.

Frisch, S.M. and Screaton, R.A. (2001). Anoikis mechanisms. Curr. Opin. Cell Biol., 13, p. 555–562.

Fuchs, .T, Stange, R., Schmidmaier, G. and Raschke, M.J. (2011). The use of gentamicin-coated nails in the tibia: Preliminary results of a prospective study. Arch. Orthop. Trauma Surg., 131, p. 1419–1425.

Garcia A.J. (2005). Get a grip: integrins in cell–biomaterial interactions. Biomaterials, 26, p. 7525–7529.

Geiger, B., Bershadsky, A., Pankov, R. and Yamada, K.M. (2001). Transmembrane extracellular matrix-cytoskeleton crosstalk. Nat. Rev. Mol. Cell Biol., 2, p. 793–805.

Gerstenfeld, L.C., Cho, T.J., Kon, T., Aizawa, T., Cruceta, J. and Graves, B.D. et al. (2001). Impaired intramembranous bone formation during bone repair in the absence of tumor necrosis factor-alpha signaling. Cells Tissues Organs, 169, p. 285–24.

Giancotti, F.G. and Ruoslahti, E. (1999). Integrin signaling. Review Signal Transduction. Science, 285, p. 1028–1033.

Gigliobianco, G., Regueros, S.R., Osman, N.I., Bissoli, J., Bullock, A.J., Chapple, C.R. and MacNeil, S. (2015). Biomaterials for pelvic floor reconstructive surgery: how can we do better? Biomed. Res. Int., 2015:968087.

Glaser, Z.R. (1993). Some unanticipated changes in implant biocompatibility produced by alteration of the surface of the implant, Pharmacopeial Forum, 19, p. 5035–5039.

Goldblum, R.M., Pelley, R.P., O'Donell, A.A., Pyron, D. and Heggers, J.P. (1992). Antibodies to silicone elastomers and reactions to ventriculoperitoneal shunts, Lancet, 340, p. 510–513.

Gong, G., Seifter, E., Lyman, W.D., Factor, S.M, Blau, S. and Frater, R.W.M. (1993). Bioprosthetic cardiac valve degeneration: Role of inflammatory and immune reactions, J. Heart Valve Disease, 2, p. 684–693.

Götze, O. (1994). The potential role of basophilic leukocytes and mast cells, Nephrol. Dial. Transplant. 9, (Suppl.2), p. 57–59.

Gould, J.A., Liebler, B., Baier, R., Benson, J., Boretos, J., Callahan, T., Canty, E., Compton, R., Marlowe, D., O'Holla, R., Page, B., Paulson, J. and Swanson, C. (1993). Biomaterials availability: Development of a characterization strategy for interchanging silicone polymers in implantable medical devices, J. App. Biomater., 4, p. 355–358.

Grammer, L.C., and Patterson, R. (1987). IgE against ethylene-oxide-altered human serum albumin (ETO-HSA) as an etiologic agent in allergic reactions of hemodialysis patients, Artif. Organs, 11, 97–99.

Greenhalgh, T. (2014). How to Read a Paper: the Basics of Evidence Based Medicine. London: John Wiley & Sons.

Havelin, L.I., Fenstad, A.M., Salomonsson, R., Mehnert, F., Furnes, O., Overgaard, S. et al. (2009): The Nordic Arthroplasty Register Association: A unique collaboration between 3 national hip arthroplasty registries with 280,201 THRs. Acta Orthop., 80 (4), p. 393–401.

Healy, K.E. and Ducheyne, P. (1991). A physical model for the titanium-tissue interface. ASAIO Trans., 37(3),M150-1.

Healy, K.E. and Ducheyn, e P. (1992a). The mechanisms of passive dissolution of titanium in a model physiological environment. J. Biomed. Mater. Res., 26(3), p. 319–338.

Healy, K.E. and Ducheyne, P. (1992b). Hydration and preferential molecular adsorption on titanium in vitro. Biomaterials, 13(8), p. 553–561.

Hench, L.L. and Paschall, H.A. (1973). Direct chemical bond of bioactive glass-ceramic materials to bone and muscle. J. Biomed. Mater. Res., 7, p. 25–42.

Hench, L.L. (1998). Bioceramics. J. Am. Ceram. Soc., 81, p. 1705–1728.

Hench, L,L, and Best, S. (2004). Ceramics, glasses and glass-ceramics. In: Ratner, B., Hoffman, A., Schoen, F.J. and Lemons, J.E. (eds), Biomaterials Science. 2nd ed. San Diego: Elsevier.

Henning, W., Bohn, W., Nebe, B., Knopp, A., Rychly, J. and Strauss, M. (1994). Local increase of beta 1-integrin expression in cocultures of immortalized hepatocytes and sinusoidal endothelial cells. Eur J Cell Biol., 65(1), p. 189–199.

Henson, P.M. (1971). The immunologic release of constituents from neutrophil leukocytes: II. Mechanisms of release during phagocytosis, and adherence to nonphagocytosable surfaces, J. Immunol., 107, p. 1547–1557.

Hill, D. (1995). Material selection, Med. Dev. Technol., 6, p. 14–21.

Hippler, R., Kersten, H., Schmidt, M. and Schoenbach, K.H. (2008). Low temperature plasma physics: Fundamental aspects and applications. Weinheim: Wiley-VCH.

Hirsch, J. A., Nicola, G., McGinty, G., Liu, R. W., Barr, R. M., Chittle, M.D. and Manchikanti, L. (2016). ICD-10: History and Context. Am. J. Neurorad., 37 (4), p. 596–599.

Hoang, Q.Q., Siceri, F., Howard, A.J. and Yang, D.S.C. (2003). Bone recognition mechanism of porcine osteocalcin from crystal structure. Nature, 425, p. 977–980.

Hoene, A., Walschus, U., Patrzyk, M., Finke, B., Lucke, S., Nebe, B., Schroeder, K., Ohl, A. and Schlosser, M. (2010). In vivo investigation of the inflammatory response against allylamine plasma polymer coated titanium implants in a rat model, Acta Biomater., 6, p. 676–683.

Hoene, A., Prinz, C., Walschus, U., Lucke, S., Patrzyk, M., Wilhelm, L., Neumann, H.G. and Schlosser, M. (2013a) In vivo evaluation of copper release and acute local tissue reactions after implantation of copper-coated titanium implants in rats, Biomed. Mater., 8, 035009.

Hoene, A., Patrzyk, M., Walschus, U., Straňák, V., Hippler, R., Testrich, H., Meichsner. J., Finke, B., Rebl, H., Nebe, B., Zietz, C., Bader, R., Podbielski, A. and Schlosser, M. (2013b). In vivo examination of the local inflammatory response after implantation of Ti6Al4V samples with a combined low-temperature plasma treatment using pulsed magnetron sputtering of copper and plasma-polymerized ethylenediamine, J. Mater. Sci. Mater. Med., 24, p. 761–771.

Hoene, A., Patrzyk, M., Walschus, U., Finke, B., Lucke, S., Nebe, B., Schröder, K. and Schlosser, M. (2015). Systemic IFNγ predicts local implant macrophage response, J. Mater. Sci. Mater. Med., 26, 131.

Holgers, K.M. Thomsen, P., Tjellström, A., and Bjursten, L.M. (1995). Immunohistochemical study of the soft tissue around long-term skin-penetrating titanium implants, Biomaterials, 16, p. 611–616.

Howlett, C.R., Chen, N., Zhang, X., Akin, F.A., Haynes, D., Hanley, L., Revell, P., Evans, P., Zhou, H. and Zreiqat, H. (2000). The effect of biomaterial chemistries on the osteoblastic molecular phenotype and osteogenesis: In vitro and in vivo studies. In: Davies, J.E. (ed.). Bone Engineering. Toronto, Canada: em squared Inc. p 240–255.

Huang, Z.M., Zhang, Y.Z., Kotaki, M. and Ramakrishna, S. (2003). A review on polymer nanofibers by electrospinning and their applications in nanocomposites, Comp. Sci. Technol., 63, p. 2223–2253.

Hunt, J.A., McLaughlin, P.J. and Flannagan, B.F. (1997). Techniques to investigate cellular and molecular interactions in the host response to implanted biomaterials. Biomaterials, 18, p. 1449–1459.

Huss. R.S., Huddleston, J.I., Goodman, S.B., Butcher, E.C. and Zabel, B.A. (2010). Synovial tissue-infiltrating natural killer cells in osteoarthritis and periprosthetic inflammation. Arthrit. Rheumatol., 62(12), p. 3799–3805.

Hynes, R.O. (1992). Integrins: versatility, modulation and signaling in cell adhesion. Cell, 69, p. 11–25.

Hynes, R.O. (1999). The dynamic dialogue between cells and matrices: implications of fibronectin's elasticity. Proc. Natl. Acad. Sci. USA, 96, p. 2588–2590.

Ideker, T. (2004). A system approach to discovering signaling and regulatory pathways – or, how to digest large interaction networks into relevant pieces. Adv. Exp. Med. Biol., 547, p. 21–30.

Ideker, T. and Lauffenburger, D. (2003). Building with a scaffold: emerging strategies for high- to low-level cellular modeling. Trends Biotechnol., 21, p. 255–262.

Ingber, D.E. (2003a). Tensegrity II. How structural networks influence cellular information processing networks. J. Cell Sci., 116, p. 1397–1408.

Ingber, D.E. (2003b). Mechanobiology and diseases of mechanotransduction. Ann. Med., 35, p. 564–577.

Inoue, T., Cox, J.E., Pilliar, R.M., and Melcher A.H. (1987). Effect of the surface geometry of smooth and porous-coated titanium alloy on the orientation of fibroblasts in vitro, J. Biomed. Mater. Res., 21, p. 107–126.

Jayakumar, R., Prabaharan, M., Sudheesh Kumar, P.T, Nair, S.V. and Tamura, H. (2011). Biomaterials based on chitin and chitosan in wound dressing applications. Biotechnol. Adv., 29, p. 322–337.

Kaafarani, H.M.A. and Rosen, A.K. (2009). Using administrative data to identify surgical adverse events: an introduction to the Patient Safety Indicators. Am. J. Surg., 198 (5 Suppl), p. S63-S68.

Kaiser, W. (2009). Silicone and autoimmunity, Autoimmunity, 14, p. 341–342.

Kalteis, T., Lüring, C., Gugler, G., Zysk, S., Caro, W., Handel, M. and Grifka, J. (2004). Acute tissue toxicity of PMMA bone cements. Z. Orthop. Grenzgeb., 142(6), p. 666–672.

Kanchanawong, P., Shtengel, G., Pasapera, A.M., Ramko, E.B., Davidson, M.W., Hess, H.F. and Waterman, C.M. (2010). Nanoscale architecture of integrin-based cell adhesions. Nature, 468, p. 580–586.

Kandzari, D.E., Smits, P.C., Love, M.P. and 20 additional authors (2017). Randomized comparison of ridaforolimus- and zotarolimus-eluting coronary stents in patients with coronary artery disease: Primary results from the BIONICS trial (BioNIR Ridaforolimus-Eluting Coronary Stent System in Coronary Stenosis). Circulation. Oct 3; 136(14). P. 1304–1314.

Keegan, A.D. (2001). IL-4. In: Oppenheim, J.J., Feldmann, M. (eds). Cytokine reference: a compendium of cytokines and other mediators of host defense. San Diego: Academic Press; p. 127–135.

Keselowsky, B.G., Collard, D.M. and Garcia, A.J. (2005). Integrin binding specificity regulates biomaterial surface chemistry effects on cell differentiation. Proc. Natl. Acad. Sci. USA, 102(17), p. 5953–5957.

Kim. J., Berger, G., Gildenhaar, R., Shapiro, I,. Ducheyne, P. and Knabe, C. (2010). Survival and apoptosis in vitro of osteoblastic cells in contact with synthetic bone grafts. Trans. 36th Ann. Meet. Soc. Biomater. U.S.A., April 21-24, 2010, Seattle, Washington, USA, p. 830.

Kiplinger, A. (1991). In The Kiplinger Washington Letter, 68(20), p. 4.

Kircher, T. (1980). Silicone lymphadenopathy. A complication with silicone elastomer finger joint prostheses, Hum. Pathol., 11, p. 240–244.

Knabe, C., Ostapowicz, W., Radlanski, R.J., Gildenhaar, R., Berger, G., Fitzner, R. and Gross, U. (1998). In vitro investigation of novel calcium phosphates using osteogenic cultures. J . Mater. Sc.i Mater. Med., 9. p. 337.

Knabe, C., Houshmand, A., Berger, G., Ducheyne, P., Gildenhaar, R., Kranz, I., and Stiller, M. (2008). Effect of rapidly resorbable bone substitute materials on the temporal expression of the osteoblastic phenotype in vitro. J. Biomed. Mater. Res. A, 84, p. 8568–68.

Knabe, C. and Ducheyne, P. (2008). Chapter 6 – Cellular response to bioactive ceramics, In: Kokubo, T. (ed.). Handbook of Bioceramics and their Applications. Cambridge: Woodhead Publishing Inc. p. 133–164.

Knabe, C., Berger, G., Gildenhaar, R., Koch, C., Axmann, I., Jonscher, S., Rack, A., Ducheyne, P. and Stiller, M. (2009). Effect of rapidly resorbable calcium-alkali-orthophosphate bone grafting materials on osteogenesis after sinus floor augmentation in sheep. Trans. 35th Ann. Meet. Soc. Biomater. U.S.A., April 22-25, 2009, San Antonio, Texas, USA, p. 29.

Knabe, C., Berger, G., Gildenhaar, R., Houshmand, A., Müller-Mai, C., Bednarek, A., Koch, C., Jörn, D. and Stiller, M. (2010). Effect of resorbable calcium-alkali-orthophosphate bone substitute cements on osteogenesis after implantation in the rabbit femur. Trans. 36th Ann. Meet. Soc. Biomater. U.S.A., April 21-24, 2010, Seattle, Washington, USA, p. 10.

Knabe, C., Stiller, M. and Ducheyne, P. (2011a). Dental graft materials. In: Ducheyne, P., Healy, K., Hutmacher, D., Grainger, D. and Kirkpatrick, J.P. (eds). Comprehensive Biomaterials. Oxford: Elsevier, 6, p. 305–324.

Knabe, C. and Ducheyne, P. (2011b). Bioactivity – mechanisms. In: Ducheyne, P., Healy, K., Hutmacher, D., Grainger, D. and Kirkpatrick, J.P. (eds). Comprehensive Biomaterials. Oxford: Elsevier, 6, p. 245–258.

Knabe, C., Stiller, M., Adel-Khattab, D. and Ducheyne, P. (2017a). Dental graft materials. In: Ducheyne, P. Healy, K., Hutmacher, D., Grainger, D. and Kirkpatrick, J.P. (eds). Comprehensive Biomaterials II. Oxford: Elsevier, 7, p. 378–405.

Knabe, C., Adel-Khattab, D. and Ducheyne, P. (2017b). 1.12 Bioactivity: Mechanisms. In: Ducheyne, P., Healy, K., Hutmacher, D., Grainger, D. and Kirkpatrick, J.P. (eds). Comprehensive Biomaterials II, Oxford: Elsevier, 1, p. 291–310.

Knabe, C., Mele, A., Peleska, B., Kann, P.H., Adel-Khattab, D., Renz, H., Reuss, A., Bohner, M. and Stiller, M. (2017c). Effect of sex-hormone levels, sex, body mass index and other host factors on human craniofacial bone regeneration with bioactive tricalcium phosphate grafts. Biomaterials, 123, p. 48–62.

Knabe, C., Adel-Khattab, D., Kluk, E., Struck, R. and Stiller, M. (2017d). Effect of a particulate and a putty-like tricalcium phosphate-based bone-grafting material on bone formation, J. Funct. Biomater., 8(3), 31.

Knabe, C., Knauf, T., Adel-Khattab, D., Peleska, B., Hübner, W.D., Peters, F., Rack, A,. Gildenhaar, R., Berger, G., Günster, J., Houshmand, A. and Stiller, M. (2017e). Effect of a rapidly resorbable calcium alkali phosphate bone grafting material on osteogenesis after sinus floor augmentation in humans. Key Eng. Mater., 758, p 239–244.

Knabe, C., Adel-Khattab, D., Hübner, W.D., Peters, F., Knauf, T., Peleska, B., Barnewitz, D., Genzel, A., Kusserow, R., Sterzik, F., Stiller, M. and Müller-Mai, M. (2019). Effect of silicon-doped calcium phosphate bone grafting materials on bone regeneration and osteogenic marker expression after implantation in the bovine scapula J. Biomed. Mater. Res. B Appl Biomater., 107(3), p. 594–614.

Knahr, K. (ed.) (2013). Total Hip Arthroplasty – Tribological Considerations and Clinical Consequences. Heidelberg: Springer Verlag.

Kochanowski, A., Hoene, A., Patrzyk, M., Walschus, U., Finke, B., Luthringer, B., Feyerabend, F., Willumeit, R., Lucke, S. and Schlosser, M. (2011). Examination of the inflammatory response following implantation of titanium plates coated with phospholipids in rats, J. Mater. Sci. Mater. Med., 22, p. 1015–1026.

Kohn, J., Welsh, W.J. and Knight, D. (2007). A new approach to the rationale discovery of polymeric biomaterials. Biomaterials, 28(29), p. 4171–4177.

Kosobrodova, E., Kondyurin, A., Chrzanowski, W., Theodoropoulos, C., Morganti, E., Hutmacher, D. and Bilek, M.M.M. (2018). Effect of plasma immersion ion implantation on polycaprolactone with various molecular weights and crystallinity. J. Mater Sci: Mater. Med., 29: 5.

Kossovsky, N., Heggers, J.P. and Robson, M.C. (1987). Experimental demonstration of the immunogenicity of silicone-protein complexes, J. Biomed. Mater. Res,. 21, p. 1125–1133.

Kowalczewski, C.J. and Saul, J.M. (2018). Biomaterials for the delivery of growth factors and other therapeutic agents in tissue engineering approaches to bone regeneration. Front. Pharmacol., 9:513.

Kunz, F., (2005). Chemistry effects on cell differentiation. Proc. Natl. Acad. Sci. USA, 102, p. 5953–5957.

Kunz, F., Bergemann, C., Klinkenberg, E.-D., Weidmann, A., Lange, R., Beck, U. and Nebe, J.B. (2010). A novel modular device for 3-D bone cell culture and non-destructive cell analysis. Acta Biomater., 6, p. 3798–3807.

Kurtz, S. M., Lau, E., Ong, K., Zhao, K., Kelly, M. and Bozic, K.J. (2009). Future young patient demand for primary and revision joint replacement: National projections from 2010 to 2030. Clin. Orthop. Rel. Res., 467 (10), p. 2606–2612.

Lange, R., Lüthen, F., Beck, U., Rychly, J., Baumann, A. and Nebe, B. (2002). Cell-extracellular matrix interaction and physico-chemical characteristics of titanium surfaces depend on the roughness of the material. Biomol. Eng., 19, p. 255–261.

Lauffenburger, D.A. and Horwitz, A.F. (1996). Cell migration: a physically integrated molecular process. Cell, 84(3), p. 359–369.

Lee, J.Y., Nam, S.H., Im, S.Y., Park, Y.J., Lee, Y.M., Seol, Y.J., Chung, C. and Lee, S,J. (2002). Enhanced bone formation by controlled growth factor delivery from chitosan-based biomaterials. J. Control. Release, 78, p. 187–197.

Liang, R., Knight, K., Abramowitch, S. and Moalli, P.A. (2016). Exploring the basic science of prolapse meshes. Curr. Opin, Obstet. Gynecol., 28(5), p. 413–419.

Linkow, L.I. (1972). An honest evaluation of blade type implants. Bull. Hudson Cty. Dent. Soc., 41(6), p. 22–24.

Linkow L.I., Giauque, F., Ghalili, R. and Ghalili, M. (1995). Levels of osseointegration of blade-/plate-form implants. J. Oral Implantol., 21(1), p. 23–34.

Lobel, K.D. and Hench, L.L. (1998). In vitro adsorption and activity of enzymes on reaction layers of bioactive glass substrates. J. Biomed. Mater. Res., 39(4), p. 575–579.

Lopes, H.B., Pereira Freitas, G., Maciely D., Fantacini, C., Picanço-Castro, V., Covas, D.T., Luiz Rosa, A. and Beloti M.M. (2019). Titanium with nanotopography induces osteoblast differentiation through regulation of integrin αV. J. Cell Biochem., 120(6), p. 1-10.

Lübbeke, A., Silman, A. J., Barea, C., Prieto-Alhambra, D. and Carr, A. J. (2018). Mapping existing hip and knee replacement registries in Europe. Health Policy, 122 (5), p. 548–557.

Lübbeke, A., Silman, A.J., Prieto-Alhambra, D., Adler, A.I., Barea, C. and Carr, A. J. (2017). The role of national registries in improving patient safety for hip and knee replacements. BMC Musculoskeletal Disorders, 18 (1), p. 414.

Lucke, S., Hoene, A., Walschus, U., Kob, A., Pissarek, J.W. and Schlosser, M. (2015). Acute and chronic local inflammatory reaction after implantation of different extracellular porcine dermis collagen matrices in rats, BioMed. Res. Internat., 938059.

Lucke, S., Walschus, U., Hoene, A., Schnabelrauch, M., Nebe, J.B., Finke, B. and Schlosser, M. (2018). The in vivo inflammatory and foreign body giant cell response against different poly(l-lactide-co-d/l-lactide) implants is primarily determined by material morphology rather than surface chemistry, J. Biomed. Mater. Res. A, 106, p. 2726–2734.

Lüthen, F., Lange, R., Becker, P., Rychly, J., Beck, U. and Nebe, B. (2005). The influence of surface roughness of titanium on β1- and β3-integrin adhesion and the organization of fibronectin in human osteoblastic cells. Biomaterials, 26, p. 2423–2440.

Luo, H., Xiong, G., Zhang, C., Li, D., Zhu, Y., Guo, R. and Wan, Y. (2015). Surface controlled calcium phosphate formation on three-dimensional bacterial cellulose-based nanofibers. Mater. Sci. Eng. C, 49, p. 526–533.

Malchau, H., Herberts, P., Eisler, T., Garellick, G. and Soderman, P. (2002). The Swedish Total Hip Replacement Register. J. Bone Joint Surg. (Am. Volume), 84-A (Suppl 2), p. 2–20.

Malchau, H., Garellick, G., Berry, D., Harris, W.H., Robertson, O., Karrlholm, J. et al. (2018). Arthroplasty implant registries over the past five decades: Development, current, and future impact. J. Orthop. Res., 36 (9), p. 2319–2330.

Maloney, W. J. (2001). National Joint Replacement Registries: has the time come? J. Bone Joint Surg. (Am. Volume), 83 (10), p. 1582–1585.

Mandracci, P., Mossano, F., Rivolo, P. and Carossa, S. (2016). Surface treatments and functional coatings for biocompatibility improvement and bacterial adhesion reduction in dental implantology. Coatings, 6(1), 7.

Marchant, R., Hiltner, A., Hamlin, C., Rabinovitch, A., Slobodkin, R. and Anderson, J.M. (1983). In vivo biocompatibility studies: I. The cage implant system and a biodegradable hydrogel, J. Biomed. Mater. Res., 17, p. 301–325.

Martinez, F.O., Helming, L. and Gordon S. (2009). Alternative activation of macrophages: an immunologic functional perspective. Ann. Rev. Immunol., 27, p. 451–483,

Matsuda, T. and Hirano T. (2001). IL-6. In: Oppenheim, J.J. and Feldmann, M. (eds). Cytokine reference: a compendium of cytokines and other mediators of host defense. San Diego: Academic Press; p. 537–563.

Mattila, P.K. and Lappalainen, P. (2008). Filopodia: molecular architecture and cellular functions. Nat. Rev. Mol. Cell Biol., 9 (6), p. 446–454.

Matschegewski, C., Staehlke, S., Loeffler, R., Lange, R., Chai, F., Kern, D.P., Beck, U. and Nebe, B.J. (2010), Cell architecture-cell function dependencies on titanium arrays with regular geometry, Biomaterials, 31(22), p. 5729–5740.

McGee, M.A., Howie, D.W., Neale, S.D., Haynes, D.R. and Pearcy, M.J. (1997). The role of polyethylene wear in joint replacement failure. Proc. Inst. Mech. Eng. H., 211(1):65

McKenzie, A.N.J. and Matthews, D.J. (2001). IL-13. In: Oppenheim, J.J. and Feldmann, M. (eds). Cytokine reference: a compendium of cytokines and other mediators of host defense. San Diego: Academic Press, p. 203–211.

MDR (2017). Medical Devices Regulation of the European Union. https://eur-lex.europa.eu/eli/reg/2017/745/oj (accessed January 29, 2020).

Meyer, U., Joos, U., Mythili, J., Stamm, T., Hohoff, A., Fillies, T, et al. (2004). Ultrastructural characterization of the implant/bone interface of immediately loaded dental implants. Biomaterials, 25, p. 1959–1967.

Moerke, C., Müller, P. and Nebe, B. (2016). Attempted caveolae-mediated phagocytosis of surface-fixed micro-pillars by human osteoblasts. Biomaterials, 76, p. 102–114.

Moerke, C., Rebl, H., Finke, B., Dubs, M., Nestler, P., Airoudj, A., Roucoules, V., Schnabelrauch, M., Koertge, A., Anselme, K., Helm, C.A. and Nebe, J.B. (2017). Abrogated cell contact guidance on amino-functionalized micro-grooves. ACS Applied Materials & Interfaces, ACS Appl. Mater. Interfaces, 9(12), p. 10461–10471.

Mofrad, M.R.K. and Kamm, R.D. (2006). Cytoskeletal mechanics. Cambridge, New York, Melbourne, Madrid, Cape Town, Singapore, Sao Paulo: Cambridge University Press. ISBN-13 978-0-521-84637-0. p.11.

Molnar, C. and Gair, J. (2015). Concepts of Biology, 1st Canadian edition. Chapter 3.3 Eukaryotic cells. p. 89.

Mosser, D.M. and Zhang X. (2008). Interleukin-10: New perspectives on an old cytokine, Immunol. Rev., 226, p. 205–218.

Müller-Mai, C.M., Stupp, S.I., Voigt, C. and Gross, U. (1995). Nanoapatite and organoapatite implants in bone: histology and ultrastructure of the interface. J. Biomed. Mater. Res., 29(1), p. 9–18.

Müller-Mai, C.M. (2003). Bioaktive Granulate in der Unfallchirurgie. München: VNM Science Publishing.

Müller-Mai, M., Berger, G., Stiller, M., Gildenhaar, R., Jörn, D., Ploska, U., Houshmand, A., Bednarek, A., Koch, C. and Knabe, C. (2010). Evaluation of degradable bone cements for percutaneous augmentation of bone defects. Mat.wiss. Werkstofftech., 41(12), p. 1–8.

Nair, L.S. and Laurencin, C.T. (2006). Polymeric applications as biomaterials in the areas of tissue engineering and controlled drug delivery. Adv. Biochem. Engng/Biotechnol. Special Issue: Tissue Engineering, 102, p. 47–90.

Nakamura, A., Kawasaki, Y., Takada, K. and 10 additional authors (1992). Difference in tumor incidence and other tissue responses to polyetherurethanes in long-term subcutaneous implantation in rats. J. Biomed. Mater. Res., 26, p. 631–650.

National Joint Registry (2015). 11th annual report 2014: National Joint Registry for England, Wales, Northern Ireland and the Isle of Man. Hemel Hempstead (UK).

Nebe, J.B., Finke, B., Hippler, R., Meichsner, J., Podbielski, A., Schlosser, M. and Bader, R. (2013). Physical plasma processes for surface functionalization of implants in orthopedic surgery, Hyg. Med., 38, p.192–197.

Nebe, B., Lüthen, F., Lange, R., Becker, P., Beck, U. and Rychly, J. (2004). Topography-induced alterations in adhesion structures affect mineralization in human osteoblasts on titanium. Mater. Science Eng. C, 24, p. 619–624.

Nebe, B., Lüthen, F., Finke, B., Bergemann, C., Schröder, K., Rychly, J., Liefeith, K. and Ohl, A. (2007). Improved initial osteoblast's functions on amino-functionalized titanium surfaces. Biomol Eng., 24(5), p. 447–454.

Nebe, J.B. and Lüthen, F. (2008). Integrin- and Hyaluronan-mediated cell adhesion on titanium – Hyaluronan-mediated adhesion. In: Breme, J., Kirkpatrick, C.J. and Thull, R. (eds). Metallic Biomaterial Interfaces. Weinheim: Wiley-VCH. ISBN 978-3-527-31860-5, 2008, p. 179–182.

Nebe, J.B., Rebl, H., Schlosser, M., Staehlke, S., Gruening, M., Weltmann, K.D., Walschus, U. and Finke, B. (2019). Plasma polymerized allylamine – the unique cell-attractive nanolayer for dental implant materials. Polymers, 11(6), 1004 (19 pp.).

Neumann, H.-G. and Klinkenberg, E.-D. (Hrsg.) (2014) Knöcherne Geweberegeneration. Zellphysiologie im Dreidimensionalen [Bony tissue regeneration. Cell physiology in 3D]. Aachen: Shaker Verlag. ISBN 978-3-8440-3049-5, p.139.

Niepel, M.S., Mano, J.F. and Groth, T. (2016). Effect of polyelectrolyte multilayers assembled on ordered nanostructures on adhesion of human fibroblasts. ACS Appl. Mater. Interfaces, 8, p. 25142–25151.

Nolfi, A.L., Brown, B.N., Liang, R., Palcsey, S.L., Bonidie, M.J., Abramowitch, S.D, and Moalli, .PA. (2016). Host response to synthetic mesh in women with mesh complications. Am. J. Obstet. Gynecol., 5(2):206.e1-8.

Ochsner, M., Textor, M., Vogel, V. and Smith, M.L. (2010) Dimensionality controls cytoskeleton assembly and metabolism of fibroblast cells in response to rigidity and shape. PLoSONE, 5(3), e9445.

OECD (1997). Test No. 471: Bacterial Reverse Mutation Test, OECD Guidelines for the Testing of Chemicals, Section 4, OECD Publishing, Paris, https://doi.org/10.1787/9789264071247-en.

Olavarria, O.A., Shah, P., Bernardi, K., Lyons, N.B., Holihan, J.L., Ko, T.C., Kao, L.S and Liang, M.K. (2019). Lack of regulations and conflict of interest transparency of new hernia surgery technologies. J. Surg. Res., https://doi.org/10.1016/j.jss.2019.09.061.

Onuma, K., Oyane, A., Kokubo, T., Treboux, G., Kanzaki, N. and Ito, A. (2000). Precipitation kinetics of hydroxylapatite revealed by the continuous-angle laser light-scattering technique. J. Phys. Chem. B, 104, p.11950–11956.

Palsson, B. (2000). The challenges of in silico biology. Nature Biotechnol., 18, p. 1147–1150.

Patrzyk, M., Hoene, A., Jarchow, R., Wilhelm, L., Walschus, U., Zippel, R. and Schlosser, M. (2010). Time course of fibronectin in the peri-implant tissue and neointima formation after functional implantation of polyester-based vascular prostheses with different porosity in pigs, Biomed. Mater., 5, 055003.

Peng, P., Kumar, S., Voelcker, N.H., Szili, E., Smart, R.S.C. and Griesser, H.J. (2006). Thin calcium phosphate coatings on titanium by electrochemical deposition in modified simulated body fluid. J .Biomed. Mater. Res. A, 76(2), p. 347–355.

Pereira, B.J.G. and Dinarello, C.A. (1994). Production of cytokines and inhibitory proteins in patients on dialysis, Nephrol. Dial. Transplant., 9 (Suppl. 2), p. 60–71.

Peters, F. and Reif, D, (2004). Functional materials for bone regeneration from β-tricalcium phosphate', Materialwiss. Werkstofftechn., 35, p. 203–207.

Piehler, J.M. and Crichlow, R.W. (1978). Primary carcinoma of the gallbladder. Surg. Gynecol. Obstet., 147(6), p. 929–942.

Qu, H., Knabe, C., Burke, M., Radin, S., Garino, J., Schaer, T. and Ducheyne, P. (2014). Bactericidal micron-thin sol-gel films prevent pin tract and periprosthetic infection. Mil. Med., 9(8 Suppl), p. 29–33.

Qu, H., Knabe, C., Radin, S., Garino, J. and Ducheyne, P. (2015). Percutaneous external fixator pins with bactericidal micron-thin sol-gel films for the prevention of pin tract infection. Biomaterials, 62, p. 95–105.

Piskin, E., Bölgen, N., Egri, S. and Isoglu, I.A. (2007). Electrospun matrices made of poly(α hydroxy acids) for medical use, Nanomedicine, 2, p. 441–457.

Rahbek, O., Overgaard, S., Lind, M., Bendix, K., Buenger, C. and Søballe, K. (2001). Sealing effect of hydroxylapatite coating on peri-implant migration of particles. J. Bone Joint Surg. [Br], 83, p. 441–448.

Ratner, B.D. (2015). Healing with medical implants: The body battles back. Sci. Transl. Med.,7 (272):272fs4.

Ratner, B.D. (2016). A pore way to heal and regenerate: 21st century thinking on biocompatibility. Regen. Biomater., 3(2), p. 107–110.

Ratner, B.D. (2019). Biomaterials: Been there, done that, and evolving into the future. Ann. Rev. Biomed. Eng., 21, p. 171–191.

Rau, J.V., Curcio, M., Raucci, M.G., Barbaro, K., Fasolino, I., Teghil, R., Ambrosio, L., De Bonis, A. and Boccaccini, A.R. (2019). Cu-releasing bioactive glass coatings and their in vitro properties. ACS Appl. Mater. Interfaces, 11(6), p. 5812–5820.

Rebl, H., Finke, B., Schmidt, J., Mohamad, H.S., Ihrke, R., Helm, C.A. and Nebe, J.B. (2016). Accelerated cell-surface interlocking on plasma polymer-modified porous ceramics. Mat. Sci. Eng. C, 69, p. 1116–1124.

Reneker, D.H., Yarin, E., Zussman, A.L. and Xu, H. (2007). Electrospinning of nanofibers from polymer solutions and melts, Adv. Appl. Mech., 41, p. 43–46.

Revell, P., Braden, M. and Freeman, M. (1998). Review of the biological response to a novel bone cement containing poly(ethyl methacrylate) and n-butyl methacrylate. Biomaterials, 19, p. 1579–1586.

Revell, P.A., Damien, E., Zhang, X.S., Evans, P. and Howlett, C.R. (2003). The Effect of magnesium ions on bone bonding to hydroxyapatite coating on titanium alloy implants. Key Eng. Mater., 254-256, p. 447–450.

Romano, P.S., Chan, B.K., Schembri, M.E. and Rainwater, J.A. (2002). Can administrative data be used to compare postoperative complication rates across hospitals? Med. Care, 40(10), p. 856–867.

Roth, J.R. (1995). Industrial Plasma Engineering. Volume 1: Principles. Bristol: Institute of Physics Publishing,.

Roth, J.R. (2001). Industrial Plasma Engineering. Volume 2: Applications to Nonthermal Plasma Processing. Bristol: Institute of Physics Publishing.

Rowley, M.J., Cook, A.D., Teuber, S. and Gershwin, M.E. (1994). Antibodies to collagen: Mapping in women with silicone breast implants, systemic lupus erythematosus and rheumatoid arthritis, J. Autoimm. 7, p. 775–789.

Ruyter, I.E. (1995). The importance of materials characterisation as it relates to polymeric materials and their applications. In: Ellerton, A. (ed). Biocompatibility Meeting the Challenge, Confer. Proc., p. 13–29, Medical Device Technology, Advanstar Commun.

Schlosser, M., Ziegler, B., Abel, P., Fischer, U. and Ziegler, M. (1994a). Implantation of non-toxic materials from glucose sensors: Evidence for specific antibodies detected by ELISA. Horm. Metab. Res., 26, p. 534–537.

Schlosser, M., Ziegler, B., Abel, P., Fischer, U. and Ziegler, M. (1994b). Humoral immune response as an additional parameter of biocompatibility testing? In: Goh, J.C.H., Nather, A. (eds). Proc. Eighth Intern. Conf. Biomed. Eng. p. 482–484. Singapore: BAC Printers.

Schlosser, M., Wilhelm, L., Urban, G., Ziegler, B., Ziegler, M. and Zippel, R. (2002). Immunogenicity of polymeric implants: Long-term antibody response against polyester (Dacron) following the implantation of vascular prostheses into LEW.1A rats, J. Biomed. Mater. Res., 61, p. 450–457.

Schlosser, M., Zippel, R., Hoene, A., Urban, G., Ueberrueck, T., Marusch, F., Koch, A., Meyer, L. and Wilhelm, L. (2005). Antibody response to collagen after functional implantation of different polyester vascular prosthesis in pigs. J. Biomed. Mater. Res. A, 72, p. 17–25.

Schnabelrauch, M., Wyrwa, R., Rebl, H., Bergemann, C., Finke, B., Schlosser, M., Walschus, U., Lucke, S., Weltmann, K.-D. and Nebe, J.B. (2014). Surface-coated polylactide fiber meshes as matrices for soft and vascular tissue engineering, Int. J. Polym. Sci., 6, 439784.

Schofield, P.F., Valsami-Jones, E., Sneddon, I.R., Wilson, J., Kirk, C.A., Terrill, N.J., Martin, C.M., Lammie, D. and Wess, T.J. (2005). Nucleation and growth of nano-apatite: applications to biomineralisation. Geochim. Cosmochim. Acta, 69 (10),p. A72-A72, Suppl. S.

Schutte, R.J., Xie, L., Klitzman, B. and Reichert, W.M. (2009). In vivo cytokine associated responses to biomaterials, Biomaterials, 30, p. 160–168.

Sedrakyan, A., Paxton, E.W., Phillips, C., Namba, R., Funahashi, T., Barber, T. et al. (2011). The International Consortium of Orthopaedic Registries: Overview and summary. J. Bone Joint Surg. (Am.volume), 93(Suppl 3), p. 1–12.

Shirakawa, T., Kusaka, Y. and Morimoto, K. (1992). Specific IgE antibodies to nickel in workers known reactivity to cobalt, Clin. Exp. Allergy, 22, p. 213–218.

Shiwaku, Y., Anada, T., Yamazaki, H., Honda, Y., Morimoto, S., Sasaki, K. and Suzuki, O. (2012). Structural, morphological and surface characteristics of two types of octacalcium phosphate-derived fluoride-containing apatitic calcium phosphates. Acta Biomater., 8(12), p. 4417–4425.

Sill, T.J. and von Recum, A.F. (2008). Electrospinning: Applications in drug delivery and tissue engineering, Biomaterials, 29, p. 1989–2006.

Silverman, E. and Skinner, J. (2004). Medicare upcoding and hospital ownership. J. Health Econ., 23(2), p. 369–389.

Silverman, W. A. (1985). Human Experimentation: A Guided Step into the Unknown. Oxford: Oxford University Press.

Sjaastad, M.D. and Nelson, W.J. (1997). Integrin-mediated calcium signaling and regulation of cell adhesion by intracellular calcium. Bioessays, 19, p. 47–55.

Smith, H.R. (1995). Do silicone breast implants cause autoimmune rheumatic diseases? J. Biomater. Sci. Polymer Edn., 7, p. 115–121.

Smith, K.A. (2001). IL-2. In: Oppenheim, J.J., Feldmann, M. (eds). Cytokine reference: a compendium of cytokines and other mediators of host defense. San Diego: Academic Press. p. 113–125.

Snow, R.B. and Kossovsky, N. (1989). Hypersensitivity reaction associated with sterile ventriculoperitoneal shunt malfunction, Surg. Neurol., 31, p. 209–214.

Staehlke, S., Koertge, A. and Nebe, B. (2015). Intracellular calcium dynamics in dependence on topographical features of titanium. Biomaterials, 46, p. 48–57.

Steinemann, .SG. and Straumann, F. (1984). Ankylotic anchorage of implants. Schweiz. Monatsschr. Zahnmed., 94(8), p. 682–687.

Steinemann, S.G. (1996). Metal implants and surface reactions. Injury, 27(3). p. 16–22.

Steinemann, S.G. (1998). Titanium–the material of choice? Periodontol 2000, 17(1), p. 7–21.

Stokes, K. (1993). Biodegradation. Cardiovasc. Pathol., 2, p. 111–119.

Stupp, S.I. and Braun, P.V. (1997). Molecular manipulation of microstructures: biomaterials, ceramics, and semiconductors. Science, 277 (29 August), p.1242–1248.

Svitkina, T.M. (2018). Ultrastructure of the actin cytoskeleton, Curr. Opin. Cell Biol., 54, p. 1–8.

Szentivanyi, A., Chakradeo, T., Zernetsch, H. and Glasmacher, B. (2011). Electrospun cellular microenvironments: understanding controlled release and scaffold structure, Adv. Drug Deliv. Rev., 63, p. 209–220.

Szycher, M. (1992). Szycher's Dictionary of Biomaterials and Medical Devices. p. 22, Lancaster, PA: Technomic Publishing Comp., Inc.

Tang, L., Lucas, A.H., and Eaton, J.W. (1993a). Inflammatory responses to implanted biomaterials. Role of surface-adsorbed immunoglobulin G, J. Lab. Clin. Med., 122, p. 292–300.

Tang, L., and Eaton, J.W. (1993b). Fibrin(ogen) mediates acute inflammatory responses to biomaterials, J. Exp. Med., 178, p. 2147–2156.

Tang, L. and Eaton, J.W. (1995). Inflammatory responses to biomaterials, Am. J. Clin. Pathol., 103, p. 466–471.

Thomsen, J.L., Christensen, L., Nielsen, M., Brandt, B., Breiting, V.B., Felby, S. and Nielsen, E. (1990). Histologic changes and silicone concentrations in human breast tissue surrounding silicone breast prostheses, Plast. Reconstr. Surg., 85, p. 38–41.

Tomita, M. (2001). Whole-cell simulation: a grand challenge of the 21st century. Trends Biotechnol., 19, p. 205–210.

Tsuchiya, T., Takahara, A., Cooper, S.L. and Nakamura, A. (1995). Studies on the tumor-promoting activity of polyurethanes: Depletion of inhibitory action of metabolic cooperation on the surface of a polyalkyleneurethane but not a polyetheruethane, J. Biomed. Mater. Res., 29, p. 835–841.

Tzafriri, A.R., Garcia-Polite, F., Li, X., Keating, J., Balaguer, J.M., Zani, B., Bailey, L., Markham, P., Kiorpes, T.C., Carlyle, W. and Edelman, E.R. (2018). Defining drug and target protein distributions after stent-based drug release: Durable versus deployable coatings. J. Control. Release, 274, p. 102–108.

Unanue, E.R. and Allen P.M. (1987). The basis for the immunoregulatory role of macrophages and other accessory cells, Science, 236, p. 551–557.

Varga, J., Schumacher, H.R. and Jimenez, S.A. (1989). Systemic sclerosis after augmentation mammoplasty with silicone implants. Ann. Intern. Med., 111, p. 377–383.

Vince, D.G., Hunt, J.A. and Williams, D.F. (1991). Quantitative assessment of the tissue response to implanted biomaterials, Biomaterials, 12, p. 731–736.

Vivier, E., Tomasello, E., Baratin, M., Walzer, T. and Ugolini, S. (2008). Functions of natural killer cells, Nat. Immunol., 9, p. 503–510.

Vroman, L. (1988). The life of an artificial device in contact with blood: initial events and their effect on its final state, Bull. NY Acad. Med., 64, p. 352–357.

Walenkamp, G.H.I.M., Murray, D.W., Henze, U. and Kock, H.J. (eds) (2001). Bone Cements and Cementing Technique. Heidelberg: Springer Verlag.

Wallin, R.F. (1995a). Global biocompatibiliy, Med. Dev. Technol., 6, p. 34–38.

Wallin, R.F. (1995b).Biocompatibility testing. In: Ellerton, A. (ed.). Biocompatibility Meeting the Challenge, Confer. Proceed. p. 7–19, Medical Device Technology, Advanstar Commun.

Walschus, U., Goldmann, H., Ueberrueck, T., Hoene, A., Wilhelm, L. and Schlosser, M. (2008). Evaluation of the biocompatibility of a new vascular prosthesis coating by detection of prosthesis-specific antibodies, J. Mater. Sci. Mater. Med., 19, p. 1595–1600.

Walschus, U., Hoene, A., Neumann, H.-G., Wilhelm, L., Lucke, S., Lüthen, F., Rychly, J. and Schlosser, M. (2009). Morphometric immunohostochemical examination of the inflammatory tissue reaction after implantation of calcium-phosphate-coated titanium-plates in rats, Acta Biomater., 5, p. 776–784.

Walschus, U., Hoene, A., Kochanowski, A., Neukirch, B., Patrzyk, M., Wilhelm, L., Schroeder, K. and Schlosser, M. (2011). Quantitative immunohistochemical examination of the local cellular reactions following implantation of biomaterials, J. Microsc., 242, p. 94–99.

Walschus, U., Hoene, A., Patrzyk, M., Finke, B., Polak, M., Lucke, S., Nebe, B., Schröder, K., Podbielski, A., Wilhelm, L. and Schlosser, M. (2012). Serum profile of pro- and anti-inflammatory cytokines in rats following implantation of low-temperature plasma-modified titanium plates, J. Mater. Sci. Mater. Med., 23, p. 1299–1307.

Walschus, U., Hoene, A., Patrzyk, M. and 9 additional authors (2017). A cell-adhesive plasma polymerized allylamine coating reduces the in vivo inflammatory response induced by Ti6Al4V modified with plasma immersion implantation of copper. J. Funct. Biomater., 8, p. 1–13.

Wang, L., Chen, L., Yan, Z. and Fu, W. (2010). Optical emission spectroscopy studies of discharge mechanism and plasma characteristics during plasma electrolytic oxidation of magnesium in different electrolytes. Surf. Coat. Technol., 205, p. 1651–1658.

Ward, R.A. (1994). Phagocytic cell function as an index of biocompatibility, Nephrol. Dial. Transplant. 9, (Suppl.2), p. 46–56.

Weiner, S., Dove, P.M. (2003). An overview of biomineralization processes and the problem of the Vital Effect. In: P.M. Dove, J.J. De Yorco, S. Weiner (eds), Biomineralization. Rev. Min. Geochem., 54, 1–29, Min. Soc. Am., Washington, D.C.

Weiss, S.J. (1989). Tissue destruction by neutrophils, N. Engl. J. Med., 320, p. 365–376.

Wilhelm, L., Zippel, R., von Woedtke, T., Kenk, H., Hoene, A., Patrzyk, M. and Schlosser, M. (2007). Immune response against polyester implants is influenced by the coating substances, J. Biomed. Mater. Res. A, 83, p. 104–113.

Williams, D.F. (1989). A model for biocompatibility and its evaluation, J. Biomed. Eng., 11, p. 185–191.

Williams, D.F. (1991). Objectivity in the evaluation of biological safety of medical devices and biomaterials, Med. Dev. Technol., 1, p. 44–48.

Williams, D.F. (1993). Diamond-like carbons and other thin films, Med. Dev. Technol., 4, p. 8–13.

Williams, D.F. (1994a). Biodegradation and stability: The inevitability of ageing, Med. Dev. Technol., 5, p. 8–12.

Williams, D.F. (1994b). The capricious nature of biocompatibility, Med. Dev. Technol., 5, p. 8–11.

Williams, D.F. (2008). On the mechanisms of biocompatibility, Biomaterials, 29, p. 2941–2953.

Williams, D.F. (2014). There is no such thing as a biocompatible material, Biomaterials, 35, 10009–14.

Williams, D. (2014). Essential Biomaterials Science. Cambridge Texts in Biomedical Engineering. ISBN 978-0-521-89908-6, p. 62.

Williams, D.F. (2015). Regulatory biocompatibility requirements for biomaterials used in regenerative medicine. J. MaterSci. Mater. Med., 26(2): 89.

Willumeit, R., Schuster, A., Iliev, P., Linser, S. and Feyerabend, F. (2007a). Phospholipids as implant coatings, J. Mater. Sci. Mater. Med., 18, p. 3673–80.

Willumeit, R., Schossig, M., Clemens, H. and Feyerabend, F. (2007b). In-vitro interactions of human chondrocytes and mesenchymal stem cells, and of mouse macrophages with phospholipid-covered metallic implant materials, Eur. Cell. Mater., 13, p. 11–25.

Wolf, B.R., Lu, X., Li, Y., Callaghan, J.J. and Cram, P. (2012). Adverse outcomes in hip arthroplasty: long-term trends. J. Bone Joint Surg. (Am. volume), 94(14), p. e103.

Wood, N.K., Kaminski, E.J. and Oglesby, R.J. (1970). The significance of implant shape in experimental testing of biological materials: disc vs. rod, J. Biomed. Mater. Res., 4, p. 1–11.

Wopenka, B. and Pasteris, J.D. (2005). A mineralogical perspective on the apatite in bone. Mater. Sci. Eng. C, 25, p. 131–143.

Xynos, I.D., Edgar, A.J., Buttery, L.D., Hench, L.L. and Polak, J.M. (2000a). Ionic products of bioactive glass dissolution increase proliferation of human osteoblasts and induce insulin-like growth factor II mRNA expression and protein synthesis. Biochem. Biophys. Res. Commun., 276, p. 461–465.

Xynos, I.D., Hukkanen, M.V., Batten, J.J., Buttery, L.D., Hench, L.L. and Polak, J.M. (2000b). Bioglass 45S5 stimulates osteoblast turnover and enhances bone formation In vitro: implications and applications for bone tissue engineering. Calcif. Tissue Int., 67, p. 321–329.

Xynos, I.D., Edgar, A.J., Buttery, L.D., Hench, L.L. and Polak, J.M. (2001). Gene-expression profiling of human osteoblasts following treatment with the ionic products of Bioglass 45S5 dissolution. J. Biomed. Mater. Res., 55, p. 151–157.

Yamada, K.M. and Geiger, B. (1997). Molecular interactions in cell adhesion complexes. Curr. Opin. Cell Biol., 9, p. 76–85.

Yamada, K.M., Pankov, R. and Cukierman, E. (2003). Dimensions and dynamics in integrin function. Braz. J. Med. Biol. Res., 36, p. 959–966.

Yang, J. and Merritt, K. (1994). Detection of antibodies against corrosion products in patients after Co-Cr total joint replacements, J. Biomed. Mater. Res., 28, p. 1249–1258.

Yao, J., Radin, S., Reilly, G., Leboy, P.S. and Ducheyne, P. (2015). Solution-mediated effect of bioactive glass in poly (lactic-co-glycolic acid)-bioactive glass composites on osteogenesis of marrow stromal cells. J. Biomed. Mater. Res. A., 75(4), p. 794–801.

Yerokhin, A.L., Nie, X., Leyland, A., Matthews, A. and Dowey, S.J. (1999). Plasma electrolysis for surface engineering. Surf. Coat. Technol., 122, p. 73–93.

Zaidel-Bar, R., Itzkovitz, S., Maayan, A., Iyengar, R. and Geiger, B. (2007). Functional atlas of the integrin adhesome. Nat. Cell Biol., 9, p. 858–867.

Zamir, E. and Geiger, B. (2001). Components of cell-matrix adhesions. J. Cell Sci., 114, p. 3577–3579.

Zhang, Y., Bindra, D.S., Barrau, M.-B. and Wilson, G.S. (1991). Application of cell culture toxicity tests to the development of implantable biosensors. Biosens. Bioelectron., 6, p. 653–661.

Zhang, Z., Li, G. and Shi. B. (2006). Physicochemical properties of collagen, gelatin and collagen hydrolysate derived from bovine limed split wastes. J. Soc. Leather Technol. Chem., 90, p. 23–28.

Zhao, J.H., Reiske, H. and Guan, J.L. (1998). Regulation of the cell cycle by focal adhesion kinase. J. Cell Biol., 143, p. 1997–2008.

Ziats, N.P., Pankowsky, D.A., Tierney, B.P., Ratnoff, O.D. and Anderson, J.M. (1990). Adsorption of hageman factor (factor XII) and other human plasma proteins to biomedical polymers, J. Lab. Clin. Med., 116, p. 687–696.

Ziegler, M., Schlosser, M., Abel, P. and Ziegler, B. (1994). Antibody response in rats against nontoxic glucose sensor membranes tested in cell culture, Biomaterials, 15, p. 859–864.

Zimmermann, E., Geiger, B. and Addadi, L. (2002). Initial stages of cell-matrix adhesion can be mediated and modulated by cell-surface hyaluronan. Biophys. J., 82, p. 1848–1857.

Zinger, O., Zhao, G. Schwartz, Z., Simpson, J. Wieland, M., Landolt, D. and Boyan, B. (2005). Differential regulation of osteoblasts by substrate microstructural features. Biomaterials, 26, p. 1837–1847.

Zippel, R., Wilhelm, L., Marusch, F., Koch, A., Urban, G. and Schlosser, M. (2001). Antigenicity of polyester (Dacron) vascular prostheses in an animal model, Eur. J. Vasc. Endovasc. Surg., 21, p. 202–207.

Zippel, R., Hoene, A., Walschus, U., Jarchow, R., Ueberrueck, T., Patrzyk, M., Schlosser, M. and Wilhelm, L. (2006). Digital image analysis for morphometric evaluation of tissue response after implanting alloplastic vascular prostheses, Microsc. Microanal., 12, p. 366–375.

Zippel, R., Wilhelm, L., Hoene, A., Walschus, U., Ueberrueck, T. and Schlosser, M. (2008). Local tissue reaction and differentiation of the prosthesis-specific antibody response following functional implantation of vascular grafts in pigs, J. Biomed. Mater. Res. B Appl. Biomater., 85, p. 334–353.

Chapter 5

Amato, S.F. (2015a). Regulatory strategies for biomaterials and medical devices in the USA: classification, design, and risk analysis. In: Amato, S.F. and Ezzell, R.M. (eds). Regulatory Affairs for Biomaterials and Medical Devices. Woodhead Publ. Series in Biomaterials. ISBN 978-0-85709-542-8. p. 27–46.

Amato, S.F. (2015b). Supply chain controls for biomaterials and medical devices in the USA. In: Amato, S.F. and Ezzell, R.M. (eds). Regulatory Affairs for Biomaterials and Medical Devices. Woodhead Publ. Series in Biomaterials. ISBN 978-0-85709-542-8. p. 79–92.

Amato, S.F. (2015c). Clinical development and endpoint strategies for biomaterials and medical devices. In: Amato, S.F. and Ezzell, R.M. (eds). Regulatory Affairs for Biomaterials and Medical Devices. Woodhead Publ. Series in Biomaterials. ISBN 978-0-85709-542-8. p. 47–66.

Andre, C. and Velasquez, M. (2014). Of cures and creatures great and small. https://www.scu.edu/ethics/publications/iie/v1n3/cures.html (accessed Jan 29, 2019).

DIN EN ISO/IEC 17025:2018-03. General requirements for the competence of testing and calibration laboratories.

FDA, PMA Review Process. https://www.fda.gov/medical-devices/premarket-approval-pma/pma-review-process. Accessed July 14, 2019.

FDA News (2019). Guidance on transitioning from MDD to MDR. Newsletter, Dec 3, 2019.

Fernando, D., Adhikari, A., Nanyakkara, C., Dilip de Silva, E., Wijesundera, R.L.C. and Soysa, P. (2016). Cytotoxic effects of ergone, a compound isolated from Fulviformes fastuosus. BMC Compl. Altern. Med., 16 (1):484.

Geysels, Y., Bamford, C.A. and Corr, R.H. (2017). The new European Union Regulation for clinical trials. Clinical Researcher, Feb 1, 2017.

Gross, K.A. (2014). Calcium phosphate research in Australia. J. Austral. Ceram. Soc., 50(1), p. 74–85.

Guidance Good manufacturing practice and good distribution practice (2019). Medicines and Healthcare products Regulating Agency and Department of Health and Social Care, Feb 12, 2019. https://www.gov.uk/guidance/good-manufacturing-practice-and-good-distribution-practice (accessed Feb 24, 2019).

Hamid, R., Rotshteyn, Y., Rabadi, L., Parikh, R. and Bullock, P. (2004). Comparison of Alamar blue and MTT assays for high through-put screening, Toxicol. In Vitro, 18(5), p. 703–710.

Hoang, Q.Q., Sicheri, F., Howard, A.J. and Yang, D.S.C. (2003). Bone recognition mechanism of porcine osteocalcin from crystal structure, Nature, 425, p. 977–980.

ISO 13485:2016. Medical devices-Quality management systems-Requirements for regulatory purposes. International Organization for Standardization. www.iso/org (accessed Feb 24, 2019).

ISO 9001:2015. Quality management system-requirements. International Organization for Standardization. www.iso/org (accessed Feb 24, 2019).

ITA (2016). ITA Medical Devices Top Markets Report. International Trade Administration. https://www.trade.gov/topmarkets/pdf/Medical_Devices_Executive_Summary.pdf (accessed March 2, 2019).

Kammula, R.G. and Morris, J.M. (2001). Consideration for the biocompatibility evaluation of medical devices. Newsletter Medical Device and Diagnostic Industry (MDDI), May 1, 2001.

Kari, G., Rodeck, U. and Dicker, A. P. (2007). Zebrafish: an emerging model system for human disease and drug discovery. Clin. Pharmacol. Therapeutics, 82 (1), p. 70–80.

Kashi, A. and Saha, S. (2017). Ethical issues in biomedical research. In: Bose, S. and Bandyopydhyay, A. (eds). Materials for Bone Disorder. Academic Press. ISBN 978-0-12-802792-9. p. 493–503.

Marchant-Forde, J.N. and Herskin, M.S. (2018). Pigs as laboratory animals. In: Advances in Pig Welfare. Herd and Flock Welfare. Woodhead Publ., p. 445–475.

Merryman, W.D. (2008). Development of a tissue engineered heart valve for pediatrics: A case study in bioengineering ethics. Sci. Eng. Ethics, 14, p. 93–101.

Mill, C.A. (1988). The Quality Audit: A Management Evaluation Tool. McGraw-Hill. ISBN 978-0-0704-2428-9. 309 pp.

Mishra, S. (2011). Social and ethical concerns of biomedical engineering research and practice. In: Shukla, A. and Tiwari, R. (eds). Biomedical Engineering and Information Systems. Technologies, Tools and Applications. Hershey, PA: IGI Global Publ. ISBN 978-1-6169-2005-0. p. 54–80.

Monzon, J.E. (1999). Teaching ethical issues in biomedical engineering. Int. J. Engng. Educ., 15(4), p. 276–281.

Poundarik, A.A., Diab, T., Sroga, G.E., Ural, A., Boskey, A.L., Gundberg, C.M. and Vashishth, D., (2012). Dilatational band formation in bone, PNAS, 109(47), p. 19178–19183.

Rajab, T., Rivard, A. L., Wasiluk, K. R., Gallegos, R. P. and Bianco, R. W. (2013). Ethical issues in biomaterials and medical devices. In: Ratner, B., Hofman, A., Schoen, F. and Lemons, J. (eds). Biomaterials Science: An Introduction to Materials, 3rd ed. Academic Press. ISBN 978-0-1237-4626-9. p. 1425–1431.

Ratner, B.D., Hoffman, A.S., Schoen, F.J. and Lemons, J.E. (2013). Biomaterials Science. An Introduction to Materials in Medicine. Elsevier, ISBN 978-0-12-374626-9.

Rollin, B.E. (1990). The Experimental Animal in Biomedical Research: A Survey of Scientific and Ethical Issues for Investigators, Volume I, CRC Press. ISBN 0-8493-4981-8.

Rozé, J., Hoornaert, A. and Layrolle, P. (2014). Correlation between primary stability and bone healing of surface treated titanium implants in the femoral epiphyses of rabbits. J. Mater. Sci.: Mater. Med., 25(8), p. 1941–1951.

Saha, S. and Saha, P. (1987). Bioethics and applied biomaterials. J. Biomed. Mater. Res., 21 (Suppl. 2A), p. 181–190.

Sastri, V.R. (2014). Regulations for medical devices and applications to plastic suppliers: History and Overview. In: Modjarrad, K. and Ebnesajjad, S. (eds). Handbook of Polymer Applications in Medicine and Medical Devices. Elsevier. ISBN 978-0-323-22805-3. p. 337–346.

Sun, L. (2018). Thermal spray coatings on orthopedic devices: When and how the FDA reviews your coatings. J. Thermal Spray Technol., 27, p. 1280–1290.

Tarabah, F. (2015). Good manufacturing practice (GMP) for biomaterials and medical devices in the EU and the USA. In: Amato, S.F. and Ezzell, R.M. (eds). Regulatory Affairs for Biomaterials and Medical Devices. Woodhead Publ. Series in Biomaterials. ISBN 978-0-85709-542-8. p. 115–143.

Tas, A.C. (2014). The use of physiological solutions or media in calcium phosphate synthesis and processing, Acta Biomater., 10(5), p. 1771–1792.

Vaquete, C., Ivanovski, S., Hamlet, S.M. and Hutmacher, D.W. (2013). Effect of culture conditions and calcium phosphate coating on ectopic bone formation. Biomaterials, 34, p. 5538–5551.

Wang, B.Y., Liao, M.L., Hong, G.C., Chang, W.W. and Chu, C.C. (2017). Near infrared-triggered photodynamic therapy toward breast cancer cells using dendrimer-functionalized upconversion nanoparticles. Nanomaterials, 7(9):269.

Zadpoor, A.A. (2014). Relationship between in vitro apatite-forming ability measured using simulated body fluid and in vivo bioactivity of biomaterials. Mater. Sci. Eng. C, 35, p. 134–143.

Zenios, S., Makower, J., Yock, P., Kumar, U.N., Denend, L. and Krummel, T.M. (2010). Biodesign: The Process of Innovating Medical Technologies. Cambridge: Cambridge University Press. ISBN 978-0-5115-1742-3.

Zhang, B.G.X., Myers, D.E., Wallace, G.G., Brandt, M. and Choong, P.F.M. (2014). Bioactive coatings for orthopaedic implants: Recent trends in development of implant coatings. Int. J. Mol. Sci., 15, p. 11878–11921.

Glossary

A

Acetabulum Cavity on the lateral surface of the hip bone. The hip bone is composed of three parts: *ilium*, *ischium*, and *pubis*, and provides the socket into which the head of the *femur* fits.

Akt A cellular homolog of murine thymoma virus akt8 oncogene is an essential component of the PI3K (phosphatidylinositol 3-kinase) pathway.

Alveolar ridge A ridge that forms the borders of the upper and lower jaws and contains the sockets of the teeth.

Amphiphilic Organic compounds combining both hydrophilic ("water-loving") and lipophilic ("fat-loving") properties. Phospholipids, a class of amphiphilic molecules, are the main components of biological membranes. The amphiphilic nature of these molecules defines the way in which they form membranes. They arrange themselves into bilayers by positioning their hydrophilic polar groups ("heads") toward the surrounding aqueous medium such as the extracellular fluid (ECF), and their lipophilic chains ("tails") toward the inside of the bilayer.

Anastomosis The surgical joining of two structures that are not typically connected. It is performed during surgery and may be done to repair a defect, to make the anatomy functional again after tissue is removed, or to make treatment possible.

Animal model An animal sufficiently similar to humans in its anatomy, physiology, or response to a pathogen or drug to be used in medical research to obtain results that can be extrapolated to human medicine.

Angiogenesis Formation of new blood vessels

Anterior cruciate ligament A ligament of each knee that attaches the front of the *tibia* with the back of the *femur* and functions especially to prevent hyperextension of the knee and is subject to injury especially by tearing during excessive sportive activities.

Apoptosis A form of programmed cell death that occurs in multicellular organisms.

Arthrodesis The artificial induction by surgery of joint ossification between two bones to relieve intractable pain in a joint that cannot be managed by pain medication, splints, or other normally indicated treatments.

Arthroplasty Surgery to relieve pain and restore range of motion by realigning or reconstructing a joint, for example by implanting an artificial hip or knee joint.

B

Biocompatibility The ability of a material to perform with an appropriate host response and confined to a specific application. Biocompatibility is neither a single specific property of a material nor a single phenomenon but is meant to describe a collection of properties and processes that involve different but interdependent interaction mechanisms between inorganic materials and living human or animal tissue.

https://doi.org/10.1515/9783110619249-011

Biointegration The ability of an implant material to form with the surrounding living tissues a continuum with stable, low-grade interaction.

Biopassivation Meaning that a biomaterial is not recognized as foreign by surrounding ECF or tissue, and that its clandestine behavior can persist for clinically meaningful periods of use.

C

Cancellous bone Porous ("*trabecular*") bone

Cardioverter A medical device for the administration of an electric shock in cardioversion

Cartilage A usually translucent, somewhat elastic tissue that forms most of the skeleton of vertebrate embryos and except for a small number of structures (such as some joints, respiratory passages, and the external ear) is replaced by bone during ossification in the higher vertebrates.

Cell proliferation An increase in the number of cells

Cell viability The ability of cells to live, develop, grow, and spread

Central composite design (CCD) A statistical experimental design, useful in response surface methodology, for building a second-order (quadratic) model for the response variables without needing to use a complete three-level factorial design.

Cerebroperitoneal shunt A surgical passage created to divert cerebrospinal fluid from the brain to the *peritoneal cavity*.

Chemical vapor deposition (CVD) In typical CVD, the substrate is exposed to one or more volatile precursors that react and/or decompose on the substrate surface to produce the desired deposit.

Chondroid cells Cartilage cells

Chirality A geometric property of some molecules and ions. A chiral molecule/ion is nonsuperposable on its mirror image. The presence of an asymmetric carbon center is one of several structural features that induce chirality in organic and inorganic molecules.

Complement protein A substance that is produced by a predecessor protein or in response to the presence of foreign material in the body and that triggers or participates in a complement reaction.

Conductive hearing loss Inability to transferring sound waves anywhere along the pathway through the outer ear, tympanic membrane (eardrum), or middle ear (*ossicles*).

Connective tissue A tissue of mesodermal origin that consists of various cells (such as fibroblasts and macrophages) and interlacing protein fibers (as of collagen) embedded in a chiefly carbohydrate ground substance, that supports and binds together other tissues, and that includes loose and dense forms (such as adipose tissue, tendons, ligaments, and aponeuroses) and specialized forms (such as cartilage and bone).

Cortical bone The hard, rigid form of connective tissue constituting most of the skeleton of vertebrates composed chiefly of hydroxylapatite–collagen I composite. It consists of multiple microscopic columns, each called an *osteon*. Each column is defined by multiple layers of osteoblasts and osteocytes around a central canal called the *Haversian canal*. The *Volkmann canaliculi* at right angles to the former connect the osteons together. Cortical bone is covered by the *periosteum* on its outer surface, and the *endosteum* on its inner surface.

Covalent bond A chemical bond that involves the sharing of electron pairs between atoms.

Crevice corrosion Refers to corrosion occurring in confined spaces to which the access of the working fluid from the environment is restricted.

CRISPR/Cas Clustered regularly interspaced short palindromic repeats/CRISPR-associated nucleases.

Cytocompatibility Ability to be in contact with a living system without producing an adverse effect on living cells.

Cytokine Any of a class of immune-regulatory proteins that are secreted by cells especially of the immune system. Cytokines are classified as lymphokines, interleukins, and chemokines, based on their presumed function, cellular locus of secretion, or target of action. Recombinant cytokines being used as drugs include bone morphogenetic proteins (BMPs), used to treat bone-related conditions.

Cytoskeleton The network of protein filaments and microtubules in the cytoplasm that controls cell shape, maintains intracellular organization, and is involved in cell movement.

Cytotoxicity The degree to which an agent has specific destructive action on certain cells.

D

Deep brain stimulation A neurosurgical procedure involving the placement of a medical device (neurostimulator) that sends electrical impulses, through implanted electrodes, to specific targets in the brain nuclei for treating movement disorders, including Parkinson's disease and *dystonia*.

Deoxyribonucleic acid (DNA) Any of various nucleic acids that are usually the molecular basis of heredity, are constructed of a double helix held together by hydrogen bonds between purine and pyrimidine bases that project inward from two chains containing alternate links of deoxyribose and phosphate, and that in eukaryotes are localized chiefly in cell nuclei.

Dialysis osteomalacia A disease in adults that is characterized by softening of the bones and is analogous to rickets in young people.

Diaphyseal fractures Fracture of the shaft of a long bone

Diaphysis The main or midsection (shaft) of a long bone. It is made up of cortical bone and usually contains bone marrow and adipose tissue (fat).

DNA origami The nanoscale folding of DNA to create nonarbitrary two- and three-dimensional shapes at the nanoscale. The specificity of the interactions between complementary Watson–Crick base pairs makes DNA a useful construction material. Future applications may include enzyme immobilization, drug delivery systems, nanorobots, and nanotechnological self-assembly of materials.

Drug-eluting stent (DES) A peripheral or coronary stent (a scaffold) placed into narrowed, diseased peripheral or coronary arteries that slowly releases a drug to block cell proliferation.

E

Ebner cement line A line visible in microscopic examination of bone in cross section, marking the boundary of an osteon.

Elastomers A polymer with viscoelasticity and very weak intermolecular forces, and generally low Young's modulus and high failure strain compared with other materials.

Electrophoresis The motion of dispersed particles relative to a fluid under the influence of a spatially uniform electric field. This is caused by the presence of a charged interface between the particle surface and the surrounding fluid. It is the basis for analytical techniques used in chemistry for separating molecules by size, charge, or binding affinity, as well as deposition of thin coatings

ELISA Enzyme-linked immuno sorbent assay. It is an analytical biochemistry assay using a solid-phase *enzyme immunoassay (EIA)* to detect the presence of a ligand (commonly a protein) in a liquid sample using antibodies directed against the protein to be measured.

Encephalopathy A disease of the brain, specifically a syndrome of overall brain dysfunction that can have many different organic and inorganic causes.

Endoalveolar infibulation A hollow spiral screw in stainless steel wire or tantalum used as endosseous implants.

Endosseous implant A dental implant that is inserted into the alveolar and/or basal bone and protrudes through the *mucoperiosteum*.

Endosteal implant See endosseous implant

Environmental impact statement (EIS) A document required by national environmental authorities for certain actions that significantly affect the quality of the human environment. An EIS is a tool for decision making. It describes the positive and negative environmental effects of a proposed action, and it usually also lists one or more alternative actions that may be chosen instead of the action described in the EIS.

Epiphysis The rounded end of a long bone, at its joint with adjacent bone(s).

Epitope The part of an antigen that is recognized by the immune system, specifically by antibodies, B cells, or T cells.

EPR effect Enhanced permeability and retention; the (controversial) concept by which molecules of certain sizes such as liposomes, nanoparticles, and macromolecular drugs tend to accumulate in tumor tissue much more than they do in normal tissues

Extracellular fluid (ECF) The internal environment of all multicellular animals, and in animals with a blood circulatory system a proportion of this fluid is blood plasma. Plasma and interstitial fluid are the two components that make up at least 97% of the ECF.

Extracellular matrix (ECM) A 3D network of macromolecules, such as collagen, enzymes, and glycoproteins that provide structural and biochemical support of surrounding cells.

F

Fatigue resistance Resistance to the weakening of a material caused by repeatedly applied loads

Foreign body reaction (FBR) FBR to biomaterials is composed of foreign body giant cells and the components of granulation tissue (e.g., macrophages, fibroblasts, and capillaries in varying amounts), depending on the form and topography of the implanted material

Fibrillogenesis The assembly of collagen fibrils by a self-assembly process, the details of which are subject to much speculation about the specifics of the mechanism through which the body produces collagen fibrils.

Fibroblast A connective tissue cell of mesenchymal origin that secretes proteins and especially molecular collagen from which the extracellular fibrillar matrix of connective tissue forms. Fibroblasts are the most common cells of connective tissue in animals.

Fractography The study of the fracture surfaces of materials. Fractographic methods are routinely used to determine the cause of failure in engineering structures, especially in product failure and the practice of forensic engineering or failure analysis.

Fracture toughness A property that describes the ability of a material to resist fracture, and is one of the most important properties of any material for many design applications. The linear-elastic fracture toughness of a material is determined from the stress intensity factor K at which a thin incipient crack in the material begins to grow.

Freund's adjuvant A solution of antigen emulsified in mineral oil and used as an immunopotentiator

Friction coefficient A value that shows the relationship between the force of friction between two objects and the normal reaction between the objects that are involved.

G

Gene delivery The process of introducing foreign genetic material, such as DNA or ribonucleic acid (RNA), into host cells. Gene delivery is a necessary step in gene therapy to introduce or silence a gene to promote a therapeutic outcome in patients.

Genotoxicity Describes the property of chemical agents that damages the genetic information within a cell, causing mutations.

Glycosaminoglycans Long unbranched polysaccharides consisting of a repeating disaccharide unit. The repeating unit consists of an amino sugar (*N*-acetylglucosamine or *N*-acetylgalactosamine) along with an uronic sugar (glucuronic acid or iduronic acid) or galactose. Prominent glycosaminoglycans are hyaluronic acid, chondroitin sulfate, and heparin.

Grignard reaction An organometallic chemical reaction in which alkyl, vinyl, or aryl-magnesium halides add to a carbonyl group in an aldehyde or ketone. This reaction is important to form carbon–carbon bonds.

H

Hemocompatibility Ability to be in contact with blood without producing an adverse effect.

Heterotopic ossification The process by which bone tissue forms outside (ektopic) of the skeleton, for example, in soft tissue.

Hip arthroplasty A surgical procedure in which the hip joint is replaced by an endoprosthetic implant.

Hot isostatic pressing (HIP) A technique to improve characteristics relevant to fracture mechanics in ceramic implants. HIPing reduces porosity, increases density and purity, and thus leads to a relevant increase in long-term life expectancy and a decrease in the tendency to subcritical crack growth.

Homeostasis The state of steady internal physical and chemical conditions maintained by living systems. This dynamic state of equilibrium is the condition of optimal functioning for the organism and includes many variables, such as body temperature and fluid balance, being kept within certain preset limits (homeostatic range).

Human mesenchymal stem cells (hMSCs) Pluripotent stromal cells that can differentiate into a variety of other cell types, including osteoblasts (bone cells), chondrocytes (cartilage cells), myocytes (muscle cells), and adipocytes (fat cells).

Human umbilical vein endothelial cells (HUVECs) Cells derived from the endothelium of veins from the cord. They are used as a laboratory model system for the study of the function and pathology of endothelial cells.

Humoral Relating to the body fluids or humors

Hydrogen bond A primarily electrostatic force of attraction between a hydrogen (H) atom which is covalently bound to a more electronegative atom or group, particularly the second-row elements nitrogen (N), oxygen (O), or fluorine (F) as the hydrogen bond donor (Dn), and another electronegative atom bearing a lone pair of electrons as the hydrogen bond acceptor (Ac).

I

In silico Research performed on computer or via computer simulation in reference to biological experiments, in contrast to *in vitro* or *in vivo*.

Intercranial angioplasty A minimally invasive, endovascular procedure to widen narrowed or obstructed arteries or veins, typically to treat arterial atherosclerosis. A deflated balloon attached to a catheter is passed over a guide wire into the narrowed vessel and then inflated to a fixed size.

Intergranular corrosion A form of corrosion in which the boundaries of crystallites of the material are more susceptible to corrosion than their insides.

Integrins Transmembrane receptors that facilitate cell–extracellular matrix (ECM) adhesion. Upon ligand binding, integrins activate signal transduction pathways that mediate cellular signals such as regulation of the cell cycle, organization of the intracellular cytoskeleton, and movement of new receptors to the cell membrane.

Intermetallic phase Solid phases containing two or more metallic elements, with optionally one or more nonmetallic elements, whose crystal structure and composition differ from that of the other constituents.

Interstitial defect A variety of crystallographic defects where atoms assume a normally unoccupied site in the crystal structure. In interstitial defects, three or more atoms may share one lattice site, thereby increasing its total energy. Alternatively, small atoms may occupy interstitial sites in energetically favorable configurations, such as hydrogen in palladium.

Intervertebral disk Lies between adjacent vertebrae in the spine. Each disk forms a fibrocartilaginous joint (*symphysis*) to allow slight movement of the vertebrae, to act as a ligament to hold the vertebrae together, and to function as a shock absorber for the spine.

Intramedullary nail (IM) A metal rod forced into the medullary cavity of a bone. IM nails have long been used to treat fractures of long bones of the body.

Intramembranous ossification One of the two essential processes during fetal development of the skeletal system by which rudimentary bone tissue is created. Intramembranous ossification is an essential process during the natural healing of bone fractures and the rudimentary formation of bones of the head. Unlike endochondral ossification, which is the other process by which bone tissue is created during fetal development, cartilage is not present during intramembranous ossification.

Intraspinal implant Intraspinal decompression implants comprise two plate attachments to a titanium metal implant that fits between the spinous processes of the vertebrae in the lower back, decompressing the neuroelements. They are implanted without fixation to the bone or ligament to preserve physiological spinal motion. These devices are particularly useful for patients who suffer from degenerative disk disease, spinal stenosis, or lateral recess syndrome.

Ion channel Pore-forming membrane proteins that allow ions to pass through the channel pore located in a cell membrane. Their functions include establishing arresting membrane potential, shaping action potentials, and other electrical signals by gating the flow of ions across the cell membrane, controlling the flow of ions across secretory and epithelial cells, and regulating cell volume.

Isomorphous A close similarity in the crystalline structure of two or more substances of similar chemical composition.

K

Kinine Biologically active polypeptides that function as mediators for inflammatory responses by triggering the immune system.

Knee arthroplasty A surgical procedure to replace the weight-bearing surfaces of the knee joint to relieve pain and disability. It is most commonly performed for osteoarthritis and other knee diseases such as rheumatoid arthritis and psoriatric arthritis.

L

Lactate dehydrogenase An enzyme found in nearly all living cells (animals, plants, and prokaryotes). LDH catalyzes the conversion of lactate to pyruvate and back, as it converts NAD^+ (nicotinamide adenine dinucleotide) to NADH and back. NAD^+ and NADH are important coenzymes in redox metabolism. Among others, they link the citric cycle and oxidative phosphorylation to transform ADP to ATP.

Langmuir–Blodgett (LB) film A nanostructured system formed when monolayers are transferred from the liquid–gas interface to solid supports during the vertical passage of the support through the monolayers, thus compressing them.

Life cycle assessment (LCA) The goal of LCA is to compare the full range of environmental effects assignable to products and services by quantifying all inputs and outputs of material flows and assessing how these material flows affect the environment. This information is used to improve processes, support policy, and provide a sound basis for informed decisions.

Lipid A biomolecule that is soluble in nonpolar solvents. The functions of lipids include storing energy, signaling, and acting as structural components of cell membranes. Lipids can be defined as hydrophobic or amphiphilic small molecules; the amphiphilic nature of some lipids allows them to form structures such as vesicles, liposomes (see below), or membranes in an aqueous environment.

Liposome A spherical vesicle formed by at least one lipid bilayer. Liposomes are most often composed of phospholipids, especially phosphatidylcholine, but may also include other lipids.

Lotus effect Refers to self-cleaning properties of a surface as a result of superhydrophobicity. The self-cleaning qualities of superhydrophobic surfaces come from physicochemical properties at the microscopic to nanoscopic scale rather than from the specific chemical properties of the surface.

M

Macrophage A type of white blood cells of the immune system that engulfs and digests cellular debris, foreign substances, microbes, cancer cells, and anything else that does not have the type of proteins specific to healthy body cells on its surface in a process called phagocytosis.

Magnetron sputtering A film deposition technology involving a gaseous plasma that is generated and confined to a space containing the material to be deposited, the "target." The surface of the target is eroded by high-energy ions within the plasma, and the liberated atoms travel through the vacuum environment and deposit onto a substrate to form a thin film.

Matrix vesicle An extracellular membrane-bound body measuring 0.1–0.2 μm in diameter, which may form the nidus for calcification initiated by precipitation of hydroxylapatite.

Medullary canal The central cavity of (long) bone shafts where red bone marrow and/or yellow bone marrow (adipose tissue) are stored. Located in the main shaft of a long bone (*diaphysis*) (consisting mostly of compact, i.e., cortical bone), the medullary cavity has walls composed of spongy bone (cancellous bone) and is lined with a thin, vascular membrane (endosteum).

Metaphyseal fracture A fracture that involves the metaphysis of tubular bones. They may occur in pediatric or adult patients. Examples of metaphyseal fractures in adults are neck of *humerus* fracture, distal radial fractures or transtrochanteric fracture. In children, distal radial buckle fractures.

Microindentation hardness A method for measuring the hardness of a material on a microscopic scale. A precision diamond indenter is impressed into the material at loads from a few grams to 1 kg. The impression length, measured microscopically, and the test load are used to calculate a hardness (Vickers or Knoop) value.

MTT assay A colorimetric assay for measuring cell metabolic activity. It is based on the ability of nicotinamide adenine dinucleotide phosphate (NADPH)-dependent cellular oxidoreductase enzymes to reduce the 3-[4,5-dimethylthiazole-2-yl]-2,5-diphenyltetrazolium bromide (MTT) dye to its insoluble purple formazan.

Mucoperiosteum Mucous membrane and periosteum intimately united as to form practically a single membrane.

N

Neointimal hyperplasia Refers to proliferation and migration of vascular smooth muscle cells primarily in the *tunica intima*, resulting in the thickening of arterial walls and decreased arterial lumen space. It is the major cause of restenosis after percutaneous coronary interventions such as stenting or angioplasty.

Neuropeptide A small protein-like molecule (peptide) used by neurons to communicate with each other. The neuronal signaling molecules influence the activity of the brain and the body in specific ways. Neuropeptides such as oxytocin and vasopressin are involved in a wide range of brain functions, including analgesia, reward, food intake, metabolism, reproduction, social behaviors, learning, and memory.

Noncollagenous proteins (NCPs) A heterogeneous group, unique to the bone and probably playing a role in mineralization. NCPs are formed exogenously or locally, and include osteocalcin, osteonectin, proteoglycans, bone sialoprotein, matrix glutamic acid (Gla) protein, and serum proteins.

Nosocomial infection An infection that is acquired in a hospital or other healthcare facility.

Nucleotide Organic molecules that serve as the monomer units to form the nucleic acid polymers DNA and RNA, both of which are essential biomolecules within all life forms on the Earth. Nucleotides are the building blocks of nucleic acids; they are composed of three subunit molecules: a nitrogenous base (nucleobase), a five-carbon sugar (deoxyribose or ribose), and at least one phosphate group.

O

Occlusal Pertaining to the contacting surfaces of opposing occlusal units (teeth or occlusion rims), or the masticating surfaces of the posterior teeth.

Opsonin Any molecule that enhances phagocytosis by marking an antigen for an immune response or marking dead cells for recycling

Osseoconductivity The ability of biomaterials to support the in-growth of bone cells, blood capillaries, and perivascular tissue into the operation-induced gap between implant body and existing (cortical) bone bed.

Osseoinductivity The ability to transform undifferentiated mesenchymal precursor stem cells into osseoprogenitor cells that precede endochondral ossification. The process relies crucially on the osseoinductive action of noncollagenous proteins (NCPs).

Osseointegration Involves a FBR whereby interfacial bone between an inorganic implant and living bone is formed as a defense reaction to shield off the implant from tissue.

Osseoprogenitor cell A precursor cell that arises from mesenchymal stem cells (MSC) in the bone marrow. It has the ability to differentiate into osteoblasts or chondrocytes depending on the signaling molecules it is exposed to, giving rise to either bone and cartilage cells, respectively.

Ossicular chain Three bones (*malleus*, *incus*, and *stapes*) in the middle ear that are among the smallest bones in the human body. They serve to transmit sound from the air to the fluid-filled labyrinth (*cochlea*).

Osteoblast A specialized, terminally differentiated product of MSCs with a single nucleus that synthesize bone. However, in the process of bone formation, osteoblasts function in groups of connected cells. Individual cells cannot make bone. A group of organized osteoblasts together with the bone made by a unit of cells is called the osteon.

Osteoclast A type of bone cell that breaks down the bone tissue. This function is critical in the maintenance, repair, and remodeling of bone of the vertebral skeleton. The osteoclast disassembles and digests the composite of hydrated protein and mineral at a molecular level by secreting acid and a collagenase, a process known as bone resorption. This process also helps regulate the level of blood calcium.

Osteolysis Active resorption of bone matrix by osteoclasts, the reverse of ossification. Although osteoclasts are active during the natural formation of healthy bone, the term osteolysis specifically refers to a pathological process. Osteolysis often occurs in the proximity of a prosthesis that causes either an immunological response or changes in the bone's structural load. Osteolysis may also be caused by pathologies like bone tumors, cysts, or chronic inflammation.

Osteonecrosis A disease caused by reduced blood flow to bones in the joints. In people with healthy bones, new bone is always replacing old bone. In osteonecrosis, the lack of blood causes the bone to break down faster than the body can make enough new bone. Hence, the bone starts to die.

Osteoporosis A pathological imbalance between bone resorption and bone formation. In normal bone, matrix remodeling of bone is constant; up to 10% of all bone mass may be undergoing remodeling at any point in time. The process takes place in bone multicellular units. Osteoclasts are assisted by transcription factor PU.1 to degrade the bone matrix, while osteoblasts rebuild the bone matrix. Low bone mass density can then occur when osteoclasts are degrading the bone matrix faster than the osteoblasts are able to rebuilding the bone.

Osteosarcoma A cancerous tumor in a bone. Specifically, it is an aggressive malignant neoplasm that arises from primitive transformed cells of mesenchymal origin and that exhibits osteoblastic differentiation and produces malignant osteoid.

Osteosynthesis An operation in orthopedics that involves the surgical implementation of implants for the purpose of repairing a bone, a concept that dates to the mid-nineteenth century and was made applicable for routine treatment in the mid-twentieth century. An internal fixator may be made of stainless steel or titanium. Bioresorbable metals such as magnesium, zinc, or iron are presently explored as alternative materials that do not require a second operation as they are resorbed by the body.

Osteotomy A surgical operation whereby a bone is cut to shorten or lengthen it, or to change its alignment such as in *Halux valgus* operations.

P

Pelvic bone Is formed posteriorly (in the area of the back) by the *sacrum* and the *coccyx*, and laterally and anteriorly (forward and to the sides) by a pair of hip bones. Each hip bone consists of three sections: *ilium*, *ischium*, and *pubis*.

Peptide Short chains of amino acid monomers linked by peptide (amide) bonds. The covalent chemical bonds are formed when the carboxyl group of one amino acid reacts with the amino group of another.

Percutaneous coronary intervention A nonsurgical procedure used to treat narrowing (*stenosis*) of the coronary arteries of the heart. After accessing the blood stream through the femoral or radial artery, the procedure uses coronary catheterization to visualize the blood vessels on X-ray imaging. After this, a deflated balloon is advanced into the obstructed artery and inflated to relieve the narrowing; certain devices such as stents can be deployed to keep the blood vessel open.

Pericytes Multifunctional mural cells of the microcirculation that wrap around the endothelial cells lining the capillaries and *venules* throughout the body.

Periodontal pocket The presence of an abnormal depth of the gingival *sulcus* near the point at which the gingival tissue contacts the tooth.

Periprosthetic tissue reactions A complex set of inflammatory reactions near an implant site, including necrosis, lymphocyte infiltration, histiocytosis, pseudotumor, hypersensitivity to metal ions, and aseptic lesions.

Phagocyte A cell that protects the body by ingesting harmful foreign particles, bacteria, and dead or dying cells. Phagocytes are crucial in fighting infections, as well as in maintaining healthy tissues by removing dead and dying cells that have reached the end of their lifespan.

Photocatalysis The acceleration of a photoreaction in the presence of a catalyst. In photogenerated catalysis, the photocatalytic activity depends on the ability of the catalyst to create electron–hole pairs that generate free radicals (e.g., hydroxyl radicals: •OH) able to undergo secondary reactions.

Photothermal therapy Efforts to use electromagnetic radiation (most often in infrared wavelengths) to treat various medical conditions, including cancer. This approach is an extension of photodynamic therapy, in which a photosensitizer is excited with specific band light. This activation brings the sensitizer to an excited state where it then releases vibrational energy (heat) to kill the targeted cells.

Pitting corrosion A form of extremely localized corrosion that leads to the creation of small holes in the metal. The driving power for pitting corrosion is the depassivation of a small area that becomes anodic while an unknown but potentially vast area becomes cathodic, leading to very localized galvanic corrosion.

Pleiotropy Pleiotropy occurs when one gene influences two or more seemingly unrelated phenotypic traits.

Polyamide A macromolecule with repeating units linked by amide bonds.

Polyester A polymer that contains the ester functional group in their main chain. As a specific material, it most commonly refers to a type called poly(ethylene terephthalate).

Polymersomes A class of artificial vesicles, forming tiny hollow spheres that enclose a solution and are made using amphiphilic synthetic block copolymers to form the vesicle membrane.

Polysaccharide Polymeric carbohydrate molecules composed of long chains of monosaccharide units bound together by glycosidic linkages, They range in structure from linear to highly branched.

Examples include storage polysaccharides such as starch and glycogen, and structural polysaccharides such as cellulose and chitin.

Protease An enzyme that helps proteolysis, that is, protein catabolism by hydrolysis of peptide bonds.

Protein Large biomolecules consisting of one or more long chains of amino acid residues. Proteins perform a vast array of functions, including catalyzing metabolic reactions, DNA replication, responding to stimuli, providing structure to cells, and transporting molecules from one location to another.

Pseudoarthrosis Permanent failure of healing following a broken bone (nonunion) unless intervention (such as surgery) is performed.

Pulsed laser deposition (PLD) A physical vapor deposition technique where a high-power pulsed laser beam is focused inside a vacuum chamber to strike a target of the material that is to be deposited. This material is vaporized from the target (in a plasma plume), which deposits it as a thin film on a substrate.

R

Radiopacity Opacity to the radio wave and X-ray portion of the electromagnetic spectrum, that is, the relative inability of those kinds of electromagnetic radiation to pass through a particular material. Medical devices often contain a radiopacifier to enhance visualization during implantation for temporary implantation devices, such as catheters or guide wires, or for monitoring the position of permanently implanted medical devices, such as stents, hip and knee implants, and screws. Metal implants usually have sufficient radiocontrast that additional radiopacifier is not necessary. Polymer-based devices, however, usually incorporate materials with high electron density contrast compared to the surrounding tissue.

Regenerative medicine (RM) A branch of medicine aimed to find solutions to regenerate damaged tissues.

Reocclusion The reoccurrence of occlusion in an artery after it has been treated (as by balloon angioplasty) with apparent success (see *restenosis*).

Restenosis The recurrence of *stenosis,* a narrowing of a blood vessel, leading to restricted blood flow. Restenosis usually pertains to an artery that has become narrowed, received treatment to clear the blockage, and subsequently become renarrowed.

Revascularization The restoration of perfusion to a body part or organ that has suffered ischemia. It is typically accomplished by surgical means. Vascular bypass and angioplasty are the two primary means of revascularization.

Ribonucleic acid (RNA) A polymeric single-strand molecule essential in various biological roles in coding, decoding, regulation, and expression of genes.

S

Sham operation Sham surgery is a faked surgical intervention that omits the step thought to be therapeutically necessary. In clinical trials, sham surgery is an important scientific control as it isolates the specific effects of the treatment as opposed to the incidental effects caused by anesthesia, the incisional trauma, pre- and postoperative care, and the patient's perception of having had a regular operation. Thus, sham surgery serves an analogous purpose to placebo drugs, neutralizing biases such as the placebo effect.

Simulated body fluid (SBF) A solution with an ionic concentration close to that of human blood plasma kept under mild conditions of pH and identical physiological temperature. There exist many formulations, all intended to mimic the composition of the ECF. An SBF can be used as an *in vitro* testing method to study the formation of a hydroxylapatite layer on the surface of implants so as to predict *in vivo* bioactivity. It ought to be emphasized that apatite formation is a necessary but not a sufficient proof of bioactivity.

Solid solution Solid-state solution of one or more solutes in a solvent. Such a multicomponent system is considered a solution rather than a compound when the crystal structure of the solvent remains unchanged by addition of the solutes, and when the chemical components remain in a single homogeneous phase.

Solution treatment (ST) Heating of a material to temperatures sufficiently high for dissolving its soluble phases, that is, forming solid solutions.

Somatic gene therapy The therapeutic delivery of nucleic acid into a patient's cells as a drug to treat disease.

Sprague–Dawley rat A rat of the species *Rattus norvegicus domesticus* that is bred and kept for scientific research. While less commonly used for research than mice, rats have served as an important animal model for research in psychology and biomedical science.

Standard operating procedures (SOP) A set of step-by-step instructions compiled by an organization to help workers carry out complex routine operations. SOPs aim to achieve efficiency, quality output, and uniformity of performance, while reducing miscommunication and failure to comply with industry regulations.

Statistical design of experiments (SDE) The design of any task that aims to describe or explain the variation of information under conditions that are hypothesized to reflect the variation. The term is generally associated with experiments in which the design introduces conditions that directly affect the variation, but may also refer to the design of thought experiments, in which natural conditions that influence the variation are selected for observation.

Stem cell A cell that can differentiate into other types of cells and can also divide in self-renewal to produce more of the same type of stem cells. Adult stem cells are frequently used in various medical therapies (e.g., bone marrow transplantation). Stem cells can now be artificially grown and transformed (differentiated) into specialized cell types with characteristics consistent with cells of various tissues such as muscles or nerves.

Stenosis An abnormal narrowing or contraction of a body passage or opening.

Stress corrosion cracking (SCC) The growth of crack formation in a corrosive environment. It can lead to unexpected sudden failure of normally ductile metals subjected to a tensile stress, especially at elevated temperature. SCC is highly chemically specific in that certain alloys are likely to undergo SCC only when exposed to a small number of chemical environments.

Subchondral bone The *epiphysis*, that is, the rounded end of a long bone, is covered with articular cartilage below that covering is a zone similar to the epiphyseal plate, known as subchondral bone lying under articular cartilage and containing marrow.

Superhydrophilicity The phenomenon of excess hydrophilicity or attraction to water. In superhydrophilic materials, the contact angle of water is around zero degrees.

Synovial fluid A viscous, non-Newtonian fluid found in the cavities of synovial joints. The principal role of synovial fluid is to reduce friction between the articulating cartilage of synovial joints during movement. Synovial fluid is composed of hyaluronic acid and proteoglycan 4 (lubricin), proteinases, and collagenases.

T

Tacticity The relative stereochemistry of adjacent chiral centers within a polymeric macromolecule

Teratogenicity Relating to, or causing developmental malformations in embryos

Thermochemical processing (TCP) Use of hydrogen as a temporary alloying element in titanium alloys to enhance both processability and final mechanical properties.

Thermoplastics A plastic polymer material that becomes pliable or moldable at a certain elevated temperature and solidifies upon cooling. The polymer chains associate by intermolecular forces that weaken rapidly with increased temperature, yielding a viscous liquid. In this state, thermoplastics may be reshaped and are typically used to produce parts by various polymer processing techniques such as injection molding, compression molding, calendering, and extrusion.

Transmembrane protein A protein that spans the entire length of the cell membrane. It is embedded between the phospholipids, providing a channel through which molecules and ions can pass into the cell. Transmembrane proteins also facilitate communication between cells by interacting with chemical messengers

Thrombogenicity The tendency of a material in contact with the blood to produce a thrombus, or clot. A thrombogenic implant will eventually be covered by a fibrous capsule, the thickness of this capsule can be considered one measure of thrombogenicity, and if extreme can lead to the failure of the implant.

Tissue engineering (TE) The use of a combination of cells, engineering methods and materials design, and appropriate signaling factors to improve or replace biological functions.

Total hip arthroplasty (THA) A surgical procedure in which the hip joint is replaced by a prosthetic implant. Such surgery is generally conducted to relieve arthritic pain or in some hip fractures. A total hip replacement (total hip arthroplasty) consists of replacing both the *acetabulum* and the femoral head.

Total knee arthroplasty (TKA) A surgical procedure to replace the weight-bearing surfaces of the knee joint to relieve pain and disability. It is most commonly performed for osteoarthritis, and also for other knee diseases such as rheumatoid arthritis and psoriatic arthritis.

Trabecular bone Porous bone composed of spongy bone tissue. It can be found at the ends of long bones like the *femur*, where the bone is actually not solid but is full of holes connected by thin rods and plates of bone tissue.

Transfection The process of deliberately introducing naked or purified nucleic acids into eukaryotic cells. Transfection of animal cells typically involves opening transient pores or "holes" in the cell membrane to allow the uptake of material. Transfection can be carried out using calcium phosphate (i.e., tricalcium phosphate), by electroporation, by cell squeezing or by mixing a cationic lipid with the material to produce liposomes that fuse with the cell membrane and deposit their cargo inside.

Transformation toughening A way of drastically improving the fracture toughness of an inherently brittle material such as certain ceramics. This is accomplished by introducing a mechanism (structural phase transformation) for triggering a uniform distribution of low-level internal strain within the bulk of the material. This strain acts locally to oppose the stress field associated with crack propagation, thus annihilating it at the approaching crack tip front.

Tribochemical reactions Chemical reactions that occur between the lubricant/environment and the surfaces under boundary lubrication conditions. The precise nature of the chemical reactions is not well understood. What causes the reactions to take place is also a subject of speculation.

Tribocorrosion A material degradation process due to the combined effect of corrosion and wear. Biotribocorrosion results from surface transformations by the interaction of mechanical loading and chemical/electrochemical reactions that occur between elements of a tribological system exposed to biological environments. It has been studied for artificial joint prostheses. It is important to understand material degradation processes for joint implants to achieve longer service life and better safety issues for such devices.

Tribofilm Films that form on tribologically stressed surfaces. While there exists no universal definition of the term, it is mostly used to refer to solid surface films that result from a chemical reaction of lubricant components and/or tribological surfaces. Tribofilms play an important role in reducing friction and wear in lubricated systems.

Tunica intima The innermost layer of an artery or vein. It is made up of one layer of endothelial cells and is supported by an internal elastic lamina. The endothelial cells are in direct contact with the blood flow.

V

van der Waals force A distance-dependent weak interaction between atoms or molecules. Unlike ionic or covalent bonds, these attractions do not result from a chemical electronic bond; they are comparatively weak and therefore more susceptible to disturbance. The van der Waals force quickly vanishes at longer distances between interacting molecules.

Vascular graft An arteriovenous fistula consisting of a venous autograft or xenograft or a synthetic tube grafted onto the artery and vein.

X

Xanthate A salt with the formula $ROCS-2M^+$ (R = alkyl; $M^+ = Na^+, K^+$), thus O-esters of dithiocarbonate. These organosulfur compounds are important in two areas: the production of cellophane and related polymers from cellulose and (in mining) for extraction of certain ores by flotation.

X-ray mapping An image showing the spatial distribution of elements in a sample. The image is produced by progressively rastering an electron beam point-by-point over an area of interest and recording the characteristic X-ray radiation given off by individual atoms.

Z

Zeta potential The electric potential in the interfacial double layer (DL) at the location of the slipping plane relative to a point in the bulk fluid away from the interface.

Zwitterionic A molecule with two or more functional groups, of which at least one has a positive and one has a negative electrical charge and the net charge of the entire molecule is zero.

Index

www.ingramcontent.com/pod-product-compliance
Lightning Source LLC
Chambersburg PA
CBHW060937210326
41598CB00031B/4651